SOIL-PLANT RELATIONSHIPS

SECOND EDITION

C. A. BLACK
Department of Agronomy
Iowa State University
Ames, Iowa

JOHN WILEY & SONS, INC. New York · London · Sydney

6 7 8 9 10

Library of Congress Catalog Card Number: 67-28946
Printed in the United States of America
ISBN 0 471 07723 2

Preface

This second edition is a completely rewritten account of the subjects included in the first edition. About half of the literature cited and examples have been published since the first edition was prepared.

Two general shifts in emphasis from the first edition are worthy of mention. First, the treatment of subject matter has moved further in the direction of showing the operation of the scientific method in experiments on soil-plant relationships. Second, more attention has been given to the significance of measurements of soil properties. Both these changes should benefit primarily the serious student. They are intended to give him an understanding of the basis for current views and at the same time to provide some appreciation for their limitations. The reader who is searching for only a specific fact or dictum may find his subject matter somewhat diluted, but hopefully the presence of the background material will caution that all is not so simple or so final as we might wish to think. Concepts of soil-plant relationships have changed in the past and are likely to change further in the future.

As in the first edition, the emphasis is on soil as a substrate for plant growth. A special attempt is made to set forth the pertinent facts and ideas about soil behavior as a basis for understanding but to limit technical details and use of jargon. Most of the concepts are illustrated by examples from specific locations. Except for purposes of illustration, however, little or no attempt is made to interpret the principles in terms of practice because frequently the proper application must take into account other factors that are not considered here.

The principal scientific problems in the area of soil-plant relationships are those of measurement and integration. The basic conditions that affect the growth of plants must be measured, and the measurements then must be integrated in the proper manner to obtain an index of plant growth. For the present, the major concern is measurement. Integration has not progressed far. Nevertheless, because of their importance in the long-range objectives of the study of soil-plant relationships, the broad concepts of measurement and integration are woven into the background throughout the text.

Limitation of coverage to the subjects that form the nine chapters is not intended to imply that these are the only aspects of soil-plant relationships of significance. Other matters should be treated but have been omitted for lack of time to prepare them for inclusion.

The subject matter is covered in part by the author in a course taught at Iowa State University. The level of presentation is intended to be suitable for students who have attained some knowledge of soils, plant physiology, and the basic sciences at the undergraduate level. Review papers are cited wherever applicable because these are the most useful source material to a student who wishes to pursue further any particular subject. Other literature citations are included primarily to give credit for material cited.

The author wishes to acknowledge indebtedness to Dr. K. L. Babcock, of the University of California at Berkeley, for many hours of profitable discussion of problems of soil chemistry during the period of literature study and revision of lecture notes that preceded writing of the second edition. For photographs or other illustrative material, the author is indebted to J. G. Cady, Roy D. Bond, R. V. Ruhe, Arthur H. Lachenbruch, Troy L. Péwé, Roy W. Simonson, E. G. Hallsworth, J. A. Kittrick, L. E. Allison, H. M. Taylor, F. L. Duley, L. A. Richards, E. P. Whiteside, W. H. Gardner, W. R. Gardner, J. Letey, Karl Grossenbacher, S. A. Barber, C. D. Foy, Frank M. Eaton, C. D. Moodie, R. K. Schulz, J. L. Paul, L. C. Dumenil, C. R. Weber, A. J. Ohlrogge, Roscoe E. Hill, G. J. Racz, A. D. McLaren, J. Skujins, A. D. Scott, M. M. Mortland, S. J. Angstrom, G. E. Wilcox, J. C. Shickluna, and Samuel Merrill, Jr. Their contributions to the text will increase its appeal to the student.

For granting permission to use copyrighted material, I am indebted to these sources: Academic Press, Acta Agriculturae Scandinavica, American Association for the Advancement of Science, American Chemical Society, American Potash Institute, American Society of Agronomy, Australian Journal of Agricultural Research, Australian Journal of Biological Sciences, S. A. Barber, L. Joe Berry, Blackwell Scientific Publications, The Clarendon Press, R. K. Cunningham, Deutsches Kalisyndikat, Elsevier Publishing Company, E. G. Hallsworth, Her Majesty's Stationery Office, M. L. Jackson, Journal of Forestry, E. R. Lemon, Macmillan Journals, Martinus Nijhoff Publishers, McGraw-Hill Book Company, Mededlingen van de Landbouwhogeschool en de Opzoekingsstations van de Staat te Gent, Nederlandsche Botanische Vereeniging, North-Holland

Publishing Co., Oxford University Press, W. H. Patrick, Jr., Pergamon Press, J. D. Säuerlander's Verlag, United Nations Educational, Scientific and Cultural Organization, University of Chicago Press, Verlag Chemie, and Verlag Paul Parey.

C. A. Black

Ames, Iowa
December, 1967

Contents

1 *Soil Characterization*

This chapter deals with characterization of soil as a whole from an evaluation of different properties. The subject matter is intended to provide both a brief review of much basic material and an appreciation for the value and limitations of the various types of evaluations for characterizing soil as a substrate for plant growth.

Particle-Size Composition

Soils have been described for many years in terms of the proportions of particles of different sizes that they contain. This basis of characterizing soils developed because particle size is an obvious characteristic related to soil behavior and plant response. In the early 1700s, Jethro Tull proposed that plants lived on the fine particles of soil they ingested (Russell and Russell, 1950, pp. 4–5). Although this view received wide acceptance at the time, soil particles as such are no longer thought to be a primary requirement for the growth of plants. The only known direct effect of particle size is the mechanical impediment offered by large particles that deflect the passage of roots or prevent the vertical emergence and growth of the aerial parts. Effects of particle size on plant growth appear to be mainly indirect.

The properties of coarse and fine soil particles differ considerably, but there is no sharp, natural division of any kind at any particle size. For practical purposes, however, some arbitrary boundaries have been established. The size limits used in several systems of classifying soil particles are shown in Fig. 1.1.

The particle-size composition of a soil is the percentage of the mineral matter by weight in each fraction obtained by separating the mineral particles into two or more mutually exclusive size-classes. Soils ordinarily are separated into at least three size-classes, which are usually called sand, silt, and clay. The process used for measuring particle-size composition is particle-size analysis. The numerous methods of particle-size analysis are based on the fact that the rate of fall of soil particles through

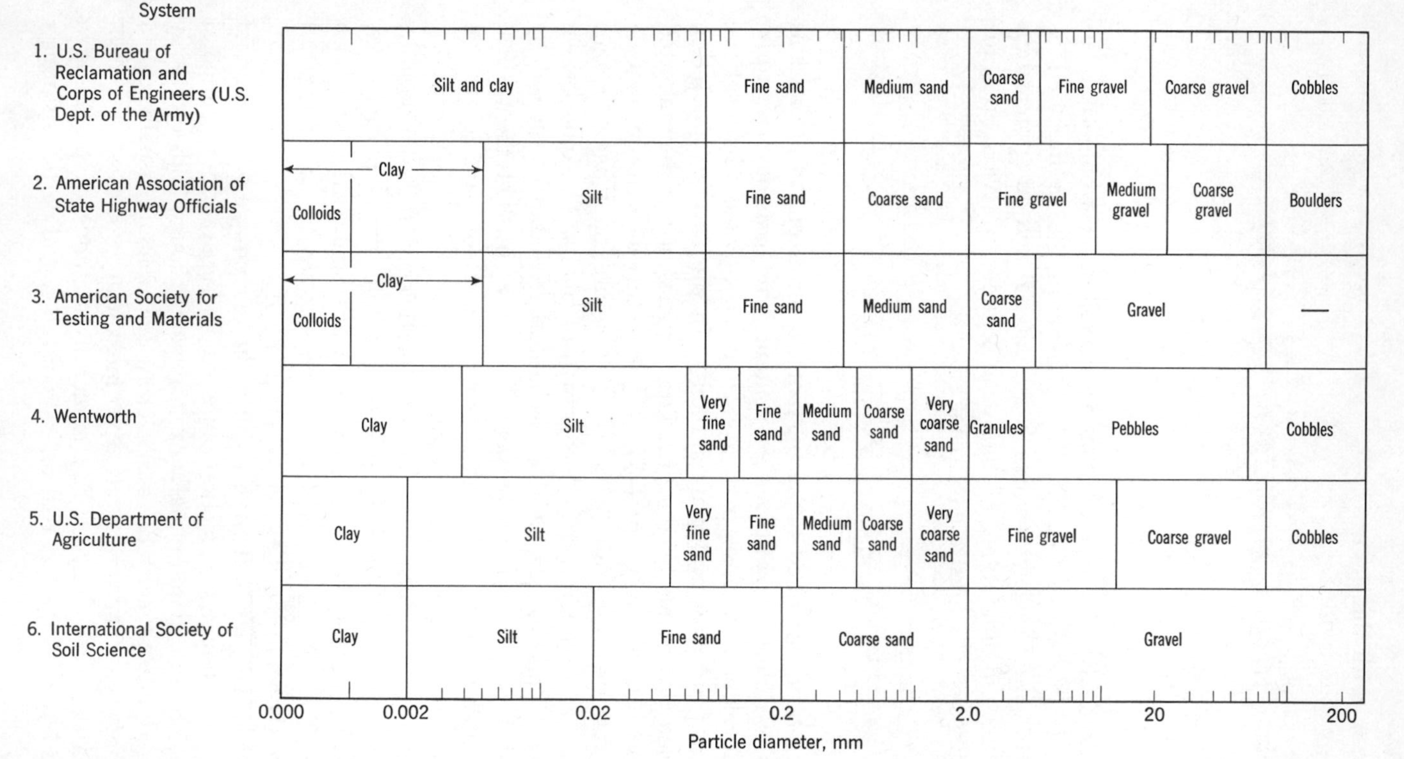

Figure 1.1. Size limits and names of particle-size classes according to six systems. The first three systems are used in soil engineering, the fourth in geology, and the fifth and sixth in soil science. In system 1, silt and clay are distinguished on plasticity instead of particle size. In the fifth system, the term "cobbles" is used for rounded particles; other terms are used for particles that are not rounded. Note the logarithmic scale. (After a similar unpublished diagram prepared by E. P. Whiteside)

water increases with the diameter of the particles or on some combination of this property with the use of sieves for separating coarse particles (Day, 1965).

The relative proportions of sand, silt, and clay in a soil determine what is called the soil class, textural class, textural grade, or texture. The number of possible combinations of sand, silt, and clay is obviously infinite. For practical purposes, however, certain arbitrary divisions are made, and a descriptive name is applied to all particle-size compositions within each arbitrary division. Figure 1.2 shows the textural classification used currently by the United States Department of Agriculture; this figure may be used to determine the name to apply when one has at hand the results of a particle-size analysis. In soil survey work, the textural class is estimated in the field by the feel of the wet soil when rubbed between finger and thumb; each textural class has particular distinguishing characteristics. As may be inferred from the relatively large range in particle-size composition classified as "clay" in Fig.

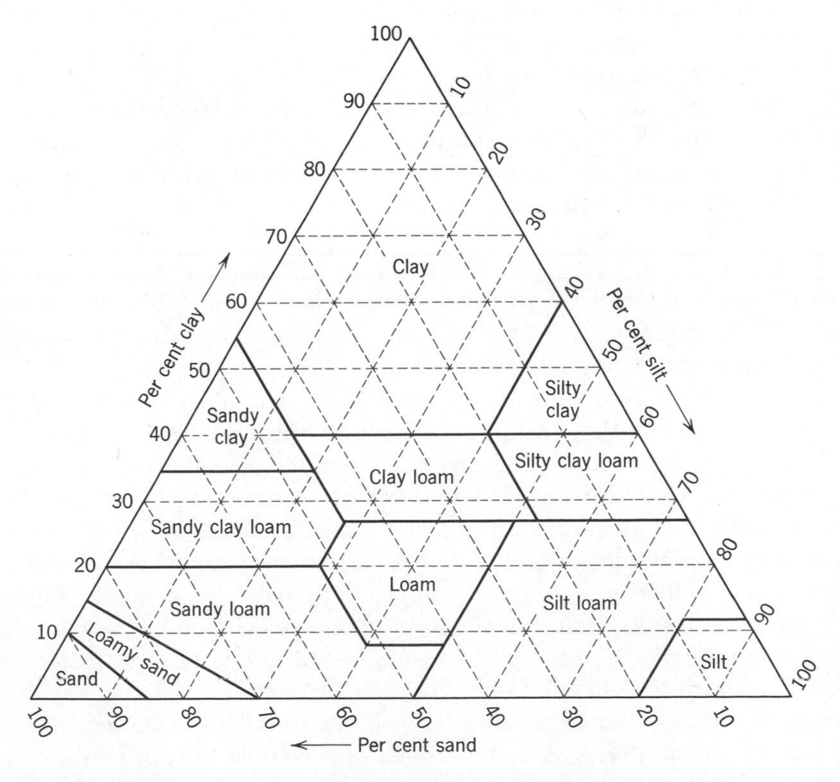

Figure 1.2. Guide for textural classification of soils. (Soil Survey Staff, 1960)

1.2, the characteristics of clay are dominant over those of sand and silt in determining the properties of the soil as a whole.

Soil texture as such is not of direct importance for most purposes. Where use is made of classification of soils according to texture, one is usually interested in the information about other properties that may be provided by knowledge of texture together with supplementary information on the relationships between texture and the properties of interest. Two philosophical aspects of this situation will be examined in turn: (1) Does texture provide a suitable basis for making inferences about other properties? (2) Do the definitions of textural classes correspond to the variation of other soil properties with particle-size composition?

Experience has shown that many important soil properties vary with texture. Nevertheless, none of the relationships applies under all circumstances. Consequently, texture does not provide an entirely suitable basis for making inferences about other properties. Two general causes of this ambiguity may be mentioned. One is that soil particles of a given size are not necessarily similar in other respects. Experiments by Odell et al. (1960) illustrate the problems in inference that arise in this way. These investigators found that they needed to know not only the percentage content of clay but also the kind of clay to estimate the plasticity of soils of Illinois. For their purpose, the montmorillonite content of the clay was important. Montmorillonite is a mineral with marked properties of plasticity, and the proportion of the clay composed of this mineral differed among the soils they examined. A second cause of ambiguity in inferences based on soil texture is that the relationship in question may depend on other conditions that are not inherent in the particles. Illustrations will be given later in connection with the relation of soil texture to plant growth.

Because there is no theoretical way of determining how classes of soils on a particle-size basis correspond to those on other bases, the only guide is experience. The system of textural classification currently used by the United States Department of Agriculture (Fig. 1.2) has not always had its present form, and other systems are now in use. No comprehensive investigation of the degree to which these classifications reflect other soil properties appears to have been made. Figure 1.3 gives the results of an investigation of the association of the particle-size composition of soils with one other soil property, the cation-exchange capacity. Although the particle-size composition of the soils evidently could be used to provide information about the cation-exchange capacity, the particle-size compositions classified as textural groups in Fig. 1.2 do not correspond to the particle-size compositions classified

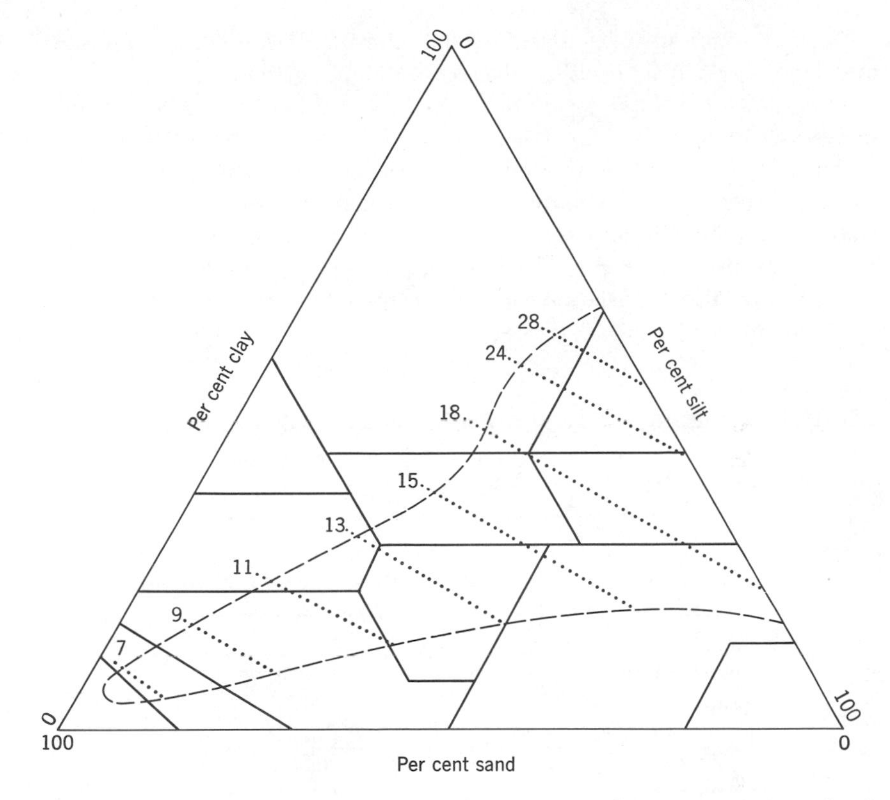

Figure 1.3. Cation-exchange capacity in relation to particle-size composition of soils from the lower Mesopotamian Plain in Iraq. The areas delineated by solid lines are named textural classes. The area delineated by a broken line includes the particle-size analysis of 98% of all the samples analyzed. The dotted lines indicate approximately the particle-size analysis associated with equal cation-exchange capacities, and the numbers give the cation-exchange capacities in milligram equivalents per 100 g of soil. (Delver, 1962)

into groups on the basis of cation-exchange capacity in Fig. 1.3. Franzmeier et al. (1960) made a similar investigation of the relationship of the particle-size composition of soils of Michigan to their readily available water capacity. Results obtained by these investigators do not agree with the classification of soils into textural groups in Fig. 1.2, nor do they agree with the trends in cation-exchange capacity in Fig. 1.3. Probably the most appropriate classification of soils into textural classes would be somewhat different for use in making inferences about each property, and the most appropriate classification might well vary with the nature of the particle-size fractions. Thus the classification in Fig.

1.2 could apparently be improved by altering it for certain properties and certain groups of soils; however, not enough is known to indicate whether any particular alteration would represent an improvement for conditions in general.

That plant growth is related to the particle-size composition of soils has been recognized for many years. A bulletin on the subject of adaptation of crops to different soil textures was published over half a century ago by Whitney (1896). Whitney accounted for the observed effects in terms of the association of soil texture with the supply of water available to plants.

Table 1.1. Available Water Capacity of Soils of Different Textural Classes in Tennessee (Longwell, Parks, and Springer, 1963)

Textural Class	Available Water Capacity,[1] cm of Water/cm of Soil Depth
Sand	0.015
Loamy sand	0.074
Sandy loam	0.121
Fine sandy loam	0.171
Very fine sandy loam	0.257
Loam	0.191
Silt loam	0.234
Silt	0.256
Sandy clay loam	0.209
Silty clay loam	0.204
Sandy clay	0.085
Silty clay	0.180
Clay	0.156

[1] As estimated by the difference between the quantities of water held at matric suctions of 0.34 and 15.2 bars. (See Chapter 2 on soil water for an explanation of the significance of this terminology for describing the condition of water in soil.)

One of the most important ways in which soil texture affects plant growth is through the influence of texture on water supply. The supply of water to plants usually is greater in soils of moderately fine texture than in those of coarse texture in humid regions such as the eastern part of the United States, where Whitney's observations were made.

The reason for this is that the capacity to hold water in a condition available to plants is greater in soils of moderately fine texture than in soils of coarse texture (Table 1.1), and usually the rainfall is sufficient to fill soils of all textures to capacity at intervals while plants are growing. Under these conditions, the amount of water available to plants increases with the capacity of the soils to hold water in available form.

Where the rainfall is not sufficient to fill soils of all textures to capacity, however, the amount of water available to plants may be determined not by the capacity of the soils but by other properties such as the rate of infiltration and the rate of evaporation. Soils of coarse texture usually are superior to those of fine texture in these respects. Thus, whereas soils of coarse texture are considered to be "drouthy" in humid regions, soils of fine texture are drouthy in dry regions. As an example of the lesser known condition in dry regions, Yankovitch and Berthelot (1948) observed that in Tunisia, where the annual rainfall is 20 to 30 cm, olive trees grow best on sandy soils. The yields decrease if the clay content of the soil exceeds 8 to 10%, and the crop fails completely if the clay content of the soil is as high as 25%. These observations illustrate how conditions other than the nature of the soil particles may affect the comparative properties of soils differing in texture.

A second important soil condition that varies with the texture is the nitrogen supply. Under similar environmental conditions, the availability of soil nitrogen to plants usually increases as the texture becomes finer. An example is shown in Fig. 1.4, which represents the yields of hemp

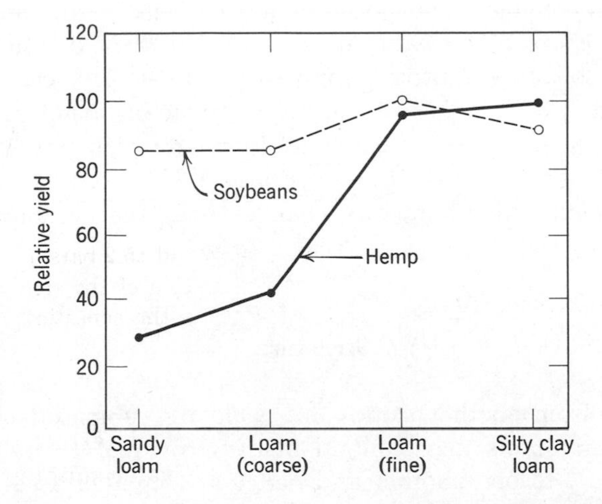

Figure 1.4. Yields of hemp and soybean on soils with increasing fineness of texture in the same field in Iowa. To facilitate the comparison of crops, the highest yield of each crop is given a value of 100. (Wilsie, Black, and Aandahl, 1944)

and soybeans on soils differing in texture. These two crops were grown side by side on four soil types in the same field in Iowa in a season in which the rainfall was ample for plants on soils of all textures. With increasing fineness of soil texture, there was a pronounced increase in the yield of hemp but not of soybeans. The similarity of yields of soybeans on the soils of different texture indicates that the soils were equally suitable for this crop, which is essentially independent of the supply of soil nitrogen if it is grown with root nodules effective in fixing elemental nitrogen from the atmosphere. Nitrogen deficiency symptoms were not evident in the soybeans but were pronounced in hemp grown on the soil of coarsest texture and became progressively less evident as the texture became finer. The evidence thus points to differences in supply of soil nitrogen as the principal cause of the marked increase in yield of hemp as the texture became finer.

The relationship of nitrogen availability to soil texture varies with the conditions. Under some circumstances of poor drainage, the nitrogen availability may even decrease as the texture becomes finer. This matter will be considered further in the chapter on nitrogen.

Water availability and nitrogen availability, which have been mentioned here, are perhaps of greatest general importance among the soil properties through which soil texture indirectly affects plants. Many other soil properties are related to soil texture, however, and in some situations the ones mentioned may not be of greatest importance. Many of the indirect effects will become apparent as the various properties are considered.

From the foregoing discussion it should be evident that an evaluation of the particle-size composition of soils provides some insight into the character of soil as a substrate for plant growth. The characterization thus obtained is not of general utility for predicting plant growth, however, because none of the fundamental factors of plant growth are measured directly and because the relationship between soil texture and the various plant growth factors changes with the circumstances that prevail.

Structure

As viewed from another standpoint, soils are composed of structural units. Structural units may contain many individual particles, but the particles adhere to one another and hence act together as a group. The over-all structure of a soil is a resultant of the arrangement and bonding of individual soil particles into structural units and the subsequent arrangement and bonding of these units. The individual particles in a structural unit are attached more strongly to each other than to adjacent particles, which may or may not be part of another structural unit. Grouping of particles into structural units occurs in all soils; however, the strength of the bonds, the size and shape of the units, and the proportion of the soil particles involved in the units differ considerably among soils.

Soil structure does not supply any of the factors essential to plant growth. Nevertheless, it may modify plant growth in various ways, and so an appreciation of the significance of soil structure is essential in the study of soil-plant relationships. The significance of soil structure and the extent of knowledge now available about it justify treatment of soil structure as a separate chapter. The subject is discussed in this chapter because the concept of soil structure includes the whole soil, and evaluation of soil structure is one way of characterizing soil as a whole.

Formation

Two distinct aspects of the process of structure formation may be recognized. The first, the development of interparticle bonds, confers the stability; the second, the separation of structural units from one another, confers the size and shape characteristic of individual units.

Interparticle bonds develop from both short-term and long-term processes. The attraction that exists between clay particles and between clay particles and larger mineral grains can cause marked bonding within a short time. Most clay particles have a platelike form. If a dilute, well-dispersed suspension of clay is poured into a bed of sand and then allowed to dry, examination of the resulting mixture shows that the clay has become oriented with planar surfaces together, forming coatings on the sand grains and bridges at points of contact (Brewer and Haldane, 1957). Bonding of large particles by oriented clay in soil is illustrated in Fig. 1.5. The planar surfaces of the clay attract each other, and the surface tension of the water aids in flattening the particles together as drying proceeds. Freezing may have an effect similar to drying because, if freezing occurs slowly, ice crystals form in the larger pores at the expense of the water in the surrounding region.

The nature of the attractive forces is not definitely known. Possibilities that may be suggested, however, include the attraction of adjacent surfaces for hydrogen (so-called hydrogen bonding) and other cations and for water molecules lying between them. There is evidence that surface forces produce orientation of water molecules in the vicinity of clay particles (Grim, 1958; Low, 1961). Surfaces of both sand and clay parti-

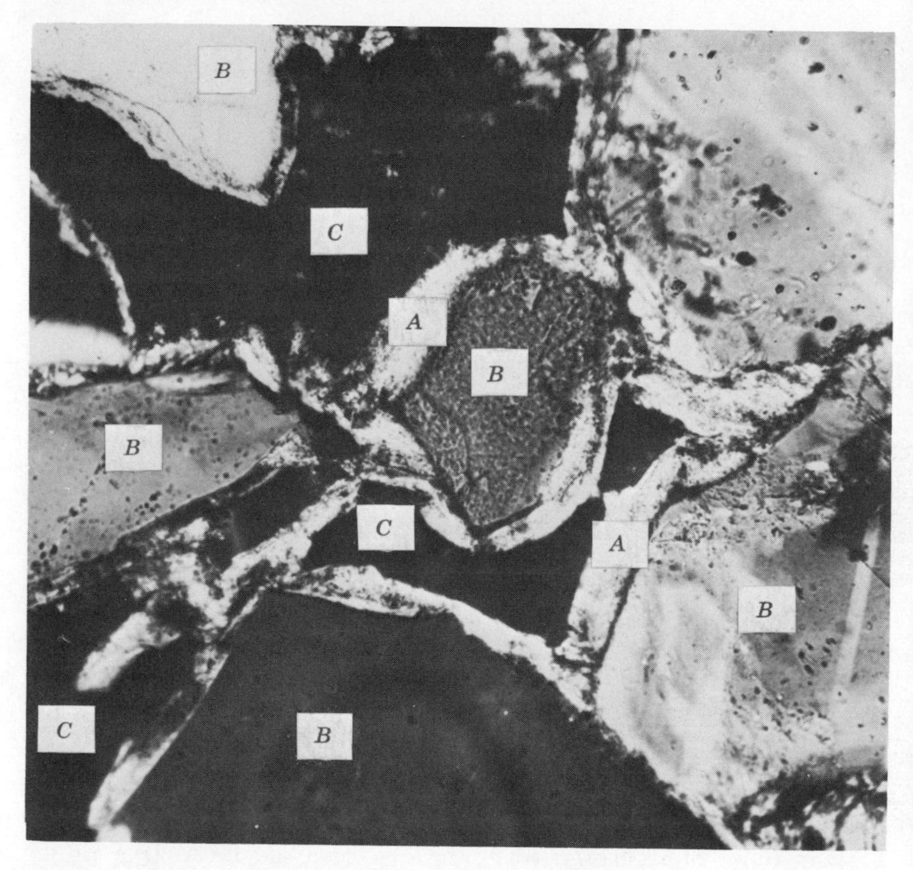

Figure 1.5. Photomicrograph of a thin section of a sandy soil showing how clay may accumulate in a sand as coatings on the grains of sand and as bridges between the grains. *A* = clay coating, *B* = sand grain, *C* = pore. (Photomicrograph courtesy of John G. Cady; Soil Survey Staff, 1960)

cles are mostly oxygen atoms with some hydroxyls, and hence hydrogen bonding and mutual attraction for intermediate molecules of water may also be responsible for bonding between the sand and clay particles. The strength with which adjacent sand particles are bound together by intervening oriented clay should increase with the content of clay as a result of the increased area of contact. Structure formation in this way appears to vary with the nature of the clay. Hagin and Bodman (1954) found that water-stable aggregates were formed by drying mixtures of sand with montmorillonite but not with kaolinite. Hydrogen bonding should be operative with kaolinite, but the attraction of surfaces due to other intermediate cations should be relatively weak because of the low cation-exchange capacity of kaolinite.

Deposition of organic matter strengthens the bonds produced among larger particles by the clay alone. Statistical computations made by Heinonen (1955) on results of analyses of many soils of Finland indicate that the percentage of the soil mass occurring in water-stable aggregates is influenced much more strongly by unit mass of soil organic matter than of clay.

Peele (1940) found that marked aggregation of soil particles into water-stable units occurred where sterile soil was treated with sucrose and inoculated with a soil suspension but not where the inoculation was omitted, thus demonstrating the importance of microbial action in producing the stability. Microbial activity appears to affect structural stability in at least three different ways. First, filaments of microbial tissue may form a network within and around groups of mineral particles. Bond (1959) examined a number of sandy soils of Australia and found that many grains were held together in this way. Figure 1.6 illustrates his observations. Second, certain soil organisms produce polysaccharides that have a mucilaginous nature and may cement mineral particles together. The composition of the polysaccharide material extracted from soil and purified by chemical methods (Whistler and Kirby, 1956) indicates that it is a mixture of polysaccharides of microbial origin.

Figure 1.6. Photomicrograph showing network of filaments of microbial tissue among sand grains. (Photomicrograph courtesy of Roy D. Bond)

Rennie et al. (1954) found that addition of only 0.02 g of the extracted and purified polysaccharide material to 100 g of soil increased the water-stable aggregates >0.1 mm in diameter from 44 g to 60 g; and statistical computations made by Chesters et al. (1957) on analyses of a large number of soils indicated that, per unit mass, the polysaccharide fraction was more effective in stabilizing the mineral particles into structural units than was the nonpolysaccharide portion of the soil organic matter. Third, microorganisms produce nonpolysaccharide organic residues of various kinds that may aid in binding mineral particles. Swaby (1950) found that treatment of extracted soil organic matter with hypoiodite, which is supposed to oxidize lignin-like compounds but not polysaccharides, decreased the stabilizing effect of the extract on subsequent addition to a soil. Experiments by Mehta et al. (1960) and Greenland et al. (1962) with chemical degradation techniques specific to polysaccharides on the one hand or to lignin and "humic" substances on the other suggest that, in some soils, the greater contribution to the stability of soil aggregates is made by the polysaccharides; and, in other soils, the greater contribution is made by lignin and humic substances.

Evans and Russell (1959) found that organic matter extracted from soil was adsorbed by clay; and their evidence indicated that the principal cause of the adsorption was interaction of the organic matter, which has cation-exchange properties, with the exchangeable aluminum and calcium present in the clay. Other evidence of chemical interaction is provided by their finding that mixtures of clay and soil organic matter had a lower cation-exchange capacity than that calculated for the sum of the individual components. They pointed out, however, that an ionic reaction between the organic matter and clay is not necessarily the sole cause of the reduction in exchange capacity because a reduction also occurs upon addition of starch, which is not ionic. Hindrance of ionic exchange by positional effects is thus a possibility. Further evidence of chemical interaction is provided by Swaby's (1950) work showing that the tendency of extracted soil organic matter to produce water-stable aggregates when added to soil was reduced markedly by certain chemical treatments designed to inactivate different kinds of exchange sites in the organic matter.

Edwards and Bremner (1967) proposed the theory that soil aggregation is primarily a consequence of chemical interaction among organic and inorganic particles. According to their theory, the basic structural units in soil are aggregates of silt and fine sand size, consisting largely of complexes of clay and organic matter bound together by polyvalent cations. The primary bonds are electrostatic, the elementary unit is clay-polyvalent cation-organic matter, and a number of elementary units may be bound together to form a single aggregate. This theory is related to both the second and third effects of microorganisms just mentioned. In support of the theory, they found that soil particles could be dispersed and would remain dispersed, even with the polyvalent cations present, if the soil was merely shaken a long enough time with water or if it was subjected to sonic or ultrasonic vibration. Energy supplied in these treatments should break the bonds in the elementary units and larger aggregates. Moreover, they found that dispersion of soil particles was promoted by addition of reagents that removed polyvalent cations from soil by precipitation or chelation and not by those that did not, as would be expected if the bonding occurred through such cations.

Effects of organic matter are partly short-term and partly long-term in nature. Stabilization of mineral particles by microbial filaments and mucilaginous polysaccharides may occur within a period of a few days in the presence of an ample supply of decomposable organic material. Long-continued stabilization by these materials probably requires continuous production because they are not highly resistant to decay. Martin et al. (1965) found that 8 to 80% of the carbon added as bacterial polysaccharides and cells to different soils was evolved as carbon dioxide in 8 weeks. The major part of the organic matter in soils, however, has been accumulated over a period of many years and has a half-life of many years. Stabilization by this portion of the organic matter is thus a long-term process.

Deposition of secondary minerals from solution on and among other particles may play a significant part in binding the particles together. Oxides and hydrous oxides of iron are thought to be the most important. These minerals usually accumulate in soils as weathering proceeds. Sometimes portions of the soil (particularly the parts some distance below the surface) may be impregnated with iron oxides and hydrous oxides to form aggregates that resist the usual chemical and physical treatments employed to produce dispersion of mineral particles preparatory to particle-size analysis. Aggregates of this sort up to a diameter of a few millimeters are usually called concretions; other terms may be used, however, particularly if the aggregates are large (Sherman, 1959). Considerable manganese oxide may accompany the iron oxide in concretions. In extreme cases, large masses of soil material may become impregnated to form what is known as laterite (Sivarajasingham et al., 1962). Considerable aluminum oxide may accompany the iron oxide in laterite. Oxides and hydrous oxides of iron are thought to be of major importance in binding soil particles in certain soils of warm, humid regions and in the B horizon of soils developed under coniferous forest vegetation in cool-temperate regions. Lutz (1937), Weldon and Hide (1942), Kroth and Page (1947), McIntyre (1956), and Walter (1965) published evidence supporting this view. Figure 1.7 shows an example.

Walter's (1965) work on formation of aggregates from dispersed silt and clay in the laboratory emphasized the importance of aging and the interaction of hydrous iron oxide with organic matter in formation of stable aggregates. On the other hand, Deshpande et al. (1964) found that removal of iron oxides by selective chemical treatment did not reduce the stability of aggregates from some soils high in iron oxide. The swelling of sodium-saturated aggregates from other soils was decreased by precipitation of aluminum hydroxide in their presence, which suggests the possible significance of bonding by aluminum; precipitation of ferric hydroxide did not reduce the swelling.

Calcium carbonate also may act as a binding agent if precipitated

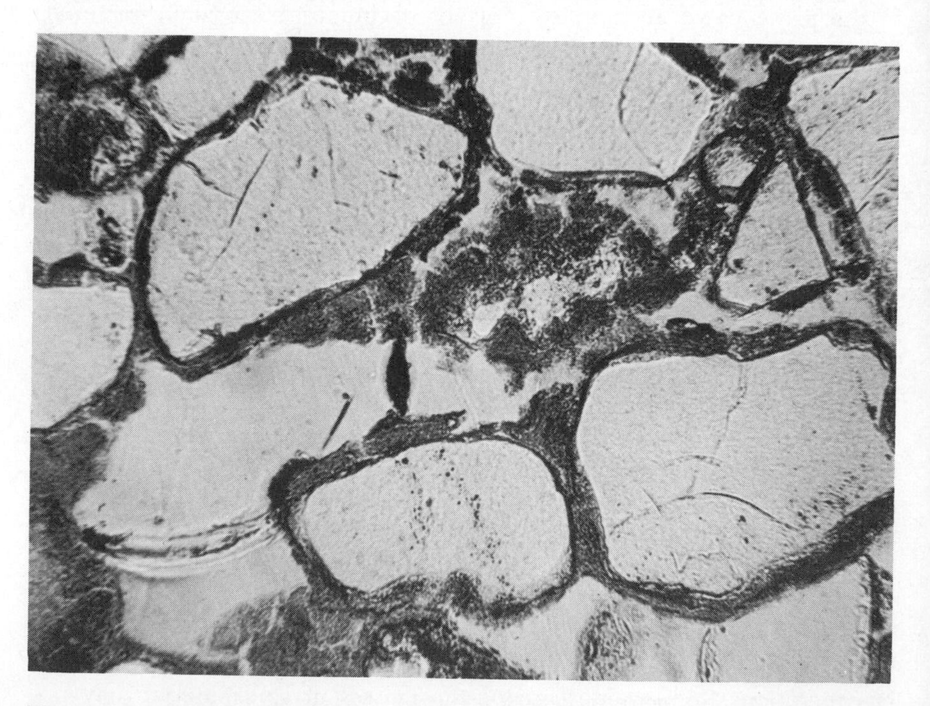

Figure 1.7. Photomicrograph of a thin section of the B horizon of a sandy soil developed under coniferous forest vegetation under cool, humid conditions. Bridges between mineral grains, formed by connecting sesquioxide-organic matter coatings, are responsible for the hardness of the horizon. Cracked coatings (upper part of micrograph) seem to be the first phase in the disintegration of the hard structure. (Photomicrograph courtesy of Klaus Flach; Soil Survey Staff, 1960)

from solution in and around soil aggregates. Concretions high in content of calcium carbonate occur in some soils. An extreme condition, in which an entire layer of soil is impregnated by precipitated calcium carbonate, is found in certain dry regions (Stuart et al., 1961). Stabilization by calcium carbonate occurs most frequently in layers of soil below the surface. An example is shown in Fig. 1.8.

Silica appears to be of importance in some instances. Breese (1960) observed bridges of secondary silica between quartz particles in a deep layer of a soil from New South Wales, Australia. Knox (1957) obtained evidence that colloidal silica is partly responsible for cementation of a hard layer found deep in some soils of New York. Silicates and amorphous silica may occur together with carbonates in "caliche" layers sometimes found in soils of dry regions (Shreve and Mallory, 1933; Brown, 1956).

Development of intergrowths by deposition of secondary minerals

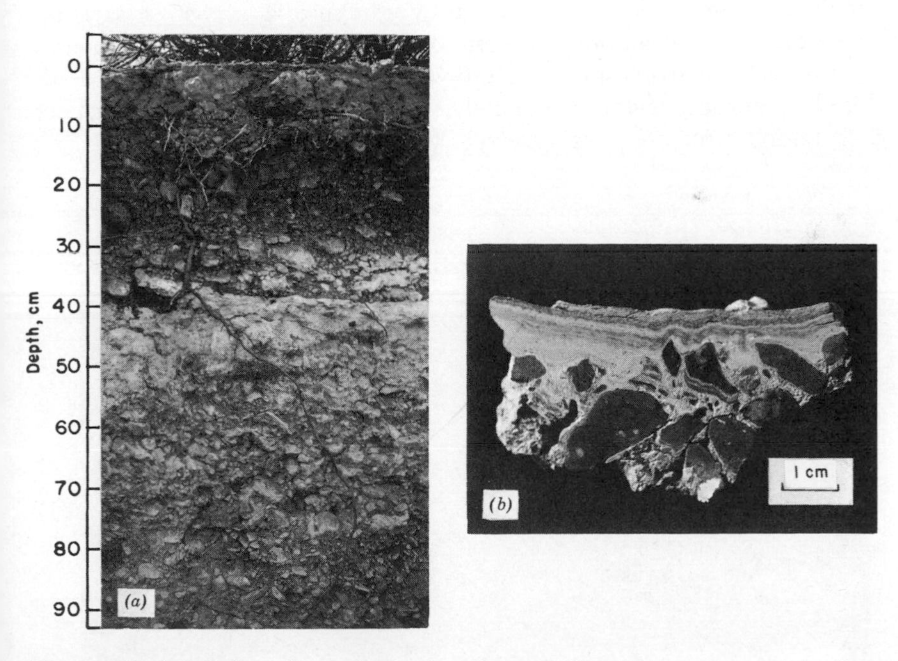

Figure 1.8. (*a*) Soil profile in Doña Ana County, New Mexico, containing a layer indurated with calcium carbonate, and (*b*) enlarged view of a vertical cross section of the top of the indurated layer at a depth of about 38 cm. The calcium carbonate content of the soil ranges from less than 1% at the surface to 47.5% in the upper part of the indurated layer. The rhyolite gravel from which the soil has developed contains less than 0.4% calcium; hence much of the calcium carbonate must have originated elsewhere. (Photographs and data courtesy of R. V. Ruhe)

from solution is a slow process, requiring many years. The occurrences of these intergrowths suggest that a small quantity of the secondary mineral produces the greatest stabilization in locations where the position of the particles with respect to each other remains fixed over long periods of time. Alternate swelling and shrinking associated with wetting and drying of clays and dislocations due to freezing and thawing or other surface disturbances do not seem to favor stabilization of aggregates of soil particles by intergrowths of secondary minerals. Thus, in contrast to the favorable effect of clay in bonding of soil particles on a short-term basis, the direct effect of clay in the long-term development of intergrowths appears to be unfavorable. The more important role of clay in the long-term process is probably that of supplying by dissolution some of the soluble components that are subsequently redeposited as intergrowths.

The first aspect of structure formation, binding soil particles into aggregates, would probably be undesirable from the standpoint of plant growth if it were not accompanied by the second aspect, separation of structural units from one another. If all the soil particles were bonded to form a single structural unit, growth of higher plants would be inhibited if not prevented completely. Nevertheless, the second aspect of structure formation has received less attention than the first.

Figure 1.9. Cracks in dried bed of intermittent lake in Doña Ana County, New Mexico. (Photograph courtesy of R. V. Ruhe)

Disruption of a solid soil mass into smaller parts occurs, it appears, because the parts representing the structural units are pulled apart by internal forces or pushed apart by external forces or because material is removed between them. Once a separation has occurred, the original zone of weakness tends to be perpetuated despite subsequent consolidation.

The shrinkage of soil associated with drying (Fig. 1.9) is one way in which structural units may be formed by forces acting internally. Buehrer and Deming (1961) implicated this mechanism in development of a condition of relatively low compaction and high permeability to water and in production of good growth of plants in certain soils of Arizona. Their investigation showed that formation of shrinkage cracks on drying was more extensive in soils high in montmorillonite clay than in illite clay. Analogous behavior was demonstrated with montmorillonite and illite.

Examples of forces acting between structural units are freezing and growth of roots. The effect of freezing perhaps is illustrated most spectacularly by the polygonal structures produced by development of

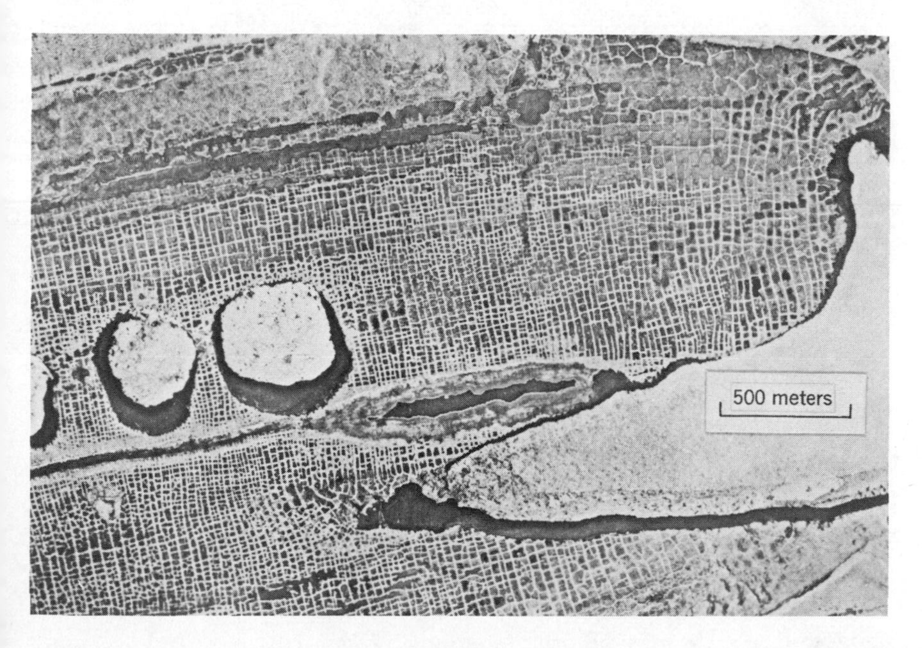

Figure 1.10. Aerial view of a rectangular pattern of soil polygons in northern Alaska. The light lines denote the location of the tops of ice wedges illustrated in Fig. 1.11. The circular objects on the left and the pointed object on the right are lakes containing ice. (Lachenbruch, 1963)

Figure 1.11. Ice wedge in perennially frozen sediments near Livengood, Alaska. (Péwé, 1963)

wedges of ice in soils of arctic regions (Figs. 1.10 and 1.11). These structures may be 20 meters or more in diameter, and the wedges of ice bounding them may have a thickness up to a meter or more. Another factor of significance is the disruption of masses of soil that occurs upon rapid wetting. The wetted soil swells and may fracture because of stresses along the wetting front. Moreover, water may seal the outer pores in masses of soil and may cause them to fracture as a result of compression of air in the inner pores as the water is drawn inward by capillarity. Thus the proportion of soil present in water-stable aggregates of a given size is found to be greater where the soil is wetted under a vacuum with de-aerated water than where it is wetted at atmospheric pressure (Kemper and Chepil, 1965).

Disruption of soil to form structural units by removal of material may occur as a result of burrowing action of animals. Earthworms are thought to be of significance in this connection (Barley, 1961). In addi-

Figure 1.12. Platelike structure in a soil developed under deciduous forest vegetation in Iowa. The view shown is a vertical cross section of part of the A2 horizon. (Photograph courtesy of Roy W. Simonson)

Figure 1.13. Exposed tops of columnar structural units in the B horizon of a sandy loam soil in New South Wales, Australia. The deposit of silica has been brushed out from between the tops of the columns. The A horizon at this location is about 46 cm in thickness. (Hallsworth and Waring, 1964)

tion to making channels through the soil, earthworms pass soil through their bodies and form it into casts, some of which are deposited on the surface of the soil. The stability of the casts as structural units may exceed that of the structural units formed in other ways in the soil. Measurements at different locations in Europe suggest a figure of

Figure 1.14. Angular blocky structure in a soil developed under grass vegetation in Iowa. The view shown is a vertical cross section of part of the B horizon.

3 or 4 kg of soil per square meter per year as the rate of deposition of worm casts on the surface. Removal of material may occur also as a result of weathering and solution. This process forms pores in laterite, for example (Sivarajasingham et al., 1962); and these pores are no doubt of significance in formation of aggregates if the laterite later disintegrates.

More extensive discussions and reviews of literature on structure formation may be found in publications by Stallings (1952), Martin et al. (1955), Baver (1956), Jacks (1963), and Griffiths (1965).

Characterization

Qualitative. Structural units of soils exist in a variety of shapes and sizes. Much can be determined about the structure of a soil from visual examination, and qualitative descriptions of structure form an important part of soil evaluation. A detailed qualitative classification of structural units was published by the Soil Survey Staff (1960).

Platelike structural units often are developed in the A2 horizon of uncultivated soils developed under deciduous forest vegetation in humid, temperate climates. Figure 1.12 shows an example. Prismlike units commonly are developed in the B horizon of soils under grass vegetation in temperate to cool semiarid regions. The tops of the prisms may become rounded in time, and the structural units are then said to be columnar (Fig. 1.13). Blocklike units often are developed in the B horizon of soils of fine texture developed under both grass and deciduous forest vegetation in humid, temperate climates (Fig. 1.14). Spheroidal

Figure 1.15. Granular structure in the A horizon of a soil developed under grass vegetation in Iowa.

units often are developed in the A horizon of uncultivated soils formed under grass vegetation in subhumid, temperate regions, as seen in Fig. 1.15.

Quantitative. A complete quantitative characterization of soil structure would include evaluation of the shapes and sizes of the structural units, the strength of the interparticle bonds within and among structural units, and the size-distribution and continuity of pore spaces within and among units. The structure of a soil as a whole represents a summation of these characteristics in the soil profile. Soil structure is a complex phenomenon that cannot be characterized adequately by a single physical measurement or by a single number derived from several measurements. Quantitative methods presently used evaluate only a portion of the overall phenomenon.

Probably the most widely used measurement of soil structure is that obtained by the wet-sieving or aggregate-analysis method (Kemper and Chepil, 1965). In this method, a sample of soil is placed on the uppermost member of a nest of sieves with successively smaller openings. By a mechanical device, the sieves are gently raised and lowered in water for an arbitrary time to cause separation of the different size fractions of aggregates. The aggregates left on the individual sieves are then dried and weighed. The size fractions reported often include aggregates with diameters of 5 to 2 mm, 2 to 1 mm, 1 to 0.2 mm, and <0.2 mm.

To provide a single index number for soil structure that will facilitate comparisons of structure with measurements of other soil properties or of plant growth, the results of wet-sieving frequently are expressed in terms of the percentage of the soil mass in size fractions larger than, say, 0.2 mm in diameter. This technique involves the assumption that equal masses of aggregates of any size larger than the arbitrary size chosen contribute equally to the effect that structure may have on soil behavior. Because large aggregates (up to 1 or 2 mm in diameter) are usually considered to be more favorable than small ones for most agricultural purposes, methods sometimes are employed in which greater weight is arbitrarily given to the larger aggregates (e.g., the mean weight-diameter method described by Van Bavel, 1950). Mazurak (1950) proposed the use of the geometric-mean diameter as a single-value index of soil structure. This parameter is the aggregate diameter above and below which 50% of the soil weight occurs. Gardner (1956) found that most aggregate size distributions can be represented to a good approximation by the geometric-mean diameter and the logarithmic standard deviation. The logarithmic standard deviation is the ratio (aggregate diameter at 84% oversize)/(aggregate diameter at 50% oversize).

A simple test to determine whether data may be characterized in this way may be made by plotting aggregate diameters on a logarithmic scale against percentage oversize on a normal probability scale on graph paper available for the purpose. In the example in Fig. 1.16, the geometric-mean aggregate diameter and the logarithmic standard deviation are

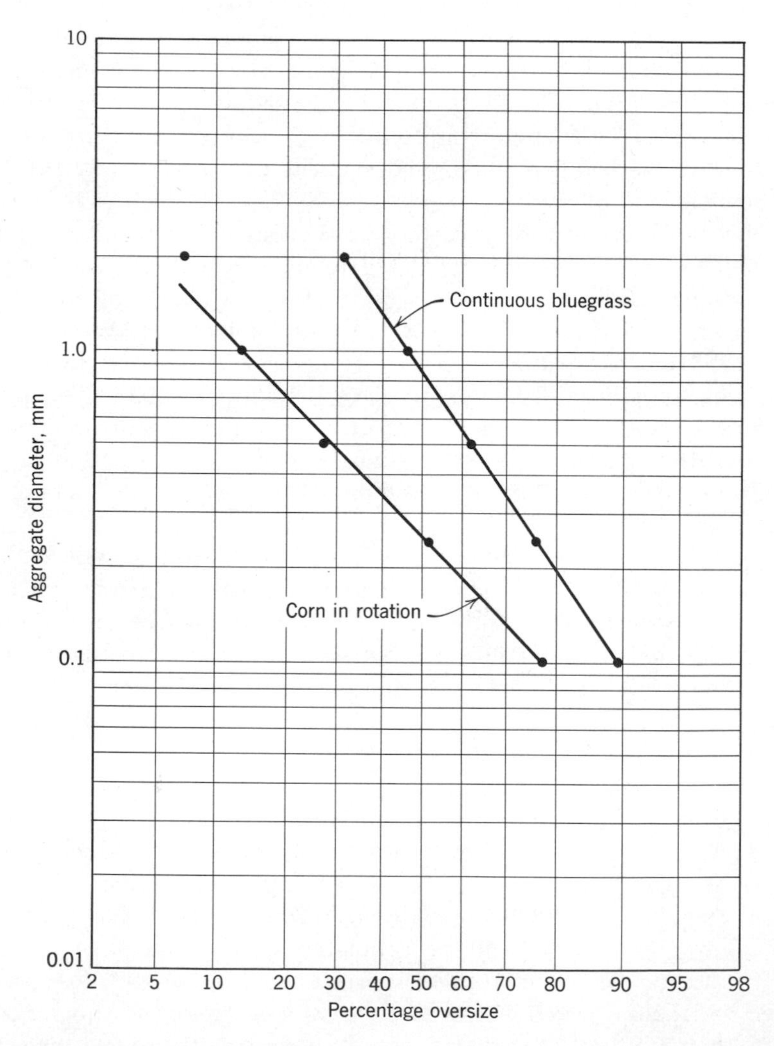

Figure 1.16. Plot of aggregate diameter on a logarithmic scale against percentage oversize on a probability scale. The aggregate distributions were found in samples of soil from continuous bluegrass and from the corn phase of a 3-year rotation on a loam soil in Iowa. (Gardner, 1956)

about 0.25 mm and 0.3 in soil under corn and about 0.85 mm and 0.2 in soil under bluegrass.

The wet-sieving or aggregate-analysis method evaluates, in part, the size-distribution of the structural units (and perhaps also the size-distribution of relatively coarse individual particles if no distinction between particles and aggregates is made by a subsequent particle-size analysis of the size-fractions of aggregates) and, in part, the stability of the units under the treatments employed. As pointed out by Kemper and Chepil (1965), the results depend on the procedure employed; and one must consider the purpose for which a particular set of measurements is to be made in selecting the most suitable procedure.

The wet-sieving method is only one of many that have been employed to give information on soil structure. Reviews by Low (1954), Cornfield (1955), and Vomocil (1957) provide a classification of methods as well as information about the methods and citations to original literature.

Effect of Soil Treatments

In general, the structure of soils deteriorates when the native vegetation is destroyed and the soils are cultivated. The results of aggregate analyses by Sokolovsky (1933), Paschall et al. (1935), Puhr and Olson (1937), and others may be cited in this connection. Although the deterioration is associated with a decrease in content of soil organic matter, this probably is not the sole cause. The passage of implements through and over the soil and the impact of raindrops on the bare soil no doubt are partly responsible.

Where comparisons of structure are made on soil growing different crops, the usual observation is that the bonding of mineral particles into water-stable aggregates is greater under close-growing, noncultivated crops than under row crops. Table 1.2 gives an example. The results of various investigations summarized by Mazurak and Ramig (1963) suggest that the increase in water-stable aggregates that occurs in soil under a temporary stand of grass may remain in evidence as long as 3 years under humid conditions and 6 years under relatively dry conditions.

Applications of organic matter, such as crop residues and manure, increase the bonding of mineral particles into water-stable aggregates. Synthetic organic soil-aggregating agents have a much more pronounced effect per unit mass than do the foregoing materials. The synthetic organic soil-aggregating agents are long-chain molecules that apparently combine chemically with mineral particles and form bridges among them. Cost is undoubtedly the major deterrent to widespread use of these substances. Application of certain inorganic substances, such as silicones and sodium and potassium silicates, also results in marked bonding of mineral particles into water-stable aggregates. These substances have not found practical application for agricultural purposes because of both possible deleterious effects on plants and cost.

Table 1.2. Aggregate Analysis of a Silt Loam Soil under Different Crops in Iowa (Browning et al., 1948)

Crop	Percentage of Soil in Aggregates >0.25 mm Diameter
Continuous corn	33
Corn in corn-oats-clover rotation	42
Oats in corn-oats-clover rotation	51
Clover in corn-oats-clover rotation	57
Continuous alfalfa	60
Continuous bluegrass	62

Soil-Plant Relationships

Many experiments have been reported in which particular soil treatments or management practices have resulted in an improvement in soil structure and an increase in crop yield. Data of Rynasiewicz (1945) in Table 1.3 may be cited as an example. The correlation between onion yields and aggregates >0.5 mm in diameter is 0.99, which means that the onion yields increased in an almost exactly linear fashion with the aggregation.

Correlation is one form of evidence that is useful in establishing cause-and-effect relationships, but it is not sufficient evidence because variables can be related in a functional manner when one is not the cause and another the effect. In the example being considered here, there is reason to question whether the differences in aggregation were the cause of the differences in onion yields. The differences in both aggregation and onion yields were associated with differences in crop rotations, and it is well known that crop rotations may affect plant growth in a variety of ways other than the structural changes they bring about. Experiments such as this, in which modifications of soil structure are produced by treatments that may have independent effects on plants, are therefore not entirely suitable as a source of evidence for the importance of soil structure to plant growth. Most soil treatments or management practices that affect soil structure may affect plants in other ways as well. To use evidence from such experiments in relating soil structure to plant growth, supplementary information is needed on the various factors involved to aid in arriving at a proper appraisal of cause-and-effect relationships.

Table 1.3. Aggregation of a Very Fine Sandy Loam Soil and Yield of Onions Obtained with Different Crop Rotations in Rhode Island (Rynasiewicz, 1945)

Crop Rotation	Percentage of Soil in Aggregates >0.5 mm Diameter	Yield of Onions per Hectare, Metric Tons
Onions, mangels, mangels	22	9.4
Onions, buckwheat, buckwheat	23	13.1
Onions, corn, corn	26	14.9
Onions, redtop, redtop	37	29.2

In experimental work, it is frequently desirable to evaluate the significance of soil structure, not only for the sake of the knowledge itself but also because the most suitable management practices may well depend on the extent to which structure affects the growth of plants. With the advent of synthetic organic soil-aggregating agents, a means was provided by which such an evaluation could be made. According to present information, some of these chemicals are capable of making profound changes in aggregation with only small direct effects on the microbiological population and nutrient status of soils.

The synthetic organic soil-aggregating agents are long-chain polymers of different types. According to Emerson (1963), the repeating units in several such chemicals are as follows: polyvinyl alcohol, $—[CH_2—CH(OH)]—$; polyacrylic acid, $—[CH_2—CH(COOH)]—$; polyacrylamide, $—[CH(CONH_2)—CH_2]—$. Polyvinyl alcohol apparently is held to clays by hydrogen bonding through the —OH groups of the polymer and the oxygen atoms of the clay surfaces. Polymers with —COOH groups appear to be bound electrostatically by aluminum on the edges of clay particles and by exchangeable aluminum and calcium (presuably by magnesium also) on the basal surfaces. The =C=O group appears to be attached by hydrogen bonding (Kohl and Taylor, 1961). The strength of the cationic bonds may cause the cation-exchange capacity of a clay-polymer mixture to be less than the sum of the exchange capacities of the polymer and clay components individually (Archibald and Erickson, 1955). Polyacrylamide probably is bound to clays in a complex manner. The

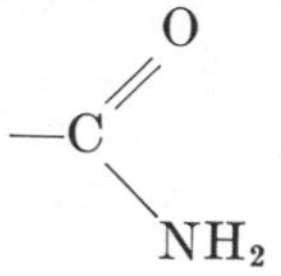

$$-\mathrm{C}\begin{matrix}\nearrow\!\!\!\!\nearrow \mathrm{O} \\ \searrow \mathrm{NH_2}\end{matrix}$$

group may take up a hydrogen ion to form the positively charged

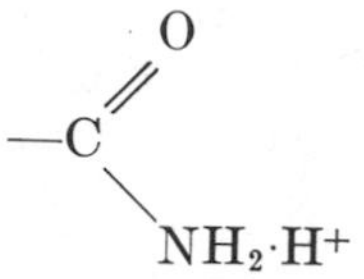

$$-\mathrm{C}\begin{matrix}\nearrow\!\!\!\!\nearrow \mathrm{O} \\ \searrow \mathrm{NH_2 \cdot H^+}\end{matrix}$$

group, which presumably could replace exchangeable cations; also, it may exist in the form

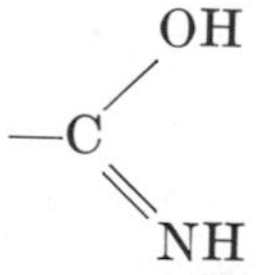

$$-\mathrm{C}\begin{matrix}\nearrow \mathrm{OH} \\ \searrow\!\!\!\!\searrow \mathrm{NH}\end{matrix}$$

which would be subject to hydrogen bonding to oxygen atoms of clay surfaces. Figure 1.17 shows an electron micrograph, illustrating the bonding of particles of kaolinite with strands of a copolymer of isobutylene and half-ammonium, half-amid salt of maleic acid. Greenland (1965) reviewed the evidence on this subject.

The most common use of synthetic organic soil-aggregating agents for experimental purposes has been qualitative testing to find whether soil structure limits plant growth in selected instances. If growth is improved upon addition of the chemical, structure is said to be limiting. Supplementary measurements are desirable to verify that structure has indeed been improved by the chemical treatments. Experiments of this type have produced various effects on crops. Increases in yield sometimes have been large, as in work by Alderfer (1954) in Pennsylvania, where lima beans yielded 3.8 metric tons per hectare on control plots and 7.0 metric tons on plots treated with a synthetic organic aggregating agent, an increase of 84%. In other instances, yields have been decreased. For example, Haise, Jensen, and Alessi (1955) in North Dakota obtained yields of 28.3 metric tons of sugar beets per hectare on control plots and 22.9 metric tons on plots treated with a synthetic organic aggregating agent, a decrease of 19%. Soil structure thus does not invariably limit crop yields. In experiments by Martin et al. (1952) on soils of fine

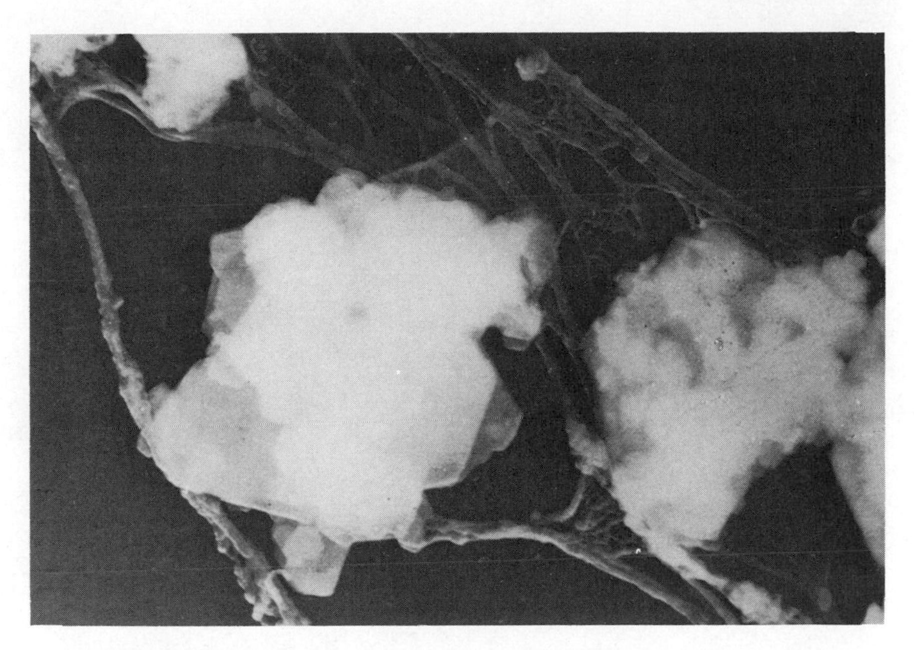

Figure 1.17. Electron micrograph of kaolinite treated with a synthetic soil-aggregating agent. The strands that appear to bond kaolinite particles to one another have been observed to do so upon rupture of the film of Formvar that supports the preparation. The diameter of the kaolinite particle just left of the center of the micrograph is about 1μ. (Micrograph by S. L. Rawlins; Kittrick, 1965)

texture with poor visual structure in Ohio, addition of a synthetic organic aggregating agent produced a marked increase in aggregation. In some instances crop yields were increased, but in others no increase was obtained. Evidently it cannot be assumed that crop yield is limited by soil structure simply because a soil appears to have poor structure.

Final yields of the harvested portion of crops do not provide complete information on plant responses to soil structure. In a number of instances [see the review by Quastel (1954)], investigators have reported favorable effects of soil-aggregating agents on growth early in the season but no increase in final yield. Boekel (1959; 1963; and private communication) described an experiment in the Netherlands in which peas were grown on a sandy clay soil. The results are shown in Fig. 1.18, where structure is represented in terms of the air content of the soil in May. Midway in the growing season, a visual index of the yield of peas increased with the air content of the soil throughout the range of measurements up to about 21%. The final yield of peas, however, increased with

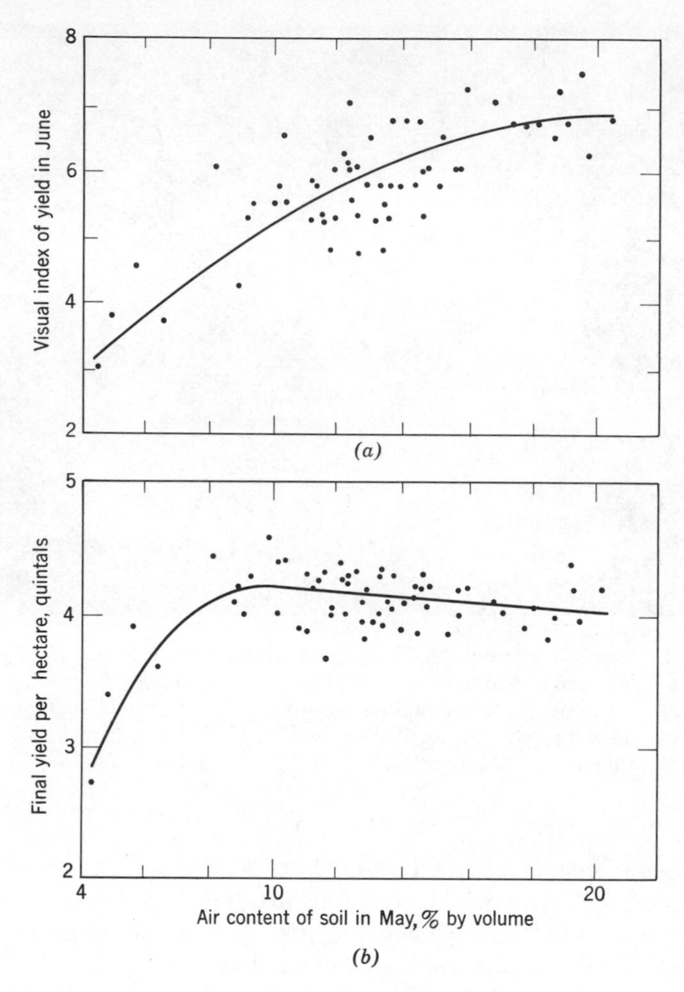

Figure 1.18. (*a*) Visual index of yield of peas in June and (*b*) final yield of peas versus air content of soil in May in an experiment in the Netherlands in which soil structure was varied by use of various synthetic organic soil-aggregating agents and a soil-structure-deteriorating agent. (Boekel, 1959)

the air content up to only 8 to 10% and decreased slightly at higher percentages. The author explained the results in the following way. The first part of the season was rainy, and the soil was continuously rather wet; consequently, the increase in growth with improved structure early in the season apparently represented a response to improved aeration. During the latter part of the season the weather was dry, and the soil was dried to the wilting point in the rather shallow zone where most of the roots occurred. The wilting point was reached sooner in the plots with good structure than in those with poor structure, apparently because of more rapid utilization of water by the larger plants on plots with good structure. Hence the slight decrease in final yield of peas growing on soil with the improved structure (air space in May greater than 8 to 10%) apparently was a result of water deficiency.

In addition to their use in determining whether plant growth or yield is limited by soil structure under a single set of circumstances, synthetic soil-aggregating agents may be used to provide similar information under each of several conditions that may have been produced experimentally by certain treatments. When used this way, synthetic soil-aggregating agents may provide an insight into the significance of structure differences associated with the treatments. Supplementary evidence on other factors may then provide information on the cause or causes of observed effects of the treatments.

Work of Clement (1961) may be cited as a first example. Clement conducted an experiment with synthetic soil-aggregating agents to determine the cause of the improvement in yield of wheat associated with increases in aggregation produced by different kinds of grass-legume vegetation in Great Britain. The experiment showed that, although soil aggregation was much increased by application of a synthetic soil-aggregating agent, the yield of wheat was not increased. Supplementary measurements were made on the quantities of mineral nitrogen produced in samples of soil taken following the various kinds of vegetation, where the samples were incubated in moist condition in a laboratory for 20 days at 30°C. The measurements on nitrogen were made to provide an index of the availability of nitrogen to wheat because differences in availability of nitrogen as well as differences in structure were known to result from the treatments. Statistical tests showed that the variations in yield of wheat following the different kinds of grass-legume vegetation were associated principally with the values for the index of nitrogen availability and not with the values for aggregation. The information derived from this work thus indicates that the differences in structure associated with the different kinds of grass-legume vegetation were incidental as far as their effect on the yield of the following crop of wheat was concerned.

Work by Boekel (1963 and private communication) may be described as a second example. Boekel reported that liming increased the soil pH and improved the structure in a long-term experiment in the Netherlands. Yields of crops were higher on limed plots than on unlimed plots. To determine whether the differences in yield were associated with differences in soil pH or structure, a synthetic soil-aggregating agent was applied to some of the unlimed plots to improve the structure; and a disaggregating agent was applied to some of the limed plots to break down the structure. A statistical analysis of the results showed that the differences in yield then obtained were associated primarily with structure and not with pH.

Once it has been decided that differences in plant growth observed in a particular situation are attributable primarily to differences in structure, the next question is the manner in which the effects on plant growth have been brought about. Soil structure may affect plant growth in a variety of ways, some direct and some indirect.

Direct Effects. Changes in soil structure may be said to have a direct effect on plant growth if differences in expansion of plant parts can be attributed to differences in mechanical impediment presented by the soils under consideration. Deciding whether these observed differences are attributable to differences in mechanical impediment is seldom easy in practice. Nevertheless, a few instances can be cited in which direct effects undoubtedly exist and apparently are primarily responsible for the effects produced.

One instance of direct effects of structure has to do with emergence of seedlings in crusted soils. There is little doubt that the bending of seedlings that sometimes occurs where they reach a dry crust from beneath is a direct result of the mechanical impediment presented by the crust. Several different types of behavior were recognized by Arndt (1965), who analyzed the phenomenon of emergence of seedlings in relation to crusting of soils. Work with synthetic organic soil-aggregating agents (Quastel, 1954) suggests that one of the major benefits from application of such substances to soils that produce surface crusts is

Figure 1.19. Stand of corn on high-sodium soil subject to crusting. Plots in the foreground are controls. Those in the background were treated with a synthetic soil-aggregating agent. (Allison, 1956)

the resulting increase in seedling emergence. An example is shown in Fig. 1.19, which represents results obtained by Allison (1956) on a high-sodium soil subject to crusting. Hemwall and Scott (1962) found that applying as little as 100 to 400 micrograms of 4-tertiary-butylpyrocatechol per gram of crust-forming soil directly over the seed row caused the soil to fracture linearly over the row, thereby permitting the seedlings to emerge. This chemical has the formula

$$\text{4-tert-butylpyrocatechol: benzene ring with —OH (top), —OH (adjacent), and } H_3C\text{—}C(CH_3)\text{—}CH_3 \text{ (bottom)}$$

The mode of action is probably that of hydrogen bonding of the —OH groups to clay surfaces. The hydrocarbon part of the molecule is oriented outward and has little attraction for the corresponding hydrocarbon part of the adsorbed layer on adjacent clay surfaces. Hence the bonding between adjacent clay particles is reduced. As the soil dries and shrinks, this zone of weakness leads to the linear fracture over the seed row.

A second, fairly clear direct effect is the restriction of root development that may occur in hard surface soils. Figure 1.20 is a photograph by Taubenhaus, Ezekiel, and Rea (1931) showing the appearance of roots of cotton plants grown in clay soils that had been compacted by rain or irrigation and then hardened by extended hot, dry weather. The normal taproot failed to develop. Instead, a constricted zone less than 1 mm in diameter and up to 25 mm in length developed in the root in the hardest layer of soil, located just below the surface. The portion of the taproot found in the less compact soil beneath was more nearly normal in size. Such malformed plants often died during the season, presumably when the rate of water movement through the restricted part became insufficient to meet plant requirements. Taylor et al. (1963) obtained similar results by placing clamps around the roots to prevent normal enlargement.

A third possible direct effect is the failure of roots to penetrate compacted or cemented layers of soil. Because various factors may contribute to such an effect under field conditions, a careful analysis is required to permit proper interpretation.

The fact that roots will penetrate limited thicknesses of soft, nonporous materials such as certain waxes (Hunter and Kelley, 1946) shows that

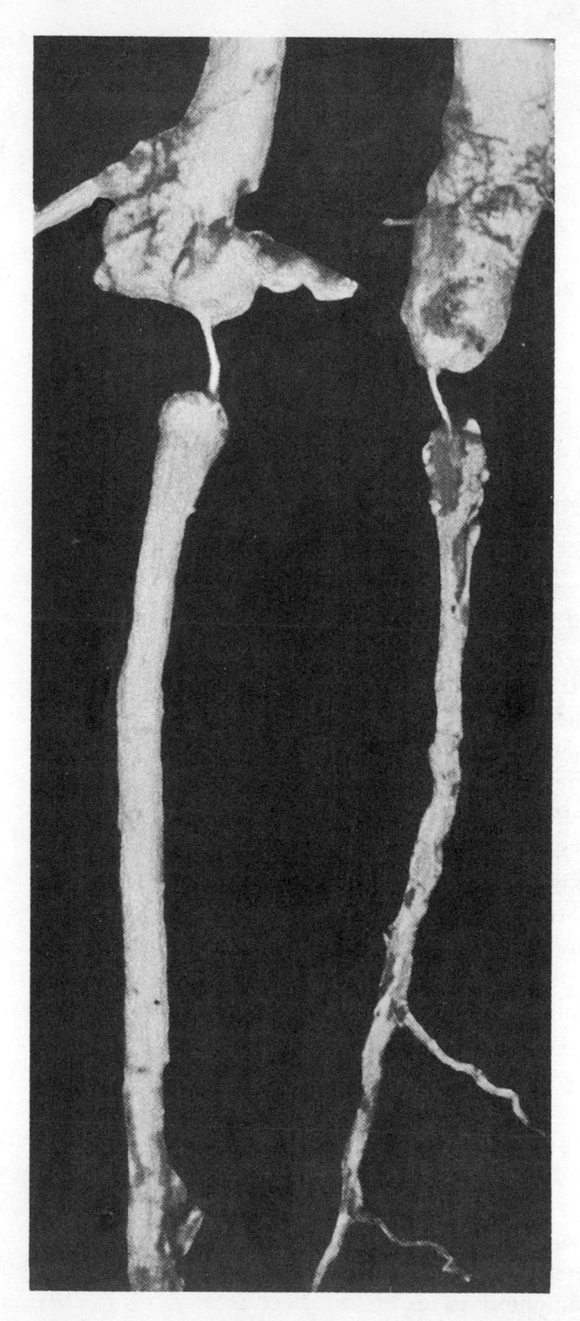

Figure 1.20. Development of taproot of cotton plants in soil in which a hard layer developed just below the surface during growth of the plants. The diameter of the roots is less than 1 mm in the portion restricted by the hard layer of soil. (Taubenhaus, Ezekiel, and Rea, 1931)

smallness of pores in the substrate is not of critical importance as such. Size of pores is of importance, however, if the pore walls are unyielding under the pressure of the developing root. Wiersum (1957) investigated the penetration of roots of a number of plants into glass filter discs in which the particles of glass were rigidly bonded by sintering. He used discs with three ranges of pore size: 0.2 to 0.5 mm, 0.15 to 0.205 mm, and 0.09 to 0.15 mm. The lower limit of pore sizes penetrated varied with the diameter of the roots. For example, *Portulaca grandiflora*, which had roots 0.125 to 0.21 mm in diameter, penetrated all three discs; but *Lathyrus odoratus*, which had roots 0.61 to 0.715 mm in diameter, penetrated only the disc with the largest pores.

To investigate the response of the roots to pore structures with different degrees of rigidity, Phillips and Kirkham (1962) grew corn seedlings in glass tubes of different diameters that were filled with sand particles 0.25 to 0.5 mm in diameter and were flushed periodically with nutrient solution. Because of the restraint provided by the rigid walls of the tubes, the sand particles were held more firmly in place in tubes of small diameter than in those of large diameter; the depth of penetration of roots into the sand was greater in tubes of large diameter than in those of small diameter; and the depth of root penetration increased linearly with the depth of penetration of a metal probe applied to the surface of the sand with a given force, thus verifying the dependence of root penetration on the force required to displace the sand particles from their original position.

With the foregoing as background, an experiment with soil may now be described. Taylor and Gardner (1963) planted cotton seeds on cylinders of a fine sandy loam soil that had been compacted to different degrees and in which the water content had been adjusted to different matric suctions. Each cylinder was covered with a layer of the same soil in which the water content was adjusted to the same matric suction as the underlying cylinder of soil and was firmed with a pressure that was the same in all instances. The percentage of the tap roots that penetrated the compacted layer beneath the seeds was then determined. Figure 1.21*a* shows that the root penetration decreased with increasing bulk density (grams of dry soil per cubic centimeter) of the compacted layer beneath the seed. The drier the soil (higher matric suction of water—see the chapter on water for an explanation of this term), the lower was the penetration at a given bulk density. The differences in penetration with water content of the soil evidently were not primarily a result of differences in aeration because the highest penetration was obtained at the highest water content where the air space was least. In Fig. 1.21*b*, the values for root penetration are plotted against the

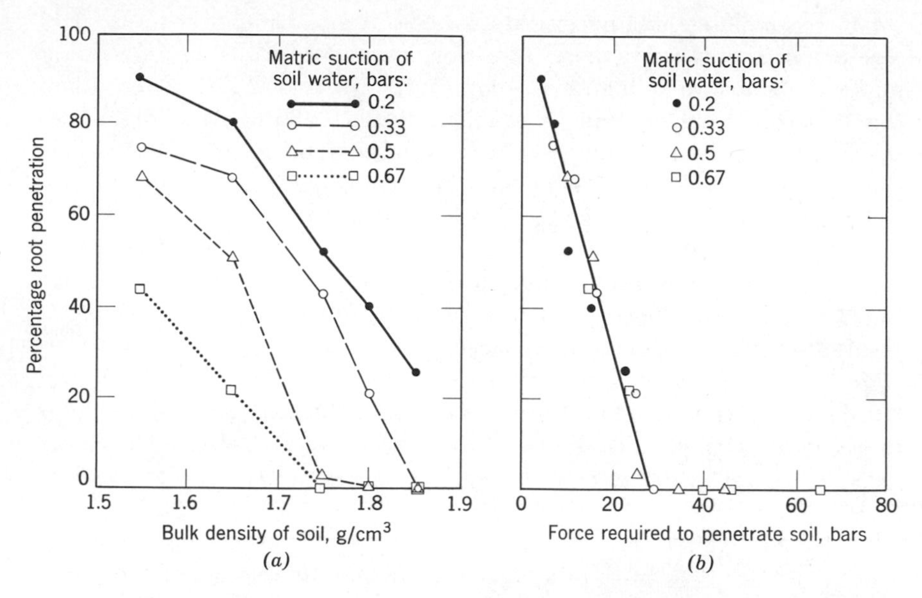

Figure 1.21. Percentage penetration of taproots of cotton through a 2.5-cm layer of a fine sandy loam soil, compacted and moistened to various matric suctions, in relation to (*a*) bulk density and (*b*) force required for penetration of the soil with a cylindrical metal probe. (Taylor and Gardner, 1963)

maximum force required to push a cylindrical metal probe 0.48 cm in diameter a distance 0.5 cm into the surface of the compacted soil. This force provides an index of the resistance of the soil particles to displacement. The various root-penetration values, which are the same ones plotted in the left portion of the figure, may be seen to fall approximately along a single line, with zero penetration at a probe-penetration force just under 30 bars. Apparently, therefore, the different degrees of root penetration resulted primarily from differences in resistance of the compacted soil to displacement by the roots and not from differences in bulk density, water content, or aeration.

With the aid of a somewhat similar experimental technique, but with provision for controlled aeration, Tackett and Pearson (1964) found that the oxygen percentage in the soil air required to obtain maximum root growth into a sandy loam subsoil increased as the bulk density was increased above 1.3 with the water content at a matric suction of 0.33 bar; however, within the range of water content in which oxygen was not limiting, the growth still decreased with an increase in bulk density. Because the water content was ample for good plant growth and because no greater growth was obtained in the presence of a higher oxygen percentage, the poor penetration apparently was due to resistance of the compacted soil to displacement by the roots.

It is difficult to obtain the type of experimental evidence needed to verify that failure of roots to penetrate a particular zone of soil in the field is caused by resistance of the soil particles to displacement, particularly in view of the fact that water content, aeration, and resistance to displacement fluctuate continually. Deficient aeration at intervals could inhibit root penetration even though aeration is adequate at other times. In some instances, the aeration might be poor enough and the resistance to displacement great enough so that either factor alone would inhibit or prevent penetration. Nevertheless, it is reasonable to suspect from observations of soils and plants that resistance of soil particles to displacement sometimes has a direct effect on expansion of plant parts. For example, Taylor and Burnett (1964) obtained differences in growth of cotton roots of the type shown in Fig. 1.22 as a result

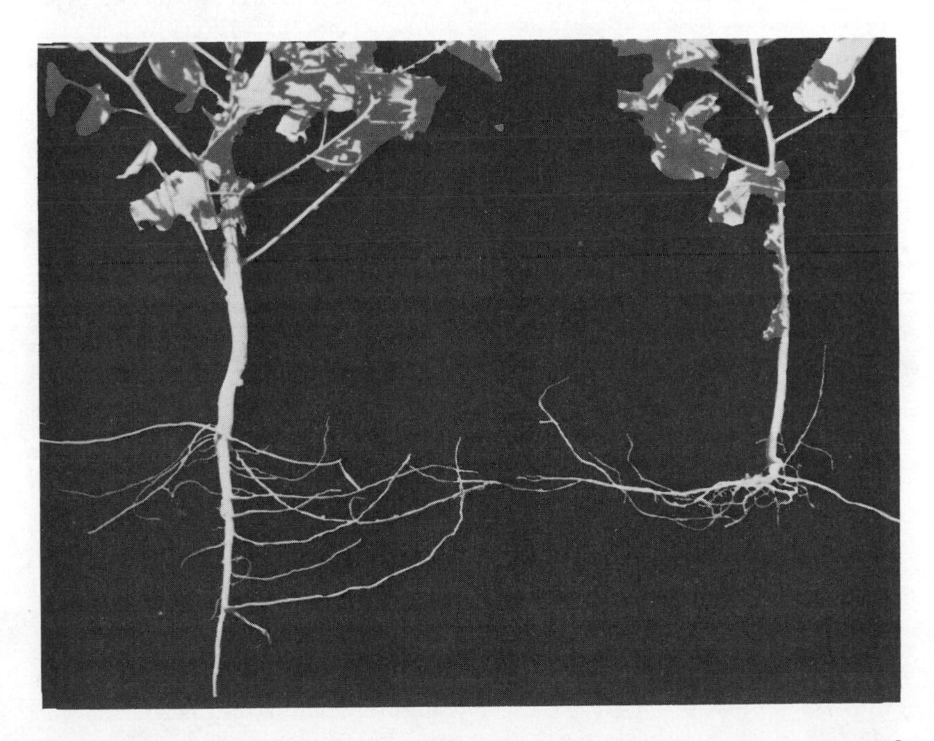

Figure 1.22. Growth of cotton plants in a fine sandy loam that was uncompacted (left) or compacted (right) by passage of a road roller over the moist soil. Soil in both areas was loosened by sweep tillage at a depth of 10 cm before cotton was planted. (Photograph courtesy of H. M. Taylor)

of a difference in compaction of a fine sandy loam soil. Here, poor aeration probably was not responsible because the air content of the soil at field capacity was about 15% of the soil volume, and the experiment was conducted in western Texas under low rainfall where irrigation was necessary. Moreover, root penetration was restricted more frequently and more severely in dry soil than in moist soil, which indicates that physical resistance of the compacted soil to displacement by the roots was responsible for the horizontal growth habit of roots illustrated in Fig. 1.22.

Indirect Effects. In addition to the direct effects of soil structure on expansion of plant parts, various indirect effects may be of significance. Because growth of roots commonly occurs between structural units and not within them, the size of structural units may affect the rate of delivery of nutrients and water from interior parts to the roots at the surfaces. Although work of Wiersum (1962) and Tepe and Leidenfrost (1958) indicates that plants respond to such indirect effects of structure, attention has been focused for the most part on the air-water relationships associated with differences in structure. Air-water relationships are sensitive to the size and continuity of pores.

Structural modifications most easily effected in soils result in changes in volume percentage of large or "noncapillary" pores. This principle may be illustrated and the implications in terms of air-water relationships perhaps best understood with the aid of some experimental data shown in Fig. 1.23. These measurements were made by Rubin (1949) on aggregates 1 to 2 mm in diameter from a silty clay loam soil without preliminary compression and after compression with a given force at 25.7% water; similar measurements were made on a sand consisting of particles the same size as the soil aggregates. Because 1 g of water at ordinary temperatures occupies a volume of 1 cm^3 to a close approximation, the values for water percentages may be taken as the cubic centimeters of water-filled pores per 100 g of dry soil or sand. The values at zero matric suction or saturation thus represent the total volume of pore space, which was 82 cm^3 in the uncompressed soil and 48 cm^3 in the compressed soil. At matric suctions of 0.5 and 1 bar, the volume of water-filled pores was almost the same, 19 $cm^3/100$ g, in both uncompressed and compressed soil; but only 2 cm^3 of water remained in the sand. The volume of air-filled pores at a matric suction of 0.5 bar is thus $82 - 19 = 63$ $cm^3/100$ g of uncompressed soil and $48 - 19 = 29$ $cm^3/100$ g of compressed soil. Compression thus reduced the volume of air-filled (large) pores at a matric suction of 0.5 bar by about one-half, but it did not reduce the volume of water-filled (small) pores at this suction. The sand has a volume of large pores of the same order of

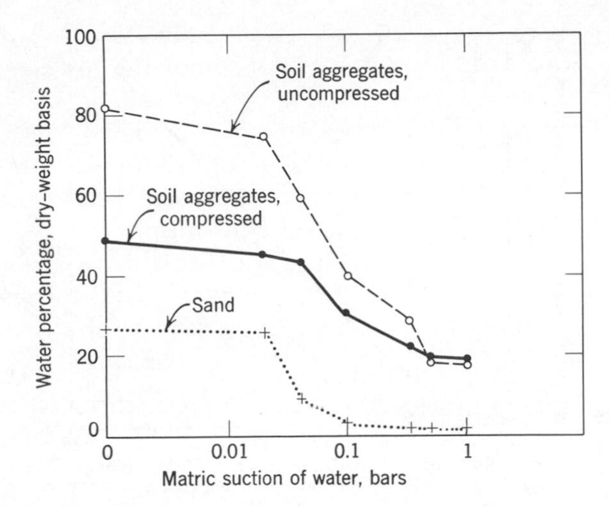

Figure 1.23. Water content of beds of soil aggregates 1 to 2 mm in diameter, with and without preliminary compression, and of sand, also 1 to 2 mm in diameter, versus matric suction of water in the beds. (Rubin, 1949)

magnitude as the soil; however, in contrast to the soil, which has considerable pore space that is drained by increasing the suction from 0.1 to 0.5 bar and still contains considerable pore space filled with water at a suction of 0.5 bar, the sand contains only a little pore space in either category. The difference no doubt is caused mainly by the presence of small pores within the soil aggregates that retain water after the large pores among the aggregates have been drained; the sand has large pores among the particles but little pore space within the particles.

The air-water relationships described in the preceding paragraph prevail under natural conditions as the structure is modified by tillage and as the soil is dried to different degrees. As water is removed from soil by evaporation, transpiration, or drainage, the large pores are the first to become filled with air; the small pores are the last. The volume of large pores is important in determining the rate of flow of water through saturated soil and the rate of infiltration of water into the soil. The volume of large pores is important in providing aeration, and the volume of both large and small pores affects the capacity of the soil for water retention.

Data of Peele (1950) may be cited as an example of the importance of large pores in determining the rate of flow of water through saturated soil. Figure 1.24 shows the averages of measurements of the rate of flow of water through a large number of saturated cores of soil 9.4 cm

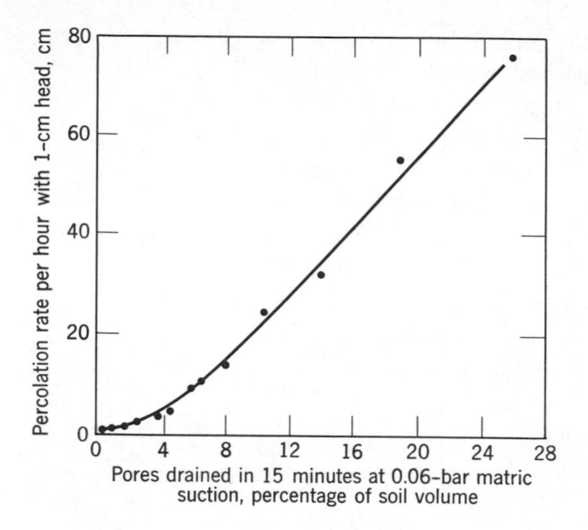

Figure 1.24. Rate of percolation of water versus pores drained in 15 minutes at a matric suction of 0.06 bar in undisturbed cores from different horizons of soils from Virginia, North Carolina, South Carolina, and Georgia. (Each point represents the average of from 16 to 65 measurements.) (Peele, 1950)

in length. The samples were collected as undisturbed cores from different layers of soils from Virginia, North Carolina, South Carolina, and Georgia. The rate of flow evidently is related not only to the rate of drainage of excess water but also to aeration and erosion.

The importance of soil structure in relation to infiltration of water is basically the same as it is for saturated flow within the soil. Infiltration is mentioned separately, however, because of the special significance of structural stability to infiltration. Structural units that are not strongly bonded internally disintegrate readily under bombardment by raindrops. The fine soil particles released are filtered out of the water as it enters the soil, thereby closing the large pores and reducing the subsequent rate of infiltration. Figure 1.25 illustrates this effect. The rate of water infiltration into saturated columns of a sandy loam soil remained essentially constant with time as long as clear water was applied without disturbance to the soil surface, but the rate dropped sharply when muddy water was applied. Figure 1.26 shows the appearance of a vertical cross section of the surface layer of a soil after it had been sprinkled with water. Runoff and erosion, of course, are increased as a result of disintegration of surface structure. Experiments demonstrating the action of soil-aggregating agents in increasing infiltration and decreasing erosion were reviewed by Quastel (1954). Mulches of organic materials produce a similar effect in a different way. Although organic mulches may improve the structure of the immediate surface of the soil, they act primarily by breaking the fall of raindrops. Literature on this subject was reviewed by Stallings (1950).

Through the control it exerts on air-water relationships, soil structure may have important indirect effects on plant growth. Unpublished observations by J. C. Russel in Iraq may be mentioned as an example. Russel noted that roots of alfalfa, barley, and wheat grown under irrigation on a certain alluvial silty clay soil fail to penetrate below a depth of about 45 cm. No hard layer is present at the depth where root penetration ceases. The cause of the root behavior apparently is poor aeration, which is associated with a lack of large pores. The soil has a moisture equivalent of 28% but pore space of only 25%, so that after irrigation the soil is saturated and poorly aerated. A native plant, *Prosopis stephtheniana,* which can endure poor soil aeration, produces roots that penetrate several times as deeply as those of alfalfa, barley, and wheat. Instances in which particular soil layers exhibit properties similar to those of the soil described by Russel have been reported by Woodruff (1940) and Veihmeyer and Hendrickson (1946). Aeration will be con-

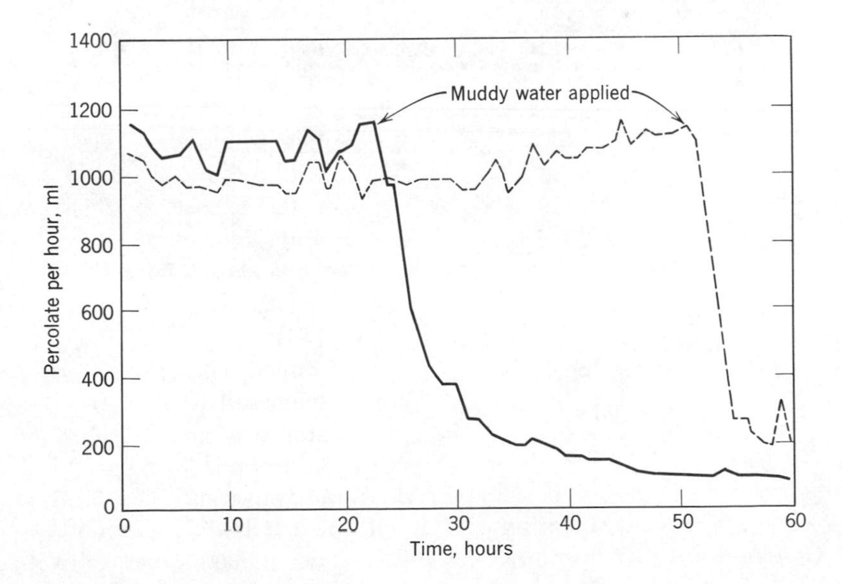

Figure 1.25. Rate of percolation of water through two columns of a sandy loam soil to which clear water was applied initially and muddy water was applied after 24 or 51 hours. (Lowdermilk, 1930)

Figure 1.26. Photomicrograph of a vertical cross section of the surface layer of a sandy loam soil after the soil had been sprinkled with water. Note the dense layer at the immediate surface and the relatively large open spaces below. (Duley, 1940)

sidered in more detail in Chapter 3. Aeration in turn has important effects on nitrogen availability, which will be discussed in Chapter 7.

Conclusion. From the foregoing consideration of soil structure in relation to plant growth, it is evident that structure is not one of the fundamental factors of plant growth and that its effects, like those of particle-size distribution, depend to a considerable extent on the circumstances.

For example, if rainfall is of borderline sufficiency, poor structure may result in decreased plant growth because of loss of water by runoff if the soil is sloping but not if it is flat. If the water table is close to the surface of the soil, poor structure may result in poor aeration and reduced plant growth; but these effects may not occur if the water table is deep in the soil. Thus, although structure involves the whole soil and has significant effects on plants, it is clearly insufficient as a basis for characterizing soil as a substrate for plant growth.

Mineralogical Composition

From the mineralogical standpoint, the inorganic portion of the soil solids is composed of minerals, of which the primary source is igneous rocks. As shown in Table 1.4, a total of 92% by weight of igneous rocks is composed of only a few types of minerals. The remaining 8% is composed of a large variety of accessory minerals, each present in small

Table 1.4. Average Mineralogical Composition of Igneous and Sedimentary Rocks

	Proportion of Mineral Present, %	
Mineral Group	Igneous Rocks (Clarke, 1924)	Sedimentary Rocks (Jeffries, 1947)
Feldspar	59	7
Amphibole	17	—
Quartz	12	38
Mica	4	20
Carbonate	—	20
Clay	—	9
Limonite	—	3
Accessory	8	3

proportion. According to Jeffries (1947), 25% of the surface of the earth is underlain by igneous rocks. The remaining 75% is underlain by sedimentary rocks, the average composition of which is shown in Table 1.4. An average mineralogical composition of soils has not been worked out; but, if this were done, the values for the various minerals probably would bear greater similarity to the average for sedimentary than for igneous rocks. The reason for this is that igneous rocks undergo alteration by weathering, whether the igneous rocks are altered to soil in place or after transportation and perhaps consolidation into sedimentary rocks.

The data in Table 1.4 illustrate the principle that the ratio of resistant to easily weathered minerals increases as the easily weathered species are selectively removed. For example, amphiboles are more readily weathered than feldspars and disappear more rapidly than feldspars as weathering proceeds; hence the ratio of feldspars to amphiboles increases. Feldspars are more readily weathered than quartz and disappear more rapidly than quartz as weathering proceeds. The net result is that quartz, which is a relatively minor constituent of igneous rocks, becomes a major constituent of sedimentary rocks and soils. Some of the minerals in sedimentary rocks and soils are not present in igneous rocks but are newly formed; this is true of carbonates, limonite, and part of the clay and mica.

Usually the mineralogical composition of the sand and silt fractions is considered separately from that of the clay fraction; only rarely is the total mineralogical composition of a soil determined. This distinction probably results partly from the fact that the mineralogical composition differs among particle-size fractions and partly from the fact that different scientific instruments are used for determining the mineralogical composition of sand and silt, on the one hand, and of clay, on the other. The petrographic microscope is the most useful instrument for determining the mineralogical composition of the sand and silt fraction (Cady, 1965), but it is not suitable for identifying single grains of clay size. The techniques most useful for examination of the clay fraction are X-ray diffraction (Whittig, 1965), thermal analysis (Barshad, 1965), and electron microscopy (Kittrick, 1965). In preparation for mineralogical analysis, soil usually is given special chemical treatments to remove salts, organic matter, and mineral materials that may encrust and cement the primary particles; then it is separated into particle-size fractions (Kunze, 1965).

Sand and Silt Fractions

McCaughey and Fry (1913) made a mineralogical examination of the sand and silt fractions of 25 soils from locations scattered widely over the United States and reported their results as shown in Table 1.5. Evidently many minerals may be identified in the sand and silt fractions of soils, but only a few species are commonly found in abundance. This table illustrates in a different way from Table 1.4 the alteration of ratios of different minerals that accompanies weathering. Quartz, which makes up only 12% of igneous rocks, is the only mineral found to be abundant in all the 25 soils examined. The ratio of quartz to

Table 1.5. Occurrence and Abundance of Certain Minerals in the Sand and Silt Fractions of Twenty-Five Soils (McCaughey and Fry, 1913)

	Number of Soils			Number of Soils	
Mineral	Present	Abundant	Mineral	Present	Abundant
Quartz	25	25	Zircon	22	1
Orthoclase	20	14	Tourmaline	21	1
Hornblende	23	12	Magnetite	6	1
Microcline	20	10	Ilmenite	2	1
Epidote	24	8	Calcite	2	1
Biotite	21	8	Sericite	1	1
Muscovite	20	6	Hematite	1	1
Plagioclase	13	5	Rutile	17	0
Andesine	6	4	Apatite	12	0
Oligoclase	7	3	Fluorite	4	0
Chlorite	21	2	Titanite	2	0
Garnet	10	2	Hypersthene	2	0
Sillimanite	10	2	Phlogopite	1	0
Pyroxene	6	2	Serpentine	1	0
Albite	5	2	Actinolite	1	0
Labradorite	3	2	Iddingsite	1	0
Augite	2	2	Staurolite	1	0

feldspar in the sand and silt or in some particle-size range less broad than this is sometimes used as an index of the degree of weathering of soils believed to be developed from similar parent materials.

Clay Fraction

As mineral particles become smaller, the ratio of surface area to mass increases greatly. Because processes of chemical weathering are imposed from the outside, subdivision of minerals to small particle sizes increases their susceptibility to alteration. Consequently, differences in chemical stability of minerals are an important factor in determining the change in mineralogical composition of soils with particle size. Changes in mineralogical composition with particle size are particularly marked as the particle size approaches that of clay. Table 1.6 illustrates the situation in one soil. Albite and quartz are representative of primary minerals, residual from particles that originally were of larger size. These two minerals occur in the coarser clay but not in the finest fraction. Quartz is the only primary mineral commonly present in large quantities in

Table 1.6. Mineralogical Composition of Different Particle-Size Fractions of the B Horizon of Houston Clay Loam Developed from Calcareous Marl under Grassland Vegetation in Mississippi (Pennington and Jackson, 1948)

Mineral	Proportion of Mineral in Particle-size Fraction Indicated, %			
	5 to 2 μ	2 to 0.2 μ	0.2 to 0.08 μ	<0.08 μ
Albite	10	5	0	0
Quartz	50	20	5	0
Illite	10	10	5	0
Mica-intermediate	10	20	30	25
Montmorillonite	5	5	50	75
Kaolinite	15	40	10	0

the clay fraction of soils. Where present, quartz is most abundant in the coarse clay. Montmorillonite and kaolinite, the last two minerals in the table, are entirely new minerals that have formed as crystals from weathering products of other minerals that were present in solution. Montmorillonite and kaolinite appear to be formed in soils; they are not components of unaltered igneous rocks. These minerals are stable in particles of clay size but occur only to a limited extent in particles of larger size, perhaps because large particles would be physically unstable. Both minerals tend to crystallize in thin sheets rather than as particles with three approximately equal dimensions. Of the two remaining minerals in the table, illite is thought to be formed from pre-existing clays in sedimentary deposits and to be inherited by soils from such deposits; there is some doubt that formation in place is responsible for the major part of the illite in soils. Mica-intermediate, a term used at one time to describe minerals having in some molecular layers the character of mica and in others the character of other structurally related minerals, may originate from alteration of mica of larger particle sizes.

Because the chemical processes leading to formation of clay in soils take place in aqueous solutions, the degree of weathering (and hence the amount of clay formed) would be expected to increase with the effective amount of water supplied to the soil. In confirmation of this supposition, Jenny and Leonard (1934) found that at constant annual temperature the clay content of soils increases from locations in eastern Colorado through Kansas and into western Missouri (Fig. 1.27). How much higher the clay content would have become had it been possible

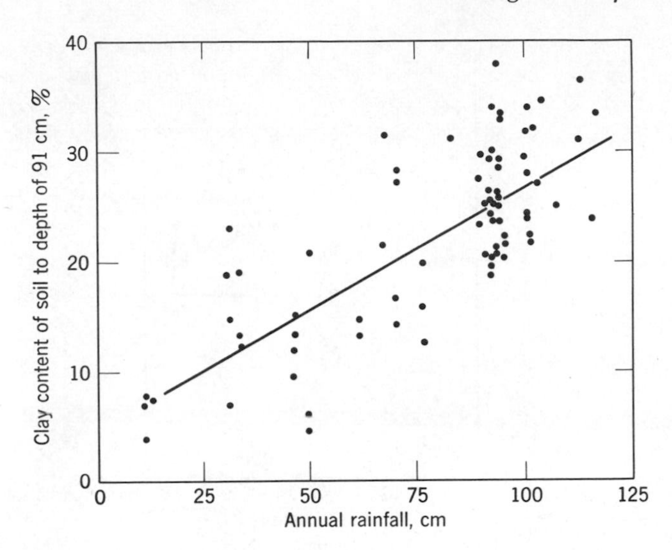

Figure 1.27. Clay content of soils versus annual rainfall along a traverse from eastern Colorado through Kansas and into western Missouri. The samples of soil were from a belt having an annual temperature of 11.1 to 13.3°C. (Jenny and Leonard, 1934)

to extend the traverse to areas of progressively higher rainfall is not known. Table 1.7 gives an instance in which the clay content decreases with increasing rainfall. This evidence, together with that in Fig. 1.27, suggests that under some conditions a maximum content of clay may be associated with a particular rainfall and that less clay would be present with either more or less rainfall. The results in Table 1.7 do

Table 1.7. Particle-Size Composition of Soils of Mauritius Developed with Different Annual Rainfall (Craig and Halais, 1934)

Average Annual Rainfall, cm	Sand and Silt, 2–0.002 mm Diameter, %	Clay, <0.002 mm Diameter, %
64–127	27	73
127–191	33	67
191–254	44	56
254–318	51	49
318–381	55	45

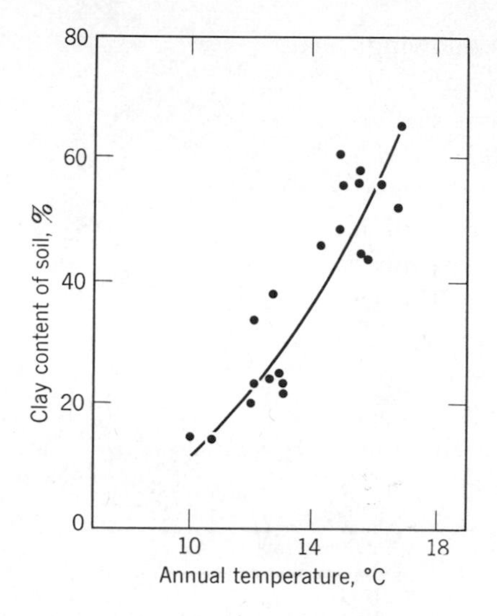

Figure 1.28. Clay content of soils versus annual temperature at constant water supply expressed as NS quotient = 400. (Jenny, 1935)

not mean that chemical weathering became less severe as the rainfall increased. Under the conditions of intense weathering that prevail in tropical Mauritius, where the data were obtained, the material of sand and silt size in the soils consists mostly of concretions and not of particles of primary minerals that have not yet been decomposed by weathering.

Because formation of clay is a result of the action of predominantly chemical processes on materials of larger particle sizes, and because the temperature coefficient of the rate of chemical reactions generally falls in the range of 1.4 to 7 for each increase of 10°C, the clay content of soils would be expected to increase with temperature. That such is the case, where due account is taken of the effect of temperature on water supply, is indicated by the data of Jenny (1935) in Fig. 1.28. Jenny found that in soils derived from basic rocks in eastern United States the clay content increased about 7% with each centigrade degree increase in mean annual temperature.

Many different minerals may be found in soil clays. For convenience they may be classified as silicates and nonsilicates. The silicates have received the most attention. This group includes the kaolins, montmorillonites, illites, vermiculites, and allophanes. Each name applies to a class of minerals based on molecular structure and chemical composition. A schematic representation of the structure of kaolinite and montmorillonite may be found in Chapter 4 on exchangeable bases. The nonsilicates include the following: quartz and other forms of silica (SiO_2); oxides and hydrous oxides of iron, such as hematite (Fe_2O_3) and goethite [$FeO(OH)$]; hydrous oxides of aluminum, such as diaspore [$AlO(OH)$] and gibbsite [$Al(OH)_3$]; and calcium carbonate ($CaCO_3$). Part of the mineral matter in soil clays is crystalline and part noncrystalline. The silicates may exist as individual minerals in discrete particles, or they may exist in mixed crystals containing some molecular layers of each of two or more species. The nonsilicates may exist in discrete particles, or they may exist as coatings on other minerals. In general, the clay minerals in soils appear to be less well crystallized than those in clay deposits.

The clay minerals of soils, being products of weathering, reflect the factors involved in their origin. Sometimes the predominant factor of origin is the parent material. For example, Peterson (1946) found that kaolinite typically is either absent or present in only small amount in the clay fraction of soils of Iowa. All soils showing this characteristic have developed from glacial drift or loess deposited during Pleistocene (glacial epoch) geologic time. In contrast, kaolinite is the dominant clay mineral in the Gosport soil. This soil has developed on recently exposed sedimentary deposits that were laid down originally during the much earlier Pennsylvanian geologic age. Kaolinite is dominant in the underlying material as well as in the soil. Therefore the Gosport soil apparently merely inherits a characteristic already present in the material from which it developed. According to Jackson (1959), inherited minerals are most abundant in soils of dry regions and cold regions and in soils where soil-forming factors have not had much influence on the parent material. Parent material may affect the mineralogical composition of soil clays indirectly through the relative susceptibility of its minerals to weathering, the concentration of magnesium it maintains in solution, and its effect on leaching. High concentrations of magnesium and conditions of little leaching favor formation of montmorillonite and secondary magnesium chlorite. Magnesium is a component of both of these minerals.

In some instances, the parent material is clearly not the deciding factor. For example, Hosking (1940) found that in Australia the same kind of parent rock gave rise to soils containing kaolinite under one set of conditions and to soils containing montmorillonite under other conditions. Conversely, Humbert and Marshall (1943) found that in Missouri parent rocks of different compositions gave rise to soils having similar clay mineral composition where the soils were developed under similar conditions. Factors of climate, time, and vegetation have significant combined effects on mineralogical composition of clays through their effect on the degree of weathering. According to Jackson (1959), gypsum, carbonates, ferromagnesians, feldspars, quartz, and illite are most abundant in clays under conditions of little weathering; and vermiculite, secondary chlorite, montmorillonite, kaolinite, and halloysite are most abundant under conditions of moderate weathering. In general, weathering appears to have been more pronounced where the dominant silicate clay mineral is kaolinite than where it is montmorillonite or illite. Kaolinite, however, is not the end product of weathering. Under conditions of strong weathering, the predominant components of soil clays are (1) secondary minerals, such as hematite, goethite, allophane, gibbsite, and anatase, and (2) resistant primary minerals, such as ilmenite and magnetite.

One may infer from the foregoing that the occurrence of different mineral components in the clay fraction of soils is not determined by chance but that some system is involved. As a means of rationalizing the distribution of minerals in the clay fraction of different soils, Jackson, Tyler, et al. (1948) placed the various minerals in 13 groups according to the rate at which they were thought to disappear if they were present in particles of clay size in soils. Arranged in order from most easily to most difficulty weathered, the type members of their 13 weathering stages are gypsum, calcite, hornblende, biotite, albite, quartz, illite, mica-intermediate [later classified as interstratified 2:1 layer silicates and vermiculite by Jackson and Sherman (1953)], montmorillonite, ka-

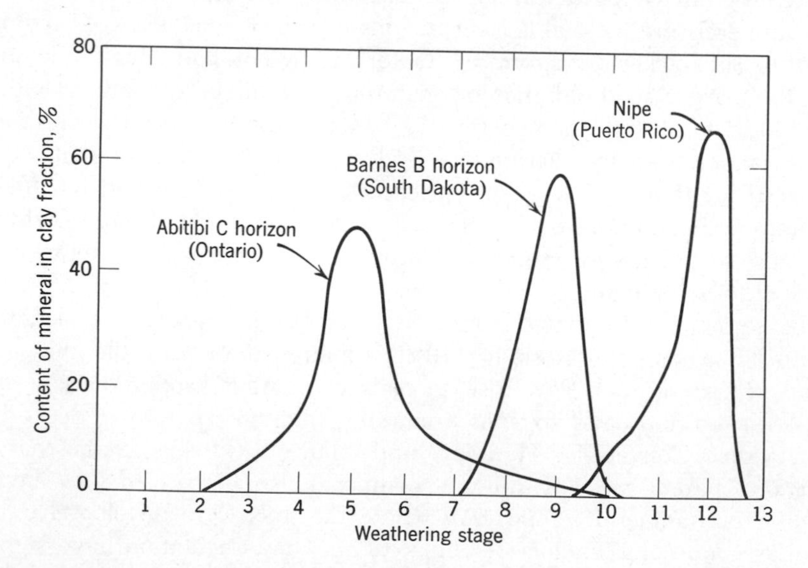

Figure 1.29. Percentage content of minerals in clay fraction of three soils versus weathering stage of minerals. (Jackson, Tyler, et al., 1948)

olinite, gibbsite, hematite, and anatase. According to these investigators, from three to five minerals of the weathering sequence usually occur in the clay of a given soil horizon. If the percentage content of the minerals at each weathering stage is plotted against the weathering stage, the distribution curve obtained is dominated by one or two members of the weathering sequence, with the percentages of the other minerals decreasing with the distance in either direction from the dominant mineral. Examples of the distribution curves obtained with the clay fraction of three soils are shown in Fig. 1.29. The degree of weathering of the clay increases as the maximum of the distribution approaches the right in the diagram.

Soil-Plant Relationships

Except for water, soil minerals occur as discrete particles or portions of particles and have no known direct effects on plants beyond the physical effect discussed previously in connection with particle-size composition. Through decomposition, solution, and ion-exchange, however, soil minerals release plant nutrients and hence are of vital indirect importance to the welfare of plants.

Mineralogical analysis has been used to a relatively small extent to provide information on the proportions of individual minerals or groups of minerals containing different plant nutrients and hence on the long-term capabilities of soils for supplying nutrients to plants. As an example of one such use, Table 1.8 gives summarized data on mineralogical analyses of 25 soils of the United States classified according to occurrence of the soils in arid, prairie, or timbered-humid regions. According to the table, the weight percentage of the sand and silt fractions composed of minerals other than quartz decreased in the order arid > prairie > tim-

Table 1.8. Average Mineralogical and Chemical Composition of Soils from Different Regions of the United States (Coffey, 1912)

	Minerals other than Quartz in Indicated Soil Fraction, %		Content of Indicated Elements in Soils, %		
Region	Sand	Silt	Calcium	Magnesium	Potassium
Arid	37	39	1.9	0.7	0.6
Prairie	20	29	0.8	0.3	0.4
Timbered-humid	8	12	0.3	0.2	0.3

bered-humid. Over equal intervals of time, alteration of uniform parent minerals by weathering would be expected to be least in the first and greatest in the last of these regions. A significant implication of the classification of minerals employed in the table is that quartz furnishes no plant nutrients on dissolution, but the minerals other than quartz include species that contain such nutrients as calcium, magnesium, and potassium. On decomposition of the minerals, these nutrients are liberated in soluble form and may be used by plants. The difference between the two mineral groups in content of these nutrients is indicated in Table 1.8 by the percentage content of calcium, magnesium, and potassium present in the same three classes of soils. The content of these nutrients varies directly with the content of minerals other than quartz.

Another example, from Australia, is shown in Table 1.9. In this work, the fertility of the soils was found to increase with the percentage of minerals other than quartz in the fine sand fraction.

Table 1.9. Mineralogy and Fertility of Soils in the Mount Gellibrand Area, Western Australia (Nichols, 1939)

	Average Content of Minerals in the Fine Sand, %		
	Brown Loam	Black Clay	Gray Loam
Rock fragments	7.6	0.3	Trace
Augite	12.3	1.0	Trace
Plagioclase	19.3	8.7	3.9
Olivine	2.2	0.8	0.5
Iron oxide	8.4	1.7	1.2
Quartz	50.3	80.8	93.7
Relative fertility	High	Medium	Low

Despite the importance of the character and quantity of mineral constituents of soil, mineralogical analysis is not used in a practical way in characterizing soil as a substrate for growth of plants. There are perhaps three reasons: (1) Obtaining a mineralogical analysis of a soil is a tedious process. (2) The methods leave much to be desired, particularly where clay is concerned. This deficiency in methods is especially important because clay is the most reactive portion of the inorganic soil solids, and the nutrients held by the clay are of more immediate concern in plant nutrition than are those of sand and silt. (3) Chemical methods can be used to obtain an index of availability of soil nutrients with less effort than is needed in a mineralogical analysis.

Although mineralogical analysis may not find much application for the purpose just mentioned, the trend in current basic research on phosphorus and potassium in soils is toward increased use of mineralogical analysis. Knowledge of the presence, absence, or alteration of specific mineral entities may be a valuable aid in interpreting other data obtained from chemical or biological measurements. Some of this work will be mentioned in Chapters 8 and 9 on phosphorus and potassium. Mineralogical analysis is useful also in studies of soil formation and classification, in which it may provide important auxiliary information on the origin, alteration, and character of soils.

Elemental Composition

From the chemical standpoint, soils contain at least traces of many chemical elements, present mainly in compounds. The elemental composition varies according to the nature of the parent rock material and the changes brought about by weathering, accumulation of organic matter, and management practices. The differences in susceptibility of certain elements to loss by weathering are illustrated by the comparative chemical composition of the lithosphere and river water shown in Table 1.10. The ratio of the percentage content of a constituent in the dissolved matter of river water to the percentage content in the lithosphere provides an indication of the comparative susceptibility of different constituents to loss by weathering. Arranged in order of increasing ratio, the constituents in the table are $(Al_2O_3 + Fe_2O_3) < Si < K < Mg < Na < Ca$. This method of comparison thus gives results in agreement with

Table 1.10. Comparative Elemental Composition of the Lithosphere and of the Dissolved Constituents in River Waters of North America (Clarke, 1924)

	Proportion of Total Weight, %					
	$Al_2O_3 + Fe_2O_3$	Si	K	Mg	Na	Ca
River water (*W*)	0.64	4.02	1.77	4.87	7.46	19.36
Lithosphere (*L*)	22.47	27.60	2.58	2.08	2.75	3.64
W/L	0.03	0.15	0.69	2.34	2.71	5.32

the weathering sequence proposed by Jackson, Tyler, et al. (1948) in which gibbsite [$Al(OH)_3$] and hematite (Fe_2O_3) are listed as types of the most resistant minerals except for anatase (TiO_2). Nevertheless, the validity of this method of comparing different elements suffers from the fact that the dissolved matter is derived from many locations, which differ in composition, and from the fact that soluble substances derived from one location may be withdrawn to form secondary minerals at other locations before the water reaches the rivers.

The chemical composition of igneous rocks and various soils is given in Table 1.11. Only a few points about these analyses will be mentioned. First, the results are expressed in oxide form. Analytical values expressed in this way will add to 100% if an analysis has been made for all constituents present in significant quantity, if the analyses are correct, and if due account has been taken of the quantities of halides present (these are usually small enough to be neglected). It may be noted that in each case the analyses in the table yield a sum close to 100%. Second, SiO_2 is the most abundant constituent in igneous rocks and in all but the Columbiana soil, which is much more strongly weathered than the other soils listed in the table. Third, the CaO, MgO, Na_2O, and K_2O percentages are much lower in the soils than in igneous rocks, indicating preferential removal of these constituents by weathering. Fourth, the soils contain organic matter and nitrogen. Unaltered igneous rocks do not contain organic matter; however, they may contain nitrogen to the extent of 80 $\mu g/g$ (Ingols and Navarre, 1952; Adams and Stevenson, 1964). The nitrogen in igneous rocks is ammoniacal, whereas in soils it is mostly organic. Fifth, the loss of weight on ignition is much greater from soils than from igneous rocks because of the presence in soils of organic matter and additional water of hydration and constitution.

The particle-size fractions of soil are not uniform in chemical composition, as would be expected from the differences that occur in mineralogical composition. Failyer, Smith, and Wade (1908) analyzed particle-size fractions of a number of soils for calcium, magnesium, potassium, and phosphorus and found that, with all these elements in most soils, the percentage content was highest in the clay fraction and lowest in the sand fraction. The difference in elemental composition among separates was most pronounced with phosphorus and magnesium.

Knowledge of the total amount of individual nutrients in soils has only limited value in predicting the adequacy of supply of nutrients for plant growth. The reason is that the availability, or effective amount, of each plant nutrient in soil is less than the total for that nutrient and often is poorly correlated with the total. Therefore, in efforts to characterize soils chemically as regards the supply of nutrients or other

Table 1.11. Elemental Composition of Igneous Rocks[1] and Soils[2]

Constituent	Igneous Rocks (Average)	Barnes Loam (South Dakota)	Caribou Loam (Maine)	Cecil Sandy Clay Loam (North Carolina)	Columbiana Clay (Costa Rica)
	Proportion of Constituents by Weight in Indicated Rocks and Soils, %				
SiO_2	59.1	69.3	57.5	74.7	19.8
Al_2O_3	15.3	11.4	7.8	12.3	37.1
Fe_2O_3	7.3	3.8	2.5	4.9	15.6
TiO_2	1.0	0.5	0.7	1.3	2.0
MnO	0.1	0.2	0.2	0.3	0.3
CaO	5.1	1.6	1.2	0.2	0.2
MgO	3.5	0.9	0.6	<0.1	0.5
K_2O	3.1	1.8	0.9	0.6	0.1
Na_2O	3.8	1.1	1.0	0.2	0.2
P_2O_5	0.3	0.2	0.2	0.2	0.3
SO_3	0.1	0.1	0.3	—	0.2
Ignition loss	1.2[4]	9.5	27.2	7.1	24.1
Total	99.9	100.4	100.1	101.9	100.4
Organic matter[3]	—	6.0	25.3	2.4	6.0
Nitrogen[3]	—	—	0.9	—	—

[1] Clarke (1924).

[2] Byers, Alexander, and Holmes (1935). Soil analyses are of the surface horizon, which had a thickness of 23 cm in the Barnes soil, 2 cm in the Caribou soil, 15 cm in the Cecil soil, and 25 cm in the Columbiana soil.

[3] Included in ignition loss. The nitrogen percentage is included in the organic matter percentage. All soils contained nitrogen although a figure for nitrogen was reported for only one soil.

[4] Includes the H_2O and CO_2 reported in the analyses.

substances that may affect plant growth, the usual objective is to obtain an index of the availability and not the total amount. Data of Semb and Uhlen (1955) may be cited as an example. These investigators conducted 92 field experiments in Norway to determine the degree to which crop yields were limited by a deficiency of soil phosphorus. On samples of the soil taken from the individual experiments, they made analyses for both total inorganic phosphorus and inorganic phosphorus extracted by a mildly acid lactate solution. They found that, although the lactate-soluble phosphorus was only a small part of the total, measurements of this fraction provided a much better estimate of the degree of phosphorus deficiency observed experimentally than did measurements of the total inorganic phosphorus.

Measurements of the total content of plant nutrients in soils are seldom made in attempts to characterize soils for plant growth. Occasionally such analyses are made to find the changes in soil composition that may occur during long periods of time in field experiments or to find differences between virgin and cropped soils. Total analysis has its greatest application as a source of auxiliary information in research work in soil mineralogy, formation, and classification.

Soil Profile

From what sometimes is called the morphological standpoint, soils are composed of a series of "horizons." Soil horizons are layers of varying thickness, distinctness, and character, oriented approximately parallel to the soil surface. Collectively, the horizons make up what is called the soil "profile." Figure 1.30 gives a diagram of a hypothetical soil profile having all the principal horizons and some general statements about the character of the different horizons. Individual soils have one or more of these horizons but do not necessarily contain all of them. Figure 1.31 gives a picture of a specific soil profile, that of Edina silt loam, with a description of the profile in terms used in soil classification work.

A soil profile is the resultant of all the factors involved in its formation. These may be classified under the headings of climate, organisms, topography, parent material, and time. A great number of different soil profiles may be produced by variations in these five factors. Certain kinds of differences in soil profiles may occur within a space of a few meters, depending on local factors. For example, the thickness of the A1 horizon may decrease as the topography becomes steeper; or the depth to the Cca horizon may increase as the soil becomes more sandy. Other kinds of differences, attributable primarily to climate, are found in profiles that usually are separated widely. For example, climate is the primary factor responsible for the difference between light-colored, relatively shallow soil profiles formed under dry conditions in Wyoming and dark-colored, deeper soil profiles formed on similar topography and parent material in eastern North Dakota.

Soil profile descriptions of the type illustrated in Fig. 1.31 are made primarily to record properties observable in the field for the benefit of individuals involved in mapping and classifying soils. If properly interpreted with the aid of knowledge of soil-plant relationships, such descriptions convey much information about the value of soils for growing plants; but they do not, in themselves, represent characterizations of soils in terms of their suitability for plant growth.

From the generalized concept of soil profiles in Fig. 1.30 and the specific example in Fig. 1.31, it is clear that horizons within a single profile may differ in qualities that are visually evident; and it may be inferred that the horizons differ in other properties as well. The environment provided for plant roots may differ significantly among horizons, as sometimes may be inferred from observations of the distribution of plant roots in soil in place. Another way to demonstrate a difference in environment is to grow plants on samples of soil taken from the various horizons and to observe the differences in growth. Lutwick and Hobbs (1964) found, for example, that yields of alfalfa were 13.5, 3.0, 1.0, and 2.5 g per culture on samples of recognizable horizons from the surface downward in a virgin soil developed under grassland vegetation in southern Alberta.

Perhaps the simplest adaptation of the profile concept in quantitative characterization of soil as a substrate for plant growth is provided by a measurement of the quantity of soil present. Figure 1.32 is an example taken from the work of Låg (1961), who classified soils in a particular portion of Norway according to the soil thickness over bedrock and found that the thickness provided an index of the annual forest production.

This adaptation of the profile concept applies to a model in which each profile in the group in question consists of two layers. The first layer affects plants as if it had the same thickness, properties, and locations in all soil profiles. The second layer affects plants as if it had the same properties and location in all profiles, but the thickness may differ among profiles. The properties of the second layer may or may not be the same as those of the first.

The success with which Låg was able to apply this adaptation of the soil profile concept may be attributed to two principal conditions. First, the thickness of soil available for the roots of the forest vegetation is so limited in the area where the data were collected that thickness (more properly, volume) is of greater importance than other differences in character of the soils. Second, the soils are located within a limited region where the other differences are not great. The circumstances under which the thickness of the soil profile may be used satisfactorily as an index of the suitability of soils for plant growth are very limited. In fact, Låg (1961) found that in areas of Norway where the thickness of the soil profile was greater than in the area with results as represented

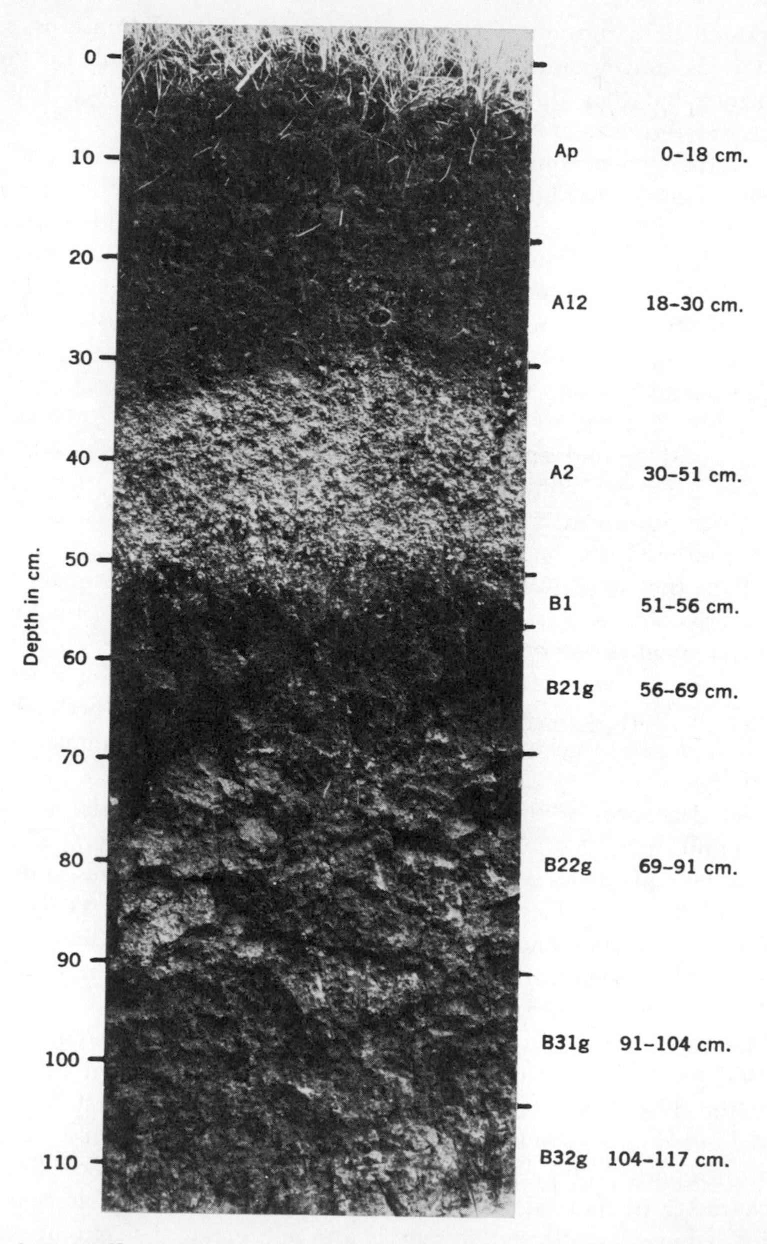

Figure 1.31. Profile of Edina silt loam with profile description. The soil is developed Ap refers to the plowed portion of the A horizon. The symbol g used with the by gleying (chemical reduction). The symbols in parentheses in the description moist soil except as noted. (Photograph courtesy of Roy W. Simonson)

Ap 0–18 cm

Very dark gray (10YR 3/1) friable silt loam; moderate medium granular structure; clear boundary.

A12 18–30 cm

Very dark gray (10YR 3/1) friable silt loam; gray (10YR 5/1, dry); moderate medium granular to weak medium platy structure; common light gray (10YR 7/1, dry) silt coats; clear boundary.

A2 30–51 cm

Gray and dark gray (10YR 5/1 and 4/1) friable silt loam, light gray (10YR 7/1, dry); few fine yellowish brown mottles; moderate medium and fine platy structure; common fine Fe-Mn nodules; clear boundary.

B1 51–56 cm

Mixed gray and very dark gray (10YR 5/1 and 3/1) slightly firm silty clay loam; moderate medium and fine subangular blocky structure; thick nearly continuous light gray (10YR 7/1, dry) silt coats; thick discontinuous clay coats on fine peds; abrupt boundary.

B21g 56–69 cm

Very dark gray and black (10YR 3/1 and 2/1) very firm silty clay; moderate medium prismatic breaking to strong medium and fine subangular structure; few medium yellowish brown mottles on ped interiors; thick continuous clay coats; many hard Fe-Mn nodules; gradual boundary.

B22g 69–91 cm

Dark gray (10YR-2.5Y 4/1) firm to very firm silty clay; weak medium prismatic breaking to moderate fine and medium subangular blocky structure; common fine yellowish brown mottles; thin continuous clay coats; few very dark gray organic coats on prism faces; many fine hard Fe-Mn nodules; gradual boundary.

B31g 91–104 cm

Dark gray to olive gray (2.5Y 4/1 to 5Y 5/2) firm silty clay; weak medium prismatic breaking to moderate fine subangular blocky structure; common fine yellowish brown and few fine yellowish red mottles; thin discontinuous clay coats; few very dark gray organic coats on prism faces; few black organic-clay coats in pores; fine hard Fe-Mn nodules common; gradual boundary.

B32g 104–117 cm

Olive gray (5Y 5/2) slightly firm silty clay loam; some vertical cleavage; common coarse yellowish brown and few fine yellowish red mottles; few black to very dark gray organic-clay coats in pores; few thin dark gray (2.5Y-5Y 4/1) clay coats on cleavage faces; few fine hard Fe-Mn nodules; diffuse boundary.

under grass vegetation on loess on nearly flat topography in Iowa. The symbol lower portions of the B horizon indicates that the portions so labeled are affected of individual horizons are color notations in the Munsell system and refer to the

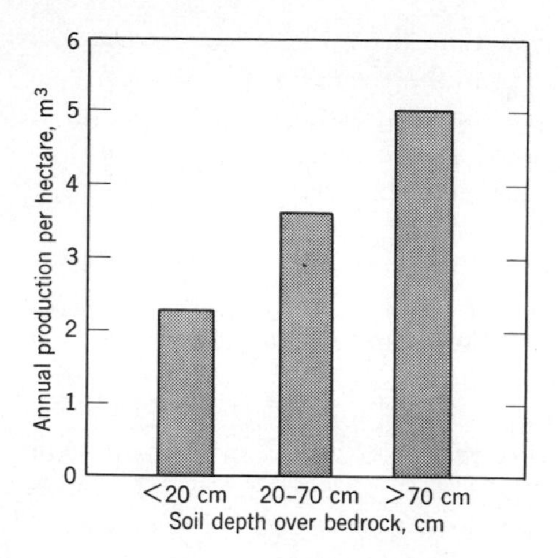

Figure 1.32. Annual forest production on soils classified according to depth over bedrock in the Agder Counties, Norway. (Låg, 1961)

in Fig. 1.32, profile characteristics other than thickness were of considerable importance.

A more complex adaptation of the profile concept in quantitative characterization of soil as a substrate for plant growth is measurement of the properties by depth increments and integration of the results by simple addition over all depth increments downward to the depth of root penetration. Clarke (1951) used this approach for relating the texture profile to the yield of wheat. He made all yield measurements in a single field having extremely heterogeneous soil. After harvesting the wheat from each of a number of small areas scattered over the field, he dug a pit in the center of each area and determined the texture and thickness of each distinct horizon above a depth of 76 cm or above the gley (chemically reduced) horizon, whichever was the more shallow (the 76-cm depth was selected on the basis of observations that few visible roots occurred below that depth). The "profile texture value" was determined by multiplying the thickness of each horizon in centimeters by an arbitrary "texture value" and summing these products. The arbitrary texture values ranged from 8, for soil containing less than 30% silt plus clay, to 20, for soil containing 45 to 55% silt plus clay, to 5, for soil containing more than 80% silt plus clay. These texture values provide an index of the yield of wheat on soil of the different textures under the conditions in the particular circumstances involved. Figure 1.33 shows a plot of yields of wheat against the profile texture values. The relatively small scatter of points around the line representing the general trend shows that the yield of wheat could be estimated to a first approximation from the profile texture values.

The process of adding the measurements on individual soil layers in the foregoing model involves the assumption that the contribution of a layer with particular properties to the over-all value of the soil for supporting plants is independent of the location of the layer in the profile within the depth limits that may be adopted. Consideration of the way in which plants grow shows that the validity of this assumption is open to question. For example, the concentration of roots in the soil changes with depth; and the maximum concentration of roots in individual layers is reached at different times, at least where annual plants are concerned. This limitation leads to a third model for soil profile evaluation.

The third adaptation of the profile concept in quantitative characterization of soil as a substrate for plant growth employs a model that alleviates to some extent the criticism of nonadditivity mentioned in the preceding paragraph. In this model, proposed by Black (1955), the properties measured in individual layers are assumed to be additive after provision has been made for multiplication of the measurements by a coefficient that is allowed to take on different values according to the depth. The procedure for determining the appropriate values involves growing plants on different soils to obtain measurements of

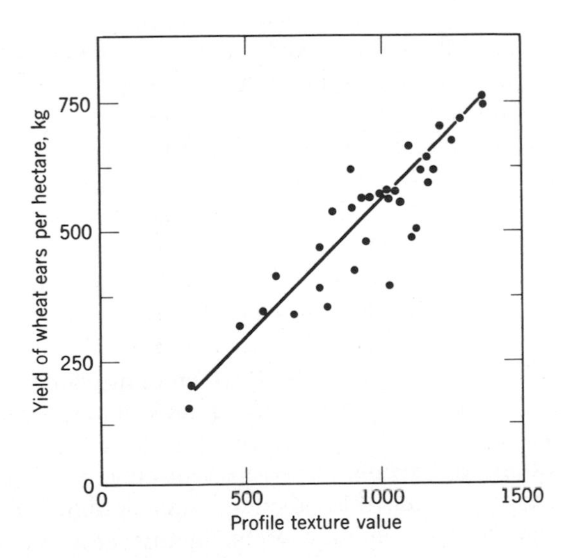

Figure 1.33. Yield of wheat versus soil profile texture value on areas in a field having heterogeneous soil. (Clarke, 1951)

yield or other characteristics, measuring the properties on all the layers in all soils, and using a statistical technique to estimate the values of the layer coefficients that lead to the best estimation of the plant measurements on the total group of soils.

In using this procedure, one must first decide on the definition of the layers. Conceivably one could define the layers in terms of horizon designations, for example, A1, A2, A3, and so on, inserting the value of zero for the property for layers that are not present. In practice, this approach has not been used because plants respond to properties and their location rather than to horizon designations, and because horizon designations do not classify soil layers in terms of properties that affect plant growth. In individual soil profiles, horizons are named partly by their location and properties in relation to each other and not entirely according to their absolute location or properties. Thus the similarity in absolute properties might be greater between the A2 horizon of soil *X* and the A1 horizon of soil *Y* than between the A1 horizons of the two soils. In practice thus far, the layers have been defined in terms of equal thicknesses in all soils without reference to horizon designations. If the layers are thick, horizons with rather different properties may occur within a single layer, which is undesirable. If the layers are thin, however, the vertical distribution of properties can be represented to a satisfactory degree of precision even though the depth and thickness of individual named horizons may differ among soils.

Work of Hanway et al. (1961) may be cited as an example. These investigators conducted 51 field experiments with alfalfa in the north-central United States and in Alaska and related various biological indexes of availability of soil potassium for the alfalfa to the exchangeable potassium content of the surface 90 cm of soil in 15-cm increments. They found that the biological indexes could be estimated with greatest precision where values for exchangeable potassium in all layers were included but that the gain in precision was small from inclusion of exchangeable potassium in layers below the first two 15-cm increments. This example, of course, represents an attempt to evaluate only one characteristic of soil on a profile basis. A total evaluation of soil profiles according to this model would require measurements of many properties in the same general way as was done with potassium and, finally, the integration of all these measurements.

In addition to being incomplete in respects that will be mentioned in following paragraphs, the third model for soil profile evaluation provides an inadequate representation of the nature of soil-plant relationships because of two basic assumptions inherent in the mathematical process used to integrate the measurements. The first assumption is that the contribution of a particular layer to the support of plants is determined by its location and by the magnitude of each property in the layer, independently of the characteristics of the other layers. The second assumption is that the coefficient or weighting factor for a particular layer is the same in different soils. Both these assumptions are known to be generally invalid, although they may provide a first approximation under certain circumstances. Both these deficiencies theoretically could be alleviated if all significant properties were measured and if the mathematical model were expanded in the proper way to take into account the interplay of effects of properties within and among the soil layers.

The adaptations of the profile concept described thus far for characterizing soil quantitatively as a substrate for plant growth have dealt with properties that may be said to be inherent, in a sense, in the soil profile. Certain other significant qualities of soils are not inherent in the soil. An example of such a quality is the water content. Water is supplied to soils from external sources. The water content of soil profiles varies continually while plants are growing, and these variations may have important consequences in terms of plant behavior. Moreover, the water content of soil profiles in practical agriculture can usually be neither controlled nor predicted with precision. Therefore the value of even the most extensive knowledge of the inherent properties of soil profiles is evidently of limited value in soil evaluation in the absence of knowledge of the fluctuations in water content and other significant properties that are not inherent in the soil. Properties of this kind may be incorporated into soil profile evaluations by making measurements over a period of years sufficient to provide an estimate of the average conditions or to permit assignment of probabilities with the aid of supplementary information on the properties in question.

Another deficiency of the models discussed is that they take no account of the qualities of plants that may cause a given set of conditions to be more suitable for one kind of plant than for another. Soil-plant relationships vary with the nature of the plant and must be determined empirically for each kind of plant.

Because of lack of knowledge, the appropriate measurements of the significant properties of soils cannot be made and integrated in a form of general validity. Research on the basic aspects of soil-plant relationships is slowly providing the desired knowledge. From the standpoint of practical agriculture, however, soil evaluation is so important that a great deal of effort is justifiably devoted to development of incomplete evaluations that may be useful in limited geographical areas. These incomplete evaluations may range all the way from relatively simple schemes in which observable profile characteristics are combined in a subjective manner for ranking different soils to relatively sophisticated schemes in which measurements are made of various soil properties, environmental factors, and crop yields at many sites for a number of years, the final results being obtained by development of empirical statistical relationships with the aid of high-speed computers. With time, the gradual incorporation of new developments in basic knowledge and techniques into the strictly empirical evaluations will bring about improvements in precision of the estimates. Adequate evaluation of soil profiles for plant growth can never be expected to become an easy matter, however, because of the many significant properties, their variation with both depth and time, and the tremendous complexity of the interrelationships.

Literature Cited

Adams, R. S., Jr., and F. J. Stevenson. (1964) Ammonium sorption and release from rocks and minerals. *Soil Sci. Soc. Amer. Proc.*, **28**:345–351.

Alderfer, R. B. (1954) Physical condition of the soil affects fertilizer utilization. *Better Crops with Plant Food* **38**, No. 10:24, 44–45.

Allison, L. E. (1956) Soil and plant responses to VAMA and HPAN soil conditioners in the presence of high exchangeable sodium. *Soil Sci. Soc. Amer. Proc.* **20**:147–151.

Archibald, J. A., and A. E. Erickson. (1955) Cation-exchange properties of a number of clay-conditioner systems. *Soil Sci. Soc. Amer. Proc.* **19**:444–446.

Arndt, W. (1965) The nature of the mechanical impedance to seedlings by soil surface seals. *Australian Jour. Soil Res.* **3**:45–54.

Barley, K. P. (1961) The abundance of earthworms in agricultural land and their possible significance in agriculture. *Adv. Agron.* **13**:249–268.

Barshad, I. (1965) Thermal analysis techniques for mineral identification and mineralogical composition. *Agronomy* **9**:699–742.

Baver, L. D. (1956) *Soil Physics.* Third edition. John Wiley and Sons, New York.

Black, C. A. (1955) Evaluation of nutrient availability in soils, and prediction of yield response to fertilization. *Iowa State College Jour. Sci.* **30**:1–11.

Boekel, P. (1959) Some remarks on the influence of soil structure on plant growth. Meded. *Landbouwhogeschool Opzoekingsstat. Gent.* **24**:52–57.

Boekel, P. (1963) Soil structure and plant growth. *Netherlands Jour. Agr. Sci.* **11**:120–127.

Bond, R. D. (1959) Occurrence of microbiological filaments in soils. *Nature* **184**:744–745.

Breese, B. F. (1960) Quartz overgrowths as evidence of silica deposition in soils. *Australian Jour. Sci.* **23**:18–20.

Brewer, R., and A. D. Haldane. (1957) Preliminary experiments in the development of clay orientation in soils. *Soil Sci.* **84**:301–309.

Brown, C. N. (1956) The origin of caliche on the northeastern Llano Estacado, Texas. *Jour. Geol.* **64**:1–15.

Browning, G. M., R. A. Norton, A. G. McCall, and F. G. Bell. (1948) Investigation in erosion control and the reclamation of eroded land at the Missouri Valley

Loess Conservation Experiment Station, Clarinda, Iowa, 1931–42. *U.S. Dept. Agr. Tech. Bul. 959.*
Buehrer, T. F., and J. M. Deming. (1961) Factors affecting aggregation and permeability of hardspot soils. *Soil Sci.* **92**:248–262.
Byers, H. G., L. T. Alexander, and R. S. Holmes. (1935) The composition and constitution of the colloids of certain of the great groups of soils. *U.S. Dept. Agr. Tech. Bul. 484.*
Cady, J. G. (1965) Petrographic microscope techniques. *Agronomy* **9**:604–631.
Chesters, G., O. J. Attoe, and O. N. Allen. (1957) Soil aggregation in relation to various soil constituents. *Soil Sci. Soc. Amer. Proc.* **21**:272–277.
Clarke, F. W. (1924) The data of geochemistry. *U.S. Geol. Surv. Bul. 770,* 5th Ed.
Clarke, G. R. (1951) The evaluation of soils and the definition of quality classes from studies of the physical properties of the soil profile in the field. *Jour. Soil Sci.* **2**:50–60.
Clement, C. R. (1961) Benefit of leys—structural improvement or nitrogen reserves. *Jour. British Grassland Soc.* **16**:194–200.
Coffey, G. N. (1912) A study of the soils of the United States. *U.S. Dept. Agr., Bur. Soils Bul. 85.*
Cornfield, A. H. (1955) The measurement of soil structure and factors affecting it; a review. *Jour. Sci. Food Agr.* **6**:356–360.
Craig, N., and P. Halais. (1934) Influence of maturity and rainfall on the properties of lateritic soils in Mauritius. *Empire Jour. Exptl. Agr.* **2**:349–358.
Day, P. R. (1965) Particle fractionation and particle-size analysis. *Agronomy* **9**:545–567.
Delver, P. (1962) Properties of saline soils in Iraq. *Netherlands Jour. Agr. Sci.* **10**:194–210.
Deshpande, T. L., D. J. Greenland, and J. P. Quirk. (1964) Role of iron oxides in the bonding of soil particles. *Nature* **201**:107–108.
Duley, F. L. (1940) Surface factors affecting the rate of intake of water by soils. *Soil Sci. Soc. Amer. Proc.* (1939) **4**:60–64.
Edwards, A. P., and J. M. Bremner. (1967) Microaggregates in soils. *Jour. Soil Sci.* **18**:64–73.
Emerson, W. W. (1963) The effect of polymers on the swelling of montmorillonite. *Jour. Soil Sci.* **14**:52–63.
Evans, L. T., and E. W. Russell. (1959) The adsorption of humic and fulvic acids by clays. *Jour. Soil Sci.* **10**:119–132.
Failyer, G. H., J. G. Smith, and H. R. Wade. (1908) The mineral composition of soil particles. *U.S. Dept. Agr., Bur. Soils Bul. 54.*
Franzmeier, D. P., E. P. Whiteside, and A. E. Erickson. (1960) Relationship of texture classes of fine earth to readily available water. *Trans. 7th Internat. Congr. Soil Sci.* **1**:354–363.
Gardner, W. R. (1956) Representation of soil aggregate-size distribution by a logarithmic-normal distribution. *Soil Sci. Soc. Amer. Proc.* **20**:151–153.
Greenland, D. J. (1965) Interaction between clays and organic compounds in soils. Part I. Mechanisms of interaction between clays and defined organic compounds. *Soils and Fertilizers* **28**:415–425.
Greenland, D. J., G. R. Lindstrom, and J. P. Quirk. (1962) Organic materials which stabilize natural soil aggregates. *Soil Sci. Soc. Amer. Proc.* **26**:366–371.
Griffiths, E. (1965) Micro-organisms and soil structure. *Biol. Rev.* **40**:129–142.

Grim, R. E. (1958) Organization of water on clay mineral surfaces and its implications for the properties of clay-water systems. *In* H. F. Winterkorn, ed. Water and its conduction in soils. *Highway Res. Board Spec. Rept. 40. National Acad. Sci.-National Res. Coun. Publ. 629, pp.* 17–23.

Hagin, J., and G. B. Bodman. (1954) Influence of the polyelectrolyte CRD-186 on aggregation and other physical properties of some California and Israeli soils and some clay minerals. *Soil Sci.* **78**:367–378.

Haise, H. R., L. R. Jensen, and J. Alessi. (1955) The effect of synthetic soil conditioners on soil structure and production of sugar beets. *Soil Sci. Soc. Amer. Proc.* **19**:17–19.

Hallsworth, E. G., and H. D. Waring. (1964) Studies in pedogenesis in New South Wales. VIII. An alternative hypothesis for the formation of the solodized-solonetz of the Pilliga district. *Jour. Soil Sci.* **15**:158–177.

Hanks, R. J., and F. C. Thorp. (1957) Seedling emergence of wheat, grain sorghum, and soybeans as influenced by soil crust strength and moisture content. *Soil Sci. Soc. Amer. Proc.* **21**:357–359.

Hanway, J. J., S. A. Barber, R. H. Bray, A. C. Caldwell, L. E. Engelbert, R. L. Fox, M. Fried, D. Hovland, J. W. Ketcheson, W. M. Laughlin, K. Lawton, R. C. Lipps, R. A. Olson, J. T. Pesek, K. Pretty, F. W. Smith, and E. M. Stickney. (1961) North central regional potassium studies. I. Field studies with alfalfa. *Iowa Agr. Home Econ. Exp. Sta. Res. Bul. 494* (*North Central Reg. Publ. 124*).

Heinonen, R. (1955) Soil aggregation in relation to texture and organic matter. *Maatalouskoelaitoksen Maatutkimusosasto Agrogeologisia Julkaisuja* N:o. 64.

Hemwall, J. B., and H. H. Scott. (1962) Use of 4-tert-butylpyrocatechol as a fracturing aid in crusting soils. *Agron. Jour.* **54**:535–538.

Hosking, J. S. (1940) The soil clay mineralogy of some Australian soils developed on granitic and basaltic parent material. *Jour. Council Sci. Indus. Res.* (Australia). **13**:206–216.

Humbert, R. P., and C. E. Marshall. (1943) Mineralogical and chemical studies of soil formation from acid and basic igneous rocks in Missouri. *Missouri Agr. Exp. Sta. Res. Bul. 359.*

Hunter, A. S., and O. J. Kelley. (1946) A new technique for studying the absorption of moisture and nutrients from soil by plant roots. *Soil Sci.* **62**:441–450.

Ingols, R. S., and A. T. Navarre. (1952) "Polluted" water from the leaching of igneous rocks. *Science* **116**:595–596.

Jacks, G. V. (1963) The biological nature of soil productivity. *Soils and Fertilizers* **26**:147–150.

Jackson, M. L. (1959) Frequency distribution of clay minerals in major great soil groups as related to the factors of soil formation. *Clays and Clay Minerals. Proc. Sixth Natl. Conf. Clays and Clay Minerals,* pp. 133–143.

Jackson, M. L., and G. D. Sherman. (1953) Chemical weathering of minerals in soils. *Adv. Agron.* **5**:219–318.

Jackson, M. L., S. A. Tyler, A. L. Willis, G. A. Bourbeau, and R. P. Pennington. (1948) Weathering sequence of clay-size minerals in soils and sediments. I. Fundamental generalizations. *Jour. Phys. Col. Chem.* **52**:1237–1260.

Jeffries, C. D. (1947) The mineralogical approach to some soil problems. *Soil Sci.* **63**:315–320.

Jenny, H. (1935) The clay content of the soil as related to climatic factors, particularly temperature. *Soil Sci.* **40**:111–128.

Jenny, H., and C. D. Leonard. (1934) Functional relationships between soil properties and rainfall. *Soil Sci.* **38**:363–381.

Kemper, W. D., and W. S. Chepil. (1965) Size distribution of aggregates. Agronomy **9**:499–510.

Kittrick, J. A. (1965) Electron microscope techniques. *Agronomy* **9**:632–652.

Knox, E. G. (1957) Fragipan horizons in New York soils: III. The basis of rigidity. *Soil Sci. Soc. Amer. Proc.* **21**:326–330.

Kohl, R. A., and S. A. Taylor. (1961) Hydrogen bonding between the carbonyl group and Wyoming bentonite. *Soil Sci.* **91**:223–227.

Kroth, E. M., and J. B. Page. (1947) Aggregate formation in soils with special reference to cementing substances. *Soil Sci. Soc. Amer. Proc.* (1946) **11**:27–34.

Kunze, G. W. (1965) Pretreatment for mineralogical analysis. *Agronomy* **9**:568–577.

Lachenbruch, A. H. (1963) Contraction theory of ice wedge polygons: a qualitative discussion. *Proc. Permafrost Internat. Conf. Natl. Acad. Sci.—Nat. Res. Council Pub.* **1287**:63–71.

Låg, J. (1961) Some investigations on the productivity of forest soils in Norway. *Acta Agr. Scand.* **11**:82–86.

Longwell, T. J., W. L. Parks, and M. E. Springer. (1963) Moisture characteristics of representative Tennessee soils. *Tennessee Agr. Exp. Sta. Bul. 367.*

Low, A. J. (1954) The study of soil structure in the field and the laboratory. *Jour. Soil Sci.,* **5**:57–74.

Low, P. F. (1961) Physical chemistry of clay-water interaction. *Adv. Agron.* **13**:269–327.

Lowdermilk, W. C. (1930) Influence of forest litter on run-off, percolation, and erosion. *Jour. Forestry* **28**:474–491.

Lutwick, L. E., and E. H. Hobbs. (1964) Relative productivity of soil horizons, singly and in mixture. *Candian Jour. Soil Sci.* **44**:145–150.

Lutz, J. F. (1937) The relation of free iron in the soil to aggregation. *Soil Sci. Soc. Amer. Proc.* (1936) **1**:43–45.

Martin, J. P., J. O. Ervin, and R. A. Shepherd. (1965) Decomposition and binding action of polysaccharides from *Azotobacter indicus* (*Beijerinckia*) and other bacteria in soil. *Soil Sci. Soc. Amer. Proc.* **29**:397–400.

Martin, J. P., W. P. Martin, J. B. Page, W. A. Raney, and J. D. DeMent. (1955) Soil aggregation. *Adv Agron.* **7**:1–37.

Martin, W. P., G. S. Taylor, J. C. Engibous, and E. Burnett. (1952) Soil and crop responses from field applications of soil conditioners. *Soil Sci.* **73**:455–471.

Mazurak, A. P. (1950) Effect of gaseous phase on water-stable synthetic aggregates. *Soil Sci.* **69**:135–148.

Mazurak, A. P., and R. E. Ramig. (1963) Residual effects of perennial grass sod on the physical properties of a chernozem soil. *Soil Sci. Soc. Amer. Proc.* **27**:592–595.

McCaughey, W. J., and W. H. Fry. (1913) The microscopic determination of soil-forming minerals. *U.S. Dept. Agr., Bur. Soils Bul. 91.*

McIntyre, D. S. (1956) The effect of free ferric oxide on the structure of some terra rossa and rendzina soils. *Jour. Soil Sci.* **7**:302–306.

Mehta, N. C., H. Streuli, M. Müller, and H. Deuel. (1960) Rôle of polysaccharides in soil aggregation. *Jour. Sci. Food. Agr.* **11**:40–47.

Nichols, A. (1939) Some applications of mineralogy to soil studies. *Jour. Australian Inst. Agr. Sci.* **5**:218–221.

Odell, R. T., T. H. Thornburn, and L. J. McKenzie. (1960) Relationships of Atterberg limits to some other properties of Illinois soils. *Soil Sci. Soc. Amer. Proc.* **24**:297–300.

Paschall, A. H., R. T. A. Burke, and L. D. Baver. (1935) Aggregation studies on the Muskingum, Chester and Lansdale silt loams. *Amer. Soil Survey Assoc. Bul.* **16**:44–45.

Peele, T. C. (1940) Microbial activity in relation to soil aggregation. *Jour. Amer. Soc. Agron.* **32**:204–212.

Peele, T. C. (1950) Relation of percolation rates through saturated soil cores to volume of pores drained in 15 and 30 minutes under 60 centimeters tension. *Soil Sci. Soc. Amer. Proc.* (1949) **14**:359–361.

Pennington, R. P., and M. L. Jackson. (1948) Segregation of the clay minerals of polycomponent soil clays. *Soil Sci. Soc. Amer. Proc.* (1947) **12**:452–457.

Peterson, J. B. (1946) Relation of parent material and environment to the clay minerals in Iowa soils. *Soil Sci.* **61**:465–475.

Péwé, T. L. (1963) Ice-wedges in Alaska—classification, distribution, and climatic significance. *Proc. Permafrost Internat. Conf. Natl. Acad. Sci.—Natl. Res. Council Publ.* **1287**:76–81.

Phillips, R. E., and D. Kirkham. (1962) Mechanical impedance and corn seedling root growth. *Soil Sci. Soc. Amer. Proc.* **26**:319–322.

Puhr, L. F., and O. Olson. (1937) A preliminary study of the effect of cultivation on certain chemical and physical properties of some South Dakota soils. *South Dakota Agr. Exp. Sta. Bul. 314.*

Quastel, J. H. (1954) Soil conditioners. *Annual Rev. Plant Physiol.* **5**:75–92.

Rennie, D. A., E. Truog, and O. N. Allen. (1954) Soil aggregation as influenced by microbial gums, level of fertility and kind of crop. *Soil Sci. Soc. Amer. Proc.* **18**:399–403.

Rubin, J. (1949) The influence of externally applied stresses upon the structure of confined soil material. Unpublished Ph.D. Thesis, University of California, Berkeley.

Russell, E. J., and E. W. Russell. (1950) *Soil Conditions and Plant Growth.* Eighth ed. Longmans, Green and Co., London.

Rynasiewicz, J. (1945) Soil aggregation and onion yields. *Soil Sci.* **60**:387–395.

Semb, G., and G. Uhlen. (1955) A comparison of different analytical methods for the determination of potassium and phosphorus in soil based on field experiments. *Acta Agr. Scand.* **5**:44–68.

Sherman, G. D. (1959) Nature and types of secondary mineral aggregates, concretions, nodules and layers of soil. *Jour. Indian Soc. Soil Sci.* **7**:193–197.

Shreve, F., and T. D. Mallery. (1933) The relation of caliche to desert plants. *Soil Sci.* **35**:99–113.

Sivarajasingham, S., L. T. Alexander, J. G. Cady, and M. G. Cline. (1962) Laterite. *Adv. Agron.* **14**:1–60.

Soil Survey Staff. (1951) Soil Survey Manual. *U.S. Dept. Agr. Handbook 18.*

Soil Survey Staff. (1960) Soil Classification, a Comprehensive System, 7th Approximation. Soil Conservation Service, U.S. Department of Agriculture, Washington, D.C.

Sokolovsky, A. N. (1933) The problem of soil structure. *Int. Soc. Soil Sci., Trans. First Commission, Soviet Section, A,* **1**:34–110.

Stallings, J. H. (1950) Keep crop residues on surface of ground. *Better Crops with Plant Food* **34**, No. 8:9–16, 48–49.

Stallings, J. H. (1952) Soil aggregate formation. *U.S. Dept. Agr. SCS-TP-10.*

Stuart, D. M., M. A. Fosberg, and G. C. Lewis. (1961) Caliche in southwestern Idaho. *Soil Sci. Soc. Amer. Proc.* **25**:132–135.

Swaby, R. J. (1950) The influence of humus on soil aggregation. *Jour. Soil Sci.* **1**:182–194.

Tackett, J. L., and R. W. Pearson. (1964) Oxygen requirements of cotton seedling roots for penetration of compacted soil cores. *Soil Sci. Soc. Amer. Proc.* **28**:600–605.

Taubenhaus, J. J., W. N. Ezekiel, and H. E. Rea. (1931) Strangulation of cotton roots. *Plant Physiol.* **6**:161–166.

Taylor, H. M., and E. Burnett. (1964) Influence of soil strength on the root-growth habits of plants. *Soil Sci.* **98**:174–180.

Taylor, H. M., E. Burnett, and N. H. Welch. (1963) Cotton growth and yield as affected by taproot diameter within a simulated restraining soil layer. *Agron. Jour.* **55**:143–144.

Taylor, H. M., and H. R. Gardner. (1963) Penetration of cotton seedling taproots as influenced by bulk density, moisture content, and strength of soil. *Soil Sci.* **96**:153–156.

Tepe, W., and E. Leidenfrost. (1958) Ein Vergleich zwischen pflanzenphysiologischen, kinetischen und statischen Bodenuntersuchungswerten. I. Mitteilung. Die Kinetic der Bodenionen, gemessen mit Ionenaustauchern. *Landw. Forsch.* **11**:217–229.

Van Bavel, C. H. M. (1950) Mean weight-diameter of soil aggregates as a statistical index of aggregation. *Soil Sci. Soc. Amer. Proc.* (1949) 14:20–23.

Veihmeyer, F. J., and A. H. Hendrickson. (1946) Soil density as a factor in determining the permanent wilting percentage. *Soil Sci.* **62**:451–456.

Vomocil, J. A. (1957) Measurement of soil bulk density and penetrability: a review of methods. *Adv. Agron.* **9**:159–175.

Walter, B., (1965) Über Bildung und Bindung von Mikroaggregaten in Böden. II. Mitteilung. *Zeitschr. Pflanzenernähr., Düng., Bodenk.* **110**:43–49.

Weldon, T. A., and J. C. Hide. (1942) Some chemical properties of soil organic matter and of sesquioxides associated with aggregation in soils. *Soil Sci.* **54**:343–352.

Whistler, R. L., and K. W. Kirby. (1956) Composition and behavior of soil polysaccharides. *Jour. Amer. Chem. Soc.* **78**:1755–1759.

Whitney, M. (1896) Texture of some important soil formations. *U.S. Dept. Agr., Bur. Soils Bul. 5.*

Whittig, L. D. (1965) X-ray diffraction techniques for mineral identification and mineralogical composition. *Agronomy* **9**:671–698.

Wiersum, L. K. (1957) The relationship of the size and structural rigidity of pores to their penetration by roots. *Plant and Soil* **9**:75–85.

Wiersum, L. K. (1962) Uptake of nitrogen and phosphorus in relation to soil structure and nutrient mobility. *Plant and Soil* **16**:62–70.

Wilsie, C. P., C. A. Black, and A. R. Aandahl. (1944) Hemp production experiments: cultural practices and soil requirements. *Iowa Agr. Exp. Sta. Bul. P63.*

Woodruff, C. M. (1940) Soil moisture and plant growth in relation to *p*F. *Soil Sci. Soc. Amer. Proc.* **5**:36–41.

Yankovitch, L., and P. Berthelot. (1948) Enracinement de l'olivier et des autres arbres fruitiers dans le sud de la Tunisie. *Compt. Rend Acad. Agr. France* **34**:774–776.

2 *Water*

Life processes in plants take place in water, which, in actively growing herbaceous plants, is commonly present in weights four to eight times as great as those of total solids. In addition, for every kilogram of dry matter produced, several hundred kilograms of water pass through plants and are lost from their surfaces by transpiration or evaporation into the atmosphere.

Plants can absorb quantities of certain mineral nutrients beyond their immediate needs and then can grow for weeks without additional absorption, but most plants require water continually. During periods of active growth, even a day without absorption of water may cause a decrease in subsequent growth or may even be fatal.

Soil acts as an absorbent for water received from precipitation or irrigation and then serves as a source of water for plants in the intervals between additions. The water retention capacity of soils may be sufficient to supply water for an entire season, without recharge, to plants with a deep root system or a short growth cycle. For most agricultural plants, however, the capacity of soils to hold water within the root zone is not great enough to meet the requirements for an entire season; thus periodic recharging is necessary as growth proceeds.

As a factor limiting plant growth over the land surface of the earth, water probably is foremost in importance. Vast land areas have too little water, and smaller areas have too much, on the average. Variations in timing and quantity of rainfall cause most areas to experience variations in water supply that have significant consequences for plants.

Retention of water by soils and movement of water in soils constitute two major segments of the field of soil physics. The physical behavior of water in soils forms the basis for much of the subject matter of this chapter. The principal objective of this chapter, however, is not to treat the properties of the soil-water system as such, but rather to describe the soil-plant relationships in terms based on these properties. Reviews of related subject matter may be found in publications by Veihmeyer (1956), Ruhland (1956), Kramer (1959, 1963), Russell (1959), W. R. Gardner (1960a, 1965), Milthorpe (1960), Oppenheimer (1960),

Staple (1960), Stocker (1960), Taylor (1960), Vaadia et al. (1961), Slatyer (1962), Viets (1962, 1966), Rutter and Whitehead (1963), Henckel (1964), Kozlowski (1964), W. H. Gardner (1965), and Pierre et al. (1965).

Physical Condition of Water in Soil

On a volume basis, solids usually occupy 40 to 70% of the total space in soils. If the remaining space is filled with water, the soil is said to be saturated. Usually, the space not occupied by solids is filled with water and with air, which contains water in the vapor state. The soil is then said to be unsaturated.

In unsaturated soils, the water occurs in films around the solid particles (Fig. 2.1) and not as discrete droplets. Water occurs as films instead of droplets because the water is attracted strongly by the solid particles. The principal sources of the attraction are probably hydrogen bonding, which is the mutual attraction of the oxygen atoms of water and of the surface of soil particles for the hydrogen atoms of the water, and hydration of the exchangeable cations. The firmness with which the

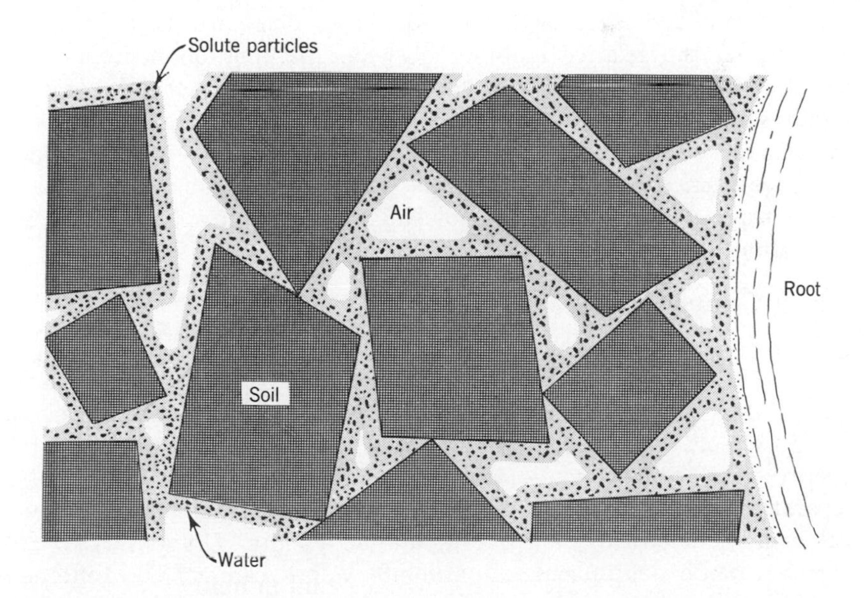

Figure 2.1. Schematic drawing of a cross section of a plant root and soil showing water film with solute particles around the root and soil particles. (After Richards, 1959)

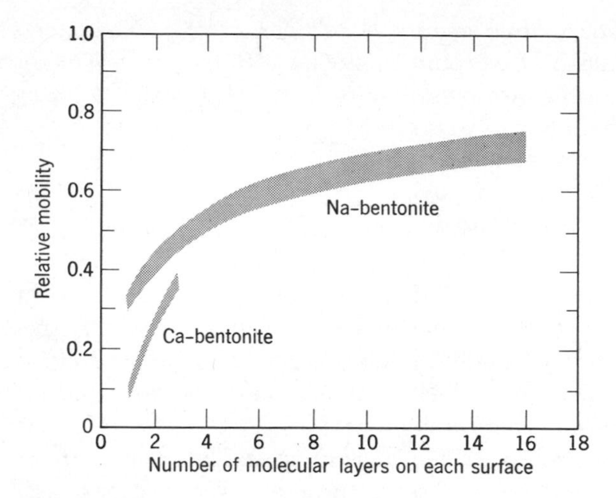

Figure 2.2. Mobility of water on surface of sodium- and calcium-saturated bentonite clay relative to that of bulk water versus number of molecular layers of water on each surface of the clay. Relative mobility values were inferred from measurements on diffusion of deuterium oxide in ordinary water. (Kemper, Maasland, and Porter, 1964)

water is held increases as the film thickness decreases, as indicated in Fig. 2.2 by the relative mobility of water in films on bentonite clay and in bulk. Removal of this film water, by plants or other means, requires application of energy sufficient to overcome the attraction of the soil for water. According to Richards (1961), most of the removal of water from soils by plants occurs while the film thickness is in the range from six or eight monomolecular layers to two or three times this value.

Energy Concept

There seems to be general agreement that the retention of water by soil and the tendency of water to move from place to place in soil, from soil to plant, and from plant to atmosphere are consequences of energy effects. The most appropriate term for this appears to be free energy. Free energy includes all types of energy that may be described separately, such as kinetic energy, potential energy, heat energy, chemical energy, electrical energy, and atomic energy. Differences in free energy can be determined but absolute values cannot. The difference in free energy between a reference state and another state is defined as the maximum amount of work (force $\times$ distance) that can be accomplished by the change from the reference state to the state in question.

The difference in free energy between water in soil and in the usual reference state of free, pure water at the same temperature, pressure, and height is thus a negative value because of the loss of energy associated with retention of water by the soil.

If the difference in free energy between two states is known, the direction (but not the rate) of spontaneous energy changes between the two states can be predicted. The tendency is for loss of free energy from the higher state and gain by the lower state. For example, because the free energy of water in relatively dry soil is lower than that in free, pure water at the same temperature, atmospheric pressure, and height, the tendency is for water to move from the reference state into the soil. Thus when a column of dry soil is placed in a container of water, water rises in the soil above the height of water in the container. Or, if quantities of dry soil and water are placed in separate beakers under a bell jar, water evaporates from the beaker of water, diffuses through the air as vapor, and condenses in the soil, eventually causing the soil to become wet. Both these changes result in lowering the total free energy of the water present originally in the free, pure state and of increasing the total free energy of the water in the soil.

Water in soil is subject to several kinds of forces that may cause its free energy to differ from that in free, pure water. Attraction of the soil solids for water has already been mentioned. Another force, negligible in many soils but important in some, is the attraction of solutes for water. Differences in height between soil water and reference water introduce gravitational effects. The atmospheric pressure may differ (such differences are not of consequence under practical circumstances, but they are of great significance in some methods of measurement). Differences in temperature also are a cause of differences in free energy between soil water and water in the reference state. All these components combine to form the total difference in free energy between soil water and reference water.

A moment's reflection on methods of measurement brings to mind that temperatures are measured by thermometers; pressures are measured by manometers or gauges; solutes are measured by chemical analysis, electrical conductivity, or other means; and so on. Conceivably, a number of measurements might need to be made independently, converted to energy units, and then integrated to obtain the total difference in free energy between soil water and reference water. Here is where the difficulty lies. Although the free-energy concept offers the most satisfying representation of the status of soil water to people interested in theory, the problems of measurement are such that strict application of the concept to conditions in the field is not yet feasible. The best and most advanced method available for determining the difference in free energy between free, pure water and soil water involves measuring the vapor pressure of water in air in equilibrium with soil (L. A. Richards, 1965). This approach is based on the principle that the free energy of water is the same in any two phases in equilibrium with each other. Although the principle is simple, the application is most difficult. In soils containing enough water to support plant growth, the air is from about 99 to 100% saturated with water, a range of only 1%. It is the smallness of this range that makes measurements so difficult. To obtain meaningful values requires precision work with complex apparatus that is not suitable for use outside the laboratory.

Practical Terminology

Measurements made by the water-vapor method do not take into account all components of free energy that possibly may differ between soil water and free water, but they do integrate two recognizable and significant classes of effects that may be measured independently by simpler techniques usable on a practical basis. These will be discussed in following paragraphs. With the measurement techniques, a nomenclature and set of units differing from those of thermodynamics have been developed. For people who are concerned with measurements, as opposed to thermodynamic theory, the practical units and nomenclature have the advantage of appeal to the senses in terms that can be understood more readily than those of thermodynamics. Although the units can be converted easily to those of thermodynamics, to do so would be to lose the advantage just mentioned and at the same time to imply, inaccurately, that they represent integrated free-energy effects. The practical terms are matric suction, solute suction, and total suction; and their significance may be explained by reference to Fig. 2.3.

Definition of Terms. In Fig. 2.3, a container of moist soil at atmospheric pressure is in contact with soil solution and pure water through appropriate membranes. Matric suction is indicated by the central manometer. The mercury in the reservoir of the manometer is at atmospheric pressure, the same as the soil. The fact that the mercury in the tube leading to the closed, rigid chamber of soil solution (which, in turn, is connected to the soil by a membrane that is permeable to water and solutes but not to air) stands above the level of mercury in the reservoir means that the pressure on the mercury in the manometer tube is less than the pressure of the atmosphere on the mercury in the reservoir. The pressure on the mercury is the pressure of the soil solution, which is in equilibrium with the soil. The pressure of the soil solution is attributable to the soil solids or the soil matrix with which

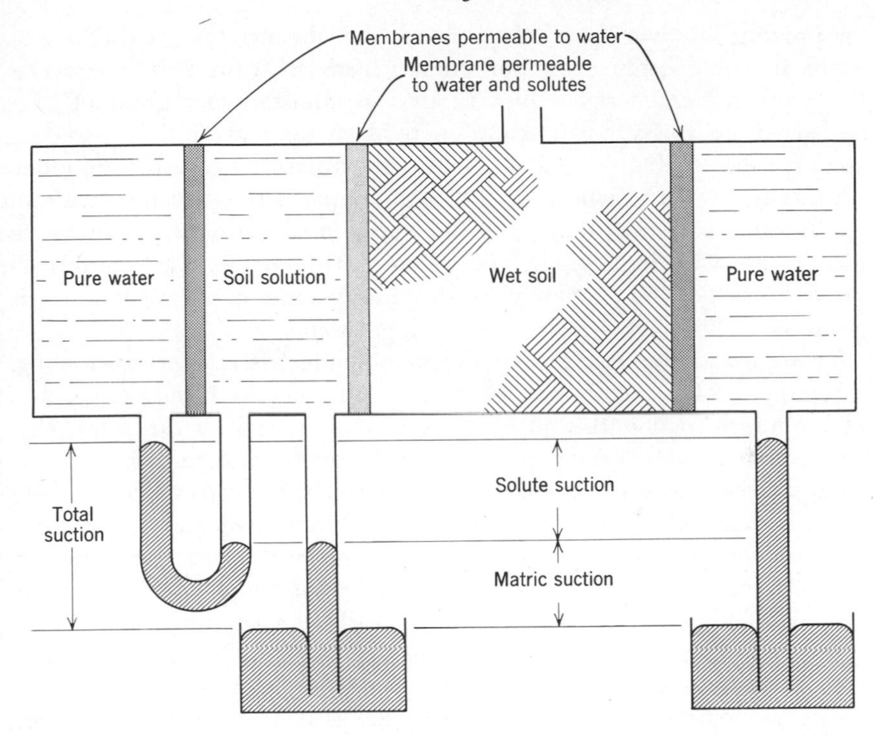

Figure 2.3. Hypothetical membrane system under isothermal equilibrium conditions illustrating the concepts of matric suction, solute suction, and total suction of soil water. See text for further explanation. (After L. A. Richards, unpublished)

the soil solution is in contact. Because the pressure is negative with respect to that of the atmosphere, the term "suction" is appropriate. The mercury manometer registers the minimum suction required to remove an infinitesimal quantity of solution from the soil solids.

The height of the column of mercury in the central manometer is an index of the matric suction of water in the soil. If desired, matric suction may be expressed in millimeters of mercury, which is common practice where atmospheric pressures are concerned. More commonly, it is expressed in terms of centimeters of water, atmospheres, or bars. These are all units of pressure (force/area). At 25°C, 1 cm of water produces a pressure equivalent to 977.8 dynes/cm.^{2} One standard atmosphere exerts a pressure equivalent to 1036 cm of water or 1.013×10^6 dynes/cm^2. The bar is a metric unit of pressure equivalent to 1023 cm of water (at 25°C) or 10^6 dynes/cm^2. The atmosphere and the bar thus are almost equivalent units. Multiplication of the numerical values for pressure in dynes per square centimeter by the specific volume of water in cubic centimeters per gram (1.003 cm^3/g at 25°C) converts the pressures to energy values of dyne-centimeters per gram or ergs per gram. In the vapor-pressure method mentioned in the preceding section, the suction value in bars is obtained from the product, $-1372 \log_e (p/p_0)$, where p is the vapor pressure of water in air in equilibrium with soil, and p_0 is the vapor pressure of water in air in equilibrium with bulk water, and where both vapor pressures are measured at 25°C. The suction value in energy units of ergs per gram is given by the product $-1.376 \times 10^9 \times \log_e (p/p_0)$.

To return to Fig. 2.3, the manometer on the left is connected across a semipermeable membrane separating soil solution from pure water. (Membranes having this property are hypothetical in the sense that no membrane yet developed will restrain solute molecules completely while permitting water molecules to pass freely.) Solute suction is indicated by the difference in height of the column of mercury in the two sides of this manometer. Solute suction thus represents the minimum suction required to remove an infinitesimal quantity of pure water from the soil solution. Solute suction is numerically equal to the osmotic pressure of the soil solution and may be expressed in the same pressure units as matric suction.

The manometer at the right in Fig. 2.3 is connected to pure water, which, in turn, is in contact with the soil solution in the moist soil through a semipermeable membrane that permits passage of water but not solutes. The mercury reservoir in this manometer is open to the atmosphere. This manometer responds to total suction (the sum of matric suction and solute suction), and the height of the column of mercury above the mercury in the reservoir is equal to the sum of the heights in the other two manometers.

Matric Suction. For measuring the matric suction of water in soil, the instrument most widely used is the tensiometer, developed by Gardner et al. (1922). A tensiometer is a fired, porous, clay vessel that is filled with water and attached to a manometer or a vacuum gauge (Fig. 2.4). The clay vessel is placed in the soil, where contact between water in the vessel and the soil solution in the soil is made through the porous clay walls. If the vessel is placed in a soil that is moist but not flooded, and if the initial suction reading on the manometer or gauge is zero (the zero value is determined previously by finding the equilibrium reading obtained when the clay vessel is immersed in water to half its depth), water will flow from the interior of the vessel through the pores in the walls and into the soil until equilibrium is established. Removal of water causes the gauge or manometer liquid to register

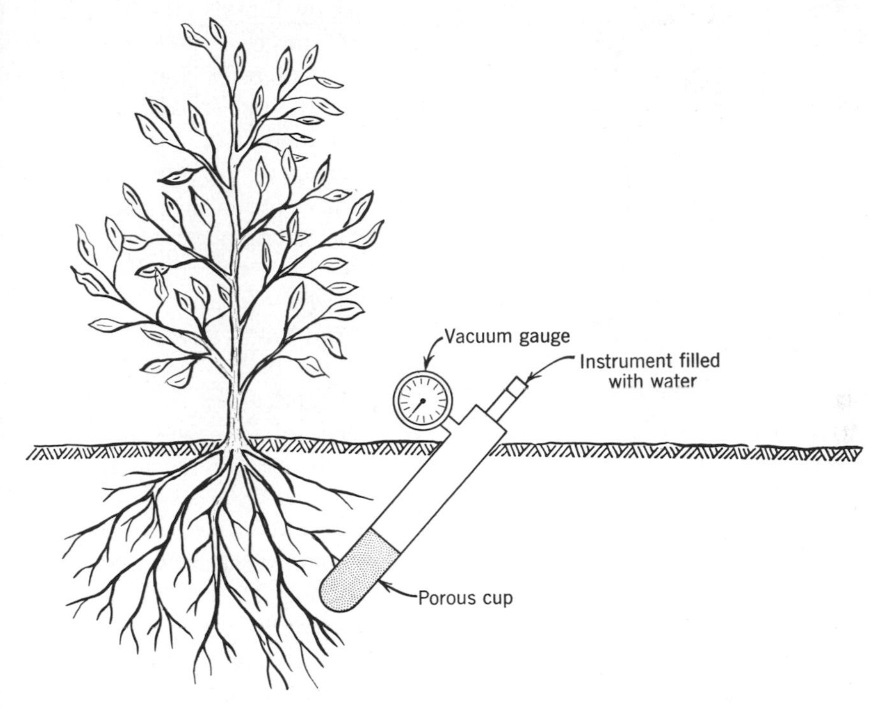

Figure 2.4. Vacuum-gauge-type tensiometer placed in soil to measure matric suction. (After Richards, 1959)

the matric suction characteristic of the soil around the porous clay vessel. The suction value registered at equilibrium is essentially unchanged if, instead of water, a tensiometer is filled with a salt solution. Thus the reading obtained does not reflect solute effects as such.

Tensiometers fail to operate at suctions greater than about 0.85 bar and thus cover less than one-tenth the range of matric suction at which plants grow well. Although the limited range places a serious restriction on the versatility of the instruments, the moist range they do cover is the most important in supplying water to plants. Tensiometers placed at one or more depths below the surface of the soil are widely used to determine when to irrigate; they may also be used to determine whether water is moving up or down in the lower part of the soil, an important matter from the standpoint of salt accumulation or removal (S. J. Richards, 1965). The importance of the range within which tensiometers operate may be inferred from Fig. 2.5, in which the water content of several soils is plotted against the matric suction. The content

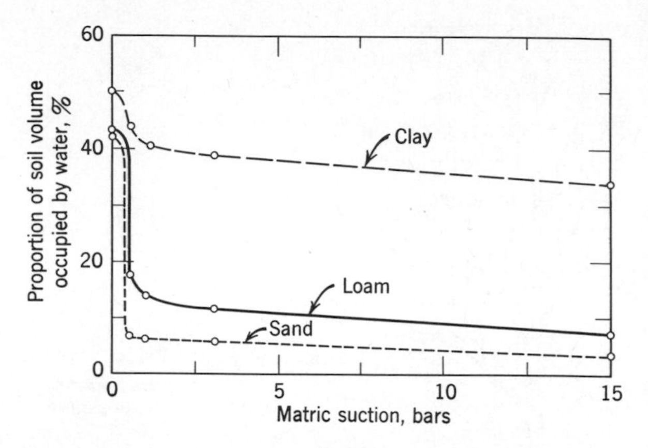

Figure 2.5. Volumetric water content of three soils at different matric suctions. (Richards, 1959)

of water in nonsaline soils at which plants wilt permanently corresponds approximately to that held at a matric suction of 15 bars. Of the water removed from the soils at matric suctions up to 15 bars, however, the greater part was removed in the range below 1 bar. Figure 2.5 illustrates also the considerable difference in water content at a given matric suction that may be found in soils differing in texture.

As yet, no single instrument has been developed for field use that will measure the complete range of matric suction of interest for the growth of plants. An innovation in the tensiometer method described by Kemper and Amemiya (1958), however, seems capable of development to provide such measurements. These investigators found that matric suction could be estimated from the permeability to air of empty porous clay vessels after equilibration with soil. The walls of the vessel are first moistened with water. Then, if the pores are large enough in relation to the matric suction, some of them will be emptied by withdrawal of water into the soil; and air will flow through them if pressure is applied. As the matric suction increases, the number of pores that will be emptied and the rate at which air will flow increase. The rate of air flow is measured by attaching the vessel to a source of compressed air and a manometer and determining the rate at which the liquid falls in the manometer. Although each porous material tested by Kemper and Amemiya had only a limited range of sensitivity, the desired range could be obtained by preparation of special ceramics with the desired size distribution of pores or by preparation of a single instrument including sections of the different types of ceramics needed to cover the range.

For laboratory work, extensive use is made of the pressure-membrane apparatus described by L. A. Richards (1965). This apparatus is suitable for producing a desired value of matric suction in the water in a sample of soil but not for measuring the matric suction. A sample of soil is placed on a ceramic or cellulose acetate membrane arranged to provide adequate support while permitting water to flow out beneath the membrane. Air pressure is then applied above the sample and membrane. The solution in the soil flows out through the fine pores in the membrane, the lower side of which is at atmospheric pressure. Flow ceases when the matric suction of the soil water is equal to the difference in air pressure above and below the membrane. With the aid of a calibration curve, pressure-membrane measurements could serve indirectly for determining the matric suction of a given soil at any particular measured content of water.

In the use of a matric-suction, water-content calibration derived from removal of water from moist soil, however, one must keep in mind that it does not apply to soil that is gaining water. The content of water corresponding to a given matric suction is lower when soil is absorbing water than when it is losing water. This behavior is indicated in Fig. 2.6. Presumably there are two causes of the difference between

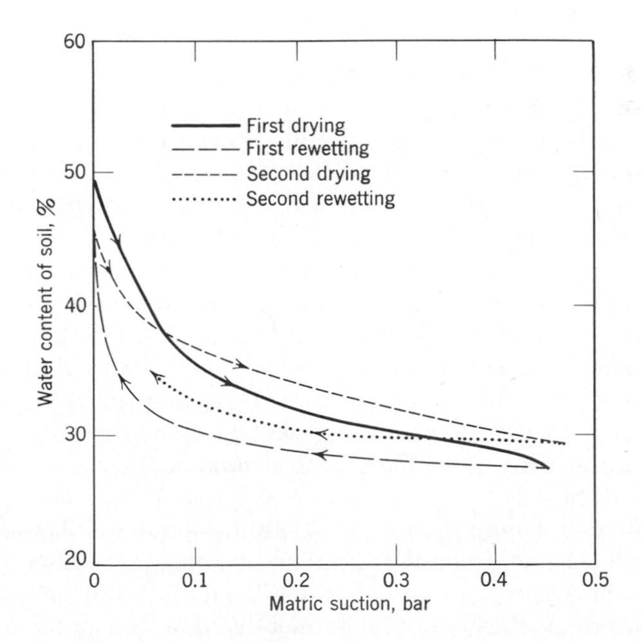

Figure 2.6. Water content in a silt loam soil versus matric suction during two successive cycles of drying and rewetting. (Richards, 1941)

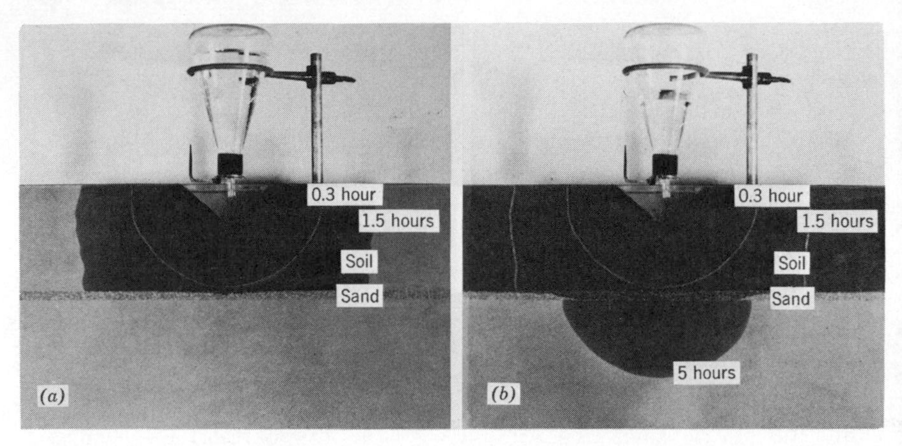

Figure 2.7. Penetration of water into a soil of fine texture containing a horizontal layer of sand, illustrating the principle that the matric suction at which pores fill with water decreases as the pore size increases. Penetration is indicated by the relatively dark color of the wet soil and by the marks and labels in the figures. (*a*) Penetration at the end of 1.5 hours. Note that the water has spread laterally in the soil from the central furrow where the water was added from the inverted flask, but that it has not penetrated the sand. (*b*) Penetration at the end of 5 hours. Note that the surface layer of soil has been wetted completely and that the water has penetrated the layer of sand into the soil below. (Photographs courtesy of W. H. Gardner)

wetting and drying curves. First, the matric suction at which a given soil pore will be drained during drying is determined by the diameter of the largest bounding neck in contact with air, but the matric suction at which the drained pore will fill with water during wetting is determined by the maximum diameter of the pore itself. The latter effect is illustrated in Fig. 2.7 by the slower penetration of water into a layer of sand during wetting than into the surrounding soil of finer texture. Second, the soil particles may undergo some incompletely reversible rearrangement to a closer packing as the soil shrinks during drying, with the result that in the presence of a given quantity of water the soil occupies a smaller volume during wetting than during the preceding drying. Evidently, therefore, there is no unique matric-suction, soil-water curve for a given soil.

Solute Suction. Solute suction results from interactions between water and the solutes and from the dilution of water by solute molecules and ions. The vapor pressure of water decreases with an increase in the concentration of solutes. Consequently, if a beaker of a solution and another beaker of water are placed under a bell jar, water will be transferred by the vapor phase from the pure water to the solution, indicating that the free energy of water in the solution is lower than that of the pure water.

Solute suction of soils is ordinarily estimated from measurements of electrical conductivity of a quantity of extracted soil solution. Conductivity measurements do not respond to nonionic solutes, but the quantities of such solutes in soils are negligible in comparison with those of ionic solutes in instances where solute suction is of consequence. Campbell, Bower, and Richards (1949) found that at 25°C the solute suction of solutions derived from soils could be estimated from the electrical conductivity by the relationship,

$$\text{Solute suction in bars} = (0.325)\,(\text{electrical conductivity in millimhos/cm})^{1.065}$$

The concept of solute suction as separate from matric suction is ambiguous in a sense because the exchangeable cations attached to soil particles produce solute suction and affect the electrical conductivity of soil; yet the fact that solute suction is measured on extracted soil solution implies that the exchangeable cations do not produce solute suction. This inconsistency is reconciled by adoption of the philosophy that the exchangeable cations are bound to the matrix, and their effect is automatically evaluated as a part of matric suction even though it may not originate in the same way as other effects attributed to the matrix. A second question with regard to the validity of evaluations of solute suction from conductivity of extracted soil solution arises from the fact that the solution extracted may not be representative of the total soil solution. The relative magnitude of the error from this cause would be expected to decrease with increasing solute suction and therefore would be of least consequence in circumstances in which one is concerned about the contribution of solute suction to total suction. Richards and Ogata (1961) found that the sum of matric suction and solute suction measured independently was approximately equal to, but frequently a little greater than, the total suction measured by the vapor-pressure method.

Movement of Water in Soil

Pore-Size Effects

The rate of flow of water in both saturated and unsaturated soils is proportional to the potential gradient, which may be expressed as centimeters of water per centimeter of distance, bars per meter, or other

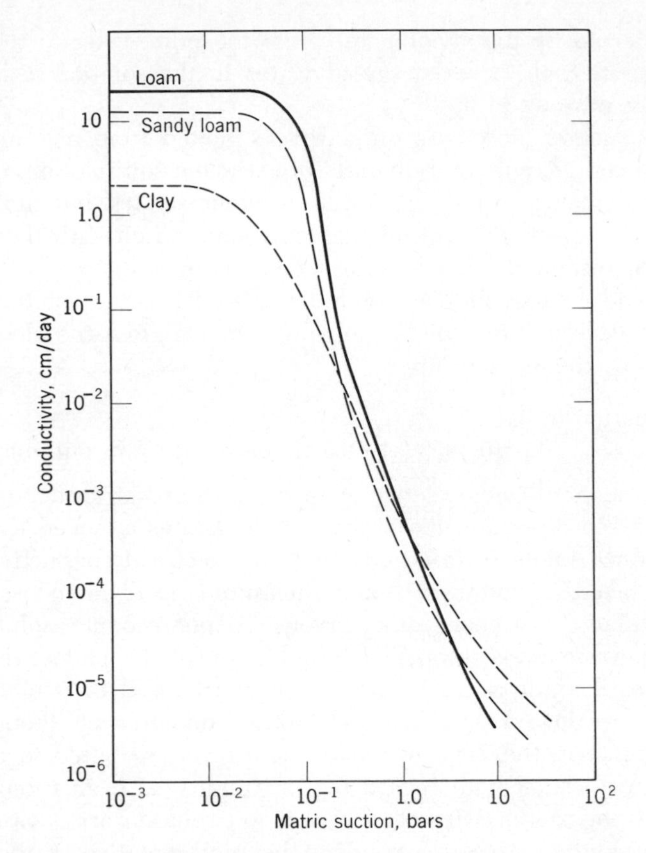

Figure 2.8. Conductivity of three soils for water versus matric suction. The conductivity in centimeters per day represents the number of cubic centimeters of water moving per day through a cross-sectional area of 1 cm^2 where the gradient is 1 cm of water per centimeter of distance in the direction of flow. (Gardner, 1960a)

units, as desired. The proportionality factor decreases with decreasing water content or increasing matric suction of water in a given soil, as illustrated in Fig. 2.8.

The behavior in Fig. 2.8 may be explained in a general way by the analogy of flow of water in capillary tubes. The quantity of water that flows through individual capillary tubes per unit time decreases rapidly as the bore becomes smaller, being proportional to the fourth power of the internal radius. Thus, if the rate of flow through a capillary tube having a radius of 1 mm is taken arbitrarily as unity, the rate of flow through the number of smaller tubes sufficient to give the same total cross-sectional area will be 0.01 if the radius is 0.1 mm and 0.0001 if the radius is 0.01 mm. The decrease in rate of flow is accounted for on the basis that an increasing proportion of the water molecules in the tube is influenced by the proximity of the wall as the tube diameter becomes smaller. Water molecules adjacent to the wall are thought to

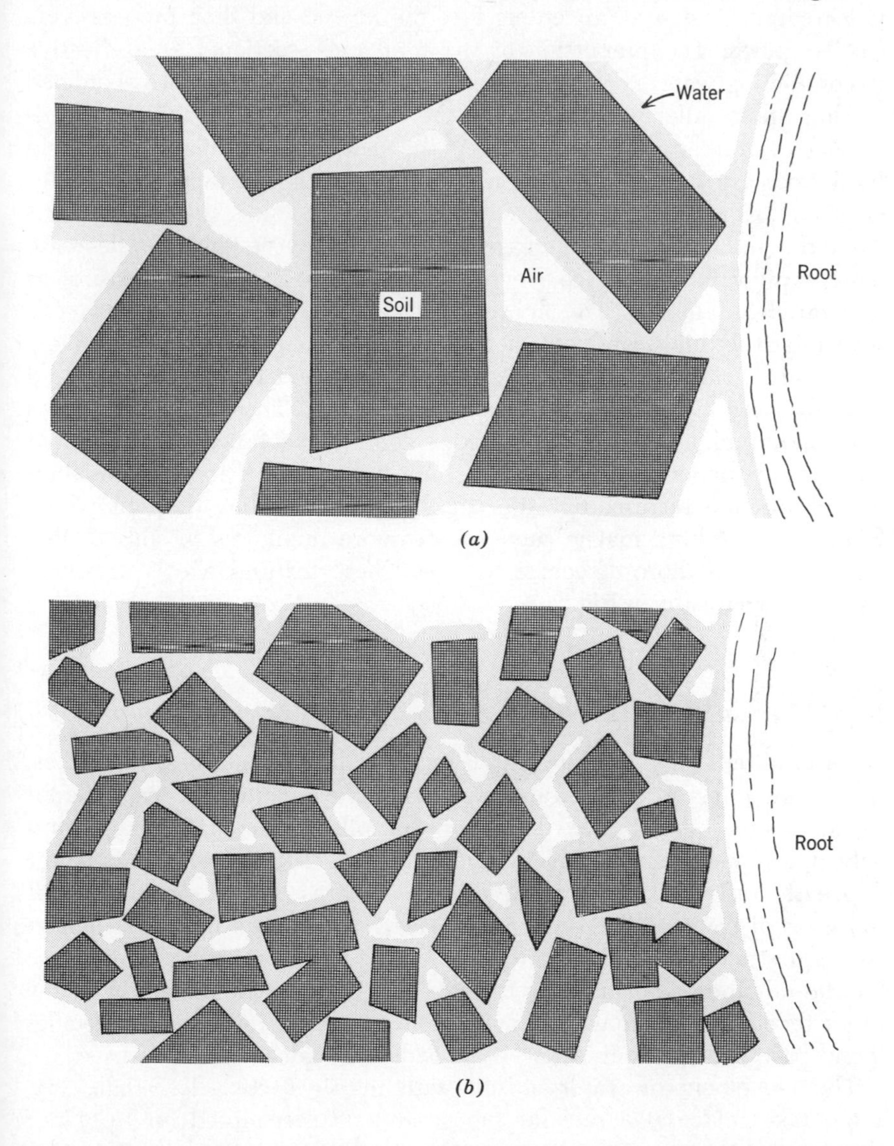

Figure 2.9. Schematic drawings of a cross section of a plant root in soil illustrating the difference between soil of (*a*) coarse texture and (*b*) fine texture at equal matric suction. (After Richards, 1959)

be held to it by hydrogen bonding and to have relatively low mobility (see Fig. 2.2). Soils are not composed of continuous capillary tubes, but they do have pores of different sizes; and water is held to the surface of soil particles as it is to glass tubes. As the matric suction is increased from zero, air enters first the largest and then progressively smaller pores. The proportion of the total cross-sectional area effective in conducting water is thereby decreased; and the movement is confined to films and smaller pores, in which the rate of movement is low according to the capillary-tube principle. At the same time, the flow is impeded by a decrease in continuity of water films among particles and an increase in tortuosity of the path.

As the matric suction increases, the conductivity tends to decrease more rapidly in soils of coarse texture than in soils of fine texture, as illustrated in Fig. 2.8. The figure illustrates also the tendency for soils with relatively high conductivity at low matric suction to have relatively low conductivity at high matric suction. This reversal no doubt is related to differences in the physical condition of water. The large pore spaces that permit high rates of flow under conditions of near saturation at low matric suctions are more numerous in soils of coarse texture than in soils of fine texture, but the small pores and films through which flow occurs at high matric suctions are more numerous in soils of fine texture than in those of coarse texture. These textural effects are illustrated schematically in Fig. 2.9.

Liquid Versus Vapor Transfer

Unsaturated soils contain water in both liquid and gaseous forms. As the quantity of liquid water decreases, the quantity of water vapor at first increases and then decreases. In soils that contain water films only a few molecular layers in thickness, the mobility of liquid water is greatly restricted; and the question arises as to the significance of transport of water in the vapor phase relative to that in the liquid phase. This question is a difficult one to investigate experimentally because the liquid and vapor phases cannot be studied independently. They occur together, their movements are associated, and individual molecules spend part of the time in each phase. Indirect evidence must be used.

The movement of water in soil with matric suction less than, say, 1 bar takes place at a rate far too great to be accounted for by vapor transfer and thus must occur predominantly in the liquid phase. Information about movement under relatively dry conditions may be derived from the comparative movement of water and solutes. Nonvolatile solutes do not move in the vapor phase. Thus, if movement of water in soil exceeds that of solutes, it is evident that part of the water moved without concomitant movement of solutes, from which one may infer that part of the water moved as the vapor. Marshall and Gurr (1954) measured the loss of water and chloride from the lower centimeter of an initially uniform 2-cm column of different soils at various water contents upon exposure of the surface to evaporation. Their results show that the proportion of the water lost from the lower centimeter of soil exceeded the proportion of the chloride lost in five of the six soils tested where the initial water content corresponded to the permanent wilting percentage or the 15-bar percentage. The difference between loss of water and chloride was small in three soils and substantial in three. In the soil with the largest relative difference, loss of water exceeded that of chloride by a factor of approximately five. Although these results cannot be interpreted quantitatively because of several complications, it does seem likely that movement of water as a vapor is of importance in some soils at the lower range of water availability to plants. From a combination of theory and experimental measurements, Jackson (1964) concluded that the movement of water occurred almost entirely in the vapor form where the relative humidity of the air in equilibrium with soil was 90% or less (corresponding to a suction value of 144 bars or more), depending on the soil.

Field Capacity

Field capacity is considered at this point, between movement of soil water and loss by evaporation, because of its relationship to both these subjects. Many soils have the property of retaining water in a metastable condition against the downward pull of gravity over periods of time so long that for practical purposes the condition is permanent. During entry of water into a dry soil, the water moves downward in a "front," across which the water content changes in a short distance from that of the moistened soil to the initial water content of the dry soil. When intake of water ceases, penetration of the front continues; but usually its movement has almost ceased within 2 or 3 days. The water content of the moistened portion of the soil, after the excess water has drained away and the rate of downward movement has decreased materially, was defined by Veihmeyer and Hendrickson (1950) as the field capacity. The water percentage in soils at the field capacity, determined in this way, ranges from less than 10 in soils of coarse texture to more than 30 in soils of fine texture. Values of water retained by soil at a matric suction of 0.3 bar are frequently used as laboratory estimates of field capacity.

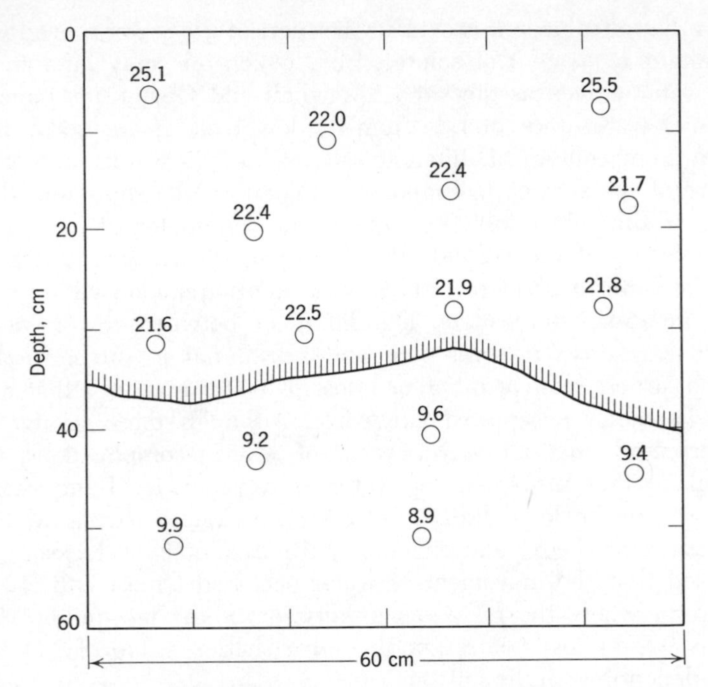

Figure 2.10. Vertical distribution of water in a loam soil in the field 48 hours after a rainfall of 5.5 cm. The circles indicate the location of the samples in the 60-cm cross section, and the numbers are the water percentages. The approximately horizontal line through the cross section represents the boundary between moistened soil above and unmoistened soil below. (Veihmeyer, 1927)

The phenomenon of field capacity is illustrated by Figs. 2.10 and 2.11. Figure 2.10 shows the distribution of water in a vertical cross section of a loam soil 2 days after a rain. In the upper or moist part of the cross section, the range of water content is from 21.6 to 25.5%. In the lower portion, not moistened by the rain, the range of values is from 8.9 to 9.9%. Figure 2.11 shows the results of an experiment in which a 61-cm column of soil moistened to the field capacity was stored for 144 days between two similar columns of dry soil. Loss of water by evaporation was prevented during the period of storage. During storage, the dry soil accumulated water at the expense of the moist soil; but no detectable increase in water content of the dry soil occurred at distances greater than about 36 cm from the initial boundary.

The phenomenon of field capacity in soil may be explained by considering two principles. First, the suction gradients in the soil water decrease with increasing time after addition of water. Second, the conductivity of the soil for water at the boundary between moist and dry soil decreases rapidly as the matric suction increases (Fig. 2.8).

The field-capacity effect is not possessed in the same degree by all soils. Water moves downward more rapidly in some soils than in others. Presumably, the difference among soils in this respect is caused by the arrangement and size-distribution of pores. For soils that allow water to flow readily under conditions of saturation, the persistence of the metastable moist zone at the surface might be expected to be most pronounced in soils of fine texture with good granular structure. Such soils should have both a relatively high numerical value for field capacity and a relatively low rate of downward movement because many fine pores within granules hold water and large unsaturated pores among granules impede transfer.

The foregoing considerations have dealt with the condition of moist

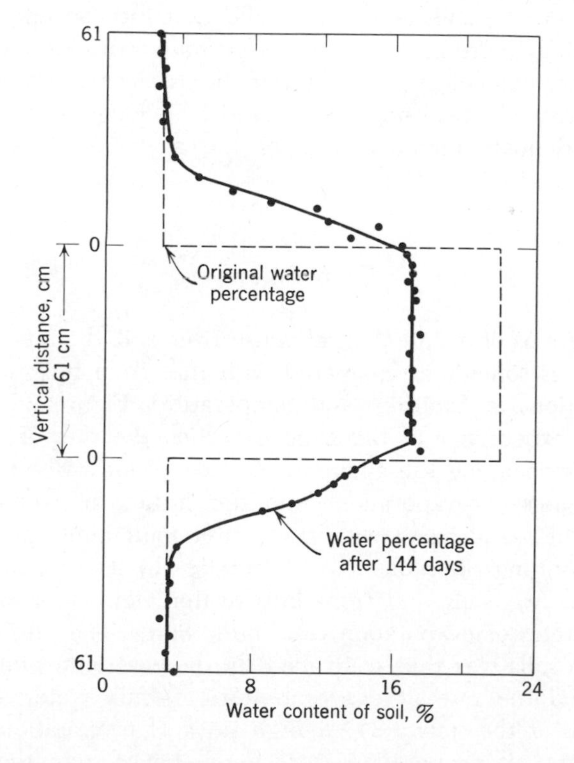

Figure 2.11. Upward and downward movement of water during 144 days in a 183-cm column of clay loam soil. The initial water percentages in the soil were 3% in the upper 61 cm, 22% in the central 61 cm, and 3.5% in the lower 61 cm. (Veihmeyer, 1927)

soil overlying dry soil. Where the underlying soil is moist at the time of application of water to the surface, the impediment to downward movement is not so great; and the metastable condition does not persist so long. Near the level of free water, moist soil overlies soil that is even more moist. Although the Veihmeyer and Hendrickson (1950) definition of field capacity does not exclude this condition, the water content of soil near the free-water level after the excess has drained away and downward movement has decreased substantially may be considerably greater than that found in moistened soil in contact with dry soil beneath. In fact, the soil near the free-water surface may be completely saturated.

Because field capacity cannot be defined precisely in physical terms, and because the field capacity of a soil may be influenced by conditions that vary with time and cannot be reproduced in the laboratory, soil physicists do not recognize the field capacity as a characteristic water-retaining property of soil. Some even hold that the concept called field capacity has done more harm than good. Although this view is justifiable from one standpoint, the metastable retention of water near the surface of soil is of great practical importance; and knowledge of it is indispensable to an adequate understanding of the behavior of water in soil in relation to plant growth.

Evaporation

In investigations of evaporation of water from soil, the rate of evaporation from soil is sometimes compared with that from bulk water under similar conditions of exposure and temperature. Figure 2.12 gives the results of an experiment of this type in which the rate of evaporation was varied by changing the environmental conditions. The figure shows steady-state rates of evaporation from the surface of columns of two soils and quartz sand, each with free water maintained at a depth of 60 cm. Two limiting situations may be noted: (1) The rates of evaporation from the two soils and from bulk water were approximately the same at low rates of evaporation from bulk water. The rate of evaporation from the soils was thus controlled by the environmental conditions that controlled the rate of evaporation from bulk water and not by the properties of the soils. (2) At high rates of evaporation from bulk water, the rates of evaporation were much lower from the soils than from bulk water, the rates became rather different for the two soils, and the rate for each soil seemed to approach an upper limit. At high rates of evaporation from bulk water, therefore, the rates of evaporation

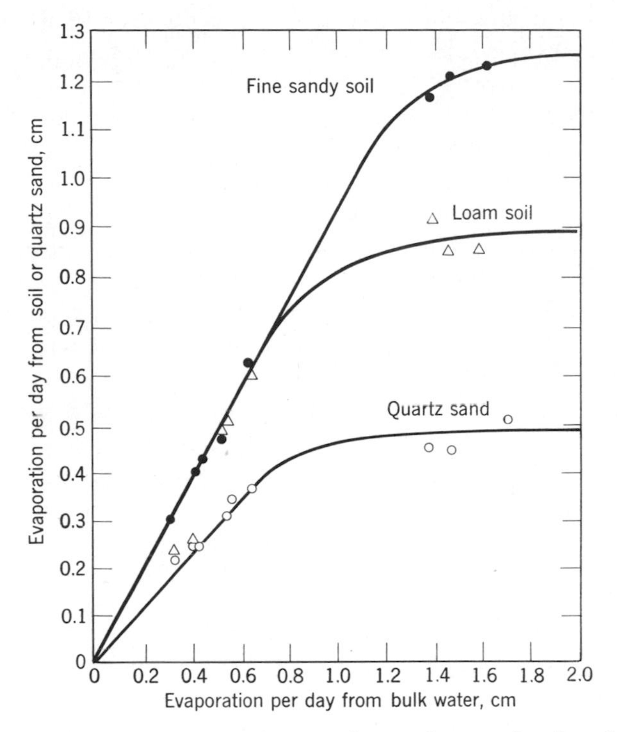

Figure 2.12. Rate of evaporation of water from columns of soil and quartz sand with free water at a depth of 60 cm versus rate of evaporation from a free-water surface. (Gardner and Fireman, 1958)

from the soils were not controlled by the environmental conditions.

The first of the foregoing two limiting situations may be explained on the basis that the source of water was close enough to the surface and the conductivity of the soils for water was great enough to keep the soils moist to the surface at low rates of evaporation. The surface of the soils was then covered with a film of water, and the film behaved like the surface of bulk water with regard to susceptibility to evaporation. The lower rate of loss of water from quartz sand than from the soils and the water in bulk at low rates of evaporation may be attributed to the failure of the relatively large sand particles to remain moist at the surface, even at low rates of evaporation. The second of the limiting situations may be explained on the basis that at high rates of evaporation from bulk water, the rate of evaporation from the surface of the soils exceeded the rate at which water could be conducted upward at a gradient of matric suction low enough to keep the surface moist. Accordingly, the surface of the soils dried, and the maximum rate of evaporation was controlled by the rate at which water was conducted through the soil to the surface.

In the experiment described in Fig. 2.12, the maximum rate of evaporation from the columns of soil occurred when the surface of the soil was driest because this was the condition under which potential evaporation was greatest. If the steady-state rate of evaporation from soil with free water at different depths below the surface is determined under uniform environmental conditions, the greater the depth to the free water, the drier the surface of the soil and the lower the rate of evaporation. Moreover, if a soil is initially moist throughout but has no steady supply of water at any depth, the rate of evaporation under uniform environmental conditions is high initially but falls with time because of exhaustion of the water from the upper part of the soil. Eventually, a dry layer is produced at the surface, and the rate of evaporation is comparatively low because most of the water that reaches the surface is then changed from the liquid phase to a vapor below the soil surface and diffuses as a vapor through the dry layer to the surface. According to Gardner (1958), the rate at which water can be lost by vapor diffusion through a dry surface layer of soil is usually less than 20% of the maximum rate of loss by evaporation from a soil that is moist at the surface.

As a result of the relatively low rate of evaporation from soil having a dry surface layer, water remains for a long time in fallow soil that is originally moist throughout. Veihmeyer (1927) described the results of an experiment in California in which a 122-cm column of soil contained in a tank open at the top but with closed sides and bottom was moistened with water to approximately the field capacity and allowed to stand out-of-doors in a fallow condition for 4 years. No rain was allowed to fall on the soil during the entire period. The tank was weighed at intervals to determine the total water loss, and samples were taken at different depths to determine the vertical distribution of water in the soil. The results in Fig. 2.13 show that the loss during the entire 4-year period amounted to only one-fourth the amount added. Nearly half the total loss occurred in the first 3 months. Water was lost rapidly from the surface but slowly from greater depths.

The long persistence of water stored in the soil is basic to the success of the alternate crop, fallow system of agriculture used widely in semiarid regions. Nevertheless, the efficiency of fallowing over a summer season in conserving water for subsequent crop production is usually rather poor. The information reviewed by Staple (1960) indicates that only about one-fourth the rainfall received during a year without a crop is present when the next crop is planted in areas of the northern

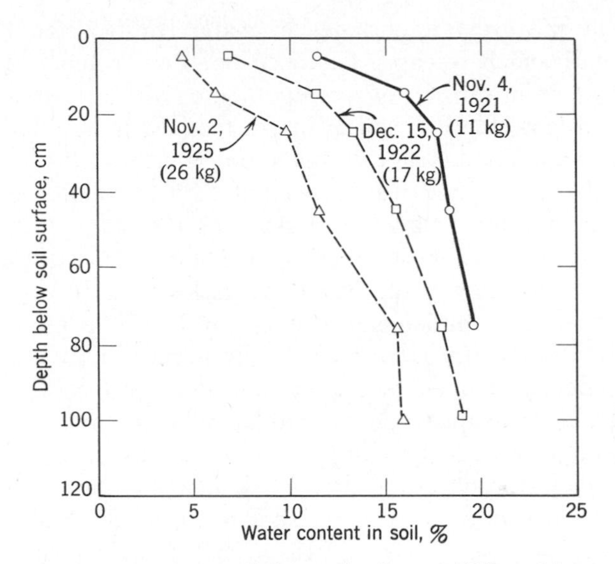

Figure 2.13. Water percentage at different depths in 454 kg of bare, uncultivated, clay loam soil in a tank 122 cm in depth at different times after addition of water on August 17, 1921. The addition of water was sufficient to moisten the soil to within 15 cm of the bottom, and the total content of water after addition was 91 kg giving an average water content of 20% for the full depth. The figures in parentheses below the dates indicate the total number of kilograms of water lost by evaporation from August 17, 1921, to the respective dates. (Veihmeyer, 1927)

Great Plains of the United States and Canada, where most of the rainfall is received in the summer season. Besides runoff, the cause of such low efficiency is evaporation. At Adelaide, Australia, Butler and Prescott (1955) noted that, in any month, the loss of water by evaporation from bare fallow was equal to approximately half the sum of the rainfall and the available water stored in the surface 61 cm of soil.

In areas where summer fallowing is practiced for water accumulation, much of the precipitation usually occurs as light rains that moisten only the surface of the soil. Little water storage results from such rains. The greatest storage per inch of rainfall results from the heavier rains that penetrate more deeply. For the weather conditions in May and June at Swift Current, Saskatchewan, Hopkins (1940) estimated that 34% of a 1-day rainfall of 2.5 cm would be lost by evaporation during the following 10 days but that 70% would be lost if the same amount of rain was received in 5 daily showers, each of 0.5 cm. The long persistence of the water in the soil illustrated in Fig. 2.13 thus may be attributed, in large part, to the initial deep penetration.

Because of the great importance of water conservation in areas of limited rainfall, much research has been done on techniques to reduce losses of water from soils by evaporation. Some of the early experiments showed that a loose layer of soil at the surface was particularly effective in reducing evaporation. Later it was appreciated that the special effectiveness of the loose layer was associated with maintenance of a free-water surface not far below the surface of the soil, a situation that does not correspond to most practical conditions in arid and semiarid regions. Moreover, most of the water that is lost readily by evaporation is lost before tillage implements can be employed to create a mulch of loose soil; and the dry layer that usually forms readily at the surface is itself an effective barrier to evaporation. Perhaps the greatest benefit from reduction of evaporation on producing a loose layer of soil on the surface under practical conditions would be obtained with soils that shrink and crack deeply during drying. Adams and Hanks (1964) found that evaporation from side walls of shallow shrinkage cracks in a clay soil in Texas varied from 35 to 91% of that from a comparable area of surface soil.

Despite the efficiency of a surface layer of dry soil as a mulch, considerable benefit may be derived from a mulch of crop residues or other material on the surface of the soil. For example, in experiments by Krantz (1949) in North Carolina, average yields of corn on unmulched and mulched plots were 3.8 and 5.1 metric tons per hectare under drouthy conditions and 6.3 and 6.7 metric tons per hectare under conditions of normal to above normal rainfall. The greater benefit was derived from the mulch, therefore, when the water supply was critical.

Organic mulches act in several ways to increase the supply of water in soil. First, they commonly increase the infiltration of water into the soil on land where runoff occurs. This effect is probably more important than any other on such soils. Second, they shade the soil, and the soil temperature is therefore lower during the heat of the day than is the temperature at a comparable depth in a bare soil. In one of the experiments by Krantz (1949), the average soil temperature at a depth of 1.3 cm on two sunny afternoons was 48°C in unmulched soil and 27°C in soil covered with a mulch of 6.7 metric tons of straw per hectare. The temperature of the air was about 35°C in both instances. The lower temperature in mulched soil decreases the evaporation because both the amount of water vapor in the soil air and the rate of diffusion decrease with the temperature. In this connection, it should be noted that the effect of a lower temperature of mulched soil during the day may be counterbalanced to some extent by a higher temperature at night (Lemon, 1956). Third, a surface mulch acts as a windbreak, in effect increasing the distance through which water vapor must diffuse between the soil water and the free atmosphere above the soil. Because of the reduction in vapor-pressure gradient, the evaporation is reduced. Russel (1940) determined the evaporation during a 9-hour period from columns of soil originally at the field capacity and found that, where evaporation from bare, exposed soil (0.48 cm) was taken as 100, the relative losses were 64 where the soil was shaded, 47 where the soil was shaded and protected from wind by a cardboard collar to give a 23-cm column of still air over the soil, 27 where the soil was protected by a straw mulch of 9 metric tons per hectare (about 3.8 cm in thickness), and 9 where the soil was mulched with a 3.8-cm layer of air-dry soil from a fine, cloddy seedbed. A surface layer of dry soil was thus more effective in reducing evaporation than was an organic mulch of equal thickness. Russel found also that the value of an organic mulch in controlling evaporation was greatest where the surface of the soil was moist and that it decreased rapidly as the surface became dry.

In some instances, organic mulches may have a detrimental effect. Zingg and Whitfield's (1957) summary of research comparing the yields of wheat under stubble-mulch (residues left on the surface and gradually worked down as the seedbed is prepared for the next crop) and moldboard-plow systems showed that the stubble-mulch system was the better under arid and semiarid conditions in western United States and the moldboard-plow system was the better under subhumid and humid conditions. Burrows and Larson (1962) found that progressively heavier applications of organic mulch in Iowa decreased the soil temperature and the early growth of corn. In an analysis of results of a number of experiments conducted in eastern United States, Allmaras et al. (1964) found that the magnitude of the reduction in early growth of corn from organic mulches became smaller as the soil temperature became higher and that the effect of mulching became negligible where the soil temperature was optimum for growth of corn. In this work, therefore, where soil temperatures were mostly below the optimum for the growth of corn and where water supply was generally sufficient, mulching aggravated a pre-existing soil-temperature problem; and water conservation was of only secondary or indirect significance.

In addition to the foregoing cultural techniques for decreasing water loss by evaporation, there has been some interest in chemical treatments of soil. Synthetic soil-aggregating agents may cause formation of a granular structure of the surface that increases infiltration and retards evaporation (Hedrick and Mowry, 1952). Substances that cause the soil particles to repel water may decrease the conduction of water to the surface of the soil and thus reduce the evaporation (Lemon, 1956). Substances that lower the surface tension of water may cause water to penetrate more deeply into the soil and thereby reduce the loss by evaporation (Lemon, 1956). Certain substances that form a monomolecular film on the surface of water may reduce the evaporation. The long-chain alcohol hexadecanol [$CH_3(CH_2)_{15}OH$] has this property and has been the subject of a number of investigations. There is evidence that this compound reduces evaporation of water from soils under some conditions (Olsen et al., 1964).

As yet, chemical treatments to suppress evaporation from soils have not reached the stage of extensive practical application; but potentialities appear to exist.

Evapotranspiration

Evapotranspiration is the sum of evaporation of water from soil and of evaporation (transpiration) of water from plant surfaces. The term is widely used because in practical agriculture the net loss of water vapor is of principal interest, and this is what is measured by the techniques employed.

The net loss of water vapor cannot be allocated accurately to evaporation and transpiration if the two processes are taking place simultaneously. Nevertheless, experiments indicate that both processes are important and that their relative importance depends on the circumstances.

Veihmeyer (1927) reported the results of an experiment in which transpiration appeared to be much more important than evaporation from the soil. In his experiment, tanks of soil 122 cm in depth were watered on August 17 and then were allowed to stand undisturbed out-of-doors in California without additional water. The tanks were covered during rains. Eleven kilograms of water had been lost by evaporation by November 4. On this date, one tank was planted to vetch. From November 4 to June 30, when the vetch was mature, the loss of water from this tank amounted to 39 kg. During the same period, only 5 kg of water were lost from the comparable control tank that was kept free of vegetation. Although there is little doubt that transpiration far exceeded evaporation in this experiment, the transpiration cannot be calculated by subtracting 5 from 39 because the evaporation from the cropped tank no doubt was different from the evaporation from the control tank.

The data in the preceding paragraph represent an extreme example because the surface of the bare soil was dry during the period of measurement and because the vetch in the planted tank was exposed to conditions of greater potential evaporation (because of greater solar radiation, greater wind movement, and lower relative humidity) than an equal area in a field of vetch. At the other extreme, evaporation from the soil surface would obviously make up most of the total evapotranspiration where plants are in the seedling stage in wide-spaced rows with moist soil between.

In recent years, comparative losses of water vapor from soil have been determined in the presence and absence of a soil cover that prevents evaporation. Polyethylene is useful for this purpose because it permits passage of oxygen and carbon dioxide while practically preventing interchange of water vapor. Peters (1960) summarized results of this type of work done with field-grown row crops in north-central United States. Loss of water from the soil was decreased by about one-half by covering the soil. Table 2.1 shows results obtained in two seasons in Illinois that illustrate the significance of the environmental conditions. In this work the crop was soybeans grown in rows spaced 102 cm. In 1958, when rainfall was sufficient to keep the surface of the soil moist most of the time, loss of water from covered soil was 33% as great as that from soil not covered. In 1959, when rainfall was lower and the surface of the soil was dry most of the season, loss of water from covered soil was 61% as great as that from soil not covered. Although this technique does not give results that can be interpreted as loss from soil and vegetation individually as they coexist naturally, it does indicate that loss from both plants and soils may contribute significantly to the total.

Table 2.1. Loss of Water from Plots of Field-Grown Soybeans in Illinois (Peters and Johnson, 1960)

			Total Water Loss, cm		
Period and Year of Measurements	Rainfall Received, cm	Open-Pan Evaporation cm	Soil Not Covered (A)	Soil Covered with Polyethylene Film and Irrigated (B)	$\frac{B}{A}$
July 2 to Sept. 26, 1958	28.2	41.0	39.0	13.0	0.33
July 1 to Sept. 20, 1959	14.2	52.1	21.4	13.1	0.61

Parallel with the research on chemical treatment of soil to reduce evaporation has been research on chemical inhibition of evaporation from plants. Abdalla and Flocker (1963) found that the alcohol hexadecanol used to control evaporation from soil may reduce the loss of water from plants independently of its effect on evaporation from soil; however, observations of this effect on plants have not been reported consistently. Oppenheimer (1960) reviewed work on the effect of naturally occurring substances with evaporation-inhibiting properties in plants. Other chemicals act more like hormones. Halevy and Kessler (1963) found that treatment of soil with 2,4-dichlorobenzyltributyl phosphonium chloride in low concentration not only exerted a water-sparing effect in soil in which plants were growing but also increased the tolerance of bean plants to drought. El Damaty et al. (1965) found that (2-chloroethyl)-trimethyl-ammonium chloride had no appreciable effect on either the yield of total dry matter or the transpiration per gram of total dry matter produced by wheat; however, the chemical increased the ratio of grain to straw and resulted in an increase in grain yield under dry conditions.

As with the chemical treatments to suppress evaporation from soil, chemical control of transpiration is still in the experimental stage. Results thus far are promising enough to indicate that practical applications are forthcoming.

Energy Relationships in Uptake of Water by Plants

Plants apparently do not expend energy directly in absorption of water from soils. Rather, it appears that metabolic energy is used in the production of cells and differentially permeable membranes within the cells, and that water then enters by passive diffusion through the membranes into the interior, where the free energy is lower on account of dissolved and suspended materials. Entry of water into cells increases the pressure of the contents on the walls (known as turgor pressure), and net entry ceases when the internal pressure balances the solute suction. In the xylem tissue, which conducts water upward in the interior of both the roots and tops of plants, there may be an actual physical suction resulting from loss of water from the transpiring surfaces and from the rigidity of the tissue.

Although the status of water in plants is usually described with terminology different from that employed for water in soils, the concept of total suction applies equally well to plants and soils. A plant should absorb water from soil if the suction in the plant exceeds that in the

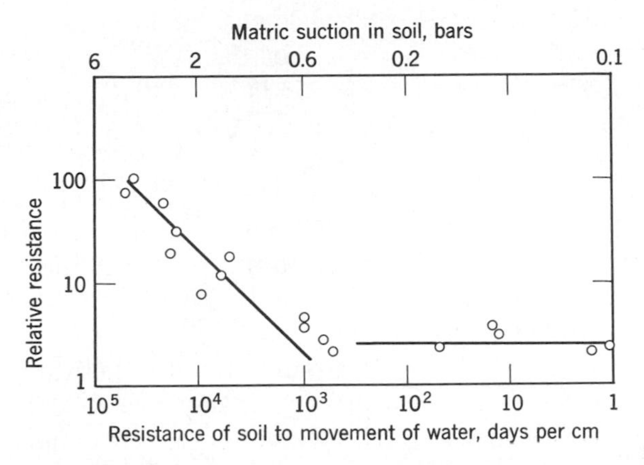

Figure 2.14. Relative resistance of plant and soil to water movement versus matric suction and resistance of soil for water. Circles represent experimental data. The sloping line on the left represents the expected relationship for negligible resistance in the plant. The horizontal line on the right indicates constant resistance in the plant and negligible resistance in the soil. The resistance of the soil for water in days per centimeter is the reciprocal of the conductivity, given by the cubic centimeters of water that flow per day across a cross-sectional area of 1 cm^2 where the gradient of matric suction corresponds to 1 cm of water per centimeter of distance in the direction of flow. (Gardner and Ehlig, 1962)

soil. Gardner and Ehlig (1962) reasoned that the rate of uptake of water by a plant from soil should be directly proportional to the suction gradient and inversely proportional to the resistance to water movement, the total resistance being partly in the soil and partly in the plant. They tested this hypothesis experimentally with the aid of measurements of suction in pepper plants and soil and measurements of resistance of soil to movement of water. The resistances in the plants were not measured directly. They then plotted their experimental observations in a manner similar to that shown in Fig. 2.14. The sloping line on the left represents the expected relationship if the resistance to movement in the plants was negligible compared with that in the soil, and the horizontal line on the right is the expected relationship if the resistance to movement in the plants was constant and the resistance to movement in the soil was negligible. The correspondence between the experimental observations and the theoretical considerations indicates that (1) the rate of uptake of water from the soil was proportional to the difference in suction between the plants and the soil, (2) the resistance to movement of water was primarily in the plants when the soil suction was below 0.6 bar, and (3) the resistance to movement of water was primarily in the soil when the suction was above 0.6 bar. In other work with a somewhat different theoretical approach, Gardner and Ehlig (1962a) obtained results indicating that resistance to flow of water in the soil exceeded that in the plants at soil matric suctions at least as low as 0.1 bar.

Despite its value, the energy concept has certain limitations as a basis for expressing the status of water in soils in relation to plants. These follow in part from difficulties related to measurement and in part from inadequacy of the concept as such. These two types of limitations are discussed in the following paragraphs.

If a sample of soil is allowed to equilibrate with the water it contains so that there is no tendency of the water to move from one place to another, the same value of total suction should be obtained on each of any number of subsamples that might conceivably be withdrawn from the larger sample. If a plant is allowed to grow in the larger sample and then is removed, re-equilibration of the soil and remeasurement will lead to a new and higher value of total suction than was obtained originally. The difference between the two measurements will represent the difference between the final and initial equilibrium states but not necessarily the difference between the final and initial states in all subsamples while the plants were growing because plants withdraw water from only the small proportion of the total soil in contact with the roots. The question, then, is whether equilibration within the soil keeps pace with removal. If it does not, the total suction of water in contact with the roots will be higher than that in soil distant from the roots and higher than the total suction measured in the sample after removal of the plant and re-equilibration of the soil.

Experimental evidence indicates that the problem of rate of equilibration is of little significance if plants are grown in relatively small quantities of moist soil. Figure 2.15 shows the matric suction in soils registered by tensiometers located at different radial distances from a central porous cylinder through which water was removed at a suction of 0.59 bar. The central cylinder may be considered to be analogous to a root with similar diameter. The fact that the values of matric suction were essentially unaffected by distance from the central absorber is evidence that the rate of equilibration was sufficient, over the distance investigated, to prevent development of any substantial difference in suction as a result of removal of water. In this experiment, the soils were all moist, as indicated by the low values of matric suction.

Movement of water is much less rapid in relatively dry soils than in moist soils, and indications are that rate of equilibration is of some significance under relatively dry conditions. Veihmeyer (1956) reviewed

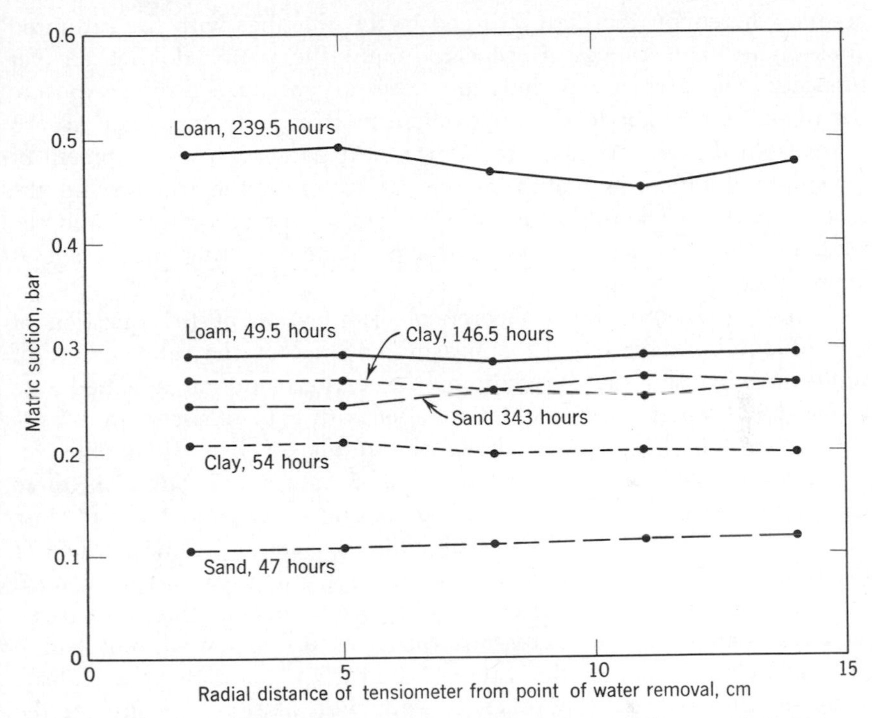

Figure 2.15. Matric suction registered by tensiometers at different radial distances from the center of beds of clay, loam, and very fine sand from which water was removed at the center through a porous cylinder under a matric suction of 0.59 bar. The times indicated are the numbers of hours during which the suction had been applied at the time the measurements of matric suction were made. (Read, 1959)

work on determination of the permanent wilting percentage that has a bearing on rate of equilibration. One of the findings has been that the permanent wilting percentage of soil is a little higher if measured with tree seedlings than with grass. This observation might result from the greater number of roots of grass than of trees per unit volume of soil and the shorter distance through which movement of water would be necessary to reach the roots of grass than of trees. Additional evidence is provided by the technique employed to measure the permanent wilting percentage (Peters, 1965). Plants are grown on the soil in containers out-of-doors or in a greenhouse until they wilt. Then they are placed in a chamber in which the atmosphere is approximately saturated with water vapor to see if they recover turgidity. If they recover, they are again exposed to evaporation and to radiation from the sun until wilting occurs, whereupon they are returned to the chamber with the saturated atmosphere. This process is repeated until the plants do not recover turgidity. The recovery periods are necessary because during exposure the plants lose water to the atmosphere more rapidly than they absorb water from the soil. As indicated in the next paragraph, development of a marked gradient in matric suction of water in the soil around the roots is partly responsible for the slowness of uptake of water and the necessity for recovery periods as the permanent wilting percentage is approached.

Gardner (1960) made a theoretical examination of the question of rate of equilibration based on measured values of the rate of flow of water through soils at different matric suctions and an assumed rate of uptake of water by a single plant root with zero diameter in a large volume of soil. Figure 2.16 shows the results of his calculations for a sandy loam soil. At a matric suction of 5 bars in the bulk soil, there was almost no gradient in soil suction leading away from the root; but, at a suction of 15 bars in the bulk soil, the suction at the surface of the root was in excess of 25 bars, and the gradient extended outward into the soil to a distance of about 3 cm. These calculations were based on an assumed rate of water absorption of 0.1 ml/cm of root length per day. Gardner chose the value of 0.1 ml/cm per day as a realistic value on the basis of experimental work with alfalfa. The greater the rate of uptake of water, the greater is the suction gradient extending outward from the root.

Considerations here have been limited to conditions in which plants grow in relatively small quantities of soil. Although these concepts are

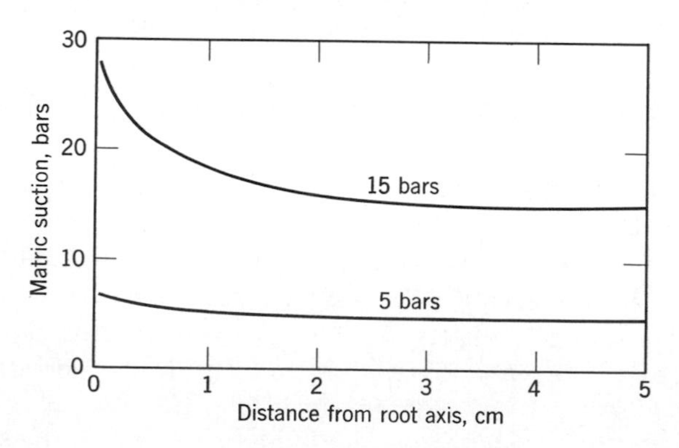

Figure 2.16. Matric suction of water in soil versus distance from a root as calculated for a root with zero diameter absorbing water from a large volume of a sandy loam soil at a rate of 0.1 ml/cm of root length per day, where the matric suction in the bulk soil is at 5 and 15 bars. (Gardner, 1960)

applicable under natural conditions, additional factors need to be considered; these will be discussed in subsequent sections.

With regard to the adequacy of the energy concept as a basis for representing the status of water in soils in relation to plant growth, two matters will be considered: the additivity of solute and matric suction, and the failure of the energy concept to provide information on quantities of water.

In the section on physical condition of water in soil, matric suction and solute suction were represented as additive components of total suction. Experimental measurements of matric, solute, and total suction by independent methods indicate that the sum of the first two does approximately equal the third. The methods of measurement leading to this conclusion are purely physical methods having nothing to do with plants. The question therefore arises regarding the additivity of the components where availability of soil water to plants is concerned. No categorical answer can be given to this question. Although some of the experimental evidence (to be considered in Chapter 6 on soil salinity and excess sodium) indicates that the two components are additive in their effect on availability of water to plants, the total evidence now supports the view that the situation is more complex than the energy concept would suggest. Both kinetic and physiological factors seem to be involved. The kinetic aspect will be discussed here, and the physiological aspect will be discussed in Chapter 6.

The rate of movement of water in response to a given difference in matric suction is high if the matric suction is low and is low if the matric suction is high. But solute suction does not have much effect on the rate of movement of water in response to differences in matric suction. Hence, at a given value of total suction, the rate of flow of water in response to a given difference in matric suction will increase with an increase in the proportion of solute suction in the total suction.

Figure 2.17 gives the results of an experimental test of the additivity of matric suction and solute suction under conditions in which movement of water was important. In the experiment, germinating corn seeds were placed in soil or perlite (an inert, porous, mineral material) after the root had emerged from the seed and was about the length of the seed. The soil was low in soluble salts and therefore had low solute suction; it had been pre-adjusted to different values of matric suction. In the perlite, the matric suction was not zero but was undoubtedly low and approximately uniform because the additions of mannitol solutions used to vary the solute suction were in all instances sufficient to supply 500 g of water/100 g of perlite. After a test period of 24 hours, the embryonic plants were detached from the remainder of the seeds and the water

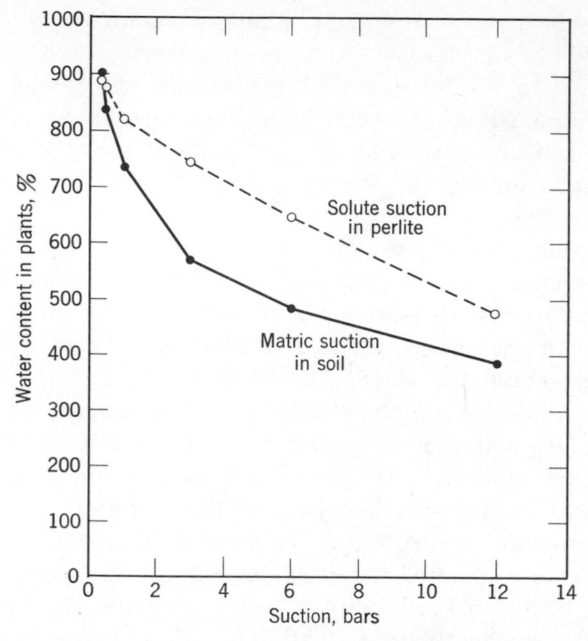

Figure 2.17. Water percentage in embryonic corn plants versus matric suction in soil and solute suction in perlite. (Danielson and Russell, 1957)

content was determined. The results in Fig. 2.17 show that the water percentages in the plants did not follow a single curve, which indicates that matric suction and solute suction did not have equivalent effects on water availability and hence were not additive.

To apply the theory about water movement to the experimental data in Fig. 2.17, the greater uptake of water by the seedlings at high values of suction from the mannitol solution in perlite than from the soil may be attributed to the greater rate of movement of water to the root in response to the difference in matric suction produced from uptake of water by the root in the perlite than in the soil. Results of other experiments by Gingrich and Russell (1957) and Collis-George and Sands (1962) support this theory. The complete theoretical explanation, however, is undoubtedly more complex. For example, if the root absorbs water selectively in relation to solutes, accumulation of solutes in the water adjacent to the root will increase the solute suction; and the increase in solute suction will impede further intake of water. Although solutes readily move to roots by transport in the moving water, they move away from roots by diffusion, a relatively slow process. This phe-

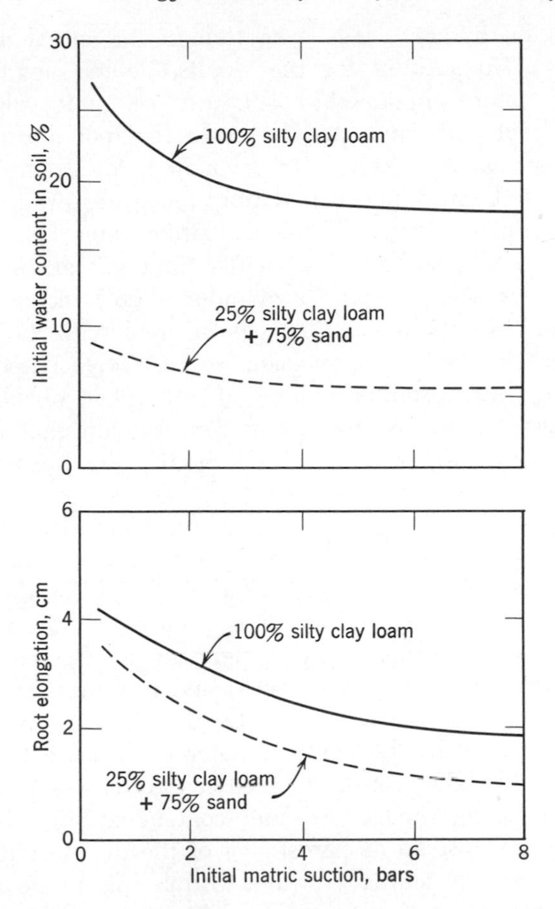

Figure 2.18. Elongation of corn roots in 24 hours and initial water percentage in two soils versus initial matric suction. (Peters, 1957)

nomenon has been little studied, but data obtained by Eaton and Bernardin (1964) verify its existence.

A second inadequacy of the energy basis for characterizing the status of soil water in relation to plant growth is its failure to reflect quantities of water. Peters (1957) investigated this matter with the aid of a technique similar to that used by Danielson and Russell (1957) in obtaining the data in Fig. 2.17. Germinating corn seeds with a root length of 8 to 12 mm were placed in different soil mixtures that had previously been adjusted to the desired value of matric suction. Figure 2.18 shows the root elongation in 24 hours corresponding to the different matric suctions for two of the soil mixtures. Evidently, the elongation was not the same in the different soils even though the initial matric suction was the same. To account for the results, Peters suggested that the soil of finer texture contained more water per unit volume adjacent to the plant root and conducted water to the root more rapidly than did the soil of coarser texture. He calculated, for example, the radius of the cylinder of soil required to supply the water needed by the root if the root were to extract all the available water from that volume of soil. The water absorbed in 24 hours from soil at an initial matric suction of 3 bars would require a cylinder of soil with a radius of 0.6 and 0.3 cm in the mixture of 75 parts of sand with 25 parts of silty clay loam and in the silty clay loam, respectively. These calculations demonstrate that a substantial volume of soil was involved in supplying the water absorbed by the roots, and they suggest that differences in capacity to supply water may greatly modify the behavior of plants on soils with the same initial suction.

The failure of the energy concept to represent the quantity factor in soil water supply is serious enough that a combination of the energy concept and the quantity concept is used commonly in applied research involving soil water. The technique is simply to determine the field capacity and the permanent wilting percentage (frequently estimated by the 0.33-bar and 15-bar percentages) and to represent the difference as the capacity of the soil to hold water in a form available to plants. [With increasing use of the neutron meter to measure the quantity of water in soil (W. H. Gardner, 1965), the tendency is increasing for expression of the values for water content in terms of percentage of the soil volume instead of percentage of the dry weight of the soil.] Differences in energy status of water within this range then may or may not be ignored, depending on the objectives of the work and the measurements made. Where tensiometers are employed in irrigation control, a common procedure is to use as an indication of need for irrigation the time when tensiometers in the major portion of the root zone have reached the limiting value of about 0.85 bar. Most of the available water has been used at this matric suction although there are still marked differences among soils in the quantity of available water remaining.

In light of this second inadequacy of the energy basis for characterizing the status of soil water in relation to plant growth, it may be worthwhile to examine the situation in more detail in an attempt to determine how a theoretical combination of the energy and quantity factors might be devised. Although the free-energy or total-suction value characterizes all the soil water, the characterization applies only where the removal of water by plants or other means is infinitesimal. Removal of a finite quantity of water causes the total suction of all the remaining water to increase. The next increment of water thus is removed at a higher suction. One may suppose, therefore, that some sort of integration would be needed to characterize the water-withholding tendency of a soil during removal of water by plants. Little has been done in this direction, perhaps because of the large amount of work required in both measurements and computations. Wadleigh (1946), Taylor (1952), and Gardner (1964) proposed methods for integrating the water supply of soil for plants over depths and times. Each method is based on different assumptions. Wadleigh and Ayers (1945) and Bahrani and Taylor (1961) described applications showing the relationship between integrated values and plant yields.

Soil-Plant Relationships in Water Availability

Upward Movement of Water in Soil and Roots

Movement of water from its original location in the soil to the aerial parts of plants takes place partly through the soil and partly through the roots. The relative importance of upward transfer by movement in the soil and in the roots is illustrated in Fig. 2.19 for two contrasting situations.

In the experiment in Fig. 2.19*a*, the water supply was relatively plentiful because of the presence of a free-water surface at a depth of 110 cm. Near the free-water surface, where the soil was wet, the conductivity of the soil for water was high, the roots were few in number, and almost all the upward movement took place in the soil. The soil became progressively drier toward the surface, which resulted in a decrease of conductivity. Furthermore, the density of roots was greater close to the surface than near the free-water surface. The net consequence of the lower conductivity and the greater density of roots was an increasing proportion of the water movement in the roots as the surface was approached.

The measurements shown in Fig. 2.19*a* were made while approximately steady-state conditions prevailed. That is, the rate of loss from the surface was equal to the rate of upward movement from the free-water surface, and the distribution of water in the soil column did not change appreciably with time. Under these conditions, therefore, one may say that the free-water surface was supplying the total needs of the plants. The total daily upward flow of nearly 1.5 cm^3 of water/cm^2 of cross-sectional area of the soil column should have been adequate to prevent any substantial limitation of plant growth from water deficiency. Under these conditions, root extension during the period of measurement may be said to have been of no significance in supplying the plants with water.

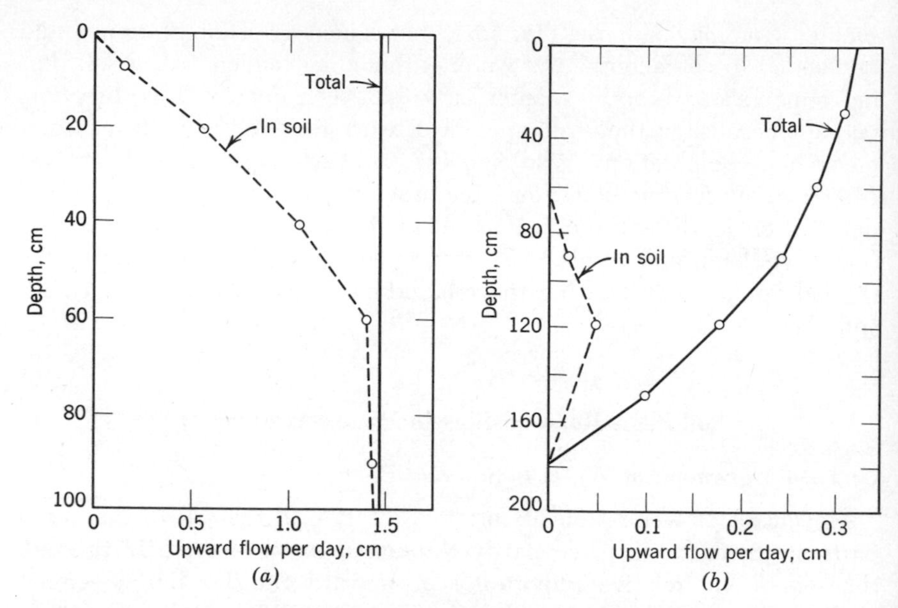

Figure 2.19. Total upward flow of water per day through a fine sandy loam soil under two conditions, and calculated upward flow through the soil as a result of the soil suction gradient. Upward flow in the roots may be represented as the difference between total upward flow and upward flow in the soil. Upward flow in centimeters per day is derived from the number of cubic centimeters of water that moved upward per day through a cross-sectional area of 1 cm^2. (*a*) Results obtained in the greenhouse with birdsfoot trefoil in cylinders 14 cm in diameter and 133 cm in depth, with a free-water surface maintained at a depth 110 cm below the surface. Measurements were made after a steady-state condition had been reached. (*b*) Results obtained outdoors on a small plot of alfalfa from the 22nd to the 50th day following irrigation. The soil was not wetted much below 180 cm. Because the plants had started to wilt before the 16th day, the rate of uptake was relatively low. Note the difference in scales between *a* and *b*. (Gardner and Ehlig, 1962a)

In Fig. 2.19*b*, the soil was relatively dry, the conductivity for water was relatively low, and most of the upward movement at all depths took place in the roots. In the experiment that provided the data in Fig. 2.19*b*, the plant roots were already extended throughout the soil depth of 180 cm at the beginning of the period of measurement. If they had not been, however, root extension would have been necessary for use of most of the water from the lower depths. Under proper conditions, root extension can be a relatively long-range process that brings the absorbing surfaces into position for the movement of water through the soil to be effective, as will be discussed subsequently.

The significance of upward movement of water from a free-water surface in supplying the water used by plants in the field is a question of long standing on which some difference of opinion still exists. The level of free water sometimes is near enough to the surface of the soil to make it appear that much of the water requirement of plants is satisfied by water rising from the free-water surface.

Wind (1955) investigated the upward movement of water in a clay soil under field conditions in the Netherlands. The water level in this soil was about 45 cm below the surface, and the roots of the grass growing on the soil were confined almost entirely to the surface 10 cm. Upward movement of water was demonstrated qualitatively by the fact that the surface 30 cm of soil became drier where this layer was separated from the underlying soil by a sheet of waterproof material than where contact with the underlying soil was maintained. Quantitative measurements made in periods without drainage from May through September indicated that upward movement of water totaled 15.3 cm in 96 days. During this time, the loss of water from the soil by evapotranspiration amounted to 32.2 cm. The rise of water through the soil thus was equivalent to 48% of the water lost by evapotranspiration.

Although further evidence is needed, the available data suggest that in a humid region with a free-water level at a depth of about 1 meter, water that is moved upward from the water level may be equivalent to a significant but not a major part of the water lost from the soil by evapotranspiration. At water levels of perhaps 2 meters or more, the upward movement is small, and perhaps negligible, in relation to loss by evapotranspiration.

Under conditions of relatively low rainfall, on the other hand, a free-water surface in the soil may provide the major part of the water lost to the atmosphere. An example of this was reported by Fox and Lipps (1955) in western Nebraska, where the natural rainfall is inadequate to produce good yields of alfalfa. In the Platte River Valley of that area, a large acreage of relatively high-yielding alfalfa is produced by "subirrigation" from the free-water surface that exists within the depth of root penetration. Once the roots reach the moist soil that is fed by water from the free-water surface beneath, the plants have an additional source of water; and their dependence on current rainfall greatly decreases. Figure 2.20 indicates that upward movement of water from the free-water surface in soils of this area may be effective over distances as great as 2 meters. Much of the water used by the alfalfa apparently is absorbed near the free-water surface, however, as indicated by the proliferation of roots within the 30 cm of soil immediately overlying the free-water surface.

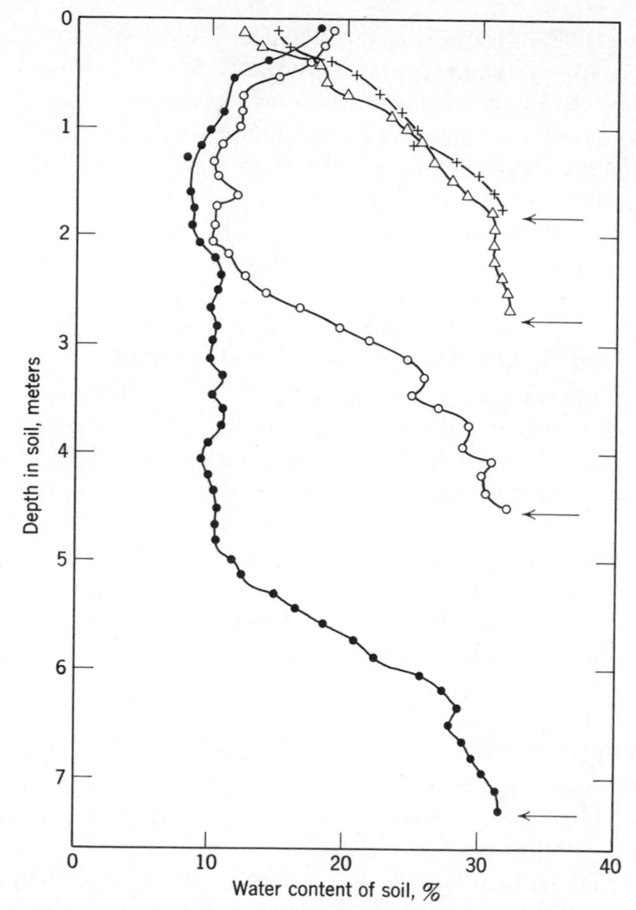

Figure 2.20. Average water percentage during the period from April through October in four soils bearing alfalfa and having a free-water surface at the depth indicated by the horizontal arrows. (Lipps and Fox, 1956)

Several reasons may be advanced for attaching less importance to upward movement of water under conditions of high rainfall than under the conditions described by Fox and Lipps (1955). First, water that falls as rain is used preferentially because it is held primarily in the upper part of the soil, which is the zone from which plants withdraw water most rapidly if the water is available. Second, water that falls as rain decreases the gradient from the free-water surface upward and thus decreases the tendency for upward movement. Third, rainfall causes a downward displacement of water that has previously risen from the free-water surface. Fourth, a temporary rise of the free-water level from the heavier rainfall that occurs in the more humid regions may cause plant roots below the water level to die because of poor aeration. Hence as the water level falls maintenance of the close proximity of active roots to the free-water level, near which upward movement is most effective, may require reinvasion of the subsoil by a new set of roots growing downward from above. (The water level remained at a fairly constant depth under the conditions described by Fox and Lipps, so that, once established, the root system of the alfalfa could operate continuously in the same soil zone.)

Optimum Depth of Free Water

Different kinds of plants do not respond uniformly to depth of free water in soil. Such was found to be the case, for example, by Ellis and Morris (1946) in Indiana. These investigators compared the performance of several crops on muck soil with the water level maintained 41, 69, or 96 cm below the surface. As shown in Table 2.2, all the crops except mint benefited appreciably from an increase in depth of free water from 41 to 69 cm; but not all crops gave an additional response to a further increase in depth of free water from 69 to 96 cm. Mint behaved singularly in that the yield of hay was unaffected by the water level, but only a 55% yield of oil was produced with the shallow water level. Apparently, a certain degree of water stress was necessary to raise the oil content.

The optimum depth of free water for accumulation of dry matter by a given species may change between seasons and within seasons. Frankena and Goedewaagen (1942) found in experiments with grass that almost invariably the highest yield of grass was obtained with the

Table 2.2. Yield of Crops on Muck Soil with Water Level Controlled at Different Depths (Ellis and Morris, 1946)

Average Depth of Water Level, cm	Relative Yield of Indicated Crop (Highest Value = 100)					
	Mint					
	Hay	Oil	Sweet Corn	Carrots	Red Beets	Onions
41	99	55	48	9	26	12
69	100	94	77	100	97	76
96	98	100	100	99	100	100

water level held at the 20-cm depth; but the degree to which the yield was decreased with the 50- and 80-cm depths varied with the season, the reduction being greatest when the water requirement was greatest. This effect of seasonal conditions on the relationship between water level and crop yield was probably caused by the plants exhausting the water from the upper part of the soil to a greater extent and drawing more of their water supply from soil near the free-water level at times when rainfall was low and evapotranspiration was high than when rainfall was ample and evapotranspiration was low. The crop yield decreased primarily because of water deficiency, which followed from the greater energy requirement for movement of water through the greater distance and through a smaller cross section of conducting tissue. The smaller cross section of conducting tissue is a consequence of the decrease in number of roots with increasing depth.

The significance to plants of the depth of free water is not limited entirely to the water supply. As the water level is raised toward the surface of the soil, the quantity of water at the disposal of the plant is increased; but the quantity of oxygen is decreased. In consequence of the complementary relationship between water level and oxygen content of the soil, the most suitable water level will depend on the oxygen supply and hence may be expected to be at a greater depth in soils of fine texture than in soils of coarse texture.

The most suitable depth of free water depends also on the nutrient supply, but not in a simple manner. If there is sufficient rainfall to prevent the surface soil from becoming dry, raising the level of free water toward the surface will reduce the nutrient availability. An instance in which this was true for nitrogen was reported by Eden et al. (1951) in England. They observed that nitrogen deficiency symptoms in the crop became more pronounced as the level of free water was raised toward the surface of the soil. The field observations were cor-

Table 2.3. Yield and Protein Content of Grass on Peat Soil with Water Level Maintained at Different Depths (Eden et al., 1951)

Depth of Water Level, cm	Yield of Dry Grass per Hectare, kg	Crude Protein in Dry Matter	
		%	Per Hectare, kg
39	3760	14.8	560
61	7030	21.1	1480
98	6890	24.6	1690

roborated by data on the yield and protein content of the crop shown in Table 2.3. On the other hand, under conditions dry enough to prevent appreciable root activity in the surface soil, raising the level of free water will moisten the surface soil. The resulting increase in nutrient availability will tend to offset the decrease in availability associated with loss of root space in the subsoil. One may therefore expect that from the standpoint of nutrient availability, the optimum depth of free water in the soil will be closer to the surface under conditions of low rainfall than under those of high rainfall. Indications from the work of Burgevin and Hénin (1943) are that the optimum depth of free water tends to approach the surface as the supply of nutrients in the surface soil increases relative to that in the subsoil.

Evidently the depth at which the water level should be maintained for maximum yields is by no means constant. It varies with the characteristics of the crop, the season, and the soil.

Root Extension

Plants themselves play an important part in influencing availability of soil water through their capability to extend roots downward into moist soil. Environmental factors have a marked effect on root extension. When the water supply is ample throughout the soil, root extension is controlled by factors such as photosynthesis, nutrients, and temperature. As the water supply is depleted from the portion of the soil occupied by roots, however, the decreased availability of water in the soil causes a deficiency of water in the roots in that area. Root extension then occurs more readily in other areas where the availability of soil water is higher because cell elongation (which requires water absorption) is less limited in these areas.

Roots may continue to live in soil layers maintained at the permanent wilting percentage if water is available in some other part of the soil. Growth of new roots into such dry soil, however, is limited (Hendrickson and Veihmeyer, 1931; Hunter and Kelley, 1946). The maximum volume of soil that can be exploited by plant roots is not much greater than the original volume of contiguous moist soil. A dry layer lying between moist soil occupied by roots and a deeper layer of moist soil acts as a barrier to use of water from the deeper layer. Observations made by Cole and Mathews (1939) (Fig. 2.21) on an area in western Nebraska planted each year to wheat may be cited as an example. These investigators found that the soil contained water in excess of the permanent wilting percentage at depths less than 0.6 meter and at depths greater than 1.5 meters in the spring of 1911. The water percentage in soil at the 0.6- to 1.5-meter depth was below the permanent wilting

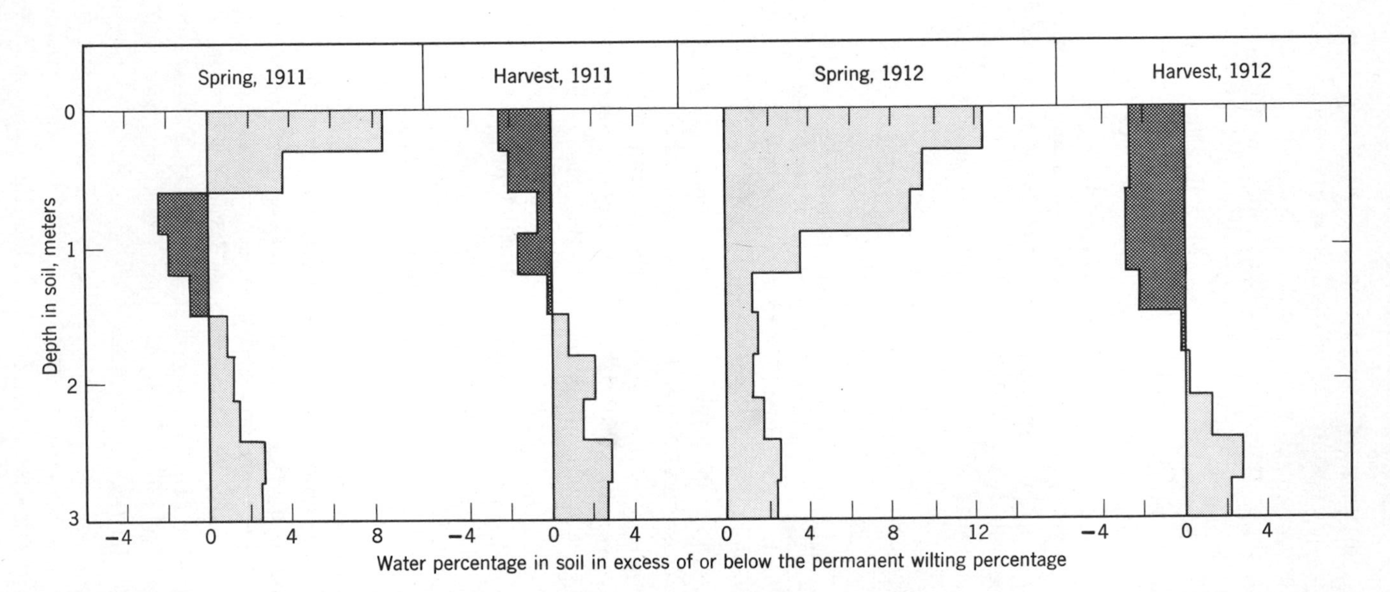

Figure 2.21. Water percentage in excess of or below the permanent wilting percentage at different depths in the spring and at harvest in two consecutive years at North Platte, Nebraska, in soil planted each year to spring wheat. (Cole and Mathews, 1939)

percentage as a result of growth of the previous crop. The 1911 crop removed all the available water in the upper part of the profile but did not use the water available below the 1.5-meter depth. By the spring of 1912, the water percentage in the profile was above the permanent wilting percentage throughout. In that year the soil was depleted of water to the permanent wilting percentage, or below, throughout the entire profile to a depth of 2.1 meters, and apparently a little water was removed from greater depths.

Pronounced differences exist among plants with regard to the maximum distance through which root extension takes place into moist soil. Such differences may have important practical consequences. For example, Burton et al. (1954) found that drought resistance of different grasses on a sandy soil in Georgia was associated with root extension. During a 2-month drought, foliage of carpetgrass turned brown and died, whereas foliage of bermudagrass showed little evidence of injury. Excavation of the roots showed that the proportion of the root system by weight in the soil below a depth of 0.6 meter was 6% with the carpetgrass and 35% with bermudagrass.

A second instance involving differences in root extension among plants was reported by Kiesselbach, Anderson, and Russel (1934), who investigated the utilization of subsoil water by crops in eastern Nebraska. They found that with a soil initially well supplied with subsoil water (as a result of accumulation over many years), 5 years of cropping to sweet clover and red clover failed to deplete the soil of water below a depth of 2.1 meters. Four years of cropping to alfalfa, however, had reduced the subsoil water nearly to the permanent wilting percentage to a depth of at least 4.6 meters (the maximum depth of sampling). The response of these crops to a supply of subsoil water was rather different, as indicated in Table 2.4 by the comparative yields obtained on land not previously in alfalfa and on land and that had been cropped

Table 2.4. Yields of Crops in Eastern Nebraska with and without Previous Cropping to Alfalfa (Kiesselbach, Anderson, and Russel, 1934)

	Yield of Hay per Hectare Annually, Metric Tons		
	Alfalfa	Sweet Clover	Red Clover
Land previously in alfalfa	6.0	6.0	6.7
Land not previously in alfalfa	8.7	6.2	7.6

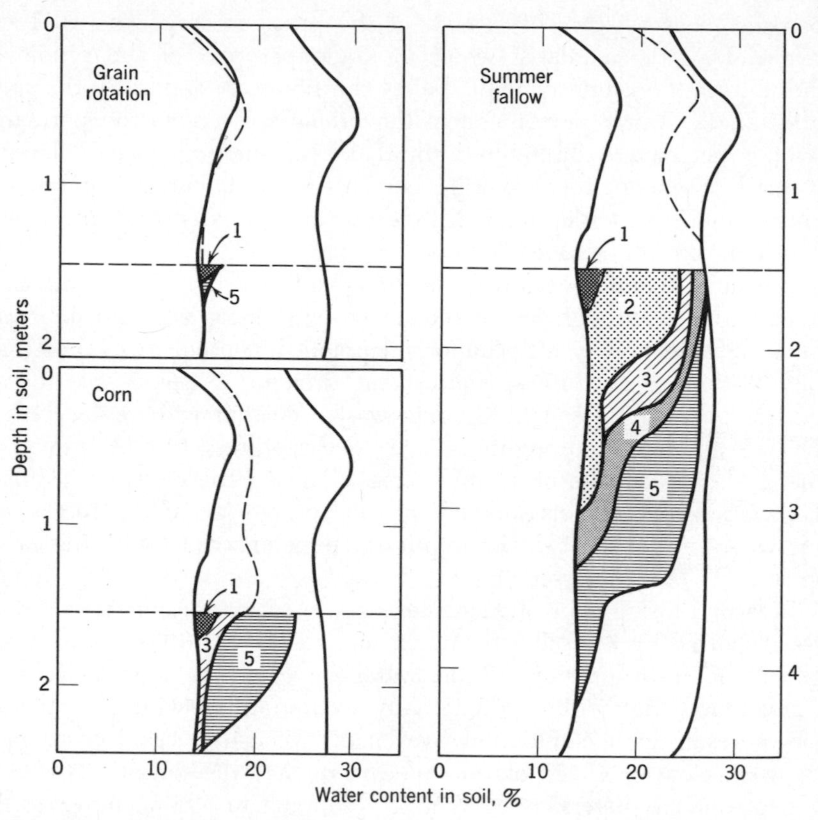

Figure 2.22. Restoration of subsoil water during 5 years (1928–1932) of grain rotation (corn, oats, wheat), continuous corn, and continuous summer fallow in soil cropped formerly to alfalfa (1922–1927). The solid line on the left in each diagram indicates the initial water content in the autumn of 1927. The heavy solid line on the right indicates the calculated field capacity. The broken line above the 1.5-meter level shows the average water content during the last three years. The shaded areas indicate the amounts of storage below the 1.5-meter level during each of the five successive yearly intervals (Nos. 1–5) between samplings. The precipitation during the 5 years was as follows: 1928—75 cm; 1929—53 cm; 1930—38 cm; 1931—68 cm; 1932—89 cm. (Kiesselbach, Anderson, and Russel, 1934)

previously to alfalfa. All crops produced higher yields on the land that had not been in alfalfa previously, but the increase was greatest with alfalfa, presumably because much more subsoil water was utilized by the alfalfa than by the red clover or sweet clover.

Figure 2.22 shows that once the subsoil water had been depleted by alfalfa, considerable time would be required for its restoration under the prevailing conditions even if the soil was kept in bare fallow. The 2-year average yields of alfalfa grown on old alfalfa land were 6.4 metric tons of hay per hectare with no fallow, 7.8 tons with 1 year of fallow, and 8.5 tons with 2 years of fallow. The yield was 9.6 metric tons per hectare on land in alfalfa for the first time. Because of the slowness with which the subsoil water was restored, Kiesselbach, Anderson, and Russel concluded that, under upland soil conditions in eastern Nebraska, the most profitable procedure after the subsoil had become depleted of water was to plow up the alfalfa and to reseed it on land that had never before grown alfalfa.

Experimental work on this same problem was conducted in Kansas (Grandfield and Metzger, 1936; Myers, 1936; Metzger and Grandfield, 1938; Hobbs, 1953), where the results were similar except for the deeper penetration of sweet clover roots (4 meters in 2 years) and the more rapid restoration of subsoil water when the land was fallowed or cropped to cereals.

Differences in distribution of roots of grasses and trees are well known and have significant implications. Because of their deeper root system, trees may remain active continuously where temperatures are favorable, whereas grasses are dormant during dry seasons. Lewis and Burgy (1964) found that oak trees contained tritium after tritiated water had been injected below the free-water level of 22 meters at a location in California, thus verifying that the trees absorbed water from a great depth. Oppenheimer (1960) reviewed the observations of many authors on the penetration of root systems of different plants.

The differences among plants with regard to maximum distance of root extension may be related to differences in plant structure that affect the energy required for movement of water through the roots. On the basis of the capillary-tube principle discussed previously, the energy requirement for movement of water at a given rate through a fixed length of root may be expected to increase with decreasing cross-sectional area of conducting tissue and of individual conducting channels within that tissue. Thus in trees the water-conducting channels are of relatively large diameter, leading to comparatively low resistance. Trees characteristically have deeply penetrating root systems. In grasses, the water-conducting channels are of relatively small diameter, leading to comparatively high resistance. Grasses are characteristically shallow-rooted. Wind (1955a) elaborated on this theme.

That energy is required for movement of water through roots is evidenced in another way by Davis' (1940) observations on uptake of water from soil by corn plants. Davis planted corn seeds in one end of a box of soil and allowed the plants to grow until the roots appeared to be distributed uniformly throughout the soil. He noted that even though the soil was moistened uniformly after this time, the soil a meter away from the base of the plants remained near the field capacity, whereas the soil near the base of the plants was exhausted of water to nearly the permanent wilting percentage. The lesser absorption of water by the more remote portion of the root system is consistent with a positive energy requirement for movement of water from these parts to the base of the stem. If the energy requirement for transportation of water through the root system had been zero, water should have been withdrawn from the soil at approximately the same rate throughout the box.

Water Supply and Plant Behavior

Water Balance in Plants

Growing plants are not in equilibrium with their environment as regards water. As may be inferred from the general upward movement of water in plants, there normally must be a free-energy gradient from high in the roots to low in the above-ground portions from which water is lost to the atmosphere. The free energy of water in the atmosphere during the day is usually far below that of water in plants and soils. The principal gradient is between the above-ground parts of plants and the atmosphere, and not between the tops of plants and the roots or between the roots and the soil.

Plants respond in a sensitive manner to their water environment in the soil and atmosphere. They do this primarily by loss and gain of water and solutes, which modifies turgor pressure and solute suction.

As the suction increases in the water in the root environment, the tendency of water to enter the roots is decreased. Transpiration continues, however, and causes the rate of loss from the above-ground portion of the plants to exceed the rate of entry from the soil. The content of water in the plants is then less than before, which causes the outward pressure of the contents on the cell walls to decrease and the solute suction to increase. With the energy status of the water thus adjusted, the plant is adapted to absorb water from the soil at the higher suction. As soil dries, therefore, there is a continuous adjustment of the energy status of the water in the plants the soil supports. This adjustment is illustrated in Fig. 2.23, which shows the progress of the increase in suction of the water in a leaf of a pepper plant with the increase in suction in the soil water as the water was used by evapotranspiration.

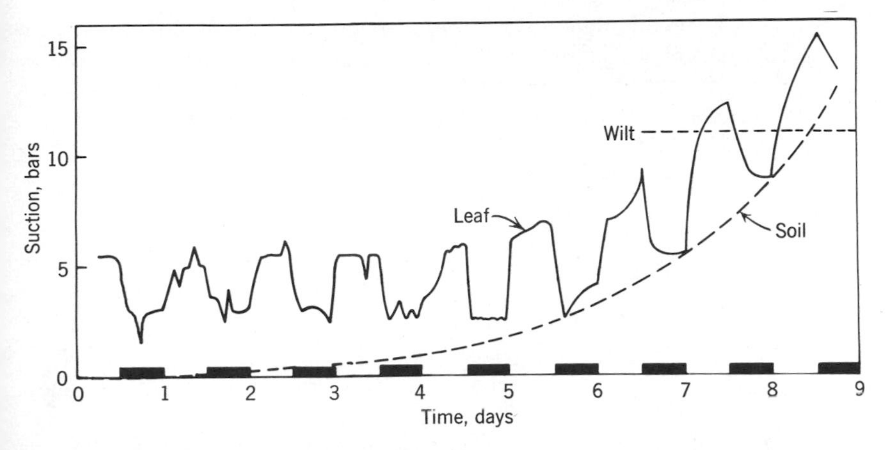

Figure 2.23. Diurnal changes in suction in the water in a leaf on a pepper plant and in the soil, with gradual exhaustion of the soil water by evapotranspiration. The horizontal line marked "wilt" indicates the suction in the leaf at which wilting occurred. The light and dark areas indicate periods of light and darkness. (Gardner and Nieman, 1964)

Superimposed on the gradual increase of suction in the leaf is a marked daily cycle resulting from net loss of water from the leaf in the daytime and gain at night.

Growth

Growth is basically an increase in volume resulting from the formation and enlargement of cells. Water plays an important direct part in cell enlargement. The walls of young cells are more plastic than those of old cells. The turgor pressure exerted against the walls of cells from the inside by the water they absorb causes stretching of cell walls, while they are in the plastic condition, and enlargement of the cells. A deficiency of water thus reduces the rate of growth because the cells do not become fully stretched. Eventually, deposition of cellulose and other constituents in the cell wall decreases the plasticity, so that full enlargement does not occur even if sufficient water is supplied. A deficiency of water has been found to result in a reduction in the rate at which glucose carbon-14 supplied to plants is incorporated into cell wall constituents (Plaut and Ordin, 1964), which suggests that loss of plasticity in the walls may be delayed by water deficiency.

The importance of full turgidity for cell enlargement provides an explanation for the limited penetration of roots into dry soil. If the water needed by a root tip for cell enlargement is not absorbed from the

Table 2.5. Rate of Increase in Height of Corn Plants In Soil at Different Water Percentages (Davis, 1940)

Water Content of Soil, %	Number of Observations	Rate of Increase in Height per Hour, mm.
9–12	26	1.4
12–15	15	2.4
15–18	15	3.0
18–21	10	5.0
21–24	8	5.1

surrounding soil, the water must be moved in through the root system from elsewhere, which probably means transportation through the xylem elements involved in long-range water movement. The xylem elements extend upward through the plant to the leaves, and rapid transpiration prevents the water in them from exerting full pressure. In times of water deficiency, plant stems have been found to become smaller in diameter in the daytime than at night, a response that probably results from a difference in turgor pressure.

Lack of full turgidity restricts the growth of above-ground parts of plants as well as roots. As indicated in Table 2.5, this effect is of significance within the range between the field capacity and the permanent wilting percentage. Table 2.5 shows measurements made by Davis (1940) on the rate of increase in height of corn plants at different water percentages in the soil in which they were grown in the greenhouse. All the measurements were made at soil-water percentages above the permanent wilting percentage (6.8%). Because the water content of the soil at field capacity is commonly about 1.8 times that at the permanent wilting percentage, the soil evidently failed to supply water rapidly enough to maintain the same rate of elongation at all water contents between the field capacity and the permanent wilting percentage. If water is deficient, growth of above-ground parts may occur mostly at night, when transpiration is reduced and turgidity is regained (Loomis, 1934).

The word growth is commonly used in reference to the height of a plant or to the quantity of some particular plant part. Although all such measurements bear some relation to growth, expressed in terms of total volume or total weight of the green plant, the degree of sensitivity of the different characters to the supply of water is not the same. Table 2.6 gives an example taken from a field experiment on cotton.

Table 2.6. Response of Cotton to Different Applications of Irrigation Water in California (Adams, Veihmeyer, and Brown, 1942)

Water Applied, cm	Relative Plant Height	Relative Yield of Above-Ground Plant Parts	
		Total Minus Seed Cotton	Seed Cotton
33	73	51	90
63	91	81	96
109	100	100	100

In this experiment, the yield of dry matter of the vegetative parts of the plant showed the greatest degree of change with the water supply (and therefore was the most sensitive index of water supply), plant height showed an intermediate change, and seed cotton showed the least change.

Transpiration and Photosynthesis

Effect of Atmospheric and Soil Conditions. Transpiration and photosynthesis are distinctly different processes, but they are considered together here because of their physical relationship. Loss of water vapor from plants and intake of carbon dioxide by plants occur principally through the stomata in the leaves. Stomata are openings from the interior of leaves to the atmosphere; they are located between specialized cells that regulate their opening and closing. The aperture is affected by various factors, including the supply of water. The tendency is for the stomata to be open if the plant is well supplied with water and closed if it is not, as illustrated in Fig. 2.24. Water deficiency thus restricts the area through which water vapor may be lost, and this restricts transpiration. At the same time, however, the intake of carbon dioxide required in photosynthesis is restricted. Figure 2.25 illustrates the decline of both transpiration and assimilation of carbon dioxide by leaves of an apple tree in soil from which the water was gradually being depleted and the rapid increase in both transpiration and assimilation when water was again supplied.

Numerous observations of photosynthesis and transpiration in relation to the water supply in soils have been made, and considerable divergence in findings is apparent. In some instances these processes did not seem to be much affected until the water content of the soil approached the permanent wilting percentage, and in others a reduction occurred

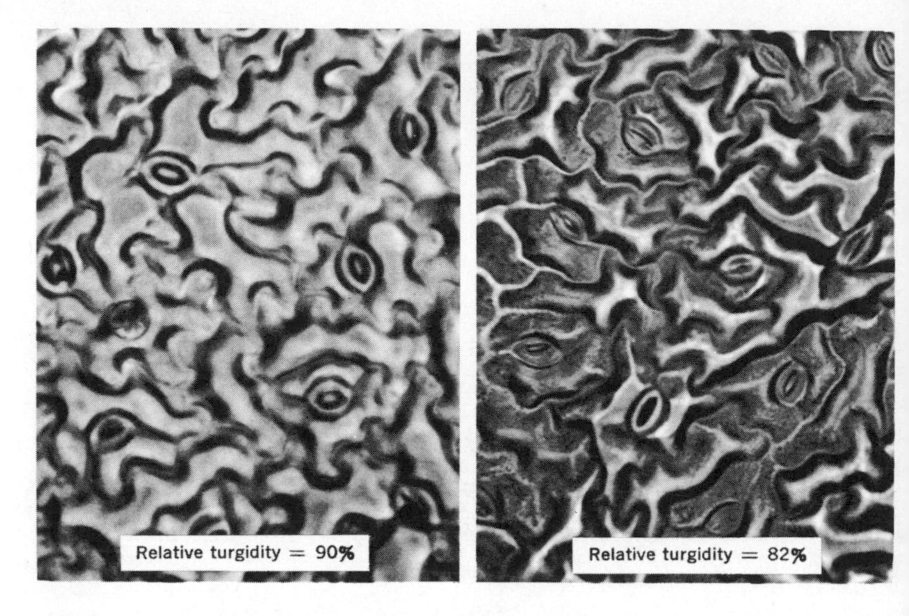

Figure 2.24. Photomicrographs of lower surface of soybean leaves showing appearance of stomata under different conditions of turgidity. Wilting symptoms became evident at about 85% relative turgidity. Relative turgidity is defined as the weight of water in a sample of leaf blade expressed as a percentage of the weight of water in the same sample after it has been floated on distilled water long enough to allow it to absorb water and reach full turgidity or maximum weight. (Laing, 1966)

while the soil still contained much water available to plants. An important reason for the divergence seems to be that the balance between the rate of supply of water to the plants from the soil and the potential rate of loss of water from the plants by transpiration was not the same under the different conditions.

The importance of the effect of potential rate of transpiration (as reflected by the transpiration of plants grown in moist soil) is suggested by the experimental data in Fig. 2.26. The results show that the relative transpiration on a given day invariably decreased within the range between the field capacity and the permanent wilting percentage. When the rate of transpiration from plants in soil at the field capacity was high, the decrease of transpiration was marked and occurred with soil matric suction less than 0.5 bar. On the other hand, when the rate of transpiration from plants in soil at the field capacity was relatively low, the decline in relative transpiration with an increase in matric suction was comparatively small.

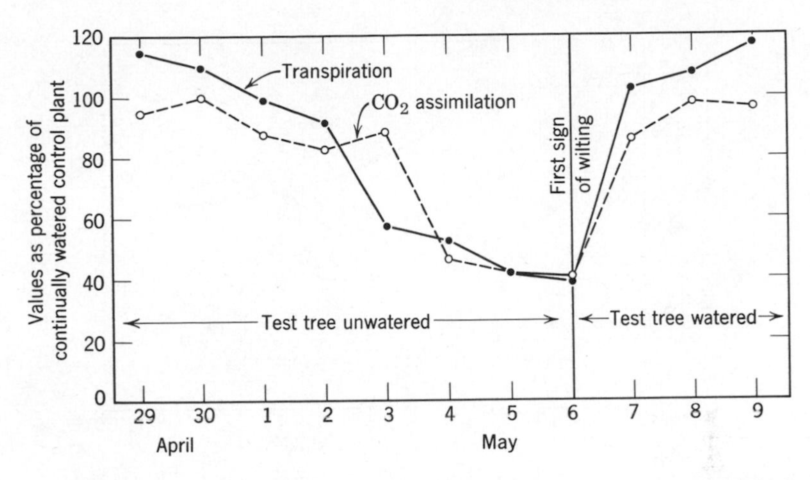

Figure 2.25. Transpiration and CO_2 assimilation per hour per 100 cm^2 of leaf surface of a McIntosh apple tree that was left unwatered from April 28 to May 6 as percentages of values for a comparable tree that was watered continually. (Schneider and Childers, 1941)

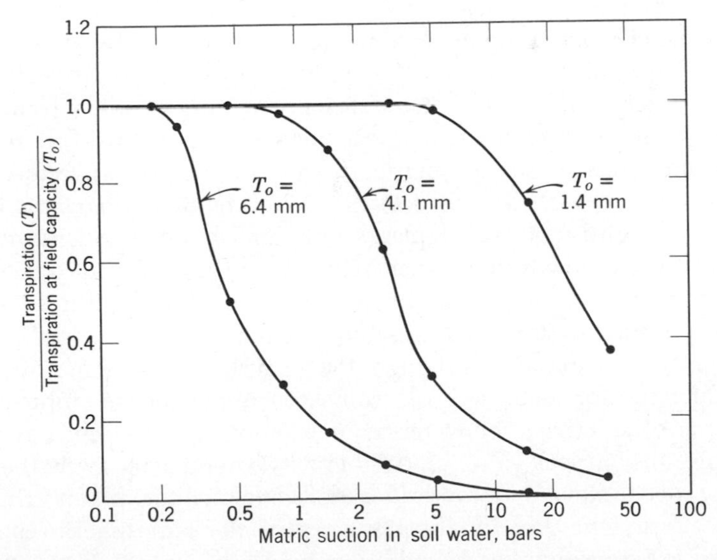

Figure 2.26. Relative rate of transpiration of corn versus soil suction on 3 days in which the rate of transpiration of corn grown in soil at the field capacity differed. The rate of transpiration of corn grown in soil at the field capacity in millimeters per 24 hours on the individual days is shown with the respective curves. (Denmead and Shaw, 1962)

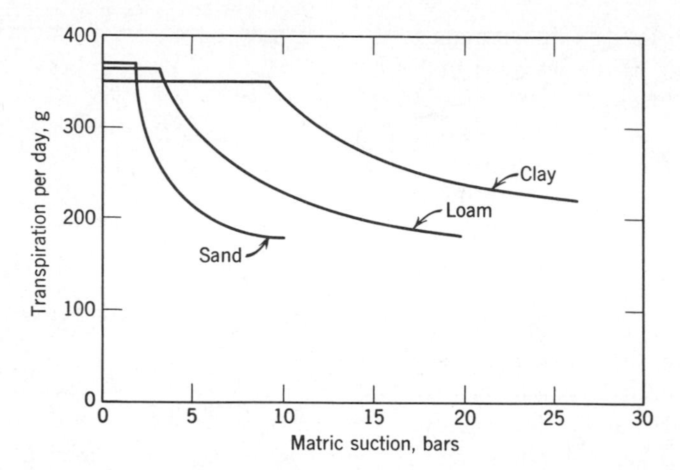

Figure 2.27. Rate of transpiration of birdsfoot trefoil versus matric suction of water in three soils. The plants were grown in glazed vessels with a diameter of 23 cm, were of the same age, and were clipped to the same height before the test. Values for the conductivity of the soil to water at the breaks in the curves were 2×10^{-5}, 2×10^{-5}, and 4×10^{-5} cm per day in the sand, loam, and clay, respectively. (Gardner and Ehlig, 1962a)

The lower relative transpiration values under conditions of greater potential transpiration in Fig. 2.26 presumably resulted from deficiency of water in the plants, and this deficiency in turn resulted from failure of the plants to take up water fast enough to maintain full turgidity. The significance of soil properties in this process is suggested by the results of an experiment shown in Fig. 2.27. The figure gives the transpiration from birdsfoot trefoil plants growing on three soils as the soils were dried. At low values of matric suction, the rate of transpiration was essentially constant and almost the same for the three soils. As the matric suction increased, a value was reached in each soil at which transpiration decreased. Although the critical value of matric suction was different for each soil, all values corresponded to approximately the same conductivity for water, namely, 2 to 4×10^{-5} cm/day. Presumably, the critical value of conductivity would vary with the extent of the root system and the water deficiency in the plants, the latter being affected in turn by the tendency of the atmospheric conditions to produce transpiration. In addition, roots are known to manifest an increase in resistance to uptake of water under conditions of low soil temperatures and poor aeration.

When plants are transpiring, a gradient of partial pressure of water vapor exists from the air in the intercellular spaces in the leaves leading outward through the stomata to the atmospheric air. Air in the intercellular spaces in the leaves remains almost saturated with water vapor. The gradient normally extends outward from the stomata for some distance (usually estimated to be of the magnitude of a few millimeters). Movement of air past the leaves replaces the moist air with drier atmospheric air, decreasing the distance through which diffusion is necessary and increasing the rate of transpiration. Martin and Clements (1935) investigated the significance of this process by comparing the rate of transpiration from sunflower plants in winds of different velocities with the rate of transpiration from comparable plants in still air. The net effects observed in short test periods are shown in Fig. 2.28. At wind velocities up to about 3 km/hour the rate of transpiration was increased from 20 to 30%; and this value was maintained throughout the entire test period of 2 to 4 hours, the stomata remaining open. At higher wind velocities, however, there was a marked initial increase in transpiration, lasting no more than 15 minutes, which caused some wilting of the plants and closure of the stomata, with consequent reduction in the rate of transpiration. The plants then gradually regained turgidity, which was associated with a gradual increase in rate of transpiration.

Plants grown in continuous wind suffer a reduction in size and leaf area, as demonstrated by Martin and Clements (1935), presumably because of the continued deficiency of water and reduction of photosynthesis. After undergoing such modification, plants transpire less water per unit area of leaf surface than do unconditioned plants subjected to similar wind velocities (Whitehead, 1962). [See the review by Stocker

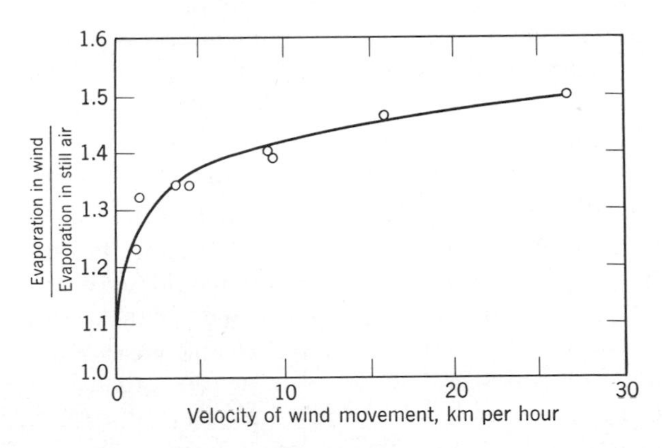

Figure 2.28. Relative rate of evaporation from sunflower plants exposed to wind and in still air versus wind velocity. The values are all averages for test periods 2 to 4 hours in length during the daytime. (Martin and Clements, 1935)

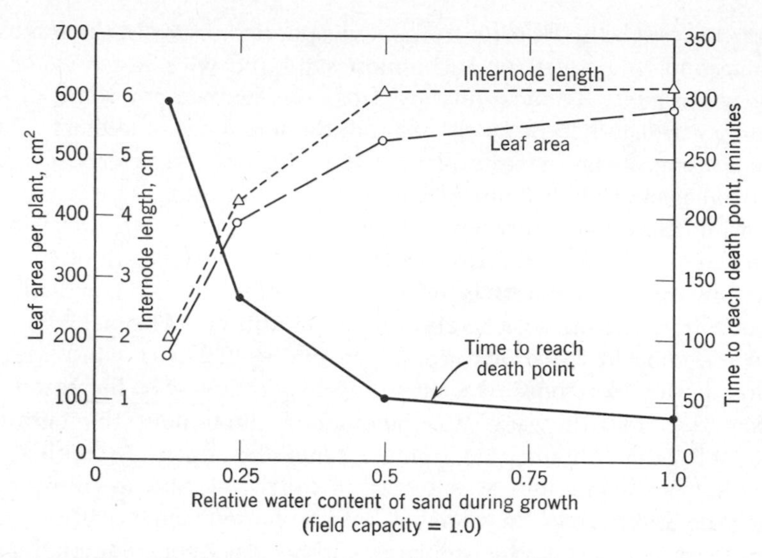

Figure 2.29. Internode length and leaf area of sunflower plants grown on sand cultures supplied with different quantities of water and time required to reach the death point when the plants were subjected to a wind of 64 km/hr. The water content of all cultures was adjusted to the field capacity before the test. (Whitehead, 1963)

(1960) for a discussion of the physiological basis for conditioning to drought.]

Whitehead (1963) conducted an experiment in which sunflowers were grown 38 days in sand cultures with different additions of water to produce different degrees of preconditioning of the plants. The water content of all cultures was then brought to the field capacity, and the cultures were exposed to a wind of 64 km/hour. A wind of this velocity is beyond the tolerance of the plants, according to Whitehead, and produces desiccation despite the protection afforded by closure of the stomata. The results of his experiment are shown in Fig. 2.29, which includes values of the time required for the foliage to turn to a darker green-black color indicating death of the plants. The plants that had been grown with an ample supply of water reached the death point much more quickly than those that had been preconditioned by continuous deficiency of water during growth. In this experiment, the supply of water to the roots was ample during exposure to the wind. In practice, dry winds may occur when the soil is relatively dry, so that plants have the advantage of some preconditioning to drought but the disadvantage of a deficiency of water in the soil.

Photosynthesis in Relation to Transpiration. Loss of water vapor in transpiration and entry of carbon dioxide gas in photosynthesis both occur in response to absorption of radiant energy by plants and take place mostly through stomatal openings in the leaves. Nevertheless, a given loss of water by transpiration is not necessarily accompanied by a fixed amount of photosynthesis.

Three general situations may be recognized in a consideration of photosynthesis in relation to transpiration or, in practical terms, plant yield in relation to evapotranspiration. First, photosynthesis is limited by factors other than water supply, and evapotranspiration is limited by water supply. The yield then increases little or none with evapotranspiration. Second, both photosynthesis and evapotranspiration are limited by water supply. The yield then increases with evapotranspiration. This situation is common and is illustrated in Fig. 2.30*a* by data obtained on wheat at different locations and in different years in the northern Great Plains region of the United States. At the time of wheat harvest, the soils in this area have usually been exhausted of water to the permanent

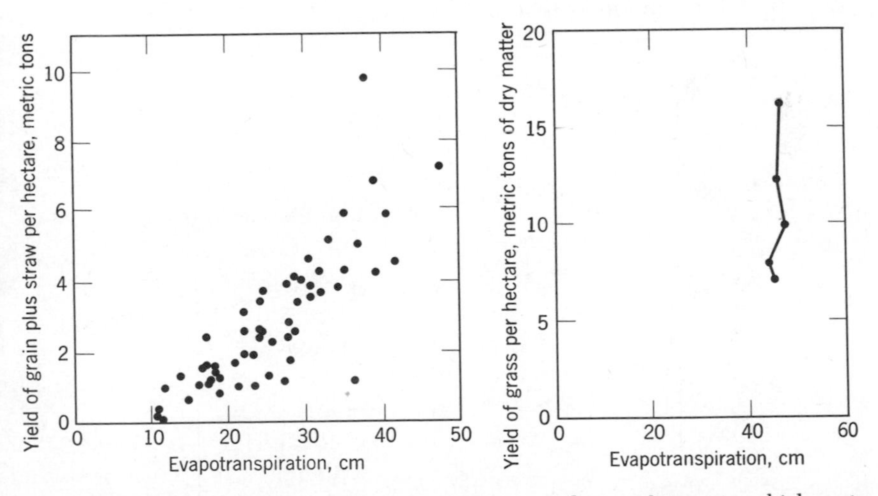

Figure 2.30. Plant yield versus evapotranspiration under conditions in which water supply limited (*a*) both evapotranspiration and photosynthesis and (*b*) neither evapotranspiration nor photosynthesis. (*a*) Yield of grain and straw of wheat versus evapotranspiration at eleven locations in the northern Great Plains of the United States in the period from 1909 to 1919 (Cole and Mathews, 1923). (*b*) Yield of grass versus evapotranspiration in an experiment in the Netherlands. The successively higher yields were obtained with applications of 0, 70, 120, 270, and 520 kg of nitrogen per hectare. (Wind, 1954)

wilting percentage. In the third situation, photosynthesis is limited by factors other than the water supply; but evapotranspiration is not limited by the water supply. The yield then may vary with conditions that affect photosynthesis without much change in evapotranspiration. This situation is illustrated in Fig. 2.30*b* by data from an experiment in the Netherlands. Yields of grass were more than doubled by application of nitrogen fertilizer, but evapotranspiration was hardly affected.

Differentiation Products

Formation of differentiation products, such as rubber, essential oils, and alkaloids, is favored by conditions under which the precursors accumulate in the plant and are not used for other purposes. Because water deficiency limits growth relatively more than photosynthesis, relatively more carbohydrate is available for purposes other than growth under conditions of water deficiency than under those of water sufficiency.

The work of Wadleigh et al. (1946) may be cited as an example of water-supply, growth, differentiation relationships in guayule, with rubber being the differentiation product. Figure 2.31 shows that an increase in total suction in the soil water resulted in a decrease in the

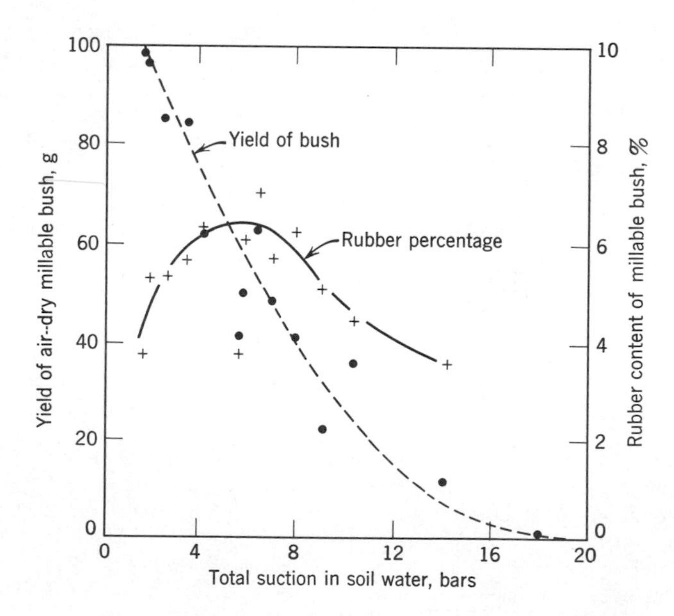

Figure 2.31. Yield and rubber content of guayule bush versus total suction in soil water. (Wadleigh et al., 1946)

yield of millable bush (the air-dry stems and main roots). The rubber content of the millable bush increased from about 4% where the suction was 1.5 bars to a maximum of about 6% where the suction was 4 to 8 bars and then decreased with further increase in the suction. (The authors accounted for the decrease in rubber percentage at high suction on the basis of loss of many leaves, which would decrease the rate of photosynthesis.) The absolute yield of rubber, that is, the product of yield of millable bush and rubber percentage, remained approximately the same as the suction in the soil water increased from 1.5 to 4 bars and decreased with further increase in suction. In some instances, the absolute yield of rubber is increased by soil dryness. Veihmeyer and Hendrickson (1961) reported such an instance. Time appears to be a factor here; that is, the rubber content increases slowly.

Proportion of Roots and Tops

If the supply of water is deficient, addition of water usually increases the yield of both the tops and roots of plants. The ratio of the weight of above-ground parts to below-ground parts increases with increasing water supply, as illustrated in Table 2.7 by Davis' (1942) data on nutgrass. Similar results have been obtained with other types of plants, as noted in a review by Stocker (1960). The relatively low ratio of tops to roots under conditions of water deficiency probably is responsible for the common belief that water deficiency stimulates root growth.

The change in proportion of roots and tops of plants in response to changes in water supply may be explained in a general way on the basis of the direct effects of water on cell enlargement and the indirect

Table 2.7. Yield of Tops and Tubers of Nutgrass in Soil with Different Minimum Water Percentages (Davis, 1942)

Minimum Water Content of Soil before Irrigation,[1] %	Fresh Weight of Plants, g		Ratio of Tops to Tubers
	Tops	Tubers	
6	4	26	0.15
9	9	50	0.18
12	16	80	0.20
15	25	116	0.22
18	34	130	0.26

[1] The moisture equivalent of the soil was 16.9%, and the permanent wilting percentage was 8.2%.

effects of water on the disposition of carbohydrates within the plant. The suction of water in the tops of plants normally exceeds the suction of water in the roots. Hence the availability of water for cell enlargement is greater in the roots than in the tops. This condition prevails whether the supply of water is ample or deficient. But with accumulation of carbohydrate unused for growth in the aerial part of the plant because of water deficiency, a greater proportion of the total carbohydrate is translocated to the roots, where it is available for root growth. In confirmation of this theory, Eaton and Ergle (1948) found that drought produced a large increase in percentage content of starch and sugars in the roots of cotton plants but not in the leaves. Thus it appears that the water relations of plants favor root extension over top extension. This tendency is expressed to a greater extent under dry than under moist conditions because of a relatively greater supply of carbohydrate to the roots and a relatively lesser supply of water to the tops.

Other considerations are involved, however. For example, during the severe drought of the 1930s, Albertson and Weaver (1945) observed that cottonwood trees growing on river flood plains died from lack of water; whereas they survived at more elevated locations. On the flood-plain locations the roots were shallow, an adaptation to the excess of water usually present; at the higher elevations the roots had grown more deeply. This observation suggests the importance of rate of response. Certainly the subsoil was initially moist in the flood-plain locations. Apparently the trees were incapable of responding rapidly enough in developing a deeper root system to use the subsoil water. Use of water from the upper part of the soil by more rapidly growing plants in the natural habitat may have reduced the capability of the cottonwoods to respond.

For some plants, an additional factor that may be significant is competition within the plants themselves. Early in the season roots grow rapidly. Later they grow less rapidly, one reason being that they are in competition with the developing fruit for carbohydrate produced by photosynthesis. The fruit is generally considered to be a stronger competitor than the roots.

There is some evidence of a change in character of the root system with water supply. Lundkvist (1955) observed that a deficiency of water in the soil decreased the weight of lateral roots of garden bean plants in relation to the weight of the tap root. This phenomenon apparently is not just a consequence of drying of soil from the surface downward because Kausch (1955) made a similar observation where the same kind of plant was grown in solutions with varied solute suction. Oppenheimer (1960) reviewed this subject with particular reference to adaptation of plants to drought.

Table 2.8. Yield of Barley Grain and Straw in Sand Cultures with Different Amounts of Water (Hellriegel, 1883)

Percentage of Water-Holding Capacity	Yield of Dry Matter per Culture, g		Ratio of Grain to Straw
	Grain	Straw	
5	0	0.1	0
10	0.7	2.3	0.3
20	7.7	6.9	1.1
30	9.7	10.0	1.0
40	10.5	11.3	0.9
60	10.0	12.8	0.8
80	8.8	10.9	0.8

Proportion of Grain and Straw

As shown in Table 2.8, Hellriegel (1883) found that where barley was grown in sand cultures supplied with different amounts of water, the entire crop consisted of straw with the smallest amount of water. With increasing amounts of water, the yields of both straw and grain passed through a maximum and then decreased; but the maximum grain yield occurred with a smaller amount of water than did the maximum straw yield. The maximum ratio of grain to straw occurred with a still smaller amount of water. The causes of the difference in behavior between grain and straw are not known.

Interpretation of the results of Hellriegel's experiment is complicated by the fact that he added only enough water to bring the water content up to the desired level at each application instead of adding enough to moisten the entire culture at each watering. As was pointed out previously, addition of a limited quantity of water to a dry soil results in moistening only the upper part of the soil to field capacity. Volume of soil and water supply thus changed together in Hellriegel's experiment. The same situation in general exists in the field, and results reported from field experiments agree with those obtained by Hellriegel. Figure 2.32 shows that the ratio of grain to straw increased with the quantity of water used by the spring wheat crop in different years at locations in North Dakota and Nebraska. The range in yield of grain per hectare was from 270 kg with the smallest water use to 2350 kg with the greatest water use, indicating the great importance of water as a limiting factor. The maximum in the ratio of grain to straw found

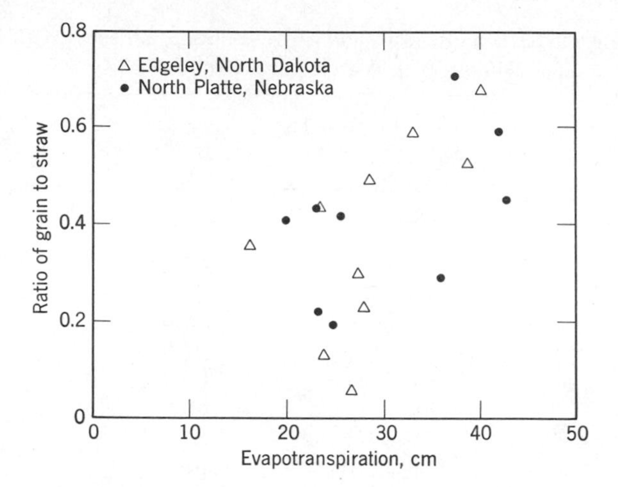

Figure 2.32. Ratio of grain to straw versus water used by spring wheat at Edgeley, North Dakota, during the years 1907 to 1917, and at North Platte, Nebraska, during the years 1908 to 1919. (Cole and Mathews, 1923)

in Hellriegel's experiment was not observed here, presumably because the range in water supply did not extend high enough. In an experiment in Utah (Table 2.9) where water was less important as a limiting factor, a decrease in ratio of grain to straw was associated with increasing water supply.

Time of Maturity

In seed-producing crops, maturation is in part a matter of translocation of organic and mineral substances to the seed and in part a matter of desiccation. The time of maturity thus is influenced by the rate of these processes. The rate of desiccation increases and the rate of translocation decreases as the water supply to the plant diminishes. Nevertheless, maturation is the last of a chain of events, and water supply may affect the time of final maturity through effects on some of the earlier growth stages.

Table 2.9. Ratio of Grain to Straw of Spring Wheat with Different Amounts of Water Available from Precipitation and Irrigation (Widtsoe, 1912)

Water available, cm	48	53	61	74	99	125	163
Ratio of grain to straw	0.80	0.76	0.75	0.69	0.63	0.60	0.49

Table 2.10. Length of Growing Period of Wheat with Water Supply Varied in Different Ways During Growth (Harris, 1914a)

Water Content of Soil During Indicated Growth Stage, %			
Planting to Five Well-Developed Leaves	Five Well-Developed Leaves to Boot Stage	Boot Stage to Maturity	Number of Days from Planting to Maturity
30	30	30	208
30	30	15	205
30	15	15	204
15	15	15	212
15	15	30	215
15	30	15	208

Table 2.10 gives the results of an experiment that shows the effect of supply of water at different stages of growth on the length of the growing period of wheat. The growing period was shortest where the soil water supply was high in the first stage or in the first two stages and low in the last stage or in the last two stages. It was longest where the soil water supply was low in the first two growth stages and high in the third. The growing period was shorter where the water supply was high throughout than where it was low throughout.

The usual situation in the field is that the water supply is greater at the beginning of the season than at the end of the season; hence additions of water to the soil when needed will have relatively less effect on the early growth of the plant than on the mid-season and late-season growth. As would be expected from Table 2.10, such water applications often are observed to delay the maturity. Thus Harris (1914) found in Utah that corn receiving 102 cm of irrigation water matured a week later than did corn receiving no supplemental irrigation. Shutt and Hamilton (1934) reported that in British Columbia the average length of the growing period of wheat was 116 days under irrigation and 104 days without irrigation. Drinkwater and Janes (1955) in Connecticut found that heavy and frequent irrigation doubled the yield of onions over no irrigation and delayed the maturity 19 days.

Field experiments by Schwalen and Wharton (1940) with lettuce in

Arizona, where normally the early season supply of soil water is low, showed that early maturity was achieved when ample irrigation water was supplied throughout the season. Still earlier maturity was obtained if the water content of the soil was lowered during harvest. On the other hand, if the lettuce was supplied with ample water after an early season deficiency of water, the growth period was prolonged; and the lettuce matured relatively late. The effect of late irrigation in lengthening the growing period of barley was noted by Harlan and Anthony (1921) in Idaho. These investigators found that the water content of the grain on August 6 was 36% where the barley had not been irrigated since June 23. On plots that had been irrigated June 25 and at later dates, the water percentages in the grain were 43, 50, and 47 where the barley was irrigated July 14, July 20, and July 29, respectively.

Plants supplied with an excess of water in the early part of the season usually mature relatively late. This behavior is often observed in areas where water has stood for some time during the early part of the season.

With some plants, the water supply has a pronounced effect on onset of flowering and hence on time of maturity. Bronchart (1964) noted that the tropical seed plant *Geophila renaris* grew vegetatively for 280 days but failed to flower even after 485 days if the soil was kept moist. But if the water content of the soil was allowed to drop temporarily after 280 days, the plant flowered even though the soil was later kept moist like that of the nonflowering control. Somewhat similar behavior has been noted in a variety of amaryllis, which is a bulbous ornamental often grown as a potted plant.

Critical Period

The effect on final yield produced by a temporary water deficiency in the soil depends on the time in the growth cycle of the plant at which the deficiency occurs. Figure 2.33 gives the results of an experiment in which soybeans were grown on a loam soil in large metal cans sunk in the soil out-of-doors in Iowa. In one series of cultures, the soil was maintained moist continuously. In other series, the water content of the soil was maintained at the same level as that in the continuously moist cultures except for a period of 1 week, when the water content was allowed to drop low enough to produce the first signs of wilting in the upper leaves for 4 days. The first 1-week period began 4 days after the first flowers were produced. The other seven 1-week drought periods followed consecutively, the last one ending 2 weeks before maturity.

The yield of soybeans per plant may be represented as the product of number of pods, number of beans per pod, and weight per bean,

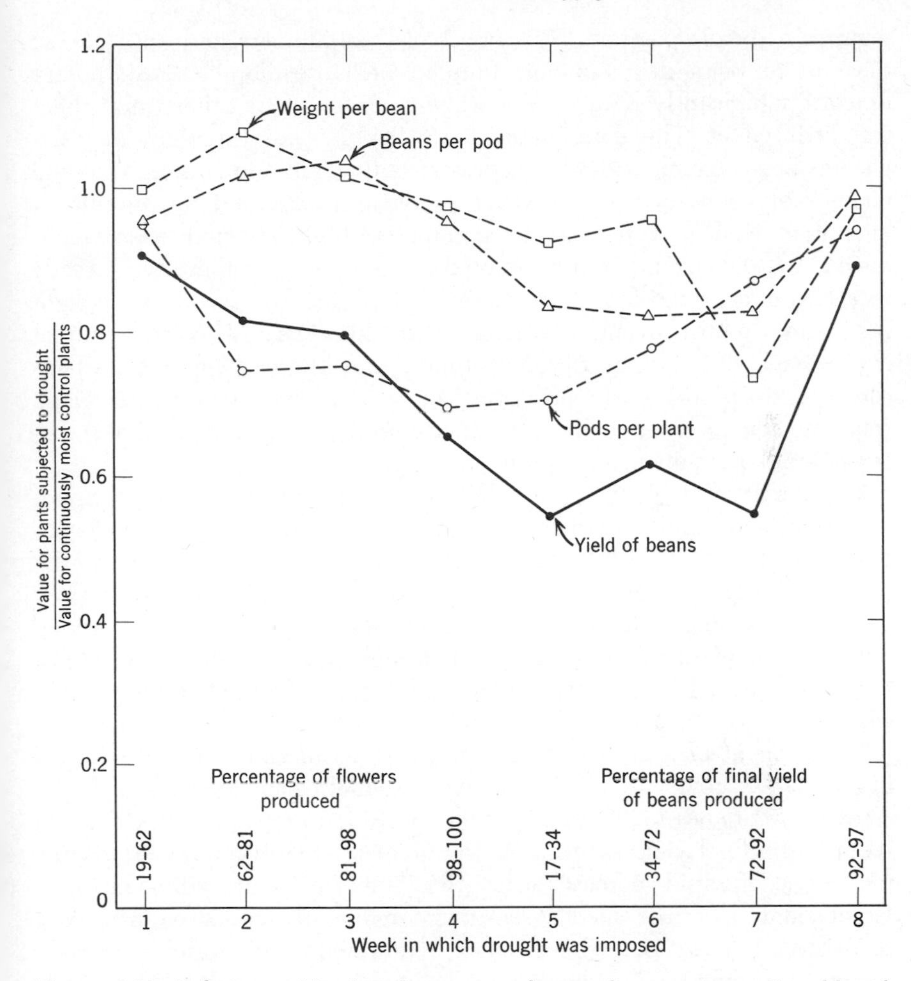

Figure 2.33. Relative yield of soybeans per plant and components of yield with temporary drought imposed in each of eight successive weeks on individual cultures. The soil was maintained in a moist condition continuously in control cultures and continuously except for the week of drought in all other cultures. Data on plant development at the bottom of the figure apply to plants on continuously moist soil. The lower value is for the beginning of the week of drought and the higher value for the end. (Laing, 1966)

values for all of which arc shown in Fig. 2.33. During flowering, the yield component most severely affected by drought was the number of pods. This effect was presumably a result of dropping of flowers, as was observed by Lambeth (1950) for lima beans. As plant development continued, the next yield component affected by drought was the number of beans per pod. This effect of drought persisted until almost the end of bean development. Late in bean development, temporary drought significantly reduced the weight per bean. At other times there was little effect. The data in the figure are for harvestable beans, defined as beans having a short axis greater than 3.5 mm in length. The total number of beans per pod was not significantly affected by drought at any time. Hence, if weights of the undersize beans had been determined, the data could be represented as showing that there were really only two effects, the first on number of pods and the second on weight per bean. In any event, temporary drought clearly affected different components of yield at different times; and these component effects combined to produce a broad critical period, in yield of beans per plant, from the fourth through the seventh drought period. Salter (1963) obtained similar results in an experiment on peas.

Van der Paauw (1949) conducted an experiment on oats with experimental design similar to that used by Laing in the work on soybeans discussed in the preceding paragraph. Van der Paauw found that temporary drought during the time the plants were heading produced greater shortening of the stems, greater barrenness of florets, and greater reduction in yield of grain than did drought during other stages. The time of heading was thus critical as far as drought damage was concerned.

The concept of a critical period of water requirement was developed in Russia. In reviewing the early work, which was done mainly with cereals, Maximov (1929, pp. 396–399) concluded that temporary drought reduces the final yield of grain to the greatest extent if it occurs during the period of rapid increase in length of internodes preceding heading. He accounted for the special sensitivity of the plant to drought at this time on the basis that the greatest requirement for water for tissue expansion occurs during the time of maximum rate of growth in length. A deficiency of water during this time decreases the size of the cells. Other work on critical periods in water requirement was reviewed by Vaadia et al. (1961).

Removal of Inhibitors

The effects of water supply described in previous sections are associated with the action of water as a substance essential to plant growth. In specific instances water may play the additional role of removing growth inhibitors. Perhaps the most familiar example of removal of inhibitors is the leaching of saline soils with water to remove excess salts. Soil salinity will be considered in Chapter 6. The purpose here is to note two less familiar types of effects, removal of inhibitors from seeds and removal of nonsaline inhibitors from soils, literature on which was reviewed by Oppenheimer (1960).

According to Went (1953), the seeds of certain desert annual plants germinate after heavy rainfall but not after light rainfall even if the soil is kept moist continuously by artificial means. Germination occurs with less water in the presence of charcoal than in its absence. Went explained these observations on the basis that the seeds contain diffusible germination-inhibitors that are removed by rain if movement of water past the seed continues long enough. The action of charcoal suggests that inhibitors of an organic nature are involved. Inorganic inhibitors may be involved also, however, because Went stated that low concentrations of salts prevent germination.

The presence of inhibitors in seeds of certain desert plants was demonstrated subsequently in other ways. For example, Koller and Negbi (1955) found that the percentage germination of seeds of the desert legume *Colutea istria*, after removal of the outer seed coat, was much reduced if they placed the outer seed coats in the dish with the inner seeds during the germination test. In another experiment they found that a water extract of the outer seed coats inhibited growth of lettuce seedlings and, in relatively high concentration, inhibited germination of lettuce seeds also. Most of the detrimental effect disappeared if the extract was heated during the concentration process, which indicates that the inhibitor was organic in nature. Water-soluble inhibitors that act in this way permit the greatest germination under conditions in which the plant is most likely to become established and to reach maturity.

Removal of growth inhibitors by water may also be important in the survival of seedlings in specific instances. Bonner and Galston (1944) found that roots of guayule, a desert shrub, exude substances that are toxic to seedlings of the same species. Cinnamic acid is one of these toxins. Went (1953) used this evidence to explain the fact that the density of the shrubs increases as the rainfall increases, the inference being that the seedlings can become established if the toxins are washed down by rain. As additional evidence, he cited observations on another desert shrub, *Larrea divaricata,* which shows the same habit of wide, regular spacing of density that increases with the rainfall. After heavy summer rains, *Larrea* seedlings were observed to develop under and between shrubs of that species. A few weeks later the seedlings under the old plants died. Competition for water apparently was not the deciding factor at this time because seedlings of other plants were not affected. With time, the radius of death of *Larrea* seedlings progressed outward from the old plants until only the seedlings farthest removed from the existing shrubs were left.

Seed Germination

The suction in the water in dry seeds is relatively high (frequently in excess of 500 bars); thus dry seeds will absorb water from soils below the permanent wilting percentage. Whitney and Cameron (1904) mixed 50 g of cowpea seeds with 50 g of soil containing 7.5 g of water and found that in 12 hours the cowpea seeds had absorbed all but 0.65 g of the soil water, leaving the soil in essentially air-dry condition. Seeds that absorb an amount of water too small to permit germination may nevertheless be moistened sufficiently to permit invasion by fungi. Fungal hyphae will develop at values of matric suction well above those at which higher plants will grow (Griffin, 1963). In time the fungi will damage the seed so that germination will not take place even if more water is added.

Under favorable conditions of water supply, temperature, and seedcoat permeability, seeds may absorb enough water to double their weight within a period of 2 days. In soils with water content suitable for plant growth, the volume of soil required to contain this amount of water is considerably greater than the volume of the seed that absorbs it. Whitney and Cameron (1904) found that seeds placed in a large mass of moist soil appeared to dry the surrounding soil to a distance of about 3 mm. Sometimes the soil next the seed appeared to be nearly air-dry. This observation suggested that large seeds, especially those that are approximately round, might not be able to absorb enough water from the soil to permit germination unless the soil were rewetted by at least one rainfall or irrigation. In confirmation of this hypothesis, they found that seeds of clover and wheat, which are relatively small, would germinate if they were placed on the surface of a moist soil in a humid atmosphere; seeds of lima bean, which are relatively large, would not germinate under these conditions.

More recent investigations by Doneen and MacGillivray (1943) on germination of seeds of different sizes in two soils, each at water contents ranging from below the permanent wilting percentage to about the field capacity, did not confirm Whitney and Cameron's hypothesis with regard to seed size. That is to say, if the germination of a given kind of seed in dry soil is expressed as a percentage of the germination in moist soil, the values obtained with large seeds did not appear to be lower than those obtained with small seeds. Moreover, some kinds of seeds germinated fairly well even in soil below the permanent wilting percentage.

In view of these results, it seems likely that the observations made by Whitney and Cameron on the failure of lima beans to germinate on the surface of soil may indeed be accounted for by failure of the large seeds to absorb enough water for germination through the limited contact zone. In their experiment, the contact zone would be less than half as great as it would if the seeds had been imbedded in the soil. The dry layer noted around the seeds, however, was probably only a temporary phenomenon that occurred soon after the dry seeds were placed in the soil.

The fact that Doneen and MacGillivray obtained good germination of some kinds of seeds in relatively dry soils indicates that seeds of some plants will germinate in soil too dry to support much plant growth; hence, at least for plants of this type, the effect of water supply is more severely limiting to growth than to germination of the seed. In general, it is probably true that water supply limits plant growth more severely than seed germination because the total amount of water required for germination is negligible compared with that required for growth. Under some circumstances, however, the water content for germination may be the more critical. For example, the seed may lie in relatively dry soil underlain by adequately moist soil, or there may be later additions of water by rainfall or irrigation that supply the plants but have little effect on germination. Another phenomenon related to the requirement of seeds for water for germination is discussed in the section on removal of inhibitors.

Relation to Soil Fertility

As illustrated by Fig. 2.34, plant yield may be increased, decreased, or unaffected by a given change in fertility level, depending on the magnitude of the change, the initial fertility level, and the water supply. The subject to be emphasized in this section is the effect on yield of increasing the soil fertility where the water supply is limited and hence where decreases in yield are a distinct possibility. Viets (1962, 1965) reviewed much of the same subject matter with emphasis on efficiency of water use.

As a point of departure, it may be helpful to recall from the section on evaporation that the rate of evaporation from a bare soil, moist at the surface, is about the same as the rate of evaporation from bulk water at the same temperature and exposure. As the density of vegetation on moist soil increases from zero up to the maximum that can be supported, the evapotranspiration remains about the same as the evaporation from bare, moist soil. Some of the water lost by evaporation from the moist soil is merely replaced by water lost by transpiration from the plants. In the presence of a dense cover of vegetation, most of the evapotranspiration would be transpiration and not evaporation from the soil surface. This behavior may be explained theoretically on the basis

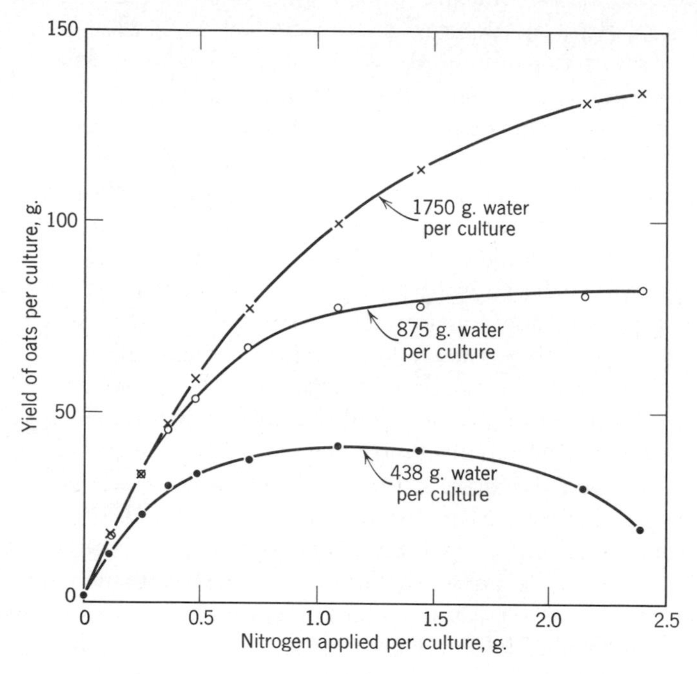

Figure 2.34. Yield of oats in sand cultures with different levels of water and nitrogen as ammonium nitrate. The values given for water content are the quantities of water maintained per culture by frequent additions. (Lange, 1938)

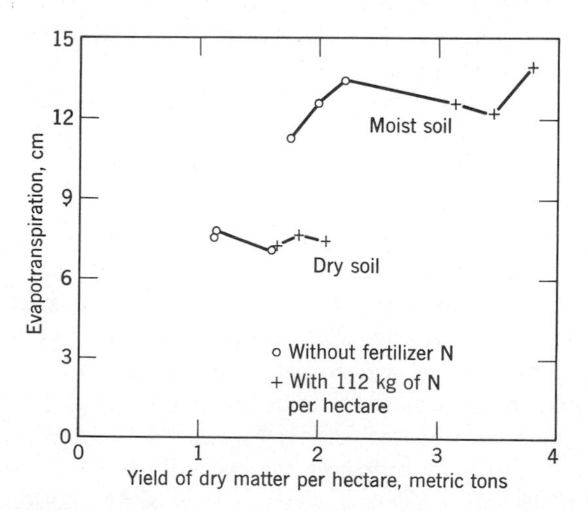

Figure 2.35. Evapotranspiration versus yield of dry matter of Sudangrass with different levels of water supply, stand density, and nitrogen fertilization in Alabama. The three observations with and without nitrogen fertilization at each level of water supply represent different densities of stand. (Weaver and Pearson, 1956)

that the availability of water for evaporation is relatively high where soil remains moist to the surface, the solar energy that brings about evaporation is independent of the character of the surface on which it falls, and the rate of removal of water vapor is about the same from the different kinds of surfaces, provided they are large and uniform. This generalization about equality of water loss from the different kinds of surfaces must be recognized as a first approximation, from which departures may occur because of differences in surface properties. Also, under conditions conducive to high rates of evapotranspiration, the approximate equality of losses from bulk water, plants, and moist soil may persist only briefly because of a decrease in turgidity of the plants and closure of the stomata and because of rapid development of a dry layer on the surface of the soil.

The situation just described may be illustrated in part by data obtained by Mitscherlich and Beutelspacher (1938) in an experiment conducted in Germany. During the growing season of oats, they recorded rainfall of 18 cm, evaporation of 26 cm from bulk water, and evapotranspiration of 26 cm from plots planted to oats.

Another illustration is found in Fig. 2.35, which gives the results of an experiment in Alabama with Sudangrass at three stand levels, each with and without an application of fertilizer nitrogen at 112 kg/hectare. Each of the six treatments was duplicated at two levels of water supply. At the high level, the soil was irrigated whenever the matric suction in the soil water at a depth of 51 cm increased to 0.5 bar. At the low level of water supply, the soil was irrigated only when wilting seemed likely. The evapotranspiration and yield of dry matter were both greater where the soil was relatively moist than where it was dry, but a pronounced upward trend of evapotranspiration with yield of dry matter was not observed with either condition. In this experiment, therefore, the evapotranspiration was largely independent of the density of vegetative canopy, within the range investigated, where evapotranspiration was definitely limited by soil dryness. Analogous results were obtained in experiments by Carlson et al. (1959) on corn in North Dakota and by Ramig (1960) in experiments on wheat in western Nebraska.

The foregoing considerations of evaporation and evapotranspiration may be synthesized in the generalized form shown in Fig. 2.36, which represents plots of evapotranspiration against density of vegetative canopy of the sort that one might expect to find under different levels of water supply in the soil but with an aerial environment that is constant except for the effects associated with the differences in supply of water. In accordance with Fig. 2.35, a range is shown under both moist and dry conditions in which the evapotranspiration does not increase with an increase in density of vegetative canopy. The strongly sloping portions

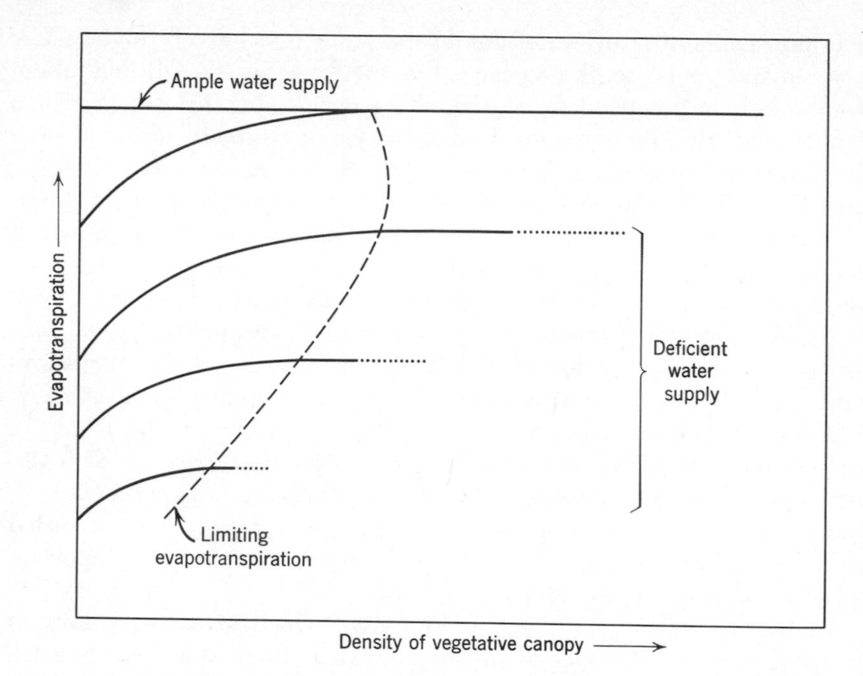

Figure 2.36. Schematic representation of evapotranspiration versus density of vegetative canopy with ample and deficient water supply.

of the curves shown in Fig. 2.36 were not found in the experiment given in Fig. 2.35, presumably because the crop densities employed experimentally were too high to show this behavior. The limiting evapotranspiration, above which the evapotranspiration is not affected by the density of vegetative canopy (or yield of dry matter, as in the experimental data described), is shown to increase with the supply of water available for evaporation and transpiration and to occur at a maximum density of vegetative canopy where the available water is only slightly deficient. The limiting evapotranspiration should be at zero density of vegetative canopy where the availability of water is unlimited.

The following theory may be suggested to account for the trend of evapotranspiration with density of vegetative canopy where the availability of soil water for evaporation and evapotranspiration is limited, as in the dry conditions in the experiment in Fig. 2.35. Where there is a deficiency of water for evaporation from bare soil containing available water at some depth below the surface, the high conductivity of plants for water substitutes in part for the loss of conductivity for water that has occurred in the surface portion of the soil. An increase in density of vegetative canopy from zero up to the maximum the soil will support then causes an increase in evapotranspiration. Concurrently with the increased use of water, however, the availability of soil water to plants decreases. Consequently, the plants lose water less rapidly than they would if the drying effect had not occurred. Because a given decrease in water content of soil has an increasingly pronounced effect on availability of the water to plants as soil dries, the decrease in availability counterbalances more effectively the greater potential loss due to extra plant surface as the supply of available water approaches exhaustion. The limiting evapotranspiration corresponds to exhaustion of available water.

According to the foregoing theory, the independence of evapotranspiration and density of vegetative canopy exists for a different reason where soil is dry than where availability of water for evaporation and evapotranspiration is unlimited. Under moist conditions, the control is in the atmosphere. Under dry conditions, the control is in the soil. Under intermediate conditions, the control may be partly in the atmosphere, partly in the soil, and partly in the plants.

The term *density of vegetative canopy* is used in Fig. 2.36 because of the association of transpiration with the expanse of transpiring surface. Differences in soil fertility affect evapotranspiration primarily by changing the expanse of transpiring surface and only in a minor way by changing the character of the surface. For a given kind of plant, the expanse of transpiring surface can be estimated to a close approximation from the yield of above-ground plant material. Accordingly, for the purpose of this section, yields of above-ground plant material are used as indexes of densities of vegetative canopy produced by a given species or variety in response to differences in soil fertility.

With regard to soil fertility and crop yields, perhaps the most significant observation to be made about Fig. 2.36 and the data on which it is based is that an increase in crop yield produced by increasing soil fertility does not produce a corresponding increase in evapotranspiration; on the contrary, wide differences in yield induced by differences in soil fertility may result in only relatively small differences in evapotranspiration.

The magnitude of the increase in yield that can be obtained by increasing the soil fertility under a given water regime may be said to depend on the fertility level of the soil and the capability of the crop to produce additional dry matter under the prevailing circumstances. An increase in production of dry matter by plants without an appreciable increase in use of water under conditions of a deficiency of water for evapotranspiration may be brought about in different ways.

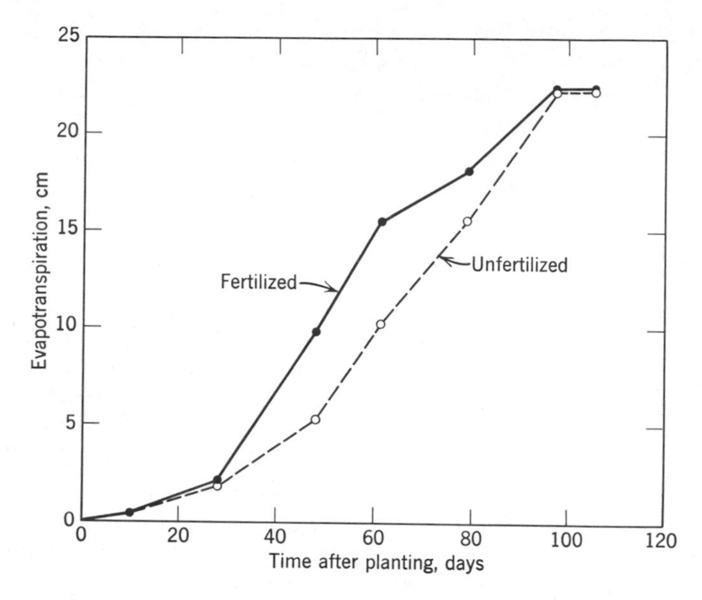

Figure 2.37. Evapotranspiration from fertilized and unfertilized corn at different times. Fertilized corn received 157 kg of nitrogen per hectare. The soil was charged with water to field capacity to a depth of 2.4 meters on the 46th day after planting and was then covered with polyethylene sheeting for the remainder of the experiment. (Linscott et al., 1962)

One possible mechanism is the capability of plants at a higher soil fertility level to carry on photosynthesis at a greater rate than those at a lower fertility level despite somewhat greater deficiency of water. The comparative sensitivity of plants to deficiency of water and to deficiency of the nutrient or nutrients in question should affect the magnitude of the increase in yield obtainable without increase in use of water.

A second way in which additional dry matter may be produced without appreciable change in total water use under conditions of water deficiency is by differences in rate of use of water at different times. Figure 2.37 illustrates the results of an experiment in which water was used more rapidly from fertilized than from unfertilized soil while the water supply was ample and more slowly after the available water was largely exhausted. The total evapotranspiration during the period of measurement was 22.2 and 22.4 cm on the unfertilized and fertilized plots, and the corresponding yields of corn grain were 1.8 and 2.5 metric tons per hectare. Sneva et al. (1958) reported analogous results in an experiment on crested wheatgrass under desert conditions in Oregon. The fertilized grass made more rapid growth early in the season, exhausted more rapidly the water supply in the soil, and matured at an earlier date than did the unfertilized grass.

The limiting evapotranspiration may be expected to be well defined in nonirrigated arid and semiarid regions. The subsoil is usually dry, and the available water is usually exhausted from the upper part of the soil by each crop; hence a fairly definite amount of available water is used completely. In humid and irrigated regions, on the other hand, the subsoil is usually moist to depths beyond the maximum extension of plant roots. An increase in soil fertility under these conditions may increase the use of subsoil water (Gliemeroth, 1951). In terms of Fig. 2.36, this greater use of subsoil water would be expected to appear as an extension of the upward-sloping portion of the evapotranspiration curves at a relatively low slope. That is, the limiting evapotranspiration is approached more gradually if the subsoil is moist than if it is dry. The extent to which subsoil water is used may be modified by the properties of the subsoil. Increased root growth is probably an important factor in increased use of subsoil water at high fertility levels (see Table 2.11). On the other hand, subsoils may have physical or chemical properties that are inhibitory to growth of roots and hence to use of subsoil water.

Thus far, attention has been focused on the yield of total above-ground dry matter. This criterion of yield may be suitable for hay and forage crops but not for grain crops. The significance of the rate of exhaustion of the water supply is not necessarily the same for forage crops, where yield is measured in terms of total dry matter, as for grain crops. The vegetative parts of grain crops are produced during the early part of the season, when the water supply is usually least limiting; but the grain is produced during the latter part of the season, when the water

Table 2.11. Distribution of Barley Roots in a Loam Soil with NPK Fertilizer Placed at Different Depths (Gliemeroth, 1955)

	Proportion of Total Weight of Roots, %		
Depth of Soil Layer, cm	Fertilizer in 0- to 18-cm Layer	Fertilizer in 18- to 36-cm Layer	Fertilizer in 36- to 54-cm Layer
0–18	50	35	27
18–36	30	45	33
36–54	19	20	40

supply is commonly most limiting. Consequently, stress of water deficiency, as modified by soil fertility, may have a different effect on production of grain than on production of total dry matter. Figure 2.38 illustrates conditions under which the behavior of grain yield is similar to and different from that of total yield. This experiment was conducted in western Nebraska with winter wheat on soil supplied with different amounts of water by preplanting irrigation. All plots received the natural rainfall, which averaged 31 cm between planting and harvest in the years the experiment was conducted. One may note from the figure that the total yield (grain plus straw) was increased by fertilization with all levels of water supply in the soil. The same was true of yield of grain at the two higher levels of water supply but not at the lowest. With the lowest water supply, the yield of grain was decreased by each increment of fertilizer after the first.

In connection with decreases in yield from fertilization under dry conditions, a brief comment on a seeming discrepancy between Figs. 2.34 and 2.38 may be worthwhile. A decrease in total yield with the highest application of nitrogen at the lowest level of water supply may be observed in Fig. 2.34 but not in Fig. 2.38. The difference is probably the result of the relatively high solute suction associated with the ammonium nitrate applied to the sand cultures in Fig. 2.34. An analogous decrease in total yield would be expected if the concentration of ammonium nitrate in soil in the field were similar; however, applications in the field are seldom sufficient to have such effects on solute suction except on a local basis in the vicinity of fertilizer bands.

Because of possible decreases in yield of grain from increasing soil fertility under dry conditions, it may be desirable to limit the rate of exhaustion of water by limiting the growth of the vegetative parts of plants, thereby saving a greater proportion of the available water for the stage of grain development. In this regard, one could theoretically increase the efficiency of water use in production of grain on nitrogen-fertilized soil by withholding the nitrogen during the first part of the season and applying it later. This practice should reduce the vegetative growth, thereby saving more of the available water for grain production. Practical application of this possibility does not seem to have been made, perhaps because of the difficulty of making a delayed application of nitrogen fertilizer to the soil so as to be effective under the dry conditions that prevail. Nitrogen sprayed on plants in the form of urea has been found to be absorbed readily, however, and this means of application might prove satisfactory for the small quantities of nitrogen that would be effective.

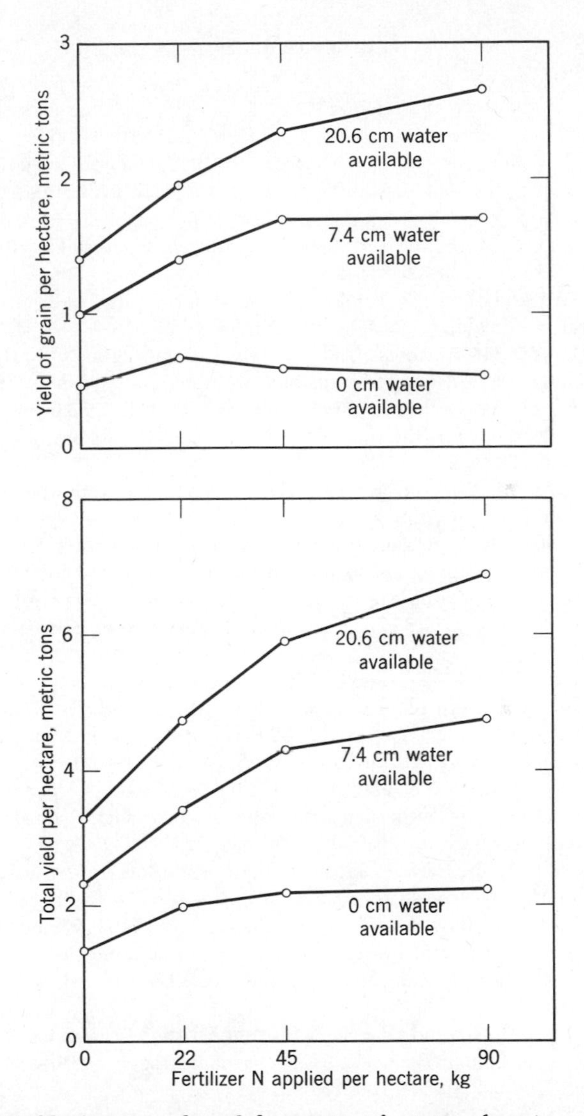

Figure 2.38. Yield of grain and total dry matter of winter wheat in western Nebraska with different applications of fertilizer nitrogen and with different quantities of available water in the soil at planting time. The figures in the graphs correspond to the centimeters of available water in the soil at planting time. The different quantities of water were supplied by preplanting irrigation and correspond to 0, 61, and 183 cm of soil wetted to the field capacity. (Ramig and Rhoades, 1963)

Literature Cited

Abdalla, A. A., and W. J. Flocker. (1963) The effect of hexadecanol on water loss from soil and plants. *Proc. Amer. Soc. Hort. Sci.* **83**:849–854.

Adams, F., F. J. Veihmeyer, and L. N. Brown. (1942) Cotton irrigation investigations in San Joaquin Valley, California, 1926 to 1935. *California Agr. Exp. Sta. Bul. 668.*

Adams, J. E., and R. J. Hanks. (1964) Evaporation from soil shrinkage cracks. *Soil Sci. Soc. Amer. Proc.* **28**:281–284.

Albertson, F. W., and J. E. Weaver. (1945) Injury and death or recovery of trees in prairie climate. *Ecol. Monogr.* **15**:393–433.

Allmaras, R. R., W. C. Burrows, and W. E. Larson. (1964) Early growth of corn as affected by soil temperature. *Soil Sci. Soc. Amer. Proc.* **28**:271–275.

Bahrani, B., and S. A. Taylor. (1961) Influence of soil moisture potential and evaporative demand on the actual evapotranspiration from an alfalfa field. *Agron. Jour.* **53**:233–237.

Bonner, J., and A. W. Galston. (1944) Toxic substances from the culture media of guayule which may inhibit growth. *Bot. Gaz.* **106**:185–198.

Bronchart, R. (1964) Research on the development of *Geophila renaris* Wild. and Th. Dur. under the ecological conditions of an equatorial forest cover. Effect of a decrease in available soil water on the onset of flowering. (Translated title) *Mém. Soc. Roy. Sci. Liège* **8**, No. 2 (*Soils and Fertilizers* **28**, Abstr. 3059, 1965).

Burgevin, H., and S. Hénin. (1943) Influence de la profondeur du plan d'eau sur le développement des plantes. *Ann. Agron.* (N.S.) **13**:288–294.

Burrows, W. C., and W. E. Larson. (1962) Effect of amount of mulch on soil temperature and early growth of corn. *Agron. Jour.* **54**:19–23.

Burton, G. W., E. H. De Vane, and R. L. Carter. (1954) Root penetration, distribution and activity in southern grasses measured by yields, drought symptoms and P^{32} uptake. *Agron. Jour.* **46**:229–233.

Butler, P. F., and J. A. Prescott. (1955) Evapotranspiration from wheat and pasture in relation to available moisture. *Australian Jour. Agr. Res.* **6**:52–61.

Campbell, R. B., C. A. Bower, and L. A. Richards. (1949) Change of electrical conductivity with temperature and the relation of osmotic pressure to electrical conductivity and ion concentration for soil extracts. *Soil Sci. Soc. Amer. Proc.* (1948) **13**:66–69.

Carlson, C. W., J. Alessi, and R. H. Mickelson. (1959) Evapotranspiration and yield of corn as influenced by moisture level, nitrogen fertilization, and plant density. *Soil Sci. Soc. Amer. Proc.* **23**:242–245.

Cole, J. S., and O. R. Mathews. (1923) Use of water by spring wheat on the Great Plains. *U.S. Dept. Agr. Dept. Bul. 1004.*

Cole, J. S., and O. R. Mathews. (1939) Subsoil moisture under semiarid conditions. *U.S. Dept. Agr. Tech. Bul. 637.*

Collis-George, N., and J. E. Sands. (1962) Comparison of the effects of the physical and chemical components of soil water energy on seed germination. *Australian Jour. Agr. Res.* **13**:575–584.

Danielson, R. E., and M. B. Russell. (1957) Ion absorption by corn roots as influenced by moisture and aeration. *Soil Sci. Soc. Amer. Proc.* **21**:3–6.

Davis, C. H. (1940) Absorption of soil moisture by maize roots. *Bot. Gaz.* **101**:791–805.

Davis, C. H. (1942) Response of *Cyperus rotundus* L. to five moisture levels. *Plant Physiol.* **17**:311–316.

Denmead, O. T., and R. H. Shaw. (1962) Availability of soil water to plants as affected by soil moisture content and meteorological conditions. *Agron. Jour.* **54**:385–390.

Doneen, L. D., and J. H. MacGillivray. (1943) Germination (emergence) of vegetable seed as affected by different soil moisture conditions. *Plant Physiol.* **18**:524–529.

Drinkwater, W. O., and B. E. Janes. (1955) Effects of irrigation and soil moisture on maturity, yield and storage of two onion hybrids. *Proc. Amer. Soc. Hort. Sci.* **66**:267–278.

Eaton, F. M., and J. E. Bernardin. (1964) Mass-flow and salt accumulations by plants on water versus soil cultures. *Soil Sci.* **97**:411–416.

Eaton, F. M., and D. R. Ergle. (1948) Carbohydrate accumulation in the cotton plant at low moisture levels. *Plant Physiol.* **23**:169–187.

Eden, A., G. Alderman, C. J. L. Baker, H. H. Nicholson, and D. H. Firth. (1951) The effect of ground water-level upon productivity and composition of fenland grass. *Jour. Agr. Sci.* **41**:191–202.

El-Damaty, A. H., H. Kühn, and H. Linser. (1965) Water relations of wheat plants under the influence of (2-chloroethyl)-trimethyl-ammonium chloride (CCC). *Physiol. Plantarum* **18**:650–657.

Ellis, N. K., and R. Morris. (1946) Preliminary observations on the relation of yield of crops grown on organic soil with controlled water table and the area of aeration in the soil and subsidence of the soil. *Soil Sci. Soc. Amer. Proc.* (1945) **10**:282–283.

Fox, R. L., and R. C. Lipps. (1955) Subirrigation and plant nutrition. I. Alfalfa root distribution and soil properties. *Soil Sci. Soc. Amer. Proc.* **19**:468–473.

Frankena, H. J., and M. A. J. Goedewaagen. (1942) Een vakkenproef over den invloed van verschillende water-standen op den grasgroei bij drie grondsoorten. *Versl. Landbouwk. Onderzoekingen* **48A**:407–461.

Gardner, W., O. W. Israelson, N. E. Edlefsen, and H. Clyde. (1922) The capillary potential function and its relation to irrigation practice. *Physical Rev.*, Ser. 2, **20**:196.

Gardner, W. H. (1965) Water content. *Agronomy* **9**:82–127.

Gardner, W. R. (1958) Some steady-state solutions of the unsaturated moisture flow equation with application to evaporation from a water table. *Soil Sci.* **85**:228–232.

Gardner, W. R. (1960) Dynamic aspects of water availability to plants. *Soil Sci.* **89**:63–73.

Gardner, W. R. (1960a) Soil water relations in arid and semi-arid conditions. *In* Plant-Water Relationships in Arid and Semi-Arid Conditions. *Arid Zone Res.* **15**:37–61.

Gardner, W. R. (1964) Relation of root distribution to water uptake and availability. *Agron. Jour.* **56**:41–45.

Gardner, W. R. (1965) Dynamic aspects of soil-water availability to plants. *Annual Rev. Plant Physiol.* **16**:323–342.

Gardner, W. R., and C. F. Ehlig. (1962) Impedance to water movement in soil and plant. *Science* **138**:522–523.

Gardner, W. R., and C. F. Ehlig. (1962a) Some observations on the movement of water to plant roots. *Agron. Jour.* **54**:453–456.

Gardner, W. R., and M. Fireman. (1958) Laboratory studies of evaporation from soil columns in the presence of a water table. *Soil Sci.* **85**:244–249.

Gardner, W. R., and R. H. Nieman. (1964) Lower limit of water availability to plants. *Science* **143**:1460–1462.

Gingrich, J. R., and M. B. Russell. (1957) A comparison of effects of soil moisture tension and osmotic stress on root growth. *Soil Sci.* **84**:185–194.

Gliemeroth, G. (1951) Der Einfluss von Düngung auf den Wasserentzug der Pflanzen aus den Unterbodentiefen. *Zeitschr. Pflanzenernähr., Düng., Bodenk.* **52**:21–41.

Gliemeroth, G. (1955) Möglichkeiten der Beeinflussung von Wurzelmasse und Wurzeldifferenzierung. *Landw. Forsch. Sonderheft* **6**:69–85.

Grandfield, C. O., and W. H. Metzger. (1936) Relation of fallow to restoration of subsoil moisture in an old alfalfa field and subsequent depletion after reseeding. *Jour. Amer. Soc. Agron.* **28**:115–123.

Griffin, D. M. (1963) Soil moisture and the ecology of soil fungi. *Biol. Rev.* 38:141–166.

Halevy, A. H., and B. Kessler. (1963) Increased tolerance of bean plants to soil drought by means of growth-retarding substances. *Nature* **197**:310–311.

Harlan, H. V., and S. Anthony. (1921) Effect of time of irrigation on kernel development of barley. *Jour. Agr. Res.* **21**:29–45.

Harris, F. S. (1914) Irrigation and manuring studies: the effect of varying quantities of irrigation water and manure on the growth and yield of corn. *Utah Agr. Exp. Sta. Bul. 133.*

Harris, F. S. (1914a) Effects of variations in moisture content on certain properties of a soil and on the growth of wheat. *New York (Cornell Univ.) Agr. Exp. Sta. Bul. 352.*

Hedrick, R. M., and D. T. Mowry. (1952) Effect of synthetic polyelectrolytes on aggregation, aeration, and water relationships of soil. *Soil Sci.* **73**:427–441.

Hellriegel, H. (1883) *Beiträge zu den naturwissenschaftlichen Grundlagen des Ackerbaus mit besonderer Berücksichtigung der agrikulturchemischen Methode der Sandkultur.* Friedrich Vieweg und Sohn, Braunschweig.

Henckel, P. A. (1964) Physiology of plants under drought. *Annual Rev. Plant Physiol.* **15**:363–386.

Hendrickson, A. H., and F. J. Veihmeyer. (1931) Influence of dry soil on root extension. *Plant Physiol.* **6**:567–576.

Hobbs, J. A. (1953) Replenishment of soil moisture supply following the growth of alfalfa. *Agron. Jour.* **45**:490–493.

Hopkins, J. W. (1940) Agricultural meteorology: a statistical study of conservation of precipitation by summer fallowed soil tanks at Swift Current, Saskatchewan. *Canadian Jour. Res.* **18C**:388–400.

Hunter, A. S., and O. J. Kelley. (1946) The extension of plant roots into dry soil. *Plant Physiol.* **21**:445–451.

Jackson, R. D. (1964) Water vapor diffusion in relatively dry soil: III. Steady-state experiments. *Soil Sci. Soc. Amer. Proc.* **28**:467–470.

Kausch, W. (1955) Saugkraft und Wassernachleitung im Boden als physiologische Faktoren. Unter besonderer Berücksichtigung des Tensiometers. *Planta* **45**: 217–263.

Kemper, W. D., and M. Amemiya. (1958) Utilization of air permeability of porous ceramics as a measure of hydraulic stress in soils. *Soil Sci.* **85**:117–124.

Kemper, W. D., D. E. L. Maasland, and L. K. Porter. (1964) Mobility of water adjacent to mineral surfaces. *Soil Sci. Soc. Amer. Proc.* **28**:164–167.

Kiesselbach, T. A., A. Anderson, and J. C. Russel. (1934) Subsoil moisture and crop sequence in relation to alfalfa production. *Jour. Amer. Soc. Agron.* **26**:422–442.

Koller, D., and M. Negbi. (1955) Germination regulating mechanism in some desert seeds. V. *Colutea istria* Mill. *Bul. Res. Council Israel, Sec. D. Bot.* **5**:73–84.

Kozlowski, T. T. (1964) *Water Metabolism in Plants.* Harper and Row, New York.

Kramer, P. J. (1959) Transpiration and the water economy of plants. *In* F. C. Steward (Ed.). *Plant Physiology.* Vol. II: *Plants in Relation to Water and Solutes,* pp. 607–726. Academic Press, New York.

Kramer, P. J. (1963) Water stress and plant growth. *Agron. Jour.* **55**:31–35.

Krantz, B. A. (1949) Fertilize corn for higher yields. *North Carolina Agr. Exp. Sta. Bul. 366.*

Laing, D. R. (1966) The water environment of soybeans. Ph.D. Thesis, Iowa State University, Ames, Iowa.

Lambeth, V. N. (1950) Some factors influencing pod set and yield of the lima bean. Ph.D. Thesis, University of Missouri, Columbia, Missouri.

Lange, A. (1938) Untersuchungen über den Wachstumsfaktor Wasser. *Landw. Jahrb.* **85**:465–499.

Lemon, E. R. (1956) The potentialities for decreasing soil moisture evaporation loss. *Soil Sci. Soc. Amer. Proc.* **20**:120–125.

Lewis, D. C., and R. H. Burgy. (1964) The relationship between oak tree roots and groundwater in fractured rocks as determined by tritium tracing. *Jour. Geophys. Res.* **69**:2579–2588.

Linscott, D. L., R. L. Fox, and R. C. Lipps. (1962) Corn root distribution and moisture extraction in relation to nitrogen fertilization and soil properties. *Agron. Jour.* **54**:185–189.

Lipps, R. C., and R. L. Fox. (1956) Subirrigation and plant nutrition: II. Utilization of phosphorus by alfalfa from the soil surface to the water table. *Soil Sci. Soc. Amer. Proc.* **20**:28–32.

Loomis, W. E. (1934) Daily growth of maize. *Amer. Jour. Bot.* **21**:1–6.

Lundkvist, L. O. (1955) Wasserüberschuss und Stickstoffmangel als Ursache gewisser Strukturveränderungen bei Mesophyten. *Svensk Bot. Tidskr.* **49**:387–418.

Marshall, T. J., and C. G. Gurr. (1954) Movement of water and chlorides in relatively dry soil. *Soil Sci.* **77**:147–152.

Martin, E. V., and F. E. Clements. (1935) Studies of the effect of artificial wind on growth and transpiration in *Helianthus annuus. Plant Physiol.* **10**:613–636.

Maximov, N. A. (1929) *The Plant in Relation to Water.* George Allen and Unwin, Ltd., London.

Metzger, W. H., and C. O. Grandfield. (1938) Extension of alfalfa roots into subsoil dried by a previous crop. *Jour. Amer. Soc. Agron.* **30**:80.

Milthorpe, F. L. (1960) The income and loss of water in arid and semi-arid zones. *In* Plant-Water Relationships in Arid and Semi-Arid Conditions. *Arid Zone Res.* **15**:9–36.

Mitscherlich, E. A., and H. Beutelspacher. (1938) Untersuchungen über den Wasserverbrauch verschiedener Kulturpflanzen und den Wasserhaushalt des natürlich gelagerten Bodens. *Bodenk. u. Pflanzenernähr.* **9/10**:337–395.

Myers, H. E. (1936) The differential influence of certain vegetative covers on deep subsoil moisture. *Jour. Amer. Soc. Agron.* **28**:106–114.

Olsen, S. R., F. S. Watanabe, F. E. Clark, and W. D. Kemper. (1964) Effect of hexadecanol on evaporation of water from soil. *Soil Sci.* **97**:13–18.

Oppenheimer, H. R. (1960) Adaptation to drought: xerophytism. *In* Plant-Water Relationships in Arid and Semi-Arid Conditions. *Arid Zone Res.* **15**:105–138.

Paauw, F. van der. (1949) Water relations of oats with special attention to the influence of periods of drought. *Plant and Soil,* **1**:303–341.

Peters, D. B. (1957) Water uptake of corn roots as influenced by soil moisture content and soil moisture tension. *Soil Sci. Soc. Amer. Proc.* **21**:481–484.

Peters, D. B. (1960) Relative magnitude of evaporation and transpiration. *Agron. Jour.* **52**:536–539.

Peters, D. B. (1965) Water availability. *Agronomy* **9**:279–285.

Peters, D. B., and L. C. Johnson. (1960) Soil moisture use by soybeans. *Agron. Jour.* **52**:687–689.

Pierre, W. H., D. Kirkham, J. Pesek, and R. Shaw (Eds.). (1965) *Plant Environment and Efficient Water Use.* American Society of Agronomy, Madison, Wisconsin.

Plaut, Z., and L. Ordin. (1964) The effect of moisture tension and nitrogen supply on cell wall metabolism of sunflower leaves. *Physiol. Plantarum* **17**:279–286.

Ramig, R. E. (1960) Relationships of soil moisture at seeding time and nitrogen fertilization to winter wheat production. Ph.D. Thesis, University of Nebraska, Lincoln.

Ramig, R. E., and H. F. Rhoades. (1963) Interrelationships of soil moisture level at planting time and nitrogen fertilization on winter wheat production. *Agron. Jour.* **55**:123–127.

Read, D. W. L. (1959) Horizontal movement of water in soil. *Canadian Jour. Soil Sci.* **39**:27–31.

Richards, L. A. (1941) Uptake and retention of water by soil as determined by distance to a water table. *Jour. Amer. Soc. Agron.* **33**:778–786.

Richards, L. A. (1959) Availability of water to crops on saline soil. *U.S. Dept. Agr., Agr. Information Bul. 210.*

Richards, L. A. (1961) Advances in soil physics. *Trans. 7th Int. Congr. Soil Sci.* **1**:67–79.

Richards, L. A. (1965) Physical condition of water in soil. *Agronomy* **9**:128–152.

Richards, L. A., and G. Ogata. (1961) Psychrometric measurements of soil samples equilibrated on pressure membranes. *Soil Sci. Soc. Amer. Proc.* **25**:456–459.

Richards, S. J. (1965) Soil suction measurements with tensiometers. *Agronomy* **9**:153–163.

Ruhland, W. (1956) *Handbuch der Pflanzenphysiologie.* III. *Pflanze und Wasser.* Springer-Verlag, Berlin.

Russel, J. C. (1940) The effect of surface cover on soil moisture losses by evaporation. *Soil Sci. Soc. Amer. Proc.* (1939) **4**:65–70.

Russell, M. B. (1959) Water and its relation to soils and crops. *Adv. Agron.* 11:1–131.

Rutter, A. J., and F. H. Whitehead (Eds.). (1963) *The Water Relations of Plants.* (British Ecol. Soc. Symp. No. 3.) Blackwell Scientific Publications, London.

Salter, P. J. (1963) The effect of wet or dry soil conditions at different growth stages on the components of yield of a pea crop. *Jour. Hort. Sci.* **38**:321–334.

Schneider, G. W., and N. F. Childers. (1941) Influence of soil moisture on photosynthesis, respiration, and transpiration of apple leaves. *Plant Physiol.* **16**:565–583.

Schwalen, H. C., and M. F. Wharton. (1940) Lettuce irrigation studies. *Arizona Agr. Exp. Sta. Bul. 133.*

Shutt, F. T., and S. N. Hamilton. (1934) The quality of wheat as influenced by environment. *Empire Jour. Exptl. Agr.* **2**:119–138.

Slatyer, R. O. (1962) Internal water relations of higher plants. *Annual Rev. Plant Physiol.* **13**:351–378.

Sneva, F. A., D. N. Hyder, and C. S. Cooper. (1958) The influence of ammonium nitrate on the growth and yield of crested wheatgrass on the Oregon high desert. *Agron. Jour.* **50**:40–44.

Staple, W. J. (1960) Significance of fallow as a management technique in continental and winter-rainfall climates. *In* Plant-Water Relationships in Arid and Semi-Arid Conditions. *Arid Zone Res.* **15**:205–214.

Stocker, O. (1960) Physiological and morphological changes in plants due to water deficiency. *In* Plant-Water Relationships in Arid and Semi-Arid Conditions. *Arid Zone Res.* **15**:63–104.

Taylor, S. A. (1952) Estimating the integrated soil moisture tension in the root zone of growing crops. *Soil Sci.* **73**:331–339.

Taylor, S. A. (1960) Principles of dry land crop management in arid and semi-arid zones. *In* Plant-Water Relationships in Arid and Semi-Arid Conditions. *Arid Zone Res.* **15**:191–203.

Vaadia, Y., F. C. Raney, and R. M. Hagan. (1961) Plant water deficits and physiological processes. *Annual Rev. Plant Physiol.* **12**:265–292.

Veihmeyer, F. J. (1927) Some factors affecting the irrigation requirements of deciduous orchards. *Hilgardia* **2**:125–291.

Veihmeyer, F. J. (1956) Soil moisture. *In* W. Ruhland (Ed.). *Handbuch der Pflanzenphysiologie.* III. *Pflanze und Wasser,* pp. 64–123. Springer-Verlag, Berlin.

Veihmeyer, F. J., and A. H. Hendrickson. (1950) Soil moisture in relation to plant growth. *Annual Rev. Plant Physiol.* **1**:285–304.

Veihmeyer, F. J., and A. H. Hendrickson. (1961) Responses of a plant to soil-moisture changes as shown by guayule. *Hilgardia* **30**:621–637.

Viets, F. G., Jr. (1962) Fertilizers and the efficient use of water. *Adv. Agron.* **14**:223–264.

Viets, F. G., Jr. (1965) Increasing water use efficiency by soil management. *In* Pierre, W. H., D. Kirkham, J. Pesek, and R. Shaw (Eds.). *Plant Environment and Efficient Water Use,* pp. 259–274. American Society of Agronomy, Madison, Wisconsin.

Wadleigh, C. H. (1946) The integrated soil moisture stress upon a root system in a large container of saline soil. *Soil Sci.* **61**:225–238.

Wadleigh, C. H., and A. D. Ayers. (1945) Growth and biochemical composition of bean plants as conditioned by soil moisture tension and salt concentration. *Plant Physiol.* **20**:106–132.

Wadleigh, C. H., H. G. Gauch, and O. C. Magistad. (1946) Growth and rubber accumulation in guayule as conditioned by soil salinity and irrigation regime. *U.S. Dept. Agr. Tech. Bul. 925.*

Weaver, H. A., and R. W. Pearson. (1956) Influence of nitrogen fertilization and plant population density on evapotranspiration by sudan grass. *Soil Sci.* **81**:443–451.

Went, F. W. (1953) The effects of rain and temperature on plant distribution in the desert. *In Desert Research. Proc. Research Council of Israel, Spec. Publ.* **2**:230–237. Jerusalem.

Whitehead, F. H. (1962) Experimental studies of the effect of wind on plant growth and anatomy. II. *Helianthus annuus. New Phytol.* **61**:59–62.

Whitehead, F. H. (1963) Experimental studies of the effect of wind on plant growth and anatomy. III. Soil moisture relations. *New Phytol.* **62**:80–85.

Whitney, M., and F. K. Cameron. (1904) Investigation in soil fertility. *U.S. Dept. Agr., Bur. Soils Bul. 23.*

Widtsoe, J. A. (1912) The production of dry matter with different quantities of irrigation water. *Utah Agr. Exp. Sta. Bul. 116.*

Wind, G. P. (1954) The influence of nitrogen fertilising on water consumption of grassland. *In European Grassland Conference,* Project 224, pp. 195–198. European Productivity Agency of the Organization for European Economic Co-operation. Paris.

Wind, G. P. (1955) A field experiment concerning capillary rise of moisture in a heavy clay soil. *Netherlands Jour. Agr. Sci.* **3**:60–69.

Wind, G. P. (1955a) Flow of water through plant roots. *Netherlands Jour. Agr. Sci.* **32**:259–264.

Zingg, A. W., and C. J. Whitfield. (1957) A summary of research experience with stubble-mulch farming in the western states. *U.S. Dept. Agr. Tech. Bul. 1166.*

3 *Aeration*

Aerobic respiration in roots of plants involves continuous absorption of oxygen and evolution of carbon dioxide. Metabolic processes of roots of plants that grow normally on well-drained soils are impaired almost immediately if this exchange of oxygen and carbon dioxide is interrupted. Inadequate gas exchange may decrease the yield of plants if continued only a day and may lead to death of roots if continued only a few days.

Aeration is the exchange of oxygen and carbon dioxide between the atmosphere and the soil and plant roots. Most of the gas exchange that is effective in aerating plant roots in well-drained soil seems to occur through the soil; in soils saturated with water, however, exchange through the plant itself may be of principal importance.

If a soil is saturated with water, the gases must move in dissolved form through the water. Under natural conditions, such movement is too slow to be effective. If the soil contains gas-filled pores, the gases dissolved in the water tend toward an equilibrium with those in the gaseous phase. If the gas-filled pores interconnect to the surface of the soil, gas exchange with the atmosphere can occur partly through the soil water and partly through the gaseous phase; but it takes place more rapidly through the gaseous phase because the rate of diffusion is much greater there than in the soil water. Although gas exchange may occur by movement of air in and out of soils with changes in wind velocity, temperature, barometric pressure, and water content, such exchange is of minor importance compared with exchange by gaseous diffusion.

Reviews of literature on aeration and allied subjects have been published by Bergman (1959), Greene (1960), Aomine (1962), Vilain (1963), Stolzy and Letey (1964), Domsch (1962), and Stotzky (1965). The last two reviews deal principally with methods of analysis.

Soil Respiration

Respiration by plant roots and microorganisms is the principal cause of oxygen absorption and carbon dioxide production by soils. The rate of respiration is controlled by such conditions as temperature, water supply, and the type and amount of respiring tissue.

Dehérain and Demoussy (1896) found that production of carbon dioxide during incubation of moist soil increased with temperature from 22°C to a maximum at 65°C and then decreased with further increase in temperature to 90°C, above which it increased markedly to values at 110°C that were three to six times greater than those at 65°C. The authors attributed the maximum at 65°C to microbial activity and the rapid increase above 90°C to chemical oxidation. Respiration of roots increases rapidly with temperature. Berry and Norris (1949) found that the rate of respiration of onion roots was more than three times as great at 30° as at 15°C.

The microbial population of soil has a wider range of adaptation than do higher plants and can carry on respiration in soils too dry to support plant growth as well as in soils saturated with water. In relatively dry soils, there is ample air space for rapid gas exchange; thus the respiration of microorganisms does not cause important competition with plants. In wet soils, gas transfer is slow, and respiration by microorganisms alone may be sufficient to exhaust the oxygen supply in the soil within a period as short as a day. Some plants and some microorganisms are adapted to the resulting anaerobic conditions, but others are not.

The relative importance of plants and soil organisms in consuming oxygen in soil under conditions favorable to plant growth is suggested by the data in Table 3.1. These measurements were made in the field with the aid of special apparatus to absorb the carbon dioxide evolved, to supply oxygen, and to measure the oxygen consumed. To judge from the results, most of the oxygen was consumed in processes going on in the soil in the absence of higher plants. From the fact that the population of microorganisms in soil is greater in the vicinity of plant roots than at a distance, one may suppose that the presence of a crop increases the consumption of oxygen in the soil and that part of the apparent effect of the crop in Table 3.1 was actually an effect of soil microorganisms associated with root surfaces. Other data obtained by use of the same apparatus were published by Brown et al. (1965).

The figures for oxygen consumption in Table 3.1 provide information on the order of magnitude of the volumes of oxygen and carbon dioxide that must be exchanged per day between soil and atmosphere. A volume of, say, 10 liters of oxygen per square meter per day corresponds to a volume of about 50 liters of air. If respiration occurs uniformly throughout the surface meter and if air-filled pores comprise 20% of the soil volume or 200 liters, the consumption of oxygen by respiration

Table 3.1. Consumption of Oxygen in Soils in the Presence and Absence of a Crop (Hawkins, 1962)

	Oxygen Consumption per Square Meter per Day, Liters		
Soil	Total in Presence of Crop	In Absence of Crop	Difference due to Crop[1]
Sandy clay loam	7.6	4.8	2.8
Peat	13.0	9.4	3.6

[1] Potatoes in the sandy clay loam and tobacco in the peat soil.

is at the rate of 25% of the amount present per day. The figures for oxygen consumption provide also a basis for some calculations of energy involved in soil respiration. Use of 10 liters of oxygen to produce 10 liters of carbon dioxide per square meter per day would result in production of 4.75×10^5 kg-calories of heat per hectare per day or power of 23 kw/hectare if the substrate being oxidized were sugar.

Characterization of Soil Aeration

Definition of soil aeration as the exchange of oxygen and carbon dioxide between soil and atmosphere is not particularly helpful in practice. One may determine the rates of oxygen uptake and carbon dioxide evolution and still not have an evaluation of soil aeration for plants. In the study of soil aeration, one of the problems is understanding the situation well enough to decide what should be measured. A second problem is devising ways to obtain the desired information. In the following subsections types of measurements that have been used to provide information on soil aeration will be discussed. None of these methods measures aeration in the sense of the definition, but most could be said to provide an index of aeration. Each method of investigation may be useful under certain conditions; but each method also has limitations, as will become evident from the text.

A third problem in characterizing soil aeration is integration. As will become clear from the following discussion, soil aeration varies with location in the soil and with time; and it involves both direct effects and indirect effects, which differ with the circumstances. The complexity of the subject and the state of knowledge about it are such that little effort has been devoted to attempts to develop a characterization of soil aeration that will integrate the aeration status of natural soils over a period of time. Erickson and Van Doren (1961) suggested that as a first approximation the length of time a soil is deficient in oxygen should provide a better measure of soil aeration for plants than the simple average over the time the plants are growing.

Verification of Aeration Status

This section has to do with determining the adequacy of soil aeration. The most direct way to investigate the adequacy of soil aeration under specific conditions is to determine the plant response to improved aeration. Aeration may be improved by increasing the volume of air-filled pores or by increasing the effectiveness of existing pores. The volume of air-filled pores may be increased in a practical way by improving soil structure or drainage. These techniques have the disadvantage of altering other conditions simultaneously, however, and so the extent to which the results may be interpreted in terms of soil aeration is uncertain.

For verification of the aeration status, the effectiveness of existing pores has been increased in two ways. One is to determine how plant growth is affected if aeration is improved by forcing air through soil by mass flow. If forced aeration improves growth, aeration is presumed to be inadequate, provided due precaution has been observed to ensure that the treatment has not changed conditions in some other way (such as evaporative drying) that is not a direct consequence of improved aeration. If no improvement in plant growth occurs, aeration may be presumed to be adequate. In this event, however, the possibility must be considered that soil aeration is inadequate but that the technique employed did not produce enough of a change to be reflected in improved plant growth. An instance in which this approach was tried in the field was reported by Melsted et al. (1949). The experiment was conducted using large tanks filled with soil having good structure. The results, given in Table 3.2, indicate that the importance of aeration as a limiting factor under the control conditions was great with corn in 1947 and small with soybeans in 1948.

A second technique to test the adequacy of aeration is to supply additional oxygen in the form of a peroxide. Melsted et al. used this technique in the same experiment with forced aeration (see Table 3.2), supplying hydrogen peroxide in the irrigation water, with results similar to those obtained with forced aeration.

If no improvement in growth results on using the method of forced aeration, oxygen may be presumed to be adequate and carbon dioxide not in excess. If growth is improved, further work is needed to find

Table 3.2. Yield of Corn and Soybeans with and without Supplemental Aeration in Illinois (Melsted et al., 1949)

	Yield per Hectare, metric tons	
Treatment	Corn (1947)	Soybeans (1948)
Control	5.9	2.8
Forced aeration	9.0	3.0
Hydrogen peroxide	8.9	3.3

the significance of a deficiency of oxygen and an excess of carbon dioxide in determining the results. The method of peroxide addition provides information on oxygen deficiency but not on carbon dioxide excess. Compared with forced aeration, use of a peroxide has the advantage that the oxygen is liberated in the soil water and the disadvantage of greater likelihood of secondary effects.

Volume Percentage of Soil Air

To obtain information on soil aeration, two kinds of measurements are made on soil air, measurements of quantity and of quality. The simpler of the two, from the standpoint of discussion, is measurement of the quantity or volume fraction of the soil occupied by air under specified conditions such as a matric suction of 0.06 or 0.1 bar in the soil water or under the conditions that may exist in the field from time to time. Methods for making such measurements were described in detail by Vomocil (1965).

The theoretical basis for the use of the volume percentage of air as an index of soil aeration is twofold. First, the rate of diffusion of gases through soil, independently of gas exchange between the liquid and gaseous phases, increases with the proportion of the soil volume occupied by gas-filled pores. Second, the rate of exchange of gases between the soil air and the respiring plant roots and soil organisms increases with the volume percentage of air because of the decreasing thickness of water films through which diffusion must occur. The significance of diffusion in the liquid and gaseous phases will be discussed in the next section.

Figure 3.1 gives an example of use of the volume percentage of soil air as an index of aeration. The figure contains results of experimental work with sugar beets on a poorly drained clay soil in northwestern

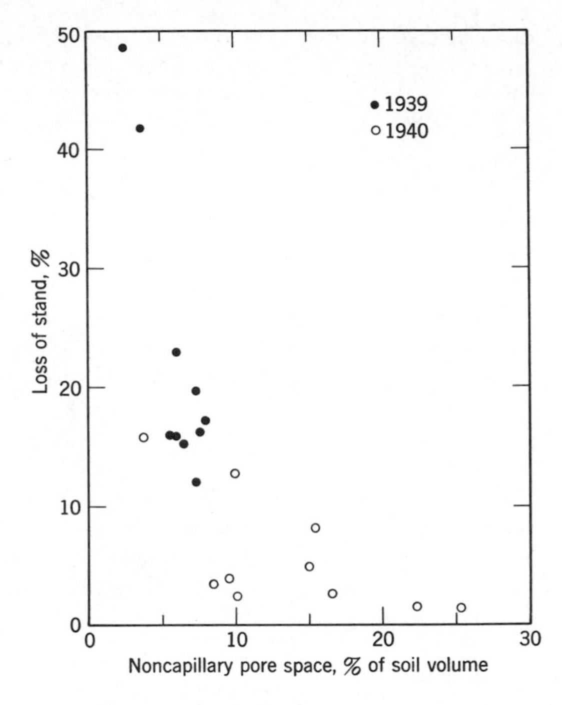

Figure 3.1. Percentage loss of stand of sugar beets between blocking and harvest versus noncapillary pore space in a clay soil in Ohio in two consecutive years. (Baver and Farnsworth, 1940)

Ohio. Chemical fertilizers were relatively ineffective compared with applications of green manure or animal manure, thus suggesting the importance of aeration as a limiting factor. As shown in the figure, differences in loss of stand of beets associated with the latter treatments were related to the volume percentage of noncapillary (large) pores in the soil. The critical value was apparently in the vicinity of 15%. Measurements of yield of beets in the same experiments suggested a critical value of about 10%. Vomocil and Flocker (1961) summarized results of work by a number of investigators showing that an appreciable reduction in growth or yield occurred if the volume of air-filled pores was as low as 5 to 15% of the soil volume, depending on the crop and other conditions.

Composition of Soil Air

The second type of measurement made on soil air is determination of the composition. Usually, the only analyses made to investigate the aeration are those for percentage content of oxygen, carbon dioxide, and nitrogen plus inert gases on a volume basis.

Measurements of composition of soil air may be made in several ways. Originally, they were made entirely by chemical analysis. Carbon dioxide was absorbed by one reagent, oxygen was absorbed by another, and the reduction in total volume of the gas sample after each treatment was measured. The unabsorbed gas was represented as nitrogen plus inert gases. In addition to chemical methods, physical methods are now being used. So-called "gas analyzers" now available may be used to determine oxygen or carbon dioxide. These analyzers, like the chemical methods, may be used in the field if desired. Another technique now being used is gas chromatography. This technique is not as simple or as adaptable to field use as the use of gas analyzers; however, it has the advantage of requiring only very small samples of gas and the capability of separating and providing a measurement of a variety of gas components. Analysis of soil air was described by van Bavel (1965).

The principal components of soil air are the same as those of the atmosphere, namely, nitrogen, oxygen, the inert gases, carbon dioxide, and water vapor. Such substances as methane and hydrogen, if present, occur in negligible quantities. From the quantitative standpoint, the main difference in composition between soil air and atmospheric air lies in the content of carbon dioxide. Carbon dioxide occurs in the atmosphere to the extent of about 0.03%. The content commonly is about 0.2 to 1% in air extracted from surface layers of soil in which aeration is adequate. Much more carbon dioxide is found under some circumstances. The atmosphere contains about 21% oxygen. Soil air contains less oxygen than atmospheric air, but the difference between the two is relatively small unless the soil air has been enriched with carbon dioxide to an extent much greater than the usual 0.2 to 1%. Under aerobic conditions the volume of carbon dioxide produced in soil is approximately equal to the volume of oxygen consumed; hence the sum of the carbon dioxide and oxygen percentages is approximately the same in soil air as in the atmosphere.

If soil is submerged, as in rice culture, circumstances are much different. Submerged soils contain no air in the usual sense of the word. Instead, gases from the atmosphere dissolve in the overlying water and are conducted through it to the soil. The dissolved oxygen is consumed in a thin surface layer of soil; and methane, carbon dioxide, and small amounts of other gases, such as hydrogen, carbon monoxide, and nitrous oxide, are produced in the soil. The amount of these gases produced exceeds that of oxygen entering the soil. Bubbles of gas thus form and rise from the surface of the soil through the overlying water. Although these bubbles may contain all the gases mentioned, they often are composed mainly of nitrogen, which is probably derived principally from the atmospheric nitrogen dissolved in the water.

The extent to which soil air differs in composition from atmospheric air is determined by the rate at which oxygen is consumed and other gases are produced and by the rate of gaseous exchange between the soil and the atmosphere. Consumption of oxygen and production of carbon dioxide are ordinarily greatest in the surface portion of the soil (Newman and Norman, 1941), and the rate of gas exchange is greatest at the surface. The net result is that the similarity of composition of soil air and atmospheric air is greatest with samples from the surface layers of soil and decreases with increasing depth of sampling. Figure 3.2 shows the departure of composition of soil air from that of the atmosphere in proceeding from the surface downward toward free water in two soils. The figure illustrates also the complementary relationship of oxygen and carbon dioxide as well as the association of air composition with the volume percentage of air in the soils at the different depths. One may note that soil *A* had both a greater volume of air and a higher percentage of oxygen in the air than did soil *B*. An extensive series of measurements of composition of air in various soil profiles under field conditions was reported by Boynton (1941). Occasionally, analyses show a maximum in carbon dioxide content and a minimum of oxygen in the soil air at some depth due to slow interchange under certain circumstances [see, for example, the paper by Linscott, Fox, and Lipps (1962)].

With a given rate of oxygen consumption and carbon dioxide production in soil, the difference in composition between soil and atmospheric air decreases with increasing rate of diffusion. The rate of diffusion increases somewhat with the temperature and is approximately proportional to the fraction of the total soil volume occupied by gas-filled pores, as noted by Penman (1940) and others. Penman found that the apparent diffusion coefficient for movement of gas through columns of soil is about 0.66 times that for gas in bulk for porosities up to about 0.6. Marshall (1959) reexamined this phenomenon and concluded that the factor 0.66 should be replaced by the square root of the porosity. The lower rate of diffusion of gas through soil pores than through the same volume in bulk is attributed primarily to the fact that in soil the gas molecules must follow the tortuous pathway provided by the soil pore spaces, so that the actual diffusion distance is greater than the measured distance. The size of gas-filled pores has little effect because it is large relative to the distance individual gas molecules move between collisions. Thus the main impediment to diffusion is collisions with other

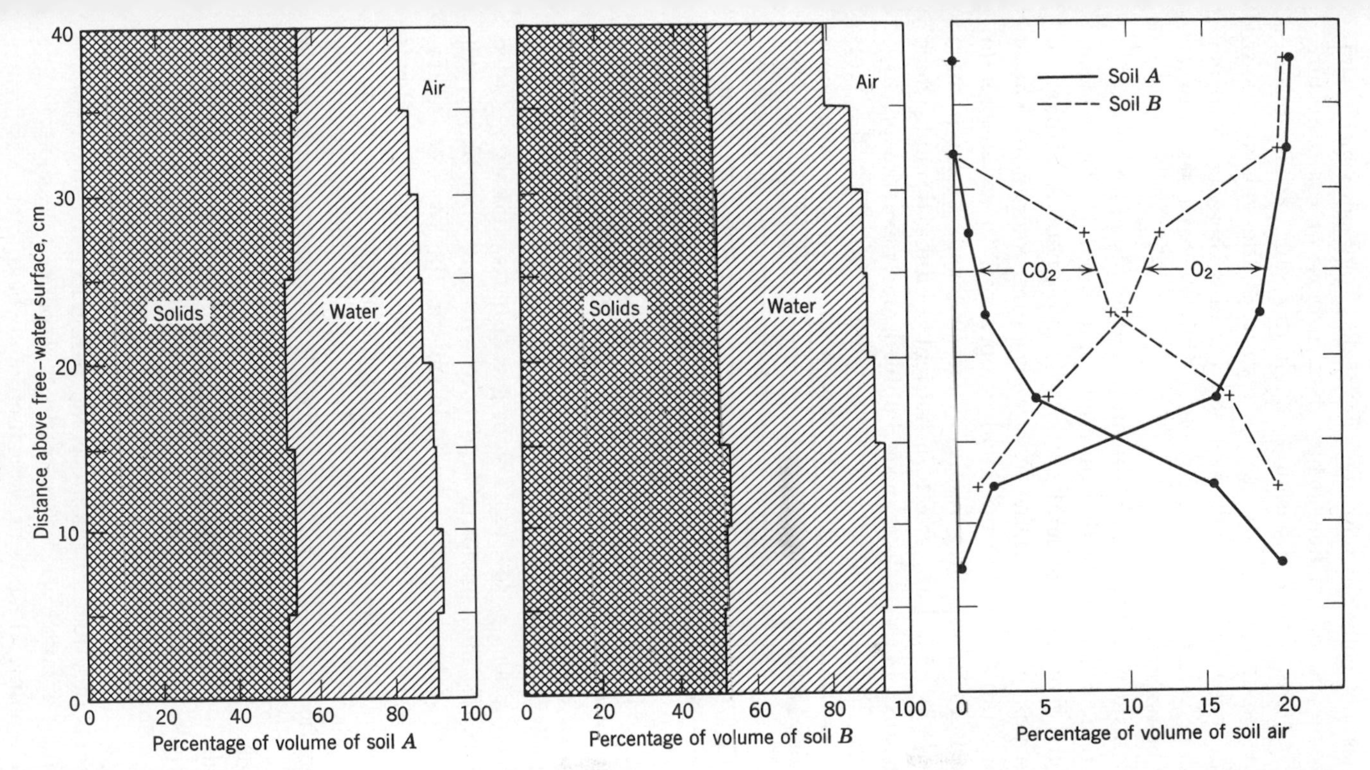

Figure 3.2. Percentage of volume of two soils occupied by solids, water, and air; and percentage of oxygen and carbon dioxide in the air at different distances above a free-water surface. Soil *A* contained 85% sand, 7% silt, 6% clay, and 2.3% organic matter; soil *B* contained 67% sand, 15% silt, 16% clay, and 2.5% organic matter. The soils were mixed individually before they were packed into the columns, and the measurements were made after a period of 1 week at 23°C. (Kristensen and Enoch, 1964)

gas molecules, not collisions with the pore walls. Total pore space commonly is greater in soils of fine texture than in those of coarse texture. Consequently, the rate of diffusion through dry soils of fine texture exceeds that through dry soils of coarse texture. If the soils are moist, however, the behavior is usually reversed because then the tendency is for a greater volume of air-filled pores in soils of coarse texture than in those of fine texture.

Where plants are grown in soil without forced aeration, use of oxygen and production of carbon dioxide proceed simultaneously. The concentrations of the two gases generally vary in complementary fashion, so that it is not possible to determine from analyses the comparative effects of the concentrations of the individual components. In an attempt to obtain a basis for interpreting results of measurements of composition of soil air, plants have been grown in nutrient solutions through which different mixtures of gases have been bubbled. Such experiments have the advantage that the condition of the solution in which the roots are bathed can be characterized and kept uniform by the continuous movement of the solution in response to the excess of gas bubbled through it. They have the disadvantage that the circumstances are not like those in soil, and the results cannot be interpreted in terms of soil aeration.

Conditions in soils have been approached more closely by passage of known gas mixtures through sand or gravel cultures in which plants were grown. In experimental work of this type, Stolwijk and Thimann (1957) used air that had been enriched to different degrees with carbon dioxide and observed some inhibition of elongation of roots from carbon dioxide concentrations as low as 1% with garden peas, 1.5% with garden beans, field beans, and sunflower, and 6.5% with oats and barley. All these values are within the range found in soil air; the lower concentrations are not uncommon. In a similar experiment with tobacco, Harris and van Bavel (1957) found that reducing the oxygen in the aerating gas mixture from 20 to 15 or 10% had no appreciable effect on the rate of respiration of the roots if carbon dioxide was absent, but inclusion of 5% carbon dioxide with the oxygen at 15% caused a definite reduction in respiration. These results indicate that the response of plant roots to poor aeration was primarily a response to toxicity of carbon dioxide.

To investigate effects of oxygen and carbon dioxide on roots in soil, Grable and Danielson (1965) determined the rate of root elongation. Results of their work on corn and soybeans indicated, in contrast to the experiments described in the preceding paragraph, that the response of roots to poor aeration was not primarily a response to carbon dioxide toxicity. For example, the rate of elongation of soybean roots was greater if the soil was aerated with a gas mixture containing 5% carbon dioxide than with a gas mixture (air) containing 0.03% carbon dioxide. The rate of elongation of corn roots was decreased less than 50% by aeration with a gas mixture containing 20% carbon dioxide. In addition, Grable and Danielson cited other work showing beneficial effects of low concentrations of carbon dioxide. The relative importance of oxygen deficiency and carbon dioxide toxicity in determining the response of plants to poor aeration in soils is thus a question to be resolved by further research.

In investigations of plant growth in relation to the composition of soil air, oxygen and carbon dioxide are usually determined; the results usually indicate that inhibition of growth or other physiological function is associated with an increase in concentration of carbon dioxide and a decrease in concentration of oxygen. Nevertheless, experimental results of the type shown in Table 3.3 were cited by Russell (1952) in support of the view that the composition of soil air does not necessarily provide a good index of soil aeration for plant growth. Both the experiments referred to in the table dealt with the effect of forced aeration on composition of soil air and growth of roses. In the first experiment, forced aeration had relatively little effect on the air composition, and analyses indicated no important deficiency of oxygen or excess of carbon dioxide; but growth was much improved by aeration. In the second experiment, forced aeration had considerable effect on the air composition but little effect on growth of roses.

Recently, the trend has been for less use of soil air composition as

Table 3.3. Growth of Roses and Composition of Soil Air in Two Investigations

Treatment	Composition of Soil Air, %		Linear Growth of Tops per Plant, cm
	Oxygen	Carbon Dioxide	
Experiment by Boicourt and Allen (1941)			
Control	18.8	1.5	95
Forced aeration	20.3	0.3	174
Experiment by Seeley (1949)			
Forced aeration			
1 liter/day	11.8	8.4	38
12 liters/day	19.7	0.8	42

an index of aeration and for greater use of measurements of the rate of diffusion of oxygen in the soil water (to be described in the next section). Aside from the convenience of the latter method and the success with which it has been applied, there may be two other important reasons for the gradual change in emphasis.

First, research by Taylor and Abrahams (1953) brought into question the validity of samples of soil air obtained by use of suction (the usual technique). These investigators found a higher percentage of oxygen in air samples obtained by suction than in samples obtained by diffusion into a sampling chamber. In an extreme case, they obtained oxygen contents of 18% by the reduced-pressure technique and 8% by the diffusion technique at a sampling depth 20 cm below the surface of the soil. Hack (1956) found that samples of soil air with a volume of about 0.01 cm^3 sometimes contained as much as 3.5% less oxygen and 2% more carbon dioxide than did samples with a volume of 10 cm^3. The obvious implication of these findings is that samples of air obtained by suction may be contaminated with air drawn down from above the orifice in the sampling device; thus the greater the volume of the sample, the greater will be the contamination. As a result of these findings, the technique of withdrawing large volumes of gas from soil under reduced pressure has been largely abandoned. Samples obtained directly by reduced pressure are relatively small; or, if larger volumes of air are used, samples are obtained by allowing soil air to diffuse into a chamber of nitrogen buried in the soil at the desired depth.

Second, more detailed analysis of the circumstances existing in soils has led to development of a concept in which soil air is not accorded the all-important position it was once thought to have in soil aeration. Considerable importance is now attached to the fact that the protoplasm of roots, where oxygen is consumed and carbon dioxide is produced, is separated from soil air by a film of water. The thickness of the film varies with the matric suction in the soil water. Also, it varies with location, being thickest where it adjoins water films around soil particles and thinnest where no soil particles are in contact (see Fig. 2.1). Aeration of soil for growth of plant roots therefore involves transfer of oxygen and carbon dioxide through the water as well as the air.

The rates of diffusion of oxygen and carbon dioxide in air are relatively high; hence pronounced gradients in concentration of these gases among contiguous soil pores at different distances from root surfaces are not to be expected. Because of solubility effects, however, the concentrations of oxygen and carbon dioxide in soil water are rather different from those in air, particularly where oxygen is concerned. At room temperature and sea-level pressure, a volume of 100 cm^3 of water will dissolve

75 cm^3 of carbon dioxide or 3 cm^3 of oxygen from atmospheres of the respective gases; or it will dissolve 1.7 cm^3 of air, including 0.6 cm^3 of oxygen. The volume of oxygen in 100 cm^3 of dry, atmospheric air is 21 cm^3. Water containing dissolved oxygen in equilibrium with air thus holds only 3% as much oxygen per unit volume as does air. The liquid phase, moreover, impedes the movement of gases to a much greater extent than does an equal thickness of gas. The diffusion coefficient of oxygen is of the order of one ten-thousandth as great in water as it is in air.

Because of these solubility and diffusion effects, the concentrations of oxygen and carbon dioxide in solution in contact with root protoplasm may be rather different from the concentrations of these gases in solution at the air-water interface. Furthermore, the proportions of oxygen and carbon dioxide in the soil air bear no definite relationship to the proportions in contact with root protoplasm. With constant concentration of oxygen in the soil air, the supply of oxygen to the root protoplasm should decrease with increasing content of water in the soil because of the increase in distance of diffusion through water. Conversely, with constant concentration of carbon dioxide in the soil air, the concentration of carbon dioxide in contact with the root protoplasm should increase with increasing content of water in the soil.

Rate of Oxygen Diffusion

The movement of gases in soil by mass transfer and diffusion was discussed by Evans (1965). The methods he described involve mostly the movement that occurs through gas-filled pores. In studies of soils in relation to plant growth, the technique most commonly used is the platinum-electrode or polarographic method introduced by Lemon and Erickson (1952) for measuring the rate of diffusion of oxygen through soil water. Research on this method was reviewed by Stolzy and Letey (1964).

Apparatus. The apparatus used to determine the rate of diffusion of oxygen through soil water is shown in schematic form in Fig. 3.3. The platinum-electrode method depends on the reduction of oxygen at the surface of the electrode under the influence of an electrical potential. The electrode used most commonly is a platinum wire that protrudes about 4 mm from the end of a piece of glass tubing into which it is sealed. The electrode is pushed into the soil to the desired location, and an electrical cell is formed by inserting into the soil nearby a so-called saturated calomel electrode. The mercury terminal in the calomel electrode is connected to the positive terminal of an external source of direct current with voltage that can be varied within the range from

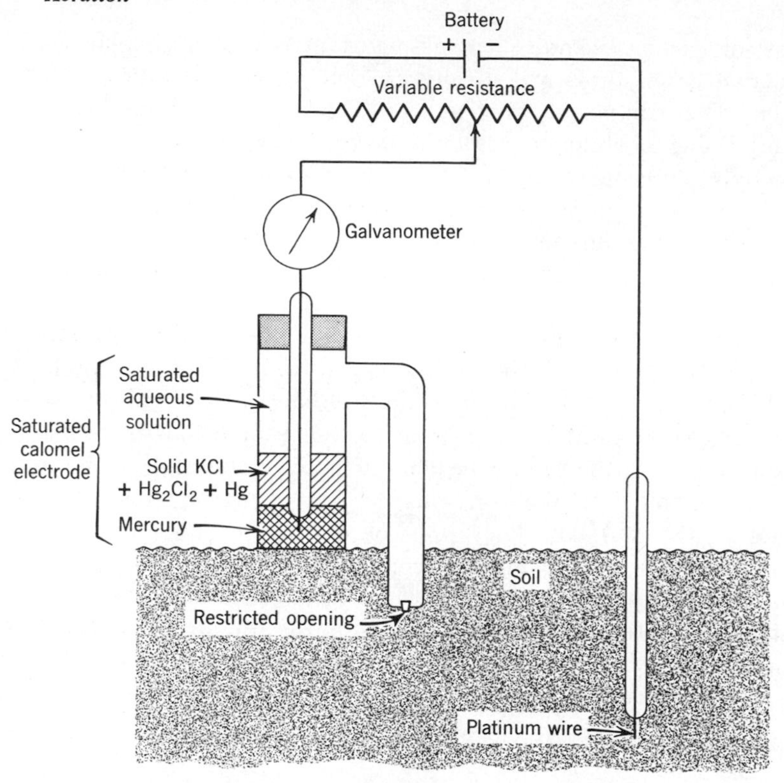

Figure 3.3. Schematic diagram of apparatus for measuring the rate of diffusion of oxygen in soil by the platinum-electrode method.

0.1 to 0.8 volt. The negative terminal of the external source is connected to the platinum wire inside the piece of glass tubing. In series with the calomel electrode is a galvanometer to register current flow of 0 to 50 μamp. Directions for preparing working equipment were published by Letey and Stolzy (1964).

Theory. If a potential of the order of 0.1 volt is applied across the cell from the external source, the current flow is negligible. As the potential is increased above about 0.2 volt, the current increases rapidly with an increase in voltage because electrons on the platinum electrode have acquired a potential sufficient to cause them to react with some of the dissolved oxygen at the surface of the electrode. Each molecule of oxygen (O_2) takes up four electrons and reacts with hydrogen ions to form water in an acid solution ($O_2 + 4H^+ + 4e^- = 2H_2O$), or it reacts with water to form hydroxyl in an alkaline solution ($O_2 + 2H_2O + 4e^-$

$= 4OH^-$). The electron transfer leads to loss of four hydrogen ions or production of four hydroxyl ions, which tends to unbalance the equality of positive and negative charges on ions in the solution at the electrode surface. This tendency is counteracted and electrical neutrality is maintained by movement of ions through the solution between the two electrodes and by an oxidation reaction at the surface of the mercury in the calomel electrode in which each of four mercury atoms loses an electron and becomes a positively charged mercury ion (Hg^+). The electrical current registered by the galvanometer is thus a measure of the rate at which oxygen is reduced at the platinum electrode.

If the voltage is maintained at a constant value in the range in which oxygen is reduced, the current falls rapidly with time because of depletion of oxygen from the solution around the electrode. In 3 to 5 minutes, however, the current becomes nearly constant, which signifies a constant rate of reduction of oxygen at the electrode. If the current then remains constant over a range of voltage, the rate of reduction of oxygen at the electrode is inferred to be controlled by the rate of diffusion of oxygen to the electrode through the surrounding solution and not by the potential at the electrode. A constant current over a range of voltage is found if measurements are made on soils saturated with water. If measurements are made on unsaturated soils, however, no range of voltage is found in which the current remains constant; instead, the current increases linearly with the voltage over a range up to about 0.75 volt and then increases more rapidly as a result of another electrode reaction. Failure to observe a range of voltage in which a constant current is obtained raises a question as to whether the current is entirely diffusion-controlled and hence whether the phrase "rate of diffusion of oxygen" should be used to describe the results of the measurements (Birkle et al., 1964). Nevertheless, by convention, the results are reported as rates of diffusion of oxygen.

The number of moles or grams of oxygen reduced at the platinum electrode per minute can be calculated readily from the measured current, the Faraday constant, and the number of electrons transferred per molecule of oxygen. The value thus obtained depends on the size and shape of the platinum electrode and on the voltage. By convention, the results are expressed as quantity of oxygen per unit area of electrode surface area per unit time. This device accounts approximately for differences in length of the electrode but not for differences in diameter or in voltage. Stolzy and Letey (1964) recommended that 22-gauge platinum wire (0.064 cm diameter) be used for electrodes to be employed in the field with a variety of soils because it is more rigid than the smaller 25-gauge wire that has been employed in some investigations.

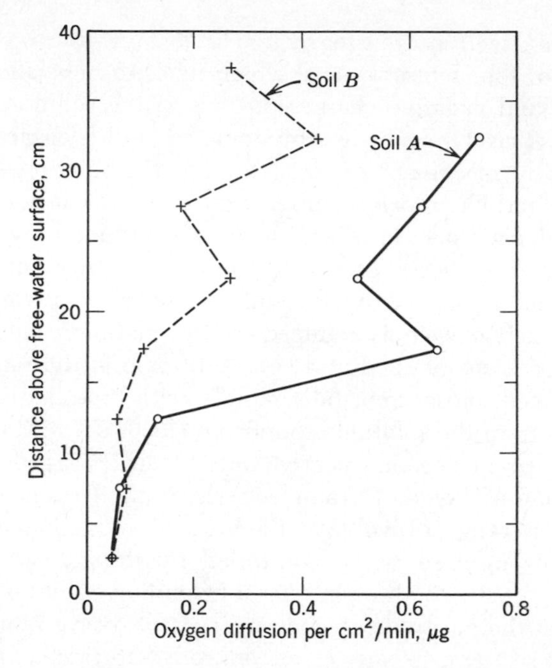

Figure 3.4. Rate of diffusion of oxygen as found by the platinum-electrode method in columns of two soils extending 40 cm above a free-water surface and in contact with air at the top. For description of the soils and other experimental measurements, see Fig. 3.2. (Kristensen and Enoch, 1964)

Birkle et al. (1964) recommended use of a potential of 0.65 volt because this potential is high enough to give relatively high readings that will reduce the relative experimental error and low enough to avoid error from other electrode reactions. Moreover, use of a given potential will make results of different investigations more comparable. Many published data have been obtained with this potential.

Figure 3.4 shows measurements of rate of oxygen diffusion by the platinum-electrode method in two soils at different distances above a free-water surface. One may note that the values decrease as the free-water surface is approached from above. This behavior is in accordance with expectations based on the greater thickness of water films near the free-water surface than at the surface of the columns of soil. The general trend of measurements of rate of diffusion of oxygen shown in Fig. 3.4 is thus in agreement with the trend of measurements of oxygen concentration in the soil in Fig. 3.2. In fact, the two sets of measurements may be compared directly because they were obtained in the same experiment.

Direct comparison of measurements of oxygen concentration in the air and rate of diffusion of oxygen in the water in Figs. 3.2 and 3.4 shows that the correlation is by no means as close as might be supposed from the similarity of the trends with depth first mentioned. Kristensen and Enoch (1964) plotted the two sets of values against each other and noted that rates of diffusion of oxygen were low where the oxygen percentage in the soil air was low; but, where the oxygen percentage in the soil air was high, the rates of diffusion of oxygen covered the full range from low to high. Further evidence of the same behavior was obtained by Lemon and Kristensen (1961). Figure 3.5, taken from their work, shows measurements of the rate of diffusion of oxygen in

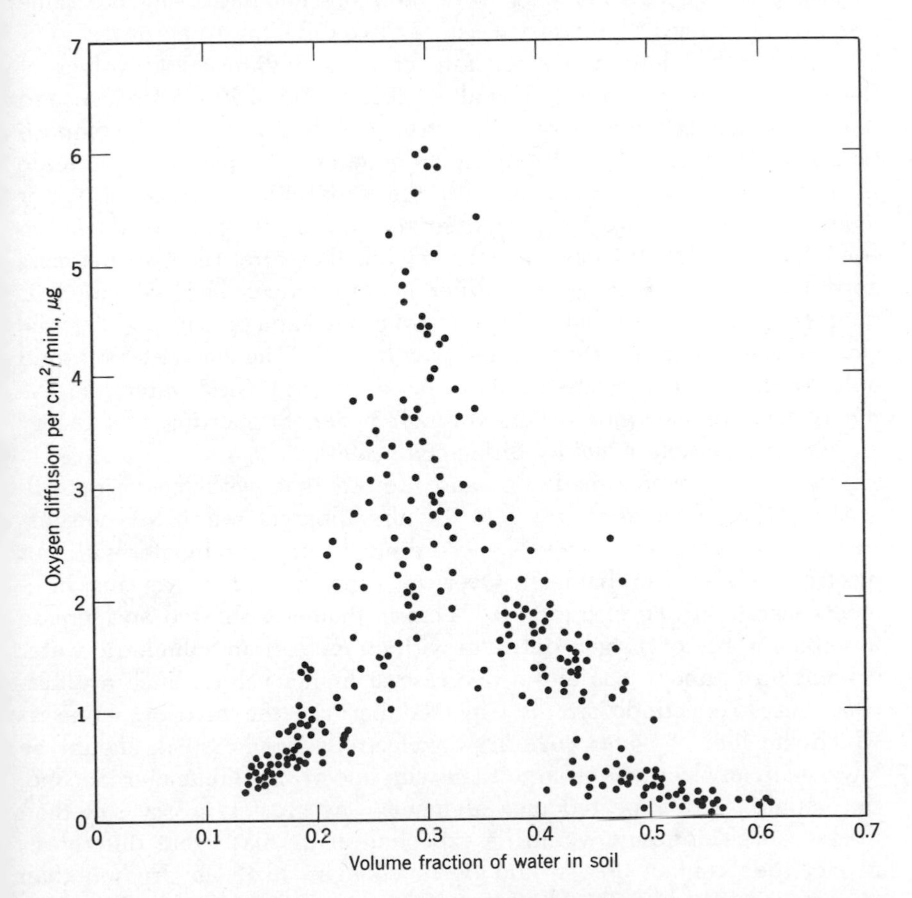

Figure 3.5. Rate of diffusion of oxygen to a platinum electrode in three soils versus volume fraction of water in the soils. (Lemon and Kristensen, 1961)

three soils that had been sieved into three particle-size groups, packed into small cylinders, and brought to different values of matric suction in the range from 0.024 to 0.39 bar. The rate of diffusion of oxygen to the platinum electrode decreased with increasing volume fraction of water above about 0.3, in accordance with the expectation that the resistance of the water films to diffusion of oxygen from the atmosphere to the electrode surface should increase with an increase in thickness of the films. Nevertheless, supplementary measurements of composition of the air in these samples of soil showed that they were at equilibrium with the oxygen content of the external atmosphere (21%). Evidently the rate of diffusion of oxygen through the soil water was sometimes low even if the oxygen content of the soil air was the same as that of the atmosphere. The two methods were therefore not measuring the same property in two ways but were measuring two different properties.

The relatively low values for rate of oxygen diffusion at values of fractional content of water below about 0.25 in Fig. 3.5 indicate a second difference between the platinum-electrode method and the air-composition method. In this case, the results obtained by the platinum-electrode method are not in agreement with the expected resistance of water films to diffusion of oxygen. Analyses of the air showed no deviation from a content of 21% oxygen. Here again, therefore, the two methods appear to be measuring two different properties. This deviation is thought to be an artifact resulting from incomplete coverage of the electrode by a film of water in relatively dry soils. The electrodes respond only to the portion of their surface area covered with water because the transfer of electrons occurs through water. Supporting this theory are two observations made by Birkle et al. (1964).

First, measurements made on sand showed that, with increasing volume fraction of water from 0.1 to 0.4, the apparent rate of oxygen diffusion decreased continuously according to measurements with an electrode 0.046 cm in diameter. Over the same range, however, measurements made with an electrode 0.064 cm in diameter showed an increase in apparent rate of oxygen diffusion with an increase in volumetric water content up to about 0.35 and a decrease at higher values. Such a difference due to electrode size is expected because the matric suction at which the film of water covering an electrode would break should be lower with an electrode of large than with one of small diameter. Second, the variability of repeated measurements was greater in dry soil than in wet soil. This result would be expected on the basis that differences in electrode contact area would be much more likely in dry soil than in wet soil. This limitation of the platinum-electrode method for use in dry soil is unfortunate but not of great practical importance because aeration is seldom if ever a problem in relatively dry soils in the field. Willey and Tanner (1963) described a flat, membrane-covered electrode which, by allowing measurement of oxygen diffusion over its entire surface even if the electrode is in air, alleviates some of the difficulty just described for the wire-type electrodes.

Relation to Oxygen Supply for Plants. If oxygen diffusion to a wire-type platinum electrode controls the reduction at the electrode surface and if the oxygen requirement of fine roots similar in shape and size to the electrode equals or exceeds the reduction at the electrode, oxygen diffusion will limit the delivery of oxygen to the surface of the roots. The rate of diffusion of oxygen to the platinum electrode will then serve as an index of the rate of diffusion of oxygen through the water films to the roots.

According to Machlis' (1944) data on the rate of respiration of barley roots at 24°C, oxygen was consumed at the rate of 0.24 $\mu g/cm^2/minute$ in the terminal 1-cm segment (if the radius of the root is assumed to be 0.05 cm). Comparison of this value with data in Fig. 3.5, for example, shows that most values in the figure were above it, as they should be if the drier samples were well aerated, and that a few values were below it, as they should be if the wettest samples were poorly aerated. Machlis' value of 0.24 μg of oxygen per square centimeter per minute turns out to be within the critical range of oxygen diffusion values at which growth of many plants is inhibited, as will be mentioned later.

Numerical values should not be taken too seriously in a comparison such as the one just made. Nevertheless, the orders of magnitude are such that they suggest the rate of diffusion of oxygen through the soil water may limit the rate of oxygen delivery to root surfaces if the water films become thick.

Closer examination of the root and electrode systems brings to light two differences that seem significant in judging the importance of oxygen diffusion to roots. First, if oxygen diffuses to a platinum electrode, the reduction occurs at the surface; and the concentration of oxygen at the surface is considered to be almost zero if the applied voltage is in the usual range. With roots, however, the oxygen is used internally rather than at the surface; hence, if the oxygen is supplied from the soil, the concentration at the surface cannot be zero but rather must be in excess of the requirements of all but the innermost cells. Otherwise, aerobic respiration of these cells will be inhibited. Consequently, the concentration of oxygen in the soil water required to supply roots with oxygen at a given rate will exceed the concentration required to supply oxygen to a platinum electrode of the same dimensions at the same rate per unit area of electrode surface.

A second difference between the root and electrode systems as regards oxygen-diffusion relationships is that of the comparative dimensions of the roots and the electrodes. The barley roots previously discussed and the wire-type platinum electrodes used for measurement have similar diameters, and so no adjustments were made for this factor in the foregoing calculations. For a given rate of oxygen consumption per unit volume of root, however, the rate at which oxygen must be absorbed per unit of surface area increases with the diameter of the root. Barley roots are comparatively thin, which suggests that the estimate of oxygen diffusion requirements based on barley roots is conservative as applied to roots of larger diameter.

These two differences between the root and electrode systems, together with the foregoing calculations and the comparison of numerical values of oxygen requirement and oxygen diffusion rate, verify that the demand of roots for oxygen can indeed be great enough to cause the rate of diffusion in the soil water to become limiting. The platinum-electrode method thus appears to provide a measurement of a process that can limit the supply of oxygen to roots in wet soils. These matters were examined in a somewhat different way and in more mathematical terms by Lemon (1962) and Lemon and Wiegand (1962).

The application of the platinum-electrode technique in an investigation of soil aeration in relation to plant growth may be illustrated by an experiment by Stolzy et al. (1961). Snapdragons were grown in cylinders of a silt loam soil that had been treated with a soil-aggregating agent. The cylinders were closed at the bottom, and the atmosphere at the top of the cylinders was controlled at different oxygen percentages by continuous passage of mixtures of nitrogen and air through an enclosed chamber at the surface. This technique resulted in differences in rate of diffusion of oxygen among depths in individual columns of soil and in differences among columns. Platinum electrodes were inserted through ports in the side of the columns, and measurements were made at different depths. Figure 3.6 shows the rates of oxygen diffusion measured in the different columns, and Fig. 3.7 shows the distribution of roots in the columns. One may note from these two figures that the depth of penetration of roots increased with the oxygen percentage in the gas mixture. The irregular, approximately vertical line in Fig. 3.6 shows that the maximum depth of root penetration in the various columns occurred where the rates of oxygen diffusion were within the range from 0.13 to 0.28 $\mu g/cm^2/minute$. Stolzy and Letey (1964) reviewed

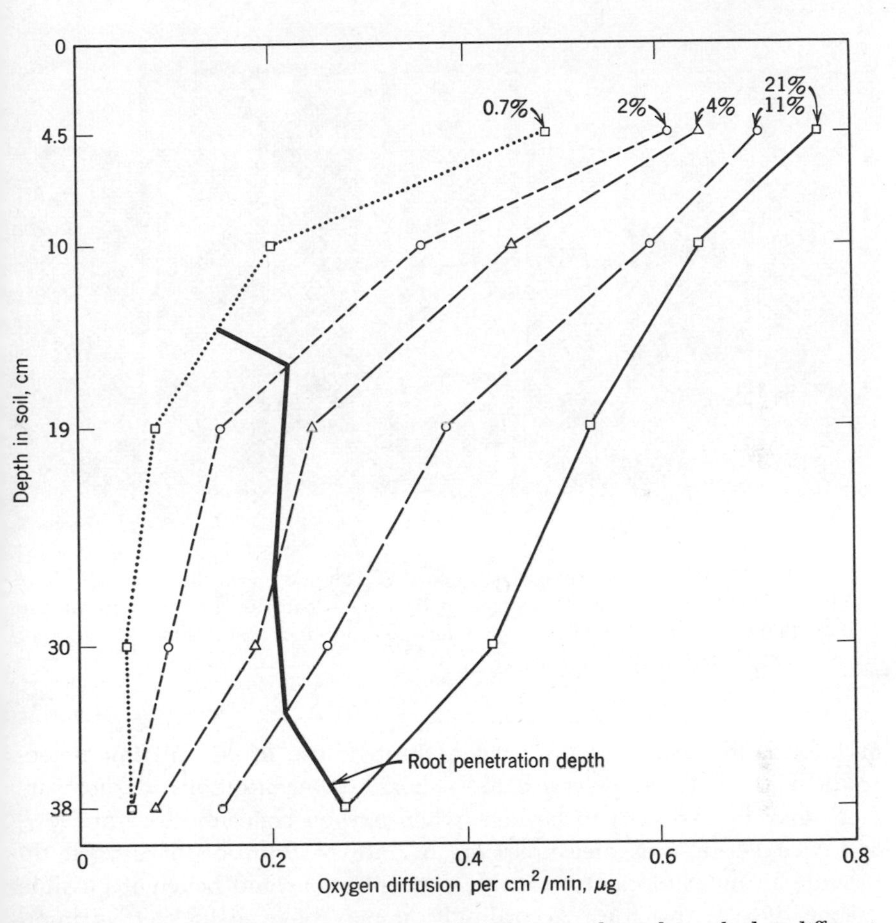

Figure 3.6. Rate of diffusion of oxygen by the platinum-electrode method at different depths in a silt loam soil. The numbers in the figure refer to oxygen percentages maintained in the air at the surface of the soil (values for 8.2 and 14.8% oxygen have been omitted for clarity). The heavy, approximately vertical line is the maximum depth of penetration of snapdragon roots. (Stolzy et al., 1961)

the work of a number of investigators and arrived at 0.2 μg of oxygen per square centimeter per minute as the rate of oxygen diffusion at which growth of roots is prevented and 0.2 to 0.3 μg as the range in which growth is affected.

Movement of carbon dioxide through water films from roots to soil air occurs simultaneously with movement of oxygen in the opposite direction. Both processes are presumably diffusion-controlled. No technique has yet been developed to provide for carbon dioxide the kind

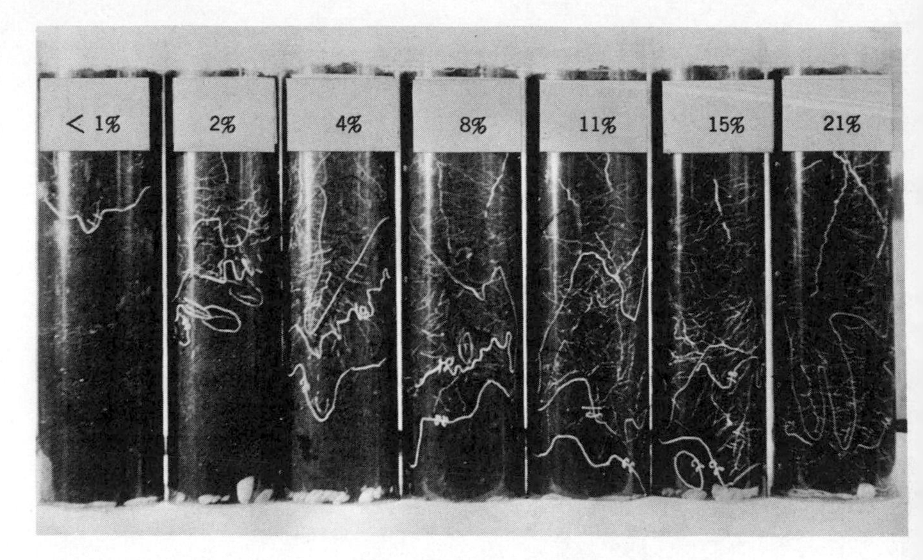

Figure 3.7. Distribution of snapdragon roots in columns of a silt loam soil. The numbers in the figure refer to oxygen percentages maintained in the air at the top of the cylinders. The crayon markings indicate root penetration at different times. (Stolzy et al., 1961)

of information available for oxygen through use of the platinum-electrode method. Nevertheless, carbon dioxide concentrations at root surfaces may be expected to be high when oxygen concentrations are low, and vice versa; thus measurements of rate of diffusion of oxygen do provide an index of carbon dioxide concentrations and hence of possible carbon dioxide toxicities. Accordingly, it may be well to keep in mind that successful correlation of measurements of rate of diffusion of oxygen with plant responses does not necessarily mean that the responses are not due in part to carbon dioxide unless proved otherwise.

Oxidation-Reduction Potential

Theory. The theory underlying oxidation-reduction potentials bears considerable similarity to the theory of the platinum-electrode method for measuring the rate of diffusion of oxygen in soil. In the method described in the preceding section, a sufficient electron potential or voltage is applied to the platinum electrode from an external source to cause transference of electrons to dissolved oxygen in contact with the electrode. In metabolic processes carried out by microorganisms and plant roots, electrons are supplied by chemical reactions in which organic substances are being oxidized. Through the catalytic action of enzymes, each molecule of oxygen takes up four electrons and is reduced, reacting with hydrogen or water to form water or hydroxyl according to the reactions given in the preceding section. Although these reactions presumably take place primarily inside of the cells of plants and microorganisms, their occurrence is reflected externally in the soil water by the disappearance of more or less of the oxygen and other substances such as ferric iron that will accept electrons, and by the appearance in the soil water of substances such as ferrous iron with reducing properties. The soil water, then, has a certain electron potential or tendency to supply electrons determined by the nature and proportion of oxidizing and reducing substances.

If a chemically inert but electrically conductive substance such as platinum is inserted into the soil water, the electron potential in the solution will be impressed on the electrode. That is, a certain tendency, characteristic of the solution, will exist for transference of electrons from the solution to the electrode. This tendency may be measured in the form of a voltage by inserting a standard electrode with known potential in the soil water nearby and measuring the voltage produced across the cell formed by the standard electrode and the platinum electrode in contact with the soil. The apparatus in principle is the same as that shown in Fig. 3.3 for measurement of the rate of diffusion of oxygen. In this application, however, the objective is to find the setting of the variable resistance that will result in imposing on the cell a voltage equal to the voltage of the cell. When the two voltages are exactly balanced, no current will flow through the galvanometer. The voltage thus found is called the oxidation-reduction potential or redox potential after due adjustment for the voltage of the standard electrode employed.

The general equation for an oxidation-reduction reaction in which a reversible equilibrium exists may be represented by

$$bB + cC + \cdots + n \text{ electrons} = gG + hH + \cdots \qquad (1)$$

where b and c are the numbers of moles of B and C reacting to produce g and h moles of G and H, and where there is a difference of n electrons between the reduced state and the oxidized state. The measured potential may be represented by

$$E_h = E_0 - \frac{RT}{nF} \ln \frac{a_G{}^g \cdot a_H{}^h \cdots}{a_B{}^b \cdot a_C{}^c \cdots} \qquad (2)$$

where a_B and so on stand for activities; E_h is the oxidation-reduction potential relative to the normal hydrogen electrode taken as zero; E_0 is a constant, characteristic of the system; $R = 8.315$ joules; $T =$ absolute temperature; $F = 96{,}500$ coulombs; and $n =$ number of electrons transferred in the reaction. The normal hydrogen electrode is, in effect, an electrode of molecular hydrogen at a pressure of 1 atmosphere adsorbed on an electrode of platinum coated with platinum black (which is finely divided metallic platinum) immersed in a solution in which the hydrogen-ion activity is unity.

To measure the oxidation-reduction potential of a simple system, for example, that represented by ferrous and ferric iron, an electrical cell is formed in the following manner:

$$H_2,\ H^+(a = 1) \,\|\, (Fe^{+++},\ Fe^{++}),\ Pt.$$

The half-cell on the left is the normal hydrogen electrode. Electrical connection to it is made through the platinum-black-coated electrode over which hydrogen gas at a pressure of 1 atmosphere is bubbled continuously. The half-cell on the right is the "iron" electrode, so-called because its potential is determined by the iron ions in solution in contact with it. When ferrous ions collide with this electrode, they tend to give up electrons to the electrode and make it negative, while the ferrous ions themselves tend to become oxidized to ferric ions. Conversely, ferric ions tend to take electrons from the electrode to form ferrous ions and leave the electrode positively charged. The activity ratio of ferric to ferrous ions around an inert electrode such as platinum determines its potential.

The vertical double line between the two half-cells indicates that the half-cells are separate physically but are joined electrically. To ensure that little or no independent potential arises from the junctions, metallic conductors are not used; instead, the connection is usually made with a solution of potassium chloride, which may be contained in a bent glass tube and stabilized in agar to prevent it from running out.

The reaction at the hydrogen electrode may be written

$$H^+ + e = \tfrac{1}{2}H_2 \tag{3}$$

and the reaction at the platinum electrode may be written

$$Fe^{++} = Fe^{+++} + e. \tag{4}$$

The sum of reactions 3 and 4 is then

$$H^+ + Fe^{++} = \tfrac{1}{2}H_2 + Fe^{+++}. \tag{5}$$

Applying Eq. 2,

$$\begin{aligned} E_h &= E_0 - \frac{RT}{nF} \ln \frac{(a_{H_2})^{1/2} \cdot a_{Fe^{+++}}}{a_{H^+} \cdot a_{Fe^{++}}} \\ &= E_0 - \frac{RT}{nF} \ln \frac{(a_{H_2})^{1/2}}{a_{H^+}} - \frac{RT}{nF} \ln \frac{a_{Fe^{+++}}}{a_{Fe^{++}}}. \end{aligned} \tag{6}$$

For the normal hydrogen electrode $a_{H^+} = 1$ and the voltage by convention is taken as zero. Accordingly, $(RT/nF) \ln (a_{H_2}^{1/2}/a_{H^+})$ is zero and $a_{H_2}^{1/2}$ must be 1. Then

$$E_h = E_0 - \frac{RT}{nF} \ln \frac{a_{Fe^{+++}}}{a_{Fe^{++}}}. \tag{7}$$

If $a_{Fe^{+++}} = a_{Fe^{++}}$, $\ln (a_{Fe^{+++}}/a_{Fe^{++}}) = 0$ and $E_h = E_0$. Therefore E_0 is the potential obtained where $a_{Fe^{+++}} = a_{Fe^{++}}$ and the measurement is made against the normal hydrogen electrode, which by definition has a voltage of zero. The conventions employed here lead to potentials that become more positive with an increase in oxidizing tendency of the solution. Strongly reducing solutions have potentials close to zero or somewhat negative.

In practice, measurements of oxidation-reduction potential of soils are not made against a normal hydrogen electrode where $a_{H^+} = 1$ (pH = 0) but against other electrodes and at hydrogen-ion activities less than 1 (pH > 0); nevertheless, the potentials are conventionally expressed with reference to the potential of the normal hydrogen electrode as zero. To express the results in this way requires an adjustment for the potential of the standard half-cell with reference to the normal hydrogen electrode. The usual reference electrode is a saturated calomel half-cell, which has a voltage that is positive with respect to the normal hydrogen electrode by 0.245 volt at 25°C. Accordingly, 0.245 volt is added to the potential measured against a saturated calomel electrode.

The oxidation-reduction potential of soils tends to decrease with increasing pH value, although there is no single characteristic relationship that applies to all soils, as indicated by Fig. 3.8. The explanation for the variation of oxidation-reduction potential with pH value is that the hydrogen ion participates in the oxidation-reduction process in some way. The participation may be direct, as in the reaction

$$MnO_2 + 4H^+ + 2e = Mn^{++} + 2H_2O$$

or it may be indirect, as in the reaction

$$Fe(OH)_3 + e = Fe^{++} + 3\ OH^-.$$

Also, the participation may be incidental, as in reactions involving competition between hydrogen ion and metallic cations such as ferric iron for combination with an organic substance that binds the iron and hence reduces its activity.

The change in oxidation-reduction potential per unit change in pH depends on the number of electrons transferred per hydrogen ion in-

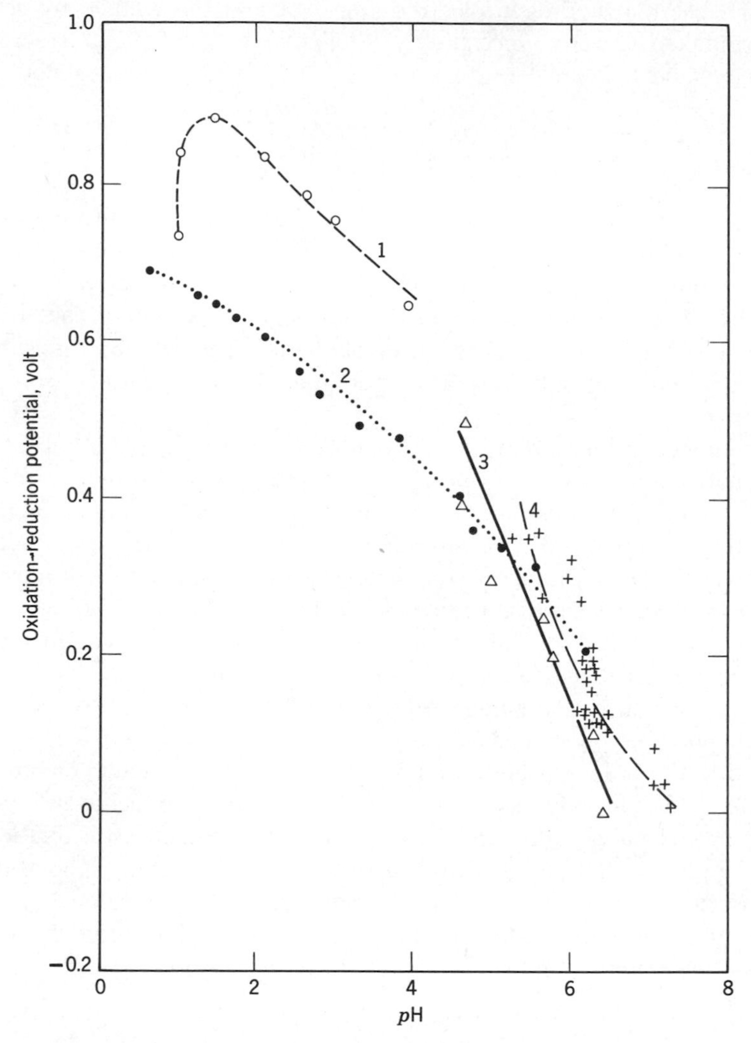

Figure 3.8. Oxidation-reduction potential versus pH of different soils. Number 1 represents a well-oxidized calcareous subsoil and 2 a reduced soil high in organic matter from a swamp. Measurements on both 1 and 2 were made on air-dry samples to which different quantities of dilute sulfuric acid had been added (Bradfield et al., 1934). Number 3 represents measurements on a suspension of a silt loam soil in water, where the oxidation-reduction potential was maintained at different values by automatic addition of oxygen (Patrick, 1961). Number 4 represents measurements after different times of incubation of a suspension of a silty clay loam soil plus ground grass roots in water. (Jeffery, 1960)

volved in the reaction. To apply Eq. 2 to the reaction involving manganese in the preceding paragraph,

$$E_h = E_0 - \frac{RT}{2F} \ln \frac{a_{Mn^{++}} \cdot a^2_{H_2O}}{a_{MnO_2} \cdot a^4_{H^+}} = E_0 - \frac{RT}{2F} \ln a^{-4}_{H^+} - \frac{RT}{2F} \ln \frac{a_{Mn^{++}} \cdot a^2_{H_2O}}{a_{MnO_2}}.$$

The term $(RT/2F) \ln a^{-4}_{H^+}$ is equal to 0.118 *p*H at 25°C, which makes the equation for the oxidation-reduction potential

$$E_h = E_0 - 0.118\ p\mathrm{H} - \frac{RT}{2F} \ln \frac{a_{Mn^{++}} \cdot a^2_{H_2O}}{a_{MnO_2}}.$$

For this reaction, therefore, the oxidation-reduction potential decreases by 0.118 volt per unit increase in *p*H value.

For reactions in which the hydrogen ion participates, the E_0 value is measured at or calculated to *p*H 0, so that the hydrogen-ion activity will be unity and E_0 will be equal to E_h (see Eq. 6). The comparative tendency for occurrence of oxidation-reduction reactions in which hydrogen ions participate thus may not be shown by the comparative E_0 values except for the condition of *p*H 0. To obtain the desired information for *p*H values of biological interest, the E_h versus *p*H relationship may be derived in the manner just described.

In chemical work with a half-cell such as that described previously for ferric and ferrous iron, oxygen is excluded to prevent partial oxidation of ferrous iron because the oxidation would change the activity ratio from the original value to some unknown value. The measured potential then could not be referred to a known activity ratio. If a platinum electrode in soil forms one of the half-cells, the situation is rather different from the simple chemical system described for ferric and ferrous iron. In natural soils, oxygen is excluded only to the extent that it has been used. Moreover, soils contain not one oxidized form and one reduced form, as in the ferric-ferrous half-cell, but a mixture of various oxidized and reduced forms, some of which are unknown. The significance of these circumstances requires special consideration.

If one wishes to make oxidation-reduction potential measurements on natural soils, the measurements should ideally represent the conditions as they exist. Hence oxygen is not excluded by artificial means. Whether or not oxygen is excluded, however, the potential present at any time and place in natural soils represents only a quasi-equilibrium condition and not a true equilibrium because of its dependence on biological processes. Moreover, techniques for measuring activity ratios of oxidized and reduced forms in soils are not well developed, so that even if oxygen and biological activities could be eliminated, the activity ratios could be measured only with difficulty or not at all. The result is that if one wishes to find the activity ratios of oxidized and reduced forms of iron or other substances corresponding to measured oxidation-reduction potentials of soils, the ratios must be calculated from relationships developed under controlled conditions on the assumption that they apply.

With regard to the mixture of oxidation-reduction systems present in soils, it may be said that some oxidation-reduction reactions proceed spontaneously because of the free transfer of electrons. The ferric-ferrous system is an example. Such reactions are said to be reversible. On the other hand, electron transfer does not take place freely in some reactions, such as $4H^+ + O_2 + 4e = 2H_2O$; and the reactions do not proceed spontaneously to an equilibrium condition. Such reactions are said to be irreversible. Many oxidation-reduction reactions involving organic substrates are of the irreversible type, although in respiration processes enzymes may facilitate the transfer of electrons, so that equilibrium conditions are approached. When irreversible and reversible systems are present in a mixture, the reversible systems interact spontaneously to yield a single oxidation-reduction potential; and this is the potential that is measured. The potential due to the reversible systems may drift with time under the influence of changes originating with the irreversible systems.

Accordingly, where soils are concerned, one may use the measured oxidation-reduction potential and the measured *p*H value, together with the E_0 value and equation for any given reversible oxidation-reduction reaction, to calculate the activity ratio of oxidized to reduced components corresponding to the reaction in question. The results of such calculations refer to the soil solution and are not applicable to irreversible reactions; and, of course, they will be hypothetical if the system in question is not present.

Because of the variation of oxidation-reduction potential with soil *p*H, some investigators prefer to express measurements on different soils or different horizons in a soil profile in terms of a constant *p*H value. The adjustment is usually made by adding to or subtracting from the oxidation-reduction potential a constant value for each *p*H unit above or below the *p*H value chosen as reference. The figure used most commonly is probably 0.059 volt/*p*H unit (at 25°C), which corresponds to transfer of one electron per hydrogen ion in the oxidation-reduction reaction. Adjustment to a common *p*H value should preferably be made by measuring the potential after removal of a quantity of the soil water and adjustment of the *p*H to the reference value, but this procedure is too difficult and tedious to have found favor in practice. Where adjustments for *p*H are not made, the conventional procedure is to report both the value of the oxidation-reduction potential as measured and the *p*H value at which it was measured.

Other complications in determining oxidation-reduction potentials of soils are those of stability and reproducibility of the potentials. The elementary apparatus shown in Fig. 3.3 would be unsuitable for practical measurements because the current flow through the soil in the process of obtaining a balance of voltages would alter the oxidation-reduction system to such an extent that the measurements would not represent the initial condition. Current flow through the soil may be practically eliminated by use of suitable electronic measuring equipment. Even with such equipment, however, the problem of instability is still troublesome because the concentrations of oxidizing and reducing substances that control the potential are usually so low that the potential is sensitive to only minor disturbances. Care is needed to prevent substantial alteration of oxidation-reduction potentials of soils in the process of measurement. Also, there is difficulty with reproducibility of potentials, apparently due to surface coatings on the platinum electrode. Care is needed in cleaning the electrode before each use.

To summarize this brief discussion of theory, it may be said that oxidation-reduction processes in soils are activated principally by metabolic processes of microorganisms and higher plants and involve reduction of elemental oxygen by organic substrates in the presence of enzymes. These changes are accompanied by reduction of certain inorganic electron acceptors as the oxygen is depleted, so that the oxidation-reduction system is a mixture of organic and inorganic oxidants and reductants. Measured oxidation-reduction potentials can be interpreted in terms of the ratios of certain oxidizing and reducing components known or assumed to be present; but not all components are known, and the oxygen and some of the organic components behave irreversibly and hence are not related in the theoretical manner to the electrode potential. Moreover, there is some difficulty in comparing potentials measured at different *p*H values. Under these circumstances, the potentials are to be regarded as an index of the intensity of reduction and their usefulness determined by the extent to which they provide information about chemical and biological conditions of interest. Oxidation-reduction potentials have the advantage of providing a measure of the intensity of reduction in soils containing no molecular oxygen; hence they may be useful in characterizing aeration conditions in the range below which the rate of diffusion of oxygen is zero. Such conditions are encountered where soils are saturated with water, and it is under these conditions that most measurements of oxidation-reduction potentials are made.

For further information on the physicochemical aspects of oxidation-reduction potentials, the works of Michaelis (1930), Hewitt (1950), and Garrels (1960) may be consulted. Reviews of theory with applications to soils were published by Ponnamperuma (1955) and Rodrigo (1963).

Relation to Soil Conditions and Plant Responses. If a sample of soil is flooded with water, the oxidation-reduction potential gradually falls to a limiting value. The more rapid the decomposition of organic matter, the lower is the limiting value (see Fig. 3.9). In the absence of oxygen, therefore, the oxidation-reduction potential falls more rapidly and to a lower level in samples of soil from surface horizons than in those from lower horizons. Under field conditions, the situation is more complex due to movement of water through the soil and a lag in change of potentials as soil conditions change. Table 3.4 illustrates the variation of oxidation-reduction potentials in a paddy soil with depth and time. The change with time is particularly pronounced in the illustration because of the intermittent flooding for production of rice. Potentials at all except the 50- to 80-cm depth were lower in the summer after flooding than in the spring before flooding. The reversal of potentials at this depth shows a lag between the time of drainage and reoxidation of the surface layer and the time of drainage and reoxidation of the 50- to 80-cm layer.

Figure 3.10 illustrates two significant aspects of the oxidation-reduction status of flooded soils that were not shown in Table 3.4. The first is

Table 3.4. Oxidation-Reduction Potential and *p*H of a Profile of Paddy Soil in Japan in the Spring Before Irrigation and in the Summer After Flooding (Aomine, 1962)

Spring (Before Irrigation)			Summer (Flooded)		
Depth, cm	*p*H	Oxidation-Reduction Potential, volt	Depth, cm	*p*H	Oxidation-Reduction Potential, volt
0–10	5.2	0.65	0–10	6.2	0.06
10–20	5.5	0.61	10–15	6.1	0.16
20–30	6.4	0.62	15–35	6.0	0.32
30–50	5.8	0.37	35–50	5.8	0.28
50–80	6.3	0.13	50–80	6.0	0.17

that the water overlying the soil may be well aerated, a condition attributed to turbulence and the growth of algae. The second is that a rapid drop in oxidation-reduction potential from oxidized to reduced conditions occurs in a thin surface layer of soil. A third aspect, significant in paddy soils, is that roots of rice may maintain the surrounding soil at a higher oxidation-reduction potential than is found in bulk soil (Alberda, 1953). This aspect may be evidenced also by formation of a crust of iron and manganese oxides on the older roots of rice in flooded soil (Sturgis, 1936).

According to Aomine (1962), oxidation-reduction potentials are correlated to some extent with observable soil conditions. The red and brown soil colors characteristic of ferric iron are generally found where the oxidation-reduction potential exceeds 0.3 volt. The gray and blue soil colors associated with reduction of ferric iron are generally found where the oxidation-reduction potential is below 0.3 volt. One or more thin layers of brown hydrous iron oxide are sometimes present just below

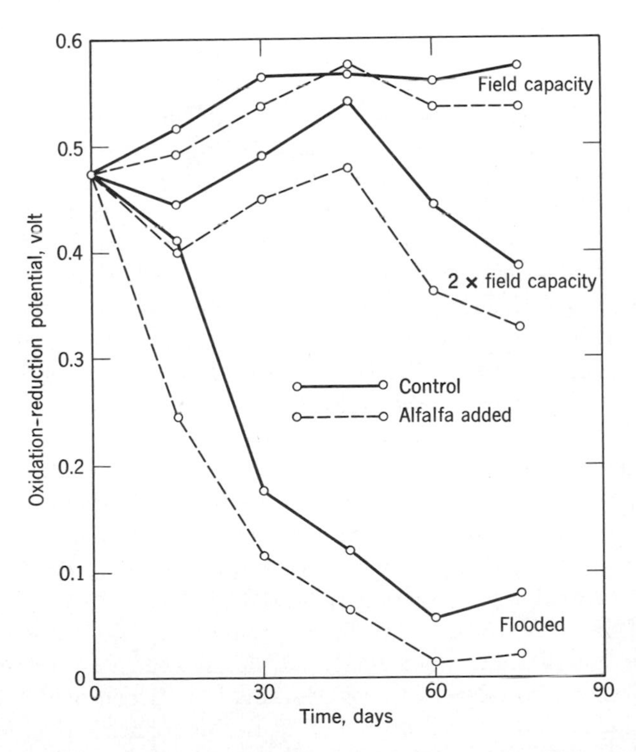

Figure 3.9. Oxidation-reduction potential of a silt loam soil at three water contents in the presence and absence of 1% ground alfalfa hay. (Savant and Ellis, 1964)

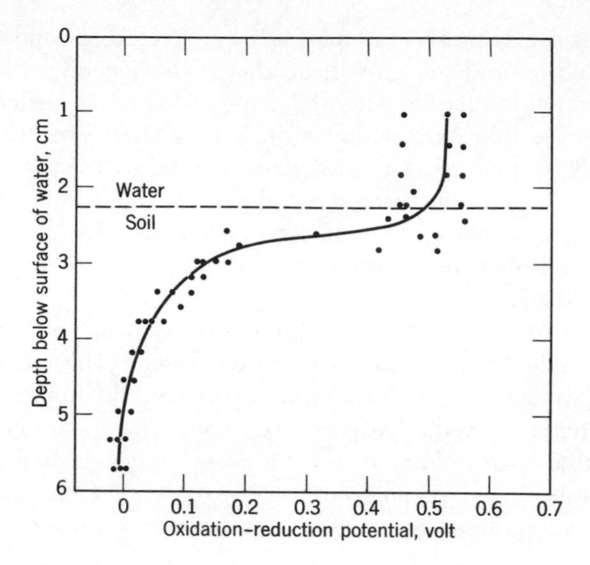

Figure 3.10. Oxidation-reduction potential at different depths in water and underlying soil. (Alberda, 1953)

the surface of flooded paddy soil, presumably marking the level where the oxidation-reduction potential is high enough to oxidize ferrous iron moving up from beneath. Yamazaki (1960) noted that the oxidation-reduction potentials in a gleyed (reduced) soil containing rusty mottlings were 0.40 volt in the rusty mottlings and 0.17 volt in the surrounding soil. Such correlations are of only general value because of a lag of visual changes relative to changes in oxidation-reduction potential and because of relatively permanent changes of appearance that may develop. For example, gray horizons may develop from reduction of iron to the ferrous form; but, if the reduced iron is removed, the gray color will remain despite subsequent rise of oxidation-reduction potential.

Oxidation-reduction potentials are related to the chemical conditions in soils, as indicated in the discussion of theory. Under conditions of precise control of oxidation-reduction potential by continual controlled addition of oxygen gas to suspensions of soil in water, Patrick (1961) found that an oxidation-reduction potential of 0.338 volt at *p*H 5.1 was the critical value for nitrate. At higher potentials, nitrate accumulated in the soil; and at lower potentials nitrate disappeared. For the most part, only general correlations have been made between oxidation-reduction potentials and the state of oxidation. Results of various investigators indicate that chemical changes in soils occur in the following order as the oxidation-reduction potential decreases: disappearance of oxygen, reduction of nitrate, reduction of manganese to divalent form, reduction of iron to divalent form, formation of sulfide, and reduction of sulfate (Aomine, 1962).

Figure 3.11 shows the results of an investigation of the association of plant response with oxidation-reduction potentials and dissolved oxygen in ground water from the gley (reduced) horizon of soils in swampy areas in northern Wisconsin and Ontario. The dissolved oxygen may be seen to decrease with decreasing oxidation-reduction potential. A substantial range in oxidation-reduction potential was encountered in which there was no significant quantity of dissolved oxygen in the water. Aspen was least tolerant and white cedar most tolerant of conditions associated with low oxidation-reduction potential. Black spruce and alder were intermediate. The limits of the range of oxidation-reduction potential for the different species are those found under conditions of natural occurrence of the species and do not necessarily represent the limits under which the species would grow in the absence of competition. Moreover, it must be recognized that the figures for oxidation-reduction potential in this application represent in a way an index of the aeration

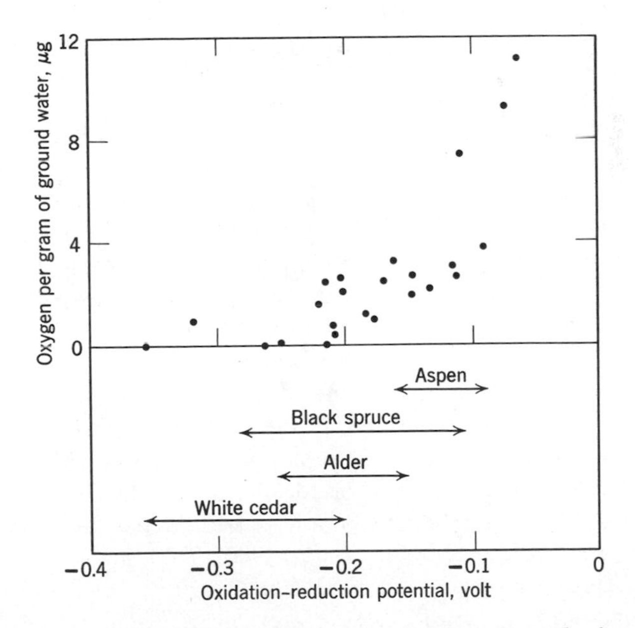

Figure 3.11. Dissolved oxygen and oxidation-reduction potential of ground water from the gley horizon in swampy areas in Wisconsin and Ontario, together with the distribution of four forest species. Values of oxidation-reduction potential have been adjusted to *p*H 7. (Pierce, 1953)

of the soil profile in which the trees were growing and not necessarily the aeration conditions to which the entire root system was exposed, because a portion of the profile occurred above the level of the gley horizon from which the water was obtained for analysis. A similar investigation of the correlation between natural vegetation and oxidation-reduction potentials was made by Pearsall (1938).

There is no evidence that the oxidation-reduction potential of soil affects plants directly. Rather, it seems that plants respond to differences in chemical environment with which the potentials are associated. In roots of plants with good internal aeration, the internal oxidation-reduction potential is largely independent of the potential of the soil; and the physiological effects are those arising indirectly from chemical changes in the soil. In plants with inadequate internal aeration, the oxidation-reduction potential of the roots increases and decreases in response to the external potential; and the physiological effects arise indirectly from external chemical changes associated with the oxidation-reduction potential of the soil and from internal chemical changes in the roots associated with their oxidation-reduction potential. The phenomenon of internal aeration is discussed in the section on plant adaptations.

Composition of Soil

Under conditions of good aeration, the supply of oxygen is sufficient to permit oxidation of organic materials to carbon dioxide and water; and the mineral constituents of soils remain almost entirely in oxidized condition. Where soils are saturated with water, the supply of oxygen is soon exhausted. Soluble organic products of metabolism and other reduced materials then accumulate. A quantitative measure of the oxygen-consuming capacity may be obtained by determining the quantity of a chemical oxidizing agent consumed by reaction with a sample of water from the soil. Measurements of this sort are standard for estimation of the oxygen-consuming capacity of sewage.

The soluble organic products that develop in flooded soils include simple forms that can be separated and identified. Considerable research on these products has been done in Japan in connection with rice culture. There is evidence for occurrence of formic, acetic, oxalic, glycolic, propionic, butyric, fumaric, succinic, malic, citric, lactic, and valeric acids and methyl mercaptan (Mitsui et al., 1960; Takijima, 1964; Takijima et al., 1961, 1963; Asami and Takai, 1963). Takijima et al. (1963) found that concentrations of formic, acetic, and butyric acid were sometimes high enough in the leachates from peaty paddy soils to inhibit growth of rice. In other work, Takijima (1961, 1964a) and Takijima and Sakuma (1962) found that organic acids accounted for only part of the inhibition of growth of rice roots observed in water extracts of paddy soils.

At low oxidation-reduction potentials, ferrous iron and sulfide sulfur are produced. These substances have received principal attention as inhibitors in flooded soils. Ota and Yamada (1960) found that the growth of rice plants decreased as the oxidation-reduction potential decreased and as the "free iron" (presumably the soluble iron) increased. Addition of iron to strongly reduced soil caused almost all plants to die from severe ferrous iron toxicity. Ponnamperuma (1955) conducted an experiment in which rice was grown on a silt loam soil at three *p*H values, each with three different quantities of organic matter and with conditions of submergence with and without drainage; and he found that the yield of rice grain plus straw decreased with increasing content of iron in leachates from the soil. The correlation between yields and iron in the leachates was closer than that between yields and total oxidizable material or oxidation-reduction potential. Concentrations of iron up to 525 μg/ml were found. Iron toxicity thus appeared to be more significant than the other factors.

Sulfide is produced under conditions of strong reduction. Osugi and Kawaguchi (1938, 1939) found that sulfide produced in certain submerged soils was responsible for a root-rot of rice. Iron is reduced to ferrous form at higher oxidation-reduction potentials than those favoring formation of sulfide, with the result that ferrous iron is present in solution in soil by the time the oxidation-reduction potential has fallen sufficiently to permit sulfide formation. Ferrous iron reacts with sulfide to produce ferrous sulfide, a compound of low solubility, thus providing protection against sulfide toxicity. Conditions of high organic matter, high sulfate, and low iron content of soils favor production of sulfide upon submergence (Tanaka et al., 1961).

Plant Response to Aeration

The response of plants to conditions of soil aeration has already been considered to some extent in connection with other topics. In this section, attention is focused more directly on the plant responses than on the soil conditions.

Respiration

In aerobic respiration, the moles of carbon dioxide evolved are approximately equal to the moles of oxygen absorbed. Carbon dioxide evolution exceeds oxygen consumption in partially anaerobic respiration and takes place without oxygen consumption in completely anaerobic respiration. In anaerobic respiration, partially oxidized organic substances accumulate; and gaseous carbon compounds other than carbon dioxide may be evolved.

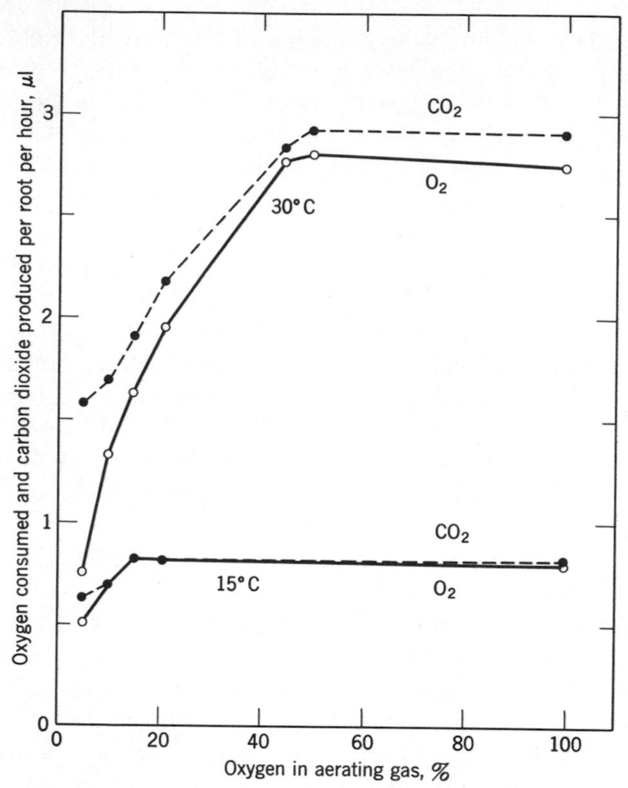

Figure 3.12. Oxygen consumed and carbon dioxide produced by the apical 5-mm segment of onion roots at two temperatures and at different oxygen percentages in the aerating gas mixture. (Berry and Norris, 1949)

Plant roots seem to be adapted to aerobic respiration. Nevertheless, partially anaerobic respiration takes place if oxygen is deficient. Figure 3.12 shows the results of measurements of respiration of onion roots at 15 and 30°C in the presence of different concentrations of oxygen in the gas mixture used for aeration. The rate of respiration, as indicated by oxygen consumption and carbon dioxide evolution, decreased at low concentrations of oxygen. Respiration at 15°C was aerobic except at the lowest concentration of oxygen, but respiration at 30°C was partially anaerobic at all concentrations of oxygen. The concentration of oxygen required to produce the maximum rate of respiration was greater at 30°C than at 15°C, presumably because metabolic use of oxygen increases more rapidly with an increase in temperature within this range than does movement of oxygen through water films around roots and through root tissue. One therefore may expect that aeration requirements of plants growing in soil should increase with the temperature [for the results of an experiment on sunflowers growing in soil at different temperatures, see the paper by Letey et al. (1962)].

Root elongation is particularly sensitive to aeration conditions, the reason presumably being that the rate of respiration (and hence the oxygen requirement) is greatest in root tips, where growth occurs, and decreases with distance from the tips. For example, in the work by Machlis (1944) cited previously, the relative rates of oxygen consumption by different 1-cm segments of barley roots were 100 in the apex segment and 73, 65, 59, 52, 47, 46, and 36 in successively older segments.

Consumption of oxygen and evolution of carbon dioxide are outward evidences of the complex of metabolic processes in roots and provide little insight into internal changes. Fulton and Erickson (1964) found that oxygen deficiency in the root medium resulted in the presence of ethyl alcohol in the sap exuded from the base of plants when the tops were removed. The concentration of alcohol increased with a decrease in the rate of diffusion of oxygen in the soil (Fig. 3.13). Appearance of ethyl alcohol is an indication of anaerobic respiration. Fulton and Erickson obtained evidence that alcohol concentrations of the order of those measured were toxic to young tomato plants, resulting in a decrease in growth even in the presence of an ample supply of oxygen. They noted further that appearance of alcohol in the exuded sap was much dependent on the depth in the soil at which the poor aeration

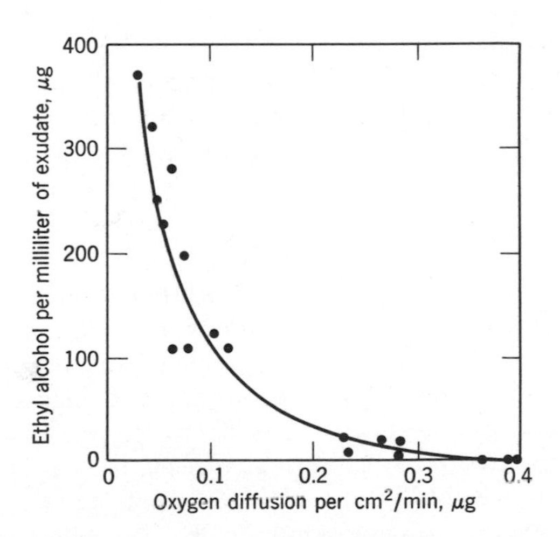

Figure 3.13. Ethyl alcohol in sap exuded from the base of tomato plants versus rate of diffusion of oxygen in the soil as measured by the platinum-electrode method at a depth of 2 cm. (Fulton and Erickson, 1964)

occurred. Exposure of only one-fifth of the total root volume to improved aeration resulted in a marked reduction of alcohol concentration, presumably because the aerated roots were metabolizing the alcohol. Further work by Fulton, Erickson, and Tolbert (1964) showed changes in other metabolic intermediates in both the roots and tops of tomato plants in response to flooding the roots with water. In work done independently, Grineva (1964) also observed the accumulation of alcohol and its exudation from roots of corn and sunflower.

The existence of alcohol toxicity, incidentally, provides evidence on the oxygen-deficiency, carbon dioxide-toxicity question discussed previously. Accumulation of alcohol is a consequence of oxygen deficiency and not of carbon dioxide excess.

Plant Adaptations

In preceding sections, emphasis has been placed on the behavior of plants adapted to growth with a supply of oxygen external to the roots. Occasional reference has been made to internal aeration of plants without elaboration of the nature of the phenomenon. Internal aeration will be considered here as one of the adaptations permitting plants to grow in poorly aerated soil.

Certain plants such as rice that are adapted to growth without an external source of oxygen for the roots have large internal air spaces. In the stem and roots, these are concentrated in the cortex. [Drawings illustrating the occurrence of internal air spaces are given in Arber's (1920) book on water plants and elsewhere.] Evidence indicates that gas exchange between the roots and the atmosphere occurs through these air spaces. For example, Vallance and Coult (1951) found that bogbean (*Menyanthes trifoliata*) plants kept in darkness with their roots in nitrogen and the stems and leaves in air contained 12.5 to 17.5% oxygen in the air spaces in the roots and slightly more in the basal part of the stem. Where the tops of the plants were in nitrogen and the roots in air, the concentration of oxygen in the roots dropped to values of 6% or less. Armstrong (1964) observed diffusion of oxygen from roots of the bog plants *Menyanthes trifoliata* and *Eriophorum angustifolium* when the roots were kept in an anaerobic medium and the tops in air. In some semisubmerged aquatic plants with internal air spaces, oxygen produced in the tops by photosynthesis seems to be an important source of oxygen used in respiration by the roots; and, conversely, carbon dioxide produced in the roots by respiration is apparently used in the tops in photosynthesis (Laing, 1940a).

Although internal air spaces are normal in roots of plants indigenous to poorly aerated environments, they are not confined to such plants.

If aeration is poor, plants that grow characteristically in aerated soil may develop internal air spaces (Bryant, 1934; McPherson, 1939). Bartlett (1961) observed that roots of such plants had an oxidizing effect on iron in flooded soil and in solutions and that the effect on solutions was greater in light than in darkness with most species. These observations suggest that oxygen was being conducted downward through the plants and outward from the roots into the external medium and that photosynthesis increased the internal content of oxygen in the roots. More direct evidence of downward conduction of oxygen in roots of plants that grow characteristically in well aerated soil has been obtained under conditions of good aeration. Jensen et al. (1964) observed that labeled oxygen entered corn roots from soil on one side of a wax seal and was exuded on the other side. Brown (1947) observed evolution of oxygen from roots of seedlings of *Cucurbita pepo* where the tops were exposed to air and the roots to nitrogen, although certain of his observations did not support the view that downward movement occurred in air spaces. Barber et al. (1962) calculated that the rate of downward movement of radioactive oxygen in roots of barley corresponded to diffusion through air spaces occupying less than 1% of the cross-sectional area.

A second type of morphological adaptation to poor aeration is development of a shallow root system in the upper part of the soil where aeration is greatest. In the case of rice Alberda (1953) noted the development of a superficial mat of roots growing horizontally above the soil in the irrigation water or vertically to the surface, and he obtained evidence that these roots (which were not present on younger plants) absorb oxygen.

A third adaptation is that of anaerobic respiration. Although roots of plants adapted to growth in aerated soils have been found to carry out some anaerobic respiration if oxygen is deficient, the effects are detrimental. The relatively low rate and energy output of anaerobic respiration and the accumulation of toxic products are perhaps the most important reasons. Rhizomes of certain aquatic plants, however, can carry out anaerobic respiration for days and perhaps much longer without noticeable injury (Laing, 1940). These organs seem to have greater tolerance than do roots of terrestrial plants to the consequences of such respiration.

Uptake of Nutrients and Water

Effects of aeration on uptake of nutrients by plants may result from changes in nutrient availability that occur in the soil in response to the aeration, or they may result from changes in metabolic status of the plants. Important among the soil effects are those connected with nitrogen. Mineralization of nitrogen (i.e., change from the organic to the inorganic form) seemingly is not much affected by the change from aerated to nonaerated conditions. After molecular oxygen has been reduced, however, nitrate-nitrogen serves as an electron acceptor. Nitrate then disappears, and part of the nitrogen is lost in gaseous forms. Tyagny-Ryadno (1960) found oxidation-reduction potentials as much as 0.07 volt lower in the interior of freshly saturated soil aggregates than on the surface, and Liu and Yu (1963) found that the oxidation-reduction potential of soil containing much easily decomposable organic matter was 0.22 volt lower in the center of clods than on the surface. Apparently, therefore, a given soil may contain, at the same time and at the same depth, locations that are aerobic, supporting nitrate production, and locations that are anaerobic, supporting reduction of nitrate and loss of some of the nitrogen in gaseous forms. Under these circumstances, the process of nitrogen loss may proceed continuously. Nitrogen transformations will be discussed in more detail in Chapter 7.

Iron and manganese behave rather similarly but unlike nitrate. These elements occur in soil in more than one valence state. The reduced forms produced in the absence of molecular oxygen are more soluble than the oxidized forms that predominate if aeration is good. Consequently, rice plants grown on submerged soils are rarely deficient in iron and manganese. On the contrary, there is more difficulty from iron toxicity than from deficiency, as mentioned previously. The aerobic conditions required by most higher plants have oxidation-reduction potentials too high to permit more than a small amount of iron and manganese to remain in the soluble, divalent forms. Nevertheless, Piper (1931) found that oats grown on aerated, manganese-deficient soil responded to previous flooding, which no doubt caused temporary reduction of manganese to the divalent form. Some of his data are shown in Table 3.5. In this experiment, the water content of the soil was kept at 60% of saturation in all treatments during growth of the plants; however, prior to planting, one portion of the soil was kept saturated with water for a week. The prior saturation treatment had about the same effect on the yield and manganese content of the plants as did application of manganous sulfate equivalent to 560 kg/hectare.

The supply of soil phosphorus for rice increases if the soil is flooded, as was found by Broeshart et al. (1965) in experimental work on soils from a number of the rice-growing countries of the world. As specific examples of the observations that have been recorded, Shapiro (1958) found that the quantity of phosphorus present in the above-ground portions of rice plants was increased from 14 mg per culture on unflooded

Table 3.5. Yield and Manganese Content of Oats Grown on a Manganese-Deficient Soil with Various Treatments (Piper, 1931)

Treatment	Yield of Dry Matter per Culture, g	Manganese per Gram of Dry Matter, μg
Control	2	5
$MnSO_4$, 125 kg/hectare	87	17
$MnSO_4$, 560 kg/hectare	77	46
Saturated with water before planting	75	42

soil to 40 mg per culture on flooded soil, whereas the corresponding yields of total dry matter were increased only from 12 g to 21 g. Analyses of the drainage water from the flooded soil showed that the concentration of inorganic phosphorus increased from 0.01 μg/ml on the fourth day after flooding to 0.03 μg/ml on the forty-fifth day after flooding. The concentrations of iron in the drainage water at the same sampling times were 0.3 and 8.3 μg/ml. Savant and Ellis (1964) found that the inorganic phosphorus extractable from soils by dilute ammonium fluoride, hydrochloric acid solution was increased by flooding, higher values being associated with lower oxidation-reduction potentials.

The improved phosphorus supply in flooded soils has been connected with dissolution of ferric phosphates in consequence of microbiological reduction of ferric iron. Shapiro (1958a) verified the role of ferric phosphate in an experiment in which uptake of phosphorus by rice from added ferric phosphate was found to be 294% greater from flooded than from unflooded soil in the absence of added organic matter and 742% greater from flooded than from unflooded soil in the presence of cellulose added as filter paper. The corresponding increases in uptake of phosphorus from aluminum phosphate were 99 and 107%.

Rice and presumably other plants that grow normally on flooded soil are in a special class in respect of their source of phosphorus. Ferric phosphates, which on reduction become available to rice grown on flooded soil, remain in ferric form during growth of plants on unflooded soil. Calcium and aluminum phosphates seem to be of greater value than ferric phosphates for plants on unflooded soil. Unpublished work by M. T. Eid and co-workers in the Ministry of Agriculture, United Arab Republic, showed that the chemical indexes of soil phosphorus availability that worked well for crops grown on unflooded soils gave unsatisfactory results for rice, where the extracting solutions were used on air-dry samples of soil. For rice, a strongly alkaline extractant, which would extract phosphorus from ferric phosphate, gave best results. Chiang et al. (1963) noted that the availability to rice plants of the phosphorus in soils of southern China increased with the amount of phosphorus present in iron phosphates.

The effects just described are chemical changes occurring in soil in response to low oxidation-reduction potentials in the absence of molecular oxygen. Except for loss of nitrate, the changes are in the direction of increased solubility and hence represent increases in availability of the nutrients in question for plants adapted to growth in the absence of a supply of oxygen external to the roots. The effects of differences in aeration on uptake of nutrients by plants adapted to growth in soil with a supply of oxygen external to the roots appear to be mostly a consequence of the effect of aeration on the physiological condition of the plants and only secondarily a consequence of chemical changes in soils. An important reason for emphasizing the physiological effects is that most of the chemical changes in soils do not occur until the aeration has become too poor for plants that require a source of oxygen external to the roots. A second reason is that the effects of aeration on uptake under conditions to which such plants are adapted are mostly in the opposite direction from solubility effects due to chemical changes in reduced soils. For example, Table 3.6 shows that the content of manganese, iron, and phosphorus in citrus leaves was increased by improving the aeration. Except for sodium, the content of all nutrients in the leaves increased with improved aeration. In this experiment, the calcium content increased markedly with improved aeration. In an experiment by Cline and Erickson (1959) with peas, the calcium content decreased almost as markedly. Perhaps the most consistent effects are increases in uptake of potassium and nitrogen with improved aeration. Experiments by Shapiro et al. (1956) indicate that an important cause of the increase in potassium content of the tops of corn plants with improved aeration is improved translocation of potassium from the roots to the tops. Stolzy and Letey (1964) reviewed work on nutrient uptake in relation to rate of oxygen diffusion in soils.

Poor aeration decreases uptake of water, as evidenced by the wilted condition of many plants after flooding. As an example of the water relations, Stolzy et al. (1963) reported for the poor and good aeration treatments in Table 3.6 that total dry weights per plant were 37 and 65 g and use of water per plant during the experiment was 1.33 and 5.07 liters. There is evidently a marked disparity between the ratios of plant weights and ratios of water use. According to Kramer (1956) the permeability of roots to water decreases under conditions of poor aeration.

Table 3.6. Mineral Content in Leaves of Sweet Orange Seedlings Grown on a Sandy Loam Soil with Different Rates of Oxygen Diffusion as Measured by the Platinum-Electrode Method (Stolzy et al., 1963)

Oxygen Diffusion per cm² per minute, μg	Mineral Content in Dry Leaves, %							
	Ca	Mg	Na	K	P	Mn	Fe	B
0.22	2.0	0.25	0.015	1.0	0.14	0.0027	0.0041	0.0050
1.04	4.4	0.27	0.006	1.5	0.16	0.0037	0.0056	0.0080

Disease Incidence

Plant roots and soil-borne disease organisms occupy the same environment and respond in characteristic ways to prevailing conditions such as aeration. Changes in aeration may alter the susceptibility of the plant, the virulence of the organism, or both, so that disease incidence may vary with conditions of soil aeration.

In experimental work on a root rot of sugarcane caused by the fungus *Pythium arrhenomanes*, Rands and Dopp (1938) investigated the effect of salicylic aldehyde, which had been reported to be present in poorly drained soil. This substance proved toxic to both sugarcane and fungus in relatively high concentration but had little effect in low concentrations. Neverthcless, the reduction in weight of plants due to inoculation of the culture with the fungus was nearly six times as great in the presence of a low concentration of salicylic aldehyde (40 μg/ml) as in its absence. In this experiment, it is not clear whether the effect was on susceptibility of the plant, virulence of the organism, or a combination of the two.

Fungi are known as aerobic organisms, but some of them are much more tolerant than higher plants to a deficiency of oxygen (Klotz et al., 1963). Their distribution in soil suggests that they may differ considerably among themselves in aeration requirement. In an investigation of fungi responsible for diseases of citrus, Klotz et al. (1966) isolated *Mucorales* species only from the surface 8 cm of soil, *Thielaviopsis basicola* mainly from the surface 15 cm, but from some samples at greater depths, *Phytophthora* species from many of the samples from the surface 15 cm but from almost all samples from greater depths, and *Pythium* species from samples throughout the profile. Rates of diffusion of oxygen decreased with depth in the soil.

Although the evidence is limited, certain fungi seem to be less tolerant than higher plants to accumulation of carbon dioxide. This difference between fungi and higher plants has been investigated in connection with the effect of aeration on the "take-all" disease of cereals caused by the fungus *Ophiobolus graminis.* In the absence of the host plant, this organism gradually disappears from the soil, presumably because it is a poor competitor with other microorganisms under these conditions. In nonsterile soil, it grows actively only along the surfaces of roots of the host, sending many short infection hyphae into the roots. Infection of plants by the organism in the field is more frequent and severe in soils of coarse texture and others that are loose than in soils of fine texture and others that are compacted, which suggests that good aeration is important for rapid growth of the hyphae. This supposition was confirmed by Garrett (1936, 1937), who found that forced aeration increased the rate of advance of the hyphae along root surfaces and that the growth rate was decreased by as little as 5% carbon dioxide in the atmosphere. These observations, together with additional evidence of a more indirect nature, led Garrett to propose that differences in concentration of carbon dioxide around the roots are primarily responsible for the apparent effect of aeration on growth rate of the hyphae observed in practice. This instance seems to be one in which the effect of aeration has to do principally with the sensitivity of the disease organism and not the plant. Cereals are moderately tolerant to poor aeration.

Attempts have been made to eliminate certain disease-producing fungi from soil by flooding the soil between crops or by growing rice under flooding in rotation with the susceptible crop. In some instances, good control has been obtained. Whether the cause of the sensitivity of the organisms to poor aeration is a deficiency of oxygen, an excess of carbon dioxide, toxicity of accumulated decomposition products, or competition with other organisms is a debated matter. In experiments on *Fusarium oxysporum* f. *cubense,* the fungus responsible for an important soil-borne disease of bananas, Newcombe (1960) found that absence of oxygen did not prohibit germination of conidia or formation of chlamydospores, which could cause subsequent infection. Carbon dioxide similarly did not prohibit germination of conidia, but it prevented formation of chlamydospores. Her view was that in flooded soil a high concentration of carbon dioxide would gradually eliminate the organism by permitting germination and destruction of the relatively short-lived conidia in the absence of a host and by preventing formation of the more resistant and longer lived chlamydospores that could otherwise cause subsequent infection upon drainage of the soil and replanting to bananas.

An extensive review of this and related subjects is found in a symposium edited by Baker and Snyder (1965). A section by Sewell in this symposium is particularly relevant.

Literature Cited

Alberda, T. (1953) Growth and root development of lowland rice and its relation to oxygen supply. *Plant and Soil* **5**:1–28.

Aomine, S. (1962) A review of research on redox potentials of paddy soils in Japan. *Soil Sci.* **94**:6–13.

Arber, A. (1920) *Water Plants.* Cambridge Univ. Press, Cambridge.

Armstrong, W. (1964) Oxygen diffusion from the roots of some British bog plants. *Nature* **204**:801–802.

Asami, T., and Y. Takai. (1963) Formation of methyl mercaptan in paddy soils (II). *Soil Sci. and Plant Nutr.* **9**, No. 2:23–27.

Baker, K. F., and W. C. Snyder (Eds.) (1965) *Ecology of Soil-Borne Plant Pathogens—Prelude to Biological Control.* University of California Press, Berkeley.

Barber, D. A., M. Ebert, and N. T. S. Evans. (1962) The movement of $^{15}O_2$ through barley and rice plants. *Jour. Exp. Bot.* **13**:397–403.

Bartlett, R. J. (1961) Iron oxidation proximate to plant roots. *Soil Sci.* **92**:372–379.

Bavel, C. H. M. van. (1965) Composition of soil atmosphere. *Agron.* **9**:315–318.

Baver, L. D., and R. B. Farnsworth. (1940) Soil structure effects in the growth of sugar beets. *Soil Sci. Soc. Amer. Proc.* **5**:45–48.

Bergman, H. F. (1959) Oxygen deficiency as a cause of disease in plants. *Bot. Rev.* **25**:417–485.

Berry, L. J., and W. E. Norris, Jr. (1949) Studies of onion root respiration: I. Velocity of oxygen consumption in different segments of root at different temperatures as a function of partial pressures of oxygen. *Biochim. Biophys. Acta* **3**:593–606.

Birkle, D. E., J. Letey, L. H. Stolzy, and T. E. Szuszkiewicz. (1964) Measurement of oxygen diffusion rates with the platinum microelectrode. II. Factors influencing the measurement. *Hilgardia* **35**:555–566.

Boicourt, A. W., and R. C. Allen. (1941) Effect of aeration on growth of hybrid tea roses. *Proc. Amer. Soc. Hort. Sci.* **39**:423–425.

Boynton, D. (1941) Soils in relation to fruit-growing in New York. Part XV. Seasonal and soil influences on oxygen and carbon-dioxide levels of New York soils. *Cornell Univ. Agr. Exp. Sta. Bul. 763.*

Bradfield, R., L. P. Batjer, and J. Oskamp. (1934) Soils in relation to fruit growing in New York. Part IV. The significance of the oxidation-reduction potential in evaluating soils for orchard purposes. *New York (Cornell Univ.) Agr. Exp. Sta. Bul. 592.*

Broeshart, H., E. Haunold, and M. Fried. (1965) The effect of water conditions and oxidation-reduction status of rice soils on the availability of soil and fertilizer phosphate. *Plant and Soil* **23**:305–313.

Brown, N. J., E. R. Fountaine, and M. R. Holden. (1965) The oxygen requirement of crop roots and soils under near field conditions. *Jour. Agr. Sci.* **64**:195–203.

Brown, R. (1947) The gaseous exchange between the root and the shoot of the seedling of *Cucurbita pepo. Ann. Bot., N.S.* **11**:417–437.

Bryant, A. E. (1934) Comparison of anatomical and histological differences between roots of barley grown in aerated and in non-aerated culture solutions. *Plant Physiol.* **9**:389–391.

Chiang, P. F., R. K. Lu, I. C. Ku, et al. (1963) The content of iron phosphates in paddy soils of southern China and their significance in the phosphorus nutrition of the rice plant. (Translated title). *Acta Pedol. Sin.* **11**:361–369. (*Soils and Fertilizers* **27**, Abstr. 1308. 1964.)

Cline, R. A., and A. E. Erickson. (1959) The effect of oxygen diffusion rate and applied fertilizer on the growth, yield, and chemical composition of peas. *Soil Sci. Soc. Amer. Proc.* **23**:333–335.

Dehérain, P. P., and E. Demoussy. (1896) Sur l'oxydation de la matière organique du sol. *Ann. Agron.* **22**:305–337.

Domsch, K. H. (1962) Bodenatmung. Sammelbericht über Methoden und Ergebnisse. *Zentbl. Bakt., Parasit. Infektionskr. u. Hygiene,* Abt. II, **116**:33–78.

Erickson, A. E., and D. M. Van Doren. (1961) The relation of plant growth and yield to soil oxygen availability. *Trans. 7th Internat. Cong. Soil Sci.* **3**:428–434.

Evans, D. D. (1965) Gas movement. *Agronomy* **9**:319–330.

Fulton, J. M., and A. E. Erickson. (1964) Relation between soil aeration and ethyl alcohol accumulation in xylem exudate of tomatoes. *Soil Sci. Soc. Amer. Proc.* **28**:610–614.

Fulton, J. M., A. E. Erickson, and N. E. Tolbert. (1964) Distribution of C^{14} among metabolites of flooded and aerobically grown tomato plants. *Agron. Jour.* **56**: 527–529.

Garrels, R. M. (1960) *Mineral Equilibria at Low Temperature and Pressure.* Harper and Row, New York.

Garrett, S. D. (1936) Soil conditions on the take-all disease of wheat. *Ann. Appl. Biol.* **23**:667–699.

Garrett, S. D. (1937) Soil conditions and the take-all disease of wheat. II. The relation between soil reaction and soil aeration. *Ann. Appl. Biol.* **24**:747–751.

Grable, A. R., and R. E. Danielson. (1965) Effect of carbon dioxide, oxygen, and soil moisture suction on germination of corn and soybeans. *Soil Sci. Soc. Amer. Proc.* **29**:12–18.

Greene, H. (1960) Paddy soils and rice production. *Nature* **186**:511–513.

Grineva, G. M. (1964) Accumulation and secretion of alcohols by plant roots supplied with insufficient oxygen. (Translated title) *Dokl. Akad. Nauk* **156**:1225–1228. (*Soils and Fertilizers* **27**, abstract 3440. 1964.)

Hack, H. R. B. (1956) An application of a method of gas microanalysis to the study of soil air. *Soil Sci.* **82**:217–231.

Harris, D. G., and C. H. M. van Bavel. (1957) Root respiration of tobacco, corn, and cotton plants. *Agron. Jour.* **49**:182–184.

Hawkins, J. C. (1962) The effects of cultivation on aeration, drainage, and other soil factors important in plant growth. *Jour. Sci. Food Agr.* **13**:386–391.

Hewitt, L. F. (1950) *Oxidation-Reduction Potentials in Bacteriology and Biochemistry.* Sixth Edition. E. and S. Livingstone, Ltd., Edinburgh.

Jeffery, J. W. O. (1960) Iron and the E_h of waterlogged soils with particular reference to paddy. *Jour. Soil Sci.* **11**:140–148.

Jensen, C. R., J. Letey, and L. H. Stolzy. (1964) Labeled oxygen: transport through growing corn roots. *Science* **144**:550–552.

Klotz, L. J., L. H. Stolzy, and T. A. DeWolfe. (1963) Oxygen requirement of three root-rotting fungi in a liquid medium. *Phytopathology* **53**:302–305.

Klotz, L. J., L. H. Stolzy, T. A. DeWolfe, and T. E. Szuszkiewicz. (1966) Distribution of root-rotting fungi in citrus orchards as affected by soil oxygen supply. *California Agr.* **20,** No. 6:15.

Kramer, P. J. (1956) Physical and physiological aspects of water absorption. *In* W. Ruhland (Ed.), *Handbuch der Pflanzenphysiologie.* Band III. *Pflanze und Wasser,* pp. 124–159. Springer-Verlag, Berlin.

Kristensen, K. J., and H. Enoch. (1964) Soil air composition and oxygen diffusion rate in soil columns at different heights above a water table. *Trans. 8th Internat. Congr. Soil Sci.* (Bucharest) **2:**159–170.

Laing, H. E. (1940) Respiration of the rhizomes of *Nuphar advenum* and other water plants. *Amer. Jour. Bot.* **27:**574–581.

Laing, H. E. (1940a) The composition of the internal atmosphere of *Nuphar advenum* and other water plants. *Amer. Jour. Bot.* **27:**861–868.

Lemon, E. R. (1962) Soil aeration and plant root relations. I. Theory. *Agron. Jour.* **54:**167–170.

Lemon, E. R., and A. E. Erickson. (1952) The measurement of oxygen diffusion in the soil with a platinum microelectrode. *Soil Sci. Soc. Amer. Proc.* **16:**160–163.

Lemon, E. R., and J. Kristensen. (1961) An edaphic expression of soil structure. *Trans. 7th Internat. Congr. Soil Sci.* **1:**232–240.

Lemon, E. R., and C. L. Wiegand. (1962) Soil aeration and plant root relations. II. Root respiration. *Agron. Jour.* **54:**171–175.

Letey, J., and L. H. Stolzy. (1964) Measurement of oxygen diffusion rates with the platinum microelectrode. I. Theory and equipment. *Hilgardia* **35:**545–554.

Letey, J., L. H. Stolzy, N. Valoras, and T. E. Szuszkiewicz. (1962) Influence of oxygen diffusion rate on sunflower growth at various soil and air temperatures. *Agron. Jour.* **54:**316–319.

Linscott, D. L., R. L. Fox, and R. C. Lipps. (1962) Corn root distribution and moisture extraction in relation to nitrogen fertilization and soil properties. *Agron. Jour.* **54:**185–189.

Liu, C. K., and T. J. Yu. (1963) Studies on the electrochemical properties of soils. II. Application of micro-electrodes to the study of soils. (Translated title) *Acta Pedol. Sin.* **11:**160–170. (*Soils and Fertilizers* **26,** Abstract 2952. 1963.)

Machlis, L. (1944) The respiratory gradient in barley roots. *Amer. Jour. Bot.* **31:**281–282.

Marshall, T. J. (1959) The diffusion of gases through porous media. *Jour. Soil Sci.* **10:**79–82.

McPherson, D. C. (1939) Cortical air spaces in the roots of *Zea mays* L. *New Phytol.* **38:**190–202.

Melsted, S. W., T. Kurtz, and R. Bray. (1949) Hydrogen peroxide as an oxygen fertilizer. *Agron. Jour.* **41:**97.

Michaelis, L. (1930) *Oxidation-Reduction Potentials.* J. B. Lippincott Co., Philadelphia, Penn.

Mitsui, S., K. Kumazawa, and T. Hishida. (1960) Dynamic studies on the nutrients uptake by crop plants. XXIII. The growth of rice plant with the use of poor drained soil as affected by the accumulation of volatile organic acids. *Soil and Plant Food* **5:**196.

Newcombe, M. (1960) Some effects of water and anaerobic conditions on *Fusarium oxysporum* f. *cubense* in soil. *Trans. British Mycol. Soc.* **43:**51–59.

Newman, A. S., and A. G. Norman. (1941) The activity of the microflora in various horizons of several soil types. *Soil Sci. Soc. Amer. Proc.* **6**:187–194.

Osugi, S., and K. Kawaguchi. (1938) On the cause of physiological disease with applied ammonium sulfate. (Translated title). *Jour. Sci. Soil Manure* **12**:298–300.

Osugi, S., and K. Kawaguchi. (1939) On the reduction of sulfate in paddy-field soils II. (Translated title). *Jour. Sci. Soil Manure* **13**:1–10.

Ota, Y., and N. Yamada. (1960) Physiological effect of iron on rice plant. (1) Effect of iron in relation to soil reduction. *Proc. Crop Sci. Soc. Japan* **28**:367–370.

Patrick, W. H., Jr. (1961) Nitrate reduction ratio in a submerged soil as affected by redox potential. *Trans. 7th Internat. Congr. Soil Sci.* **2**:494–500.

Pearsall, W. H. (1938) The soil complex in relation to plant communities. II. Characteristic woodland soils. *Jour. Ecol.* **26**:194–209.

Penman, H. L. (1940) Gas and vapour movements in the soil. I. The diffusion of vapours through porous solids. *Jour. Agr. Sci.* **30**:437–462.

Pierce, R. S. (1953) Oxidation-reduction potential and specific conductance of ground water: their influence on natural forest distribution. *Soil Sci. Soc. Amer. Proc.* **17**:61–65.

Piper, C. S. (1931) The availability of manganese in the soil. *Jour. Agr. Sci.* **21**:762–779.

Ponnamperuma, F. N. (1955) The chemistry of submerged soils in relation to the growth and yield of rice. Ph.D. Thesis, Cornell University, Ithaca, N.Y.

Rands, R. D., and E. Dopp. (1938) Influence of certain harmful soil constituents on severity of Pythium root rot of sugarcane. *Jour. Agr. Res.* **56**:53–67.

Rodrigo, D. M. (1963) Redox potential—with special reference to rice culture. *Trop. Agriculturist* **119**:85–100.

Russell, M. B. (1952) Soil aeration and plant growth. *Agronomy* **2**:253–301.

Savant, N. K., and R. Ellis, Jr. (1964) Changes in redox potential and phosphorus availability in submerged soil. *Soil Sci.* **98**:388–394.

Seeley, J. G. (1949) The response of greenhouse roses to various oxygen concentrations in the substratum. *Proc. Amer. Soc. Hort. Sci.* **53**:451–465.

Shapiro, R. E. (1958) Effect of flooding on availability of phosphorus and nitrogen. *Soil Sci.* **85**:190–197.

Shapiro, R. E. (1958a) Effect of organic matter and flooding on availability of soil and synthetic phosphates. *Soil Sci.* **85**:267–272.

Shapiro, R. E., G. S. Taylor, and G. W. Volk. (1956) Soil oxygen contents and ion uptake by corn. *Soil Sci. Soc. Amer. Proc.* **20**:193–197.

Stolwijk, J. A. J., and K. V. Thimann. (1957) On the uptake of carbon dioxide and bicarbonate by roots, and its influence on growth. *Plant Physiol.* **32**:513–520.

Stolzy, L. H., S. D. van Gundy, C. K. Laubanauskas, and T. E. Szuszkiewicz. (1963) Response of *Tylenchulus semipenetran* infected citrus seedlings to soil aeration and temperature. *Soil Sci.* **96**:292–298.

Stolzy, L. H., and J. Letey. (1964) Characterizing soil oxygen conditions with a platinum microelectrode. *Adv. Agron.* **16**:249–279.

Stolzy, L. H., J. Letey, T. E. Szuszkiewicz, and O. R. Lunt. (1961) Root growth and diffusion rates as functions of oxygen concentration. *Soil Sci. Soc. Amer. Proc.* **25**:463–467.

Stotzky, G. (1965) Microbial respiration. *Agron.* **9**:1550–1569.

Sturgis, M. B. (1936) Changes in the oxidation-reduction equilibrium in soils as related to the physical properties of the soil and the growth of rice. *Louisiana Agr. Exp. Sta. Bul. 271.*

Takijima, Y. (1961) Metabolism of organic acids in soils of paddy fields and their inhibitory effects on rice plant growth. Part 5. Growth inhibition of rice seedlings in waterlogged soil and organic acid concentration in the leachate. *Soil Sci. Plant Nutr.* **7**:167.

Takijima, Y. (1964) Studies on organic acids in paddy field soils with reference to their inhibitory effects on the growth of rice plants. Part 1. Growth inhibiting action of organic acids and absorption and decomposition of them by soils. *Soil Sci. Plant Nutr.* **10**:204–211.

Takijima, Y. (1964a) Studies on organic acids in paddy field soils with reference to their inhibitory effects on the growth of rice plants. Part 2. Relations between production of organic acids in waterlogged soils and the root growth inhibition. *Soil Sci. Plant Nutr.* **10**:212–219.

Takijima, Y., and H. Sakuma. (1962) Metabolism of organic acids in soil of peaty paddy field and their inhibitory effects on the growth of rice plant. Part VII. Production of organic acids in water-logged soil treated with green manure (Chinese milk vetch) and its relation to growth inhibition of rice seedlings. *Soil Sci. Plant Nutr.* **8**, No. **2**:41.

Takijima, Y., H. Sakuma, and M. Chiba. (1961) Metabolism of organic acids in soils of paddy fields and their inhibitory effects on rice plant growth. Part 6. Accumulation of organic acids in soil in the presence of sucrose and its relation to growth inhibition of rice seedlings. *Soil Sci. Plant Nutr.* **7**:167–168.

Takijima, Y., M. Shiojima, and K. Konno. (1963) Studies on growth inhibiting substances found in the ill-drained peaty paddy soils. (Part 1) Inhibition of plant growth and soil metabolism. *Soil Sci. Plant Nutr.* **9**:37.

Tanaka, I., K. Nojima, and Y. Uemura. (1961) Influence of the water percolation on the growth of rice plant in rice field. II. The effects of water percolation of various amounts on the growth and yield of rice, and on NH_3-N, H_2S and free CO_2 in soil. *Proc. Crop Sci. Soc. Japan* **29**:392–394.

Taylor, G. S., and J. H. Abrahams. (1953) A diffusion-equilibrium method for obtaining soil gases under field conditions. *Soil Sci. Soc. Amer. Proc.* 17:201–206.

Tyagny-Ryadno, M. G. (1958) Biophysical and chemical analysis of soil aggregates. *Soviet Soil Sci.* **1958**:1378–1387.

Vallance, K. B., and D. A. Coult. (1951) Observations on the gaseous exchanges which take place between *Menyanthes trifoliata* L. and its environment. *Jour. Exp. Bot.* **2**:212–222.

Vilain, M. (1963) L'aération du sol. Mise au point bibliographique. *Ann. Agron.* **14**:967–998.

Vomocil, J. A. (1965) Porosity. *Agronomy* **9**:299–314.

Vomocil, J. A., and W. J. Flocker. (1961) Effect of soil compaction on storage and movement of soil air and water. *Trans. Amer. Soc. Agr. Eng.* **4**:242–246.

Willey, C. R., and C. B. Tanner. (1963) Membrane-covered electrode for measurement of oxygen concentration in soil. *Soil Sci. Soc. Amer. Proc.* **27**:511–515.

Yamazaki, K. (1960) Studies on the pedogenetic classification of paddy soils in Japan. *Toyama Agr. Exp. Sta. Spec. Rpt.* **1**:1–105. (Quoted by Aomine, 1962).

4 *Exchangeable Bases*

The exchangeable cations of soils are the portions of the cations associated with soil solids that are subject to interchange with cations in solution under conditions of little or no decomposition of the remainder of the solids. The process of interchange of cations in solution with those in exchangeable form is known as cation exchange.

The principal exchangeable cations in soils from the standpoint of quantity are calcium, magnesium, potassium, sodium, aluminum, and hydrogen. Emphasis in this chapter will be on the first four of these, which are called exchangeable bases. Succeeding chapters will deal with exchangeable sodium, aluminum, and hydrogen. Ferric iron does not seem to occur in appreciable quantity as an exchangeable cation in agricultural soils, but ferrous iron may be an important exchangeable cation when reducing conditions prevail.

The exchangeable cations are chemically the most reactive part of the soil solids. Cations in solution remain nearly at equilibrium with those in exchangeable form. Therefore chemical reactions involving cations in the soil solution involve also the cations in exchangeable form. The ions in exchangeable form are usually far more numerous than those in the soil solution and constitute a source from which the bases in the soil solution are replenished. These properties are of considerable significance to the growth of plants in soils because ions occurring in soils as exchangeable bases serve as nutrients.

Distinction between Exchangeable and Nonexchangeable Bases

The definition of exchangeable cations implies that a portion of each of the cations is not in exchangeable form and hence may be termed "nonexchangeable." The distinction between exchangeable and nonexchangeable forms of a cation is not absolute and in some instances may depend to a considerable extent on the method of measurement, as will be illustrated in the paragraphs that follow. Nevertheless, both chemical and biological evidence shows that there is a significant distinction in behavior between the two classes.

The first chemical evidence to be considered is provided by observations that may be made in connection with determination of exchangeable cations. Appropriate analyses will show that (1) leaching a sample of soil with a salt solution causes removal of a much larger amount of cations than does comparable leaching with water, (2) the amount of soil cations removed by the salt solution in excess of the amount removed by water is approximately chemically equivalent to the cations retained by the soil from the salt solution, and (3) the removal of cations from the soil by the salt solution takes place rapidly at first but soon almost ceases even though relatively large quantities of the individual cations can still be detected in the soil by fusion analysis.

A second type of experiment may be conducted on the basis of the law of mass action, which states that, at equilibrium and at constant temperature, the product of the active masses on one side of a chemical equation when divided by the product of the active masses on the other side of the chemical equation is a constant, regardless of the quantities of each substance present at the beginning of the reaction. If a solution containing radioactive calcium is added to soil containing nonradioactive calcium, the following reaction may be written:

$$\mathrm{Ca}^{45}_{\mathrm{soln.}} + \mathrm{Ca}^{40}_{\mathrm{soil}} = \mathrm{Ca}^{45}_{\mathrm{soil}} + \mathrm{Ca}^{40}_{\mathrm{soln.}}$$

where Ca^{45} and Ca^{40} are the radioactive and nonradioactive forms. By putting the equation in the form of the mass-law expression.

$$\frac{(\gamma_{\mathrm{Ca}^{45}_{\mathrm{soil}}} \cdot c_{\mathrm{Ca}^{45}_{\mathrm{soil}}})(\gamma_{\mathrm{Ca}^{40}_{\mathrm{soln.}}} \cdot c_{\mathrm{Ca}^{40}_{\mathrm{soln.}}})}{(\gamma_{\mathrm{Ca}^{45}_{\mathrm{soln.}}} \cdot c_{\mathrm{Ca}^{45}_{\mathrm{soln.}}})(\gamma_{\mathrm{Ca}^{40}_{\mathrm{soil}}} \cdot c_{\mathrm{Ca}^{40}_{\mathrm{soil}}})} = K,$$

where the γ's are activity coefficients, the c's are concentrations, and K is the equilibrium constant. If the mixture of radioactive and nonradioactive calcium is uniform throughout the solution, the activity coefficients for the two forms of calcium in the solution may be expected to be the same. Similarly, if the mixture of radioactive and nonradioactive calcium in the exchangeable form is uniform throughout, these activity coefficients may be expected to be the same. The equation then simplifies to

$$\frac{(c_{\mathrm{Ca}^{45}_{\mathrm{soil}}})(c_{\mathrm{Ca}^{40}_{\mathrm{soln.}}})}{(c_{\mathrm{Ca}^{45}_{\mathrm{soln.}}})(c_{\mathrm{Ca}^{40}_{\mathrm{soil}}})} = K.$$

Borland and Reitemeier (1950) conducted experiments of this type with several clays. Results obtained with a clay extracted from a soil in Missouri may serve as an example. Table 4.1 gives the quantities of radioactive and nonradioactive calcium in solution and in the clay at equilibrium

Table 4.1. Values of Radioactive and Nonradioactive Calcium in Solution and in Clay at Equilibrium in Suspensions of a Soil Clay in Water and the Equilibrium Constant Calculated for the Reaction of Radioactive Calcium in Solution with Nonradioactive Calcium in the Clay (Borland and Reitemeier, 1950)

Reactants, mg		Products, mg		
$Ca^{45}_{soln.}$	Ca^{40}_{clay}	$Ca^{40}_{soln.}$	Ca^{45}_{clay}	K
2.10×10^{-4}	12.1	1.34	1.90×10^{-3}	1.00
9.20×10^{-4}	12.1	9.56	1.16×10^{-3}	1.00
1.96×10^{-3}	12.1	146.0	1.62×10^{-4}	1.00

for three different quantities of calcium in solution. The equilibrium constant for the reaction is unity in each case. This result verifies that calcium in solution did exchange with a quantity of calcium in the clay in accordance with the law of mass action and also that exchange between calcium in the soil solution and exchangeable form occurs continuously, a phenomenon that previously had been inferred only.

The example just given must be recognized as a special case of cation exchange. The equilibrium constant of unity indicates that the radioactive and nonradioactive calcium ions behaved so nearly alike that no difference was distinguishable. Where exchange of different ions is concerned, the situation may be far more complex, as will be explained subsequently.

Radioactive calcium may be used in another way to verify the reality of the distinction between exchangeable and nonexchangeable forms. If a solution containing a quantity of tagged calcium is allowed to equilibrate with a soil, the added calcium will start immediately to exchange with the common form of the cation present initially in the soil. Subsequent analysis of the solution therefore will show that the ratio of radioactive calcium to total calcium is lower than the ratio in the tagged source; furthermore, some of the radioactive calcium will have disappeared from the solution. When equilibrium has been attained (this ordinarily takes no more than 2 hours), the radioactive calcium will be distributed throughout the mixture of added calcium and the portion of soil calcium with which it has equilibrated. If the equilibrium constant of the reaction is unity, as was demonstrated for a soil clay in Table

4.1, the ratio of total calcium to radioactive calcium will be the same in all finite parts of the equilibrium mixture. Hence for a finite sample of the equilibrium mixture such as a quantity of the solution,

$$\frac{Ca^{40}_{sample} + Ca^{45}_{sample}}{Ca^{45}_{sample}} = \frac{Ca^{40}_{added} + Ca^{45}_{added} + Ca^{40}_{initial\ soil}}{Ca^{45}_{added}}.$$

Because all the variables are known except $Ca_{initial\ soil}$, the numerical value of the latter may be calculated by substituting the known values and solving the expression for $Ca_{initial\ soil}$:

$$Ca_{initial\ soil} = \frac{(Ca^{40}_{sample} + Ca^{45}_{sample})(Ca^{45}_{added}) - (Ca^{45}_{sample})(Ca^{40}_{added} + Ca^{45}_{added})}{Ca^{45}_{sample}}.$$

The value for soil calcium obtained in this way may be compared with the value for exchangeable soil calcium obtained by the conventional technique of leaching a sample of soil with neutral, normal ammonium acetate. Newbould (1963) made measurements of this kind with different soils, and some of his results are shown in Table 4.2. The table shows that the quantities of soil calcium that equilibrated with added radioactive calcium were about the same as those extracted with normal ammonium acetate; but the total calcium was much higher than either of these two figures, thus demonstrating the presence of considerable nonexchangeable calcium and verifying the reality of the distinction between exchangeable and nonexchangeable calcium. Blume and Smith (1954) made measurements of calcium in a number of soils by the radioisotope and ammonium acetate methods and found that the quantities of soil calcium thus measured were almost the same with most of the soils tested. In a few instances, the quantity of soil calcium found

Table 4.2. Total Calcium and Calcium Found by Equilibration with Radioactive Calcium and Extraction with Ammonium Acetate in Two Soils of the British Isles (Newbould, 1963)

	Calcium per 100 g of soil, m.e.		
Soil	By Equilibration with Ca^{45}	By Extraction with NH_4OAc	Total by Fusion with Alkali
A	10	11	27
B	15	15	70

by equilibration with radioactive calcium substantially exceeded that extracted with ammonium acetate. Analogous work by Fisher (1963) with magnesium, potassium, and sodium showed that the quantities extracted with ammonium acetate were somewhat greater than the quantities that equilibrated with the respective isotopes. Failure to obtain complete agreement, within experimental error, demonstrates that the apportionment of a cation between "exchangeable" and "nonexchangeable" forms may differ somewhat from one method of measurement to another.

Biological evidence is supplied by experiments with radioactive calcium that parallel the work described in connection with Table 4.2. It was pointed out there that after an added isotope of a soil cation has reached equilibrium with the cation in the soil, a finite sample from any part of the mixture will have the same ratio of radioactive to total calcium as will the total mixture. Instead of withdrawing a sample of solution for analysis, one may grow a plant on the soil and allow the plant to withdraw a sample of calcium. The plant can then be analyzed for its content of radioactive and total calcium. If the quantity of calcium absorbed by the plant is large in relation to the calcium present in the seed, the calcium from the seed source may be neglected; and the equation described previously for calculating the quantity of soil calcium ($Ca^{40}_{initial\ soil}$) may be applied. If the plant absorbs calcium only from the mixture of exchangeable calcium and added calcium, the proportionate quantities of the plant calcium derived from the soil and from the tagged source will be the same as the proportionate quantities of exchangeable calcium and added tagged calcium present in the soil.

An experiment of this kind was performed by Davis et al. (1953) in Tennessee. In this experiment, 1.77 g of Ca^{45}-labeled calcium were applied to a quantity of Hartsells soil containing 1.92 g of exchangeable calcium and also to a quantity of Claiborne soil containing 7.19 g of exchangeable calcium, as found by the ammonium acetate method. Hence the percentage of soil exchangeable calcium in the total of exchangeable and added tagged calcium was $(1.92)(100)/(1.92 + 1.77) = 52$ in the Hartsells soil and $(7.19)(100)/(7.19 + 1.77) = 80$ in the Claiborne soil. Analysis of ryegrass grown on the soils showed that the percentages of the calcium derived by the plants from the soil were 54 in the Hartsells soil and 80 in the Claiborne soil. The close agreement between results obtained by plant and soil analyses indicates that the calcium absorbed by the plants came from the mixture of exchangeable calcium and added calcium. If the plants had absorbed nonexchangeable calcium, the values derived from the plant analyses would have exceeded those calculated from the quantities of exchangeable and tagged calcium.

Additional experimental work of the same kind was done by Newbould (1963) with soils from the British Isles. In Newbould's work, the quantity of soil calcium estimated from measurements of radioactive and total calcium in the plants exceeded the quantity of soil calcium found by ammonium acetate. For example, the values calculated from measurements on cabbage for soils *A* and *B* in Table 4.2 were 14 and 18 mg equivalents of calcium per 100 g as compared with values of 11 and 15 mg equivalents of calcium per 100 g obtained by the ammonium acetate method. These findings, which indicate that the plants derived calcium in part from nonexchangeable forms, anticipate the topic on renewal of exchangeable bases from nonexchangeable forms, to be discussed subsequently. At this point, however, it may be mentioned that the experimental work of Davis et al. (1953) was performed with strongly weathered soils in which release of calcium from nonexchangeable forms would be expected to be low. The experiments by Newbould (1963) were on moderately weathered soils in which release of calcium from nonexchangeable forms during cropping would be expected to be more significant.

A second kind of biological evidence may be obtained by a statistical examination of data on uptake of cations by plants growing on soils. The general procedure is to grow plants on each of a group of soils, measure the content of a particular cation in the plants and the quantity of the same cation present in exchangeable and nonexchangeable forms in each of the soils, express the content of the cation in the plant as a function of the several fractions of the cation measured in the soils, and find by statistical techniques the best-fitting coefficients for each of the variables represented by the soil fractions of the cation. If a functional relationship of linear form is employed, and if the quantities of the different fractions are expressed in the same units (e.g., milligram equivalents per 100 g), the numerical values of the coefficients provide an estimate of the relative value of equal quantities of the various fractions of the cation in the soil as a source of supply for the plants. Several experiments of this kind have been performed to obtain information on potassium.

In the original experiment (Pratt, 1951), alfalfa was grown in the greenhouse on equal quantities of thirteen soils. Nine cuttings were made over a period of a year, and the total potassium absorbed by the plants during this time was determined. On samples of each of the original soils before cropping, analyses were made for exchangeable potassium, nonexchangeable potassium released from the soil to a hydrogen-saturated cation-exchange resin during incubation for 2 months, and nonexchangeable potassium not released to the exchange resin. These measurements thus divided the total soil potassium into three mutually exclusive fractions differing in resistance to extraction. The best-fitting linear coefficients of the three fractions were found to be 0.713, 0.197, and 0.0022, in the order named, which means that the estimated comparative uptakes of potassium associated with equal quantities of each fraction in the soil were as follows: exchangeable potassium = 1, nonexchangeable potassium released from the soil to the hydrogen-saturated cation-exchange resin = 0.28, and nonexchangeable potassium not released to the hydrogen-saturated cation-exchange resin = 0.003. The effectiveness of unit quantity of exchangeable potassium as a source of potassium for the alfalfa therefore was considerably greater than that of either of the two fractions of nonexchangeable potassium.

The chemical evidence thus indicates that the dividing line between exchangeable and nonexchangeable forms of a soil cation may be somewhat arbitrary but that there is nevertheless a real distinction in reactivity. The biological evidence indicates that the distinction made by chemical methods is significant as far as plant behavior is concerned.

Representation and Determination of Cation-Exchange Properties of Soils

The results of measurements of cation-exchange quantities in soils are usually expressed in terms of milligram equivalents (abbreviated m.e.) per 100 g of soil. This convention has the advantage that the quantities of different exchangeable cations are additive if expressed on this basis; also, the numbers involved are neither exceedingly large nor exceedingly small, usually lying in the range from 0.1 to 40.

The terms *exchangeable bases* and *total exchangeable bases* refer to the sum of the exchangeable bases (calcium, magnesium, potassium, and sodium) in milligram equivalents per 100 g of soil. The cation-exchange capacity (also called the total exchange capacity and the base-exchange capacity) is the total number of exchange positions or the total amount of exchangeable cations expressed as milligram equivalents per 100 g of soil. The percentage base saturation is the percentage of the cation-exchange capacity occupied by exchangeable bases.

Determination of exchangeable cations in soils generally involves two steps, extraction and analysis. Analysis of extracts is not without problems, but the principal problems are connected with extraction. The objective of extraction is to remove all the exchangeable cations without at the same time removing some nonexchangeable cations. Accordingly, care must be used to avoid selection of an extracting solution that causes some decomposition of the soil and release of nonexchangeable forms.

Due precautions must be taken also to avoid bias in the answer for exchangeable cations from inclusion of cations of soluble salts extracted with the exchangeable cations or from loss of exchangeable bases by hydrolysis.

Although the cation-exchange capacity may be determined by summing the numbers of exchange positions occupied by individual cations, as found by analysis, this method requires a separate analysis for each species of exchangeable cation. The method more commonly employed is treating soil with a salt solution to remove the exchangeable cations and concurrently saturating the exchange positions with the cation supplied by the salt solution. The excess of saturating salt is removed; and the quantity of the saturating cation retained in exchangeable form is then determined, usually by displacement and analysis.

For many years, a solution of neutral, 1-normal ammonium acetate has been used more consistently than any other for displacing exchangeable cations and for saturating exchange positions in the determination of exchangeable cations and cation-exchange capacity. This reagent has a number of advantages, as outlined by Schollenberger and Simon (1945). Moreover, it has some disadvantages for use with certain soils and for certain purposes. Aside from ammonium acetate, the reagents used most extensively have been 1-normal sodium acetate solution with a *p*H value of 8.2 for measuring the cation-exchange capacity of alkaline soils and 0.5-normal calcium acetate at *p*H 7 or barium chloride-triethanolamine solution at *p*H 8 or 8.2 for measuring exchangeable hydrogen plus aluminum in acid soils. Methods have been described by Chapman (1965, 1965a), Peech (1965), Heald (1965), Pratt (1965, 1965a), and others.

Extensive experimental work indicates that no two reagents for extracting exchangeable cations or for saturating the soil in the determination of cation-exchange capacity will give the same results with all soils. For the most part, the discrepancies are quantitatively minor; and some are evident only with particular cations or with particular cations in particular soils. One of the principal inconsistencies has been found with the exchangeable hydrogen, exchangeable aluminum complex. This matter will be discussed in Chapter 5 on soil acidity.

Exchangeable Bases in Soils

Relative Proportions

As a general rule, the relative quantities as milligram equivalents of the exchangeable bases in soils follow the order calcium > magnesium > potassium. The content of sodium may be either larger or smaller

Table 4.3. Proportions of Exchangeable Bases in Soils

		Milligram Equivalents of Indicated Base per 100 m.e. of Total Exchangeable Bases			
Soils Analyzed	Author	Ca	Mg	K	Na
Thirty-four acid soils (Finland)[1]	Marttila (1965)	79	16	4	2
Nine acid soils (U.S.)	C. A. Bower (unpublished) and Baver (1928)	69	25	3	3
Six neutral or slightly alkaline soils	Kelley and Brown (1924)	62	26	4	8
Two sodic soils	U.S. Salinity Laboratory Staff (1954)	33	36	2	29

[1] Omitting one soil with 42 m.e. of sodium per 100 m.e. of exchangeable bases.

than that of potassium. Table 4.3 contains a summary of results reported by several investigators for different classes of soils.

Soils of humid regions commonly contain substantial quantities of exchangeable aluminum, with some hydrogen, so that the degree of base saturation is less than 100%. Soils of humid regions seldom contain much exchangeable sodium except in areas affected by sea water. The B and C horizons, however, may contain relatively more magnesium than is indicated in Table 4.3, which lists analyses of surface samples. This tendency is particularly pronounced in highly weathered soils; the content of exchangeable magnesium may exceed that of calcium in the B and C horizons of such soils.

Soils of arid regions are usually base-saturated; in fact, special procedures often are needed to distinguish between exchangeable bases and the substantial amounts of bases present as soluble salts or calcium carbonate. Strongly alkaline soils are characterized by a relatively high ratio of monovalent bases (particularly sodium) to divalent bases. The two "sodic" soils in Table 4.3 are relatively high in exchangeable magnesium as well as sodium. Both samples are from western United States.

Renewal from Nonexchangeable Forms

If the exchangeable calcium, magnesium, and potassium in soils represented the total supply of the respective bases, deficiencies of these bases for plant growth in many soils would appear within a period of a few years. This is particularly true of potassium, which is found in exchangeable form in relatively small amounts and is used by plants in relatively large amounts.

Analyses of mineral soils indicate that exchangeable bases seldom constitute the bulk of the total supply and often represent only a small fraction of the total. The data obtained by Bear, Prince, and Malcolm (1945) on twenty soils of New Jersey may be cited as an illustration. A summary of their results is given in Table 4.4. Another example may be taken from the work of Anderson, Keyes, and Cromer (1942). These investigators found that the exchangeable bases calculated as a percentage of the total amounted to 47% with calcium, 16% with magnesium, 2% with potassium, and 2% with sodium as an average of three soils, one each from South Dakota, Iowa, and Indiana. In peat soils that contain little mineral matter, the situation may be rather different. Gore and Allen (1956) found that the major portions of the calcium, magnesium, potassium, and sodium in certain peat soils in England are in exchangeable form.

Although bases in nonexchangeable form are not thought to be available to plants to an important extent in that form, their gradual release replenishes the supply of exchangeable and soluble bases. Primary minerals, such as feldspars, amphiboles, and micas, form one source of nonexchangeable bases. Experiments by Graham (1940), McClelland (1951), and others have demonstrated that appreciable release of bases from such minerals to acid clays will take place if the minerals are ground to particle sizes in which they occur in soil. The nature of the cations released and the rate of release differ among minerals.

The silicate clays form a second source of nonexchangeable bases in soils. Illite and montmorillonite contain nonexchangeable magnesium, and illite contains nonexchangeable potassium. Weathering causes re-

Table 4.4. Average Content of Exchangeable, Nonexchangeable, and Total Calcium, Magnesium, and Potassium in Twenty Soils of New Jersey (Bear, Prince, and Malcolm, 1945)

	Content of Cation per 100 g of Soil, m.e.			
Cation	Exchangeable	Non-exchangeable	Total	Exchangeable as Percentage of Total
Calcium	4.0	13.3	17.3	23.1
Magnesium	1.6	40.7	42.3	3.8
Potassium	0.2	46.3	46.5	0.4

lease of magnesium and potassium from these forms. Kaolinite does not contain either magnesium or potassium in nonexchangeable form in appreciable amounts.

Calcium is not known to occur as a structural component of the silicate clays and hence presumably must be liberated from other minerals. Extensive practical use is made of limestone to supplement soil minerals as a source of nonexchangeable calcium (and magnesium). Finely ground limestone weathers rapidly in acid soils, and the originally nonexchangeable calcium is transferred to the soil as exchangeable calcium. Calcium carbonate is a natural constituent in some soils; and, in these, the rate of release or dissolution is controlled by the rate of production of acidity. Except where the calcium carbonate occurs in relatively large, isolated particles, its environment-controlled dissolution maintains soils continuously in a base-saturated condition.

The rate of release of bases from nonexchangeable forms in uniform parent material may be expected to increase with the intensity of weathering. Long-continued, intense weathering, however, depletes the supply of nonexchangeable bases, so that the rate of release is lower than it is from initially comparable soil material that has been maintained under conditions of moderate or weak weathering. If a soil becomes acid, the liberation of bases from nonexchangeable forms evidently is not proceeding rapidly enough to keep the exchange positions saturated with bases. The percentage base saturation of a natural soil is thus an index of the balance between loss of exchangeable bases from the soil and their accumulation from nonexchangeable forms. The data in Table 4.5 on cation-exchange properties of soils from different rainfall zones in Mauritius may be noted in this connection. Mauritius is a tropical island on which the soil parent material is of volcanic origin.

Table 4.5. Annual Rainfall and Cation-Exchange Properties of Soils at Different Locations in Mauritius (Craig and Halais, 1934)

Annual Rainfall, cm	Cation-Exchange Capacity per 100 g, m.e.	Exchangeable Bases per 100 g, m.e.	Base Saturation, %
64–127	29.5	24.0	81
127–191	26.2	15.9	61
191–254	22.9	8.2	36
254–318	22.3	5.4	24
318–381	20.6	4.0	19

Investigations of the release of specific bases from nonexchangeable forms in soils have been confined mainly to potassium. Release of potassium will be considered in some detail in Chapter 9. In experiments on calcium done in the greenhouse with various soils, some containing calcium carbonate, Newbould (1963) verified that release of calcium from nonexchangeable form may occur during cropping and that the rate of release may increase with the intensity of cropping. In an experiment carried out by Davis et al. (1953), no release of calcium from nonexchangeable form was measured where plants were grown on soils in the greenhouse. Some of the numerical data from these experiments were given in the section on the distinction between exchangeable and nonexchangeable bases. Salmon and Arnold (1963) investigated the behavior of magnesium in various soils of Great Britain and found that the apparent release from nonexchangeable form during a period of 11 months of intensive cropping in the greenhouse was equivalent to about one-fifth of the initial content of exchangeable magnesium.

Source of Cation-Exchange Properties

The cation-exchange positions in soils are heterogeneous in nature. Some are associated with mineral particles and some with organic matter. A number of different kinds of sites are recognized in each class. Because of the variability of the mineralogical composition and organic matter content of soils, the make-up of the cation-exchange capacity may differ from one soil to the next. Such differences may lead not only to some ambiguity in measurements of exchange properties but also to significant effects on plant nutrition.

Mineral Matter

Particle-Size Effects. The cation-exchange capacity of the mineral fraction of soils is derived from the dissociation of cations from mineral surfaces and hence increases with the surface area of a given kind exposed in the soil. Surface area and cation-exchange capacity per unit weight increase rapidly with a decrease in particle size, as indicated in Table 4.6 for seven particle-size fractions of silt and clay separated from a given soil. As would be anticipated from the trend shown in the table, the cation-exchange capacities of sand fractions are small. Also important, but not distinguishable in data such as those in Table 4.6, is the nature of the surface. Quartz, for example, does not have appreciable cation-exchange capacity in either coarse or fine particle sizes. Soils contain a mixture of various minerals that differ chemically and hence present different kinds of surfaces.

One of the problems in particle-size studies of the type shown in Table 4.6 is that of obtaining a clean separation of size fractions or of defining what is meant by a size fraction. Silt and sand fractions may include particles composed in some degree of minerals such as vermiculite that are customarily classified as clay minerals, and these inclusions may confer a relatively high exchange capacity on the fraction in which they occur. The question then is whether one should define the particles as silt or sand and say that the silt or sand has a high exchange capacity or change the method of dispersion to cause these high-exchange-capacity inclusions to disperse and thus appear predominantly in the clay fraction. This question was considered by McAleese and Mitchell (1958), who found cation-exchange capacities of silt fractions of 25 to 60 m.e./100 g and values that decreased with improved dispersion.

Because of the marked increase in cation-exchange capacity that occurs with decrease in particle size, most of the cation-exchange properties of the inorganic portion of soils are characteristically concentrated in the clay fraction. The percentage content of clay is therefore a major factor in determining the cation-exchange capacity of a soil.

Nature of Clay Minerals. The clay fractions of soils contain a number of different minerals, the cation-exchange properties of which may differ considerably. According to Grim (1953, p. 129), the cation-exchange

Table 4.6. Cation-Exchange Capacity and Specific Surface of Mineral Soil Particles Separated from a Clay Soil in Missouri (Whitt and Baver, 1930)

Soil Separate	Equivalent Diameter of Particles, mm	Calculated Surface Area per g,[1] cm²	Cation-Exchange Capacity per 100 g, m.e.
Silt	0.02 –0.005	1,800	3
	0.005 –0.002	6,200	7
Coarse clay	0.002 –0.001	16,000	22
	0.001 –0.0005	30,000	35
	0.0005–0.0001	74,000	52
Fine clay	0.0001–0.00005	320,000	56
	<0.00005	920,000	63

[1] Calculated on the basis of the average size of particles of each group.

capacity per 100 g of different clays is approximately 3 to 15 m.e. for kaolinite, 10 to 40 m.e. for illite and chlorite, 80 to 150 m.e. for montmorillonite, and 100 to 150 m.e. for vermiculite. Thus a particle-size analysis to determine the clay content will not provide much information about the comparative cation-exchange capacities of the clays unless the clays contain similar minerals in similar proportions or the proportions of the mineral components are known. The occurrence of clay minerals in soils is discussed in the section on clay mineralogy in Chapter 1. For information on the nature and properties of clays beyond that given in the following paragraphs, the books by Grim (1953) and Marshall (1964), the review paper by Rich and Thomas (1960), and the symposium edited by Rich and Kunze (1964) may be consulted.

The crystalline silicate minerals of the clay fraction of soils have a layer-type structure in which individual molecules are not distinct from each other but are joined in a repeating pattern that occurs in what

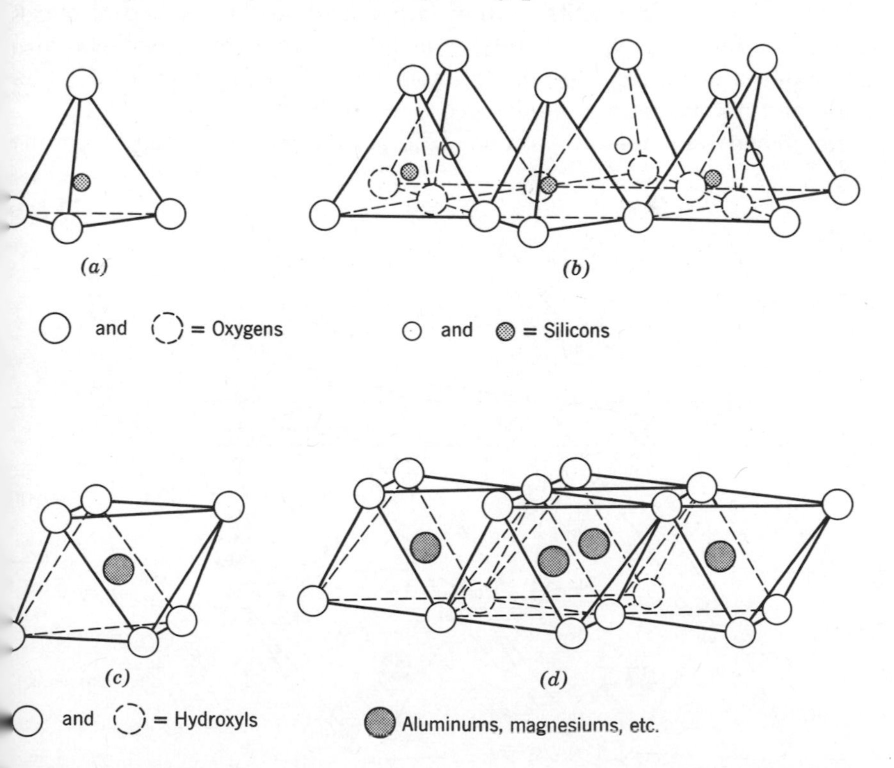

Figure 4.1. Diagrammatic sketch of basic structural units in clays. (*a*) Single silica tetrahedron. (*b*) Sheet structure of silica tetrahedrons arranged in a hexagonal network. (*c*) Single octahedral unit. (*d*) Sheet structure of octahedral units. (After Grim, 1953)

is essentially a two-dimensional layer. The crystals are composed of a number of individual layers superimposed in much the same way as the layers in crystals of mica. The crystalline silicate clay minerals commonly found in soils are of two general structural types.

The first type, frequently termed the 1-to-1 type, has in each layer a silica tetrahedral sheet and an alumina octahedral sheet tied together by oxygen atoms that are shared between the two. The silica tetrahedral sheet receives its name from the fact that each silicon atom is surrounded by four oxygen atoms to produce what may be idealized as a figure with four plane sides (a tetrahedron). The tetrahedral units may be seen in Fig. 4.1. Individual tetrahedral units are joined in a sheet in a characteristic way, as shown in the figure. The alumina octahedral sheet receives its name from the fact that each aluminum atom is surrounded by six oxygen and hydroxyl groups to give an eight-sided figure (an octahedron). Figure 4.1 shows a single octahedral unit and an assembly of octahedral units as they are joined in an octahedral sheet. There are few, if any, substitutions of other cations for the silicon and aluminum atoms in the tetrahedral and octahedral layers of 1-to-1 minerals, and the structures are balanced electrically.

As shown for kaolinite in Fig. 4.2, the surface of the layer on the

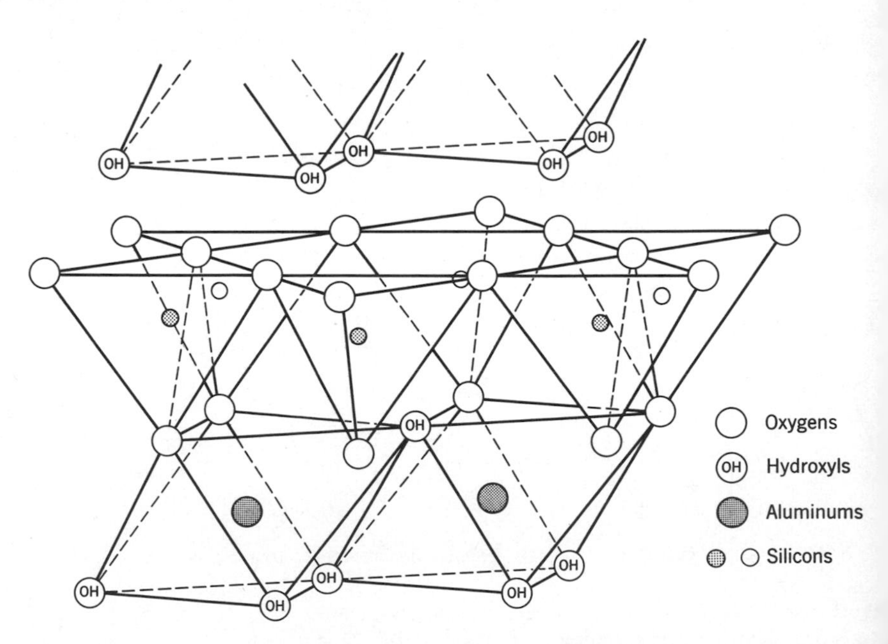

Figure 4.2. Structure of kaolinite. The linked silica tetrahedra in this structure are inverted from their orientation in Fig. 4.1. (After Grim, 1953)

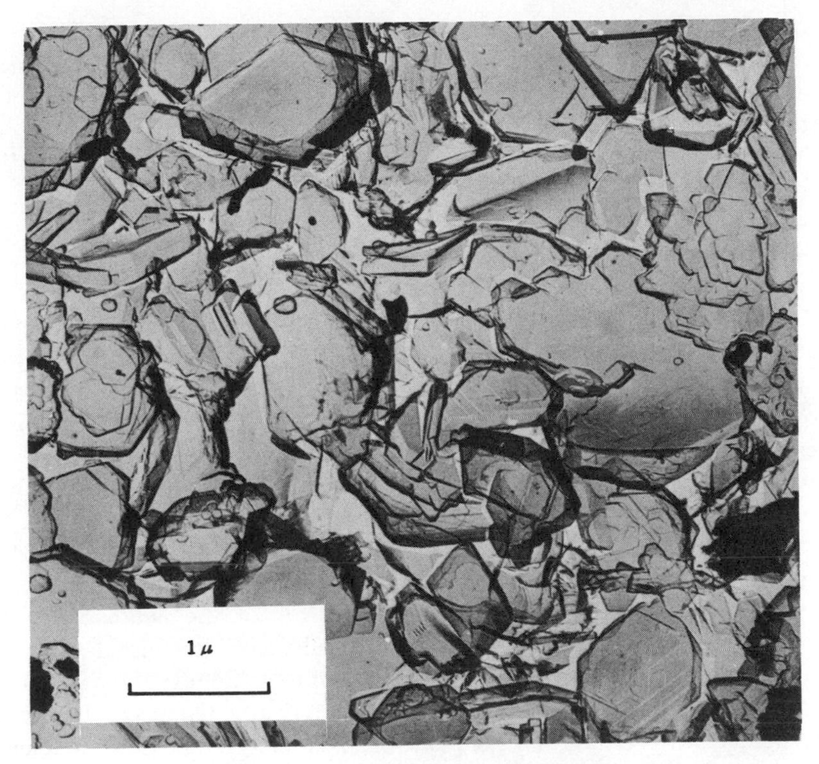

Figure 4.3. Electron micrograph of kaolinite. (Towe, 1961)

alumina side is composed of hydroxyls; and the surface on the silica side is composed of oxygens. The crystals are composed of superimposed layers with the hydroxyl and oxygen surfaces adjacent to each other. As seen with the aid of an electron microscope, crystals of kaolinite are platelike in shape with hexagonal outlines (Fig. 4.3). The layers show little tendency to separate from each other, probably because of the attraction between the adjacent oxygen and hydroxyl layers; however, the layers can be made to separate some distance if the clay is dry-ground with potassium acetate and then allowed to absorb water vapor from the atmosphere (Andrew et al., 1960).

The cation-exchange properties of kaolinite are thought to originate from ionization of hydrogen from hydroxyl groups on the outer basal plane surfaces of the particles and from the hydroxyl groups formed at the edges of the particles where the mineral structure is discontinued. The view that some of the surface hydrogens are subject to exchange with other cations is in agreement with the observation that the surface hydrogen of kaolinite comes to exchange-equilibrium with the deuterium of deuterated water (McAuliffe et al., 1948). The evidence from deuterium is equivocal, however, because the exchange could have been between OD^- and OH^- as well as between D^+ and H^+.

The molecular structure of halloysite is similar to that of kaolinite; however, successive layers are shifted in a somewhat random fashion, and a single layer of water molecules may occur between the silicate layers. The layers may be rolled into tubes, so that the appearance of particles is rather different from that of kaolinite.

In the second type of structure, frequently termed the 2-to-1 type, each layer is composed of two silica tetrahedral sheets bonded to a central octahedral sheet. The three sheets are held together by shared oxygen atoms. Individual layers are stacked to form the crystalline particles. In this type of structure, a substantial proportion of the silicon atoms in the tetrahedral layers may be replaced by aluminum. Moreover, the central atoms in the octahedral layer may be aluminum, iron, or magnesium, alone or in combination. If all the octahedral units contain a central divalent cation (principally Mg^{++} but also Fe^{++}), the structure is said to be trioctahedral; if only two-thirds of the octahedral units contain a central trivalent cation (principally Al^{+++} but also Fe^{+++}), the structure is said to be dioctahedral. In some minerals, the population of cations is such that the structure is electrically balanced. In others, the net negative charge exceeds the net positive charge. The deficiency of positive charges is then satisfied by other cations located outside the layers. For example, replacement of one tetravalent silicon in a silica tetrahedron with one trivalent aluminum would result in deficiency of one positive charge, and electrical neutrality would be preserved by the presence of one monovalent cation external to the tetrahedron. Charge-balancing cations held externally to structural layers are partly or completely exchangeable, according to circumstances that will be described subsequently. As far as is known, replacement of central atoms in tetrahedral and octahedral units is not a matter of ordinary ion exchange.

Several different clay minerals are recognized within the second structural category. In illite, which is similar to muscovite (Fig. 4.4) and biotite, about 15% of the silicon positions are occupied by aluminum. Aluminum is the dominant octahedral cation, but magnesium and iron are present also. The mineral may be dioctahedral or trioctahedral. The balancing cation is chiefly potassium, which fits in an open space surrounded by oxygen atoms in adjacent layers. The replacement of silicon by aluminum in the tetrahedral sheets provides an excess negative charge originating near the surface of the layer instead of at the center. More-

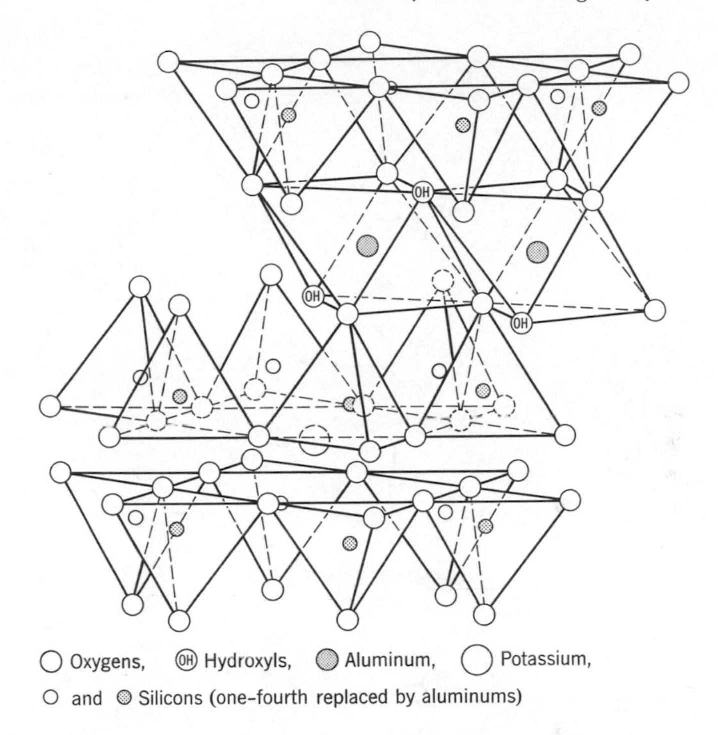

Figure 4.4. Structure of muscovite. (After Grim, 1953)

over, the number of charges requiring balancing cations per unit area of interlayer surface is relatively large. Consequently, the clay layers are held together firmly, and water molecules do not enter. The potassium ions bond the surfaces together, being attracted by negative charges from the layers on both sides. The stability of potassium ions in these locations is thought to result partly from their valence (one) but especially from their size. Potassium is of the proper size to fit snugly into openings among surrounding oxygen atoms that occur adjacent to each other in successive layers. The potassium ions are exchangeable where they occur on the surface of the particles, but those lying between the layers are not exchangeable except perhaps for a few near the edges. This phenomenon is of considerable importance in relation to potassium availability in soils and will be considered in more detail in Chapter 9. Like kaolinite, illite is thought to have exchange positions at the edges of the particles where the layers are discontinued. Illite particles are characteristically smaller than those of kaolinite.

Another important member of the second structural group is mont-

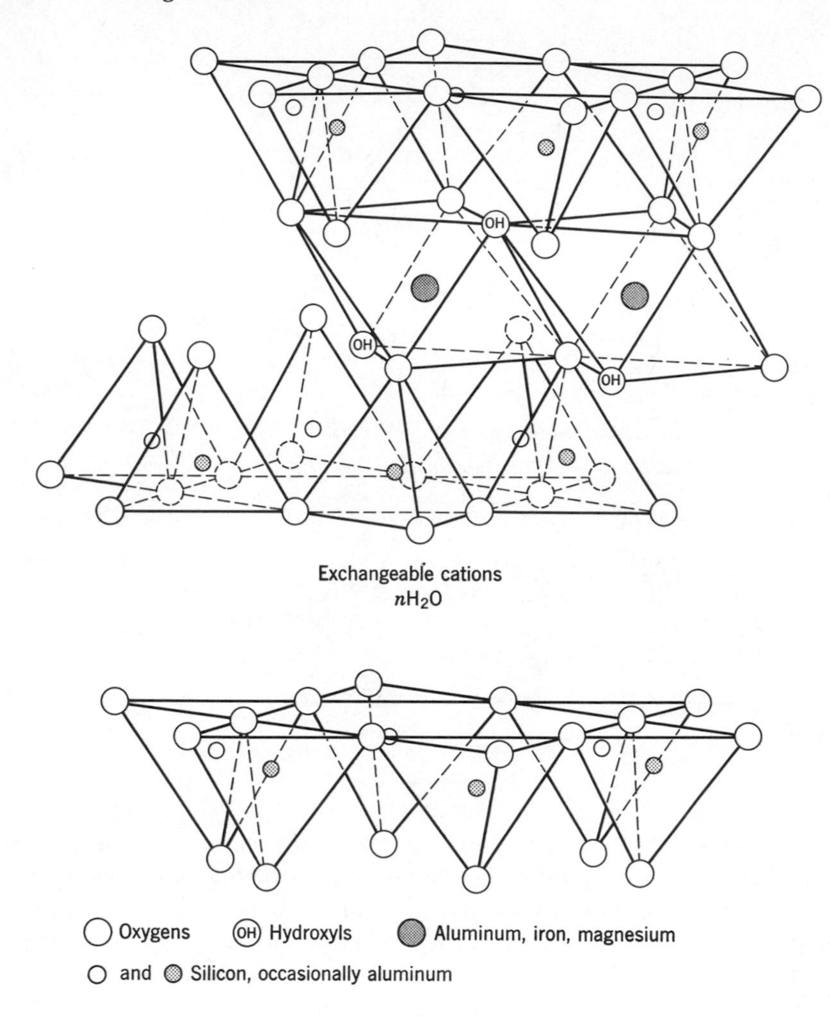

Figure 4.5. Structure of montmorillonite. (After Grim, 1953)

morillonite (Fig. 4.5). This mineral characteristically occurs in relatively thin flakes of indefinite shape and small size in comparison with kaolinite and illite. The substitutions in montmorillonite occur largely in octahedral positions, and the structure may be either dioctahedral or trioctahedral. The net negative charge results from substitutions in the octahedral layer. Replacement of approximately 20% of the aluminum by magnesium would require balancing cations to the extent of 100 m.e./100 g of clay. In montmorillonite, the source of the excess negative charge originates near the center of the layers; hence the strength of the charge at the surface is not so strong as it is with illite. Moreover, the number of charges per unit area of interlayer surface is smaller than it is with illite. Thus the layers are not held together by potassium ions even though the same openings among the surface oxygen atoms occur in montmorillonite as in illite. Because of the limited interlayer attraction, the layers of montmorillonite are forced apart readily by water molecules, which are held to the interlayer surfaces by hydrogen bonding. The distance between adjacent layers varies with the number of interposed layers of water but is sufficient to permit the balancing cations between the layers to be exchangeable. Thus montmorillonite has a higher cation-exchange capacity than illite, despite a lower number of excess negative charges, because the interlayer cations are exchangeable in montmorillonite but not in illite. In addition to the balancing cations held between the layers, exchangeable cations may be held at the edges of the particles, as in kaolinite and illite. Hendricks et al. (1940) concluded that approximately 20% of the cation-exchange positions of montmorillonite are at the edge of the flakes, and the remaining 80% are on the basal-plane surfaces.

Another member of the second structural group important in soils is vermiculite. Substitutions in this mineral are largely in the tetrahedral layers, as with illite, thus providing a source of charge near the surface of the layers. The number of charges per unit area of interlayer surface is not as great as is the case with illite, and so the tendency for strong bonding of the layers with potassium ions is less than in illite. Layers of vermiculite are forced apart by water molecules to an extent sufficient to permit the interlayer cations to be exchangeable, but not to the unlimited extent permitted in montmorillonite. When the exchange positions are occupied by potassium, however, water is excluded and the layers are held together tightly. Properties of vermiculite are thus intermediate in certain respects between those of illite and montmorillonite.

A fourth member of the second structural group important in soils is chlorite. In this mineral, there is considerable substitution of aluminum for silicon. The net negative charge is balanced by a crystalline interlayer consisting of magnesium, aluminum, or ferric iron ions, each surrounded by six hydroxyls in octahedral configuration. The number of substitutions of aluminum and iron for magnesium is sufficient to provide the excess positive charges needed to balance the excess negative charges resulting from the substitution in the clay layer. Chlorite layers are not forced apart by water, and cations in the crystalline interlayer are not exchangeable. The exchangeable cations thus occur on the edges and exposed basal-plane surfaces. This classical chlorite is unstable in acid soils. The chlorite in such soils has aluminum as the principal cation in the interlayers.

Minerals of this second group often occur in the form of mixed crystals (sometimes called interstratified or interlayered crystals) in which, within a single particle, some layers will be of one type and others of another. The arrangement may be either regular or random. Although this condition complicates the nomenclature of soil clays and certain aspects of their characterization, it probably adds little in the way of complexity to cation-exchange investigations.

The crystalline components of clay have received most attention because they can be characterized according to molecular structure and physical shape by X-ray and electron-microscope methods. Amorphous materials are present, however, and in some soils they are of significance in respect of cation-exchange properties. The general name given to amorphous aluminosilicates in clays is allophane. Allophane is an important clay mineral in certain soils derived from deposits of volcanic ash. A cation-exchange capacity of about 70 m.e./100 g was found by White (1953) for allophane in bulk deposits. In addition to allophane, soils may contain large quantities of hydrous oxides of iron and aluminum, which appear to have some cation-exchange capacity (Fieldes et al., 1952; Sumner, 1963). The amorphous components are more reactive than the crystalline components and may respond sensitively to conditions. Sherman et al. (1964) found that certain soils of Hawaii with a high content of amorphous colloidal mineral matter lose up to 91% of their cation-exchange capacity on drying and do not regain it on wetting. Initially these soils have a high content of water and a low bulk density. Drying causes an increase of bulk density and development of crystallinity in the originally amorphous hydrous oxides of iron and aluminum.

Organic Matter

Soil organic matter has a relatively high cation-exchange capacity, as may be inferred from determinations of the cation-exchange capacity of soil before and after treatment with hydrogen peroxide to remove most of the organic matter. In experiments of this type, Baver (1930) found that the loss in cation-exchange capacity due to treatment with peroxide amounted to from 112 to 252 m.e./100 g of organic matter oxidized in four soils. Mitchell (1932) determined the reduction in cation-exchange capacity of soils due to oxidation of the organic matter at low red heat. He found that the calculated cation-exchange capacity of the organic matter in fourteen soil samples was from 70 to

200 m.e./100 g of organic matter, with thirteen of the values between 113 and 200. With neither of these methods can one place great confidence in the results. In the peroxide method, the oxidized organic matter may not have the same exchange capacity as the unoxidized organic matter; moreover, the exchange capacity of the mineral matter may be affected directly or indirectly by this treatment. In the ignition method, there is some question as to whether the cation-exchange capacity of the mineral fraction is unaffected by the treatment.

The chemical nature of the functional groups in soil organic matter that hold exchangeable cations has been investigated by the technique of measuring the reduction in cation-exchange capacity resulting from treatment of preparations of soil organic matter with organic reagents that act selectively to block specific groups. From experiments with several soils, Broadbent and Bradford (1952) estimated that 54% of the exchange positions were attributable to carboxyl (—COOH) groups, 36% to phenolic and enolic hydroxyl groups, and 10% to imide nitrogen groups. The participation of different functional groups in cation-exchange reactions varies with the *p*H and the nature of the cations as will be mentioned later.

Mineral and Organic Components in Unfractionated Soils

The relative contribution of mineral and organic components to the cation-exchange capacity of soils varies greatly. With the aid of the peroxide method mentioned in the preceding section, McGeorge (1930) found that the cation-exchange properties reside almost entirely in the mineral fraction of certain low-organic-matter soils of Arizona and California, which have substantial inorganic exchange capacity. At the other extreme, nearly all the cation-exchange properties of peat and muck soils from the north-central states are attributable to organic matter. Because of the high exchange capacity of organic matter, the exchange capacity of even low-organic-matter soils may be mostly organic in nature if the soils are low enough in content of clay. Peech (1939) found this to be the case in certain sandy soils of Florida.

In attempts to obtain information about organic and inorganic components of the exchange capacities of natural soils that have not been fractionated or otherwise treated to alter the cation-exchange capacity, different statistical methods have been used. Pratt (1957) made measurements of cation-exchange capacity and organic carbon content of samples of a sandy loam soil from a field experiment in which various applications of fertilizers and organic matter had been made over a period of 28 years. As shown in Fig. 4.6, he found that the exchange capacity of 100 g of soil increased at the rate of 4.9 m.e. for each per cent of

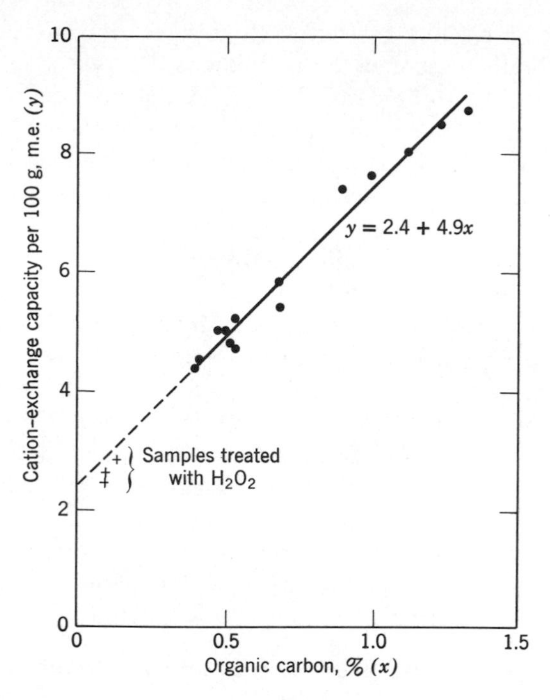

Figure 4.6. Cation-exchange capacity versus organic carbon content of the surface 15 cm of a sandy loam soil that had been variously treated with fertilizers and organic matter over a period of 28 years in southern California. (Pratt, 1957)

organic carbon. The organic carbon accumulated in the soil as a result of the treatments thus had an apparent exchange capacity of 490 m.e./100 g of organic carbon or 284 m.e./100 g of organic matter (on the basis of the conventional factor, organic matter = organic carbon $\times$ 1.724). The constant term 2.4 in the statistical equation is an estimate of the exchange capacity of the mineral portion of the soil on the assumption that the exchange capacity per unit weight of native soil organic matter is the same as that of the added organic matter. In addition to the measurements on the original samples, Pratt made measurements on three samples that were treated with hydrogen peroxide to remove most of the organic matter. These values (shown in the graph) fall somewhat below the extrapolated line but are not far from it and verify that the exchange capacity of the mineral portion of the soil is in the vicinity of the value estimated by the extrapolation procedure. One may note that the exchange capacity of the samples lowest in organic carbon is about equally divided between mineral and organic forms, according to the method of extrapolation employed, and that most of the exchange capacity is organic in the samples highest in organic carbon.

In a second statistical method, measurements are made of the cation-exchange capacity and the sand, silt, clay, and organic matter content of a number of soils. The cation-exchange capacity is then expressed as a function (usually linear) of the content of the various soil components, and the best-fitting coefficients of the individual soil fractions are used as estimates of the cation-exchange capacities of the corresponding components. Most authors neglect the coarser fractions, perhaps because the statistically derived coefficients representing the exchange capacities are usually not significantly different from zero. The cation-exchange capacities of the coarser fractions then are reflected in the other terms of the equation, which makes them slightly inaccurate for use as estimates of the exchange capacity of these fractions.

Heinonen (1960) used the statistical method just described on a group of soil samples from Finland and obtained estimates of cation-exchange capacity in milligram equivalents per 100 g equal to 234 for organic matter, 40 for clay, -7 for silt, and 2 for fine sand (20 to 60 microns effective diameter). He accounted for the apparent negative exchange capacity of the silt on the basis that the exchange capacity per 100 g of clay decreased as the silt content increased because the clay was more coarse in soils high in silt than in those low in silt. In any event, the comparatively low values for the exchange capacity of the coarser fractions are clearly evident. The value of 2 m.e./100 g of fine sand was not significantly different from zero. The constant term 2 in the equation is an estimate of the cation-exchange capacity of 100 g of material coarser than 60 μ in effective diameter.

Renger (1965) made an unusually extensive investigation, analyzing more than 1500 samples of soils of Germany. He divided the samples into groups, according to the nature of the soils, and then made a statistical analysis of data in each group. The estimated cation-exchange capacities of the various soil components (in milligram equivalents per 100 g of the component) in the different groups of soils ranged from 168 to 249 for organic matter, 38 to 51 for clay, and 2 to 22 for silt. The constant term in the equation, estimating the cation-exchange capacity of soils devoid of organic matter, clay, and silt (in other words, the cation-exchange capacity of 100 g of sand and gravel), ranged from 0.7 to 6.5.

One of the features of the work of both Heinonen (1960) and Renger (1965) is the finding that the calculated cation-exchange capacity of the organic matter decreased with an increase in content of both organic matter and clay. Two explanations have been proposed for this finding. One is that the nature of the organic matter varies with the amount of organic matter and clay present. The second is that organic matter and clay interact chemically with a resulting reduction in cation-exchange capacity. The second theory is consistent with results of experiments showing that the sum of the cation-exchange capacities of organic matter and clays measured individually exceeds the measured sum after the organic matter and clay have been allowed to interact.

Characteristics of Cation-Exchange Reactions

Rate of Reaction

According to the kinetic theory of matter, exchangeable cations are in continual motion around the point of their attachment. If an invading cation from the solution happens to penetrate within this hemispherical volume of motion at an instant when the exchangeable cation is far away, the attractive force emanating from the solid surface may be transferred to the invading cation, which then will become the exchangeable cation. The process of transference of the attractive force from one cation to another, which may be said to represent the actual exchange, undoubtedly is instantaneous.

In laboratory work on rate of exchange, cations are moved to and from the particles by rapid movement of particles in the solution in which they are suspended or by rapid movement of solution past the particles. Even with these techniques, measurements of rates of cation exchange do not show the process to be instantaneous. Some of the exchangeable cations are located in interlayer positions in individual particles, and some particles bearing exchange sites are located in inner positions in soil aggregates. Cations cannot reach all these sites by mass movement of the solution but must penetrate and leave by diffusion.

Rich and Thomas (1960) quoted unpublished work of Coleman and Craig showing that the rate of exchange in aggregated samples of a kaolinitic and a vermiculitic soil approximately doubled with each halving of aggregate size. The time required to reach 50% exchange was about twice as long with the vermiculitic soil as with the kaolinitic soil, which suggests that the slowness of diffusion of ions in the interplanar spaces impeded the exchange process. No times of exchange were quoted by Rich and Thomas, but the exchange process is generally rapid compared with processes such as absorption of cations by plants. As numerical examples that will give some conception of rates, Hissink (1925) found that shaking a 25-g sample of soil with 250 ml of 1-normal sodium chloride solution for 5 seconds, 3 minutes, 1 day, and 7 days caused release of 27, 28, 28, and 28 m.e. of calcium per 100 g of soil. Kelley (1948) quoted results of an experiment conducted by Gedroiz in which 100 g of soil were shaken with 100 ml of normal sodium chloride solution continuously or at intervals from 5 seconds to 14 months. The suspension was then filtered, and calcium was determined in the filtrate. In this experiment, the values for calcium were almost unaffected by the time. Because of the time required for filtration, the time of contact of soil and solution was, of course, longer than the time of shaking.

Some known exchange reactions are not this fast. Cernescu (1931) found, for example, that exchange equilibrium was established in 5 minutes with a clay, 10 days with permutite, and 92 days with chabazite. With the last two materials, the slowness of the over-all process probably resulted from the slowness of diffusion of ions between the solution at the surface of the particles and the points of exchange in the interior of the particles. The reaction of limestone with acid soil under field conditions is an example of another slow exchange reaction. Here the rate of the process is probably limited by a combination of slow diffusion of calcium bicarbonate through the soil and the limited solubility of the calcium carbonate, once it is surrounded by a layer of neutralized soil. Granules of unreacted limestone may be detected in soils many years after application if the original material contained large particles.

Equivalence of Exchange

If the exchange positions in a sample of soil are saturated successively with different cations, the quantities of the different ions held by the soil are sometimes chemically equivalent and sometimes not. For example, McGeorge (1930) gave the following values for the cation-exchange capacity of a peat soil in an experiment in which the exchange capacity was determined with different ions in the sequence indicated by the arrows:

$$
\begin{array}{cccccccccc}
 & & & & & & & & & Ca^{++} \; 167 \\
 & & & & & & & & \nearrow & \\
NH_4^+ & \rightarrow & Ba^{++} & \rightarrow & NH_4^+ & \rightarrow & Ba^{++} & \rightarrow & NH_4^+ & \\
 & & & & & & & & \searrow & \\
159 & & 199 & & 163 & & 177 & & 162 & K^+ \; 161
\end{array}
$$

The numerical values are milligram equivalents per 100 g. Several theories may be advanced to account for lack of equivalence in measurements such as these.

The first theory is that the usual methods for determining the cation-exchange capacity are inaccurate because they do not take due account of the effects of soluble salts and hydrolysis of exchangeable cations. The method of analysis evidently gives high results if it includes as exchangeable cations some cations that were paired in the soil with soluble anions, and it gives low results if some of the exchangeable cations have been lost by hydrolysis before analysis, for example,

$$NH_4\text{-soil} + HOH = H\text{-soil} + NH_4OH.$$

According to this theory, comparisons of cations displaced with cations retained as a measure of equivalence of cation-exchange reactions are not valid unless the salt concentration and pH are the same in both instances.

All soils contain soluble salts, but the quantities ordinarily present are so small in relation to the exchangeable cations that they are neglected; and the total cations displaced are taken as exchangeable cations. In saline soils, the proportion of soluble cations in the total displaceable cations is high enough to be of consequence; and a correction for soluble cations is made in analyses of such soils. The method most commonly used is that of subtracting the equivalents of cations in solution in a sample of soil saturated with water from the equivalents of cations displaced in a separate operation by the exchanging salt. After the exchangeable and soluble cations have been displaced, the sample of soil remains saturated with the salt solution used for displacement and hence contains a large excess of soluble cations in addition to exchangeable cations. The traditional method has been to use a 1-normal salt solution for displacement and to leach the soil with 70 or 80% methanol or ethanol in water to remove the soluble salts before displacing and determining the exchangeable cation. The amount of the cation displaced is then taken as the cation-exchange capacity for that cation, the assumption being that the soluble salt was removed quantitatively before displacement and none of the exchangeable cation was lost by hydrolysis during removal of the soluble salt. Alcohol is used to reduce dispersion of the soil and loss of exchangeable bases by hydrolysis during leaching. Isopropyl alcohol was favored by Peech et al. (1962) because it presumably dissolves less organic matter [later verified by Frink (1964)] and leads to higher values for cation-exchange capacity than do methyl and ethyl alcohol.

Experiments by Okazaki et al. (1963) indicate that loss of exchangeable cations by hydrolysis begins before removal of excess salt is complete and that the degree of washing at which the positive and negative errors balance varies with the soil. In view of these complexities, close agreement between the sum of the cations displaced and the amount of the displacing cation retained and subsequently displaced would appear coincidental if the usual methods are used.

The uncertainties connected with salt and hydrolysis errors may be avoided by standardizing the salt concentration and pH at which the cation-exchange capacity is measured. The pH is usually controlled by use of a buffered solution. To standardize the salt concentration, Okazaki et al. (1964) used the method of weighing the sample after displacement to determine the weight of displacing solution retained and hence the quantity of displacing cation present in salt form. Subsequent displacement and determination of the soluble-plus-exchangeable forms then yielded a total value from which the calculated quantity of salt present in the displacing solution retained in the soil was subtracted to obtain the quantity of the cation present in the soil in exchangeable form or the cation-exchange capacity. They found that values for cation-exchange capacity obtained in this way were the same for sodium and barium in several samples. Chapman and Kelley (1930) standardized the salt concentration similarly, using ammonium chloride for displacement and determining the chloride retained by the soil as a correction for the quantity of the cation present as a soluble salt. More recently, D. R. Keeney and J. M. Bremner (unpublished) used the same technique with ammonium nitrate as the final displacing agent, determining nitrate as a correction for the ammonium present in excess of the exchangeable ammonium.

Okazaki et al. (1964) found that the cation-exchange capacities obtained with the salt concentration standardized at 1-normal were invariably higher than those obtained by measurement of the cation retained after washing with 80% ethanol—in one instance more than twice as high—which suggests that values obtained at high salt concentrations do not necessarily represent the cation-exchange capacity as it exists in the field where the soil solution has a low concentration of salts. In unpublished work on salt-affected soils, R. K. Schulz stopped the washing process automatically by use of a relay that was actuated when the conductivity of the leachate dropped to a low, predetermined level. Frink (1964) recommended final equilibration of the soil with a 0.001-normal solution of the saturating salt and use of a small correction for the volume of 0.001-normal solution retained by the soil. The techniques of both Schulz and Frink have the advantage of permitting the determination of cation-exchange capacity at a standardized salt concentration similar to that existing in natural soil.

A second theory to account for failure of the cation-exchange capacity to be the same for different ions is that the average number of equivalents of polyvalent cations present per equivalent of exchange positions may exceed unity because in some instances the bonds of these cations are not all attached in exchangeable form. Bower and Truog (1941) first proposed this theory to account for differences in cation-exchange capacity of clays found with different cations. They obtained average values of 58, 62, and 71 m.e./100 g as the exchange capacity for ammonium, calcium, and magnesium. The cation-exchange capacity of montmorillonite was higher with various polyvalent cations than with monovalent cations where the exchange occurred in water in the usual manner. Where the exchange occurred in alcohol, however, the values for cation-exchange capacity obtained with all cations were approximately the same as those obtained with the monovalent cations in water. Bower and Truog attributed these observations to the attachment of some of the polyvalent cations in exchangeable form as basic ions (for example, $MgOH^+$) in the aqueous medium, where hydrolysis of salts could occur, but not in the alcoholic medium, where the hydroxyl ions necessary to form the basic cations would not be present. The cause of the apparent lack of equivalence in the aqueous medium thus was inferred to be the incorrectness of the usual assumption that polyvalent cations are attached entirely in polyvalent form.

In an investigation of the reactions of zinc with montmorillonite, Elgabaly and Jenny (1943) obtained evidence indicating attachment of exchangeable zinc as Zn^{++}, $ZnOH^+$, and $ZnCl^+$. More recent evidence indicates that, with iron as exchanging cation, precipitation of part of the iron may occur during removal of the excess by washing; and high values may result from solution of some of the precipitated iron during subsequent displacement (Page and Whittig, 1961).

A third theory to account for lack of equivalence of release and uptake of ions in cation-exchange reactions is that the number of certain exchange sites varies somewhat with the nature of the ion because of ion-selectivity effects. Such effects are discussed in the next section.

None of the three theories, one may note, has to do with a real lack of chemical equivalence. Each one provides an explanation for a different class of effects in which lack of chemical equivalence, as inferred from incomplete information, is only apparent.

Equilibrium Constants and Selectivity Effects

The exchange reaction between radioactive calcium in solution and nonradioactive calcium in soil clay, described previously, seems to be reversible in the classical chemical sense. The mass-action constant was found experimentally to be constant over a wide range of ratios of reactants. Exchange between isotopes is a special case of cation exchange in which the ions concerned are almost identical.

In the mass-law expression for the reaction between radioactive calcium in solution and nonradioactive calcium in exchangeable form (p. 203), the activity coefficients for the ions in exchangeable form were eliminated on the basis that they were the same for radioactive and nonradioactive forms of calcium. If the ions involved in the exchange reaction are different, however, the activity coefficients cannot be eliminated because they may not be the same for the ions undergoing exchange; moreover, no way is known to determine the activity coefficients of individual exchangeable cations.

In general, the "equilibrium constants" calculated without accounting for the activity coefficients of the ions in exchangeable form are most nearly constant in exchanges involving pairs of monovalent cations or pairs of divalent cations (Krishnamoorthy and Overstreet, 1950; Marshall, 1964, pp. 263–276). Greater variability is found for exchanges of ions differing in valence. The explanation for this situation is perhaps twofold. First, the differential selectivity of bonding of cations in exchangeable form generally differs more between monovalent and divalent cations than among monovalent ions or among divalent cations. Second, different kinds of cation-exchange positions are present, and the selectivity may vary with the nature of the exchange positions.

Much of the experimental work relative to equilibrium constants of cation-exchange reactions has been done on fairly pure clays or on synthetic organic cation-exchange resins. Although the exchange positions on such substances are not necessarily homogeneous, they are not as heterogeneous as they are in mixtures of materials. Where mixtures of exchange materials with different kinds of exchange sites are used, the variability of the "equilibrium constants" may become unmanageable because one is dealing with a constantly changing type of selectivity as the ratios of exchanging ions are varied in solution.

Some conception of the situation that exists in soils, which contain mixtures of different kinds of exchange materials, may be derived from examination of Table 4.7. This table shows the retention of calcium and ammonium in exchangeable form where different substances with exchange properties were leached with an equinormal solution of calcium and ammonium acetates. Because each material was equilibrated with a solution in which the equivalent ratio of calcium to ammonium was unity, marked differences among materials evidently existed in the selectivity of bonding of calcium and ammonium. In a mixture in which half the exchange positions are derived from "humic acid" and half from muscovite, one would expect from the data in the table that the exchangeable calcium would be present almost entirely on the humic acid with low ratios of calcium to ammonium in solution and that only after the humic acid was nearly saturated with calcium would there be much exchangeable calcium on the muscovite. Conversely, with low ratios of ammonium to calcium in solution, almost all the exchangeable

Table 4.7. Retention of Calcium and Ammonium in Exchangeable Form by Different Materials after Leaching with a Solution Containing a Mixture of 0.05-Normal Calcium Acetate and 0.05-Normal Ammonium Acetate (Schachtschabel, 1940)

	Percentage of Exchange Positions Occupied by Indicated Cation	
Material	Calcium	Ammonium
"Humic acid"	92	8
Montmorillonite	63	37
Kaolinite	54	46
Muscovite	6	94

ammonium would be present on the muscovite. Only after the muscovite was nearly saturated with ammonium would there be much exchangeable ammonium on the humic acid.

In view of the foregoing experimental difficulties with equilibrium constants, some have used the technique of preparing two-cation systems in which soil is equilibrated with solutions containing different proportions of the chosen two cations in a fixed total concentration. The proportion of one ion in the total ions present in exchangeable form is then plotted against the corresponding proportion in the solution, with all values for the ions being calculated in terms of chemical equivalents. Figure 4.7 represents an example for a soil in which the two ions employed were aluminum and calcium, potassium, or sodium. If there is no differential selectivity or preferential combination of an ion-exchange substance with either of the two ions, the ions will occur in exchangeable form in the same proportion as they occur in solution. The plot of the results will then be in the form of a straight-line diagonal from lower left to upper right in the figure. Selectivity of the ion-exchange substance for the ion in the numerator of the ratio (Al^{+++} in Fig. 4.7) will cause displacement of the experimental observations above the diagonal line, and selectivity of the ion-exchange substance for the other ion in the pair (c = Ca^{++} K^{+}, or Na^{+} in Fig. 4.7) will cause displacement below the diagonal line. The experimental observations in the figure show that the soil bonded Al^{+++} selectively over Ca^{++} and Na^{+} but bonded K^{+} selectively over Al^{+++} In a plot of this kind, the position of the line, reflecting the selectivity, may change with the total concentration of the ions in solution.

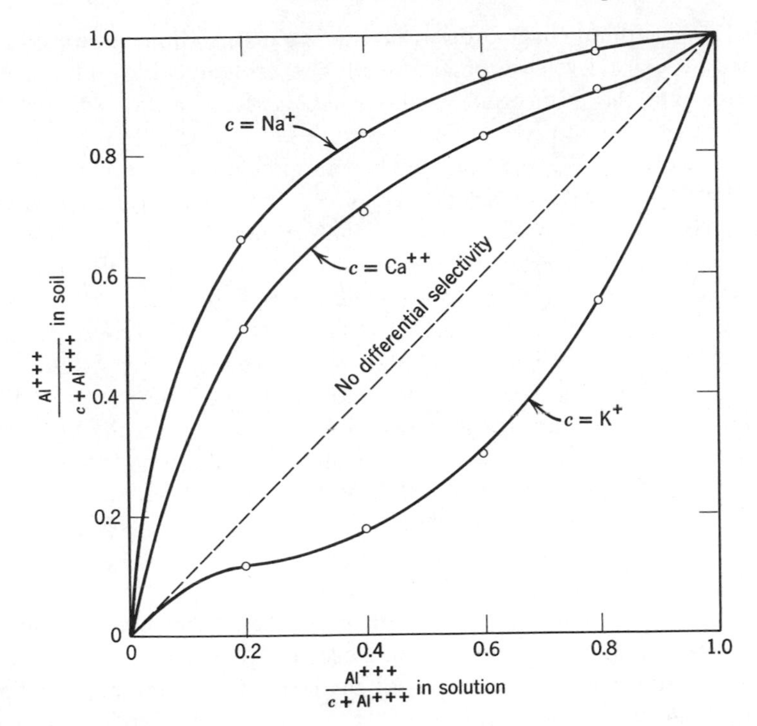

Figure 4.7. Plot of $Al^{+++}/(c + Al^{+++})$ in exchangeable form in a soil saturated with c and Al^{+++} against $Al^{+++}/(c + Al^{+++})$ in solution in equilibrium with the soil, where $c = Na^+$, Ca^{++}, or K^+, where the quantities of cations are expressed as milligram equivalents, and where the value of $(c + Al^{+++})$ in solution is one equivalent per liter. (Nye et al., 1961)

Selective bonding of cations by soils may be such that the position of the line relating the composition of exchangeable cations to the composition of the solution is not affected by the direction from which equilibrium is approached; that is, the bonding is reversible. Some selectivities are sufficient to cause a difference in position of the line according to the direction from which equilibrium is approached. This phenomenon was discussed by Kelley (1948) and Tabikh et al. (1960). The selectivities are sometimes so marked that they seem to represent irreversible reactions if investigated by the usual methods for characterizing cation-exchange properties of soils. Examples of several types of marked selectivity are discussed in the following paragraphs.

Selectivity of inorganic exchange materials for a large, organic, polyvalent cation was demonstrated by Ensminger and Gieseking (1941), who found that addition of gelatin to clays reduced the cation-exchange capacity. Addition of the equivalent of 150 g of gelatin to 100 g of montmorillonite reduced the exchange capacity from 90 to 20 m.e. McLaren et al. (1958) found that addition of lysozyme to sodium-saturated montmorillonite caused release of almost all the exchangeable sodium, which indicates that the adsorption was at least partly a cation-exchange reaction. Both gelatin and lysozyme are proteins.

Schulz et al. (1965, and personal communication) described research on a problem involving selectivity of an inorganic exchange material for an inorganic cation. These investigators noted, in measurements on soil from a field experiment on reclamation of a salt-affected soil in California, that rice was growing well on soil in which the cation-exchange capacity was approximately 70% saturated with sodium according to the methods of analysis they used. In an area showing sparse growth of rice, the exchange capacity was found to be 130% saturated with sodium. These two findings clearly indicated that the results of the measurements did not have the usual meaning on these soils—the second because a soil cannot be more than 100% saturated with a monovalent exchangeable cation, and the first and second because of a great discrepancy from the usual finding that crops fail completely at high values of sodium saturation.

Further experiments showed that the quantity of soil sodium that equilibrated with radioactive sodium exceeded the quantity replaceable by ammonium acetate and that the quantity of sodium replaceable by calcium chloride was relatively small. Eventually the discovery was made that soils which exhibited this behavior contained the uncommon mineral analcime, soils which did not show the unusual behavior did not contain analcime, and the pure mineral analcime had the sodium-exchange properties of the soils containing it. Analcime was thus evidently responsible for the behavior. Analcime shows special selectivity for sodium, presumably because the small channels that readily accommodate sodium do not allow free access of all kinds of cations to the exchange sites.

A second example of marked selectivity of an inorganic exchange material for inorganic ions is the selectivity of vermiculite-type minerals for ammonium and potassium. Data in Table 4.8 show the great differences in cation-exchange capacity found with use of different sequences of ions in vermiculite and in a particle-size fraction from a soil containing vermiculite. This phenomenon will be discussed in more detail in Chapter 9.

Another type of limited reversibility is that represented by combination of certain inorganic cations with certain organic exchange sites. DeMumbrum and Jackson (1956) found that, even in the presence of 0.5-normal calcium acetate solution, calcium-saturated peat would take up additional cupric ions from the low concentration of these ions supported in solution by solid cupric phosphate. These ions did not displace

Table 4.8. Cation-Exchange Capacity of Particle-Size Fractions as Found with the Use of Different Cations for Saturation and Displacement (Sawhney et al., 1959)

		Cation-Exchange Capacity per 100 g, m.e.	
Cation Used for Exchange Saturation	Cation in Neutral, Normal Acetate Displacing Solution	Soil Fraction, 5–20μ	Vermiculite Fraction, 0.2–2μ
Potassium	Ammonium	3	44
Calcium	Ammonium	16	71
Calcium	Sodium	21	140

exchangeable calcium but were held in excess of the exchangeable calcium in a form extractable by acid ammonium acetate solution.

Although the reactive groups responsible for the excess of exchangeable copper over exchangeable calcium are not definitely known, the suggestion has been made that a type of bonding called chelation is involved. Metal chelates are substances in which the metal is combined with two or more electron donors so that one or more rings are formed. Many chelating substances are known in organic chemistry. The general concept may be illustrated by ethylendiaminediacetic acid (left) and its copper salt (right):

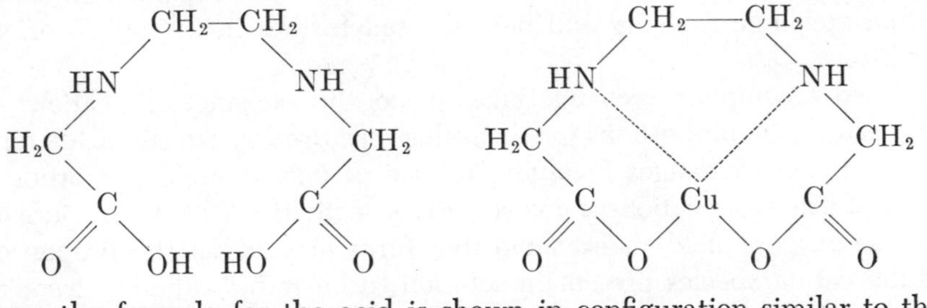

```
        CH2—CH2                          CH2——CH2
       /       \                        /        \
    HN          NH                   HN            NH
    /             \                  /  ·        ·  \
 H2C               CH2            H2C     ·    ·     CH2
    \             /                  \      ·  ·     /
     C           C                    C      Cu     C
   //  \       /  \\                //  \   /  \  /  \\
  O     OH   HO     O              O      O      O     O
```

where the formula for the acid is shown in configuration similar to the copper salt to facilitate recognition of the relationship. In this chelate, copper occurs in two five-membered rings. The primary electron donors are the two carboxyl

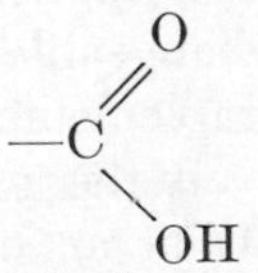

```
     O
    //
  —C
    \
     OH
```

groups. The nitrogen atoms serve as a secondary source of electrons. Each nitrogen atom has an unshared pair of electrons and tends to attract positive charges. The broken lines between the copper and the nitrogen atoms in the copper salt indicate this type of attraction. The copper is therefore held in place more firmly than would be the case with simply the electrostatic charges originating with the carboxyl groups. Calcium and other divalent cations such as manganese and zinc are similarly held, but the strength of bonding differs among cations.

Copper, manganese, and zinc usually occur in only minute quantities in ammonium acetate extracts of soils, but larger quantities of each ion are usually found to be exchangeable if the extracting solution includes one of the other ions in the group. For example, Heintze and Mann (1949) found that 1-normal ammonium acetate containing 0.02-normal copper sulfate extracted as much as forty times more manganese from soil than did 1-normal ammonium acetate alone.

One of the particularly marked selectivities is that for hydrogen ions. Hydrogen ions tend to associate strongly with certain oxygens as the hydrogen-ion activity in the solution increases, with the result that the hydrogen ions no longer act as exchangeable cations. The cation-exchange capacity of substances with exchange positions of this type thus increases as the hydrogen-ion activity in the solution decreases (or as the *p*H increases). Conversely, in displacement of metallic cations from exchangeable form with solutions of different *p*H value, the degree of release of the metallic cations increases with an increase in tendency of the exchange positions to become inactivated by hydrogen ions. Randhawa and Broadbent (1965) investigated this effect in the case of combination of zinc with soil organic matter. This *p*H-dependency of the cation-exchange capacity will be discussed further in Chapter 5 on soil acidity.

When a complete exchange takes place, the exchangeable cations are released in the proportions in which they occurred in exchangeable form. In fractional exchanges involving release of only a small proportion of the exchangeable cations, however, the selectivities with which individual cations are held in exchangeable form may cause the proportions of the cation species present in solution to be rather different from the proportions in exchangeable form. The comparative behavior of individual cations in this respect is described by the complementary-ion principle, which may be stated as follows: In a fractional exchange, the release of a cation of given species from exchangeable form increases with increasing strength of bonding of the complementary exchangeable cations.

The effect of complementary ions on the release of a given ion is illustrated in Table 4.9 by the results of an experiment published by Jarusov (1937). In this experiment, 5 m.e. of ammonium chloride in 200 ml of solution were added to 1.91-g samples of soil. Each sample

Table 4.9. Release of Exchangeable Calcium in a Fractional Exchange with Ammonium Ions in the Presence of Exchangeable Hydrogen, Magnesium, and Sodium as Complementary Cations (Jarusov, 1937)

Exchangeable Cations in Soil Sample	Quantity of Exchangeable Calcium Displaced by Ammonium Chloride	
	Milligram Equivalents	Per Cent of Total
0.5 m.e. Ca + 0.5 m.e. H	0.30	60
0.5 m.e. Ca + 0.5 m.e. Mg	0.18	36
0.5 m.e. Ca + 0.5 m.e. Na	0.09	19

of soil contained 0.5 m.e. each of calcium and a complementary ion. The displacement of the exchangeable calcium was greatest with hydrogen, intermediate with magnesium, and least with sodium as the complementary cation. Among the complementary cations, by inference, hydrogen was attached most strongly and sodium least strongly. As explained previously, the complementary-ion effect observed in an experiment with soil, which contains a variety of kinds of exchange sites, will represent the summation of the selectivity effects in all the sites and not necessarily the behavior of any one type of site.

The usual over-all complementary-ion effects in natural soils containing a variety of cations may be summarized by the series Na > K > Mg > Ca, where sodium is released most readily and calcium least readily in a fractional exchange. Thus sodium is released more

Table 4.10. Bases in Exchangeable Form and in Solution in a Loam Soil from South Dakota (Anderson et al., 1942)

Base	Content of Bases in Indicated Form per 100 g of Soil, m.e.		Bases in Solution as Percentage of Total in Exchangeable and Solution Forms
	Exchangeable	Solution	
Calcium	34.4	0.175	0.5
Magnesium	7.8	0.074	0.9
Potassium	0.9	0.011	1.2
Sodium	0.5	0.016	3.1

Table 4.11. Yield and Calcium Content of Wheat Seedlings Grown on Soil Saturated with Calcium or with Calcium and Different Complementary Cations (Ratner, 1938)

Exchangeable Cations in Soil Sample	Air-Dry Weight of Seedlings, g	Calcium in Seedlings, mg
100% Ca	1.8	9.7
60% Ca + 40% H	1.7	8.6
60% Ca + 40% Mg	1.7	8.1
60% Ca + 40% Na	1.6	5.2
Control (sand without soil)	1.7	5.2

readily if it is accompanied by a high proportion of calcium than by a high proportion of potassium; magnesium is released more readily if it is accompanied by a high proportion of calcium than of potassium; and so on. The magnitude of the spread among members of the series is indicated in Table 4.10, which shows the quantities of individual bases in solution expressed as a percentage of the sum of the exchangeable and soluble bases in a particular sample of soil.

Table 4.11 shows the uptake of calcium by plants from soil in which a fixed quantity of exchangeable calcium was accompanied by different complementary ions. The uptake of calcium was greater with hydrogen than with magnesium as the complementary cation, and greater with magnesium than sodium. The same relative effects were shown for release of calcium in a fractional exchange in Table 4.9. Because of such similarities, it appears that the same principles are involved in removal of exchangeable cations from soils by plants as by a fractional exchange with a displacing electrolyte. Application of this concept will be discussed next.

Exchangeable Bases and Plant Nutrition

Over-All Relationship

Under natural conditions, net release of bases from exchangeable form in soil occurs primarily by exchange with hydrogen ions. The hydrogen ions may be derived in varying proportions from sources external to the soil, from processes in the soil, and from activities of plants. Generally, an accumulation of exchangeable hydrogen in soils is associated with removal of bases by plants.

Jenny and Cowan (1933) grew 30 soybean seedlings 8 days in a suspension of calcium-saturated clay and found that uptake of calcium by the plants amounted to 1.02 m.e. Analysis of the residual clay suspension showed that the *p*H had dropped from the initial value of 6.3 to a final value of 4.3 and that 0.95 m.e. of calcium hydroxide was needed to raise the *p*H again to 6.3. Growth of the soybeans had thus depleted the clay of exchangeable calcium and at the same time had enriched the clay with an almost equivalent quantity of titratable acidity. The finding of an almost equivalent quantity of titratable acidity in this case may be attributed to the fact that the plants were the principal source of hydrogen ions and only a short period of time was involved in the experiment. Over a longer time, secondary reactions involving release of nonexchangeable bases, formation of exchangeable aluminum, and alteration of the exchange capacity may result in disappearance of most of the hydrogen ions from exchangeable form.

The experiment by Jenny and Cowan gives the initial condition and the final condition, the net change being found by subtraction. The mechanisms involved have been the subject of much research, of which various aspects will be considered in the following sections.

The Physical System

Understanding of the current state of knowledge of the subject of cation uptake by plants from soils may be facilitated by consideration of the physical nature of the system involved. Plant cells are encased by a wall consisting of cellulose and other polysaccharides and including a small amount of protein (Preston, 1961, pp. 229–253). The wall may become further encrusted with lignin as the cell ages. Little lignification exists in the young cells in the region in roots in which absorption of ions occurs most actively. Between cells is a layer of calcium pectate. External to the wall of the outermost cells of young roots may occur a diffuse gel-like substance of thickness several times exceeding that of the cell wall. All these structures are nonliving but are formed by the internal, living portion of cells.

Inside the cell wall is the plasma membrane, generally considered to be composed of two layers of lipid molecules, with the hydrocarbon ends pointed toward each other and the ionic ends oriented outward in both directions, the double layer of lipid molecules being sandwiched between two layers of protein molecules bounding the outer and inner surfaces of the membrane (Giese, 1962, pp. 276–279). This membrane is considered to be the site of selectivity in controlling the intake and outgo of solutes; it is the outer boundary of the living portion of the cell. Inside the plasma membrane is the cytoplasm, which surrounds the nucleus and contains other smaller organized bodies; and inside the cytoplasm are one or more vacuoles, which contain organic and inorganic solutes.

A section across a portion of the outermost cell of a barley root to the external solution is shown in Fig. 4.8. Figure 4.9 shows another section across a portion of the outermost cell of a barley root, illustrating the type of physical association that may occur between the surface of a root and adjacent particles of soil.

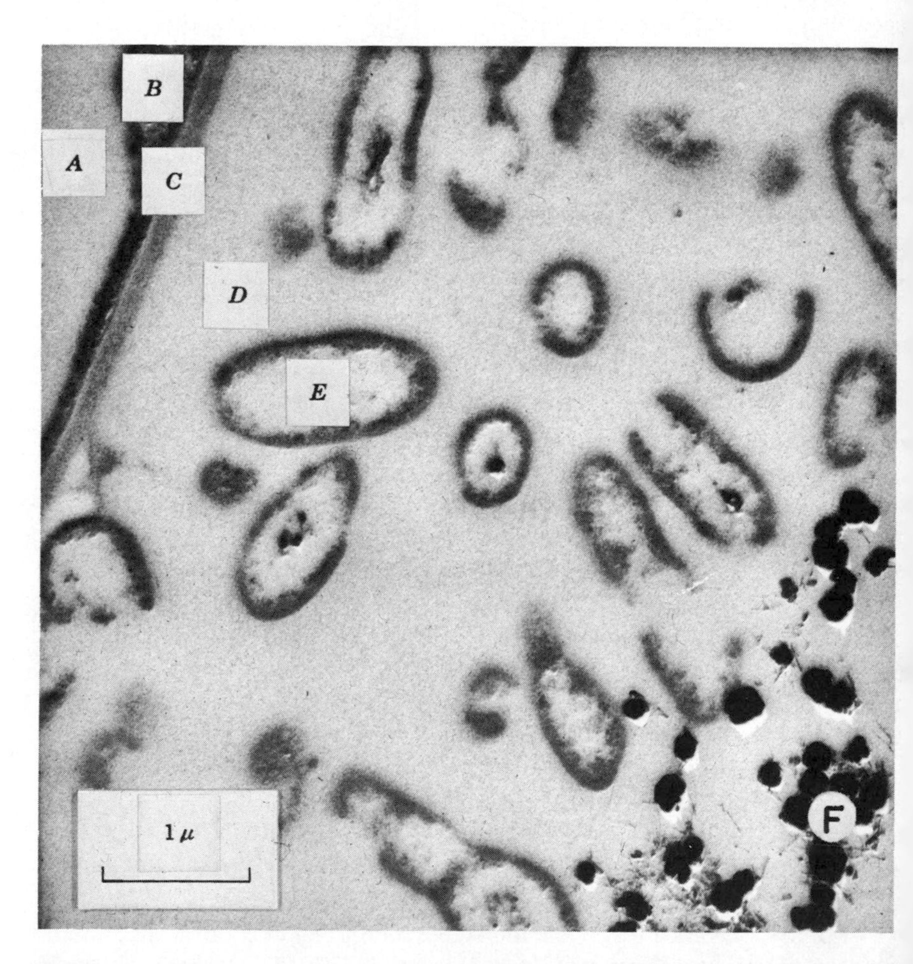

Figure 4.8. Electron micrograph of cross section of surface of root of barley: *A* = vacuole, *B* = cytoplasm, *C* = epidermal cell wall, *D* = layer of mucilage, *E* = bacterial cell in mucilage, *F* = iron oxide particles marking the outer boundary of the mucilage. (Micrograph courtesy of Karl Grossenbacher)

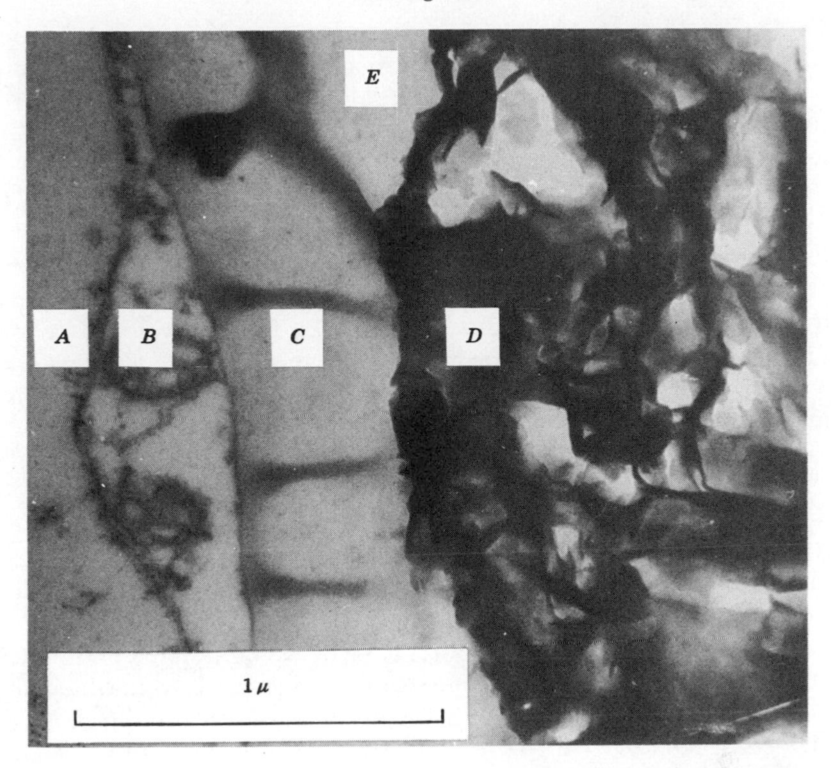

Figure 4.9. Electron micrograph of contact zone between surface of barley root and clay matrix: *A* = vacuole, *B* = cytoplasm, *C* = epidermal cell wall, *D* = clay, *E* = mucilage. (Jenny and Grossenbacher, 1963)

Forms of Absorption

Freely Diffusible Form. By use of appropriate experimental methods, several forms of cation uptake by roots may be distinguished and measured quantitatively. If roots that have been allowed to remain for some time in a dilute salt solution are blotted, centrifuged, or washed rapidly with distilled water to remove the external solution and are then transferred to distilled water, a quantity of the salt will appear in the water. The outward movement of the solute is most rapid at first and soon stops. This experimental observation gives rise to the concept that some of the salt taken up by roots is present in a freely diffusible form and hence that roots have a certain volume of "free space," representing an extension of the soil solution, into and out of which solutes diffuse freely. Absorption of cations in freely diffusible form is considered to be a passive process because similar movement of solute would occur into any volume available for diffusion, whether in a root or not.

If roots are allowed to remain in a salt solution long enough to permit the concentration of solute in the free space to come to equilibrium with the external solution before transference to water to observe the outgo of solute, the volume of free space in the root may be calculated by dividing the quantity of solute that diffuses out into the water by the concentration of the solute in the initial salt solution. Numerous experiments of this type have been performed, and the results point toward a value of approximately 10% of the root volume as free space (Epstein, 1960). Although the concept as presented here is simple, there are important complications in measurements, definitions, and nomenclature that will not be considered. Reviews by Briggs and Robertson (1957) and Marschner and Mengel (1962) deal with these matters.

The exact location of the free space is not known. Knowledge of the nature and behavior of roots, however, suggests that the free space is in the parts external to the plasma membranes. It may extend in the cell walls and calcium pectate interlayers completely around individual cells in the outer portion of the root.

Exchangeable Form. If roots that have been equilibrated with a dilute salt solution are placed in water to allow the freely diffusible salt to diffuse out of the free space, subsequent placement of the roots in a dilute salt solution with a second cation species will cause release of a quantity of the first cation species, which appears in the solution and can be determined by analysis. Release of the first cation species in the presence of the salt solution takes place rapidly and soon stops without involving the total content of that species in the roots. These observations give rise to the concept that roots absorb cations in exchangeable form and that the exchangeable cations are located in the free space.

Considerable work has been done on the cation-exchange properties of roots. The exchange properties appear to be attributable mainly to carboxyl groups present in the pectin substances. Keller and Deuel (1957) estimated that from 70 to 90% of the exchange capacity of roots of various plants could be accounted for by such sites. Heintze (1961) found that the exchange capacities were about the same, whether measured on living or dead roots, which indicates that the exchange positions are in sites external to the cell cytoplasm. Uptake in exchangeable form, like that in freely diffusible form, thus is considered to be a passive process that is distinct from intake into the interior, living portion of cells.

The exchangeable cations of roots are conceived to occur within the same physical volume as the freely diffusible cations. The two types of cation have mutual effects that give rise to most of the problems of definition and nomenclature of the free space.

Nondiffusible and Nonexchangeable Form. If roots are placed in a dilute salt solution, uptake of the cation in freely diffusible and exchangeable forms ceases in a matter of minutes; but uptake in nondiffusible and nonexchangeable form continues. Uptake in nondiffusible and nonexchangeable form may be found by determining total uptake and uptake in freely diffusible and exchangeable forms and by subtracting the sum of the last two from the first. Alternatively, the linear relationship between total uptake and time obtained after initial saturation of the free space (including exchange positions) with the ion in question is extrapolated to zero time. The intercept is an estimate of uptake by diffusion and exchange, and the difference between the intercept and the total uptake at any subsequent time is an estimate of uptake in nondiffusible and nonexchangeable form.

For brevity, uptake in nondiffusible and nonexchangeable form will be referred to here as uptake in nonexchangeable form, the argument for such usage being that the operation of removing the exchangeable

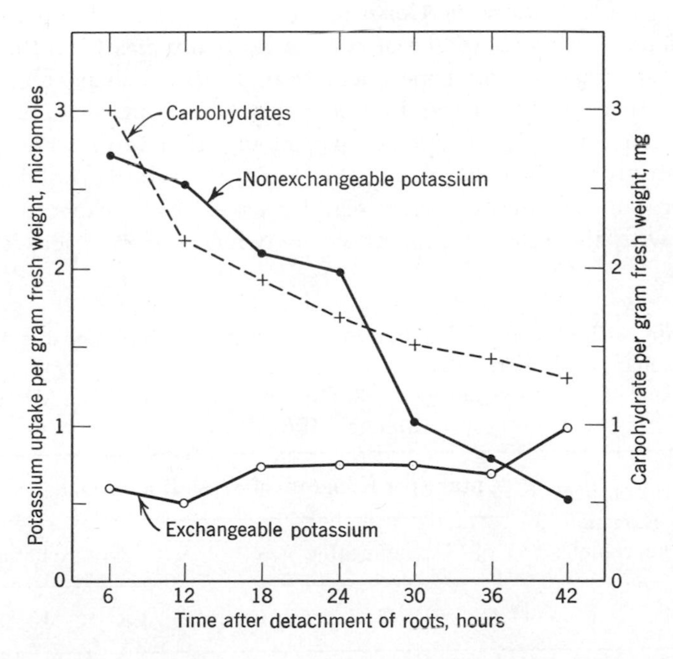

Figure 4.10. Uptake of potassium in exchangeable and nonexchangeable forms from solution by detached barley roots and hot-water-soluble carbohydrates in the roots, as measured different lengths of time after detachment. Hot-water-soluble carbohydrates include starch, sucrose, glucose, and fructose. (Mengel, 1962)

form will remove also the freely diffusible form. The terms active uptake and active transport are sometimes used in the same sense as uptake in nondiffusible and nonexchangeable form because disappearance of cations from the free space into the interior is a result of the activity of cells as living entities in contrast to passive diffusion and exchange processes.

The association of uptake of cations in nonexchangeable form with operation of roots as living systems is illustrated in Fig. 4.10, in which a decrease in uptake of potassium in nonexchangeable form with depletion of hot-water-soluble carbohydrate in plant roots is evident. Hot-water-soluble carbohydrate is an index of metabolic activity of the roots because it is the principal source of energy. Uptake of potassium in exchangeable form did not decrease with depletion of carbohydrate, indicating that this process was not connected directly with metabolism. In further work not shown in the figure, the same author found that uptake of potassium in nonexchangeable form was increased upon addition of glucose to plants depleted of carbohydrate.

Significance of Exchangeable Cations of Roots in Absorption

Exchangeable Cations in General. The significance of the exchangeable cations of roots to uptake of cations by plants has been the subject of much speculation and experimentation. If the exchangeable cations of roots are causally related to uptake by plants in nonexchangeable form, as, for example, if uptake in exchangeable form were the first step in absorption in nonexchangeable form, the relative uptake of two cations in nonexchangeable form could reasonably be expected to vary directly with the relative proportions in which the two ions occur in

Table 4.12. Uptake of Calcium and Potassium in Exchangeable and Nonexchangeable Forms from Solutions of Different Concentrations by Sections of Detached Barley Roots During a Period of Three Minutes (Mengel, 1961)

Concentration per 25 ml of External Solution, micromoles		Uptake per Kilogram of Fresh Root Weight, micromoles					
		Exchangeable			Nonexchangeable		
Ca	K	Ca	K	Ca/K	Ca	K	Ca/K
10	10	440	8.06	54.5	21.4	773	0.036
30	30	405	34.1	11.9	27.5	670	0.041
90	90	463	49.3	9.4	41.7	760	0.055

exchangeable form. Mengel (1961) conducted an experiment to test this consequence of the supposed essentiality of the exchangeable form, with calcium and potassium as the pair of cations. The results in Table 4.12 show that the ratio of calcium to potassium taken up by barley roots in nonexchangeable form increased with a decrease in the ratio of calcium to potassium taken up in exchangeable form, a behavior that does not indicate essentiality of uptake in exchangeable form.

Similarly, if uptake in exchangeable form is prerequisite to uptake in nonexchangeable form, the quantities of a given cation taken up in the two forms should be correlated. Figure 4.11 gives the results of an experiment with detached roots of corn in which radioactive potassium nitrate was added to the external solution alone or with equal molar quantities of other cations. The plot of relative uptake of potassium in nonexchangeable form against relative uptake in exchangeable form gives no evidence of a close relationship between the two. The data therefore do not support the view that a causal relationship exists between uptake of cations in exchangeable and nonexchangeable forms.

Carrier Theory. The cation-exchange properties of roots and their

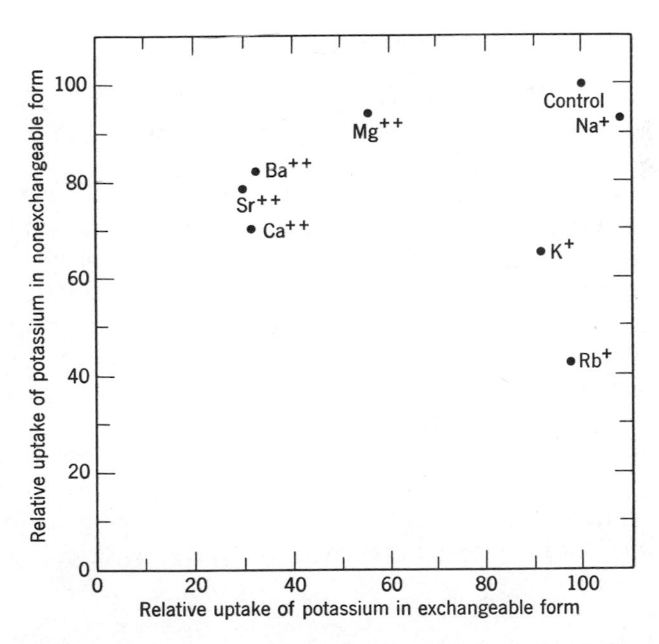

Figure 4.11. Relative uptake of radioactive potassium in exchangeable and nonexchangeable form during a period of two hours by detached corn roots from solutions containing labeled potassium nitrate alone (control) or with equal molar concentrations of the chlorides of nonlabeled sodium, potassium, rubidium, magnesium, calcium, strontium, or barium. (Mengel, 1963)

significance in plant nutrition have been investigated for many years. More recently, another closely related concept called the carrier theory has been developed. Reviews dealing with this subject include those by Laties (1959) and Fried and Shapiro (1961).

The carrier theory is an attempt to take into account both the discrimination that living cells exert in the accumulation of ions and the association of ion uptake in nonexchangeable form with metabolism. According to this theory, absorption of a cation into nonexchangeable form by roots involves its combination with a carrier, transport into some internal location in combination with the carrier, and release from the carrier internally. Metabolic energy is involved in producing the carrier and in carrying on the cyclic process of transport, ion release, and repositioning of the same or another carrier in condition suitable for absorption of another cation. The discrimination among ions in uptake in nonexchangeable form is accounted for by the postulate that carrier sites are of different kinds and have selectivities for particular ions.

The carrier theory may be expressed in the form of the equations

$$M_{\text{outside}} + R \rightleftharpoons MR$$

and

$$MR \rightleftharpoons M_{\text{inside}} + R'$$

where M is the ion, R and R' are the chemical states of the carrier before combination with the ion and after release of the ion, and MR is the ion-carrier complex.

Equations of the same form given here to express the carrier theory are used to express enzyme reactions, in which M_{outside} is replaced by substrate and R by enzyme. According to the theory of enzyme reactions, the substrate combines with the enzyme to form an unstable complex, which then breaks down to yield free enzyme and one or more products of the reaction. On the basis of such chemical equations representing the reaction, mathematical equations have been developed to express the velocity at which an enzymatic reaction will occur. The same mathematical equations have been used to express the velocity of uptake of ions by plant roots. The simplest case is that in which an ion reacts with the carrier and is absorbed in the absence of an inhibiting ion. The equation for the velocity of such a reaction is

$$v = \frac{V(M)}{K + (M)}$$

where v is the observed velocity of the reaction at the concentration (M) of the ion, V is the maximum velocity of the reaction, attained where the concentration of the ion is great enough to cause all the carrier sites to be combined with the ion, and K is the concentration of M at which $v = 0.5\,V$. The reciprocal of this equation is

$$\frac{1}{v} = \frac{K}{V(M)} + \frac{1}{V}.$$

If $1/v$ is then plotted against $1/(M)$, a straight line results. The intercept on the ordinate is $1/V$, and the slope is K/V. (M) and v can be measured experimentally, and so a plot of $1/v$ against $1/(M)$ provides a way to obtain values for V and K.

Three different kinds of inhibitory effects of one ion on absorption of another are recognized: competitive, noncompetitive, and uncompetitive. Competitive inhibition occurs if two ions combine with the same carrier site. Under these circumstances, the concentration of MR, and hence the rate of uptake of M, depends on (M) and also on the concentration of the competing ion. Competitive inhibition occurs to the greatest extent between ions that are chemically similar. Competitive inhibition has been found, for example, between potassium and rubidium and between calcium and strontium. Noncompetitive inhibition occurs if the inhibiting ion combines with the carrier at a different site than the ion M independently of the presence of M. Noncompetitive inhibition has been found between magnesium and strontium. Uncompetitive inhibition occurs if the inhibiting ion does not combine with the free carrier R but combines only with MR at a site different from that with which M combines. Each of these types of inhibition requires a different modification of the equation given for the velocity of uptake of an ion in the absence of an inhibiting ion. These modifications result in a change of intercept, slope, or both in a plot of $1/v$ against $1/(M)$; and it is by such changes that the different kinds of inhibition are recognized. Further information on the various equations and their use may be found in papers by Epstein and Hagen (1952) and Yoshida (1964).

The carrier theory has been found to represent satisfactorily the uptake of ions in nonexchangeable form in a variety of short-term experiments, mostly with detached roots or lower plants. That is to say, a plot of $1/v$ against $1/(M)$ is usually a straight line, and the behavior of the plots agrees with concepts of competition based on chemical similarity of the ions involved. The fact that the equations employed were developed initially for enzyme reactions does not necessarily mean that the supposed carrier sites are on enzymes, although they may be. The equations merely describe hypothetical processes that might be carried out by enzymes as well as other substances.

In criticism of the carrier theory, one may say that it is not readily subjected to experimental test. Carrier sites have not been identified and cannot as yet be manipulated experimentally in a useful way. Hence, in the absence of restrictions and experimental verification, a wide variety of experimental results may be accounted for by postulating at will new carrier sites, combinations of different competitive behaviors, and fast and slow reactions. In time, methods of test may be devised; currently, the theory is providing a useful frame of reference and is the one most favored by plant physiologists.

Work by Fried et al. (1958) may be cited as an example of the type of inferences developed from application of the carrier theory. These investigators determined the total uptake of calcium by detached roots of barley from a solution of 10^{-4}-molar calcium chloride over different lengths of time and obtained the results shown in the upper line in Fig. 4.12. In a companion experiment, under similar conditions, roots were allowed to absorb radioactive calcium. At the end of each absorption period, the exchangeable and freely diffusible radioactive calcium were removed by placing the roots in 0.01-molar nonradioactive calcium chloride for 30 minutes. The radioactive calcium absorbed in nonexchangeable form by the roots was then determined. Results of these measurements are shown in the lower line in Fig. 4.12. The absorption from radioactive calcium chloride was carried out in two ways. In one series (solid points), the roots were left in the solution of radioactive calcium chloride for the full absorption period indicated on the abscissa. In the second series (crosses), the roots were removed from the radioactive calcium chloride solution after the first 15 minutes, washed with deionized water for 1 minute, and then placed in a saturated atmosphere at the same temperature for the remainder of the absorption period, at the end of which the exchangeable and remaining freely diffusible radioactive calcium were removed by placing the roots in a 0.01-molar calcium chloride solution like the others. The values for nonexchangeable radioactive calcium for the two series are almost identical, and the single lower line in the figure adequately represents both series.

The results shown in the figure demonstrate that the total uptake of calcium by the roots far exceeded the uptake in nonexchangeable form, and the rate of uptake in nonexchangeable form was not diminished by removal of the roots from the source of supply and by washing out most of the freely diffusible calcium that could have served as a source to replace the calcium absorbed in nonexchangeable form. From the last observation, one may infer that the decrease in concentration of calcium in sites from which absorption in nonexchangeable form was occurring was relatively small during the test period. The authors

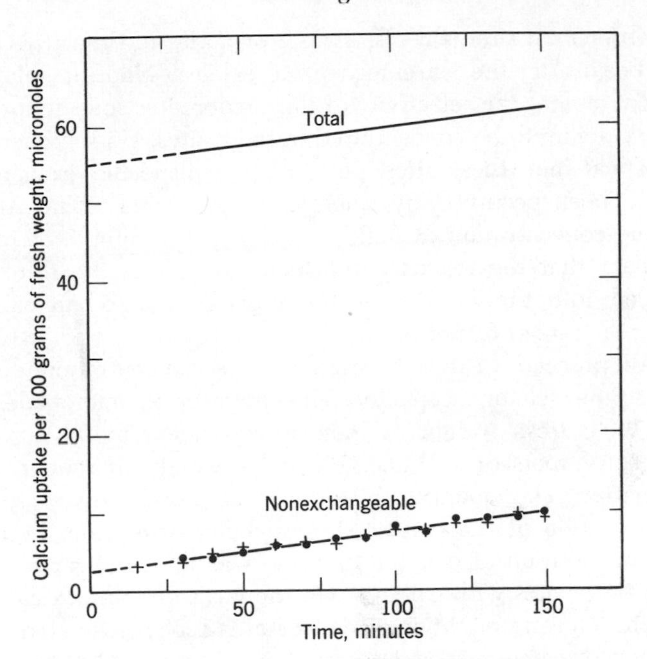

Figure 4.12. Total and nonexchangeable calcium taken up by detached barley roots during different lengths of time from a 10^{-4}-molar solution of calcium chloride at 25°C. See text for details. (Fried et al., 1958)

pointed out that analyses by their technique should have detected the effect of a decrease in concentration of calcium in active absorbing sites amounting to 10 to 15%. If, for purposes of calculation, the figure of 15% is taken as the decrease that did occur in concentration of calcium in active absorbing sites, the concentration of calcium in active absorbing sites must have been at least $(100/15) \times 6.6$ or 44 micromoles/100 g of roots, where the value 6.6 is the number of micromoles of calcium absorbed in 135 minutes (i.e., 150 minutes total time minus 15 minutes of time the roots were in the radioactive calcium chloride solution). If a 10% reduction in concentration of calcium in active absorption sites is assumed, the concentration of active absorption sites is estimated in like manner as 66 micromoles/100 g of roots. This value is unrealistic because the sum of the concentrations of calcium in active and inactive exchange sites and the calcium in freely diffusible form in the free space amounted to only 52 micromoles (the difference in intercepts of the two lines in Fig. 4.12). Consequently, it appears that almost all the calcium represented by the difference in intercepts was present in active absorption sites, that is, as *MR* or Ca*R*, and that this situation prevailed because in the extremely dilute calcium chloride solution employed (10^{-4} molar) the selectivity of the carrier sites specific for calcium caused the calcium to be concentrated in these sites. With higher concentrations of calcium, these sites plus other nonspecific exchange sites would have been occupied by calcium. If the value 52 micromoles is taken as the concentration of *MR* (=Ca*R*) in the sample of roots, one may calculate that the reaction in which calcium in the form of Ca*R* was released into the interior of the roots in nonexchangeable form occurred at the rate of 6% per hour.

One may proceed further to estimate the concentration of Ca*R* in terms of cation-exchange capacity. The quantity 52 micromoles of calcium per 100 g fresh weight of roots is equivalent to 0.104 m.e./100 g fresh weight of roots or 1.04 m.e./100 g dry weight of roots containing 10% dry matter. The quantity 1.04 does not necessarily represent the total concentration of the carrier sites specific for calcium because the external concentration of calcium chloride was not necessarily sufficient to saturate these sites with calcium. On the basis of complex calculations involving the enzyme equation, Fried et al. (1958) estimated the total concentration of calcium-specific carrier sites at 4 m.e./100 g dry weight of barley roots. Similar calculations for sodium and potassium yielded values of 2 m.e. for each. These values may be compared with a cation-exchange capacity of 12 to 24 m.e./100 g said by Fried et al. (1958) to be the range of values reported for the cation-exchange capacity of barley roots.

The carrier theory and the calculations described in the preceding paragraph provide an explanation for some of the difficulties that have been encountered in establishing a clear relationship between cation-exchange capacity of roots and uptake of cations in nonexchangeable form. The carrier sites account for only part of the exchange sites. Much of the exchange capacity consists of sites that appear to play no direct part in cation uptake. The latter exchange positions probably include those identified with carboxyl groups of pectates in the cell walls. The number and behavior of carrier sites specific for uptake of individual cations thus may not be reflected adequately in the number and behavior of the exchange sites as a whole.

Rate-Limiting Factors. In the experiment by Fried et al. (1958) discussed in the preceding section, a period of only 15 minutes or less was needed for accumulation of enough calcium in the free space to permit absorption of calcium in nonexchangeable form to proceed for 135 minutes without perceptible diminution in rate. The implication of this observation in terms of rate-limiting factors is that the roots took up calcium in freely diffusible and exchangeable form, including carrier sites, far more rapidly than they took up calcium in nonexchangeable form from the carrier sites. According to the carrier theory, therefore, the rate-limiting process in uptake of calcium in nonexchangeable form was the reaction

$$\mathrm{Ca}R = \mathrm{Ca}_{\mathrm{inside}} + R'$$

and not the formation of Ca*R* by combination of external calcium with the carrier *R*. According to one view, the process of release of cations into the interior in nonexchangeable form after combination with carrier sites is generally the slowest or rate-limiting step in absorption of cations by plants.

From another point of view, one may argue that the rate of uptake is frequently limited primarily by the soil and not by the plant. Circumstantial evidence is provided by the fact that the supply of soil potassium is frequently inadequate, to judge from the increases in uptake and growth associated with application of potassium fertilizer. Because of the obvious conflict between these points of view and the importance of a clear understanding, further examination of the situation seems worthwhile.

To place the roles of soil and plant in perspective, one must distinguish between rate-limiting processes and rate-limiting conditions. The second reaction,

$$MR = M_{\mathrm{inside}} + R',$$

represents a continuous change with time and hence is a process. If it is the slowest of all processes involved in cation uptake, it is properly designated the rate-limiting process. Certainly the evidence indicates that this process was much slower than the ones that preceded it in the experiment by Fried et al.

Now, according to the carrier theory, release of cations internally requires their pre-existence in the form *MR*; and conventionally the number of cations released internally from metal-carrier form per unit time is assumed to be proportional to the concentration of *MR*. But, according to the first reaction,

$$M_{\mathrm{outside}} + R = MR,$$

the concentration of *MR* depends on the concentration of M_{outside}. By controlling the concentration of M_{outside}, therefore, the soil may control the number of cations released internally per unit time. If the rate at which cations arrive at the carrier sites through the soil is great enough so that the concentration of M_{outside} and MR remains constant within the range in which R is not zero, the second reaction may be designated a rate-limiting process and the concentration of M_{outside} may be designated a rate-limiting condition. The rate-limiting process is then in the plant, and the rate-limiting condition is in the soil.

If, however, the uptake of a cation from soil by a plant root occurs rapidly enough to produce a local reduction in concentration of M_{outside} at the carrier sites, one rate-limiting condition (the concentration of M_{outside}) and two rate-limiting processes (the rate of transfer of cations from their source in the soil to the carrier sites and the rate of release of cations internally from carrier sites) are in effect. The importance of rate-limiting soil processes in uptake of cations by plants is not well established; nevertheless, the results of the experiment illustrated in Fig. 4.13 verify that such limitation may exist.

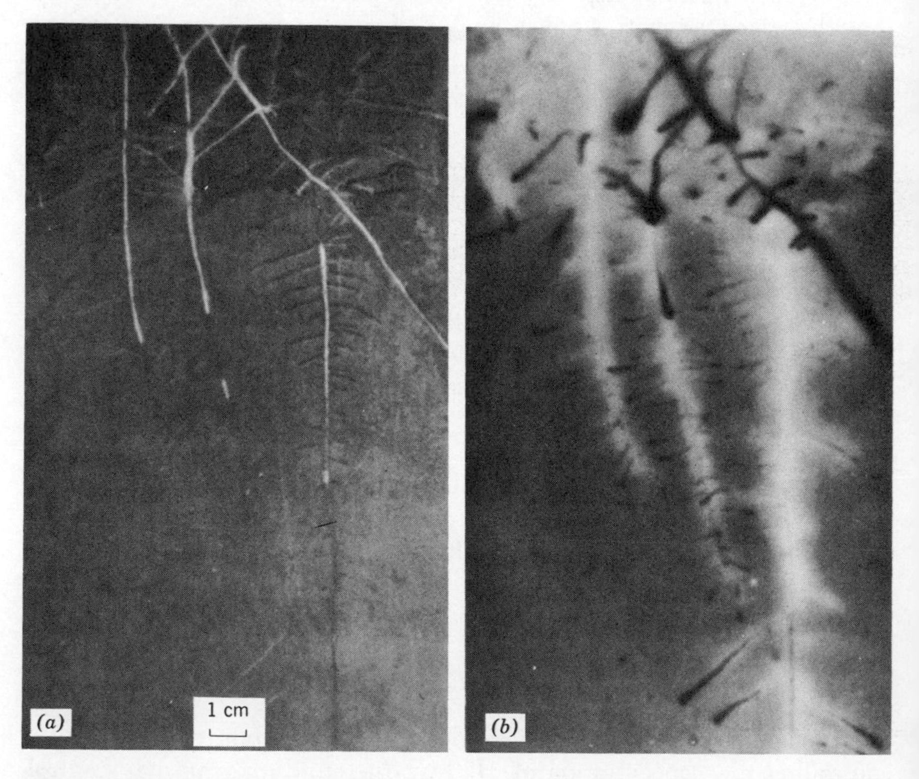

Figure 4.13. (*a*) Photograph of corn roots in soil and (*b*) autoradiograph of the same area showing distribution of rubidium-86. In (*b*), dark areas represent high rubidium content. Light areas represent zones from which rubidium has been removed. (Barber et al., 1963)

Control of the rate of uptake of cations by plants thus seems to reside partly in the plant and partly in the soil. To understand the over-all process requires appreciation of both plant and soil aspects. Both aspects need further investigation and elaboration.

Transport of Cations from Soil to Roots

Movement in Solution. Culture solutions are ordinarily stirred by the continuous bubbling of air through them; this keeps the composition of the bulk solution uniform despite local removal of ions by plant roots. In soils, mass movement of water carrying dissolved salts occurs far more slowly. The most rapid movement takes place at intervals during and after rainfall or irrigation. Much slower movement takes place during evaporation. The principal mass movement of significance in transporting ions to roots, however, is no doubt the slow movement of water to roots that occurs as plants withdraw water from the soil. Estimates made by Fried and Shapiro (1961) on the basis of cation uptake and water use by a high-yielding crop of corn indicate that, if transport of cations to plant root surfaces occurred only by mass movement of the ions in solution in the water used in transpiration, cation requirements of the plants could be met even if the concentrations of calcium, magnesium, and potassium were within the lower range of concentrations of these ions in soil solutions. Apparently, therefore, transport by mass movement can be significant. Barber (1962) drew a similar conclusion. A comparison of Jenny's (1966) calculated rates of transfer of cations by diffusion with published data on rates of uptake of cations indicates that diffusion also is an effective process in transport of cations through short distances such as those in cell walls and gelatinous coatings that may exist on roots.

Cation-Exchange Reactions. In the section on the over-all relationship in uptake of exchangeable bases by plants, an experiment by Jenny and Cowan (1933) was described in which there was evidence for an exchange of hydrogen ions derived from plants for exchangeable calcium taken up by the plants from the clay suspension on which the plants were grown. The background information about related matters set forth in succeeding pages now permits consideration of the mechanism of the exchange.

The cation-specific carrier sites involved in uptake of bases in nonexchangeable form may well have hydrogen as the counter-ion, as they are positioned prior to combination with incoming bases, so that the process of combination involves ion exchange. That such is the case is indicated by the inhibitory effects of hydrogen ions on uptake of bases in nonexchangeable form by plants, a subject that will be considered further in Chapter 5.

In addition to possible external release of hydrogen ions by exchange from cation-specific carrier sites, other hydrogen ions are derived from carbon dioxide released in root respiration. Carbon dioxide moves outward in the form of carbonic acid through the free space and into the soil. In the process, the hydrogen ions may be expected to equilibrate with carboxyl groups in root pectates. Some of these sites will therefore be present in the —COOH form. Most of the carbonic acid eventually decomposes to water and carbon dioxide, with loss of carbon dioxide to the atmosphere.

Observations on hydrogen-ion excretion by plants and the accumulation of titratable hydrogen in soils in consequence of cation uptake have given rise to two different theories of ion exchange in the nutrition of plants growing in soils: the carbonic acid theory and the contact-exchange theory. According to the carbonic acid theory (Burd, 1947; Jenny, 1957, pp. 107–132), hydrogen ions derived from carbonic acid excreted from roots exchange with exchangeable bases in the soil; the bases thus released as bicarbonates are then freely diffusible and may be absorbed by plant roots, with the hydrogen ions remaining in the soil in exchangeable form.

A saturated solution of carbonic acid in water has a *p*H of 4.0, which means that the maximum hydrogen-ion concentration obtainable from this source is about 0.0001-normal. Williams and Jenny (1952) found that the potassium released when they leached columns of soil with a saturated solution of carbonic acid depended on the volume of leachate and was independent of the rate of flow and the thickness of the soil column (25 g of soil were used in each case), which means that the soil potassium equilibrated rapidly with the carbonic acid. Moreover, the concentration of potassium in the carbonic acid leachate from 25 g of soil exceeded 0.1 μg/ml in the first 4.6 liters of leachate. Williams (1961) found that barley plants grew well and showed no potassium deficiency symptoms in culture solutions in which the potassium concentration was maintained at 0.01 μg/ml. This evidence verifies the effectiveness of the carbonic acid mechanism in releasing potassium from exchangeable form in concentrations sufficient for plant growth.

According to the contact-exchange theory (Overstreet and Jenny, 1939; Jenny, 1957, pp. 107–132), a direct exchange of cations between plant roots and soil particles occurs on contact because of the overlapping of atmospheres of exchangeable cations around roots and soil particles. The contact-exchange theory is not intended to imply that cation transfer must occur in this way, but rather that such transfer does occur as a normal process in nutrition of plants growing in soils. This theory has been supported by evidence of two principal kinds: (1) experiments with plants in which cation accumulation by roots or loss from roots took place more rapidly in suspensions of clays than in their equilibrium dialysates; and (2) experiments on nonbiological systems in which transfer of cations between solid adsorbents in the absence of appreciable quantities of diffusible anions took place more rapidly as a result of contact between adsorbents than of mass movement of cations in solution.

Of the two types of evidence, the second appears to be more nearly unequivocal. Lopez-Gonzalez and Jenny (1959) conducted an experiment with a cation-exchange resin that was prepared in the form of sheets with a thickness of 0.8 mm. A disk of such a membrane, with a diameter of 2.08 cm, containing 390 microequivalents of radioactive strontium and 42 microequivalents of exchangeable hydrogen, was placed in 150 ml of water. Initially the solution contained 0.54 microequivalent of strontium. A similar disk of hydrogen-saturated membrane was then placed in the same solution, with provision for prevention of direct physical contact between the two. The solution was agitated vigorously to provide effective transfer of ions between the membranes by mass movement. At intervals the membrane initially saturated with hydrogen was removed, and the radioactivity was determined to provide a measure of the quantity of strontium that had been accumulated. Two analogous disks were pressed together to determine the transfer that occurred when the disks were in close contact without provision for mass movement of solution between them. The results of the measurements in Table 4.13 show that accumulation of strontium by the disk initially saturated with hydrogen was far more rapid where the transfer occurred by direct contact than where it occurred by mass transfer of ions through the solution. Within 2 days, 87% of the equilibrium distribution of strontium between the two disks had been reached where they were in close contact. Many years would evidently have been required to achieve the same transfer through the solution.

Experiments with plants have produced conflicting results. In some experiments, uptake has been greater from a clay suspension than from the equilibrium dialysate (a solution in equilibrium with the suspension but containing no clay particles); in others there has been no difference. Papers illustrating different experimental techniques and findings and different viewpoints include those by Lagerwerff (1960), Olsen and Peech (1960), Scheuring and Overstreet (1961), Helmy and Oliver (1961), Franklin and McLean (1963), Werkhoven and Ohlrogge (1963), and Frere and Axley (1964). This work will not be reviewed here in detail.

In view of the negative results and the experimental difficulties, further

Table 4.13. Transfer of Strontium from a Disk of Strontium-Hydrogen Cation-Exchange Membrane to a Similar Disk of Hydrogen Membrane by Mass Movement of a Vigorously Agitated Solution and by Direct Contact Between the Membranes (Lopez-Gonzalez and Jenny, 1959)

	Strontium Transferred to the Disk of H-Saturated Membrane, microequivalents	
Time, days	By Mass Movement of the Solution	By Contact
1	—	130
2	0.090	170
7	0.126	192
23	0.140	—
65	0.175	—
82	0.180	—
(Equilibrium)	195	195

development of technique and new experiments will be needed before contact-exchange can be said to be established as a significant mode of transfer of cations between soil and plant roots. Experimental difficulties include the following: (1) Adherence of clay particles to roots may give high values for cation uptake from clay suspensions. (2) Greater depletion of cations from dialysates than from clay suspensions during experiments may give low values for cation uptake from dialysates. (3) Greater loss of calcium from roots to clay suspensions containing little or no exchangeable calcium than to dialysates may result in greater uptake of monovalent cations from clay suspensions than from dialysates because of the differential effects on root permeability. The third difficulty has been recognized only recently. Calcium seems to have a special function in regulating root permeability. Jacobson et al. (1961) published a paper on this subject.

Replacement of exchangeable bases from soils by hydrogen ions from carbonic acid can readily be demonstrated experimentally in the absence of plants. Occurrence of similar replacement from the carbonic acid excreted from plant roots seems reasonable. Replacement of exchangeable bases in soil by hydrogen ions by contact exchange with plant roots also seems reasonable in principle, even though there are experimental difficulties in demonstrating unequivocally the occurrence and magnitude of such exchange. The direct contact-exchange may well involve only exchange positions in the free space if the hypothesis that carrier sites are located at plasma membranes should be well founded; however, exchangeable cations may migrate from one position to another along surfaces or presumably also through cell walls. Further comments on the significance of these exchange mechanisms will be found in the last section of this chapter.

Release of Bases from Exchangeable Form in Relation to Uptake by Plants

Plants grown in culture solutions rarely absorb ions in the proportions in which the ions are supplied; rather, they act selectively, absorbing greater proportions of some ions than others. The selectivity varies with the kind of plant. Therefore there is no reason to expect that plants grown in soil will absorb ions in the proportion in which they occur in the soil solution, in the exchangeable form, or in any other form. The proportionate quantities of the bases absorbed by plants can be estimated, nevertheless, on the basis of empirical procedures that take into account both the selectivity of plants for ions and the tendency of soils to release bases from the exchangeable form to the soluble form. Two different approaches will be considered here.

The first approach, measurement of bases released from soil in a fractional exchange, was used by Mehlich and co-workers in North Carolina. Much of this work was reviewed by Mehlich and Coleman (1952). Profiting from Bray's (1942) earlier work, these investigators determined the quantities of the various bases in solution after a fractional exchange brought about by addition of a small quantity of hydrochloric acid.

An experiment by Milam and Mehlich (1954) may be cited to illustrate the method and the type of results obtained. *Crotolaria striata* was grown on five different soils in the greenhouse. Each soil had received various treatments to provide four or more different proportions of exchangeable calcium, magnesium, and potassium. The proportionate content of the various bases in the plants was estimated from the soil measurements by means of an equation that may be simplified to the following form for calcium:

$$\frac{Ca_{plant}}{Ca_{plant} + Mg_{plant} + K_{plant}} = \frac{Ca_{HCl}}{Ca_{HCl} + \alpha Mg_{HCl} + \beta K_{HCl}}$$

where Ca_{plant}, Mg_{plant}, and K_{plant} represent the milligram equivalents of the respective bases per 100 g of plant material, where Ca_{HCl}, Mg_{HCl}, and K_{HCl} represent the milligram equivalents of the respective bases in

Table 4.14. Correlation of Proportionate Content of Calcium in *Crotolaria striata* with Different Measures of Soil Calcium, Based on Data of Milam and Mehlich (1954)

Variable Correlated with $Ca_{plant}/(Ca_{plant} + Mg_{plant} + K_{plant})$	Correlation Coefficient (r)
Exchangeable calcium ($Ca_{exch.}$)	0.49
Calcium released in fractional exchange with hydrochloric acid (Ca_{HCl})	0.89
$Ca_{exch.}/(Ca_{exch.} + 2.31\ Mg_{exch.} + 3.21\ K_{exch.})$[1]	0.94
$Ca_{HCl}/(Ca_{HCl} + 0.97\ Mg_{HCl} + 3.41\ K_{HCl})$[1]	0.95

[1] Constants fitted by the method of least squares.

the extract obtained by shaking 100 g of soil for 15 minutes with 500 ml of water containing 1 m.e. of hydrochloric acid, and where α and β are constants, representing the selectivity of the plants for magnesium and potassium relative to calcium in the solution. An analogous equation was used to estimate the proportionate content of the other bases in plants.

The comparative precision with which the proportionate content of calcium in the plants was estimated by the method mentioned and by several others is indicated in Table 4.14 by correlation coefficients calculated by Ethel Tyler of Cornell University. Evidently the precision of estimation (as indicated by the magnitude of the correlation coefficients) was greater if all ions were considered than if only calcium was considered, as would be expected from the existence of complementary-ion effects. Also, the precision of estimation was greater where estimates were based on ions released in a fractional exchange than on the total exchangeable ions, as would be expected if the ions available for absorption by the plants were those released in a fractional exchange.

The second approach to estimation of the proportionate content of bases in plants is of a more physicochemical nature than the first. A certain amount of background explanation is needed to provide an understanding of the reasoning behind the procedure.

According to theory, the activity of an electrolyte such as calcium chloride must be the same at equilibrium in the bulk portion of a solution in contact with soil as in the portion of the solution lying within the atmospheres of exchangeable cations around individual soil particles. The activity of an electrolyte, each mole of which dissociates into $v+$ cations and $v-$ anions, is given by $(a_{cation})^{v+}\ (a_{anion})^{v-}$, where the *a*'s are activities representing the product of the molar concentration and an activity coefficient. The chloride ions nominally associated with calcium are equally associated with every other cation species; hence, in calculation of the activity of each chloride salt in the bulk solution, the total concentration of chloride is used. Similarly, the chloride ions in solution in the ionic atmospheres around the soil particles are equally associated with calcium and with all other cation species whether nominally in solution or in exchangeable form; hence, in calculation of the activity of each chloride salt in the ionic atmospheres, the total chloride again, in theory, may be used. Because of the relatively high concentration of exchangeable cations in the ionic atmospheres, the activity of chloride ions within the ion atmospheres must be much less than it is in the bulk solution if the activity of calcium chloride, for example, is to be the same in the solution within the atmospheres of exchangeable cations as in the bulk solution. This prediction from theory has been well verified in a qualitative way by measurements on soils [see, for example, a paper by Bower and Goertzen (1955)] showing that the concentration of soluble anions is lower within the ionic atmospheres than in the bulk solution. However, it is not yet possible to determine the activity of anions within the ionic atmospheres; and the activity of the cations is not known.

The problem of the unknown cation and anion activities in the ionic atmospheres may be circumvented in part. If a bulk solution is in equilibrium with soil, one may write equalities for activities of the calcium, magnesium, and potassium salts of any common anion as follows, where the common anion is taken as chloride:

$$\begin{matrix}\text{Activity of salt in} \\ \text{bulk solution}\end{matrix} = \begin{matrix}\text{Activity of salt in} \\ \text{ionic atmosphere}\end{matrix}$$

$$(a_{\mathrm{Ca}_b^{++}})(a_{\mathrm{Cl}_b^-})^2 = (a_{\mathrm{Ca}_i^{++}})(a_{\mathrm{Cl}_i^-})^2$$

or

$$(a_{\mathrm{Ca}_b^{++}})^{1/2}(a_{\mathrm{Cl}_b^-}) = (a_{\mathrm{Ca}_i^{++}})^{1/2}(a_{\mathrm{Cl}_i^-});$$

$$(a_{\mathrm{Mg}_b^{++}})(a_{\mathrm{Cl}_b^-})^2 = (a_{\mathrm{Mg}_i^{++}})(a_{\mathrm{Cl}_i^-})^2$$

or

$$(a_{\mathrm{Mg}_b^{++}})^{1/2}(a_{\mathrm{Cl}_b^-}) = (a_{\mathrm{Mg}_i^{++}})^{1/2}(a_{\mathrm{Cl}_i^-});$$

and

$$(a_{\mathrm{K}_b^+})(a_{\mathrm{Cl}_b^-}) = (a_{\mathrm{K}_i^+})(a_{\mathrm{Cl}_i^-});$$

where the subscripts b and i refer to the solution in bulk and the solution within the ionic atmospheres (including the exchangeable cations). Because the total chloride and its activity coefficient are the same for all species of chloride salts in the bulk solution and for all species of chloride salts in the ionic atmospheres, the second and third equations may be divided by the first, yielding ratios that do not contain chloride:

$$\frac{(a_{\text{Mg}_b^{++}})}{(a_{\text{Ca}_b^{++}})} = \frac{(a_{\text{Mg}_i^{++}})}{(a_{\text{Ca}_i^{++}})}$$

or

$$\frac{(a_{\text{Mg}_b^{++}})^{1/2}}{(a_{\text{Ca}_b^{++}})^{1/2}} = \frac{(a_{\text{Mg}_i^{++}})^{1/2}}{(a_{\text{Ca}_i^{++}})^{1/2}};$$

and

$$\frac{(a_{\text{K}_b^{+}})}{(a_{\text{Ca}_b^{++}})^{1/2}} = \frac{(a_{\text{K}_i^{+}})}{(a_{\text{Ca}_i^{++}})^{1/2}}$$

The last two equations show the manner in which activity ratios of ions in a bulk solution in equilibrium with a soil are related to the activity ratios of the corresponding ions in the ionic atmospheres around the soil particles. (The cation valences have been shown to make clear the connection between the valences and the exponents. Usually the valences are omitted for simplicity of representation.) From these equations, it may be perceived that suitable measurements in solutions in equilibrium with soils can provide information about ratios of activities of cations in the ionic atmospheres. The next step is to consider the way in which such measurements may be made.

If a salt solution is added to soil, an exchange of cations between solution and soil will occur. From analysis of the solution after equilibration with the soil, one can calculate the activity ratios of cations present; these presumably are equal to the activity ratios of the same cations in the ion atmospheres if formulated as just described. If much salt has been added, the activity ratios found for the cation species that occur in exchangeable form in soils will not represent the activity ratios initially existing in the ion atmospheres because of the extensive ion exchange that occurred during equilibration. On the other hand, if relatively little salt is added, and if the ratio of solution to soil is not too great, the activity ratios found in solution will provide an estimate of the initial activity-ratios of cations in the ion atmospheres in the soil. Schofield and Taylor (1955) verified the constancy of the ratios for dilute solutions differing in concentration.

To determine the activity ratios of exchangeable cations, Taylor

(1958) used the technique of equilibrating samples of soil with solutions containing different concentrations of the chlorides of calcium and potassium and of determining the activities of calcium, magnesium, and potassium in the equilibrium solutions. His analytical method measured the sum of calcium and magnesium in the equilibrium solutions, which means that, in effect, differences in behavior between magnesium and calcium were neglected. The initial ratios of calcium to potassium in the solution added were chosen to avoid an extensive exchange of calcium for potassium in the soil. The equilibrium ratio in the solution after contact with soil may then be plotted against the initial ratio in the solution added. The value of the initial ratio that yields an equal activity-ratio in the equilibrium solution, found by interpolation, is taken as an estimate of the initial activity-ratio of exchangeable cations in the soil. Figure 4.14 illustrates the process for two experimental series

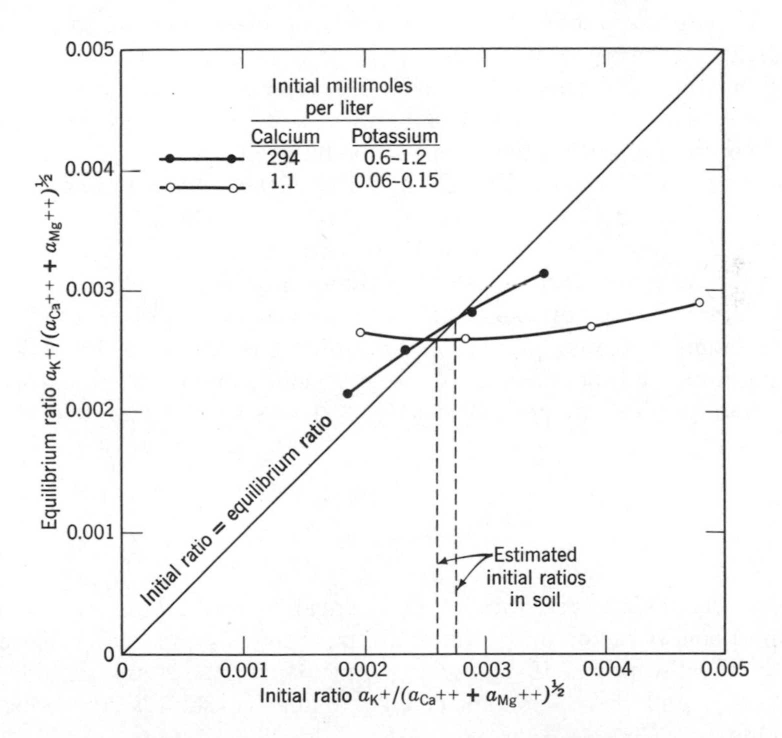

Figure 4.14. Plot of equilibrium values against initial values of $a_{K^+}/(a_{Ca^{++}} + a_{Mg^{++}})^{1/2}$ in solutions of the chlorides of calcium and potassium added to a soil in the ratio of two parts of solution to one of soil by weight, illustrating for two experimental series the technique of finding the composition of the initial solution in equilibrium with the soil. (Taylor, 1958)

with one soil. Ideally, all such trials should yield the same intercept. The intercepts obtained with six different experimental series were similar for this soil. The variation was greater with some soils.

This theoretical and experimental background provides a basis for testing the suitability of measurements of activity ratios of exchangeable bases in estimation of the proportionate uptake of different bases from soil by plants. The activity ratios of bases in exchangeable form should be equal to the activity ratios of bases released to the soil solution in an infinitesimal exchange with hydrogen ions derived from carbonic acid excreted by plants, from production of nitric acid in soil by nitrification, or from contact exchange, or in an infinitesimal exchange involving other ions. As explained previously, plants absorb ions selectively from the medium in which they grow. It is not to be anticipated, therefore, that the process of estimating activity ratios of cations in soil will in any way affect this property. Conceivably, however, activity ratios may reflect more accurately than other kinds of measurements the relative tendency of bases to be given up by soil in a nonselective removal. If, then, due allowance is made for the selective qualities of different kinds of plants, activity ratios of bases in soils may provide useful estimates of the proportionate content of the bases in plants.

Salmon (1964) reported results of a test of the activity-ratio approach. He used the technique just described for estimating the ratios $a_K/(a_{Ca+Mg})^{1/2}$ and $(a_{Mg})^{1/2}/(a_{Ca+Mg})^{1/2}$ in two soils, each adjusted to four different ratios of exchangeable calcium to exchangeable magnesium, with four values of exchangeable potassium at each ratio of calcium to magnesium. Ryegrass grown on the soils was analyzed for calcium, magnesium, and potassium. The magnesium content of the grass in milligram equivalents per 100 g (Mg_{grass}) was then represented by the equation

$$Mg_{grass} = \frac{k'(a_{Mg})^{1/2}}{\alpha(a_{Ca+Mg})^{1/2} + \beta a_K}$$

where the a's are activities of exchangeable ions in the soil, k' is a proportionality factor, and α and β are parameters representing the selectivity of absorption. If numerator and denominator are divided by $\alpha(a_{Ca+Mg})$, and if k'/α is replaced by a new constant k, the equation becomes

$$Mg_{grass} = k\left[\frac{(a_{Mg})^{1/2}}{(a_{Ca+Mg})^{1/2}}\right]\left[\frac{1}{1 + \dfrac{\beta a_K}{\alpha(a_{Ca+Mg})^{1/2}}}\right].$$

If the ratio of the selectivity factors β/α is defined by

$$B = \frac{\beta}{\alpha} = \left(\frac{K}{Ca + Mg}\right)_{grass} \Bigg/ \frac{a_K}{(a_{Ca+Mg})^{1/2}},$$

a value for this ratio may be determined by plotting $[K/(Ca + Mg)]_{grass}$ against $[a_K/_{Ca+Mg})^{1/2}]$ and taking the slope of the resulting line as B. The equation then becomes

$$Mg_{grass} = k\left[\frac{(a_{Mg})^{1/2}}{(a_{Ca+Mg})^{1/2}}\right]\left[\frac{1}{1 + \dfrac{Ba_K}{(a_{Ca+Mg})^{1/2}}}\right].$$

The equation in this form shows the ion activity ratios as measured. Rearrangement yields an equation in a form similar to the original:

$$Mg_{grass} = \frac{k(a_{Mg})^{1/2}}{(a_{Ca+Mg})^{1/2} + Ba_K}$$

A plot of the magnesium content of the grass against the values calculated for the right side of the equation (excluding k) yielded the results shown in Fig. 4.15. A close relationship was obtained for each soil, indicating good precision of estimation from the measurements of activity ratios.

The two approaches to estimation of bases in plants from soil analyses may now be placed in perspective by a few general remarks. The over-all process of interest involves release of exchangeable cations from soils and their uptake by plants. The first part of the process is affected by the proportions in which the exchangeable cations occur and by the selectivities operating in the soil. The fractional exchange approach is one method of taking into account these two factors, and the activity-ratio approach is another. Both methods have been shown to provide more precise estimates of cations in plants than do measurements of the total quantities of individual exchangeable cations. It is not yet clear which method is the better.

There are essentially two differences between the two methods. First, the activity-ratio method employs activity coefficients, whereas the fractional exchange method does not, as it has been used in the past. Activity coefficients could be used with the fractional exchange method if desired. In connection with the use of activity coefficients, it should be noted that it has not been established whether activities are more appropriate than concentrations for characterizing solutions as sources of ions for plant uptake. Unpublished work of K. L. Babcock at the University of California has shown that absorption of a given cation by barley

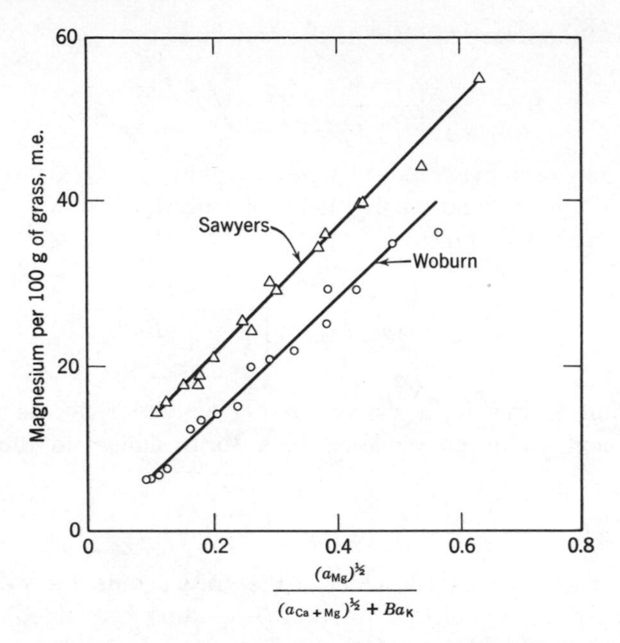

Figure 4.15. Magnesium content of ryegrass grown on two soils from Great Britain versus values of $(a_{Mg})^{1/2}/[(a_{Ca+Mg})^{1/2} + Ba_K]$. (Salmon, 1964)

roots from different salt solutions is not the same even though the activity of the cation is the same in the various solutions. This has long been known to be true where concentrations are concerned.

The second essential difference between the two approaches is that the activity-ratio method uses different exponents than the fractional exchange method. Any desired exponents could be used with the fractional exchange method, but the comparative value of using different ones has not been determined. In connection with the appropriateness of different exponents, two experiments by Broeshart (1962) may be mentioned. In the first experiment on uptake of sodium and calcium in exchangeable form by plant roots, he found that the proportionate uptake of the two ions could be described fairly well by the ratio of the concentration of sodium to the square root of the concentration of calcium. In the second experiment, he found that the proportionate contents of potassium, calcium, and magnesium in the tops of oats grown to maturity in sand cultures supplied continuously with various nutrient solutions were related to the ratio of the concentrations of the cations in the solutions and not to the ratio of the concentration of potassium to the square root of the concentration of calcium or magnesium. These results suggest that use of the ratio of the concentration or activity of monovalent ions to the square root of the concentration or activity of divalent ions is useful for estimating the relative uptake of the ions in exchangeable form but that it is not the best way of estimating the relative uptake in nonexchangeable form.

A limitation common to both approaches is the fact that as plants remove bases from the soil, changes are to be expected in both activities and concentrations and in both activity ratios and concentration ratios. Hence the removal of a substantial proportion of one or more of the exchangeable bases during growth may be expected to produce changes in soils that are not reflected adequately by a single set of measurements with either method. Salmon (1964) discussed this matter and pointed out that the finding of a different relationship for the two soils in Fig. 4.15 may have resulted from such changes in the soils subsequent to measurement. Ideally, the relationship for a given kind of plant at a given stage of growth should be the same with different soils.

Ion Ratios versus Ion Concentrations

Both the methods in the preceding section are based on ratios of cations in solution in equilibrium with soils on which plants are to be grown. Use of ratios is peculiar to research involving exchangeable bases. Concentrations are used in most soil tests for the supply of individual nutrients, including potassium, in soils. One may reasonably inquire, therefore, to what extent the nutritional environment of plants with regard to cationic nutrients is defined by the ratios of the cations without reference to the absolute concentrations or activities of these ions in the soil solution. This question probably cannot be answered satisfactorily on the basis of information now available, but certain aspects of the subject may be presented for consideration.

In support of the use of ratios in the two methods described in the last section, one may argue from the following line of reasoning that ratios in equilibrium solutions have more significance for cations than for anions. If a soil is leached thoroughly with water to remove soluble salts and is then kept sterile with no plants growing in it, the concentration of soluble cations remains exceedingly low for an indefinite length of time. If microbial activity is introduced, however, mineral acids (carbonic, nitric, sulfuric) are produced; these bring cations into solution by exchange. The number of equivalents of the cations in solution is determined almost completely by the number of equivalents of nitrate, sulfate, and bicarbonate anions resulting from microbial activity. Nitrate is quantitatively the most important. Because of this process, it is reasonable to say that under natural conditions and in nonsaline soils the concentration of nutrient anions largely controls the concentration of soluble cations. Moreover, as the salt concentration varies under the influence of microbial activity and removal of ions by plants and leaching, the concentrations of all the cations increase and decrease together (although the ratios are not constant) because of continuous equilibration of the ions in the soil solution with those in exchangeable form.

To carry the argument further, one may perceive that as long as the supply of nutrient anions in the soil at root surfaces is adequate to meet plant needs, the supply of total cations present at root surfaces is adequate also because anions are accompanied by an equal number of equivalents of cations and the number of equivalents of anionic nutrients taken up by plants exceeds the number of equivalents of cationic nutrients taken up. If the concentration of soluble salts is low enough to result in deficiency of anionic nutrients, the cation requirements will be reduced; but plants will still be assured of an adequate total supply of cations through release from exchangeable form by the carbonic acid and contact-exchange mechanisms. At this point, the experiment by Jenny and Cowan (1933) demonstrating accumulation of titratable acidity in clay approximately equivalent to the calcium absorbed by soybean seedlings from the clay may be placed in perspective more clearly than heretofore. In this experiment, externally derived anions were not of significance to the plants because the concentrations were too low. (The clay used had been electrodialyzed and then saturated with calcium by addition of calcium hydroxide. No nutrients other than calcium were added.) But the hydrogen ions supplied by the plants served as exchanging ions that released calcium and permitted the plants to take up calcium without taking up nutrient anions. Because of the conditions in this experiment, the titratable acidity that developed in the clay during the test period may be attributed to these plant-derived hydrogen ions. On the other hand, if cations are released from exchangeable form to the soil solution by microbially derived acids, titratable acidity develops in the soil from this cause. The normal uptake of more anions than cations by plants should then reduce the titratable acidity produced by the microorganisms.

One final aspect deserving of comment is that competition for internal position in plants is greater among cationic than among anionic nutrients. The proportion in which cations occur externally affects this competition. Anionic nutrients in plants remain in part as anions but are metabolized to a considerable degree to organic forms that no longer enter into the ionic balance. Although individual cations have specific functions, they exist in plants for the most part in soluble form, balancing organic and inorganic anions. Considerable mutual substitution of cations in plants is possible without much effect on metabolism, growth, and yield and without much variation in total number of cation equivalents; a paper on this subject was published by van Itallie (1938). If a particular cation makes up a high proportion of the cations externally, its rate of absorption relative to the other cations is increased, and its competitive ability relative to the other cations for the fairly constant number of internal cation equivalents is increased. The consequence is that the ion in high proportion externally tends to be in high proportion internally and that the proportion of other cations in the plant is reduced.

Literature Cited

Anderson, M. S., M. G. Keyes, and G. W. Cromer. (1942) Soluble material of soils in relation to their classification and general fertility. *U.S. Dept. Agr. Tech. Bul. 813.*

Andrew, R. W., M. L. Jackson, and K. Wada. (1960) Intersalation as a technique for differentiation of kaolinite from chloritic minerals by X-ray diffraction. *Soil Sci. Soc. Amer. Proc.* **24**:422–424.

Barber, S. A. (1962) A diffusion and mass-flow concept of soil nutrient availability. *Soil Sci.* **93**:39–49.

Barber, S. A., J. M. Walker, and E. H. Vasey. (1963) Mechanisms for the movement of plant nutrients from the soil and fertilizer to the plant root. *Jour. Agr. Food Chem.* **11**:204–207.

Baver, L. D. (1928) The relation of exchangeable cations to the physical properties of soils. *Jour. Amer. Soc. Agron.* **20**:921–941.

Baver, L. D. (1930) The effect of organic matter upon several physical properties of soils. *Jour. Amer. Soc. Agron.* **22**:703–708.

Bear, F. E., A. L. Prince, and J. L. Malcolm. (1945) Potassium needs of New Jersey soils. *New Jersey Agr. Exp. Sta. Bul. 721.*

Blume, J. M., and D. Smith. (1954) Determination of exchangeable calcium and cation-exchange capacity by equilibration with Ca^{45}. *Soil Sci.* **77**:9–17.

Borland, J. W., and R. F. Reitemeier. (1950) Kinetic exchange studies on clays with radioactive calcium. *Soil Sci.* **69**:251–260.

Bower, C. A., and J. O. Goertzen. (1955) Negative adsorption of salts by soil. *Soil Sci. Soc. Amer. Proc.* **19**:147–151.

Bower, C. A., and E. Truog. (1941) Base exchange capacity determination as influenced by nature of cation employed and formation of basic exchange salts. *Soil Sci. Soc. Amer. Proc.* (1940) **5**:86–89.

Bray, R. H. (1942) Ionic competition in base-exchange reactions. *Jour. Amer. Chem. Soc.* **64**:954–963.

Briggs, G. E., and R. N. Robertson. (1957) Apparent free space. *Annual Rev. Plant Physiol.* **8**:11–30.

Broadbent, F. E., and G. R. Bradford. (1952) Cation-exchange groupings in the soil organic fraction. *Soil Sci.* **74**:447–457.

Broeshart, H. (1962) Cation adsorption and absorption by plants. *In Radioisotopes in Soil-Plant Nutrition Studies,* pp. 303–313. International Atomic Energy Agency, Vienna.

Burd, J. S. (1947) Mechanism of release of ions from soil particles to plants. *Soil Sci.* **64**:222–225.

Cernescu, N. C. (1931) Cation exchange and structure. Comparative studies with clay, permutite and chabazite. *Annuar. Inst. Geol. Romaniei* **16**:777–859. [*Chem. Abstr.* **29**:3897 (1935).]

Chapman, H. D. (1965) Cation-exchange capacity. *Agronomy* **9**:891–901.

Chapman, H. D. (1965a) Total exchangeable bases. *Agronomy* **9**:902–904.

Chapman, H. D., and W. P. Kelley. (1930) The determination of the replaceable bases and the base-exchange capacity of soils. *Soil Sci.* **30**:391–406.

Craig, N., and P. Halais. (1934) The influence of maturity and rainfall on the properties of lateritic soils in Mauritius. *Empire Jour. Exptl. Agr.* **2**:349–358.

Davis, D. E., W. H. MacIntire, C. L. Comar, W. M. Shaw, S. H. Winterberg, and H. C. Harris. (1953) Use of Ca^{45} labeled quenched calcium silicate slag in determination of proportions of native and additive calcium in lysimeter leachings and in plant uptake. *Soil Sci.* **76**:153–163.

DeMumbrum, L. E., and M. L. Jackson. (1956) Copper and zinc exchange from dilute neutral solutions by soil colloidal electrolytes. *Soil Sci.* **81**:353–357.

Elgabaly, M. M., and H. Jenny. (1943) Cation and anion interchange with zinc montmorillonite clays. *Jour. Physical Chem.* **47**:399–408.

Ensminger, L. E., and J. E. Gieseking. (1941) The absorption of proteins by montmorillonitic clays and its effect on base-exchange capacity. *Soil Sci.* **51**:125–132.

Epstein, E. (1960) Spaces, barriers, and ion carriers: ion absorption by plants. *Amer. Jour. Bot.* **47**:393–399.

Epstein, E., and C. E. Hagen. (1952) A kinetic study of the absorption of alkali cations by barley roots. *Plant Physiol.* **27**:457–474.

Fieldes, M., L. D. Swindale, and J. P. Richardson. (1952) Relation of colloidal hydrous oxides to the high cation-exchange capacity of some tropical soils of the Cook Islands. *Soil Sci.* **74**:197–205.

Fisher, T. R. (1963) Measurements of adsorbed cations in soils employing radioactive isotope equilibration methods. *Diss. Abstr.* **23**:2284.

Franklin, R. E., Jr., and E. O. McLean. (1963) Effect of electrolyte concentration on Donnan systems and the resulting uptake of cations by plants. *Soil Sci. Soc. Amer. Proc.* **27**:137–141.

Frere, M. H., and J. H. Axley. (1964) Cation uptake by excised barley roots from solutions and suspensions. *Soil Sci.* **97**:209–213.

Fried, M., J. C. Noggle, and C. E. Hagen. (1958) The relationship between adsorption and absorption of cations. *Soil Sci. Soc. Amer. Proc.* **22**:495–499.

Fried, M., and R. E. Shapiro. (1961) Soil-plant relationships in ion uptake. *Annual Rev. Plant Physiol.* **12**:91–112.

Frink, C. R. (1964) The effects of wash solvents on cation-exchange capacity measurements. *Soil Sci. Soc. Amer. Proc.* **28**:506–511.

Giese, A. C. (1962) *Cell Physiology.* Second ed. W. B. Saunders Co., Philadelphia.

Gore, A. J. P., and S. E. Allen. (1956) Measurement of exchangeable and total cation content for H^+, Na^+, K^+, Mg^{++}, Ca^{++}, and iron, in high level blanket peat. *Oikos* **7**:48–55.

Graham, E. R. (1940) Primary minerals of the silt fractions as contributors to the exchangeable-base level of acid soils. *Soil Sci.* **49**:277–281.

Grim, R. E. (1953) *Clay Mineralogy.* McGraw-Hill Book Co., New York.

Heald, W. R. (1965) Calcium and magnesium. *Agronomy* **9**:999–1010.

Heinonen, R. (1960) Über die Umtauschkapazität des Bodens und verschiedener Bodenbestandteile in Finnland. *Zeitschr. Pflanzenernähr., Düng., Bodenk.* **88**: 49–59.

Heintze, S. G. (1961) Studies on cation-exchange capacities of roots. *Plant and Soil.* **13**:365–383.

Heintze, S. G., and P. J. G. Mann. (1949) Studies on soil manganese. *Jour. Agr. Sci.* **39**:80–95.

Helmy, A. K., and S. Oliver. (1961) Cation absorption by excised barley roots from soil suspensions and their equilibrium true solutions at different time intervals. *Soil Sci.* **91**:339–340.

Hendricks, S. B., R. A. Nelson, and L. T. Alexander. (1940) Hydration mechanism of the clay mineral montmorillonite saturated with various cations. *Jour. Amer. Chem. Soc.* **62**:1457–1464.

Hissink, D. J. (1925) Base exchange in soils. *Trans Faraday Soc.* **20**:551–566.

Itallie, T. B. van. (1938) Cation equilibria in plants in relation to the soil. *Soil Sci.* **46**:175–186.

Jacobson, L., R. J. Hannapel, M. Schaedle, and D. P. Moore. (1961) Effect of root to solution ratio in ion absorption experiments. *Plant Physiol.* **36**:62–65.

Jarusov, S. S. (1937) On the mobility of exchangeable cations in the soil. *Soil Sci.* **43**:285–303.

Jenny, H. (1957) Contact phenomena between adsorbents and their significance in plant nutrition. *In* Emil Truog (Ed.) *Mineral Nutrition of Plants.* University of Wisconsin Press, Madison, Wisc.

Jenny, H. (1966) Pathways of ions from soil into root according to diffusion models. *Plant and Soil* **25**:265–289.

Jenny, H., and E. W. Cowan. (1933) Über die Bedeutung der im Boden adsorbierten Kationen für das Pflanzenwachstum. *Zeitschr. Pflanzenernähr., Düng. Bodenk.* **31A**:57–76.

Jenny, H., and K. Grossenbacher. (1963) Root-soil boundary zones as seen in the electron microscope. *Soil Sci. Soc. Amer. Proc.* **27**:273–277.

Keller, P., and H. Deuel. (1957) Kationenaustauschkapazität und Pektingehalt von Pflanzenwurzeln. *Zeitschr. Pflanzenernähr., Düng., Bodenk.* **79**:119–131.

Kelley, W. P. (1948) *Cation Exchange in Soils.* A.C.S. Monograph No. 109. Reinhold Publishing Corp., New York.

Kelley, W. P., and S. M. Brown. (1924) Replaceable bases in soils. *California Agr. Exp. Sta. Tech. Paper 15.*

Krishnamoorthy, C., and R. Overstreet. (1950) An experimental evaluation of ion-exchange relationships. *Soil Sci.* **69**:41–53.

Lagerwerff, J. V. (1960) The contact-exchange theory amended. *Plant and Soil* **13**:253–264.

Laties, G. C. (1959) Active transport of salt into plant tissue. *Annual Rev. Plant Physiol.* **10**:87–112.

Lopez-Gonzalez, J. de, and H. Jenny. (1959) Diffusion of strontium in ion-exchange membranes. *Jour. Colloid Sci.* **14**:533–542.

Marschner, H., and K. Mengel. (1962) Apparent free space (AFS). *Zeitschr. Pflanzenernähr., Düng., Bodenk.* **98**:30–44.

Marshall, C. E. (1964) *The Physical Chemistry and Mineralogy of Soils. Vol. 1: Soil Materials.* John Wiley and Sons, New York.

Marttila, U. (1965) Exchangeable cations in Finnish soils. *Maat. Aikak.* **37**:148–161.

McAleese, D. M., and W. A. Mitchell. (1958) Studies on the basaltic soils of northern Ireland. V. Cation-exchange capacities and mineralogy of the silt separates (2–20μ). *Jour. Soil Sci.* **9**:81–88.

McAuliffe, C. D., N. S. Hall, L. A. Dean, and S. B. Hendricks. (1948) Exchange reactions between phosphates and soils: hydroxylic surfaces of soil minerals. *Soil Sci. Soc. Amer. Proc.* (1947) **12**:119–123.

McClelland, J. E. (1951) The effect of time, temperature, and particle size on the release of bases from some common soil-forming minerals of different crystal structure. *Soil Sci. Soc. Amer. Proc.* (1950) **15**:301–307.

McGeorge, W. T. (1930) The base exchange property of organic matter in soils. *Arizona Agr. Exp. Sta. Tech. Bul. 30.*

McLaren, A. D., G. H. Peterson, and I. Barshad. (1958) The adsorption and reactions of enzymes and proteins on clay minerals: IV. Kaolinite and montmorillonite. *Soil. Sci. Amer. Proc.* **22**:239–244.

Mehlich, A., and N. T. Coleman. (1952) Type of soil colloid and the mineral nutrition of plants. *Adv. Agron.* **4**:67–99.

Mengel, K. (1961) Die Donnan-Verteilung der Kationen im Freien Raum der Pflanzenwurzel und ihrer Bedeutung für die aktive Kationenaufnahme. *Zeitschr. Pflanzenernähr., Düng., Bodenk.* **95**:240–253.

Mengel, K. (1962) Die K- und Ca-Aufnahme der Pflanze in Abhängigkeit vom Kohlenhydratgehalt ihrer Wurzel. *Zeitschr. Pflanzenernähr., Düng., Bodenk.* **98**: 44–54.

Mengel, K. (1963) Die Bedeutung von Kationenkonkurrenzen im Free Space der Pflanzenwurzel für die aktive Kationenaufnahme. *Agrochimica* **7**:236–257.

Milam, F. M., and A. Mehlich. (1954) Effect of soil-root ionic environment on growth and mineral content of *Crotolaria striata. Soil Sci.* **77**:227–236.

Mitchell, J. (1932) The origin, nature and importance of soil organic constituents having base exchange properties. *Jour. Amer. Soc. Agron.* **24**:256–275.

Newbould, P. (1963) Relationship between isotopically exchangeable calcium and absorption by plants. *Jour. Sci. Food Agr.* **14**:311–319.

Nye, P., D. Craig, N. T. Coleman, and J. L. Ragland. (1961) Ion exchange equilibria involving aluminum. *Soil Sci. Soc. Amer. Proc.* **25**:14–17.

Okazaki, R., H. W. Smith, and C. D. Moodie. (1963) Hydrolysis and salt retention errors in conventional cation-exchange-capacity procedures. *Soil Sci.* **96**:205–209.

Okazaki, R., H. W. Smith, and C. D. Moodie. (1964) Some problems in interpreting cation-exchange-capacity data. *Soil Sci.* **97**:202–208.

Olsen, R. A., and M. Peech. (1960) The significance of the suspension effect in the uptake of cations by plants from soil-water systems. *Soil Sci. Soc. Amer. Proc.* **24**:257–261.

Overstreet, R., and H. Jenny. (1939) Studies pertaining to the cation absorption mechanism of plants in soil. *Soil Sci. Soc. Amer. Proc.* **4**:125–130.

Page, A. L., and L. D. Whittig. (1961) Iron adsorption by montmorillonite systems: II. Determination of adsorbed iron. *Soil Sci. Soc. Amer. Proc.* **25**:282–286.

Peech, M. (1939) Chemical studies on soils from Florida citrus groves. *Florida Agr. Exp. Sta. Bul. 340.*

Peech, M. (1965) Exchange acidity. *Agronomy* **9**:905–913.

Peech, M., R. L. Cowan, and J. H. Baker. (1962) A critical study of the $BaCl_2$-triethanolamine and the ammonium acetate methods for determining the exchangeable hydrogen content of soils. *Soil Sci. Soc. Amer. Proc.* **26**:37–40.

Pratt, P. F. (1951) Potassium removal from Iowa soils by greenhouse and laboratory procedures. *Soil Sci.* **72**:107–117.

Pratt, P. F. (1957) Effect of fertilizers and organic materials on the cation-exchange capacity of an irrigated soil. *Soil Sci.* **83**:85–89.

Pratt, P. F. (1965) Potassium. *Agronomy* **9**:1022–1030.

Pratt, P. F. (1965a) Sodium. *Agronomy* **9**:1031–1034.

Preston, R. D. (1961) Cellulose-protein complexes in plant cell walls. *In* M. V. Edds, Jr. (Ed.) *Macromolecular Complexes. Soc. General Physiol., Sixth Annual Symp.* Ronald Press Co., New York.

Randhawa, N. S., and F. E. Broadbent. (1965) Soil organic matter-metal complexes: 5. Reactions of zinc with model compounds and humic acid. *Soil Sci.* **99**:295–300.

Ratner, E. I. (1938) The availability for plants of exchangeable cations in connection with chemical amelioration of soils. *Bul. Acad. Sci. U.R.S.S., Classe Sci. Math. Nat., Sér. Biol.* **1938**:1153–1183.

Renger, M. (1965) Berechnung der Austauschkapazität der organischen und anorganischen Anteile der Böden. *Zeitschr. Pflanzenernähr., Düng., Bodenk.* **110**: 10–26.

Rich, C. I., and G. W. Kunze (Eds.). (1964) *Soil Clay Mineralogy.* University of North Carolina Press, Chapel Hill.

Rich, C. I., and G. W. Thomas. (1960) The clay fraction of soils. *Adv. Agron.* **12**:1–39.

Salmon, R. C. (1964) Cation-activity ratios in equilibrium soil solutions and the availability of magnesium. *Soil Sci.* **98**:213–221.

Salmon, R. C., and P. W. Arnold. (1963) The uptake of magnesium under exhaustive cropping. *Jour. Agr. Sci.* **61**:421–425.

Sawhney, B. L., M. L. Jackson, and R. B. Corey. (1959) Cation-exchange capacity determination of soils as influenced by the cation species. *Soil Sci.* **87**:243–248.

Schachtschabel, P. (1940) Untersuchungen über die Sorption der Tonmineralien und organischen Bodenkolloide, und die Bestimmung des Anteils dieser Kolloide an der Sorption im Boden. *Kolloid-Beihefte* **51**:199–276.

Scheuring, D. G., and R. Overstreet. (1961) Sodium uptake by excised barley roots from sodium bentonite suspensions and from their equilibrium filtrates. *Soil Sci.* **92**:166–171.

Schofield, R. K., and A. W. Taylor. (1955) Measurements of the activities of bases in soils. *Jour. Soil Sci.* **6**:137–146.

Schollenberger, C. J., and R. H. Simon. (1945) Determination of exchange capacity and exchangeable bases in soil—ammonium acetate method. *Soil Sci.* **59**:13–24.

Schulz, R. K., R. Overstreet, and I. Barshad. (1965) Some unusual ionic exchange properties of sodium in certain salt-affected soils. *Soil Sci.* **99**:161–165.

Sherman, G. D., Y. Matsusaka, H. Ikawa, and G. Uehara. (1964) The role of the amorphous fraction in the properties of tropical soils. *Agrochimica* **8**:146–163.

Sumner, M. E. (1963) Effect of iron oxides on positive and negative charges in clays and soils. *Clay Min. Bul.* **5**:218–226.

Tabikh, A. A., I. Barshad, and R. Overstreet. (1960) Cation-exchange hysteresis in clay minerals. *Soil Sci.* **90**:219–226.

Taylor, A. W. (1958) Some equilibrium solution studies on Rothamsted soils. *Soil Sci. Soc. Amer. Proc.* **22**:511–513.

Towe, K. McC. (1961) Lateral variations in clay mineralogy across major facies boundaries in the Middle Devonian (Ludlowville), New York. Ph.D. Thesis, University of Illinois, Urbana.

United States Salinity Laboratory Staff. (1954) Diagnosis and improvement of saline and alkali soils. *U.S. Dept. Agr. Handbook 60.*

Werkhoven, C. H. E., and A. J. Ohlrogge. (1963) Uptake of rubidium by plants from cation-exchange resin suspensions and their equilibrium solutions. *Soil Sci. Soc. Amer. Proc.* **27**:523–526.

White, W. A. (1953) Allophanes from Lawrence County, Indiana. *Amer. Mineral.* **38**:634–642.

Whitt, D. M., and L. D. Baver. (1930) Particle size in relation to base exchange capacity and hydration properties of Putnam clay. *Jour. Amer. Soc. Agron.* **29**:703–708.

Williams, D. E. (1961) The absorption of potassium as influenced by its concentration in the nutrient medium. *Plant and Soil* **15**:387–399.

Williams, D. E., and H. Jenny. (1952) The replacement of nonexchangeable potassium by various acids and salts. *Soil Sci. Soc. Amer. Proc.* **16**:216–221.

Yoshida, F. (1964) Interrelationships between potassium and magnesium absorption by oats (*Avena sativa* L.). *Versl. Landbouwk. Onderz. 642.*

5 *Soil Acidity*

Acidity of soils is associated with the presence of hydrogen and aluminum in exchangeable form. For this reason, the subjects of acidity and exchangeable bases might be classified under the general heading of exchangeable cations. The many indirect chemical, mineralogical, and biological effects associated with soil acidity, however, justify separate consideration of the subject.

Soil acidity involves intensity and quantity aspects. The intensity aspect is universally characterized by measurements of hydrogen-ion activity, expressed as pH. The quantity aspect is characterized, directly or indirectly, by the quantity of alkali required to titrate soil to some arbitrarily established endpoint. Neither of these techniques, as applied to soils, is as simple and satisfactory as with solutions of simple electrolytes. Different methods of measurement yield different answers.

Meaningful correlation of soil acidity with its effects, including biological responses, is accomplished through experimental measurements and appreciation of their significance. Accordingly, the first part of this chapter will be concerned with the nature of soil acidity and the way in which it is related to experimental measurements. The significance will be considered at the end.

Nature of Soil Acidity

Titratable Acidity

The titratable acidity of soils will be considered first because it involves consideration of the nature of soil acidity. The term titratable acidity is used here as a general operational description of the measurement of the quantity of soil acidity. In one way or another, methods for measuring the quantity of soil acidity determine the quantity of base consumed in neutralizing the acidity to some arbitrary degree.

For information on methods of analysis, papers by Peech (1965, 1965b) may be consulted. Jenny (1961) traced the development of theories of the nature of soil acidity.

The T-S Concept. Research on soil acidity in the early 1900s led to the concept that the quantity factor in soil acidity is properly expressed by the difference between the total capacity of a soil to adsorb bases (T) and the exchangeable bases already present (S), where both are expressed in chemical equivalents. The terms T and S were used by D. J. Hissink (1924) in the Netherlands, and he defined them in terms of methods he used.

Neither Hissink's terms T and S nor his methods are currently in general use, but the concept of titratable acidity as the difference between the cation-exchange capacity and the exchangeable bases already present has remained and has been found useful. The titratable acidity (T-S) was defined by Hissink as the capacity of the soil to take up barium in exchangeable form when barium hydroxide was added in excess. In effect, therefore, the cation-exchange capacity of the soil was defined as the current exchangeable bases plus the capacity of the soil to take up barium from a solution made strongly alkaline with barium hydroxide.

For many years since Hissink's time, the most common practice has been to define the cation-exchange capacity of soil as the ammonium retained in exchangeable form when a sample of the soil is leached exhaustively with a neutral, 1-normal solution of ammonium acetate. Because neutral, 1-normal ammonium acetate is strongly buffered against acid, the last portion of the solution remains at *p*H 7 after contact with the soil. Although there has been no overt titration, the soil has been neutralized to *p*H 7; and the titratable acidity is defined, in effect, by the difference between the cation-exchange capacity and the exchangeable bases removed from the soil by the ammonium acetate solution.

Effect of pH on Cation-Exchange Capacity. Adoption of the practice of determining the cation-exchange capacity by use of a buffered solution at *p*H 7 tended to inhibit due consideration of the significance of variations in cation-exchange capacity (and hence in quantity of soil acidity) with different displacing solutions. The quantity of ammonium retained in exchangeable form by acid soil is lower if, for example, the displacing solution is neutral, 1-normal ammonium chloride than if it is neutral, 1-normal ammonium acetate. Although this has been known for many years, for a long time there was some question as to whether one should say that the cation-exchange capacity of acid soil is lower if measured by ammonium chloride than ammonium acetate or whether one should say that ammonium chloride does not measure the cation-exchange capacity. Because neutral ammonium chloride solution is unbuffered against acidity and becomes acid when passed through an acid soil, one might argue that ammonium ions have relatively low replacing power for hydrogen ions and will not replace all the exchangeable hydrogen ions when the solution is acid.

The problem has recently been reexamined, and the outcome has been that the cation-exchange capacity is now considered to vary with the *p*H value. The reasoning and rationale of the definitive experiment that resulted in a change in way of thinking about the cation-exchange capacity are described in the following paragraph.

If samples of soil are treated repeatedly with ammonium chloride solutions having different *p*H values, the exchangeable cations will be removed; and the soil exchange sites will reach an equilibrium with the solution. That is to say, no more ammonium ions will be retained in exchangeable form if the displacement process is continued for a longer period of time. If the treated samples of soil are weighed to determine the quantity of solution retained, the quantity of ammonium chloride in solution in the sample can be calculated. If the samples then are treated repeatedly with potassium nitrate solution, the combined potassium nitrate extracts will contain the exchangeable ammonium plus the ammonium in solution plus the hydrogen ions in the ammonium chloride residual in the soil plus all hydrogen ions displaced from exchangeable form by the potassium nitrate. The hydrogen ions displaced from exchangeable form by potassium nitrate can be determined by titrating the extract to, say, *p*H 6 and subtracting the quantity of alkali required to titrate the calculated quantity of ammonium chloride solution retained by the soil to the same *p*H value. The exchangeable cations are then given by the sum of the milligram equivalents of ammonium and hydrogen in the potassium nitrate extract minus the sum of the milligram equivalents of ammonium and hydrogen calculated to be present in the quantity of solution residual in the soil after the ammonium chloride treatment. Schofield (1950) conducted an experiment in this way with a sample of a subsoil from England containing mainly illite in the clay fraction and found that the total exchangeable cations remained constant at 23.3 to 23.4 m.e./100 g where the pH of the ammonium chloride solutions varied from 2.6 to 3.8. The total exchangeable cations increased to 24.5 m.e. at *p*H 5.5 and further to values of 25.7, 27.0, and 28.2 m.e. at *p*H 6.2, 7.15, and 7.4.

From these results, it seems reasonable to conclude that the cation-exchange capacity varied with the *p*H. Two critical aspects of technique in this experiment were continuation of the treatment until equilibrium had been obtained and determination of ammonium retention by soil in equilibrium with the same concentration of ammonium at all *p*H values.

On further examination of the results of Schofield's experiment, one may note that the cation-exchange capacity decreased somewhat with decreasing *p*H; but, when the logarithmic nature of the *p*H scale is taken into account, there was a wide range of hydrogen-ion activities in the solution in the range below about *p*H 5 in which the exchange capacity decreased relatively slowly with an increase in hydrogen-ion activity in the solution. The existence of the nearly constant exchange capacity under strongly acid conditions, extending to *p*H values below those of usual practical interest, gave rise to the concept that the cation-exchange capacity of soils includes some "permanent-charge" positions that hold cations in exchangeable form under acid conditions as well as neutral and alkaline conditions. The increase in exchange capacity with increasing *p*H gave rise to the concept that part of the exchange positions are "*p*H-dependent," the proportion of these present as operative exchange positions being dependent on the *p*H of the solution with which the soil is equilibrated.

As conceived originally, the permanent-charge exchange positions were thought to include sites that hold cations by electronic valency in which the valence electron or electrons are transferred from the cation to an oxygen, so that the cation exists in ionic form. The exchange positions in aluminosilicate clays that arise from charge unbalance within the clay layers (see Chapter 4 on exchangeable bases) were thought to hold cations in this way. The *p*H-dependent exchange positions were thought to be sites in which the bonding of hydrogen is intermediate between electrovalency and covalency. (In covalent bonding, the electron of the hydrogen is not transferred to the oxygen; this electron and one from the oxygen are shared, and the hydrogen remains attached to the oxygen). At low *p*H values, the numerous hydrogen ions in solution suppress the dissociation of hydrogen from most of the *p*H-dependent positions; thus no exchange between these hydrogens and the displacing ions occurs. At higher *p*H values, there are fewer hydrogen ions in solution; more of the hydrogen dissociates from the electrovalent-covalent sites in the soil, and the displacing cations can then be attached in exchangeable form.

The concept of permanent-charge and *p*H-dependent cation-exchange positions, as explained in the preceding paragraphs, seems now to be an oversimplification. For example, different answers for permanent-charge and *p*H-dependent cation-exchange positions may be obtained on untreated acid soil and on soil that has been held for a time at a high *p*H value before analysis. These effects, which will be discussed subsequently, are thought to result from blocking of some exchange positions by aluminum at low *p*H values and from partial removal of the aluminum as insoluble aluminum hydroxide under approximately neutral conditions or as soluble aluminate under alkaline conditions.

The *p*H-dependency of the cation-exchange capacity in five soils of California is illustrated in Fig. 5.1. In all these samples, the *p*H-dependency of the exchange capacity is considerably greater, as a percentage of the total exchange capacity, than was the case in Schofield's (1950) experiment described previously. The explanation for the difference appears to be that the sample used in Schofield's experiment was a subsoil, low in organic matter, whereas the samples employed in the work shown in Fig. 5.1 were all surface soils, containing more organic matter. Different forms of evidence indicate that the *p*H-dependent exchange positions are principally in the organic matter. McLean et al. (1965) found that most of the *p*H-dependent exchange capacity of acid soil high in organic matter disappeared when the soil was treated with hydrogen peroxide to remove most of the organic matter. From a statistical examination of analyses made on a number of soils, Pratt and Bair (1962) estimated that the difference in cation-exchange capacity between *p*H 3 and *p*H 8 was 15.8 m.e./100 g of clay and 370 m.e./100 g of organic matter. Helling et al. (1964) made a similar investigation of the cation-exchange capacity of samples of 60 soils of Wisconsin. Their results indicated that the increases in exchange capacity in milligram equivalents per 100 g were 18 for clay and 140 for organic matter in the *p*H range from 3.5 to 8.0. Further discussion of the concept of permanent-charge

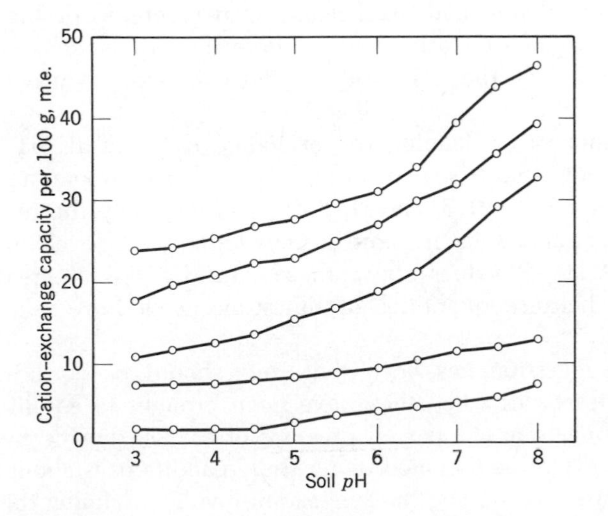

Figure 5.1. Cation-exchange capacity versus *p*H of five soils of California. (Pratt and Bair, 1962)

and *p*H-dependent cation-exchange positions may be found in a review paper by Coleman et al. (1958).

According to the foregoing considerations, the cation-exchange capacity of soils found by use of neutral, normal ammonium acetate is arbitrarily defined; and the titratable hydrogen calculated as the difference between the total exchangeable bases and the cation-exchange capacity of the soil at *p*H 7 thus is also arbitrarily defined. The figure for the exchange capacity of an acid soil at *p*H 7 includes the exchange capacity operative at the *p*H of the soil plus the portion of the *p*H-dependent exchange capacity that becomes active at *p*H values up to 7. The *p*H-dependent portion of the exchange capacity is fictitious as far as the soil in its natural state is concerned.

As in many matters of this kind, there are differences in point of view on the proper way to define the exchange capacity. If one wishes to determine the exchange capacity of soil as it exists in the field, the measurement should be made at the *p*H of the soil. Some work of this kind has been done with acid soils using normal potassium chloride as a displacing agent. This solution is only slightly buffered, and its *p*H is determined by the soil to which it is applied. Another technique has been to use for displacement a solution of barium chloride buffered at the *p*H of the soil. Measurement of the cation-exchange capacity at the *p*H of the soil may be helpful also in investigations of nutrition of plants on different soils. Martin and Page (1965) found that the correlation of calcium and magnesium percentages in citrus leaves with exchangeable calcium and magnesium as percentages of the cation-exchange capacity of soils was greater where the cation-exchange capacity was determined at the *p*H of the soil than where it was determined at pH 7.

The advantages of defining the exchange capacity at *p*H 7 are that *p*H 7 represents acid-base neutrality, activities of hydrogen and hydroxyl ions being equal at *p*H 7, and that *p*H 7 is generally the highest value attained when acid soils are limed. Knowledge of the exchange capacity of acid soils at *p*H values above their natural *p*H is of great practical significance because quantities of limestone needed are related to this property.

Another suggestion has been that soils should be considered to be saturated with bases when they have been brought to equilibrium with calcium carbonate at the partial pressure of carbon dioxide in the atmosphere. The *p*H value obtained under these conditions is about 8.2 except for sodic soils; accordingly, the suggestion involves defining the exchange capacity at *p*H 8.2. This definition has the advantage that calcium carbonate is a naturally occurring constituent of many soils. Probably most measurements of cation-exchange capacity of calcareous soils are now made with a displacing solution of *p*H 8.2.

Further discussion of these matters and a description of methods of measurement may be found in papers by Peech (1965, 1965b). Because of the differences among soils and among objectives of people who make measurements on soils, it does not seem likely that a single *p*H value will be decided upon as the proper one for all soils and all purposes. Where results of various investigators are being compared, it is therefore important to keep in mind the consequences of the methods they use. Figure 5.2 provides an example. It gives a plot of two sets of values of percentage base saturation of the same group of acid soils against the *p*H value of the soils. The indicated values of percentage base saturation are much lower where the cation-exchange capacity was measured at *p*H 8.2 than at the *p*H of the soil, the reason being that the cation-exchange capacity at *p*H 8.2 includes the titratable acidity between the soil *p*H and *p*H 8.2 and hence resulted in larger numbers

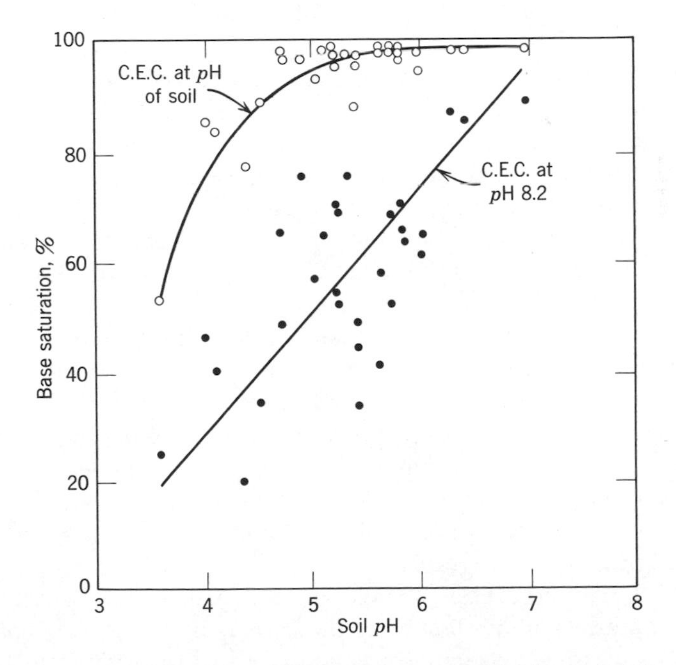

Figure 5.2. Percentage base saturation of various acid soils of California versus soil *p*H, where the cation-exchange capacity (C.E.C.) was measured at the *p*H of the soil by treatment with normal potassium chloride or at *p*H 8.2 by treatment with barium chloride, triethanolamine solution. (Pratt and Bair, 1962)

for cation-exchange capacity in the denominator of the ratio used to calculate the percentage base saturation.

Exchangeable Aluminum. If an acid soil is treated with a solution of a neutral, unbuffered salt, such as sodium chloride or potassium chloride, the extract is acid and contains aluminum in addition to the conventional exchangeable bases (calcium, magnesium, potassium, and sodium). The origin of the aluminum in neutral-salt extracts of soils was a matter of debate for many years. As Jenny (1961) pointed out in his review of the development of theories of the nature of soil acidity, the original view was that the aluminum was present in exchangeable form and was displaced by the salt solutions. Later, as *p*H measurements became common, it was evident that soils from which substantial quantities of aluminum were extracted were already acid before addition of the salt solution. This and other observations led to the theory that acid soils contain exchangeable hydrogen. The occurrence of aluminum in the extracts was then accounted for on the basis that the cation of the chloride salts usually employed exchanged with the hydrogen in the soil to produce hydrochloric acid, which reacted with aluminum hydroxide in the soil to produce the aluminum chloride that appeared in the extract.

Within recent years, new experiments have been conducted on this old problem; the results have led back to the early theory that acid soils contain exchangeable aluminum. Perhaps the most important evidence has been obtained in experiments on clays acidified artificially in a way that permits differentiation of exchangeable hydrogen and exchangeable aluminum, a capability previously lacking. The technique of acidification is simply to pass a dilute suspension of clay in water through a column of a hydrogen-saturated cation-exchange resin. The cation-exchange capacity of the resin is many times that of the clay. As the clay moves past the stationary resin particles, metallic exchangeable cations are continuously removed from the clay by the resin and are replaced by hydrogen ions. At the moment of emergence from the bottom of the exchange column, therefore, the cation-exchange sites on the clay should be occupied by hydrogen ions, although whether they are cannot be determined without additional work such as that described in the next paragraph.

Figure 5.3 gives the results of an experiment in which bentonite clay (which contains montmorillonite as the principal silicate clay mineral), prepared by passing it through a column of hydrogen-saturated cation-exchange resin, was titrated with sodium hydroxide immediately or after heating for different lengths of time at 95°C. The titration curve for the unheated sample that was titrated immediately shows a definite strong-acid character, that is, consumption of considerable alkali without much rise in *p*H value, followed by a rapid increase in *p*H. The titration curves obtained with successively longer periods of heating show successively less strong-acid character, indicating that heating produced a change in the nature of the acidity that was being titrated. Separate samples of each of the suspensions were extracted with normal barium chloride solution, and the extracts were analyzed for aluminum. The quantities found, in milligram equivalents per 100 g of clay, were 8 in the suspension that was titrated immediately without heating and 34, 40, 51, and 59 in the suspensions titrated after they had been heated for 1, 2, 4, and 12 hours.

The combination of evidence provided by the method of preparing the acid clay, the change in nature of the titration curves, and the aluminum displaced from the different samples indicates that the clay

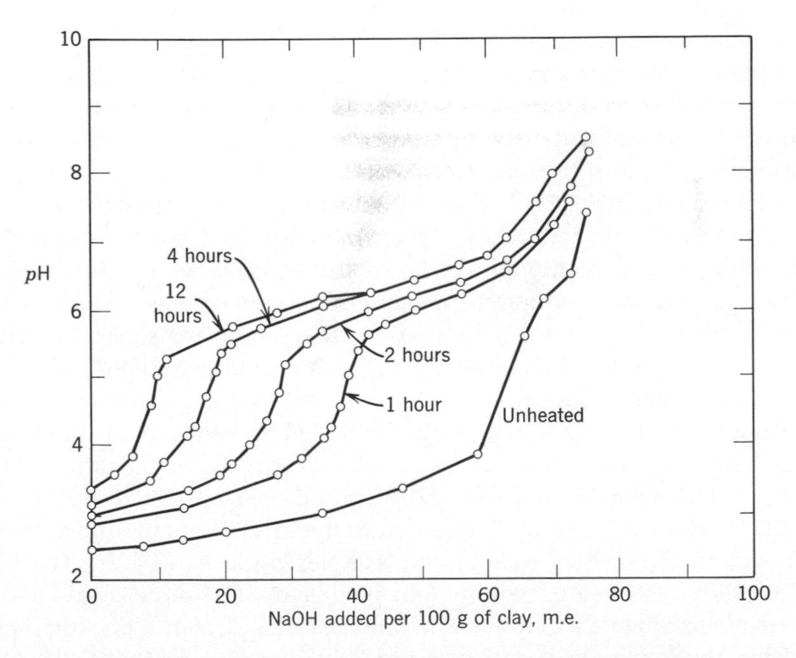

Figure 5.3. Titration curves for a suspension of bentonite clay determined immediately after emergence of the suspension from a column of hydrogen-saturated cation-exchange resin and after heating for 1, 2, 4, or 12 hours at 95°C. (Davis et al., 1962)

was mostly a hydrogen clay as it emerged from the column of hydrogen-saturated cation-exchange resin. The aluminum in the extracts did not result from dissolution of aluminum hydroxide by hydrochloric acid produced by exchange because the most hydrochloric acid should have been formed upon addition of barium chloride to the original unheated suspension of clay, and this sample yielded the least aluminum. The aluminum evidently appeared in extractable form during the reaction period before displacement with barium chloride. The hydrogen-saturated clay thus must have been unstable, decomposing spontaneously with release of aluminum.

Experiments similar to the one just described have been done by various investigators. This work has shown that magnesium as well as aluminum is released from nonexchangeable form from hydrogen-saturated clays that contain magnesium as a structural ion.

The rate of decomposition is slower at ordinary temperatures than at the temperature of 95°C used in the experiment described in Fig. 5.3, but it is still relatively rapid in terms of field behavior of soils. For example, Coleman and Craig (1961) found that at 30°C a period of 18 hours was required for half of the exchangeable hydrogen to be replaced by aluminum in a sample of bentonite clay that had a degree of hydrogen saturation of 95% or more as originally prepared.

In the era of exchangeable hydrogen, it was not clear whether aluminum could exist in soil as an exchangeable cation. Results of the foregoing experiments indicate that both hydrogen and aluminum may exist as exchangeable cations. They indicate further that early experiments done with what were thought to be hydrogen clays were actually done with clays that were saturated mostly with aluminum because the clays were frequently "hydrogen saturated" by a slow process such as electrodialysis and usually were allowed to stand for some time before use.

Further evidence has now been obtained to verify that aluminum may indeed exist as an exchangeable cation. The behavior of aluminum in exchange positions was investigated by Lin and Coleman (1960). They leached samples of montmorillonite and various soils with aluminum chloride solution and washed out the excess aluminum chloride with water. They then treated the samples with normal solutions of the chlorides of calcium, sodium, and potassium. The three reagents displaced almost identical numbers of milligram equivalents of aluminum, and the aluminum displaced was almost equivalent to the number of milligram equivalents of calcium retained in exchangeable form upon leaching the samples with calcium chloride. Similar quantities of aluminum were displaced with 3-normal and 1-normal potassium chloride. In other tests with samples of acid soils, the number of milligram equivalents of aluminum plus calcium plus magnesium displaced by normal potassium chloride was almost the same as the number of milligram equivalents of calcium retained as an exchangeable cation when the samples were treated with calcium chloride.

These experimental results indicate that aluminum behaves as an exchangeable cation in a manner similar to the exchangeable bases and that little hydrogen was replaceable in the soils analyzed. Two other significant points about the findings are the following: (1) The salt solutions used for displacement were of the unbuffered type and hence would not replace hydrogen or aluminum from *p*H-dependent exchange positions at *p*H values above that of the soil. (2) The milligram equivalents of aluminum were calculated by multiplying the millimoles of aluminum by three. The implication of the findings is therefore that all three valences of the aluminum ions displaced were attached to exchange sites and none to hydroxyl.

Figure 5.4 shows the quantity of exchangeable aluminum found in a number of naturally acid soils of North Carolina. Each point represents a different soil. The quantity of exchangeable aluminum was less than 1 m.e./100 g at *p*H 5.5 and above but increased rapidly at *p*H values below 5.5. With one exception, aluminum was less than 10% of the exchangeable cations at *p*H 5.5 or above. At *p*H values below 5, aluminum comprised more than 40% of the exchangeable cations. The total exchangeable cations in this case include the hydrogen displaced by barium chloride, triethanolamine solution at *p*H 8.2.

Although the experimental work reported by Coleman et al. (1959) on acid soils of North Carolina indicates that the titratable acidity of normal potassium chloride extracts is attributable entirely to aluminum, Yuan (1963) found that hydrogen sometimes exceeds aluminum in similar extracts from acid sandy soils of Florida. The largest quantity of exchangeable hydrogen found by Yuan was only 0.82 m.e./100 g, but this represented 15% of the cation-exchange capacity. The same soil contained 0.78 m.e. of exchangeable aluminum. Schwertmann (1961) reported that in his analyses of soils of Germany only soils high in organic matter released significant amounts of exchangeable hydrogen to potassium chloride solution. In all other samples, the acidity of the potassium chloride extracts was due almost entirely to aluminum. These observations suggest that Yuan's (1963) finding of exchangeable hydrogen as well as aluminum may be attributed to the existence of most of the exchange capacity of his soils in the organic fraction.

The extractant most commonly used for exchangeable aluminum is a 1-normal solution of potassium chloride. Potassium is held in exchangeable form preferentially to aluminum at this concentration, at least in

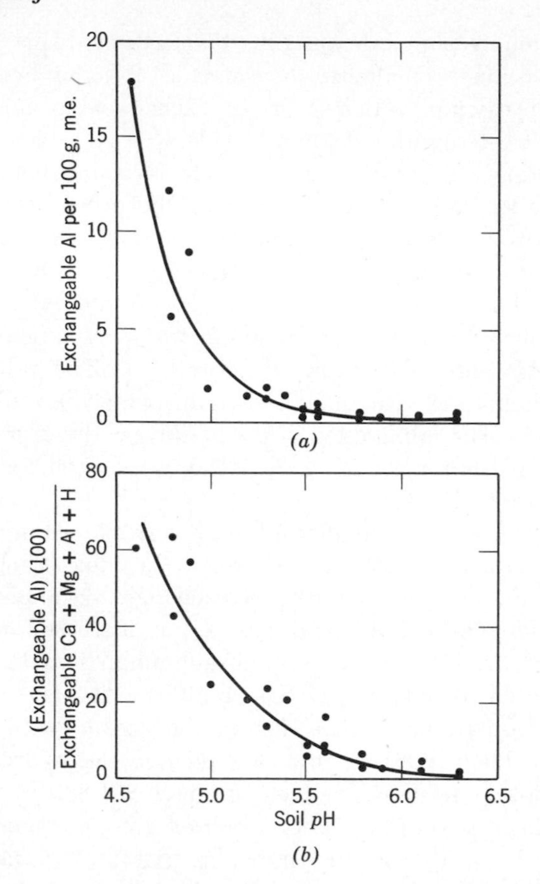

Figure 5.4. (*a*) Exchangeable aluminum at different *p*H values and (*b*) exchangeable aluminum as a percentage of the principal exchangeable cations at different *p*H values in soils of North Carolina. (Coleman et al., 1959)

some soils. Normal solutions of calcium and sodium chlorides are less effective than potassium chloride in extraction. For further information on methods of extraction and analysis, the paper by McLean (1965) should be consulted.

Polymeric Hydroxy-Aluminum Interlayers. In the preceding sections, reference to the reactions of hydrogen and aluminum ions has implied that the hydrogen ion is H^+ and the aluminum ion is Al^{+++}. Whereas this implication was of no consequence there, evidence points to the existence of these species as $H_2O \cdot H^+$ or H_3O^+, called hydronium ions, and $Al(H_2O)_6^{+++}$ or $Al(OH_2)_6^{+++}$. The aluminum species is a coordination complex in which a central aluminum atom is bonded to the oxygen of six surrounding molecules of water in octahedral configuration:

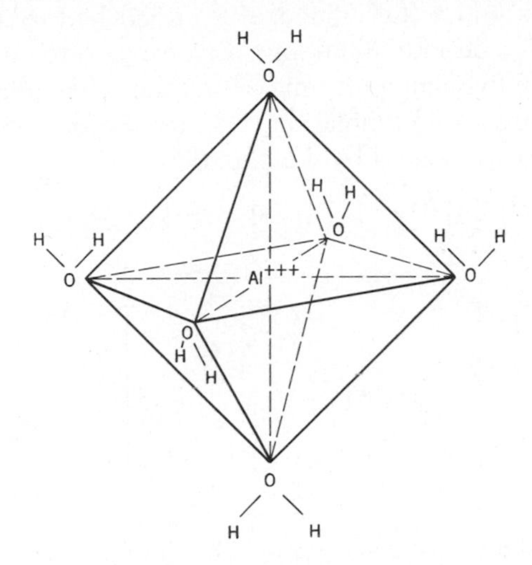

Pokras (1956) used the term "aquo-aluminum" to describe the hydrated aluminum ion. The species has been viewed also as a substituted form of hydronium and termed "aluminohexahydronium" (Jackson, 1960) and "hexaluminohydronium" (Jackson, 1963).

The hydrated aluminum ion is an acid in the general sense in that it contains removable protons (hydrogen ions). In this acid, the protons are removed from the water molecules surrounding the aluminum. These are not water molecules in the ordinary sense because the oxygen of each one is bonded to aluminum with an average charge of 0.5. Each hydrogen is then bonded to oxygen with an average charge of 0.75, leaving an average positive charge of 0.25 for each hydrogen or a total of $12 \times 0.25 = 3$ per hydrated aluminum ion. The steps in the dissociation of protons from the hydrated aluminum ion in water may be represented by the following equations:

$$[Al(H_2O)_6]^{+++} + H_2O = [Al(H_2O)_5OH]^{++} + H_3O^+,$$

$$[Al(H_2O)_5OH]^{++} + H_2O = [Al(H_2O)_4(OH)_2]^{+} + H_3O^+,$$

and

$$[Al(H_2O)_4(OH)_2]^{+} + H_2O = [Al(H_2O)_3(OH)_3] + H_3O^+.$$

The second and third equations are hypothetical in that the process leading to final formation of aluminum hydroxide involves poorly defined interactions among individual ions. Hydrated aluminum ions are thought to be stable as single ions in solutions that are sufficiently acid because the individual positively charged units repel one another. As the first reaction proceeds under the influence of added hydroxyl ions, however, the positive charge on individual ions diminishes. A tendency then exists for bonding of individual units into larger units containing two, three, or more aluminums with elimination of water. Formation of the simplest such unit may be represented by the equation

$$2[Al(H_2O)_5(OH)]^{++} = [Al_2(H_2O)_8(OH)_2]^{++++} + 2H_2O.$$

The structure

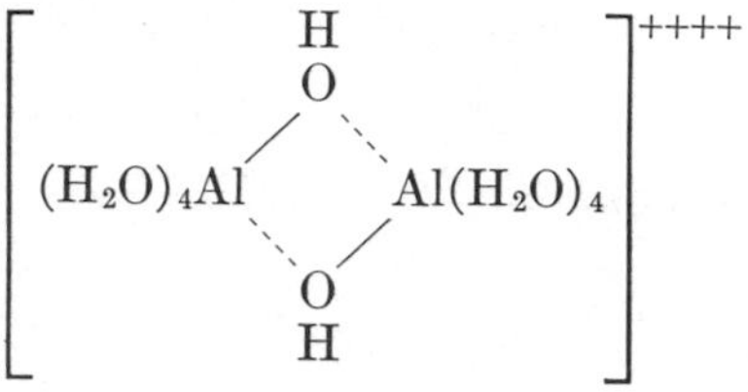

$$\left[\begin{array}{ccc} & H & \\ & O & \\ & \diagup \quad \diagdown & \\ (H_2O)_4Al & & Al(H_2O)_4 \\ & \diagdown \quad \diagup & \\ & O & \\ & H & \end{array}\right]^{++++}$$

has been proposed for the two-unit ion, indicating that the bonding occurs through the hydroxyl groups. Dissociation of additional protons from this two-unit ion would provide additional hydroxyl groups which could join with other units to form a polymer containing many units. Evidence indicates existence of polymers in solution, but there is difficulty in describing them exactly because they may be of various sizes and of various degrees of stability, depending on the age of the preparation and other conditions, such as the nature of the associated anions.

If a solution of aluminum chloride containing some excess hydrochloric acid is titrated with an alkali such as sodium hydroxide, the alkali reacts first with the hydrochloric acid because of its complete dissociation. The dissociation constant for the first hydrogen of the hydrated aluminum ion is 1.4×10^{-5} (Gilreath, 1958, p. 234), which makes the hydrated aluminum ion a weak acid somewhat like acetic acid. Accordingly, the neutralization of the hydrated aluminum ion produces a titration curve in which the *p*H increases slowly with addition of alkali within the range from *p*H 4 to *p*H 8. As titration of the hydrated aluminum ion proceeds, successive protons are removed from individual ions and from the polymers formed during the process. Precipitation of polymers occurs before the final aluminum hydroxide stage is reached. The shape of the titration curve varies with the age of the sol before titration, reflecting, no doubt, the effect of the slow increase in stability of the polymers. The quantity of alkali required to reach a given *p*H value decreases with an increase in age of the preparation (Schwertmann and Jackson, 1964).

Evidence from many sources shows that aluminum occurs also in interlayer positions in clays, often forming complete layers to which the term chlorite is sometimes applied. The aluminum in such layers is not in the form of single hydrated aluminum ions but rather as the polymeric ions that occur in solution during titration of hydrated aluminum ions, or as some modification of these polymers. Lodging of the polymers in interlayer positions in montmorillonite and vermiculite minerals is promoted not only by their size and tendency to grow by addition of more aluminums laterally between the layers but also by the fact that the polymers are charged and are attached to exchange positions on both the layers. Jackson (1963) discussed at length the evidence for the marked stability of the polymers in interlayers in minerals, pointing out that aluminum released in soils during chemical weathering tends to accumulate in the form of interlayers in preference to gibbsite (crystalline aluminum hydroxide).

The aluminum in polymeric form in interlayers in silicate clays does not appear to be exchangeable to any significant degree. Barnhisel and Rich (1963) carried out a laboratory experiment in which polymeric aluminum was precipitated in interlayer positions in montmorillonite at different *p*H values and found that the exchange capacity was markedly reduced by the treatment. Interlayers produced where the molar ratio of hydroxyl to aluminum added was 0.35, 0.75, and 1.50 were stable during the 6-month test period, and the clay did not gain in exchange capacity. Where the OH/Al ratio was 3 (corresponding to aluminum hydroxide and gibbsite), the cation-exchange capacity was initially low; but the original exchange capacity was regained by the end of 3 months. The increase was associated with formation of gibbsite. Hsu and Bates (1964) conducted a similar experiment with vermiculite and found that aluminum interlayers did not form where the OH/Al ratio was three. Gibbsite crystallized outside the clay particles within a few weeks. This evidence suggests that the blocking of exchange positions by polymeric interlayers is not necessarily permanent under the proper conditions.

The suggestion has been made that the interlayer polymers may contain some iron as well as aluminum; however, the hydrated ferric ion $[Fe(H_2O)_6]^{+++}$ (analogous to hydrated aluminum) dissociates protons to form ferric hydroxide at a lower *p*H value than does the hydrated aluminum ion, and this property probably limits extensive accumulation of hydroxy-iron polymers during chemical weathering of soils. In short-term laboratory experiments, Thomas and Coleman (1964) found that ferric iron added artificially to soils and clays tended to disappear from exchangeable form, being replaced by hydrogen (hydronium), aluminum, and magnesium, the last two ions being released from nonexchangeable form. The treatment caused some reduction in cation-exchange capacity, which could have been due to formation of polymeric hydroxy-iron interlayers. Clark (1964) found that saturation with either ferric iron or aluminum reduced the cation-exchange capacity of soil clays and of an ion-exchange resin carrying carboxyl groups. Repeated extraction of the aluminum-saturated soil and the aluminum- and iron-saturated resin with neutral, normal potassium acetate effected a gradual increase in exchange capacity. These findings suggest not only the formation of interlayers of hydroxy-iron polymers but also a sluggish reaction of iron in carboxyl-type exchange positions in soil.

Sawhney (1960) found that chemical removal of interlayers by treatment of soils with neutral sodium citrate solution at 100°C resulted in an increase in cation-exchange capacity. Frink (1965) used an adaptation of this method on a group of natural soil clays and found that the increase in cation-exchange capacity due to the sodium citrate treatment was highly correlated with the aluminum extracted but not with the iron or magnesium extracted. These findings verify that the principal metallic component of the interlayers was aluminum and not iron or magnesium.

Polymeric hydroxy-aluminum interlayers appear to be, in effect, incompletely neutralized, hydrated-aluminum ions that are stabilized against further neutralization and against release by ion exchange by their polymerization, physical location, and chemical combination with the adjacent layers of clay. If the potassium chloride solutions usually used to remove exchangeable aluminum replaced aluminum from hydroxy-aluminum interlayers, the sum of the cations released (counting one mole of aluminum as three equivalents) would exceed the equivalents of potassium or other cations retained in exchangeable form.

Occurrence of hydroxy-aluminum interlayers provides a reasonable explanation for the well-known difficulty in reaching a stable endpoint when an acid soil is titrated with alkali. With an increase in time, the *p*H value associated with a given quantity of alkali gradually decreases, as would be expected if the hydroxy-aluminum polymers were slowly being neutralized to aluminum hydroxide with concurrent unblocking of the exchange positions they occupy. This phenomenon has not yet received the attention it deserves in relation to the performance of laboratory methods for determining titratable acidity for estimation of limestone needs.

From the standpoint of the nature of soil acidity, one may perceive that the exchange positions from which blocking was removed during neutralization of polymeric hydroxy-aluminum interlayers may be classified as part of the *p*H-dependent exchange capacity. Analysis of an originally acid soil after treatment with alkali and return to the initial *p*H would be expected to show that most of the exchange positions uncovered in interlayer locations are no longer *p*H-dependent exchange positions but rather add to the cation-exchange capacity at the initial soil *p*H.

Forms in Different pH Ranges. The forms of soil acidity titrated in different *p*H ranges may be summarized in the following classification, which is somewhat modified from the one given in a paper by Jackson (1963): (1) Strong acids, soil *p*H 4.2 and below. Sulfuric acid and basic ferric sulfate are the principal acids, occurring under special conditions to be described later. (2) Weak acids, soil *p*H 5 or 5.2 and below. The hydrated aluminum ion appears to be the only acid that accumulates in large quantity in most instances. Small quantities of hydronium may be present. Possibly some carboxyl groups in the organic matter may contribute. Basic ferric sulfate and basic aluminum sulfate may be important under special conditions. (3) Very weak acids, soil *p*H 5.2 to 6.5 or 7. Carboxyl groups of organic matter are important. Edge groups of hydroxy-aluminum polymers in interlayer positions and edges of silicate clay particles contribute. Carbonic acid contributes trace quantities. Basic aluminum sulfate may be significant under special conditions. (4) Very, very weak acids, soil *p*H 6.5 or 7 to 9.5. Phenolic groups of organic matter may be significant. Edge groups of hydroxy-aluminum polymers in interlayer positions and edges of silicate clay particles contribute. Bicarbonates of calcium and sodium in the soil solution are usually unimportant quantitatively. (5) Extremely weak acids, soil *p*H above 9.5. These include alcoholic groups in organic matter, silicic acid, and gibbsite. This classification does not include the slow release of protons from inner portions of hydroxy-aluminum polymers. Presumably this process takes place in ranges 3, 4, and 5, depending on the conditions.

*p*H Values

The total acidity titratable to *p*H 7 or some other arbitrarily chosen *p*H value is an appropriate measurement to make for purposes such as estimating the quantity of limestone to apply to different acid soils to provide conditions suitable for growing an acid-sensitive crop like alfalfa. For determining whether or not limestone is needed, however, the total, titratable acidity is not the appropriate measurement because the response of plants is determined primarily by the intensity of soil acidity and not by the quantity. The intensity and the quantity are closely related for a given soil or similar soils but not for dissimilar soils. The concept of *p*H and the methods of measurement underlying it were developed in an attempt to evaluate the intensity of acidity to which biological systems respond. The *p*H value has long been used to evaluate the intensity of soil acidity and has long been accepted as one of the standard criteria for soil characterization.

The *p*H is defined as the negative logarithm of the hydrogen-ion activity, where activity is understood to mean the effective concentration. The activity is the product of the concentration and an activity coefficient. Concentrations may be expressed as moles per liter of solution or moles per 1000 g of water, which are essentially equal as far as *p*H values are concerned. According to this definition, *p*H values of 1, 3, 5, 7, 9, and 11, for example, correspond to hydrogen-ion activities of 10^{-1}, 10^{-3}, 10^{-5}, 10^{-7}, 10^{-9}, and 10^{-11}. Activities of hydrogen and hydroxyl are equal at *p*H 7, and solutions having a *p*H of 7 are said to be neutral. Solutions with *p*H values below 7 are acid, and those with *p*H values above 7 are alkaline. Although this concept is simple enough to be comprehended readily, complications exist with regard to measurement of *p*H, in both solutions and soils, and with regard to interpretation of the measurements in soils.

Theory of pH Measurement. If an electrical cell is formed by

$$\mathrm{Pt};\ \mathrm{H_2}(\mathrm{gas},\ p_1),\ \mathrm{H^+}(a_1)\ \|\ \mathrm{H^+}(a_2),\ \mathrm{H_2}(\mathrm{gas},\ p_2);\ \mathrm{Pt},$$

an electrical potential may be measured between the two pieces of platinum. The pieces of platinum are coated with finely divided platinum (platinum black) which adsorbs gaseous hydrogen, catalyzes the reversible conversion of hydrogen gas to hydrogen ions, and causes the electrodes to behave as if they were hydrogen electrodes. Hydrogen gas at pressure p_1 on the left and p_2 on the right is bubbled past the electrodes continuously to maintain the layer of adsorbed hydrogen gas. Surrounding each electrode is an aqueous solution containing hydrogen ions at activity a_1 on the left and a_2 on the right. The pair of vertical lines between the two half-cells symbolizes an electrolytic bridge of saturated potassium chloride that makes an electrical connection between the two solutions. A saturated solution of potassium chloride is used in the bridge because the potassium and chloride ions move at almost the same velocity (if there are no complicating interactions with one or the other in the solutions) and hence do not cause an appreciable potential at the junction between the bridge and the solutions. Moreover, with such a concentrated solution, the current is carried almost entirely by the potassium and chloride ions, thus limiting the potential that might arise at the junction from differences in velocity of movement

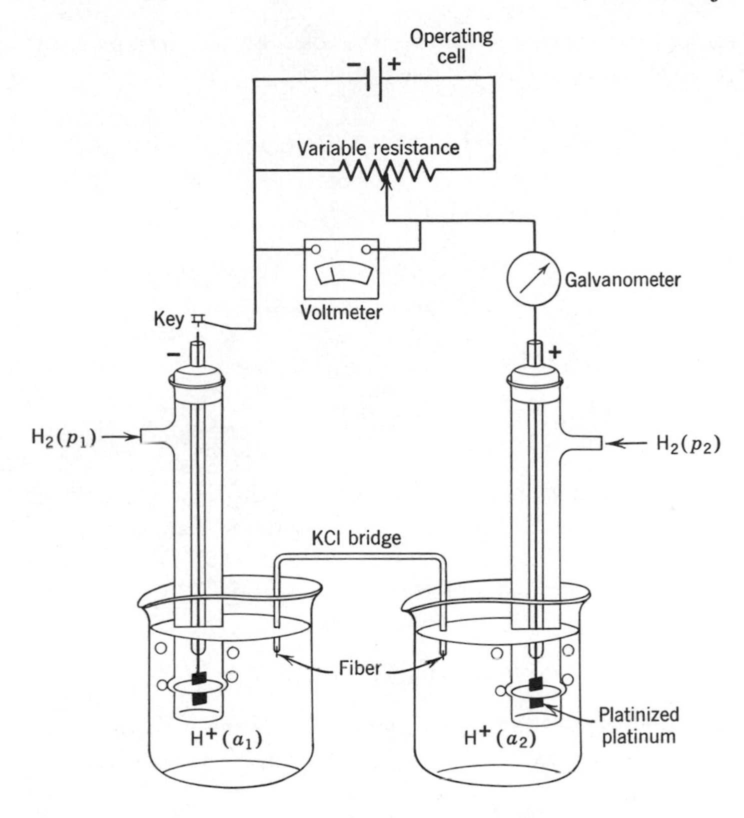

Figure 5.5. Schematic diagram of hydrogen electrodes and potentiometer for measuring voltages.

of the other positive and negative ions in the solutions. The apparatus, including a potentiometer for measuring the voltage, is shown schematically in Fig. 5.5.

The electrical potential is derived from the reactions

$$\frac{1}{2}\,H_2(\text{gas},\ p_1) = H^+(a_1) + \text{electron}$$

at the left and

$$H^+(a_2) + \text{electron} = \frac{1}{2}\,H_2(\text{gas},\ p_2)$$

at the right. If the hydrogen-ion activity in the solution at the right is greater than that at the left ($a_2 > a_1$), the electrode at the right will be positive with respect to the one at the left as indicated in the figure. The sum of the two reactions is

$$\frac{1}{2}\,H_2(\text{gas},\,p_1) + H^+(a_2) = \frac{1}{2}\,H_2(\text{gas},\,p_2) + H^+(a_1).$$

The equilibrium constant K for the reaction is

$$K = \frac{[H_2(\text{gas},\,p_2)]^{1/2} \times [H^+(a_1)]}{[H_2(\text{gas},\,p_1)]^{1/2} \times [H^+(a_2)]}$$

and the voltage E of the cell is

$$E = -\frac{RT}{F}\ln\frac{[H_2(\text{gas},\,p_2]^{1/2}}{[H_2(\text{gas},\,p_1)]^{1/2} \times [H^+(a_2)]} - \frac{RT}{F}\ln\,[H^+(a_1)].$$

If $p_2 = p_1 = 1$ atmosphere, and if $a_2 = 1$, the first term on the right side of the equation is equal to 0. If the numerical values $R = 8.3144$ joules/degree/mole, $T = 298$ degrees absolute (25°C), and $F = 96{,}500$ coulombs are used in place of the symbols, and if the natural logarithm is changed to the base 10, the equation becomes

$$E = -0.0591 \log H^+(a_1).$$

The pH is defined as the negative logarithm of the hydrogen-ion activity, that is, $p\text{H} = -\log a_{H^+} = -\log\,[H^+(a_1)]$, and so the equation may be written as

$$E = 0.0591\,p\text{H}.$$

Although the theory and basic standard for pH measurements are dependent on the hydrogen electrode, this electrode is inconvenient for practical use. For practical pH measurements, the left hydrogen electrode is usually replaced by a glass electrode, the right hydrogen electrode by a calomel half-cell, and the slide-wire potentiometer by a vacuum-tube voltmeter. The calomel half-cell was described in Chapter 3. This half-cell provides a standard voltage in a more convenient form than the hydrogen electrode. A vacuum-tube voltmeter serves the same purpose as the slide-wire potentiometer in Fig. 5.5 and is required because glass-electrode circuits have high electrical resistance. If the voltage in such a circuit is to remain unchanged during measurement, the voltmeter must have even higher resistance. Vacuum-tube voltmeters can be made with the necessary high resistance.

The glass electrode consists of a half-cell inside a bulb of a special kind of glass. The glass bulb is dipped in the test solution to make the measurement. The half-cell inside the glass is usually silver that is coated with silver chloride and immersed in 0.1-normal hydrochloric acid. The potential depends on the nature of the half-cell and on the nature and condition of the glass. Metallic cations pass through the glass to only a very small extent, and the selectivity of the glass electrode for hydrogen ions appears to depend on this property.

The glass-electrode, calomel-electrode assembly used in practical measurements must be standardized in terms of the response of the hydrogen electrode. Standardization is accomplished by use of buffer solutions that have been found to produce certain voltages (that is, to have certain *p*H values) by use of the hydrogen electrode. Standardization is carried out by setting the vacuum-tube voltmeter to read the same voltage or *p*H value with the glass-electrode assembly as it would with the hydrogen electrode (the standard potential). The measured voltage or *p*H value then increases or decreases to the same degree as it does with the hydrogen electrode in solutions of different *p*H if the sodium-ion activity is not too high at high *p*H values, if the time of contact with the solution is not too long, and if the temperature is not too high. Usual practice is to use two buffer solutions differing in *p*H to see whether the standard *p*H value of the second is obtained where the instrument is standardized against the first. In practical measurements, the *p*H of a test solution is calculated from the relationship

$$pH = \frac{E - E_b}{0.0591} + pH_b$$

where E is the observed voltage with the test solution at 25°C, E_b is the standard voltage set with a buffer solution at the same temperature, and pH_b is the standard *p*H value of the buffer solution. Setting the instrument with the buffer solution makes allowance for the potential of the saturated calomel reference electrode, the potential of the internal glass electrode, the difference in potential between the inside and outside of the glass (known as the asymmetry potential), and the potential that may exist at the junction between the potassium chloride bridge and the test solution (if this is the same as the potential at the junction between the potassium chloride bridge and the buffer solution). Commercial *p*H meters are calibrated in terms of *p*H units instead of volts, and so no calculations are required in their use.

In the device shown schematically in Fig. 5.5, only one potential is measured between the two electrodes; but the measured value is the algebraic sum of the potential between the platinum electrode on the left and the solution, the potential between the solution on the left and the potassium chloride bridge, the potential between the potassium chloride bridge and the solution on the right, and the potential between the solution on the right and the platinum electrode on the right. An analogous situation exists with regard to the glass-electrode, calomel-electrode assembly used in practical measurements. The potassium chloride bridge may be eliminated by use of another type of cell in which one electrode is sensitive to hydrogen ions and the other to chloride ions, for example, but what one then measures is the activity of hydrochloric acid ($a_{H^+} \times a_{Cl^-}$) and not the activity of either ion independently of the other. Consequently, a given hydrochloric acid activity may be obtained with either a high activity of hydrogen ion and a low activity of chloride ion or the reverse.

The outcome of this situation is that no way is known to measure hydrogen-ion activities unambiguously. Problems connected with this matter are still under active consideration [see, for example, the book by Bates (1964)]. Nevertheless, various indirect means have been investigated to determine whether the values obtained do, in fact, represent hydrogen-ion activities. The results of these studies, described by Bates (1964), indicate that the degree of uncertainty in interpreting the values as hydrogen-ion activities is too small to be of consequence in measurements of practical significance provided the standard potential remains the same when the buffer solution is replaced by the test solution. This condition is approximately fulfilled if the test solution is a dilute aqueous solution of simple solutes and if the acidity of the test solution is not much different from that of the buffer solution. A further limitation is that *p*H measurements at different temperatures are not strictly comparable.

Factors Affecting Soil pH Values. The precise numerical values commonly reported for *p*H values of soils tend to convey the erroneous impression that soils have characteristic *p*H values. As illustrated in Table 5.1, however, a variety of *p*H values may be found for a given soil.

Three methods are in common use for measuring soil *p*H values. Each gives a different answer. In the first and most commonly used method, a quantity of dry soil is mixed with water and allowed to stand for perhaps 30 minutes. Then the mixture is stirred, the electrodes are inserted, and the measurement is made on the soil suspension. According to this technique, the *p*H values of most soils lie in the range between *p*H 4 and *p*H 8.5; however, values have been recorded as low as *p*H 1.2 (Chenery, 1954) and as high as *p*H 10.7 (Fireman and Wadleigh, 1951). Bailey (1944, 1945) tabulated *p*H values of a number of soils of the United States. The second method for measuring soil *p*H, investigated thus far mostly from the research standpoint, is similar

Table 5.1. *p*H Values of a Soil Measured under Different Conditions (Coleman et al., 1951)

	*p*H Values		
	Supernatant Solution	Suspension	Sediment
Natural soil	6.2	5.8	4.7
Soil leached to remove soluble salts	6.5	5.9	5.2
Soil in 1 *N* KCl	5.1	5.1	5.1

to the first except that 0.01-molar calcium chloride is used in place of water. This method gives lower *p*H values than the first. The third method again is similar to the first except that a normal solution of potassium chloride is used in place of water. This method gives lower values than either of the first two. The potassium chloride method is used commonly in Europe, where the values obtained have been used to some extent as a basis for estimating lime requirement. The values obtained are not supposed to represent the actual *p*H of the soil.

Extensive experience with measurements of *p*H by the first method has shown that the *p*H is not a constant and characteristic soil property even where a given method is used. One of the most important causes of differences in *p*H values measured on a given soil by the first method is the electrolyte concentration. Soil *p*H decreases with an increase in electrolyte concentration. Figure 5.6 provides an example of the effect of natural variations in electrolyte content on *p*H values of soil in water measured in a certain way. The individual data points are *p*H values and nitrate-nitrogen contents of samples of soil taken from the field at different times during a summer season. The *p*H values may be seen to decrease with an increase in content of nitrate-nitrogen. Addition of different quantities of nitric acid to samples of the soil in the laboratory produced a curve showing the same trend as the samples taken from the field. A different trend was obtained with additions of calcium nitrate. Apparently, therefore, the *p*H value of the soil was affected by the nitric acid produced by nitrification.

Another group of electrolytes that may have significant effects on soil *p*H values is carbonic acid and the bicarbonates produced as a result of interaction of carbonic acid with exchangeable bases. Because of the limited dissociation of carbonic acid, its greatest effect is found at *p*H values above 7. Whitney and Gardner (1943) found that soil

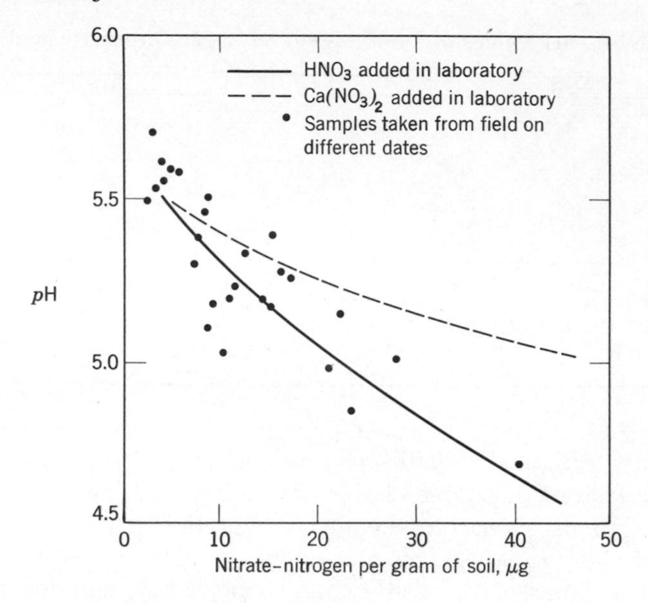

Figure 5.6. Nitrate content and *p*H of a sandy soil. (Lehr, 1950)

*p*H decreased linearly with an increase in the logarithm of the partial pressure of carbon dioxide in the atmosphere. In the most extreme case they reported, the *p*H value of a soil high in exchangeable sodium decreased from 9.2 to 6.4 as the partial pressure of carbon dioxide was increased from 0.0003 atmosphere to 0.77 atmosphere. Nichol and Turner (1957) felt that the carbon dioxide effect is significant enough in noncalcareous soils near neutrality to require specification of the partial pressure of carbon dioxide at which the measurement is made.

Although the nitric acid and carbonic acid produced biologically influence soil *p*H by their acidic nature, the electrolyte effect is not merely the result of addition of hydrogen ions in the electrolyte as such; it involves also the interaction of the electrolyte with the exchangeable cations in the soil. For example, addition of calcium nitrate to the soil in Fig. 5.6 produced a definite lowering of the measured *p*H value into the acid range even though calcium nitrate solution is nearly neutral. Conversely, leaching soil with water to remove the electrolytes raises the measured *p*H value.

The electrolyte effect is involved to a significant degree in the results of *p*H measurements made at different dilutions of soil with water. As soil is diluted with water, the concentration of electrolytes decreases; and net exchanges of cations occur between electrolytes and exchangeable cations, with the consequence that the dilution effect is a complex process. In tests on soils of Finland, Ryti (1965) found that the *p*H usually increased by about 0.15 *p*H unit when the ratio of soil to water was decreased from 1/2.5 to 1/5 and that there was a further increase of similar magnitude when the ratio was decreased from 1/5 to 1/10. Puri and Ashgar (1938) found that the *p*H of soil treated to remove soluble salts, however, was substantially constant over the range of soil-to-water ratios from 1/5 to 1/25.

In recognition of the effect of dilution, and with a view to convenience in performance of the measurements, the International Society of Soil Science (Anonymous, 1927) adopted a ratio of 1 part of soil to 2.5 parts of water by weight as standard. Such standardization has the value of making results of different investigators comparable, in a sense; but it provides no way of estimating the values at different dilutions such as the comparatively limited dilution that is normal in the field. This deficiency is an important one because the change of *p*H with dilution is not always the same. More recently, some investigators have used a 1-to-1 ratio. Some have used only enough water to bring the soil to a pastelike consistency, and a few have used even less water in an attempt to measure the *p*H of the soil under conditions approaching those in the field.

Soil *p*H values may change markedly under conditions of flooding. Soils that are strongly acid under aerobic conditions (*p*H 5 or below) may have *p*H values near 7 under anaerobic conditions (Romanoff, 1945). Although electrolyte concentration may be involved, the principal cause is apparently neutralization of the soil by hydroxyl groups activated by reduction of ferric oxide or hydroxide to the more soluble ferrous form

$$Fe(OH)_3 + \text{electron} = Fe^{++} + 3OH^-.$$

Table 5.2 illustrates the disappearance of exchangeable aluminum and its replacement with ferrous iron upon flooding of an acid soil of Guyana (Cate and Sukhai, 1964). The authors proposed the following neutralization reaction:

$$2\text{Al-clay} + 3Fe(OH)_2 = 3\text{Fe-clay} + 2Al(OH)_3.$$

As aerobic conditions return, the ferrous iron is oxidized, the hydrated ferric ion thus formed serves as a proton donor, and exchangeable hydrogen accumulates in the soil, to be followed by aluminum on dissolution of aluminum hydroxide or decomposition of other aluminum-bearing minerals; and the acid condition is regained. An associated process that

Table 5.2. Exchangeable and Soluble Ions in an Acid Soil Before and After Flooding (Cate and Sukhai, 1964)

	Ions per 100 g of Soil, m.e.			
	Exchangeable		Water-Soluble	
	Before Flooding	After Flooding for 6 Months	Before Flooding	After Flooding for 6 Months
Ferrous iron	0.0	20.0	0.0	1.0
Aluminum	15.0	1.0	5.0	0.0
Calcium + magnesium	1.5	4.5	4.5	1.5
Sulfate			10.0	1.5

takes place during flooding is the reduction of sulfate to sulfide and the accumulation of part of the ferrous iron as ferrous sulfide. This phenomenon will be discussed at some length in the section on development of soil acidity.

Source of Apparent Hydrogen-Ion Activities in Soils. As Bates (1964) pointed out, the *p*H theory applies to dilute aqueous solutions of simple salts. Readings on *p*H meters may readily be obtained on systems to which the theory does not apply, and the values obtained may provide useful correlations with other properties of interest; however, there is some question as to the significance of the measurements in terms of hydrogen-ion activities.

If a sample of soil is shaken with water and is then allowed to settle or is filtered, the clear solution probably corresponds to the conditions for interpreting the results as hydrogen-ion activities. If, then, the *p*H values obtained on the clear solutions agreed closely with the *p*H values obtained on insertion of the electrodes into the suspension of soil, one could infer that the latter values also are valid measurements of hydrogen-ion activity. Some of the earliest measurements on soils, however, showed that the *p*H values measured in the filtrates are usually different from those measured in the suspensions. In measurements on a large number of soils of Denmark, Christensen and Jensen (1924) found that the *p*H of the suspension usually exceeded that of the filtrate at *p*H values above 7 and was lower than the *p*H of the filtrate in soils with *p*H values below 6. The difference in *p*H between filtrate and suspension was usually less than 0.2 *p*H unit, but differences as great as 0.77 *p*H unit were recorded. Existence of such differences has been confirmed in other experiments. Recent work on this subject was reported by Ryti (1965).

Further research on measurement of *p*H values of supernatant solutions and suspensions showed (Fig. 5.7) that the difference in *p*H due to electrode locations is associated with the location of the junction between the calomel electrode and the medium. The position of the glass electrode may be changed at will from the solution to the suspension without affecting the measured *p*H value. These observations may be explained by alternative theories about the source and sign of individual potentials included in the over-all potential that is measured, and there seems to be no direct way to determine how the total potential is, in fact, divided among its possible components. Indirect evidence therefore must be used.

Jenny et al. (1950) and Coleman et al. (1951) proposed the theory that the difference in *p*H between the suspension and the solution is a result of a junction potential arising at the boundary between the suspension and the potassium chloride diffusing out of the potassium chloride bridge. According to this theory, the equality of mobilities of potassium and chloride ions in water and dilute solutions where there are no complicating interactions is disturbed as the potassium chloride diffuses into soil suspensions. In soil suspensions, the potassium ions move through the ionic atmospheres as well as the solution, but chloride ions are largely excluded from the ionic atmospheres and must move mostly through the outer solution. In consequence of these interactions, the potassium ions tend to move faster than the chloride ions, making

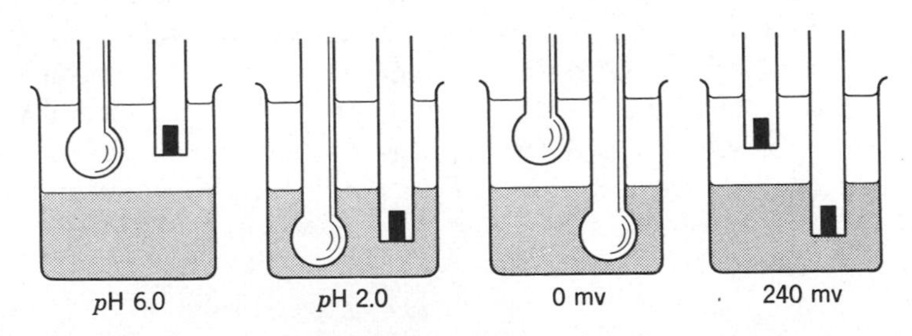

Figure 5.7. Illustration of the effect of location of electrodes on *p*H values and voltages measured in a heterogeneous system consisting of supernatant solution in equilibrium with a suspension of material with high cation-exchange capacity. The glass electrode is symbolized by the spheres, the calomel electrode by the large rectangles, and the potassium chloride bridge by the small black rectangles. The difference in potential corresponding to 4 *p*H units on the left is equivalent to the difference of 240 mv in the potential on the right. (After Jenny et al., 1950)

the advancing front of potassium chloride positive with respect to the potassium chloride in the bridge leading to the reference calomel electrode. These investigators were able to calculate the difference in voltage measured between electrodes in the solution and in suspensions of ion-exchange resins, clay, and soils (all potassium-saturated) from measurements of the degree to which movement of chloride was inhibited, thus verifying the theory that differences in *p*H measured between solutions and suspensions are attributable to a difference in junction potential between the potassium chloride bridge in the solution and the potassium chloride bridge in the suspension. The junction potentials calculated for potassium-saturated soils were highly correlated with the difference in measured *p*H between the solution and the sediment of twelve soils.

The foregoing theory does not account for the fact that the *p*H value of the suspension is higher than that of the solution with some soils and lower with others. To account for this behavior, Peech et al. (1953) and Raupach (1957) proposed a modified version to the effect that the junction potential in soils may be produced by the salt formed by exchange at the interface between the advancing potassium chloride and the soil. If the cations released are predominantly divalent, the *p*H of the suspension will exceed that of the solution; and, if the cations released are predominantly monovalent, the *p*H of the suspension will be below that of the solution. Experimental verification of this theory has been obtained with soils saturated with either monovalent or divalent cations and with a soil clay saturated with monovalent and divalent cations in different proportions. Presumably, the rate of diffusion of divalent cations is sufficiently below that of monovalent cations to permit the associated chloride to move ahead of the divalent cations and produce a junction that has a negative potential with respect to the potassium chloride in the bridge. The manner in which this accounts for suspension *p*H values below the solution *p*H values for naturally acid soils is still not clear, however, because the ions released from such soils on exchange with potassium are predominantly divalent and trivalent. The mobility of hydrogen ions far exceeds that of other cations, and perhaps this is the explanation.

The evidence that the difference in *p*H between solution and suspension is due to a junction potential and the fact that the measured *p*H value is the same whether the hydrogen-ion-sensitive glass electrode is located in the solution or in the suspension brings into question the hydrogen-ion activity in the ionic atmospheres around the soil particles. Theoretical considerations and experimental evidence described in Chapter 4 lead to the view that the activity of hydrogen ions should be greater in the ion atmospheres than it is in the bulk solution. There is independent experimental evidence that hydrogen-ion activity is indeed greater in the ionic atmospheres than in the bulk solution. Enzymes have a characteristic *p*H optimum at which the reaction they catalyze takes place most rapidly. McLaren and Estermann (1957) determined the rate at which the enzyme chymotrypsin digested denatured lysozyme protein where the enzyme was present in solution and where it was adsorbed on kaolinite. Their results in Fig. 5.8 show that the *p*H optimum for the reaction was about two units higher in the suspension than in the solution. (In the dilute suspensions employed, the *p*H of the suspensions and solutions agreed within 0.01 *p*H unit, indicating that junction potentials were not of significance.) Because of the logarithmic nature of the *p*H scale, these results indicate that the hydrogen-ion activity in the ionic atmospheres where the adsorbed enzyme was located was about one-hundred times as great as that in the solution.

The failure of the glass electrode to register any difference in *p*H where it is in the solution or suspension thus does not appear to mean that the hydrogen-ion activity is the same in the ion atmospheres around the soil particles as in the solution. The theoretical aspects of this phenomenon are not of concern here, but it may be mentioned that the

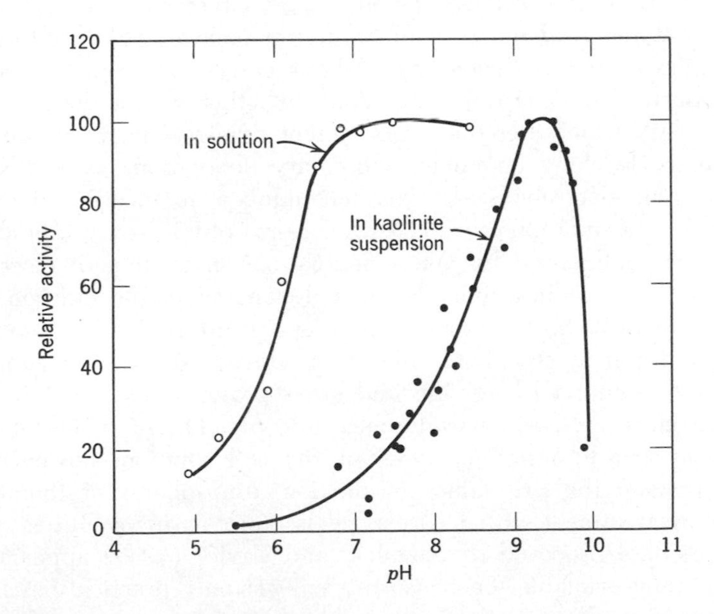

Figure 5.8. Relative activity of chymotrypsin enzyme in digesting heat-denatured lysozyme protein at different *p*H values in solution and in a suspension of kaolinite. (McLaren and Estermann, 1957)

most appropriate explanation appears to be that the glass electrode does respond to the difference in hydrogen-ion activity between solution and suspension but that it responds also to a "membrane potential" of magnitude equal to, and sign opposite to, the difference in potential associated with the hydrogen ions in the ionic atmospheres.

In summary, evidence now available indicates that the apparent *p*H of soil obtained by use of a glass-electrode *p*H meter reflects hydrogen-ion activities in the bulk solution surrounding the electrodes and not hydrogen-ion activities in the ion atmospheres around the soil particles. In addition, the apparent *p*H values may be biased to some extent by another voltage at the junction between the soil and the potassium chloride bridge leading to the calomel reference electrode.

Measurement of Soil pH. If measurements of *p*H made on soil suspensions are biased to some degree by junction potentials, as there is reason to think, the usual practice is evidently unsatisfactory. More accurate methods must be sought. The first possibility that comes to mind is that of equilibrating soil with water and then measuring the *p*H of the supernatant solution or the *p*H of a filtrate. There are two general objections to this technique. First, as may be inferred from the section on factors affecting soil *p*H, the value obtained cannot be said to represent the *p*H of the soil. Second, the supernatant solutions are usually poorly buffered, and reproducible values are not always easy to obtain. The values may be affected by a slight carry-over from one solution to the next. The same objections apply whether or not the soil is first washed with water to reduce the content of electrolytes, although the preliminary leaching eliminates temporary fluctuations associated with variations in electrolytes. Another technique is to measure the *p*H in 1-normal potassium chloride. The results thus obtained are reproducible and are not influenced by minor fluctuations in content of electrolytes in the soil. The values obtained are independent of the location of the electrodes (Table 5.1), and junction potentials are probably of no importance. Addition of potassium chloride produces extensive exchange and brings into solution hydronium and other proton donors that influence the glass electrode and lower the measured *p*H. The *p*H values obtained do not appear to represent those of the soil solution; however, they may approach the *p*H values in the ion atmospheres of the original soil before treatment with 1-normal potassium chloride solution.

The method proposed by Schofield and Taylor (1955) appears to be the best now available for measuring soil *p*H on a practical basis if the objective is to obtain an estimate of the *p*H of the soil solution where the water content of the soil corresponds to field conditions. These investigators proposed that, where C represents a cation having the valence

v, the ratio $a_H/(a_C)^{1/v}$ is constant in dilute solutions in equilibrium with a given soil. The general concept is the same as that outlined in Chapter 4. Table 5.3 contains some of their data verifying the proposal. The ratio $a_H/(a_{Ca+Mg})^{1/2}$ is seen to be substantially constant despite differences in weight of soil and concentration of calcium chloride solution added. Schofield and Taylor preferred to express the result in terms of the negative logarithm of the ion activity ratio [pH − ½p(Ca + Mg)], as shown in the last column of the table. They called this value the "lime potential." They argued that the lime potential is a more characteristic soil property than the pH value and is more desirable for this reason. Justification for their point of view may be seen by noting that in Table 5.3 the pH value was affected by the concentration of calcium chloride, but the lime potential was not. Hence one may perceive that small fluctuations of electrolyte content which would cause changes in the pH value would have little or no influence on the lime potential. More recent measurements by Raupach (1957) and Ryding and Salmon (1964) have shown that the lime potential is not always as constant as was found in some of Schofield and Taylor's work.

Although Schofield and Taylor placed emphasis on the lime potential, other investigators for the most part have viewed the procedure from the standpoint of measuring pH values. Peech (1965a) recommended that pH measurements on soils be made by equilibrating 10 g of soil with 20 ml of 0.01-molar calcium chloride solution and then inserting the calomel electrode into the supernatant solution and the glass electrode into the partly settled suspension. The results should then be reported as "soil pH measured in 0.01-molar calcium chloride." Junction

Table 5.3. Experimental Test of Constancy of $\frac{a_H}{(a_{Ca+Mg})^{1/2}}$ Employing Rothamsted Soil (Schofield and Taylor, 1955)

Weight of Soil in 50 ml of $CaCl_2$ Solution, g	pH	a_H	Concentration of Ca + Mg per Liter, mole	$a_{(Ca+Mg)}$	$\frac{a_H}{a_{(Ca+Mg)}^{1/2}}$	$p\text{H} - \frac{1}{2}p(\text{Ca} + \text{Mg})$
30	4.04	0.0000913	0.029	0.0112	0.00086	3.06
15	4.06	0.0000871	0.029	0.0112	0.00082	3.09
30	4.52	0.0000302	0.0018	0.00134	0.00082	3.09
15	4.52	0.0000302	0.0015	0.00115	0.00089	3.05

potentials apparently are not eliminated in suspensions of soil in 0.01-molar calcium chloride (Ryti, 1965), and so Peech's method of placing the calomel electrode in the supernatant solution serves as a safety precaution. The advantage in inserting the glass electrode in the suspension is that the suspension is more strongly buffered than the supernatant solution and hence will be less influenced than the supernatant solution by minor disturbances such as carry-over from one sample to the next.

Peech pointed out that the *p*H of a soil measured in 0.01-molar calcium chloride solution is independent of dilution over a wide range of soil-to-solution ratios. The proportion of the total electrolyte contributed by a nonsaline soil to a suspension of one part of soil in two parts of 0.01-molar calcium chloride is relatively small, so that the observed *p*H is a better index of base saturation than is the *p*H of the soil measured in water. And, perhaps most important, Peech noted that the electrolyte concentration of the soil solution in nonsaline soils at optimum water content for plant growth is similar to that in 0.01-molar calcium chloride. The *p*H values obtained in 0.01-molar calcium chloride according to this method are lower than those obtained in water under similar conditions. Ryti (1965) found an average difference of about 0.5 *p*H unit in measurements on many soils of Finland, with extreme differences of 0 and 1.1 *p*H unit in individual samples. As indicated in Fig. 5.9, the difference between the two measurements was found to decrease with an increase in the concentration of electrolytes (as estimated from the conductivity) in the soil.

Measurement of the *p*H of soil in 0.01-molar calcium chloride solution approaches the ideal of diluting the soil solution with enough additional soil solution to permit measurement or that of direct measurement of the *p*H of displaced soil solution. Munns (1965b) found that *p*H values of soil solutions displaced from acid soils were 0.1 to 0.2 unit lower than those obtained according to Peech's method using 0.01-molar calcium chloride, and both measurements were about 0.8 unit lower than the *p*H values found for suspensions of one part of soil in five of water. Measurement of the *p*H of soil in 0.01-molar calcium chloride solution thus provides a better basis for comparing *p*H values of soil solutions with *p*H values of other solutions (culture solutions for example) than do measurements of soil *p*H values in water.

Nevertheless, as long as measurements are made consistently in the same manner, a variety of methods may be of almost equal value for comparative purposes; this is evidently the case in the data shown in Fig. 5.9. For some purposes, such as the experimental work described in Fig. 5.8, the *p*H measured in water is appropriate. The same is true

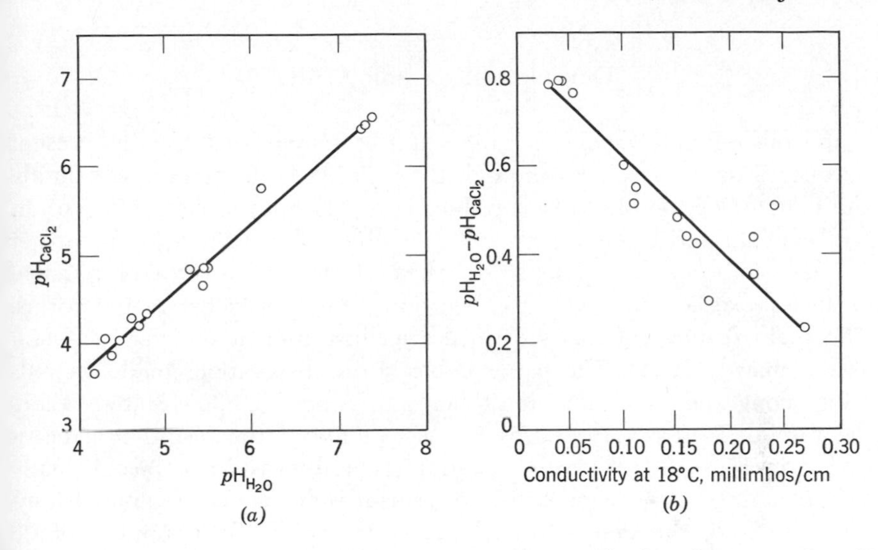

Figure 5.9. (*a*) *p*H of soils in 0.01-molar calcium chloride solution versus *p*H in water, and (*b*) difference between *p*H of soils in water and 0.01-molar calcium chloride solution versus conductivity of aqueous suspension. The same soils are represented in both parts of the figure. (Ryti, 1965)

in the quick test for detecting saline-sodic soils by the magnitude of the increase in *p*H that occurs on dilution of the soil with water. For purposes of estimating the appropriate quantities of limestone to apply to different soils, considerable practical use is made of measurements of the *p*H of soil suspended in certain buffer solutions. In general, therefore, the technique may be chosen to fit the purpose.

To conclude this section, a brief comment on terminology related to *p*H measurements seems appropriate. Solutions are said to be neutral at *p*H 7, acid at *p*H values below 7, and alkaline at *p*H values above 7; the same convention is applied to soils. From the fact that the *p*H values of soils depend on the method of measurement, however, it is evident that there is some doubt as to whether soils having *p*H values near 7 are properly described as neutral, slightly acid, or slightly alkaline. This ambiguity may be of importance if measurements by methods yielding rather different *p*H values are being compared; however, it is not of particular concern if all measurements being compared are made in the same way because correlations of *p*H with most other properties are strictly empirical, and small differences cannot be interpreted with certainty.

Development of Soil Acidity

In soils of dry regions, a large supply of bases is usually present because little water passes through the soil. With an increase in rainfall, the content of soluble salts is reduced to a low level, and any gypsum and calcium carbonate are removed, in the order named. With further increase in rainfall, a point is reached at which the rate of removal of bases exceeds the rate of liberation from nonexchangeable forms. The main features of this sequence are illustrated in a paper by Jenny and Leonard (1934). The paper is based on observations made on soils lying along the 11°C annual isotherm in central United States, where the rainfall increases from west to east. Some of the results are reproduced in Fig. 5.10. Where the annual rainfall was 35 to 50 cm, many soils contained carbonates within the surface 50 cm. As the annual rainfall increased, the depth to carbonates increased. At a rainfall of 90 to 100 cm, the soils were free of carbonates to a depth of 100 cm or more. The soil reaction changed from about *p*H 8 at a rainfall below 40 cm in eastern Colorado to about *p*H 6.7 to 6.8 at a rainfall of about 65 cm in east central Kansas, where titratable hydrogen first appeared. With further increase in rainfall through eastern Kansas and into Missouri, soil *p*H decreased continuously and titratable hydrogen increased.

The titratable hydrogen of soils is derived from several sources. One of these is water, which ionizes to a slight extent, producing hydrogen ions that exchange with the exchangeable bases. A second source is contact exchange between exchangeable hydrogen on root surfaces and the bases in exchangeable form on soil particles. The other principal source of titratable hydrogen in soils is the exchange that occurs with soluble acids. These acids arise in soils in several different ways. Large quantities of carbonic acid are produced in soils by microorganisms and higher plants. The effect is relatively small, however, because most of the carbonic acid decomposes and is lost to the atmosphere as carbon dioxide. Where leaching is absent or limited, practically all the carbonic acid is volatilized because chemical equilibria soon limit the accumulation of carbonic acid and bicarbonates in soil. Under these conditions microbial production of more stable nitric and sulfuric acids appears to be the deciding factor. For example, Desai and Subbiah (1951) found that where they allowed various organic and inorganic fertilizers to incubate in moist soil, the quantities of water-soluble calcium, magnesium, and potassium present at the end of incubation were correlated with the quantities of nitrate nitrogen (that is, nitric acid) produced during incubation. No correlation was found between the oxidized car-

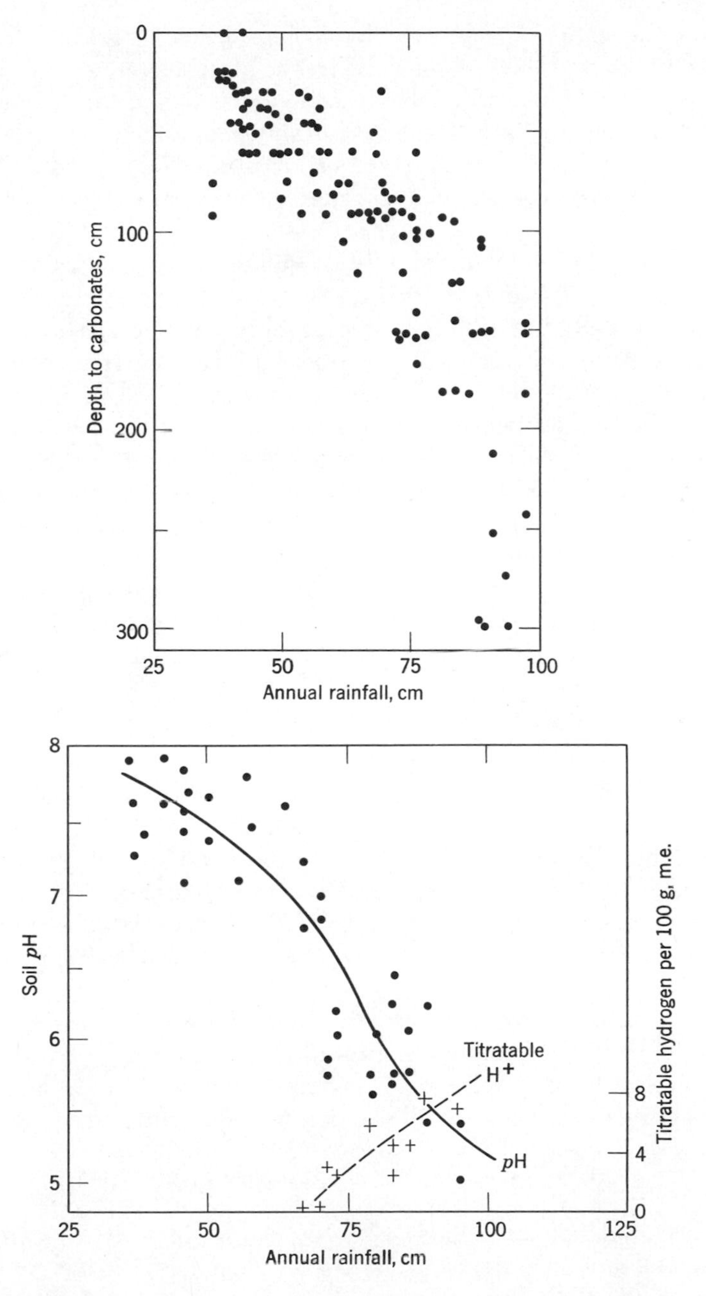

'igure 5.10. Depth to carbonates in soil profiles, and *p*H and titratable hydrogen in ıe surface portion of soils versus annual rainfall along the 11°C annual isotherm in entral United States. (Jenny and Leonard, 1934) bon and the cations made soluble. Where leaching occurs, however, a substantial part of the bases may be lost as bicarbonates.

The foregoing sources of soluble acids are common to all soils; however, others may be important in certain instances. One such source is the acidity produced from oxidation of iron sulfides. The effects are sometimes devastating. A review paper on this subject was published by Beers (1962).

Iron sulfides accumulate as ferrous sulfide (FeS) or as iron polysulfides (the most common are iron pyrite and marcasite, which have the formula FeS_2 but differ in crystal form) under anaerobic conditions as a result of reduction of iron and sulfate. Accumulation of significant quantities of iron sulfide requires a source of more sulfate than is found in most poorly drained soils. Thus the principal occurrences of sulfide-bearing soils are along the seacoast where the sulfate in the sea water (about 2 g of SO_4^{--} per liter) serves as the source. In inland locations, occurrences of sulfides in spoil materials deposited at the surface in coal mining operations have received considerable attention. These accumulations probably originated in earlier geologic times in the same way as those of modern origin along seacoasts. The general nature of the over-all reduction may be represented by the equation

$$4Fe(OH)_3 + 4CaSO_4 + 9CH_2O = 4FeS + 4Ca(HCO_3)_2 + CO_2 + 11H_2O$$

where CH_2O represents the organic matter used as an energy source by the microorganisms that are responsible for the reduction. The FeS formed originally is altered in time to polysulfides, which are more stable. The calcium bicarbonate and other bicarbonates formed in the reaction are moved out in the water if the carbon dioxide content is high, but calcium carbonate is precipitated if the carbon dioxide content is low.

These sulfide-bearing soils usually have *p*H values near neutrality as long as anaerobic conditions prevail. Upon drainage, however, the sulfides are oxidized to sulfates by a combination of chemical and microbiological processes. The over-all reaction may be written as

$$4FeS_2 + 15O_2 + 2H_2O = 2Fe_2(SO_4)_3 + 2H_2SO_4.$$

If the accumulation of sulfides is accompanied by calcium carbonate the reaction products are neutralized by calcium carbonate. According to Beers (1962), the *p*H of the soil upon oxidation usually remains above 5 if the sum of the calcium and magnesium present in carbonate and exchangeable form exceeds the sulfide (calculated as sulfate) by

50 m.e./100 g of soil. The result then is production of calcium sulfate, magnesium sulfate, and ferric hydroxide. But if the excess of calcium and magnesium is not present, acidification may be extreme; and the soil may support only the most acid-tolerant vegetation or none at all. Soil *p*H values of 2 to 4 are frequently obtained. Chenery (1954) reported an extreme value of *p*H 1.2 in a soil profile in Uganda.

The ferric sulfate produced in the oxidation reaction has a yellow color and commonly accumulates in visible streaks in the soil profile and as an efflorescence on the surface of clods upon drying. Fleming and Alexander (1961) identified the mineral species jarosite [$KFe_3(OH)_6(SO_4)_2$] and coquimbite [$Fe_2(SO_4)_3 \cdot 9H_2O$] in the dried efflorescence on clods from soils of the tidal marsh area of South Carolina. If the ferric sulfate is mixed with considerable aluminum sulfate, the streaks and efflorescence have a whitish-yellow color. The ferric ion coordinates water molecules like aluminum (discussed previously) and behaves as an acid. The first dissociation constant is 6×10^{-3}. The hydrated ferric ion is thus a considerably stronger acid than the hydrated aluminum ion, which has a first dissociation constant of 1.4×10^{-5} (Gilreath, 1958, p. 234). Neutralization of the hydrated ferric ion occurs by reactions analogous to those given previously for the hydrated aluminum ion, but the transformation to ferric hydroxide occurs at lower *p*H values than is true for conversion of the hydrated aluminum ion to aluminum hydroxide. The over-all result of the oxidation of FeS_2 is the production of 1 mole of sulfuric acid from 1 mole of sulfide sulfur. In view of the fact that sulfide-bearing soils sometimes contain more than 5% total sulfur, it is evident that great quantities of acid may be produced. Application of enough limestone to neutralize the acidity produced on oxidation of sulfides may be uneconomic.

A second source of acidity is that produced by fertilizers, of which nitrogenous fertilizers are of greatest importance. Nitrification of the ammonium in 1 mole of ammonium nitrate results in 2 moles of nitric acid. Nitrification of the ammonium in 1 mole of ammoniun sulfate results in 2 moles of nitric acid and 1 mole of sulfuric acid. Even ammonium hydroxide is acid-forming. Upon nitrification, 1 mole of ammonium hydroxide produces 1 mole of nitric acid.

Table 5.4 shows the soil *p*H values obtained in an experiment with ammonium sulfate. The fertilized plot received ammonium sulfate equivalent to 840 kg/hectare in 1925. Soil samples were taken for analysis in the fall of 1926. Within this short period of time, the *p*H of the soil from the ammonium sulfate-treated plot was lowered about 0.5 unit to a depth of 46 cm. The marked effect in this particular case can be attributed in part to the low buffer capacity of the sandy soil employed.

At one time, the soil acidification resulting from application of acid-forming fertilizers was of major importance in areas of heavy and sustained fertilizer use. Considerable research was done on the problem in the 1920s and early 1930s. Pierre (1928) devised a theory to account for the effect of different fertilizers on soil acidity and eventually proposed a laboratory method that would permit a quantitative estimation of the acid-forming tendency of fertilizers (Pierre, 1933). Efforts were then made to produce fertilizers that were not acid-forming. More recently, the trend has been toward manufacture of fertilizer with higher analysis and less "filler," and the use of large quantities of nitrogenous fertilizers is widespread. Consequently, the problem of soil acidification is being intensified. In the absence of neutralizing material as filler in the fertilizer, applications of limestone must be increased to avoid undue acidification. One associated phenomenon that eventually may prove

Table 5.4. *p*H Values at Different Depths in a Control Plot and a Plot Treated with Ammonium Sulfate Equivalent to 840 Kilograms per Hectare (Pierre, 1927)

	Soil *p*H	
Depth of Sampling, cm	Control Plot	Ammonium Sulfate Plot
0–15	5.5	4.9
15–30	5.5	5.0
30–46	5.4	5.0
46–61	5.5	5.4

troublesome is that nitrogenous fertilizers may acidify the subsoil to a considerable depth, but applications of limestone to the surface of the soil affect mostly the surface and neutralize the subsoil relatively slowly (Wander, 1954).

The acidity from nitrogenous fertilizers is not present in the fertilizer but is developed in the soil as a result of microbiological transformations. Of the common fertilizer materials, superphosphate is the only one that is inherently strongly acid. The solution that moves out into the soil from particles of superphosphate has a *p*H value between 1 and 2 (Lindsay and Stephenson, 1959). Although this solution does strongly acidify small volumes of soil around the particles on a temporary basis, the long-term effect does not appear to be of much significance because of the tendency of phosphate to react with hydrous oxides of aluminum and iron, with release of hydroxyl ions that react with the hydrogen ions present initially. The comparatively innocuous influence of superphosphate over a period of time is illustrated by the results of a field experiment conducted by Haylett and Theron (1955) on the application of ammonium sulfate and superphosphate to grass. At the end of 7 years, the soil *p*H values were 5.36, 5.36, and 5.35 with annual applications of 0, 318, and 635 kg of superphosphate per hectare and were 5.36, 4.83, 4.60, and 4.17 with annual applications of 0, 265, 530, and 1060 kg of ammonium sulfate per hectare. Superphosphate does decompose calcium carbonate, but the effect has not been considered significant in acidification of calcareous soils because of their high buffering capacity.

Finally, note may be taken of acids added from the air. Rainfall is characteristically acid. Riehm (1961) recorded *p*H values ranging from 4.3 to 6.4 at various locations in Europe. Quantities of acids added in precipitation, however, usually amount to calcium carbonate equivalents of only a few kilograms per hectare annually. Eriksson (1952) published a comprehensive review of the composition of atmospheric precipitation. Under special conditions, as in locations adjacent to smelters that release sulfur dioxide into the atmosphere, the acids derived from the air may be of great importance; much of the acid is then probably taken up directly by the soil from the atmosphere.

Soil Acidity and Plant Growth

Where neutral soils are acidified or acid soils are neutralized, many factors of the soil environment are changed simultaneously. Some of these changes may have no significant effect on plants and others may be critical. Factors that are critical in one soil may have no significant effect in another because of differences between the soils concerned. Although plant species and varieties have much in common in terms of their response, important differences may exist in individual instances. Thus both the magnitude of the effect and the importance of the various components may vary from one case to another. Occasionally a diagnosis can be made from visual examination of the plants, but usually the evaluation of soil-plant relationships under conditions of differing soil acidity is a research problem. There is no single, simple interpretation of titratable acidity or *p*H in terms of plant response to differing degrees of soil acidity. Accordingly, in this section emphasis will be placed on the principles involved in the three classes of factors that have received most emphasis in investigations of the soil-plant relationships in soil acidity: toxic substances, nutrient availability, and microbial activity.

Toxic Substances

Soil Acids. The value of limestone as a soil amendment was known long before the discovery that some soils are acid and others are not. The knowledge that the beneficial effects of liming are associated with the application of limestone to acid soils naturally led to the theory that the acids in soils are detrimental to plants and that limestone benefits plants by neutralizing these acids.

The validity of the acid-toxicity theory became doubtful as experiments involving *p*H measurements were conducted. Plants grown in nutrient solutions were found to make satisfactory growth at *p*H values lower than those tolerated in soils. For example, barley grown in the field failed at a *p*H value of 5 or below (Ohio Agricultural Experiment Station, 1938) but grew fairly well in nutrient solutions at *p*H 4.5 (Ligon and Pierre, 1932). The acid-toxicity theory was discredited further by investigations of the *p*H of tissue fluids of plants. Truog and Meacham (1919) expressed the sap from the roots of a number of plants and found *p*H values as low as 4. Pierre and Pohlman (1933) found that the sap exuded from the stump of several different plants, after removal of the tops, was strongly acid by standards used for evaluating soils and was unaffected by liming the soil. According to Small (1954), the *p*H of plant tissue fluids is usually in the range of soil *p*H values within which liming is beneficial. The evidence indicates that plant tissues tolerate acidity of the same or greater intensity than that commonly found in acid soils and leads to the reasonable inference that the usual hydrogen-ion activities in acid soils are not specifically toxic to plants.

The same type of evidence supports the view that the hydrogen-ion activity is great enough to be specifically toxic to plants under the conditions of extreme soil acidity that result from oxidation of sulfides. Arnon and Johnson (1942) found that lettuce, tomato, and bermudagrass made little or no growth in nutrient solutions at *p*H 3.

The evidence that the hydrogen-ion activity in most acid soils is not specifically toxic to plants is based on comparisons of *p*H values measured in nutrient solutions, soils, and plants. In view of the complications relative to *p*H measurements in soils, the possibility must be considered that such measurements do not reflect the hydrogen-ion activities to which plants respond. In particular, one may inquire whether plants respond to the relatively high hydrogen-ion activities in the ionic atmospheres around soil particles instead of the hydrogen-ion activities measured by a glass-electrode *p*H meter. The data in Fig. 5.8 indicate that the hydrogen-ion activity around particles of kaolinite was of the order of one-hundred times greater than that measured by a *p*H meter.

Vlamis (1953) conducted experiments that provide evidence on the *p*H condition in soils that is significant to plants. In one of these, he arranged the apparatus shown schematically in Fig. 5.11 to permit con-

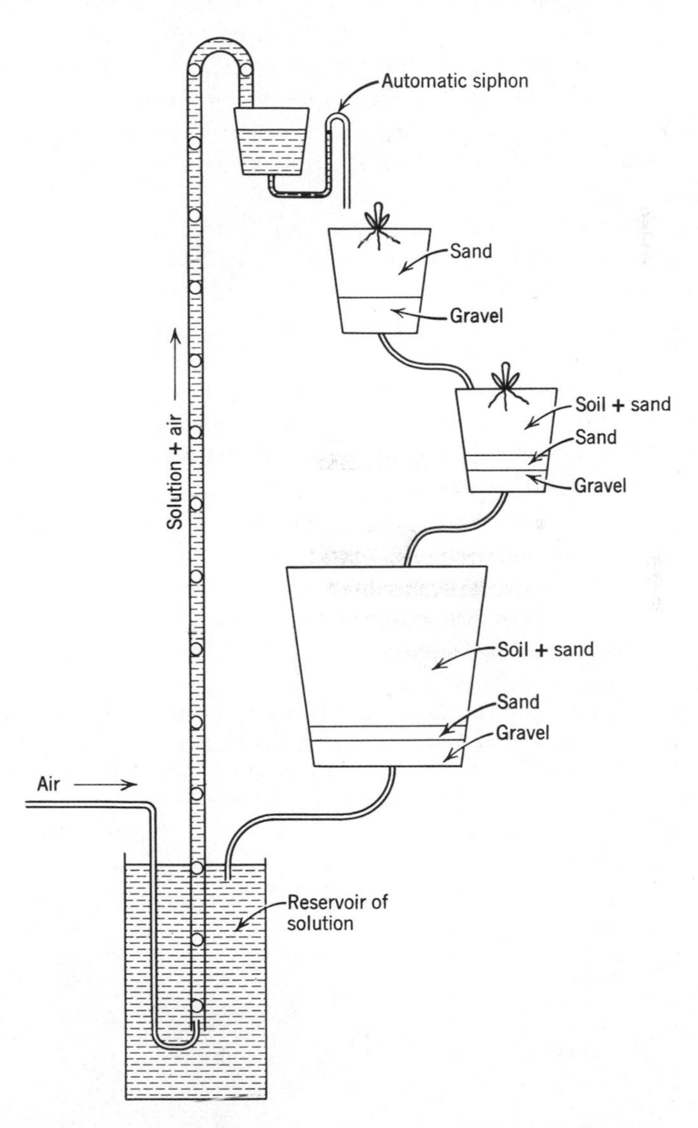

Figure 5.11. Schematic diagram of apparatus for circulating a single solution through a sand culture and a soil-plus-sand culture containing plants. (Vlamis, 1953)

tinual circulation of solution through an upper sand culture, then through a culture of the same size containing a mixture of soil and sand, then through a large container of soil and sand, and then to a reservoir of solution. An air-lift pump elevated the solution from the reservoir to an automatic siphon at the top, which periodically filled and started the siphon, draining the entire contents into the upper container of sand. Plants were grown in the upper container of sand and in the following container of soil plus sand. The large lower container of soil plus sand was used as a reservoir to permit equilibration of the solution with the soil before the solution was pumped back into the upper container of sand. Several acid soils were used, each being limed and unlimed.

Some of the data obtained are given in Table 5.5, where the results shown for lettuce and barley are averages of two and four separate trials. If plants growing in soil respond to the hydrogen-ion activity of the ion atmospheres around the soil particles instead of the hydrogen-ion activity of the solution, the yields would be expected to be lower in the soil-plus-sand cultures than in the sand cultures. The results show, however, that the yields were almost the same, from which one may infer that the plants were not responding to the hydrogen-ion activities around the soil particles and that the *p*H value of the solution does have meaning for plants growing in soils. The similarity of yields obtained in the sand and soil-plus-sand cultures permits the additional inference that the low yields on the unlimed soils and the high yields on the limed soils were attributable to substances in solution.

Table 5.5. Yield of Tops of Barley and Lettuce in Cultures of Sand Alone and of Sand plus Limed or Unlimed Soil in the Device Shown in Fig. 5.11 for Continual Circulation of Solution Through the Cultures (Vlamis, 1953)

	Unlimed Soil			Limed Soil		
		Yield of Dry Matter, g			Yield of Dry Matter, g	
Test Plant	*p*H of Solution	Sand	Soil plus Sand	*p*H of Solution	Sand	Soil plus Sand
Barley	4.2	0.50	0.55	5.2	2.32	2.20
Lettuce	4.4	0.25	0.25	5.2	2.30	2.55

The acid-toxicity theory includes possible toxicities of acid anions as well as hydrogen ions. In Chapter 3, evidence was cited for the occasional occurrence and toxicity of simple organic acids accumulated in soil under anaerobic conditions. Kaurichev et al. (1963) reported the occurrence of soluble organic acids in soils under forest vegetation. Oxalic, fumaric, and citric acids were detected, oxalic acid being found most commonly. These acids were presumably derived from the surface layer of organic material. Quantities up to 44 μg of organic acids (calculated as oxalic acid) per milliliter were found in the water that passed downward through the A0 or surface organic horizon into the underlying soil. The carbon of organic acids (calculated as oxalic acid) comprised 2 to 26% of the total carbon of the soluble organic matter collected on columns of aluminum oxide placed beneath the A0 and other soil horizons. Kononova (1962) reported occurrence of a variety of low-molecular organic acids in solutions pressed out of peat and soil. The acids found in this work on water-soluble organic material in soils are normal constituents of most, if not all, plants; and the quantities present do not appear to be sufficient to be a cause of specific toxicity.

Toxic acids that have been identified in plant residues and soils were summarized as follows by W. Moje (quoted by Chapman, 1965): benzoic acid (from soil); trans-cinnamic acid (from guayule roots); p-coumaric acid (from sugar-beet roots and fungi); threo-9,10-dihydroxystearic acid (from soil and oak roots); ferulic acid (from plant residues); p-hydroxybenzoic acid (from soil and sugar-beet roots); and 2-methylisonicotinic acid (from soil). These acids are mostly connected with particular conditions that are probably of limited extent and are not as yet to be regarded as toxic constituents in soils in general. The significance of organic acids as toxins in soils in general is yet to be determined. Whitehead (1963) reviewed evidence on this and related matters.

Aluminum. The hydrated aluminum ion is an acid in the general sense, and so aluminum toxicity might be considered under the acid-toxicity theory. Aluminum has other effects, however, and these justify separate consideration.

The nature of the evidence for aluminum toxicity in acid soils is essentially as follows. First, addition of aluminum to culture solutions depresses the growth of plants. Figure 5.12 shows that the growth of some species may be markedly reduced by aluminum concentrations less than 1 μg/ml of culture solution. Second, concentrations of aluminum in displaced soil solutions of soils with *p*H values below 5 are frequently in the range in which aluminum toxicity is found in culture solutions. For example, Fig. 5.13 shows data of Magistad (1925) on the concentration of aluminum in solution at different *p*H values (after different

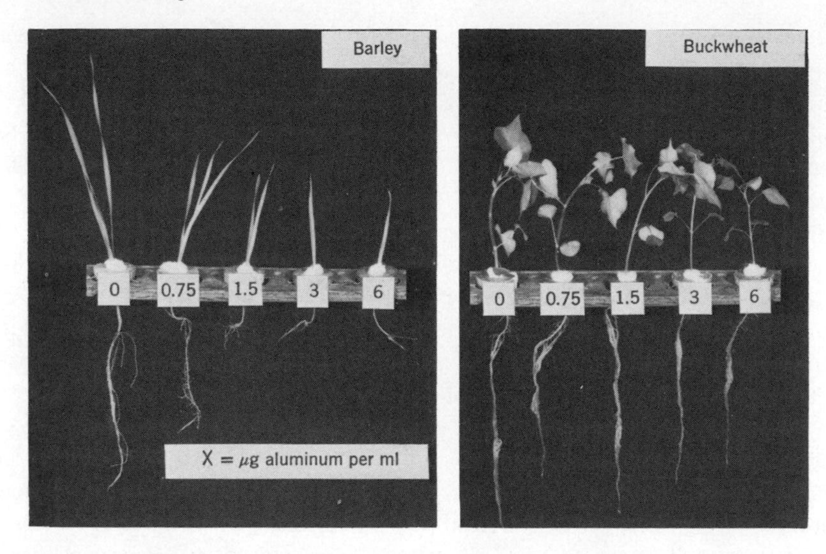

Figure 5.12. Response of barley and buckwheat to increasing concentrations of aluminum in culture solutions. Initial concentrations of aluminum from left to right with each species were 0, 0.75, 1.5, 3, and 6 μg/ml. Concentrations of aluminum in filtered solutions at the end of the growth period ranged from 0.2 to 0.8 μg/ml. (Photographs courtesy of C. D. Foy)

quantities of sodium hydroxide were added to aluminum sulfate), together with data of Magistad (1925) and Pierre et al. (1932) on the concentration of aluminum in the soil solution displaced from different soils.

Third, displaced soil solutions have been used as culture solutions for growing plants, and the toxicity of their aluminum content has been verified. Vlamis (1953) displaced the solution from an unlimed soil and from the corresponding limed soil and grew barley on each, with and without various additions. The data in Table 5.6 show that the major changes in yield of the plants were associated with changes in the aluminum content of the solutions and not with *p*H, calcium content, or manganese content. The results leave little doubt that aluminum toxicity existed in the solution from the unlimed soil. The experiment described in Table 5.6 was accompanied by the experiment described in the preceding section (Fig. 5.11 and Table 5.5) in which the growth of plants was found to be equally poor in acid soils or in solutions in equilibrium therewith, and equally good in limed soils or in solutions in equilibrium therewith. These results indicate that the cause of poor growth in the acid soil and of good growth in the limed soil was the same as that

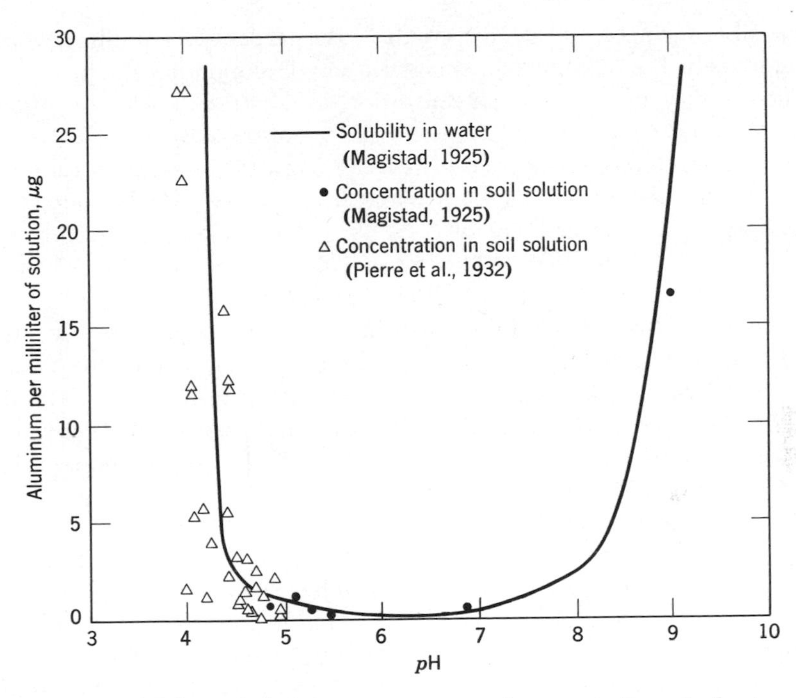

Figure 5.13. Solubility of aluminum in water, and concentration of aluminum in displaced soil solutions at different *p*H values. (Magistad, 1925; Pierre et al., 1932)

Table 5.6. Growth of Barley in Solutions Displaced from Corresponding Unlimed and Limed Samples of Soil (Vlamis, 1953)

Substrate for Growth of Barley	*p*H	Concentration per ml of Solution, μg		Yield of Barley, mg
		Al	Mn	
Solution displaced from unlimed soil				
Untreated	4.2	1.8	16.0	139
Treated with $Ca(OH)_2$	5.8	0.4	12.8	340
Treated with NaOH	5.8	0.4	12.0	286
Solution displaced from limed soil				
Untreated	5.8	0.3	7.6	353
Treated with H_2SO_4	4.2	0.35	7.4	315
Treated with $H_2SO_4 + Al_2(SO_4)_3$	4.2	1.8	7.7	176
Treated with $H_2SO_4 + MnSO_4$	4.2	0.3	16.0	341

responsible for poor and good growth in the respective equilibrium solutions, namely, the difference in concentration of aluminum.

The toxicity of a given concentration of aluminum may be greatly influenced by the accompanying cations. In culture solutions, the toxicity of aluminum tends to decrease with an increase in concentration of other cations. For example, Dios Vidal and Broyer (1962) found that aluminum at a concentration of 4 to 17 μg/ml of culture solution was toxic to corn where the concentration of magnesium was 0.1 millimole/liter but not where it was 10 millimoles/liter (calcium was present at a concentration of 4.8 millimoles/liter). In acid soils, on the other hand, addition of other cations in the form of neutral, unbuffered salts such as potassium chloride or calçium chloride may increase aluminum toxicity. The reason presumably is that the salts increase the concentration of aluminum in solution and lower the *p*H value, while part of the added cations disappear by exchange.

The salt effect in soils is illustrated in Table 5.7 by the results of an experiment in which sweet corn was grown in the greenhouse on an acid soil that had been treated with different quantities of potassium chloride. The applications of potassium chloride were greater than those commonly used for fertilizer purposes, which exaggerated the aluminum effect in comparison with practical conditions. Nevertheless, the quanti-

Table 5.7. Growth of Roots of Sweet Corn in an Acid Soil with Various Treatments, and Concentrations of Aluminum and Calcium plus Magnesium in Saturation Extracts Made after Cropping in the Greenhouse (Ragland and Coleman, 1959)

	Original Acid Soil			Acid Soil plus 2 m.e. of $Ca(OH)_2$ and 1 m.e. of MgO per 100 g		
		Ions per Liter of Saturation Extract			Ions per Liter of Saturation Extract	
Potassium Chloride Added per 100 g, m.e.	Relative Yield of Roots	Al, mg	Ca + Mg, m.e.	Relative Yield of Roots	Al, mg	Ca + Mg, m.e.
0	41	0.8	0.5	23	0.2	1.6
0.5	36	18	3.0	100	0.2	4.0
1.0	23	46	4.0	98	0.8	8.0

ties of potassium chloride were not excessive as such because they produced increases in yield where the soil had been treated to alleviate the acid condition. From the opposite effects produced by potassium chloride at the two levels of acidity, one may infer that the effect of the potassium chloride addition on the plant response represented a balance between the primary nutritive effect of the potassium and the secondary toxic effect of the aluminum. The nutritive effect was the more important where the soil acidity had been partly neutralized, and the toxic effect was the more important in the original acid soil.

Special staining tests indicate that aluminum accumulates on the root surface and in the cortex under conditions of toxicity. In the cortex, the staining occurs mostly in the cell protoplasm and especially in the nuclei; little or no staining due to aluminum is found in the conducting tissue inside the endodermis (McLean and Gilbert, 1927; Wright and Donahue, 1953). McLean and Gilbert (1927) noted that redtop, which was noticeably more resistant to aluminum toxicity than corn and cabbage, did not accumulate much aluminum in its root system, as indicated by the staining test. In work with two varieties of wheat that had been found to differ in tolerance to aluminum toxicity, Foy et al. (1965) found that the more tolerant variety tended to maintain a higher *p*H around the roots than did the less tolerant variety, thus presumably promoting precipitation of aluminum in the soil instead of the roots.

Aluminum toxicity represents a combination of effects, of which inhibition of root growth is perhaps the most obvious visually. Microscopic examination of plants grown with toxic concentrations of aluminum has shown an abnormally large number of cells with two nuclei in the meristematic region of the root tip, indicating inhibition of cell division (Rios and Pearson, 1964). Rorison (1958, pp. 43–61) proposed that the inhibition of branch-root development might be due to reaction of aluminum with the pectic substances of the young cell walls, causing them to lose their plasticity prematurely and inhibiting elongation.

Aluminum toxicity has a number of effects on plant nutrition. Foy and Brown (1963) reported that excess aluminum in nutrient solution caused decreased uptake of phosphorus, calcium, potassium, manganese, iron, sodium, and boron by cotton plants. Moreover, excess aluminum caused plants to wilt. McLean and Gilbert (1927) observed lower loss of water per unit leaf area from aluminum-treated than from control plants, and they obtained evidence for lesser movement of dyes in and lesser uptake of nitrate by aluminum-treated plants than by control plants. These observations suggest that aluminum toxicity causes a general decrease in permeability of protoplasm of root cells.

The effect of aluminum on phosphorus nutrition has received more attention than any other nutritional effect. Foy and Brown (1964) reported that foliar symptoms of aluminum toxicity in plants grown in nutrient solutions were similar to those of severe phosphorus deficiency. Munns (1965b) reported a similar finding for plants suffering from aluminum toxicity in acid soil. The aluminum-phosphorus interaction may be illustrated by Fig. 5.14, which shows the growth of cotton plants with increasing concentrations of phosphorus in culture solutions containing either 2 or 10 μg of aluminum per milliliter. At the higher concentration of aluminum, branch roots failed to develop; the growth of tops was stunted except at the highest concentration of phosphorus. At the lower concentration of aluminum, a lower concentration of phosphorus sufficed to induce formation of branch roots and growth of tops.

The aluminum-phosphorus interaction in plant nutrition may be explained in part on the basis that aluminum and phosphate ions interact chemically to form a sparingly soluble salt. If the concentration of either component of this salt in the solution is high, the solubility-product principle allows the prediction that the concentration of the other component must be low.

Applicability of the solubility-product principle was verified in a qualitative way by Munns (1965b), who found that the concentration of aluminum in 0.01-molar calcium chloride extracts and displaced soil solutions from acid soils was indeed decreased upon addition of soluble phosphate salts. Application of 708 μg of phosphorus per gram of one soil reduced the concentration of aluminum in the calcium chloride extract from about 2.7 to 0.3 μg/ml. Alfalfa grown on this soil at low levels of phosphorus had stunted roots, symptomatic of aluminum toxicity. The tops of the plants showed symptoms of phosphorus deficiency. Theory and experiment thus both indicate that plants in soils may be affected by a high concentration of aluminum and a low concentration of phosphorus at the same time. Furthermore, they may be affected by a low concentration of aluminum and a high concentration of phosphorus at the same time.

The solubility-product principle of aluminum-phosphorus interaction applies over a wide range of soil *p*H values, but it becomes significant in relation to aluminum toxicity only at low *p*H values. The reason is that formation of aluminum hydroxide and hydroxy-aluminum interlayers in clays limits the solubility of aluminum to nontoxic levels even in the absence of phosphate except at low *p*H values. Consequently, plants may be affected by phosphorus deficiency at high soil *p*H values without at the same time being affected by aluminum toxicity.

Munns (1965a) found that nutrient solutions could be prepared at *p*H 4.5 with enough aluminum (2.7 μg/ml) to be toxic to alfalfa and

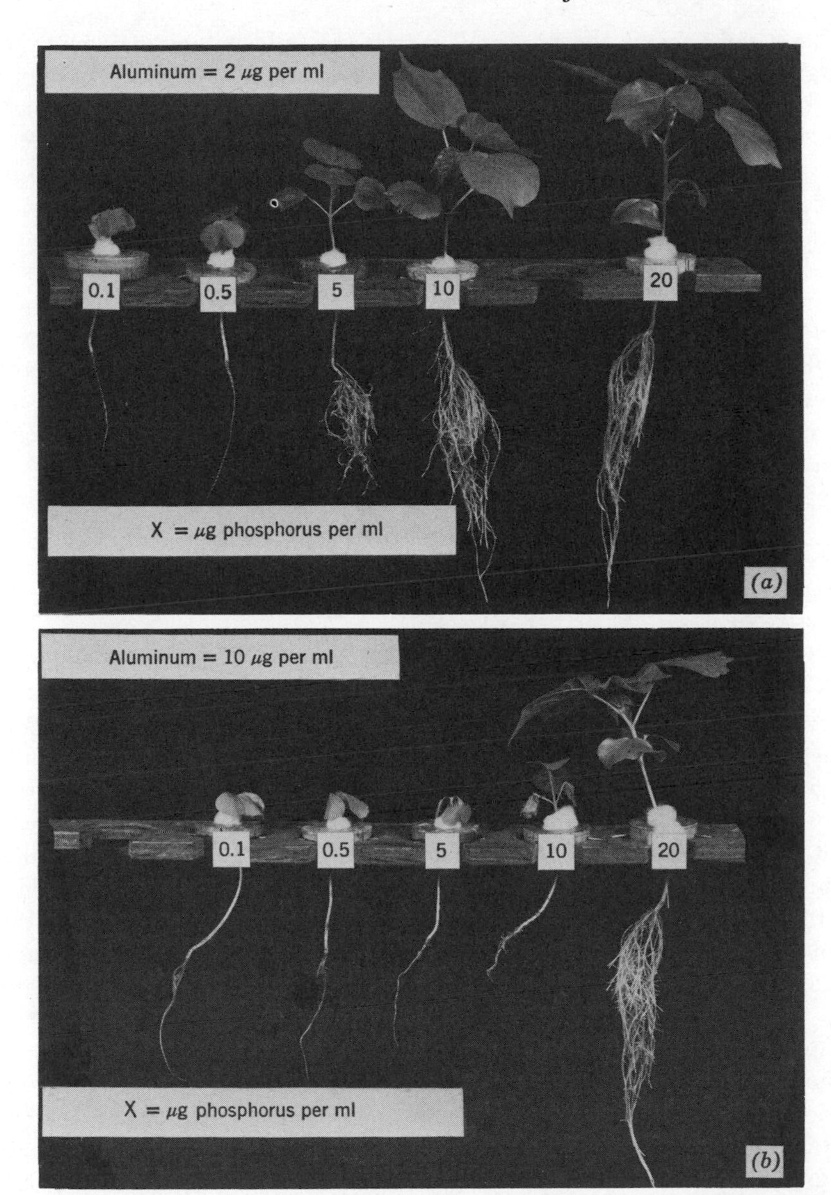

Figure 5.14. Appearance of cotton plants grown in nutrient solutions at *p*H 4 with different additions of phosphorus and aluminum. Additions of aluminum were (*a*) 2 μg/ml and (*b*) 10 μg/ml. Additions of phosphorus from left to right were 0.1, 0.5, 5, 10, and 20 μg/ml. (Foy and Brown, 1963)

still not deficient in phosphorus (up to 0.3 μg/ml) without precipitation of aluminum phosphate. In subsequent work, Munns (1965b) obtained evidence for existence of similar concentrations of aluminum and phosphorus in displaced soil solutions. These observations provide evidence for existence of a condition of aluminum toxicity without phosphorus deficiency for plants that are sufficiently sensitive to aluminum toxicity.

In practice, therefore, the increase in plant yield produced upon application of phosphate to a soil may result from the extra supply of phosphorus without aluminum toxicity if the soil has a high *p*H. If the soil has a low *p*H, the increase in yield may result from (1) an increase in the supply of phosphorus without effect of aluminum toxicity if the plant species is sufficiently tolerant to aluminum, (2) a decrease in toxicity of aluminum if the plant species is sensitive to aluminum toxicity and if the soil is not too low in phosphorus, or (3) an increase in supply of phosphorus and a decrease in aluminum toxicity if the soil is low in phosphorus and the plant species is sensitive to aluminum toxicity. No limits can be specified for *p*H values because of insufficient information and the fact that different plant species are not equally sensitive to aluminum toxicity.

Thus far, no technique has been developed by which phosphorus and aluminum can be varied independently in soil to investigate the relative significance of aluminum toxicity and phosphorus deficiency. Raising the *p*H effectively eliminates aluminum toxicity and then permits an investigation of phosphorus deficiency. Raising the soil *p*H, of course, may have effects other than reduction of aluminum concentration even if it should happen to have no significant effects on phosphorus; this limits the effectiveness of the technique for diagnostic purposes. In view of this situation, perhaps the best approach is chemical analysis of the above-ground parts of plants for aluminum and phosphorus. Although most of the aluminum usually stays in the roots, analyses of roots have the disadvantage that there is some question as to how much of the aluminum found is on the outside of the roots as a precipitate or as adhering particles of soil. The problem of adhering particles of soil is significant also in analyses of above-ground parts because the concentration of aluminum in soil particles is of the order of ten-thousand times as great as that in above-ground parts of some plants. Fox et al. (1962) and Munns (1965b) made use of this technique in investigations of aluminum toxicity and phosphorus deficiency in acid soils.

Certain complications in aluminum-phosphorus relations in plants have been discovered that will need further clarification. On first thought, one would expect from the solubility-product theory that a high concentration of either aluminum or phosphorus in the soil solution would result in a low concentration of the other in the plant because of formation of aluminum phosphate in the soil. This expectation seems to work out if phosphorus is present in excess but not if aluminum is present in excess. Experiments on plants grown in nutrient solutions show that addition of aluminum to the nutrient solution causes an increase in uptake of phosphorus by plant roots (Randall and Vose, 1963). This observation, coupled with the additional observation that a high concentration of aluminum externally usually results in a low concentration of phosphorus in the above-ground parts of plants, has led to the suggestion that the precipitation of aluminum phosphate called for by the solubility-product theory takes place partially on or in the roots. This suggestion seems reasonable in view of the concurrent accumulation of aluminum and phosphorus by the roots, but it has never been proved that the aluminum and phosphorus are combined as aluminum phosphate.

The barrier that keeps aluminum mostly confined to the roots in some plant species seems to be lacking in others. As much as 0.38% aluminum in dry, above-ground plant tissue has been reported (Hutchinson, 1945). *Pinus radiata* is one species that accumulates high concentrations of aluminum in the leaves if grown on strongly acid soils. Humphreys and Truman (1964) found that this species accumulated phosphorus as well as aluminum in the tops if aluminum was present in excess. The concentration of phosphorus in the aerial parts increased from 0.16% in plants grown on a solution lacking aluminum to 0.46% in plants grown on a solution containing 20 μg of aluminum per milliliter.

The tendency of the excess of phosphorus to be localized with the aluminum suggests that in some way the phosphorus is inactivated nutritionally. Hence one may suspect that the accumulation of phosphorus in the presence of excess aluminum does not mean that aluminum has the effect of alleviating the nutritional deficiency of phosphorus but rather that the extra phosphorus is taken up because high-aluminum plants have a high phosphorus requirement. This aluminum-phosphorus interaction has the implication that interpretation of the phosphorus content of plants in terms of the degree of nutritional sufficiency of phosphorus must take into account the content of aluminum if the plant part analyzed accumulates appreciable concentrations of aluminum.

Manganese. In acid soil, manganese behaves similarly to aluminum in that its concentration in the soil solution increases as the *p*H decreases. The concentration of manganese frequently exceeds that of aluminum, particularly at higher *p*H values.

Evidence for toxicity of manganese in acid soils is similar to that for aluminum. Experiments on plants in nutrient solutions show that low concentrations of manganese may depress the growth. As little as

1 to 4 μg of manganese per milliliter of solution may depress the yields of lespedeza, soybeans, and barley (Morris and Pierre, 1949; Olsen, 1936), whereas corn may tolerate over 15 μg and *Deschampsia flexuosa* over 60 μg of manganese per milliliter without yield depression (Olsen, 1936). Experiments with soils show that the manganese concentration in displaced soil solutions may be in the range in which toxicity occurs in nutrient solutions. Figure 5.15 summarizes data on manganese concentrations in solutions displaced from soils having different *p*H values. Manganese toxicity symptoms were observed on plants growing on the three soils represented by the two solid points and the one open point representing *p*H values of 4.78 to 4.87 and manganese concentrations of 2.2 to 4.3 μg/ml.

Manganese toxicity is an important problem in acid soils that have been steamed, a common practice in management of soils for production of crops under glass. Steaming causes an unusually high concentration of soluble and exchangeable manganese, perhaps as a result of reduction by the heated soil organic matter. Control of *p*H in steamed soils thus is more important than it is in ordinary soils. Davies (1957) published a paper dealing with this subject.

In contrast to aluminum, excess manganese produces characteristic toxicity symptoms that are especially prominent on young plants. The symptoms differ among species, but a brown spotting of leaves is frequently observed. Sometimes excess manganese produces symptoms of

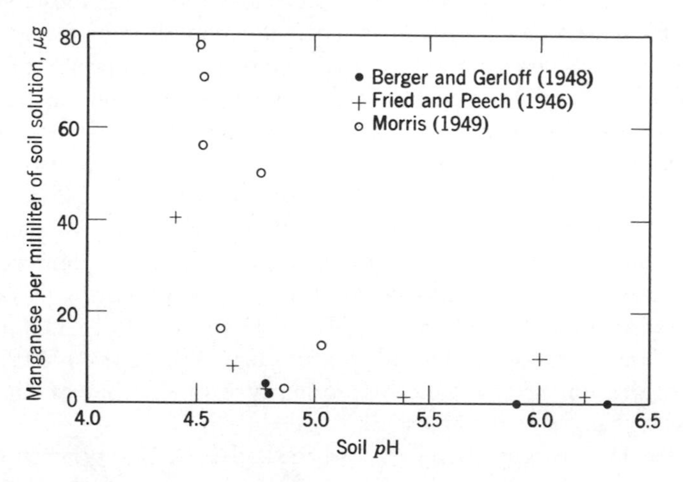

Figure 5.15. Concentration of manganese in displaced soil solutions versus soil *p*H.

iron deficiency. Aluminum does not accumulate in high concentrations in the above-ground parts of most plants even where aluminum toxicity is severe, but manganese is taken up more readily. Plants grown in soils usually do not contain more than 100 μg of manganese per gram of dry matter, but the concentration may be much higher under conditions of manganese toxicity. Hale and Heintze (1946) reported an extreme case in which leaves of field-grown potato plants contained 11,300 μg of manganese per gram of dry matter. In this connection, Vlamis (1953) observed some brown spotting of leaves indicative of manganese toxicity in plants grown in the acidified, manganese-treated solution mentioned in Table 5.6, but the yield of the plants did not differ significantly from that of plants in the control solution.

Aluminum toxicity and manganese toxicity differ also in respect of their relation to phosphorus. Application of phosphate causes precipitation of aluminum and alleviation of aluminum toxicity but does not have an analogous effect on manganese. The effect on manganese seems to be incidental and not the result of a direct chemical interaction between manganese and phosphate. Messing (1965) observed that application of superphosphate to soil with *p*H values below 5 generally raised the soil *p*H and decreased the water-soluble and exchangeable manganese; and, at *p*H values above 5.5, the effect of addition of superphosphate was generally to lower the soil *p*H and to increase the concentration of water-soluble and exchangeable manganese.

Nutrient Availability

Effects of soil acidity on nutrient availability may be designated as nonspecific or specific according to their nature. Perhaps the most important nonspecific effects are those associated with inhibition of root growth. Restriction of root growth into acid subsoil layers may be demonstrated easily, and a number of experiments have shown that the magnitude of the effect may range from little to almost complete inhibition. Figure 5.16 shows the results of one experiment in which sorghum was grown in the greenhouse on equal quantities of a common surface soil underlain by equal quantities of different acid subsoils. This experiment was done with disturbed samples, and so compaction was not a factor of significance. Limited root penetration due to subsoil acidity would evidently limit uptake of nutrients (including water) from the subsoil and would increase the dependency of the plants on the nutrients in the surface soil and on current rainfall. The effect on water availability is probably more important than that on nutrient availability in most instances. Some specific effects of acidity on nutrient availability are discussed in the following paragraphs.

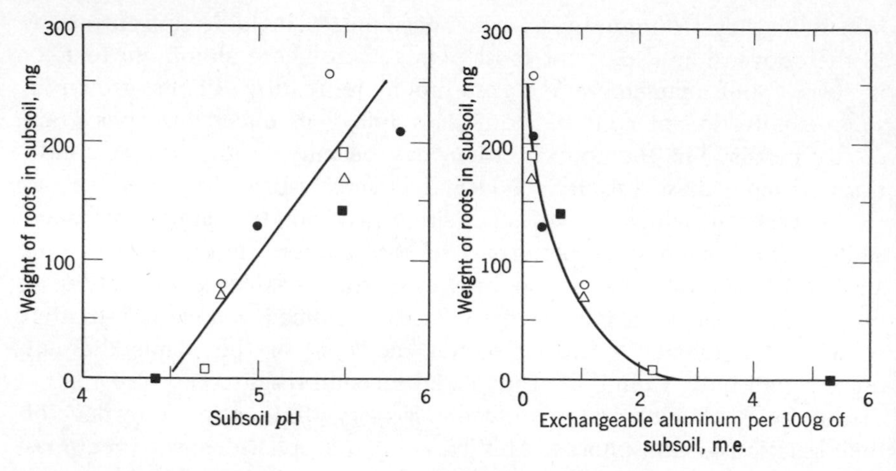

Figure 5.16. Weight of sorghum roots in different subsoils beneath a common surface soil versus subsoil *p*H and exchangeable aluminum. Legend for subsoils: ● Norfolk, ○ Goldsboro, △ Lynchburg, □ Rains, ■ Portsmouth. Two values are given for each subsoil. The sample corresponding to the lower *p*H and higher content of exchangeable aluminum was unlimed; and the sample corresponding to the higher *p*H and lower content of exchangeable aluminum received enough calcium oxide and magnesium carbonate in a 2-to-1 equivalent ratio to equal the exchange acidity displaced by normal potassium chloride solution. (Ragland and Coleman, 1959)

Exchangeable Bases. Two aspects of availability of exchangeable bases as they are related to soil acidity will be discussed in this section: the ion-uptake process and the release of bases from exchangeable form. Differences among plants will be discussed in a subsequent section.

Many experiments have shown that uptake of cations by plants is impaired by acidity of both soils and nutrient solutions. If the *p*H of nutrient solutions is low enough, there is no absorption and cations previously absorbed tend to leak out of the plants into the solutions. Two theories have been proposed to account for the effects of acidity. According to the first theory, hydrogen ions decrease uptake of cations by a competitive process. In terms of the carrier theory of ion uptake (introduced in Chapter 4), the cations probably exchange for hydrogen ions in the cation-selective sites; hence the higher the hydrogen-ion activity in the external solution, the greater will be the competition offered by hydrogen ions to each of the individual bases at the respective selective sites. According to the second theory, hydrogen ions damage the uptake mechanism by irreversible or only partly reversible processes. The evidence now indicates that the two theories are to a considerable degree complementary and that an important aspect of the unspecified damage referred to in the second theory involves the behavior of calcium.

Some appreciation for the importance of the interaction of hydrogen ions and calcium ions may be obtained from Fig. 5.17, which shows that the concentration of calcium required to obtain the maximum yield of lettuce in culture solutions increased with decreasing *p*H of the solution. Stated another way, an increase in the concentration of calcium alleviated to some degree the detrimental effect of acidity. (Because this experiment was done with nutrient solutions, toxicities due to increasing concentrations of aluminum and manganese at low *p*H values were not a factor.) The view now is that calcium plays an important part in maintaining the integrity of the absorption and selectivity mechanisms involved in uptake of cations in general and that hydrogen ions tend to displace the calcium from the sites where this control is exerted.

As an example of the competition between hydrogen ions and cations for uptake, Fig. 5.18 shows the results of an experiment on uptake of radioactive rubidium into nonexchangeable form at two pH values. (Radioactive rubidium was used in place of potassium because the two ions apparently are taken up via the same carrier site, low concentrations were used in the experiment to ensure that the uptake occurred via this specific site, and rubidium has a more convenient radioisotope than potassium.) Figure 5.18*a* shows that the quantity of rubidium absorbed by barley roots in 10 minutes was greater at *p*H 5.7 than at *p*H 4.1. Figure 5.18*b* shows a plot of the reciprocal of the velocity of rubidium

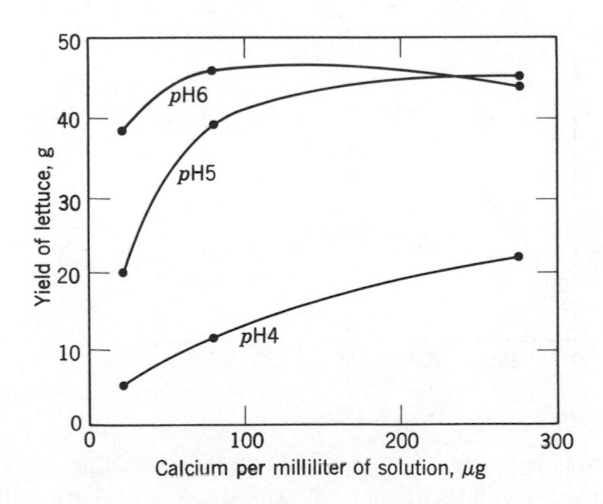

Figure 5.17. Yield of lettuce in culture solutions maintained at different *p*H values and calcium concentrations. (Arnon and Johnson, 1942)

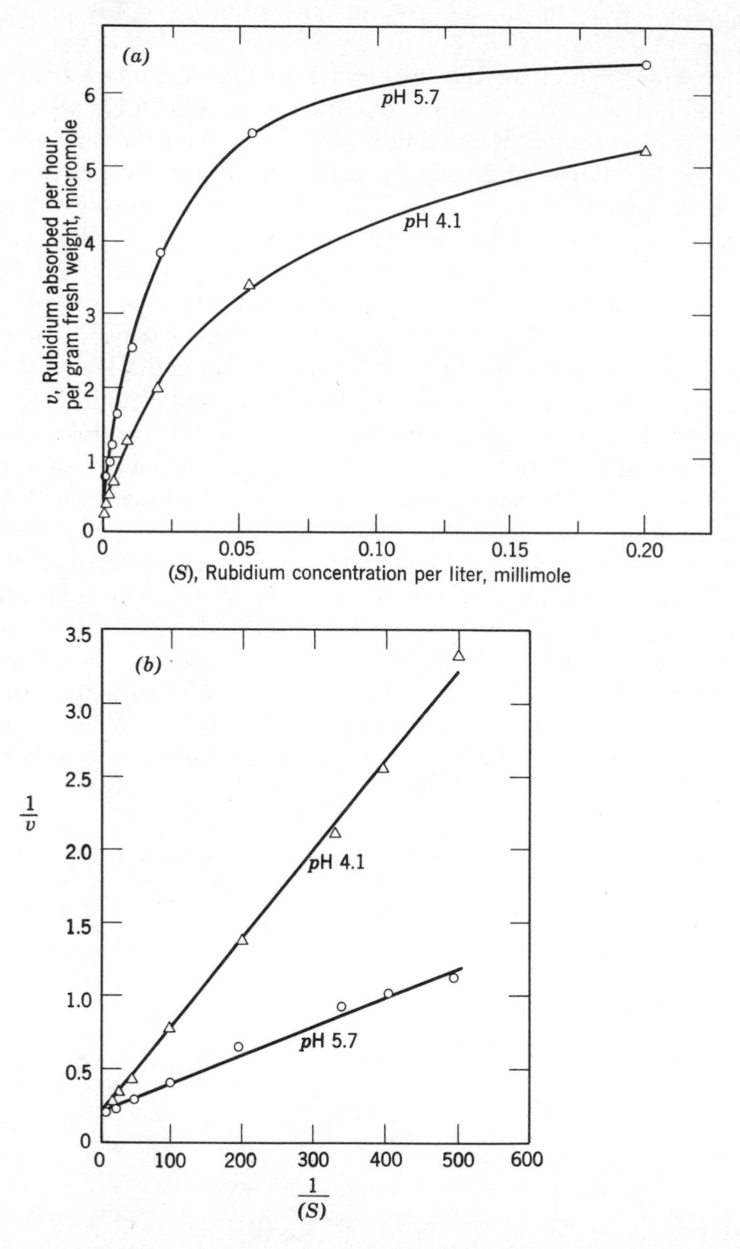

Figure 5.18. Rubidium absorption by detached barley roots versus concentration of rubidium chloride at *p*H 4.1 and 5.7. The tissue was in 0.5 millimolar calcium chloride before and during the absorption period of 10 minutes. (*a*) A conventional plot of the data. (*b*) A double-reciprocal plot of the same data. (Rains et al., 1964)

uptake against the reciprocal of the concentration of rubidium in the solution. This plot is based on the equation

$$v = \frac{V(S)K_I}{K_M K_I + K_M(I) + K_I(S)},$$

where v is the velocity of absorption of rubidium at the concentration (S) of substrate (rubidium) and (I) of inhibitor (hydrogen ion) in the solution, V is the maximum velocity of absorption attainable as the concentration of rubidium increases indefinitely, K_M is the concentration of rubidium at which the velocity of absorption of rubidium is equal to $V/2$ in the absence of inhibition by hydrogen ions, and K_I is the concentration of hydrogen ions at which the velocity of absorption of rubidium is $V/3$ where $(S) = K_M$.

This equation represents a modification of the fundamental enzyme equation (see Chapter 4 on exchangeable bases) to take into account the effect of various concentrations of a competitive inhibitor. A competitive inhibitor is one that combines with the enzyme in the same way as the substrate with which it competes, thereby reducing the concentration of enzyme-substrate molecules and the velocity with which the enzyme carries out the reaction with the substrate. In competitive inhibition, the limiting velocities at zero and infinitely high concentrations of substrate are not affected by the presence of inhibitor.

The reciprocal of the foregoing expression may be written as

$$\frac{1}{v} = \left(\frac{K_M}{V}\right)\left[1 + \frac{(I)}{K_I}\right]\left[\frac{1}{(S)}\right] + \frac{1}{V},$$

from which it may be seen that a plot of $1/v$ against $1/(S)$ should be a straight line with intercept $1/V$ and slope $(K_M/V)[1 + (I)/K_I]$ at a given concentration (I) of inhibitor. If the concentration of inhibitor increases, the slope of the line will increase; but the intercept should remain the same. This is the behavior shown in Fig. 5.18b, where the hydrogen ion is the inhibitor. As the concentration of the hydrogen ion (I) increases, the slope increases by the factor $1 + (I)/K_I$. The value of K_M may be found by determining the value of the slope at values of (I) so small that $1 + (I)K_I$ does not differ appreciably from unity, and the value of K_I may be found by determining the increase in the slope associated with a known value of (I). Such an analysis yielded the following values: $V = 7.2$ micromoles/hour/g frcsh weight, $K_M = 0.016$ millimole of rubidium per liter, and $K_I = 0.038$ millimole of hydrogen per liter. The pH value corresponding to K_I is 4.4. At pH 5.7, the value of $(I)/K_I$ is 0.05, which means that $1 + (I)/K_I = 1.05$ and that

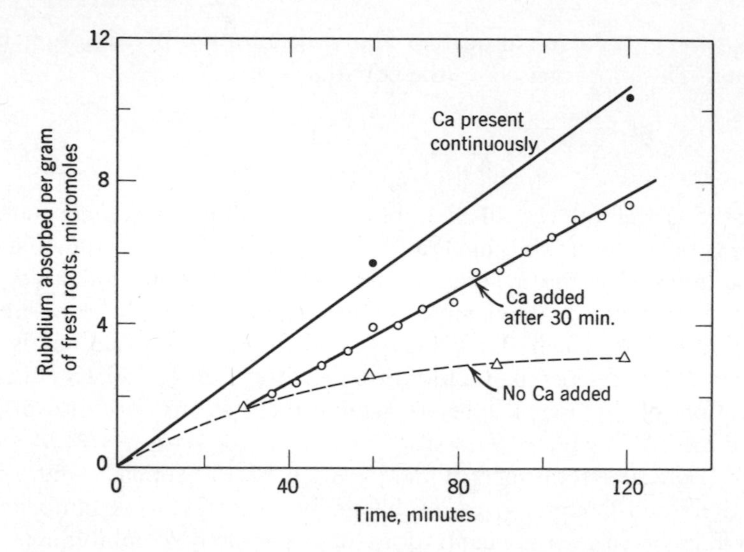

Figure 5.19. Absorption of rubidium by barley roots over different lengths of time at *p*H 3.9 from solutions containing 0.1 millimole of rubidium per liter with no added calcium, calcium present continuously at 0.5 millimole/liter, or calcium present after 30 minutes at 0.5 millimole/liter. (Rains et al., 1964)

the inhibition due to hydrogen ions should be almost nonexistent. At higher *p*H values, the inhibition should be even smaller.

The experimental data in Fig. 5.18 were obtained with a solution containing calcium at a concentration of 0.5 millimole/liter. Figure 5.19, which represents the results of another experiment reported in the same paper, shows the importance of calcium. This experiment was conducted at *p*H 3.9 in the presence and absence of calcium at a concentration of 0.5 millimole/liter and with addition of calcium at that concentration after a preliminary absorption period in which no added calcium was present. Absorption of rubidium took place at a considerably greater rate in the presence of calcium than in its absence. Furthermore, the decrease in the rate of absorption with time was much less in the presence of calcium than in its absence. Where calcium was added after an initial 30-minute period without calcium, the rate of absorption of rubidium was much greater than in the continued absence of calcium but did not return to the rate obtained where calcium was present continuously. These results indicate that calcium functions in reducing or preventing injury to the selective absorption mechanism by hydrogen ions.

Release of bases from exchangeable form also is affected by soil acidity. With an increase in acidity, the absolute amount of exchangeable bases decreases. At the same time, however, changes in two other conditions tend to oppose a decrease in release of bases in a fractional exchange. (1) The cation-exchange capacity decreases due to loss of *p*H-dependent exchange positions and accumulation of polymeric hydroxy-aluminum interlayers in expanding clays. This effect tends to maintain the degree of base saturation. (2) Aluminum accumulates as an exchangeable cation. Aluminum is more strongly bound in exchangeable form than are the exchangeable bases; hence, by virtue of the complementary-ion effect, exchangeable bases are released preferentially in a fractional exchange.

A numerical example of the degree-of-saturation effects may be quoted from results of an experiment by Mehlich (1946). He determined the release of calcium, magnesium, and potassium from exchangeable form by addition of dilute hydrochloric acid in the ratio of 500 ml of solution containing 1 m.e. of acid to 100 g of soil that had different levels of exchangeable bases and a cation-exchange capacity of 4 m.e. at *p*H 8.1. In two samples with total exchangeable bases of 3.9 and 1.4 m.e./100 g, the release of total bases in the fractional exchange was 0.9 and 0.7 m.e.

Figure 5.20 represents an example of the over-all effect of soil acidity on the concentration of exchangeable bases in soil, their release from exchangeable form in a fractional exchange, and their uptake by plants. The figure gives a plot of the total bases in leaves of sweet-orange seedlings against the total exchangeable bases in a soil that had been treated to obtain different proportions and amounts of exchangeable bases. One may note that the concentration of bases in the leaves showed no particular downward trend until the level of exchangeable bases had decreased to about half the level present at full saturation of the cation-exchange capacity at *p*H 7.6.

The three lowest values for exchangeable bases and for base concentration in the leaves in Fig. 5.20 correspond to the lowest soil *p*H values (all below 4.5) and to the highest contents of exchangeable aluminum in the soil. It has occurred to several workers that the techniques discussed in Chapter 4 for relating the bases in soil to their uptake by plants might be extended to account for effects of soil acidity such as those in Fig. 5.20. Salmon (1964) used the expression

$$\frac{(a_{\mathrm{Mg}})^{1/2}}{(a_{\mathrm{Ca+Mg}})^{1/2} + Ba_{\mathrm{K}} + Ca_{\mathrm{H}}}$$

as a soil characterization for estimating the magnesium content of plants. The *a*'s stand for activities of the different ions in exchangeable form

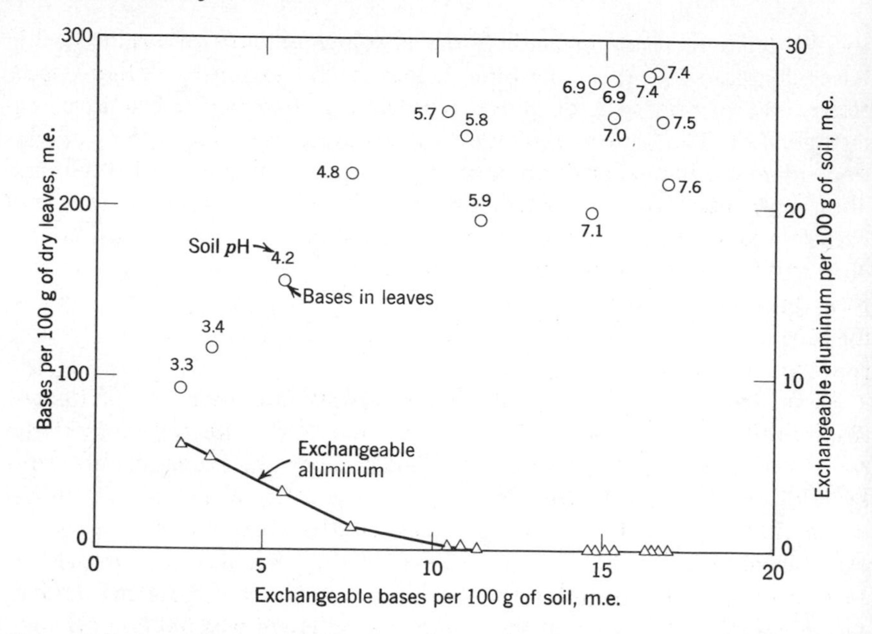

Figure 5.20. Sum of bases (calcium, magnesium, and potassium) in leaves of sweet orange seedlings versus sum of exchangeable bases in a sandy loam soil with different levels and proportions of exchangeable bases and exchangeable aluminum. The cation-exchange capacity of the soil varied from 9.1 m.e./100 g at *p*H 3.3 to 16.9 m.e. at *p*H 7.6. (Martin and Page, 1965)

in the soil, and *B* and *C* are parameters. Tinker (1964) used the activity-ratio approach in estimating the response of oil palm to potassium fertilization in southern Nigeria and tested the value of the formulations

$$\frac{a_{\mathrm{K}}}{(a_{\mathrm{Ca+Mg}})^{1/2}}$$

and

$$\frac{a_{\mathrm{K}}}{(a_{\mathrm{Ca+Mg}})^{1/2} + P(a_{\mathrm{Al}})^{1/3}}$$

where *P* is a constant. The second formulation, which provides an accounting for soil acidity, proved to be considerably better than the first.

The expression used by Milam and Mehlich (1954) (also discussed in Chapter 4) might similarly be modified to

$$\frac{\mathrm{Ca_{HCl}}}{\mathrm{Ca_{HCl}} + \alpha\mathrm{Mg_{HCl}} + \beta\mathrm{K_{HCl}} + \gamma H}$$

or

$$\frac{Ca_{HCl}}{Ca_{HCl} + \alpha Mg_{HCl} + \beta K_{HCl} + \gamma H + \delta Al_{HCl}}$$

where Ca_{HCl}, Mg_{HCl}, K_{HCl}, and Al_{HCl} are the milligram equivalents of the respective ions released in a fractional exchange with a small quantity of hydrochloric acid; H is the hydrogen-ion activity, preferably measured in 0.01-molar calcium chloride; and α, β, γ, and δ, are parameters. The last expression provides a more comprehensive accounting for the ionic effects in acid soils than the others. Further addition of sodium and manganese would not be expected to provide much improvement in estimation because of the generally small quantities involved; if, however, the objective were to account for variations in the content of sodium or manganese in the plants, use of an expression including the other ions would no doubt be advantageous.

To conclude this section on the effect of soil acidity on availability of exchangeable bases, specific attention will be directed to calcium because of the theory that soil acidity is primarily a deficiency of calcium. This theory is a reasonable inference from the fact that the content of exchangeable calcium decreases as soils become increasingly acid and calcium carbonate in the form of limestone is the most common treatment for decreasing soil acidity.

Foliar symptoms of calcium deficiency are rarely seen in the field. Diagnosis of instances of suspected calcium deficiency must usually be made by more tedious methods. From the standpoint of the experimental method, research done on calcium deficiency in peanuts in North Carolina provides a valuable example. Evidence from two kinds of experiments indicates that the observed responses of peanuts to liming were primarily an effect of the increase in supply of calcium and not a direct or indirect effect of the increase of soil *p*H associated with application of calcium carbonate.

First, Colwell and Brady (1945) conducted a number of field experiments on soils with low cation-exchange capacity and exchangeable calcium contents ranging from 0.2 to 2.2 m.e./100 g. In these experiments, the ratio of the yield obtained on the control plots to the yield on the limed plots was closely related to the exchangeable calcium but not to the *p*H of the unlimed soil. Liming did not increase the yield if the soil contained in excess of 1.4 m.e. of exchangeable calcium per 100 g.

Second, an experiment was conducted (Brady and Colwell, 1945) in which calcium, potassium, and magnesium sulfates were added separately to the soil in the rooting and fruiting zones. The particular value of these treatments from the standpoint of verification of calcium deficiency is that they provide a way of testing the effect of an additional supply of calcium without at the same time increasing the *p*H value (as is done automatically if calcium is supplied as calcium carbonate). In this experiment, the yield of peanuts was approximately doubled where calcium sulfate was applied to the fruiting zone. All other treatments reduced the yield. The combination of these forms of evidence leaves little doubt that the phenomenon was one of calcium deficiency. The results of the last experiment indicate that the calcium requirement is much higher for fruiting of the peanut than for vegetative growth. Fruiting of the peanut takes place by means of "gynophores" sent down into the soil from the above-ground stems. Calcium does not seem to move freely from the vegetative part of the plant into the gynophore but must be absorbed for the most part from the soil into which the gynophore grows.

One more example of calcium deficiency will be given because it represents a documented case history with implications for many other areas. According to Ayres (1963), liming of soils used for sugarcane in Hawaii was a well-established practice at the beginning of the twentieth century but ceased in about 1925 after field experiments had shown that there was little or no benefit from liming. Subsequent analyses showed that in some of the soils the total calcium content was as little as 5 m.e./100 g and that the degree of calcium saturation of the exchange capacity was below 1%. Although these findings strongly suggested that the soils should be deficient in calcium, there were no concurrent field experiments against which the laboratory findings could be calibrated.

Finally, in 1949, two field experiments designed to compare phosphate sources showed that superphosphate (ordinary superphosphate is primarily a mixture of calcium sulfate and monobasic calcium phosphate) was significantly better than ammonium phosphate. When analyses of the soil from the two areas showed only 67 and 90 kg of exchangeable calcium per hectare in the surface 30 cm of soil, the results of the experiments immediately suggested that the differential response of sugarcane to the two phosphate fertilizers was due to the difference in their content of calcium and that this difference was important because the soils were deficient in calcium. Because many other soils were known to contain even less exchangeable calcium, it appeared that calcium deficiency might be widespread. A number of field trials were then established in which calcium was applied in comparatively small quantities (mostly in the range of 240 to 480 kg/hectare) as calcium sulfate or calcium carbonate to test the nutritive effect of calcium without a pronounced change in *p*H of the soil. A summary of the results of 35 such experi-

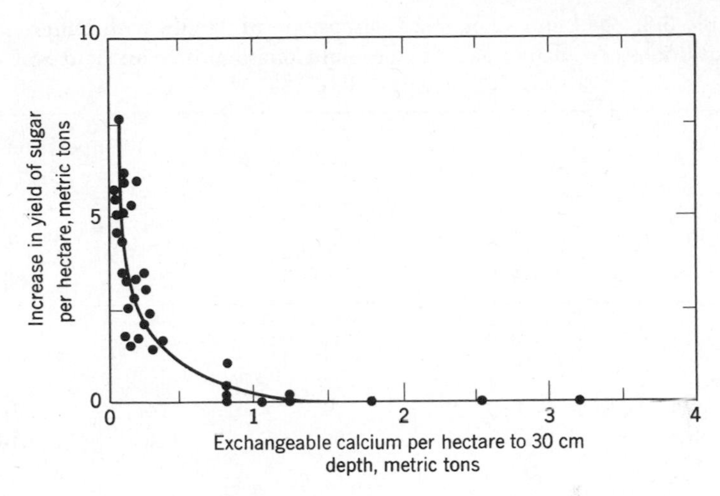

Figure 5.21. Increase in yield of sugar in sugarcane from application of calcium as the sulfate or carbonate to different soils in Hawaii. (Ayres, 1963)

ments given in Fig. 5.21 shows that large increases in yield of sugar were obtained where the soils contained less than about 500 kg of exchangeable calcium per hectare to a depth of 30 cm. This quantity is approximately equivalent to 0.7 m.e. of exchangeable calcium per 100 g of soil with a bulk density of 1.2. In explaining the results, Ayres pointed out that after the carly experiments, the combined effects of soil erosion, change of harvesting procedures to remove the calcium in the nonmillable as well as the millable portion of the crop, and change of the form of fertilizer nitrogen from sodium nitrate to ammonium salts had depleted the calcium supply of the soils.

The foregoing examples of deficiency of calcium as a nutrient are instances in which the content of exchangeable calcium is relatively low in comparison with values found in acid soils in general. If poor crop growth on acid soils is generally attributable to calcium deficiency and not to acidity or some other condition associated with it, addition of calcium in a form that does not increase the *p*H value should benefit the crop. Moreover, raising the *p*H by the use of a basic source of calcium should benefit the crop much more than an equivalent *p*H change effected in other ways. The experiment by Brady and Colwell (1945) described previously, in which the yield of peanuts was increased by application of calcium sulfate, thus provides evidence that calcium deficiency was involved in that instance because application of calcium sulfate usually decreases the soil *p*H somewhat.

Table 5.8. Yield and Chemical Composition of Alfalfa with Different Additions of Calcium and Magnesium Compounds to an Acid Soil (Kipps, 1947)

Treatment per Culture	Soil *p*H	Yield of Dry Matter per Culture, g	Composition of Dry Matter	
			Calcium, %	Manganese, μg/g
None	5.6	51	1.2	323
$Ca(OH)_2$				
139 m.e.	6.9	64	1.4	166
278 m.e.	7.3	69	1.5	130
$CaSO_4$				
139 m.e.	5.2	43	1.4	191
279 m.e.	5.2	42	1.4	277
MgO				
177 m.e.	6.8	66	1.1	167
353 m.e.	7.3	71	1.0	103

Other experiments on application of calcium sulfate to acid soils have not produced such results. Data from an experiment by Kipps (1947) are shown in Table 5.8 for example. The *p*H of the soil, the yield of the crop, and the calcium content of the crop all were increased by application of calcium hydroxide. Calcium sulfate increased the calcium content of the crop but decreased the soil *p*H and the yield. Because the manganese content of the crop was reduced by the calcium sulfate, the lowering of the soil *p*H apparently did not induce managanese toxicity. Application of magnesium oxide raised the soil *p*H and the yield and decreased the manganese content of the crop to essentially the same extent as did calcium hydroxide. The combination of evidence provided by these different treatments thus indicates that the increase in yield from application of calcium hydroxide did not result from the calcium it contained but from some other factor associated with the change in soil *p*H.

Experiments of a similar nature were conducted by Schmehl et al. (1950) and Heslep (1951). Schmehl et al. worked with a soil that contained 1.5 m.e. of exhangeable calcium per 100 g and had a degree of calcium saturation of 11%. Heslep used two soils containing 1.8 and

4.3 m.e. of exchangeable calcium per 100 g and having calcium saturation of 12 and 17%. Their results indicated that the soils were not deficient in calcium although Heslep produced calcium deficiency symptoms in lettuce by additions of magnesium oxide. Additional experiments have been done in connection with soils derived from serpentine. These soils are characteristically high in magnesium and low in calcium and are infertile. The proportion of magnesium and calcium in exchangeable form has been varied experimentally in different ways, and the results indicate that common plants not specifically adapted to serpentine-derived soils do not respond to calcium above a degree of saturation of 20 to 25% (Walker et al., 1955; Vlamis, 1949).

The evidence at hand indicates that calcium deficiency may properly be represented as one of the limiting factors in the soil-acidity complex. Occasionally it may be the dominant factor, but more frequently it is of only secondary importance.

Phosphorus. The general trend of response of crops to phosphate fertilization of soils differing in degree of acidity is indicated in Table 5.9 by a summary of 56 experiments conducted in Kenya. The smaller

Table 5.9. Response of Different Crops to Application of Superphosphate to Soils Differing in Degree of Base Saturation (Birch, 1951)

	Increase in Yield from Application of Superphosphate, %		
Base Saturation of Soils, %	Wheat	Grass	Millet
60 and below	111	26	125
61 to 70	44	11	27
71 and above	21	4	5

increase in yield from phosphate fertilization on soils with relatively high base saturation indicates that the soil phosphorus is more nearly adequate for crops on these soils and hence that the availability of soil phosphorus is greater in nearly neutral soils than in strongly acid soils. Similarly, the response of crops to phosphate fertilization on a given acid soil generally decreases as the *p*H is brought toward neutrality, indicating an increase in availability of soil phosphorus. The extent to which this effect is associated with aluminum toxicity remains to be determined, as was mentioned previously in connection with alumi-

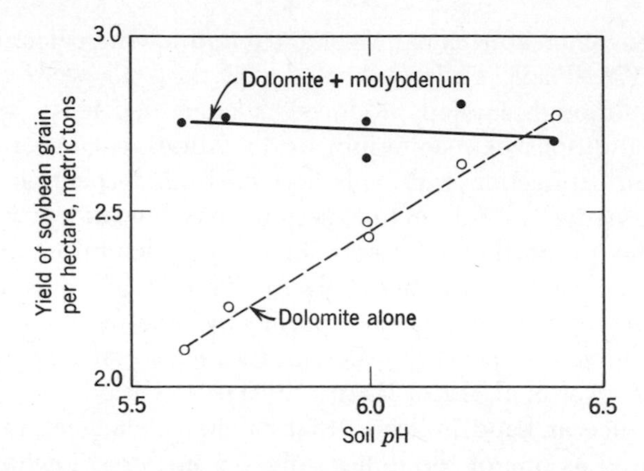

Figure 5.22. Yield of soybeans on a loam soil in Georgia with various quantities of dolomite alone and with dolomite plus an average foliar application of 336 g of molybdenum as sodium molybdate per hectare. (Parker and Harris, 1962)

num-phosphorus relations. Availability of soil phosphorus will be considered in more detail in Chapter 8 on phosphorus.

Micronutrients. Of the various micronutrients that are affected appreciably in availability by soil acidity, molybdenum is the only one that behaves like phosphorus. Availability of molybdenum is increased when acid soils are limed. In acid soils in which molybdenum deficiency is pronounced, and for crops having a relatively high molybdenum requirement, the principal effect of liming on plant growth may be associated with the increase in molybdenum availability; this was found to be the case in an experiment on soybeans in Georgia (see Fig. 5.22). The yield increased with the quantity of dolomite and with the soil *p*H in the absence of molybdenum, but there was no response to liming where molybdenum was applied. Availability of boron, iron, zinc, and manganese, on the other hand, is decreased when acid soils are limed.

An analysis of results of more than 600 experiments with copper in Sweden, where copper deficiency is common, showed no relation between copper deficiency and soil *p*H (Lundblad and Johansson, 1956). Copper toxicity from accumulated spray residues, however, has been found to increase with soil acidity (Reuther et al., 1953). The difference between these two findings may be explained in part by the marked selectivity of certain cation-exchange sites for small quantities of copper. If present in large quantities, some of the copper will occur in nonselective exchange sites, and the release to the soil solution will then be increased at low *p*H values. At high *p*H values, the excess copper will be changed to mineral forms of low solubility.

Activity of Soil Microorganisms

Organic Matter Decomposition. Laboratory experiments in which carbon dioxide evolution and nitrogen mineralization are measured on unlimed and limed soils usually indicate that the rate of decomposition of soil organic matter is increased by liming. From such experiments, the inference has been made that organic matter decomposition and nitrogen mineralization occur less rapidly in acid soils than in neutral soils.

Other evidence indicates that the prevailing view is not entirely correct. Frercks and Kosegarten (1956) limed various soils in lysimeters and measured the carbon dioxide evolution over a period of 3 years. Figure 5.23 shows the results obtained with a strongly acid peat soil that was adjusted to various *p*H values by addition of calcium oxide. In 1953, the year of the *p*H adjustment, the rate of carbon dioxide evolution was considerably increased by raising the *p*H. Carbon dioxide evolution was lowest in the unlimed soil and highest at the highest *p*H value obtained (*p*H 6.6). In 1954, the increase in carbon dioxide evolution with increase in soil *p*H was much smaller, and there were indications that the rate of carbon dioxide evolution was higher between

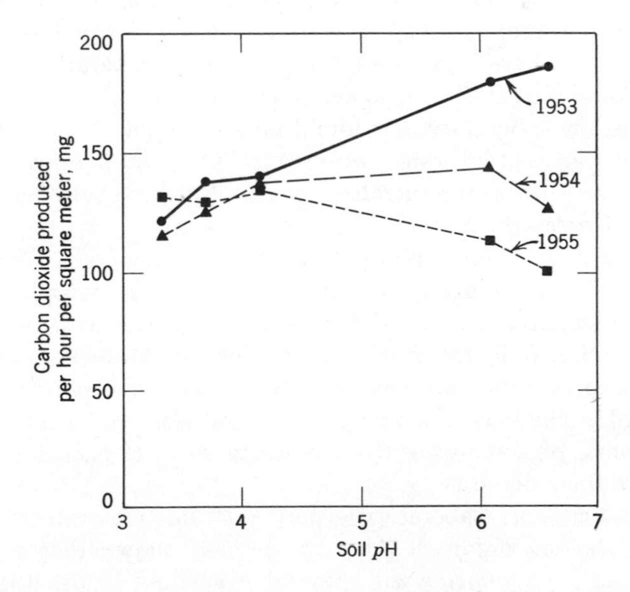

Figure 5.23. Yearly average production of carbon dioxide during three successive years by a peat soil adjusted to different *p*H values by addition of calcium oxide. (Frercks and Kosegarten, 1956)

*p*H 4 and *p*H 6 than at either the lowest or highest *p*H values. In 1955, the rate of carbon dioxide evolution was about the same from *p*H 3.4 to *p*H 4.2; at higher *p*H values, the carbon dioxide evolution was below that in the unlimed soil. Mulder (1950) conducted a field experiment in the Netherlands in which a strongly acid peat soil was limed to different *p*H values. He found that liming decreased the organic matter percentage in the soil and increased the availability of nitrogen to crops. The nitrogen-availability effect lasted about 5 years. These results indicate that the effect of raising the *p*H was only temporary and involved only a portion of the organic matter.

Thompson et al. (1954) investigated the same phenomenon in a different way. They determined the evolution of carbon dioxide, mineralization of nitrogen, and mineralization of organic phosphorus upon laboratory incubation of a group of 50 soil samples covering a range in *p*H from 5.2 to 8.1. Statistical analysis of the results showed that the carbon dioxide production per unit of organic carbon did not increase with the *p*H of the soil. Similarly, the nitrogen mineralized per unit of organic nitrogen did not increase with the *p*H of the soil. These samples had not been limed prior to their incubation in the laboratory; presumably they had been at approximately the measured *p*H values for some time in the field. These results agree with those of Frercks and Kosegarten (1956) and Mulder (1950) and indicate that the enhancement of organic matter decomposition associated with increasing the *p*H of acid soils is relatively short-lived and does not represent the general state of affairs. Further, Cornfield (1959) incubated samples of 25 unlimed soils of *p*H 4.4 to 8.1 in the laboratory and found no significant change in nitrogen mineralized per unit of total nitrogen at *p*H values above 6. Below *p*H 6, however, nitrogen mineralized per unit of total nitrogen increased as the *p*H decreased.

With regard to organic phosphorus, Thompson et al. (1954) found that the quantity mineralized per unit of total organic phosphorus during laboratory incubation increased with soil *p*H in the work with 50 soil samples mentioned in the preceding paragraph. Moreover, they found that the ratio of total nitrogen to total organic phosphorus increased with soil *p*H. These results indicate that an increase in mineralization of soil organic phosphorus with increase in soil *p*H occurs in the field as well as in the laboratory.

Nitrogen Fixation. Laboratory experiments on nonsymbiotic nitrogen fixation by the soil organism *Azotobacter* have shown that a *p*H value of 6 is critical. *Azotobacter* will grow at *p*H values below 6 if supplied with inorganic combined forms of nitrogen such as ammonium, but little or no nitrogen is fixed under these conditions. Roisin (1964) found that *Azotobacter* inoculated into acid soils did not survive for more than an hour. Although there is evidence that some soils contain acid-tolerant strains, several field surveys have shown that *Azotobacter* occurs mainly in soils having *p*H values above 6. The importance of *Azotobacter* in fixing nitrogen in such soil is not known; clearly, however, no nitrogen fixation by the organism can occur in soils that do not contain it.

Soil acidity inhibits nitrogen fixation by the symbiotic association of rhizobia with many species of host legume. One of the most sensitive is alfalfa. In experimental work to investigate factors responsible for poor growth of alfalfa on acid soils of Australia, Munns (1965) found that nitrogen fixation was the first critically limiting factor as soils became moderately acid. Evidence for this was the fact that increases in yield of alfalfa on soils with *p*H values of 5.5 to 6 were obtained from application of calcium carbonate where the nitrogen was supplied by inoculation with rhizobia and by mineralization of soil nitrogen but not where nitrogen was supplied by ammonium nitrate and by mineralization of soil nitrogen without inoculation. Earlier work on the effect of acidity on nitrogen fixation was reviewed by Wilson (1940, pp. 209–215).

Longevity of rhizobia in free-living form in the soil and infection of the host plant in soils differing in acidity are associated with the behavior of the host. If the host is sensitive to soil acidity, the tendency is for the organisms to disappear rapidly from the soil and vice versa. Nevertheless, the soil environmental requirements of the organisms and the host are not necessarily identical. For example, Loneragan and Dowling (1958) found that growth of the rhizobia that cause inoculation of subterranean clover decreased markedly at *p*H values below 5 but that the calcium supply had no appreciable influence. The organism grew vigorously in media containing only the calcium present as an impurity in the other substances added. Norris (1959) found that the calcium requirement of various strains of rhizobia, if indeed there was any requirement, could be met by traces; magnesium, however, was essential. Moreover, the requirements for nodulation may be rather different in certain respects from the requirements for independent growth of either the organisms or the host. Loneragan and Dowling (1958) found that production of nodules on roots of subterranean clover required a calcium concentration higher than that required by the host. This finding may explain Pohlman's (1946) observation that roots of alfalfa developed many nodules in limed layers of an acid soil but only a few nodules in unlimed layers. Some practical application of this phenomenon has been made by pelleting legume seeds in calcium carbonate before planting and by making a small application of limestone along rows of legumes sensitive to soil acidity.

Plant Diseases. Development of the microbial population of the soil is limited by the food supply, which leads to competition among the different organisms. Any change in conditions that affects the relative competitive ability will change the balance of the population toward the organisms whose competitive ability is favored by the change. Generally, the competitive ability of fungi relative to that of bacteria and actinomycetes increases with an increase in soil acidity (Waksman, 1952). Occurrence of plant disease caused by these different groups of organisms usually follows the general pattern for the group.

One of the early applications of soil *p*H measurements was made by Gillespie and Hurst (1918) in a survey of occurrence of potato scab in the potato-growing area of Maine. This disease is produced principally by the actinomycete *Streptomyces scabies*. Of nineteen locations with soil *p*H values between 4.5 and 5.16, eighteen fields were not infected, and one was infected slightly. In nine areas having *p*H values between 5.16 and 5.52, infection varied from none to medium. In nineteen locations having *p*H values between 5.64 and 7.21, infection was severe in all but one. The one field that grew clean potatoes (*p*H 6.2) had just been cleared and was growing potatoes for the first time. Potato scab is controlled in practice by application of sulfur, which undergoes microbiological oxidation to sulfuric acid, thus lowering the *p*H value. The effect of sulfur on potato scab, however, does not appear to be accounted for by *p*H changes alone. Vlitos and Hooker (1951) found that application of sulfur decreased the numbers of actinomycetes in soil before there was any detectable *p*H change. They attributed this effect to production of a small amount of hydrogen sulfide, which they demonstrated experimentally. Hydrogen sulfide is toxic to actinomycetes.

In contrast to diseases caused by actinomycetes, those caused by fungi usually are more prevalent in acid soils than in neutral soils. The reason for this seems to be not so much that fungi grow best under acid conditions but rather that they have a broader adaptability to hydrogen-ion activity than do bacteria and actinomycetes and hence suffer less from competition or antagonism under acid conditions. For example, Halkilahti (1964) found that infection of clover by the fungus *Sclerotinia trifoliorum* Erikss. was decreased and numbers of other organisms were increased by liming of acid soil in Finland. As shown in Table 5.10 the change in infection with soil *p*H was much more pronounced in unsterilized soil, where other organisms were present, than in sterilized soil.

The common pattern for disease incidence, explained in a general way on the basis of competition among organisms and group tolerance of the organisms to soil acidity, does not always prevail. The severity of a disease is determined not only by the numbers of organisms present in the soil as inoculum but also by conditions that affect the virulence of the organisms and the susceptibility of the host. Consequently, a full understanding of the effect of soil acidity may be most difficult to obtain because inoculum, virulence, and susceptibility may all be involved. The complexity of the situation may be inferred from Doran's (1931) observation that the severity of the root rot of tobacco caused by the fungus *Thielaviopsis basicola* was decreased where the soil was acidified by sulfuric acid (a reversal of the usual tendency for fungal diseases) but increased where the soil was acidified by phosphoric acid. Chapman (1965) reviewed observations on disease incidence in relation to soil acidity.

Growth of Different Plants

Perhaps the most comprehensive field experiment that has been conducted to determine the growth of crop plants at different soil *p*H values

Table 5.10. Infection of Clover Seedlings by the Fungus *Sclerotinia trifoliorum* in Sterilized and Unsterilized Soil at Different *p*H Values in Relation to Numbers of Other Microorganisms (Halkilahti, 1964)

	Proportion of Trials Showing Infected Seedlings, %		Millions of Microorganisms per Gram of Unsterilized Soil[1]
Soil *p*H	Sterilized Soil	Unsterilized Soil	
5.6–6.3	92	29	8.9
6.3–7.5	85	3	12.2

[1] About 82 to 93% bacteria, 1 to 6% actinomycetes, and 5 to 16% fungi.

is the legume-reaction experiment at the Ohio Agricultural Experiment Station (1938). The data from part of this experiment are shown in Table 5.11. All the crops produced the highest yield at *p*H 6.8 or 7.5. Among the grain crops, barley was the most sensitive and oats the least sensitive to soil acidity. Sweet clover and alfalfa were the most sensitive

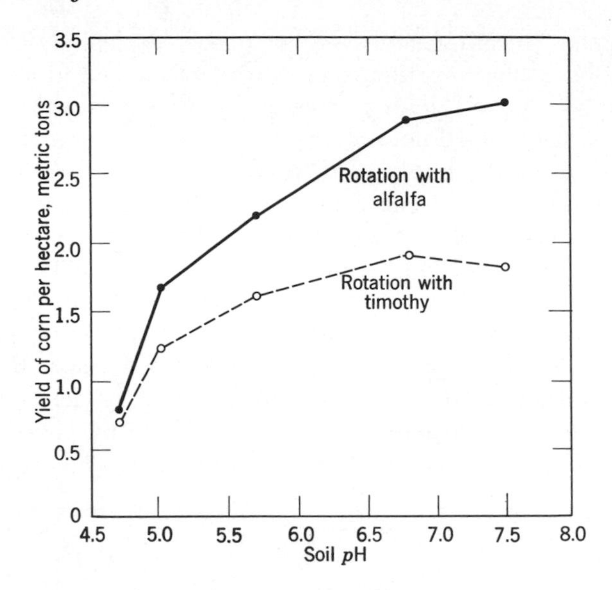

Figure 5.24. Yield of corn grown in a rotation of corn, small grain, and alfalfa, and in a rotation of corn, small grain, and timothy on soil at various *p*H values. (Ohio Agricultural Experiment Station, 1938)

of the legumes, and soybeans were least sensitive. Red, alsike, and mammoth clovers were intermediate.

Because the crops were grown in the common rotation of corn, small grain, and legume or timothy, the observed relationship between soil *p*H and yield of a particular crop may be influenced by the response of other crops in the rotation. The response of corn in Fig. 5.24 provides an example. The yield of corn was about the same following alfalfa as following timothy at low *p*H values, but the yield after alfalfa was much the higher at high *p*H values. Presumably, the extra nitrogen left in the soil at high *p*H values by the improved alfalfa crop was responsible.

With cultivated crops, the adaptation to conditions at different degrees of acidity affects the yield of a crop and sometimes the stand. Where a choice of crops is available, there is a natural tendency to plant the crops that make the best growth under the prevailing conditions. Although little scientific study has been made of the practice from the standpoint of soil acidity, the tendency to plant the best adapted crops causes the cropping pattern to vary somewhat with the acidity of the soil. The sensitivity of alfalfa to acid soil conditions, for example, is so well known that farmers seldom plant the crop without some assurance that the soil has a suitable *p*H value for growth of the crop. Arrhenius (1926) made a survey of cropping practices in relation to soil *p*H in Sweden and found that crops such as oats and timothy occupied an important part of the acreage on strongly acid soils and that barley and alfalfa were not grown on such soils. The latter two crops, however, were important on neutral and alkaline soils. These observations on cropping patterns are in accordance with the comparative sensitivity of the individual species to soil acidity as shown in Table 5.11.

Table 5.11. Yield of Crops Grown in Corn, Small Grain, Legume or Timothy Rotation at Different Soil *p*H Values (Ohio Agricultural Experiment Station, 1938)

	Relative Average Yield at *p*H Indicated				
Crop	4.7	5.0	5.7	6.8	7.5
Corn	34	73	83	100	85
Wheat	68	76	89	100	99
Oats	77	93	99	98	100
Barley	0	23	80	95	100
Alfalfa	2	9	42	100	100
Sweet clover	0	2	49	89	100
Red clover	12	21	53	98	100
Alsike clover	13	27	72	100	95
Mammoth clover	16	29	69	100	99
Soybeans	65	79	80	100	93
Timothy	31	47	66	100	95

In mixed plantings or under natural conditions, where the operator does not determine specifically the botanical composition of the vegetation, competition among species may result in a pattern of vegetation that varies with the acidity of the soil. In field experimental work in the Netherlands (Fig. 5.25), van Dobben (1956) found that barley grown alone was more sensitive to soil acidity than oats grown alone and that the sensitivity of barley was greater in mixed plantings with oats than in pure culture. In an investigation of weed associations in Germany, Hemel (1957) noted that the botanical composition varied with soil *p*H. Work of Pierre et al. (1937) on permanent pastures on acid soils in West Virginia showed that with increase in soil *p*H the prevalence of Kentucky bluegrass and white clover increased, and the proportion of weeds and bare space decreased. In a survey of pastures

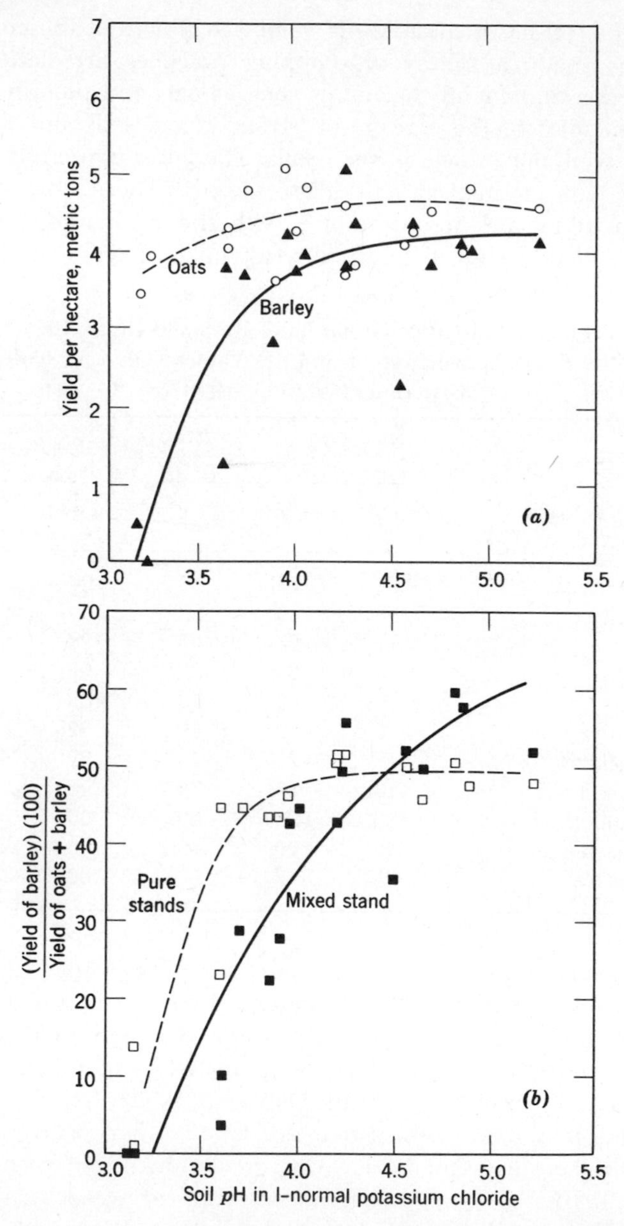

Figure 5.25. (*a*) Yield of oats and barley grown in pure stands and (*b*) percentage of barley in total yield of oats and barley in pure and mixed stands versus *p*H of soil suspended in 1-normal potassium chloride solution. (van Dobben, 1956)

in Ireland, Atkins and Fenton (1930) found that leguminous plants occurred most frequently on sites where the soil *p*H was between 5 and 8. Herbage on these sites was grazed more heavily than that on sites having *p*H values below 5 and lacking legumes.

The reasons for differences in response of individual species to conditions in acid soils are mostly unknown. The comparative sensitivity of each species to the individual conditions associated with soil acidity must be determined experimentally and then interpreted in terms of conditions in the soil in question. Hewitt (1947) grew various crops in sand cultures with different levels of aluminum, manganese, and calcium and then rated the crops according to their response to the different conditions. His observations are summarized in Table 5.12.

Table 5.12. Rating of Crops According to Their Response to High Aluminum, High Manganese, and Low Calcium in Sand Cultures (Hewitt, 1947)

Condition	Tolerant Crops	Intermediate Crops	Sensitive Crops
High aluminum	Brassicas[1] Potato	Oats	Sugar beet Barley
High manganese	Sugar beet Oats	Potato Barley	Brassicas[1]
Low calcium			Potato Sugar beet

[1] Cauliflower, marrowstem kale, and swede.

Extensive experience with leguminous plants has shown that many of them are comparatively sensitive to acidity. With subterranean clover, both calcium supply and soil *p*H appear to be important, the latter because of competitive effects (Spencer, 1950; Loneragan and Dowling, 1958). In contrast to the leguminous crop plants of temperate regions, with which most research has been done, some of the leguminous plants of tropical regions are remarkably tolerant to soil acidity. For example, Norris (1958, pp. 164–180) found that the yield of alfalfa on a base-deficient soil having only 3% base saturation was only 6% as great as the yield where the soil had received enough calcium carbonate to raise the base saturation to 57%. The corresponding figure for *Phaseolus lathyroides* (a tropical legume) was 53%. Alfalfa showed calcium deficiency symptoms in the control cultures, but *Phaseolus* did not. The numbers of nodules per gram of roots at the two levels of base saturation were 0 and 63 for alfalfa and 330 and 353 for *Phaseolus*.

Certain kinds of plants characteristically grow on acid soils and fail on neutral or alkaline soils. This behavior has been associated with iron deficiency under neutral and alkaline conditions, and there is some evidence that the iron nutrition is associated with calcium nutrition. For example, Ballinger et al. (1958) made a survey of blueberry plantations in Michigan and found that good growth was associated with a degree of calcium saturation below 10%. The growth was not associated with cation exchange capacity or exchangeable magnesium or potassium, and there was no relation between soil *p*H and percentage calcium saturation. Blueberry is susceptible to iron deficiency in neutral and alkaline soils. Tod (1956) conducted an experiment in which *Rhododendron davidsonianum* (an ornamental plant that is particularly sensitive to iron deficiency in neutral and alkaline soils) was grown on a mixture of peat and acid soil that contained less than 0.5 m.e. of exchangeable calcium per 100 g. The soil mixture was treated with enough magnesium carbonate to raise the *p*H value from 4.7 in the control cultures to *p*H 6.8 and 8.4. Although the growth of the plants over a period of several years was somewhat poorer where magnesium carbonate had been added, no iron deficiency appeared. Tod pointed out that the plants would have failed completely if calcium carbonate had been used to obtain the higher *p*H values.

Literature Cited

Anonymous. (1927) The study and determination of soil acidity. *Internat. Soc. Soil Sci., Trans. 2nd Comm.* **B**:80–82.

Arnon, D. I., and C. M. Johnson. (1942) Influence of hydrogen ion concentration on the growth of higher plants under controlled conditions. *Plant Physiol.* **17**:525–539.

Arrhenius, O. (1926) *Kalkfrage Bodenreaktion und Pflanzenwachstum.* Akademische Verlagsgesellschaft M.B.H., Leipzig.

Atkins, W. R. G., and E. W. Fenton. (1930) The distribution of pasture plants in relation to soil acidity and other factors. *Sci. Proc. Royal Dublin Soc.* (New Series) **19**:533–547.

Ayres, A. S. (1963) The utility of soil analysis in determining the need for applying calcium to sugar cane. *Proc. Internat. Soc. Sugar Cane Technol., 11th Cong.* (1962), pp. 162–170.

Bailey, E. H. (1944) Hydrogen-ion concentration of the important soils of the United States in relation to other profile characteristics: I. Pedocal soils. *Soil Sci.* **57**:443–474.

Bailey, E. H. (1945) Hydrogen-ion concentration of the important soils of the

United States in relation to other profile characteristics: II. Pedalfers and soils transitional between pedocals and pedalfers. *Soil Sci.* **59**:239–262.

Ballinger, W. E., A. L. Kenworthy, H. K. Bell, E. J. Benne, and S. T. Bass. (1958) Production in Michigan blueberry plantations in relation to nutrient-element content of the fruiting-shoot leaves and soil. *Michigan Agr. Exp. Sta. Quart. Bul.* **40**:896–905.

Barnhisel, R. I., and C. I. Rich. (1963) Gibbsite formation from aluminum-interlayers in montmorillonite. *Soil Sci. Soc. Amer. Proc.* **27**:632–635.

Bates, R. G. (1964) *Determination of pH: Theory and Practice.* John Wiley and Sons, New York.

Beers, W. F. J. van. (1962) Acid sulphate soils. *Internat. Inst. Land Recl. Imp. Bul. 3.*

Berger, K. C., and G. C. Gerloff. (1948) Manganese toxicity of potatoes in relation to strong soil acidity. *Soil Sci. Soc. Amer. Proc.* (1947) **12**:310–314.

Birch, H. F. (1951) Relationship between base saturation and crop response to phosphate in acid soils. *Nature* **168**:388–389.

Brady, N. C., and W. E. Colwell. (1945) Yield and quality of large-seeded type peanuts as affected by potassium and certain combinations of potassium, magnesium, and calcium. *Jour. Amer. Soc. Agron.* **37**:429–442.

Cate, R. B., Jr., and A. P. Sukhai. (1964) A study of aluminum in rice soils. *Soil Sci.* **98**:85–93.

Chapman, H. D. (1965) Chemical factors of the soil as they affect microorganisms. *In* K. F. Baker and W. C. Snyder (Eds.) *Ecology of Soil-Borne Plant Pathogens. Prelude to Biological Control,* pp. 120–139. University of California Press, Berkeley.

Chenery, E. M. (1954). Acid sulphate soils in central Africa. *Trans. Fifth Internat. Congr. Soil Sci.* **4**:195–198.

Christensen, H. R., and S. T. Jensen. (1924) Untersuchungen bezüglich der zur Bestimmung der Bodenreaktion benutzen elektrometrischen Methoden. *Internat. Mitt. Bodenk.* **14**:1–26.

Clark, J. S. (1964) Aluminum and iron fixation in relation to exchangeable hydrogen in soils. *Soil Sci.* **98**:302–306.

Coleman, N. T., and D. Craig. (1961) The spontaneous alteration of hydrogen clay. *Soil Sci.* **91**:14–18.

Coleman, N. T., E. J. Kamprath, and S. B. Weed. (1958) Liming. *Adv. Agron.* **10**:475–522.

Coleman, N. T., S. B. Weed, and R. J. McCracken. (1959) Cation-exchange capacity and exchangeable cations in Piedmont soils of North Carolina. *Soil Sci. Soc. Amer. Proc.* **23**:146–149.

Coleman, N. T., D. E. Williams, T. R. Nielsen, and H. Jenny. (1951) On the validity of interpretations of potentiometrically measured soil pH. *Soil Sci. Soc. Amer. Proc.* (1950) **15**:106–110.

Colwell, W. E., and N. C. Brady. (1945) The effect of calcium on yield and quality of large-seeded type peanuts. *Jour. Amer. Soc. Agron.* **37**:413–428.

Cornfield, A. H. (1959) Mineralisation, during incubation, of the organic nitrogen compounds in soils as related to soil *p*H. *Jour. Sci. Food. Agr.* **10**:27–28.

Davies, J. N. (1957) Steam sterilization studies. *Glasshouse Crops Res. Inst. Ann. Rpt. 1954–1955:*70–79.

Davis, L. E., R. Turner, and L. D. Whittig. (1962) Some studies of the autotransformation of H-bentonite to Al-bentonite. *Soil Sci. Soc. Amer. Proc.* **26**:441–443.

Desai, S. V., and B. V. Subbiah. (1951) Nitrification in relation to cation absorption by plants. *Proc. Indian Acad. Sci.* **34B**:73–80.

Dios Vidal, R., and T. C. Broyer. (1962) Efecto de niveles altos de magnesio en la absorcion del aluminio y en el crecimiento del maize en soluciones de cultivo. *An. Edafol. Agrobiol.* **21**:13–30.

Dobben, W. H. van. (1956) Concurrentie tussen haver en zomergerst op een *p*H-trappenproefveld. *Cent. Inst. Landb. Onderz.* (*Wageningen*) *Verslag.* **1955**:128–131.

Doran, W. L. (1931) Increasing soil acidity as a means of controlling black root-rot of tobacco. *Massachusetts Agr. Exp. Sta. Bul. 276.*

Eriksson, E. (1952) Composition of atmospheric precipitation. I. Nitrogen compounds. II. Sulfur, chloride, iodine compounds. Bibliography. *Tellus* **4**:215–232, 280–303.

Fireman, M., and C. H. Wadleigh. (1951) A statistical study of the relation between *p*H and the exchangeable-sodium-percentage of western soils. *Soil Sci.* **71**:273–285.

Fleming, J. F., and L. T. Alexander. (1961) Sulfur acidity in South Carolina tidal marsh soils. *Soil Sci. Soc. Amer. Proc.* **25**:94–95.

Fox, R. L., S. K. De Datta, and G. D. Sherman. (1962) Phosphorus solubility and availability to plants and the aluminum status of Hawaiian soils as influenced by liming. *Trans. Internat. Soc. Soil Sci.*, Joint Meeting Comm. IV & V, New Zealand, pp. 574–583.

Foy, C. D., and J. C. Brown. (1963) Toxic factors in acid soils: I. Characterization of aluminum toxicity in cotton. *Soil Sci. Soc. Amer. Proc.* **27**:403–407.

Foy, C. D., and J. C. Brown. (1964) Toxic factors in acid soils: II. Differential aluminum tolerance of plant species. *Soil Sci. Soc. Amer. Proc.* **28**:27–32.

Foy, C. D., G. R. Burns, J. C. Brown, and A. L. Fleming. (1965) Differential aluminum tolerance of two wheat varieties associated with plant-induced *p*H changes around their roots. *Soil Sci. Soc. Amer. Proc.* **29**:64–67.

Frercks, W., and E. Kosegarten. (1956) Die Bodenatmung von Moorböden, Heidesandböden und Sandmischkulturen in Abhängigkeit vom Kalkzustand. *Zeitschr. Pflanzenernähr., Düng., Bodenk.* **75**:33–47.

Fried, M., and M. Peech. (1946) The comparative effects of lime and gypsum upon plants grown on acid soils. *Jour. Amer. Soc. Agron.* **38**: 614–623.

Frink, C. R. (1965) Characterization of aluminum interlayers in soil clays. *Soil Sci. Soc. Amer. Proc.* **29**:379–382.

Gillespie, L. J., and L. A. Hurst. (1918) Hydrogen-ion concentration—soil type—common potato scab. *Soil Sci.* **6**:219–236.

Gilreath, E. S. (1958) *Fundamental Concepts of Inorganic Chemistry.* McGraw-Hill Book Co., New York.

Hale, J. B., and S. G. Heintze. (1946) Manganese toxicity affecting crops on acid soils. *Nature* **157**:554.

Halkilahti, A.-M. (1964) The significance of soil microorganisms as a limiting factor in infection of clover by *Sclerotinia trifoliorum* Erikss. at different times of the year. *Maataloust. Aikak.* **36**:120–134.

Haylett, D. G., and J. J. Theron. (1955) Studies on the fertilisation of a grass ley. *Union of South Africa, Dept. Agr. Sci. Bul. 351.*

Helling, C. S., G. Chesters, and R. B. Corey. (1964) Contribution of organic matter and clay to soil cation-exchange capacity as affected by the *p*H of the saturating solution. *Soil Sci. Soc. Amer. Proc.* **28**:517–520.

Hemel, J. W. (1957) Vergleichende Beurteilung des Fruchtbarkeitszustandes des Bodens mit pflanzensoziologischen und chemischen Methoden. *Landw. Forsch.* **10**:88–95.

Heslep, J. M. (1951) A study of the infertility of two acid soils. *Soil Sci.* **72**:67–80.

Hewitt, E. J. (1947) The resolution of the factors in soil acidity: the relative effects of aluminum and manganese toxicities on farm and market garden crops. *Bristol Univ., Agr. Hort. Res. Sta. Ann. Rpt.* **1947**:82–96.

Hissink, D. J. (1924–1925) Base exchange in soils. *Trans. Faraday Soc.* **20**:551–566.

Hsu, P. H., and T. F. Bates. (1964) Fixation of hydroxy-aluminum polymers by vermiculite. *Soil Sci. Soc. Amer. Proc.* **28**:763–769.

Humphreys, F., and R. Truman. (1964) Aluminium and the phosphorus requirements of *Pinus radiata. Plant and Soil* **20**:131–134.

Hutchinson, G. E. (1945) Aluminum in soils, plants, and animals. *Soil Sci.* **60**:29–40.

Jackson, M. L. (1958) *Soil Chemical Analysis.* Prentice-Hall, Englewood Cliffs, N.J.

Jackson, M. L. (1960) Structural role of hydronium in layer silicates during soil genesis. *Trans. 7th Internat. Congr. Soil Sci.* **2**:445–455.

Jackson, M. L. (1963) Aluminum bonding in soils: a unifying principle in soil science. *Soil Sci. Soc. Amer. Proc.* **27**:1–10.

Jenny, H. (1961) Reflections on the soil acidity merry-go-round. *Soil Sci. Soc. Amer. Proc.* **25**:428–432.

Jenny, H., and C. D. Leonard. (1934) Functional relationships between soil properties and rainfall. *Soil Sci.* **38**:363–381.

Jenny, H., T. R. Nielsen, N. T. Coleman, and D. E. Williams. (1950) Concerning the measurement of *p*H, ion activities, and membrane potentials in colloidal systems. *Science* **112**:164–167.

Kaurichev, I. S., T. N. Ivanova, and Y. M. Nozdrunova. (1963) Low-molecular organic acid content of water-soluble organic matter in soils. *Soviet Soil Sci.* **1963**:223–229.

Kipps, E. H. (1947) The calcium/manganese ratio in relation to the growth of lucerne at Canberra, A.C.T. *Jour. Council Sci. Indust. Res.* (*Australia*) **20**:176–189.

Kononova, M. M. (1962) Relationships between humus, the plant and micro-organisms. (Translated title). *In Studies about Humus.* Symp. Humus and Plant, pp. 111–120. Praha. (*Soils and Fertilizers* **27**, Abstr. 3299. 1964.)

Lehr, J. J. (1950) Seasonal variations in the *p*H value of the soil, as influenced by nitrification. *Fourth Internat. Congr. Soil Sci. Trans.* **2**:155–157.

Ligon, W. S., and W. H. Pierre. (1932) Soluble aluminum studies: II. Minimum concentrations of aluminum found to be toxic to corn, sorghum, and barley in culture solutions. *Soil Sci.* **34**:307–321.

Lin, C., and N. T. Coleman. (1960) The measurement of exchangeable aluminum in soils and clays. *Soil Sci. Soc. Amer. Proc.* **24**:444–446.

Lindsay, W. L., and H. F. Stephenson. (1959) Nature of the reactions of monocalcium phosphate monohydrate in soils: I. The solution that reacts with the soil. *Soil Sci. Soc. Amer. Proc.* **23**:12–18.

Loneragan, J. F., and E. J. Dowling. (1958) The interaction of calcium and hydrogen ions in the nodulation of subterranean clover. *Australian Jour. Agr. Res.* **9**:464–472.

Lundblad, K., and O. Johansson. (1956) Resultat av de senaste årens svenska mikroelementförsök. I. Försök med koppar. *St. JordbrFörs. Medd.* **60/64**:39–90. (*Soils and Fertilizers* **19,** Abstr. 1701. 1956.)

Magistad, O. C. (1925) The aluminum content of the soil solution and its relation to soil reaction and plant growth. *Soil Sci.* **20**:181–225.

Martin, J. P., and A. L. Page. (1965) Influence of high and low exchangeable Mg and Ca percentages at different degrees of base saturation on growth and chemical composition of citrus plants. *Plant and Soil* **22**:65–80.

McLaren, A. D., and E. F. Estermann. (1957) Influence of *p*H on the activity of chymotrypsin at a solid-liquid interface. *Arch. Biochem. Biophysics* **68**:157–160.

McLean, E. O. (1965) Aluminum. *Agronomy* **9**:978–998.

McLean, E. O., D. C. Reicosky, and C. Lakshmanan. (1965) Aluminum in soils: VII. Interrelationships of organic matter, liming, and extractable aluminum with "permanent charge" (KCl) and *p*H-dependent cation-exchange capacity of surface soils. *Soil Sci. Soc. Amer. Proc.* **29**:374–378.

McLean, F. T., and B. E. Gilbert. (1927) The relative aluminum tolerance of crop plants. *Soil Sci.* **24**:163–175.

Mehlich, A. (1946) Soil properties affecting the proportionate amounts of calcium, magnesium, and potassium in plants and in HCl extracts. *Soil Sci.* 62:393–409.

Messing, J. H. L. (1965) The effects of lime and superphosphate on manganese toxicity in steam-sterilized soil. *Plant and Soil* **23**:1–16.

Milam, F. M., and A. Mehlich. (1954) Effects of soil-root ionic environment on growth and mineral content of *Crotolaria striata. Soil Sci.* **77**:227–236.

Morris, H. D. (1949) The soluble manganese content of acid soils and its relation to the growth and manganese content of sweet clover and lespedeza. *Soil Sci. Soc. Amer. Proc.* (1948) **13**:362–371.

Morris, H. D., and W. H. Pierre. (1949) Minimum concentrations of manganese necessary for injury to various legumes in culture solutions. *Agron. Jour.* **41**:107–112.

Mulder, E. G. (1950) Effect of liming an acid peat soil on microbial activity. *Trans. Fourth Internat. Congr. Soil Sci.* **2**:117–121.

Munns, D. N. (1965) Soil acidity and growth of a legume. I. Interactions of lime with nitrogen and phosphate on growth of *Medicago sativa* L. and *Trifolium subterraneum* L. *Australian Jour. Agr. Res.* **16**:733–741.

Munns, D. N. (1965a) Soil acidity and growth of a legume. II. Reactions of aluminium and phosphate in solution and effects of aluminium, phosphate, calcium, and *p*H on *Medicago sativa* L. and *Trifolium subterraneum* L. in solution culture. *Australian Jour. Agr. Res.* **16**:743–755.

Munns, D. N. (1965b) Soil acidity and growth of a legume. III. Interaction of lime and phosphate on growth of *Medicago sativa* L. in relation to aluminium toxicity and phosphate fixation. *Australian Jour. Agr. Res.* **16**:757–766.

Nichol, W. E., and R. C. Turner. (1957) The *p*H of non-calcareous near-neutral soils. *Canadian Jour. Soil Sci.* **37**:96–101.

Norris, D. O. (1958) Lime in relation to the nodulation of tropical legumes. *In* E. G. Hallsworth (Ed.) *Nutrition of the Legumes.* Proc. Univ. Nottingham Fifth Easter School in Agr. Sci. Butterworths Scientific Publications, London.

Norris, D. O. (1959) The role of calcium and magnesium in the nutrition of *Rhizobium. Australian Jour. Agr. Res.* **10**:651–698.

Ohio Agricultural Experiment Station. (1938) Handbook of experiments in agronomy. *Ohio Agr. Exp. Sta. Spec. Cir. 53.*

Olsen, C. (1936) Absorption of manganese by plants. II. Toxicity of manganese to various plant species. *Compt. Rend. Lab. Carlsberg, Sér. Chim.* **21**:129–145.

Parker, M. B., and H. B. Harris. (1962) Soybean response to molybdenum and lime and the relationship between yield and chemical composition. *Agron. Jour.* **54**:480–483.

Peech, M. (1965) Exchange acidity. *Agronomy* **9**:905–913.

Peech, M. (1965a) Hydrogen-ion acitivity. *Agronomy* **9**:914–926.

Peech, M. (1965b) Lime requirement. *Agronomy* **9**:927–932.

Peech, M., R. A. Olsen, and G. H. Bolt. (1953) The significance of potentiometric measurements involving liquid junction in clay and soil suspensions. *Soil Sci. Soc. Amer. Proc.* **17**:214–218.

Pierre, W. H. (1927) Buffer capacity of soils and its relation to the development of soil acidity from the use of ammonium sulfate. *Jour. Amer. Soc. Agron.* **19**:332–351.

Pierre, W. H. (1928) Nitrogenous fertilizers and soil acidity: I. Effect of various nitrogenous fertilizers on soil reaction. *Jour. Amer. Soc. Agron.* **20**:254–269.

Pierre, W. H. (1933) Determination of equivalent acidity and basicity of fertilizers. *Indust. Eng. Chem., Anal. Ed.* **5**:229–234.

Pierre, W. H., J. H. Longwell, R. R. Robinson, G. M. Browning, I. McKeever, and R. F. Copple. (1937) West Virginia pastures: type of vegetation, carrying capacity, and soil properties. *West Virginia Agr. Exp. Sta. Bul. 280.*

Pierre, W. H., and G. G. Pohlman. (1933) Preliminary studies of the exuded plant sap and the relation between the composition of the sap and the soil solution. *Jour. Amer. Soc. Agron.* **25**:144–160.

Pierre, W. H., G. G. Pohlman, and T. C. McIlvaine. (1932) Soluble aluminum studies: I. The concentration of aluminum in the displaced soil solution of naturally acid soils. *Soil Sci.* **34**:145–160.

Pohlman, G. G. (1946) Effect of liming different soil layers on yield of alfalfa and on root development and nodulation. *Soil Sci.* **62**:255–266.

Pokras, L. (1956) On the species present in aqueous solutions of "salts" of polyvalent metals. I. History and scope of discussion. II. Polymerization of aquo-bases. III. Additional experimental methods and quantitative data. *Jour. Chem. Ed.* **33**:152–161, 223–231, 282–289.

Pratt, P. F., and F. L. Bair. (1962) Cation-exchange properties of some acid soils of California. *Hilgardia* **33**:689–706.

Puri, A. N., and A. G. Ashgar. (1938) Influence of salts and soil-water ratio on *p*H value of soils. *Soil Sci.* **46**:249–257.

Ragland, J. L., and N. T. Coleman. (1959) The effect of soil solution aluminum and calcium on root growth. *Soil Sci. Soc. Amer. Proc.* **23**:355–357.

Rains, D. W., W. E. Schmid, and E. Epstein. (1964) Absorption of cations by roots. Effects of hydrogen ions and essential role of calcium. *Plant Physiol.* **39**:274–278.

Randall, P. J., and P. B. Vose. (1963) Effect of aluminum on uptake and translocation of phosphorus^{32} by perennial ryegrass. *Plant Physiol.* **38**:403–409.

Raupach, M. (1957) Investigations into the nature of soil *p*H. *Commonwealth Sci. Indust. Res. Org. (Australia), Soil Publ.* 9.

Reuther, W., P. F. Smith, and G. K. Scudder, Jr. (1953) Relation of *p*H and soil type to toxicity of copper to citrus seedlings. *Proc. Florida State Hort. Soc.* **66**:73–80.

Riehm, H. (1961) Die Bestimmung der Pflanzennährstoffe im Regenwasser und in der Luft unter besonderer Berücksichtigung der Stickstoffverbindungen. *Agrochimica* **5**:174–188.

Rios, M. A., and R. W. Pearson. (1964) The effect of some chemical environmental factors on cotton root behavior. *Soil Sci. Soc. Amer. Proc.* **28**:232–235.

Roisin, M. B. (1964) Microbe survival in the soil assayed by the thread method. (Translated title.) *Mikrobiologiya* **33**:1074–1077.

Romanoff, M. (1945) Effect of aeration on hydrogen-ion concentration of soils in relation to identification of corrosive soils. *Jour. Res. Natl. Bur. Standards* **34**:227–241.

Rorison, I. H. (1958) The effect of aluminum on legume nutrition. *In* E. G. Hallsworth (Ed.) *Nutrition of the Legumes.* Proc. Univ. Nottingham Fifth Easter School in Agr. Sci., 1958. Butterworths Scientific Publications, London.

Ryding, W. W., and R. C. Salmon. (1964) A note on the measurement of soil *p*H in calcium chloride solutions. *Rhodesian Jour. Agr. Res.* **2**:51–52.

Ryti, R. (1965) On the determination of soil *p*H. *Maat. Aikak.* **37**:51–60.

Salmon, R. C. (1964) Cation-activity ratios in equilibrium soil solutions and the availability of magnesium. *Soil Sci.* **98**:213–221.

Sawhney, B. L. (1960) Weathering and aluminum interlayers in a soil catena: Hollis-Charlton-Sutton-Leicester. *Soil Sci. Soc. Amer. Proc.* **24**:221–226.

Schmehl, W. R., M. Peech, and R. Bradfield. (1950) Causes of poor growth of plants on acid soils and beneficial effects of liming: I. Evaluation of factors responsible for acid-soil injury. *Soil Sci.* **70**:393–410.

Schofield, R. K. (1950) Effect of *p*H on electric charges carried by clay particles. *Jour. Soil Sci.* **1**:1–8.

Schofield, R. K., and A. W. Taylor. (1955) The measurement of soil *p*H. *Soil Sci. Soc. Amer. Proc.* **19**:164–167.

Schwertmann, U. (1961) Über das lösliche und austauschbare Aluminium im Boden und seine Wirkung auf die Pflanze. *Landw. Forsch.* **14**:53–59.

Schwertmann, U., and M. L. Jackson. (1964) Influence of hydroxy aluminum ions on *p*H titration curves of hydronium-aluminum clays. *Soil Sci. Soc. Amer. Proc.* **28**:179–183.

Small, J. (1954) *Modern Aspects of pH with Special Reference to Plants and Soils.* Ballière, Tindall, and Cox, London.

Spencer, D. (1950) The effect of calcium and soil *p*H on nodulation of *T. subterraneum* L. clover on a yellow podsol. *Australian Jour. Agr. Res.* **1**:374–381.

Thomas, G. W., and N. T. Coleman. (1964) The fate of exchangeable iron in acid clay systems. *Soil Sci.* **97**:229–232.

Thompson, L. M., C. A. Black, and J. A. Zoellner. (1954) Occurrence and mineralization of organic phosphorus in soils, with particular reference to associations with nitrogen, carbon, and *p*H. *Soil Sci.* **77**:185–196.

Tinker, P. B. (1964) Studies on soil potassium. IV. Equilibrium cation activity ratios and responses to potassium fertilizer of Nigerian oil palms. *Jour. Soil Sci.* **15**:35–41.

Tod, H. (1956) High calcium or high *p*H? A study of the effect of soil alkalinity on the growth of rhododendron. *Edinb. E. Scotland Coll. Agr. Misc. Publ. 164.*

Truog, E., and M. R. Meacham. (1919) Soil acidity: its relation to the acidity of the plant juice. *Soil Sci.* **7**:469–474.

Vlamis, J. (1949) Growth of lettuce and barley as influenced by degree of calcium saturation of soil. *Soil Sci.* **67**:453–463.

Vlamis, J. (1953) Acid soil infertility as related to soil-solution and solid-phase effects. *Soil Sci.* **75**:383–394.

Vlitos, A. J., and W. J. Hooker. (1951) The influence of sulfur on populations of *Streptomyces scabies* and other streptomycetes in peat soil. *Amer. Jour. Bot.* **38**:678–683.

Waksman, S. A. (1952) *Soil Microbiology.* John Wiley and Sons, New York.

Walker, R. B., H. M. Walker, and P. R. Ashworth. (1955) Calcium-magnesium nutrition with special reference to serpentine soils. *Plant Physiol.* **30**:214–221.

Wander, I. W. (1954) Sources contributing to subsoil acidity in Florida citrus groves. *Proc. Amer. Soc. Hort. Sci.* **64**:105–110.

Whitehead, D. C. (1963) Some aspects of the influence of organic matter on soil fertility. *Soils and Fertilizers* **26**:217–223.

Whitney, R. S., and R. Gardner. (1943) The effect of carbon dioxide on soil reaction. *Soil Sci.* **55**:127–141.

Wilson, P. W. (1940) *The Biochemistry of Symbiotic Nitrogen Fixation.* University of Wisconsin Press, Madison.

Wright, K. E. (1945) Aluminum toxicity: microchemical tests for inorganically and organically bound phosphorus. *Plant Physiol.* **20**:310–312.

Wright, K. E., and B. A. Donahue. (1953) Aluminum toxicity studies with radioactive phosphorus. *Plant Physiol.* **28**:674–680.

Yuan, T. L. (1963) Some relationships among hydrogen, aluminum, and *p*H in solution and soil systems. *Soil Sci.* **95**:155–163.

6 *Salinity and Excess Sodium*

Soils are said to be *saline* if they contain an excess of soluble salts and *sodic* if they contain an excess of sodium. Two classes of saline soils are recognized, *saline-nonsodic* and *saline-sodic,* according to their content of sodium. The term *nonsaline-sodic* is applied to soils with a high concentration of sodium but not of soluble salts. The term *salt-affected* is applied to saline-nonsodic, saline-sodic, and nonsaline-sodic soils collectively. Nonsaline-sodic soils are included in the group because they are usually derived from saline-sodic soils. Because of the ease with which soluble salts are removed from soil by leaching with water, occurrences of salt-affected soils are much more frequent in arid regions than in humid regions.

Excesses of soluble salts and sodium have important influences on plant growth. Agricultural production in many parts of the world is limited by detrimental effects associated with these conditions. In the United States, salinity and excess sodium are said to reduce the productivity of over one-fourth of the 33 million acres under irrigation and to prevent the farming of additional areas.

The principal management problems are associated with the accumulation, presence, and removal of soluble salts. These matters will be discussed in the first portion of this chapter. The latter portion will emphasize the significance of salinity and excess sodium to plants.

Numerous summaries of information on salt-affected soils are available. These include monographs by Kelley (1951), the United States Salinity Laboratory Staff (1954), and Strogonov (1964); Volumes 4 and 14 of *Arid Zone Research,* which represent papers presented at symposia (some of them are comprehensive reviews); papers presented at a seminar at the Indian Agricultural Research Institute (1962); review papers by Allison (1964), Bernstein and Hayward (1958), and Hayward and Bernstein (1958); and a bibliography of publications on saline and sodic soils by Carter (1962).

Soluble Salts

Accumulation in Soils

The soluble ionic constituents of salt-affected soils are derived for the most part from primary minerals and to a smaller extent from atmospheric sources through intermediary biological activity. In general, however, the rate of production of soluble salts from these sources in place in soils is so low compared with removal of salts by water that direct accumulation is probably of little or no practical significance in development of excesses. The excesses of salts found in soils are derived mostly from weathering and biological activity in other locations. Ocean water may be the immediate source of salts in coastal areas and in uplifted marine sediments. Most commonly, the salts are gradually accumulated through underground or surface movement of water from locations of higher elevation to those of lower elevation, followed by evaporation of the water. In the United States, operation of this process on a large scale has produced the Great Salt Lake in Utah and a large area of strongly saline soil around it (see Fig. 6.1). Because the areas at low elevations in which salts may accumulate are generally the most adapted to crop production from the standpoint of topography and ease of irrigation, the problems connected with soil salinity are of major importance in highly developed agriculture in dry regions.

Irrigation waters all contain soluble salts, and in some instances as much as 20 metric tons of salts are added in this way per hectare annually. Removal of excess salts by drainage is therefore essential if an undue accumulation is to be prevented.

Figure 6.1. Saline soil with salt crust at edge of Great Salt Lake, Utah.

Composition

The major cationic constituents of the soluble salts in saline soils are sodium, calcium, and magnesium. The major anionic constituents are sulfate, chloride, and bicarbonate. Minor ionic constituents include potassium, carbonate, nitrate, and others in small quantity. Two minor constituents that occasionally are of major importance because of their toxicity to plants are lithium and borate (Aldrich et al. 1951; Bingham et al., 1964; Eaton, 1935).

Although the excesses of soluble salts in soils may be derived mainly from weathering of primary minerals, the relative proportion of the different elements is by no means the same in the soluble salts as in the primary minerals. There is a tendency for the soluble salts to have the higher proportionate content of sodium and chlorine and the lower proportionate content of calcium, magnesium, potassium, and sulfur. The cause of this change in ratio among the elements is not so much a difference in rate of release from primary minerals as a difference in tendency to form secondary minerals of low solubility. Calcium, magnesium, potassium, and sulfur form a number of such minerals.

Operation of one of the processes involved in alteration of the composition of soluble salts is illustrated by the data in Table 6.1. The data are from an experiment in which a soil was incubated under anaerobic conditions with and without addition of organic matter as plant residues. In the control, little change occurred during the incubation. Where or-

Table 6.1. Anionic Constituents in a Sandy Soil before and after Incubation under Anaerobic Conditions (Abd-el-Malek and Rizk, 1963)

Time of Incubation, Months	Constituents per 100 g of Soil, m.e.				
	In Water Extract				Insoluble Carbonate
	Cl^-	SO_4^{--}	S^{--}	HCO_3^- and CO_3^{--}	
	Control, Incubated without Additions				
0	28.5	18.2	0.0	0.2	13.0
3	29.6	17.5	0.2	0.2	12.6
	Incubated with 2.5% Ground Plant Residues				
0	29.4	18.3	0.0	1.4	12.7
3	28.9	9.3	0.9	2.3	19.8

ganic matter was added, however, sulfate was reduced; and the soil turned black as a result of the ferrous sulfide formed. Insoluble sulfide probably accounts for most of the decrease in sulfate not accounted for by soluble sulfide. Reduction of sulfate to sulfide left in solution an equivalent quantity of cations. Electrical neutrality was maintained by retention of bicarbonate and carbonate in solution. Because the solution was already saturated with calcium carbonate, insoluble carbonate was produced. The increase in bicarbonate and carbonate shown in the table is approximately equivalent to the decrease in sulfate.

Analyses of waters and soils around lakes in the Libyan Desert (Abd-el-Malek and Rizk, 1963) and around a natural drainageway in the Sacramento River Valley in California (Whittig and Janitzky, 1963) indicate the occurrence of these same transformations under natural conditions where growth of plants supplies the organic matter. Precipitation of calcium and magnesium as carbonates effects an increase in the proportion of sodium in soluble and exchangeable form.

The work of Abd-el-Malek and Rizk was done in the Wadi Natrûn, so named for the natron or sodium carbonate that occurs in the water. In this case, water moves from underground sources to the lakes through the surrounding soils, carrying the sodium bicarbonate and carbonate with it.

In the situation described by Whittig and Janitzky, movement of water during the dry season takes place from the drainageway upward and outward into the adjacent soils. Analyses of successive profiles indicate that the lateral and upward movement of bicarbonate-charged water through the soil during the dry season, with associated loss of water and carbon dioxide, results in a zone in which the soil is enriched in solid calcium and magnesium carbonate; this zone is accompanied and followed by another in which sodium bicarbonate loses carbon dioxide and changes in part to sodium carbonate. These processes may result in measurable alteration of the composition of soluble salts within lateral distances of a few meters and within vertical distances of a few centimeters in a given soil profile.

Determination

The bases present as soluble salts are customarily included with the exchangeable bases for analysis where the quantities of salts are small, as in most soils of humid regions. With saline soils, an attempt is usually made to determine separately the bases present in exchangeable and soluble forms because each class of bases has significant properties that are different from those of the other. Soluble salts contribute to the solute suction of the soil water, whereas exchangeable bases contribute to the matric suction, as explained in Chapter 2. Moreover, the effect of the soluble salts may undergo marked local variations because the salts move freely in the soil water, whereas the exchangeable bases are more permanently located. Finally, the exchangeable bases that remain after removal of the soluble salts have important effects on the physical and chemical properties of the soil.

No completely satisfactory method of analysis has been developed for determining separately the quantities of exchangeable and soluble bases in soils. The primary reason is that a complex and shifting equilibrium exists among exchangeable bases, soluble bases, and sparingly soluble bases (principally calcium sulfate and carbonates of calcium and magnesium). A secondary reason is that methods of analysis may not provide accurate values for the different forms.

The equilibrium problem may be considered to have three aspects: ion exchange, dissolution, and negative adsorption. These will be described in turn.

The soluble salts are in equilibrium with the exchangeable bases in soils, but the equilibrium depends on the water content of the soil. Accordingly, one cannot determine the cations in, say, a 1-to-2 soil-to-water extract and assume that at the lower water contents found under field conditions the ions will be present in the same proportions but merely in greater concentrations. As the solution is diluted, an exchange of cations between soluble and exchangeable forms occurs, with the result that the solution is enriched in monovalent cations at the expense of divalent cations. An example is shown in Fig. 6.2*a*. In addition to these exchanges between monovalent and divalent cations, a small amount of hydrolysis occurs; this increases with dilution. Hydrolysis reactions are ordinarily represented as an exchange between hydrogen ions furnished by water and cations present in exchangeable form, but the hydroxyl ions may at the same time replace sulfate and phosphate held by hydrous oxides.

As explained in the last part of Chapter 4, the effect of these exchanges on the proportions of cations in solution may be expressed in terms of the ratio of the activity of monovalent cations to the square root of the activity of divalent cations. The ratio tends to remain constant where small fractional exchanges are produced as a result of dilution or addition of salts. Moss (1963) found, for example, that in several soils of Trinidad the ratio $a_{K^+}/(a_{Ca^{++}+Mg^{++}})^{1/2}$ in displaced soil solutions was substantially constant up to a water content of 100% and that the tendency for deviation from constancy at higher water contents was greater where the exchangeable potassium content was low than

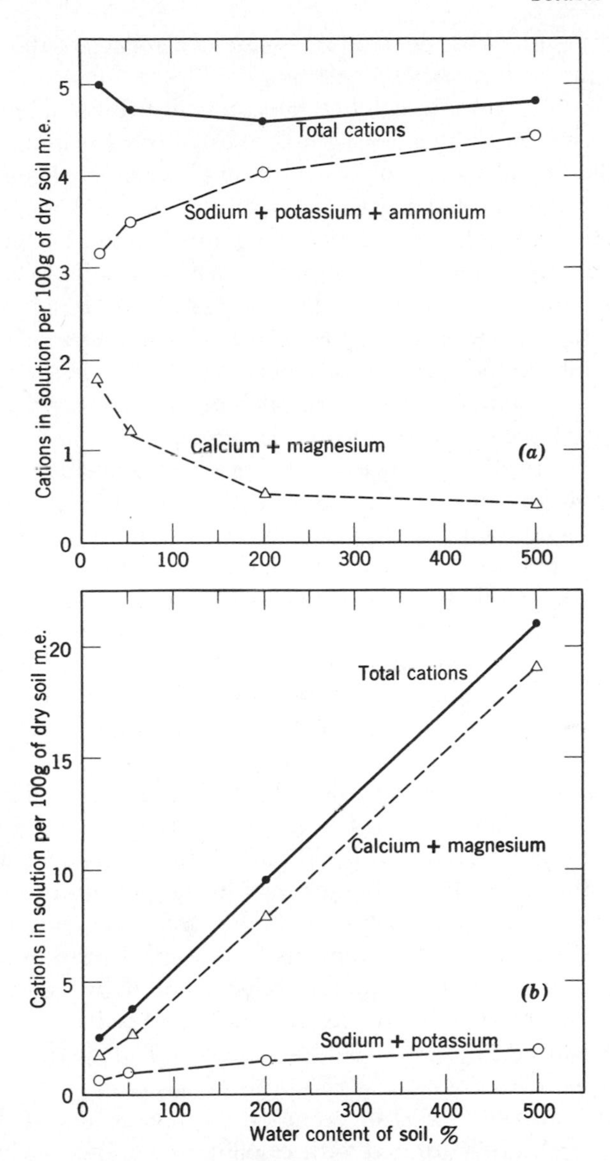

Figure 6.2. Amounts of monovalent, divalent, and total cations in solution in equilibrium with two soils where different quantities of water were present. Soil (*a*) contained no calcium carbonate or gypsum. Soil (*b*) contained 13% calcium carbonate and 3% gypsum ($CaSO_4 \cdot 2H_2O$). (Reitemeier, 1946)

where it was high. The deviations apparently reflected the degree of depletion of the exchangeable potassium.

In the case of sparingly soluble salts, the distribution between dissolved and solid forms changes with the water content of the soil. The sparingly soluble substances of greatest significance in the equilibrium are calcium sulfate and the carbonates of calcium and magnesium. In a pure solution of calcium sulfate in equilibrium with solid calcium sulfate, for example, the concentration of calcium sulfate is the same whether the quantity of water is large or small; and the quantity of calcium sulfate in solution is proportional to the quantity of water. The situation is not this simple in soils because of the presence of other salts; however, calcium sulfate and calcium carbonate do dissolve to an increasing degree as more water is added, and the calcium ions from these sources then increase the degree of exchange of divalent for monovalent ions on dilution. Figure 6.2*b* shows the sums of divalent cations and monovalent cations in solution when different quantities of water were added to a soil containing both calcium sulfate and calcium carbonate. In this soil, the monovalent, divalent, and total cations in solution all increased with the quantity of water added. This behavior contrasts with that shown in Fig. 6.2*a* and may be accounted for by dissolution of calcium sulfate (analyses for sulfate showed that the sulfate in solution was almost equivalent to the sum of calcium and magnesium.

The equilibrium between soluble salts and exchangeable cations involves partial exclusion (so-called negative adsorption) of the salts from the ionic atmospheres around the soil particles, as explained in Chapter 4. Consequently, the concentration of salts in the portion of the solution removed from the soil for analysis is higher than the concentration in the portion of the solution that remains in the soil. In the usual method of calculation, the quantity of soluble salts is found by multiplying the total volume of solution by the concentration of salts in the portion of solution removed from the soil for analysis. The quantity thus obtained is too high.

Bower and Goertzen (1955) investigated the significance of this source of error in several soils saturated with calcium or sodium and containing different quantities of added salts. They based their calculations of apparent salt content on analyses made on an extract obtained by saturating the samples of soil with water and then removing all the solution that could be extracted by vacuum filtration at a matric suction of 0.5 bar to obtain the "saturation extract." To estimate the true concentration of salts, they leached separate samples with water and determined the chloride and sulfate removed. Calculations based on their data show that the estimates of apparent salt content were 4 to 57% higher than those of true salt content. Errors were greater in sodium-saturated soil than in calcium-saturated soil and increased with an increase in the specific surface of the soil and with a decrease in the salt concentration. Analyses of solutions removed at successively greater matric suctions showed that the chloride concentration in the extract obtained from two sodium-saturated soils with fine texture between matric suctions of 1 and 4 bars was invariably less than that obtained between matric suctions of 0 and 1 bar. The same was true of calcium-saturated soils, although the difference was considerably smaller. These findings indicate that the concentration of salts increases with distance from the soil particles throughout most of the water present in soils under field conditions. Bower and Goertzen suggested that the values for apparent salt content based on analyses of saturation extracts could be made more accurate by multiplying the apparent salt content by the ratio of the true chloride content (found by leaching a comparable sample with water) to the apparent chloride content.

The monograph by the United States Salinity Laboratory Staff (1954) and the paper by Bower and Wilcox (1965) should be consulted for further information on this subject. Both sources describe specific methods of analysis for total salts and individual ions.

Classification of Salt-Affected Soils

Criteria Employed

After considerable experience with evaluation of conditions in salt-affected soils of western United States, the staff of the United States Salinity Laboratory at Riverside, California, proposed a classification of soils into four general categories. Placement of a soil in one or another of these four groups, according to its content of salts and sodium, is intended to tell whether or not the soil has an excess of salt, sodium, or both for growth of plants. The criteria are summarized in Table 6.2.

As shown in Table 6.2, soils are classified as saline or nonsaline according to the electrical conductivity of the saturation extract. The saturation extract is the solution removed from a water-saturated sample of soil by vacuum filtration, as mentioned previously. The conductivity measurements are commonly expressed as millimhos per centimeter. These units are derived from the definition of specific resistance as the resistance in ohms offered to passage of an electric current by a 1 cm cube of a substance (in this case a salt solution), where the flow of current is perpendicular to two parallel faces. Specific resistance ρ is defined

Table 6.2. Classification of Salt-Affected Soils According to Their Chemical Properties (U.S. Salinity Laboratory Staff, 1954; Bower et al., 1958)

Soil Group	Specific Conductivity of Saturation Extract at 25°C, millimhos/cm	Saturation of Cation-Exchange Capacity with Sodium, %
Saline-nonsodic soils	>4	<15
Saline-sodic soils	>4	>15
Nonsaline-sodic soils	<4	>15
Nonsaline-nonsodic (normal) soils	<4	<15

by the expression $R = \rho l/A$, where R is the resistance in ohms and l and A are the length and cross-sectional area of a conductor. If l and A are expressed as centimeters and square centimeters, the units of ρ are ohm-centimeters. Conductivity is the reciprocal of resistance, and the conductivity corresponding to a resistance of 1 ohm in 1 mho. Consequently, the specific conductivity corresponding to a specific resistance of 1 ohm-cm is 1 mho/cm. Specific conductivities of soil extracts are usually expressed in millimhos per centimeter instead of mhos per centimeter to obtain values that are small whole numbers instead of decimal fractions. A specific conductivity of 4 millimhos/cm thus means that a centimeter cube of soil extract has an electrical conductivity of 0.004 mho or a resistance of $1/0.004 = 250$ ohms.

The electrical conductivity of saturation extracts provides an index of their content of salts. From analysis of saturation extracts of samples of a number of soils, Campbell et al. (1949) found the average relationship between the milligram equivalents of salt per liter s and the specific conductivity in millimhos per centimeter c at 25°C to be

$$s = 10.37c^{1.065}$$

or

$$\log s = 1.016 + 1.065 \log c.$$

The solute suction of the water in bars may be obtained by multiplying the specific conductivity in millimhos per centimeter by the factor 0.365 (U.S. Salinity Laboratory Staff, 1954). A saturation extract with a specific conductivity of 4 millimhos/cm thus corresponds to a salt concentration of $10.37 \times 4^{1.065} = 45$ m.e./liter and to a solute suction of $4 \times 0.365 = 1.46$ bars or 1.46×10^6 dynes/cm^2.

The second criterion employed in the classification is the percentage saturation of the cation-exchange capacity with sodium, where the cation-exchange capacity is measured by use of a displacing solution having a *p*H of 8.2. The percentage sodium saturation provides an index of certain nutritional conditions and of unfavorable physical properties that either exist in the soil or may develop upon removal of the salts.

The classification of soils into the four classes shown in Table 6.2 provides a convenient frame of reference for discussion. Nevertheless, it must be recognized that the classification is arbitrary and has certain limitations.

The general intent of the classification is to group soils according to plant responses. Saline and sodic soils are supposed to contain enough salt and sodium to have adverse effects on plants, whereas nonsaline and nonsodic soils are not supposed to contain enough of these constituents to have adverse effects. Because of differences among soils, plants, and environmental conditions, the values 4 millimhos/cm for the specific conductivity of saturation extracts and 15% for the degree of saturation of the cation-exchange capacity with sodium must be regarded as general averages that are subject to revision upward or downward for specific circumstances. As an example of soil effects, it is clear that the solute suction of the soil water inferred from the specific conductivity of the saturation extract is lower than that of the soil solution to which plants are subjected under field conditions. The lower the water content of the soil, the more concentrated are the salts and the greater is the solute suction of the soil water. For many soils, the water content at the saturation percentage (at which the extract is removed for analysis) is about twice that at the field-capacity percentage and four times that at the 15-bar percentage. For such soils, the salt concentration measured in the saturation extract would be approximately doubled at the field capacity and quadrupled at the 15-bar percentage. The United States Salinity Laboratory Staff (1954) reported, however, that the water content at the saturation percentage averaged 6.4 times as great as that at the 15-bar percentage in a group of soils with coarse texture and 3.2 times as great in a group of soils with fine texture. Accordingly, a given specific conductivity of saturation extracts would correspond to higher effective salinity for plants in the group of soils with coarse texture than in the group with fine texture. As an example of differences among plants, the United States Salinity Laboratory Staff (1954) reported that the specific conductivities of saturation extracts associated with a 50% decrease in yield were 16, 10, and 4 millimhos/cm with cotton, corn, and field beans.

Saline-Nonsodic Soils

Saline-nonsodic soils contain soluble salts in quantities great enough to interfere with the growth of most crop plants. According to the United States Salinity Laboratory classification, the saturation extract of saline-nonsodic soils has a specific conductivity greater than 4 millimhos/cm at 25°C. The degree of sodium saturation of the cation-exchange capacity is less than 15%. Sodium seldom comprises more than half of the soluble cations. The *p*H value is usually below 8.5. A white crust of salts often occurs on the surface in dry weather (see Fig. 6.3).

Saline-Sodic Soils

Saline-sodic soils contain soluble salts and sodium in quantities great enough to interfere with the growth of most crop plants. The saturation extract has a specific conductivity greater than 4 millimhos/cm at 25°C. The degree of sodium saturation of the cation-exchange capacity exceeds 15%. Sodium usually comprises more than half of the total soluble cations.

Figure 6.3. Scattered native vegetation and patches of salt efflorescence on the surface of the soil in an area in the Great Salt Lake Desert of Utah.

As long as the soluble salts are present in excess, the appearance and properties of saline-sodic soils are similar to those of saline-nonsodic soils. The *p*H value is usually below 8.5.

Nonsaline-Sodic Soils

Nonsaline-sodic soils contain enough exchangeable sodium to interfere with the growth of most crop plants, but they do not contain an excess of soluble salts. The saturation extract of nonsaline-sodic soils has a specific conductivity less than 4 millimhos/cm at 25°C. The degree of sodium saturation of the cation-exchange capacity exceeds 15%. In the absence of excess salts, the soil usually has a *p*H value between 8.5 and 10. The high *p*H results from interaction of exchangeable sodium with carbonic acid formed biologically, producing a mixture of sodium bicarbonate and carbonate in the soil solution. The hydrogen ions from the carbonic acid take the place of the sodium released from exchangeable form. As loss of exchangeable sodium continues, the soil *p*H decreases; and, if the replacing ion is hydrogen, the *p*H may drop below 7 before the exchangeable-sodium-percentage is less than 15. Because of the low salt content and the relatively high degree of sodium saturation of the cation-exchange capacity, the inorganic and organic colloids tend to move with the water in the soil. Some of the organic matter may migrate to the surface of the structural units and of the soil as a result of evaporation of water. The color of the soil is darkened in this way, giving rise to the common term "black alkali" as opposed to the analogous term "white alkali" sometimes applied to saline soils that form white salts crusts when dry. (An attempt is now being made to eliminate the term "alkali" from descriptions of salt-affected soils because of its ambiguity.) Movement of organic matter seems to take place more easily and in a different direction from movement of clay, perhaps because the organic matter is more nearly in true solution than is the clay. With sufficient time, nonsaline-sodic soils may develop a B horizon having prismatic structure and a clay content higher than that of the A horizon.

Dynamics of Salt-Affected Soils

One of the significant qualities of salt-affected soils is that their salt-related properties are subject to rapid change under the proper circumstances. In the laboratory, a salt-affected soil may be changed from one of the four classes to another in less than an hour. In the field, significant changes may be made within a day, under economic conditions, although longer periods are usually involved. Maintenance of soil

Figure 6.4. Irrigated alfalfa field that became highly salinized due to inadequate drainage. The unsalinized strip in the center of the photograph is the irrigation border or dike. Salinization usually occurs in a more irregular, spotted pattern than in this field, where it is remarkably uniform. The picture was taken in the Imperial Valley, California. (Allison, 1964)

conditions suitable for growth of plants in areas where salt effects are important requires proper adaptation of management practices. Figure 6.4 shows the result of improper management at one location.

Because irrigation water always contains soluble salts, an originally nonsaline-nonsodic (normal) soil may be changed to a saline-nonsodic soil or a saline-sodic soil by irrigation. An appreciation for the ease with which this transformation can occur may be derived from some calculations based on the composition of irrigation waters. According to a summary of data on irrigation projects in western United State in which surface waters of known and reasonably constant compositio were used (U.S. Salinity Laboratory Staff, 1954), 21% of the acres were irrigated with water having a specific conductivity less than 0.25 mil limho, 46% with water of 0.25 to 0.75 millimho, 32% with water of 0.7 to 2.25 millimhos, and 1% with water of 2.25 to 5 millimhos/cm a 25°C. The most saline water contains enough salt to change a nonsalin soil to a saline condition to the full depth of wetting with a singl application. Even relatively pure water with a specific conductivity c 0.5 millimho/cm would cause rapid salinization if no drainage occurre Application of water in increments totaling 2 meters could make th surface 25 cm of soil saline. The higher the salt content of the wate the greater is the proportion of the water that must pass through th soil to remove the accumulated salt and to keep the soil in conditio suitable for plant growth.

Whether an originally nonsaline-nonsodic (normal) soil is changed to the saline-nonsodic or the saline-sodic classification on accumulation of salt depends on the composition of the irrigation water and on the proportion of the water that is removed from the soil by drainage. For estimation of the exchangeable-sodium-percentage in soil from the soluble salts in equilibrium with it, extensive use has been made of the "sodium-adsorption-ratio," $Na^+/\sqrt{(Ca^{++} + Mg^{++})/2}$, where Na^+, Ca^{++}, and Mg^{++} are the milligram equivalents of the cations per liter of solution. The utility of this ratio is derived from the fact that the exchangeable sodium and exchangeable calcium + magnesium in a given soil tend to remain constant whether the soil is in equilibrium with a concentrated solution or a dilute solution, provided the sodium-adsorption-ratio of the solution remains constant. Although the exchangeable-sodium-percentage that corresponds to a given sodium-adsorption ratio differs among soils, a fairly good estimate of the exchangeable-sodium-percentage of a number of soils was found by the United States Salinity Laboratory Staff (1954) to be provided by the empirical equation

$$\text{Exchangeable-sodium-percentage} = \frac{(1.47)(\text{sodium-adsorption-ratio}) - 1.26}{(0.0147)(\text{sodium-adsorption-ratio}) + 0.99}$$

where the measurements of sodium-adsorption-ratio were made on saturation extracts of the soils. According to this equation, an exchangeable-sodium-percentage of 15 corresponds approximately to a sodium-adsorption-ratio of 13. Saline-nonsodic soils correspond to sodium-adsorption-ratios below 13 and saline-sodic soils to ratios above 13. The numerical values of the exchangeable-sodium-percentage are about equal to those of the sodium-adsorption-ratio up to values of approximately 30.

The exchangeable-sodium-percentage may be estimated from the sodium-adsorption-ratio calculated from analyses of saturation extracts of soils, from more dilute extracts, or from water used for irrigation, provided each is in equilibrium with the soil. Irrigation water, of course, is not in equilibrium with soil before application. Inferences about exchangeable-sodium-percentages based on composition of the irrigation water must be made in the sense of a prediction of exchangeable-sodium-percentages to be expected after enough water has been applied and enough leaching has occurred to equilibrate the exchangeable cations in the soil with those added in the water. As shown in Fig. 6.5, the exchangeable-sodium-percentage of soils that have been irrigated for a long time with water of relatively constant composition is indeed related to the sodium-adsorption-ratio of the water. Most of the experimental observations, however, lie above the line representing the foregoing

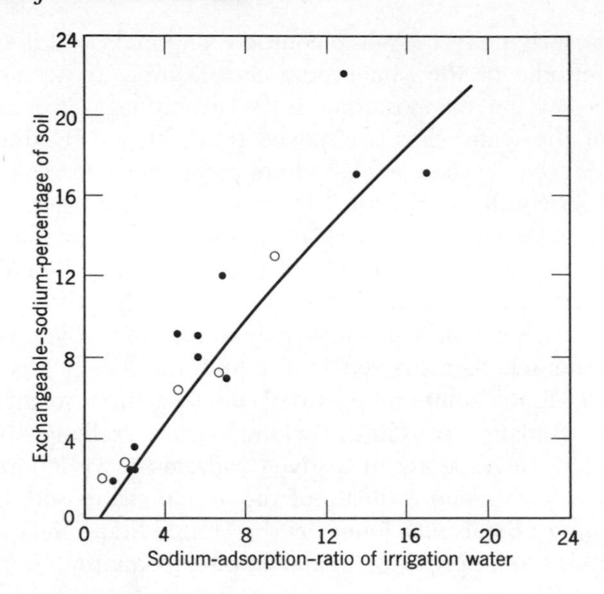

Figure 6.5. Exchangeable-sodium-percentage of samples of surface soil versus sodium-adsorption ratio of the irrigation water. Open circles represent results with small columns of soil after 42 irrigations. Solid circles represent field observations The line represents the empirical equation given in the text. (U.S. Salinity Laboratory Staff, 1954)

empirical equation. This behavior is interpreted to mean that the soils were in equilibrium with solutions having sodium-adsorption-ratios somewhat higher than those in the irrigation water, the explanation being that the salts in the irrigation water become more concentrated as a result of evapotranspiration after application to the soil. If the concentration of salts is doubled without change in composition, the sodium-adsorption-ratio is increased by the factor $\sqrt{2}$; if the concentration is quadrupled, the sodium-adsorption-ratio is increased by the factor $\sqrt{4}$.

Bowen (1961) described the application of an alternative equation developed by Eriksson (1952) for estimating the exchangeable-sodium percentage of soil from the ions in solution. This equation is apparently more suitable than the sodium-adsorption-ratio, but it requires more experimental measurements on the soil, including the cation-exchange capacity per unit surface area of the soil particles.

A second cause of uncertainty in estimating the exchangeable-sodium percentage from the composition of the irrigation water is the usual lack of information on the degree of precipitation that occurs as the irrigation water is concentrated in the soil by evapotranspiration. Depending on the composition of the water and the degree to which the water is concentrated by evaporation, from zero to almost 100% of the calcium and magnesium may be precipitated as carbonates by interaction with the bicarbonate in the water, as was pointed out by Eaton (1950). Eaton tabulated the composition of a number of irrigation waters containing enough bicarbonate and carbonate to produce a substantial increase in the sodium-adsorption-ratio upon evaporation and removal of calcium and magnesium by precipitation. Bower (1961) described a method of estimating the degree of precipitation of calcium carbonate in soil from different irrigation waters.

From the foregoing, it should be evident that nonsaline-nonsodic soils may readily be changed to saline-nonsodic or saline-sodic soils in the field, depending on the composition of the water, the quantity added and changes in composition that may occur after the water is added to the soil. Moreover, saline-nonsodic soils may readily be changed to saline-sodic soils and vice versa if the combination of conditions is changed in the appropriate way. Nonsaline-sodic soils also may be changed to saline-nonsodic or saline-sodic soils, depending on the conditions, although this process may not take place readily if the nonsaline-sodic soil is almost impermeable to water.

One of the important reasons for distinguishing between saline-nonsodic soils and saline-sodic soils is that removal of excess salts by leaching with water is followed by a reduction of the conductivity of the soil for water if excess sodium is present. This effect may not be important if the soil has sufficiently coarse texture, but the hydraulic conductivity of soils with fine texture sometimes becomes so low that the soil cannot be treated economically and must be abandoned. To avoid development of a nonsaline-sodic soil in the first place, it may be possible to change the saline-sodic soil to a saline-nonsodic soil by application of water with a lower sodium-adsorption-ratio before leaching. One way of doing this is to change the source of water. Another is to reduce the salt concentration in the water by mixing it with a gradually increasing proportion of low-salt water, making use of the principle that the sodium-adsorption-ratio of water decreases upon dilution, thus causing calcium and magnesium in the water to exchange for sodium in the soil (Reeve and Bower, 1960). Usually, however, only one source of water is available. Under these conditions, part of the salts may be removed by leaching; then some substance may be added to exchange with the sodium and facilitate its removal by further leaching. Sodium, which is less strongly held in exchangeable form than calcium and magnesium, is removed preferentially.

Three general types of substances are used to effect replacement of exchangeable sodium: (1) soluble calcium salts (calcium chloride, gypsum); (2) acids or acid-forming materials (sulfur, sulfuric acid, iron sulfate, aluminum sulfate, lime-sulfur); (3) calcium and magnesium carbonate (limestone, dolomite, by-product calcium carbonate from sugar factories). The most suitable choice among these materials depends on cost and the nature of the soil. The first two classes of amendments are suitable for use on soil containing calcium and magnesium carbonate, but the third is not. If the soil does not contain calcium and magnesium carbonate, the first two classes of amendments are effective; the third is effective if the *p*H value is below about 7.5. If the quantity of class-2 amendments required to replace the sodium would make the soil too acid (say, below *p*H 6), a mixture of class-2 materials with those of class 1 or 3 or both may be used. Soluble materials may be applied to and distributed by the irrigation water, but other materials must be applied to the soil. These principles apply to the removal of excess sodium from both saline-sodic soils and nonsaline-sodic soils; however, the time required for removal may be greater in the latter because of the slower movement of water through the soil.

Soil-Plant Relationships

Soil Salinity

Plants grown on saline soils tend to be relatively small in size, but usually there are no distinctive foliage symptoms. Figure 6.6 illustrates the type of response frequently observed. In some instances, plants grown on saline soils have a darker, more bluish-green color than similar plants grown under comparable nonsaline conditions. The foliage color results from a high content of chlorophyll and an unusually thick coating of cuticle. Occasionally, such symptoms as browning of the tip and marginal or interior portions of leaves, leaf mottling, leaf curling, and incipient chlorosis (yellowing) are seen. Internally, there may be morphological changes. In tomato, for example, the proportion of vascular or conducting tissue is reduced, and the thickness of the cell walls in the conductive tissue is increased. Frequently the thickness of the leaves increases.

Three theories have been advanced to account for different aspects of the detrimental effects of soil salinity: the water-availability theory, the osmotic-inhibition theory, and the specific toxicity theory. These will now be discussed.

Figure 6.6. Comparative growth of 13-year-old grapefruit trees on two soils, both irrigated with water from the Colorado River. The upper photograph shows an orchard on a sandy soil to which water was applied in quantities sufficient to remove excess salts. The specific conductivity of the displaced soil solution in the surface 1.8 meters of soil ranged from 2.0 to 3.2 millimhos/cm at 25°C. The trees are vigorous, and the production of fruit in the season the picture was taken was 250 kg per tree. The lower photograph shows an orchard on a silty clay soil that was allowed to become saline. The specific conductivity of the displaced soil solution in the surface 1.8 meters of soil ranged from 6.9 to 14.8 millimhos/cm at 25°C. The production of fruit in the year the picture was taken was less than 50 kg per tree. The trees and individual leaves are smaller than those in the upper photograph (note the shovel in both photographs for scale), and there are a few yellow leaves; but salt-injury symptoms of diagnostic value are otherwise lacking. The trees are heavily foliated, and the foliage color is good. Several years later, trees shown in the lower photograph had some dead wood in the tops, the leaves were small, many leaves were yellowed to varying degrees, and some browning of leaf tips was evident. (Eaton, 1942)

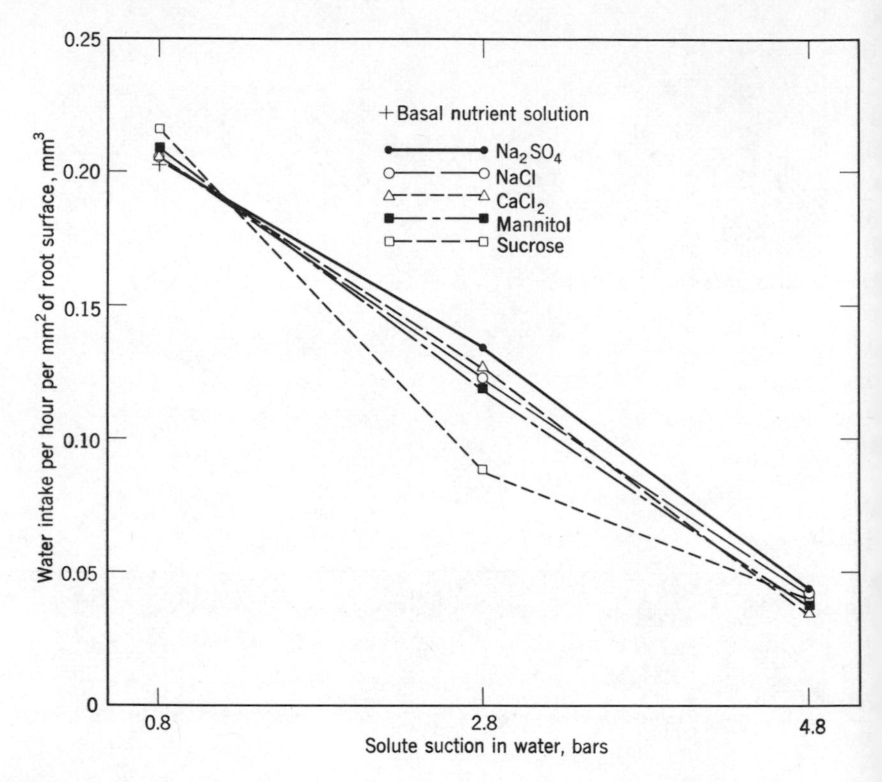

Figure 6.7. Rate of entry of water into corn roots versus solute suction of water in solutions containing different kinds and concentrations of solutes. (Hayward and Spurr, 1944)

Water-Availability Theory. According to the water-availability theory the soluble salts in saline soils increase the solute suction of the soil water. The availability of the water to plants is thereby decreased; therefore the plants suffer a deficiency of water.

An experiment conducted by Eaton (1941) provides a test of the water-availability theory. Eaton divided the roots of corn plants equally between two nutrient solutions having the same proportions of the various salts but differing in total concentration. The solute suction of the more dilute solutions was 0.3 bar, and that of the more concentrated solution was 1.8 bars. To avoid secondary effects of unequal growth of roots, the two parts of the root system were alternated between the two solutions every 2 days. Uptake of water by the plants amounted to 171 and 319 ml from the solutions having solute suctions of 1.8 and

0.3 bars, thus demonstrating a marked reduction of water uptake at the higher solute suction.

A different sort of test of the water-availability theory was made by Hayward and Spurr (1944), who grew corn plants in a nutrient solution and then transferred them to other solutions in which the kind and concentration of solutes were varied. Measurements of the rate of water absorption by the roots were made over a period of 5.5 or 6 hours beginning 30 minutes after transference of the plants to the new solutions. The rate of water intake was found to decrease with increasing concentration of each solute and to be essentially independent of the nature of the solute where the concentrations were expressed in terms of the solute suction of the solution, as shown in Fig. 6.7. The results in Fig. 6.7 are of the type to be expected if decreasing the availability of water to plants was the only effect of salts.

Investigation of the water-availability theory was carried further by Wadleigh and Ayers (1945), who grew bean plants in cultures of soil that were treated with different quantities of sodium chloride and were allowed to dry to different degrees before watering. They found that the growth of the plants decreased with an increase in both matric suction and salt content of the soil, but that it could be expressed to a good approximation as a function of the total suction of the water in the soil, as estimated by the sum of the matric- and solute-suction components (Fig. 6.8). These results indicate not only that the salt

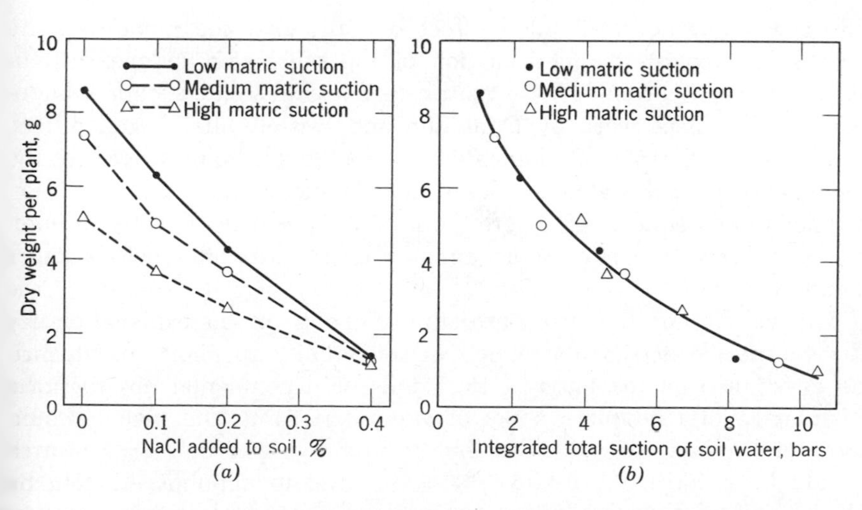

Figure 6.8. Yield of beans versus (*a*) sodium chloride added to soil and (*b*) integrated total suction of soil water. (Wadleigh and Ayers, 1945)

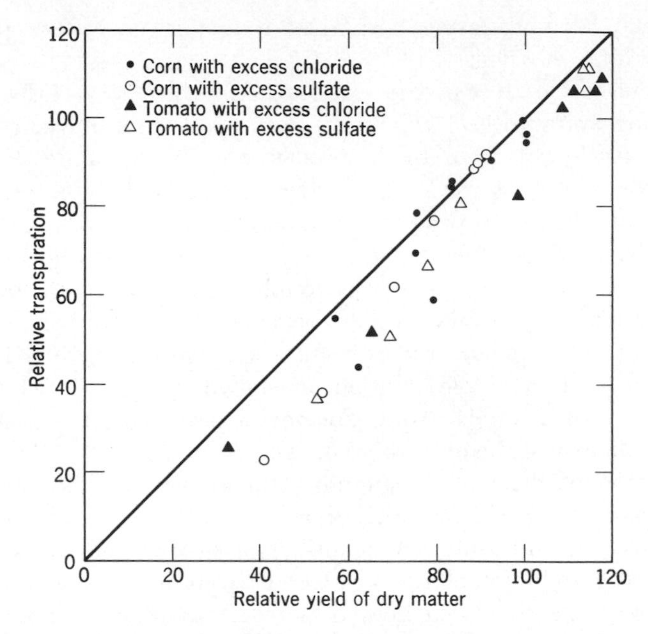

Figure 6.9. Relative yield and relative transpiration of corn and tomato in complete nutrient solutions with excess chloride and sulfate, with 50% of the excess as the sodium salt and the remaining 50% as calcium and magnesium salts. In each series the value obtained with the basal nutrient solution is represented as 100. The diagonal line represents equal relative transpiration and relative yield. (Eaton, 1942)

effect was a water-availability effect but also that the plant response was a function of the total suction of the soil water, independent of the relative proportions of the matric-suction and solute-suction components. Subsequent work by Danielson and Russell (1957) and others described in Chapter 2, indicates that although solutes reduce the availability of soil water to plants, matric suction and solute suction are not strictly equivalent, where the rate of movement of water through soil is concerned. Under some circumstances, this difference becomes significant.

The validity of the water-availability theory as an explanation for the nonspecific detrimental effects of soil salinity on plant growth may be questioned on the basis of two kinds of experimental observations. First, if salinity inhibits uptake of water by plants, the plants should lose turgor; and loss of water by transpiration per unit of leaf area should be reduced. Bernstein (1961) referred to unpublished data by W. L. Ehrler showing that transpiration per unit of leaf area was not particularly affected by salinity and to Eaton's (1942) data in Fig. 6.9

which show that, although salinity decreased the transpiration, the major part of the decrease could be accounted for by the associated decrease in weight of the plants.

The second basis for questioning the validity of the water-availability theory is that experimental fact does not support the postulate that the solute suction contributed by the salts reduces the difference between total suction of water in plant and soil and thereby reduces the tendency of water to enter plants. Figure 6.10 shows what happened to tomato plants that had been grown on a nutrient solution with solute suction of 0.7 bar and then were transferred at zero time in the figure to similar solutions supplemented with enough sodium chloride to give an additional 5 or 10 bars of solute suction. Initially, the plants lost water and wilted, in accordance with the water-availability theory. Not in accord with the water-availability theory, however, the plants recovered from wilting within 28 hours and essentially regained their original content of water. Accompanying the recovery from wilting was an intake

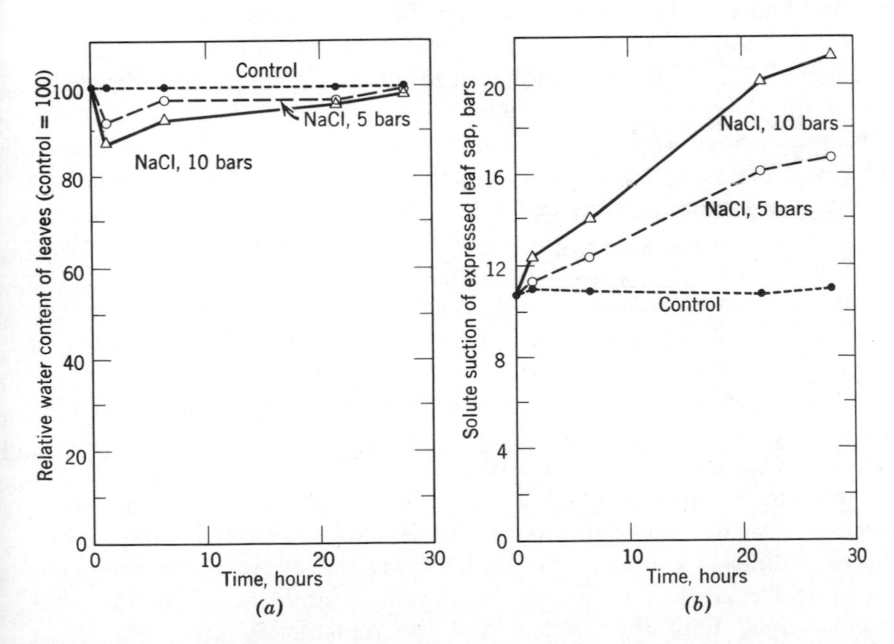

Figure 6.10. (*a*) Relative water content of leaves and (*b*) solute suction of expressed sap from leaves of tomato plants at different times after transference of the plants from a nutrient solution with a solute suction of 0.7 bar to others with solute suction increased by 5 or 10 bars by addition of sodium chloride. Control plants remained in the original solution. (Slatyer, 1961)

of solutes and an increase in solute suction in the plant sap that was approximately equal to the 5 and 10 bars of solute suction added to the external solution by sodium chloride. Adjustment made by the plants, principally in uptake of solutes, thus had eliminated the decrease in the difference in total suction between water in plants and solution that would be postulated by the water-availability theory to be responsible for a decrease in water availability to plants.

In a general sense, the water-availability theory postulates that the action of salts in saline soils is in the soil water and that this action outside of plants has indirect effects on water supply inside of plants. The foregoing evidence indicates that the water-availability theory does not apply, at least in the form originally conceived. If the water-availability theory does not account for the nonspecific detrimental effect of soil salinity on plant growth, there must be some other way of explaining this effect. The osmotic-inhibition theory discussed in the next section is an attempt in this direction.

Osmotic-Inhibition Theory. According to the osmotic-inhibition theory, plant growth is inhibited by the excess of solutes taken up from saline media. The osmotic-inhibition theory thus postulates that the salts act inside the plants, but it does not specify how the inhibition of growth is effected. The inhibition could even result in part from water deficiency in a sense different from that envisioned by the water-availability theory. The presence of excess solutes in the plant decreases the free energy of unit mass of water even though the absolute mass of water in the plant may not be reduced after the plant has adjusted to the excess of salts present externally. Papers dealing with this theory include those by Slatyer (1961) and Bernstein (1961, 1963).

Specific-Toxicity Theory. According to the specific-toxicity theory, soil salinity exerts a detrimental effect on plants through the toxicity of one or more specific ions in the salts present in excess. The clearest case for specific toxicity may be made in instances in which a highly toxic trace constituent is present in excess but salts are not. Figure 6.11 shows a view in a peach orchard in California where the saturation extract of the soil in the foreground contained 3.8 μg of boron per milliliter. This value compares with a concentration of 0.7 μg/ml considered by the United States Salinity Laboratory Staff (1954) as the approximate safe limit for sensitive crops such as peaches, apricots, and grapes. In the foreground area, both the original and the replanted peach trees either died or made little growth. The saturation extract of the soil in the background covered with trees contained only 0.19 μg of boron per milliliter. To judge from the analyses on the saturation extracts, neither the soil in the foreground nor the soil in the background contained an excess

Figure 6.11. Area of unproductive, high-boron soil in the foreground and productive, low-boron soil in the background in a peach orchard in the San Joaquin Valley of California. (Eaton, 1935)

of soluble salts. Boron toxicity has marked effects on the appearance of certain plants, as illustrated in Figs. 6.12 and 6.13.

As a point of departure in examining the specific-toxicity theory with reference to the common constituents of soluble salts in soils, it may be instructive to reconsider the experiment conducted by Hayward and Spurr (Fig. 6.7) in which the reduction in uptake of water by plants in the presence of various solutes was largely independent of the nature of the solute where the solute concentration was expressed in terms of solute suction. This experiment evidently dealt with the primary effect of solutes on water absorption. The time that elapsed during the absorption measurements was too short to permit much internal change in the plants in response to the various kinds and concentrations of solutes in the external solution. Over a somewhat longer period of time, differential absorption of solutes by the plants would have changed the internal composition of the plants; this would be a secondary effect. If the differences in internal composition eventually resulted in differences in the amount of dry matter produced, this would be another secondary effect.

If all the primary and secondary effects of the various solutes are traceable to osmotic effects measurable in the external solution, the rela-

Figure 6.12. Comparison of normal leaves of grape (center) with those showing boron deficiency (top) and boron toxicity (bottom). (Eaton, 1935)

tionship between the yield of plants and the solute suction of the water will be independent of the nature of the solute. The results of an experiment to investigate this question for the ions commonly found in excess in saline soils are shown in Fig. 6.14. In this experiment, red kidney beans were grown in a complete nutrient solution, with and without additions of various salts. The yield of dry matter per plant is plotted

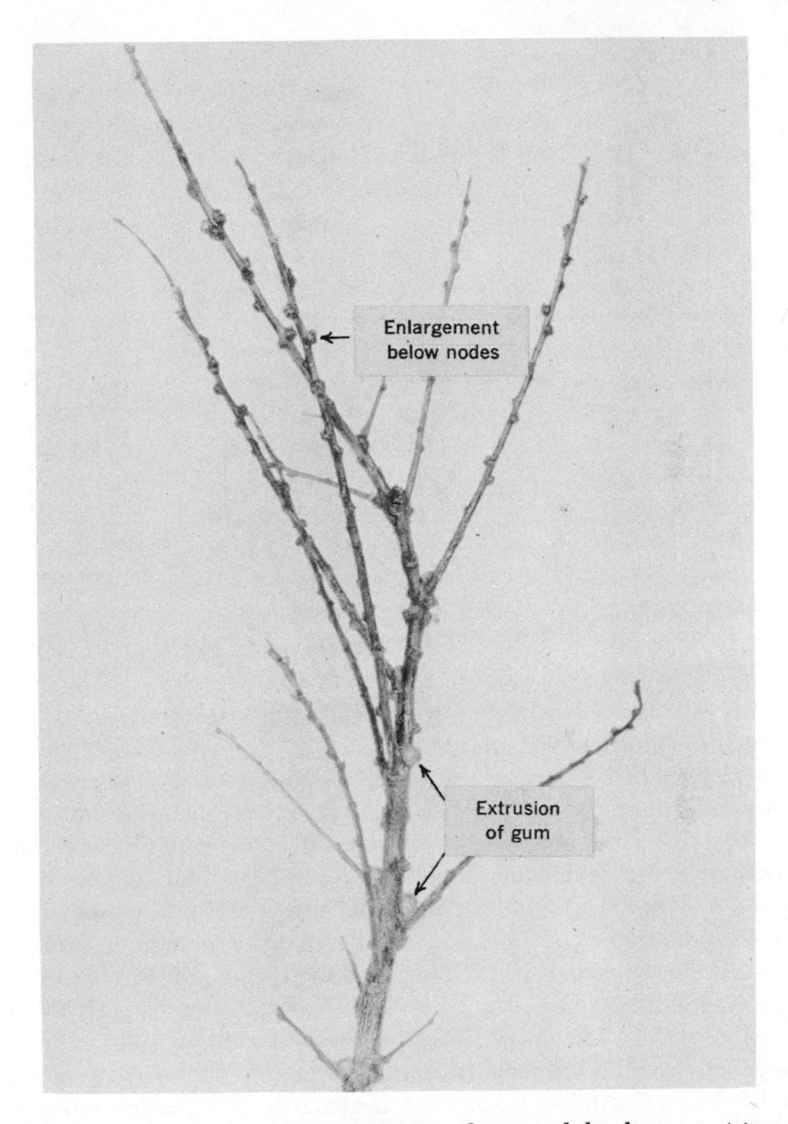

Figure 6.13. Branch of an apricot tree severely injured by boron toxicity. The most characteristic symptom of boron toxicity on apricot trees is an enlargement of the bark and wood of smaller branches just below the nodes. There is commonly an extrusion of gum above the axes of the twigs and leaves. (Eaton, 1935)

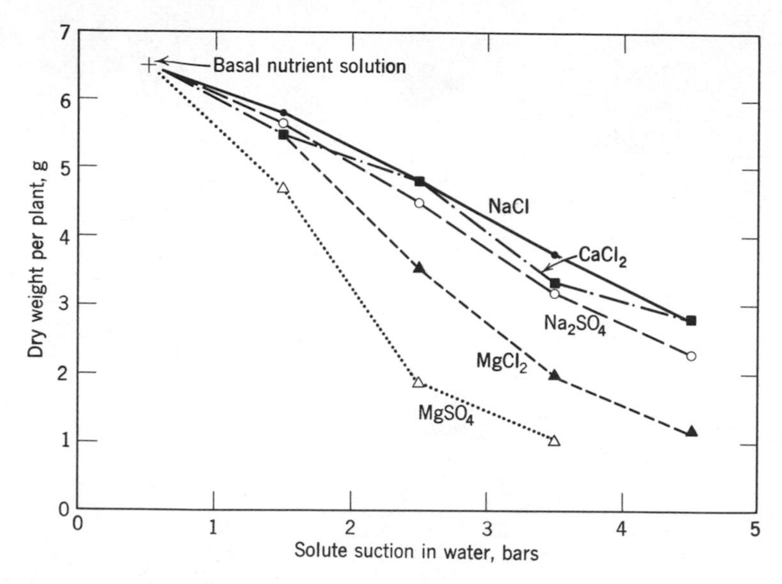

Figure 6.14. Yield of bean plants versus solute suction in water in a nutrient solution with and without addition of various salts. (Gauch and Wadleigh, 1944)

against the solute suction of the water. The response curves obtained with additions of sodium chloride, sodium sulfate, and calcium chloride are similar, although the curve for sodium sulfate is consistently a little below that for sodium chloride. These results thus provide no evidence for existence of major differences among salts, which is the type of experimental finding to be expected if the effects of the various salts were attributable entirely, or almost entirely, to their osmotic properties. The substantially lower yields obtained with magnesium chloride and magnesium sulfate, however, show that these salts had effects that differed in some way from those of the other three salts. Thus this experiment provides evidence that the effects of excesses of ions abundant in salts of saline soils are not necessarily nonspecific and merely a function of their osmotic properties.

Perhaps the simplest explanation for the lower yields of the plants with excess magnesium chloride and magnesium sulfate than of plants with excess calcium chloride, sodium chloride, or sodium sulfate is that magnesium had some specific detrimental effect that was not shared equally by sodium and calcium. The reasoning behind this interpretation is that magnesium was present in the salts that produced the lower yields and was absent from the salts that produced the higher yields.

The reverse was true of calcium and sodium. Chloride and sulfate were present in both groups of salts.

The foregoing interpretation involves the assumption that all factors and conditions operate in the same way in the presence as in the absence of magnesium. This assumption may be incorrect. For example, Fig. 6.14 shows that the difference in yield between cultures supplied with excess chloride and excess sulfate was greater where the associated cation was magnesium than where it was sodium. According to Hayward and Wadleigh (1949), the apparent specific effect of magnesium may be merely a deficiency of calcium induced by partial exclusion of calcium from the plants in the presence of excess magnesium. They cited unpublished evidence that the magnesium toxicity may be alleviated by a higher concentration of calcium. Calcium deficiency may have been involved also in the differential action of chloride and sulfate observed in Fig. 6.14. Because calcium sulfate is of limited solubility, an excess of sulfate would reduce the concentration of calcium in solution. If the effects are indeed a consequence of induced calcium deficiency, there is some question as to whether the phenomenon may properly be described as magnesium toxicity.

Experimental work by Ayers (1950) and Ayers et al. (1951) may be cited as an example of an investigation of sodium and chlorine toxicity in which there is no evidence for existence of induced nutrient deficiencies. "Leaf-burn" symptoms were relatively severe in avocado orchards in southern California during a period of dry years, and this suggested the possibility that an increased accumulation of salts in the soils might have been the cause or at least a contributing factor. Analyses of avocado leaves from a number of affected areas showed an accumulation of chlorine and occasionally of sodium. The more severe leaf injury was frequently found in areas containing appreciable concentrations of salts, particularly chlorides. Only in a few instances, however, was the specific conductivity of the saturation extract as great as 4 millimhos/cm. These observations suggested that ionic effects were important and led to an investigation of the response of avocados to individual salts in culture solutions.

To a complete basal nutrient solution with a solute suction of 0.4 bar, sodium chloride, calcium chloride, or sodium sulfate was added, each to give solutions with a solute suction of 0.9, 1.4, and 1.9 bar. The value 0.9 bar corresponds to a specific conductivity lower than 4 millimhos/cm for all three salts. All salt treatments had a marked effect. After 5 months, the diameter of the trunk 2.5 cm above the graft union was 2.0 cm with the basal solution, 1.4 cm with calcium chloride at 0.9 bar, and 1.1 cm with sodium chloride at 0.9 bar. The comparable plants grown with sodium sulfate at 0.9 bar were dead. Two distinct leaf symptoms were observed. One was a "tip-burn," so-called because the tips of the leaves died first. With increasing severity of this symptom, the dead area at the leaf tip increased in size and progressed along the leaf margins and finally into the areas between the leaf veins. The symptom occurred on plants grown with additional chloride salts, and the tendency for penetration of injury into the areas between the veins was more noticeable with sodium chloride than with calcium chloride. This type of injury was associated with relatively high chlorine content of the leaves. The second leaf symptom was a "spot-burn," so-called because of development of dead areas between the veins and about midway between the leaf margin and the midrib. This symptom developed on plants receiving additional sodium sulfate. Analyses showed that the leaves were relatively high in sodium content. These leaves had a higher percentage content of sodium than did leaves grown with comparable excesses of sodium chloride. No analyses were reported for sulfate, probably because sulfate generally shows little tendency to accumulate and is regarded as a relatively innocuous substance.

To complete the investigation, an additional survey of commercial orchards was made to collect samples of leaves for analysis. Obervations were recorded on the relative severity of each of the two types of leaf injury, and analyses for sodium and chlorine were made on the leaves. The results of this survey are shown in Fig. 6.15, where the severity of each type of leaf damage is plotted against the percentage content of sodium and chlorine in the leaves. The results show that leaf-tip damage was indeed associated with chlorine content of the leaves and leaf-spot damage with sodium content. The unusual sensitivity of avocado to soil salinity thus appears to be a consequence of specific effects of sodium and chloride, both of which are commonly found in excess in salts present in saline soils.

Ehlig (1964) explained in a similar manner the sensitivity of raspberries, blackberries, and boysenberries to salinity. He found that foliar symptoms of excess salinity were associated with high concentrations of sodium and particularly of chlorine in the leaves. The concentrations of chlorine required to produce visible injury decreased as the air temperature increased.

In some instances, differential plant response to salts having a common ion may be attributable to the common ion. Such was the case with sodium chloride and sodium sulfate in the work just described with avocados. Another instance was reported by Brown et al. (1953), who grew various stone-fruit trees in sand cultures to which were added

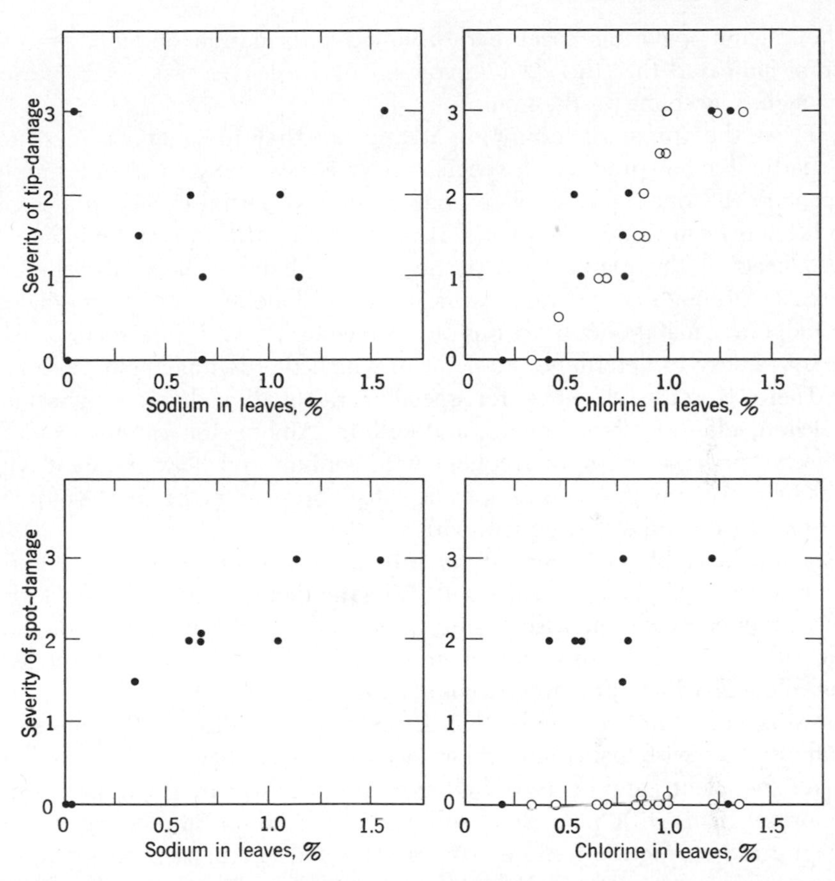

Figure 6.15. Severity of tip-damage (above) and spot-damage (below) of leaves from avocado orchards in southern California versus sodium and chlorine percentages in the dry leaves. Both sodium and chlorine analyses were made in the instances represented by solid circles. Only chlorine analyses were made in the instances represented by open circles. Leaf damage is rated on a scale from 0 for none to 3 for severe. (Ayers et al., 1951)

calcium chloride or sodium chloride in concentrations to produce a solute suction of 2 bars. The growth was much poorer in the cultures salinized with calcium chloride than in those salinized with sodium chloride, from which one might expect a toxic effect of calcium. The chloride content of the leaves, however, was much greater in the calcium chloride cultures than in the sodium chloride cultures. The excess uptake of chloride was presumed to be largely responsible for the unfavorable effect of the calcium chloride because the limited information obtained on calcium indicated that the calcium content of the leaves was not increased enough to account for the injury.

From the foregoing examples, it appears that the manner in which a particular ion produces a specific effect is not known without supplementary information. The effect may be due to partial exclusion of some other ion from uptake by plants. It may be due also to metabolic effects produced in the plants by a particular ion that has been absorbed in excess. Although considerable work has been done from the physiological standpoint, metabolic processes are so complex and interrelated that it is not easy to determine the point at which the derangement begins.

There is some evidence for specific effects of sodium, magnesium, calcium, chloride, bicarbonate, and sulfate. Among the cations, specific effects appear to be most frequent with sodium and least frequent with calcium. Among the anions, specific effects appear to be most frequent with chloride and least frequent with sulfate.

Chloride is blamed more frequently than any other specific ion for poor growth of plants on salty soils. This tendency carries over to fertilizer practice, where potassium sulfate is usually preferred over potassium chloride from the standpoint of crop quality. Chloride often increases succulence, which probably accounts for the claim of some truck-crop growers that tomatoes and other fruits are more firm with potassium sulfate than with potassium chloride and that cabbage and lettuce are more succulent (and hence of higher quality in this case) with potassium chloride than with potassium sulfate. The general impression of the magnitude of specific ionic effects in salty soils perhaps is exaggerated, however, because in experimental work single salts are usually added in increasing quantities, whereas in practice the excesses of soluble salts are mixtures.

Tolerance to Salinity. According to a review by Hayward and Wadleigh (1949), the tolerance of individual plant species and varieties to soil salinity increases with their capacity to adjust to relatively high internal solute suction and decreases with their sensitivity to this adjustment. Plants native to a saline environment have both a marked capacity for adjustment and limited sensitivity. Some of these plants, known as halophytes, normally develop internal solute suctions of the order of 30 to 50 bars; and they may grow better on saline than on nonsaline soil. Crop plants have a lesser but still considerable capacity for upward adjustment of internal solute suction, but they are more sensitive to the adjustment.

A number of classifications of crops as regards tolerance to salinity have been developed in different parts of the world. For citations to literature on this subject, reviews by Hayward and Bernstein (1958) and Bernstein (1962) should be consulted.

Various criteria for appraising the tolerance of plants to soil salinity may be employed. The United States Salinity Laboratory has chosen to use the specific conductivity of the saturation extract associated with a 50% decrement in yield from the yield on nonsaline soil. This criterion facilitates comparison of diverse crops. Table 6.3 gives a partial listing

Table 6.3. Grouping of Crops According to the Conductivity of the Saturation Extract of Saline Soil Corresponding to a 50% Decrement of the Yield on Saline Soil from That on Nonsaline Soil (U.S. Salinity Laboratory Staff, 1954)

High Salt Tolerance	Medium Salt Tolerance	Low Salt Tolerance
	Vegetable Crops	
EC[1] = 12 to 10	*EC* = 10 to 4	EC = 4 to 3
Garden beets	Tomato	Radish
Kale	Cabbage	Celery
Asparagus	Lettuce	Green beans
Spinach	Potatoes	
	Cucumber	
	Forage Crops	
EC = 18 to 12	*EC* = 12 to 4	*EC* = 4 to 2
Saltgrass	White sweetclover	White Dutch clover
Bermudagrass	Perennial ryegrass	Meadow foxtail
Western wheatgrass	Sudangrass	Alsike clover
Birdsfoot trefoil	Alfalfa	Red clover
	Orchardgrass	Ladino clover
	Bromegrass	
	Field Crops	
EC = 16 to 10	*EC* = 10 to 6	*EC* = 4
Barley	Wheat	Field beans
Sugar beet	Oats	
Rape	Rice	
Cotton	Corn	
	Castor beans	

[1] *EC* = specific electrical conductivity of the saturation extract in millimhos per centimeter at 25°C. Crops in each group are arranged in order of decreasing tolerance. The first and second numbers for *EC* in each group apply approximately to the first and last members of the group.

of the ratings included in the monograph by the U.S. Salinity Laboratory Staff (1954).

Tolerance of plants to soil salinity is not a fixed characteristic of each species or variety but may vary with the environmental conditions. Table 6.4 illustrates the effect of climatic conditions in the presence of a given saline environment. The data were obtained by growing plants in large outdoor sand cultures supplied with complete nutrient solutions adjusted to different degrees of salinity. The basal nutrient solution had a solute suction of 0.4 bar. The level of salinity at which the reduction in yield below that of the control amounted to 25% of the control yield was determined graphically. All three species showed greater tolerance

Table 6.4. Response of Three Crops to Salinity in Sand Cultures at Two Locations (Magistad et al., 1943)

	Salinity of Solution Corresponding to a 25% Reduction in Yield at Indicated Location, bars	
Crop	Torrey Pines	Indio
Bean pods	1.4	1.1
Garden beet roots	4.0	2.4
Onion bulbs	4.5	1.2

to salinity at Torrey Pines, where the environment is cool and humid, than at Indio, where the environment is hot and dry. Moreover, the order of tolerance changed from onions $>$ beets $>$ beans at Torrey Pines to beets $>$ onions $>$ beans at Indio.

A soil factor of considerable importance in relation to tolerance of plants to salinity is the location of the salts. Salts usually are distributed nonuniformly within a given soil profile. After a period of dry weather there may be a marked concentration at the surface of the soil. Observations of plant growth in relation to the salinity of the surface layer of soil at such a time would suggest that a given species is relatively tolerant to salinity. On the other hand, after rainfall or irrigation, a soil may be saline throughout the root zone with the exception of the surface layer. Observations of plant growth in relation to the salinity of the surface layer at such a time would suggest that the same species is relatively sensitive to salinity. Where furrow irrigation is practiced, the furrow is often placed midway between crop rows, which are located in the ridges. Movement of water from the furrow then carries salts toward the crop rows. The concentration of salts after furrow irrigation thus is often highest in the ridges where, from the standpoint of crop welfare, it should preferably be lowest. An example of distribution of soluble salts in a cross section of soil following furrow irrigation is shown in Fig. 6.16. In Washington, Heald et al. (1950) found that germination of sugar beet seed was much reduced by such movement of salts in saline soils that must be irrigated after planting to produce emergence. They found that germination could be improved by opening the irrigation furrow near the row and making the ridge about midway between rows. The irrigation water then carried the salt away from the seed and into the ridge. The effect of irrigation in this way is indicated by the close-up view in Fig. 6.17 showing a marked efflorescence of salt on the soil in the ridge area and the absence of an efflorecence from the soil around the row.

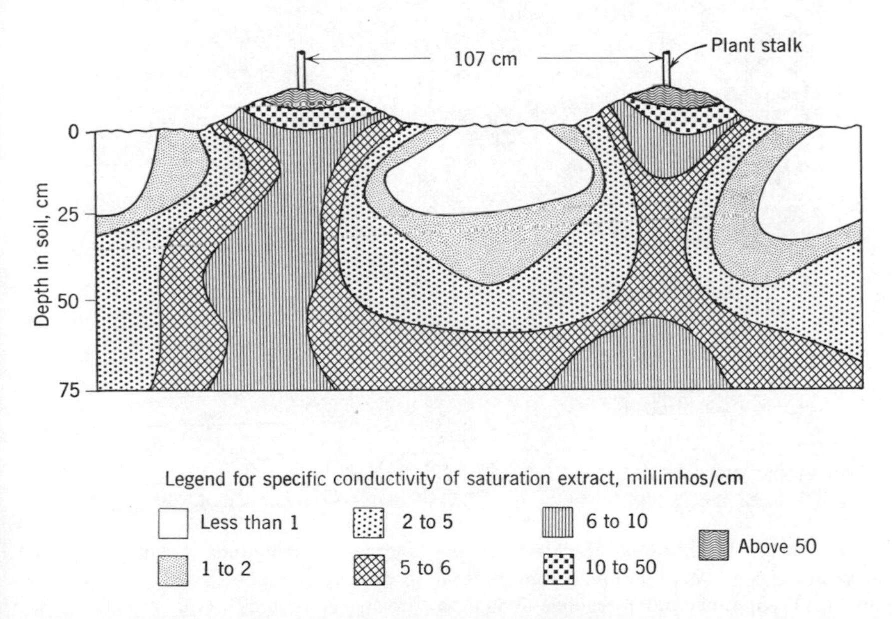

Figure 6.16. Variation of specific conductivity of the saturation extract in a vertical cross section of a sandy loam soil at right angles to rows of cotton plants grown at Riverside, California. The soil was originally salinized artificially and mixed to promote uniformity. During growth of the plants, nonsaline irrigation water was applied as needed to the furrows in increments equivalent to 7.6 surface centimeters at each irrigation. (Wadleigh and Fireman, 1949)

Figure 6.17. Distribution of salt after pre-emergence irrigation of sugar beets on a saline soil in Washington. From bottom to top in the photograph, the features are (*A*) edge of pre-emergence irrigation furrow (desalted), (*B*) pre-emergence irrigation furrow (desalted), (*C*) row of sugar beet seedlings in desalted soil at edge of furrow, (*D*) flat area of desalted soil including a pencil for scale, (*E*) flat area of soil with efflorescence of salt on the surface, and (*F*) part of ridge associated with the furrow for the next row with efflorescence of salt on the surface. (Heald et al., 1950)

Excess Sodium

Following the terminology of Bower et al. (1958), a sodic soil contains enough exchangeable sodium to interfere with growth of most crop plants. For purposes of definition, sodic soils are said to have a degree of sodium saturation equivalent to 15% or more of the cation-exchange capacity.

Excess exchangeable sodium affects both the physical and chemical properties of soils, and there is evidence that both classes of effects have important consequences in terms of the suitability of soils as a substrate for plant growth. The critical conditions vary with the nature of the soil and with the characteristics of plants.

Physical Effects. The marked influence of exchangeable sodium on the over-all physical properties of soils is associated principally with the behavior of the clay and organic matter, in which most of the cation-exchange capacity is concentrated. Table 6.5 illustrates the relative effects of exchangeable sodium and calcium on certain physical properties of the colloidal material extracted from soil. The third column shows the volume attained by 1 g of the dry material after it had been allowed to absorb water. The fourth column shows the percentages of the material that remained in suspension after it had been moistened, shaken with 100 ml of water, and allowed to stand 31 hours in cylinders in which the depth of the suspension was 10 cm.

The greater tendency for separation of individual particles from one another in the presence of exchangeable sodium than calcium presumably results from a combination of differences in forces of attraction and repulsion associated with the two ions. Interparticle bonding through distribution of charges of cations between adjacent particles should be less with monovalent sodium than with divalent calcium, and repulsion of particles due to mutual repulsion of exchangeable cations should be greater with sodium than with calcium because of the greater thickness of atmospheres of the former ion around the particles.

The last column in Table 6.5 shows the rate of horizontal movement of individual particles in an extremely dilute suspension in water under the influence of an electrical potential gradient of 1 volt/cm of horizontal distance in the suspension. Uncharged particles show no horizontal movement under these conditions, but charged particles move at a rate that depends on their size and on the difference in electrical potential between the particle and the solution. This potential, known as the zeta potential, is greater if the exchangeable cations are sodium than calcium. With addition of salt to a dilute suspension of clay or to soils, the ionic atmospheres contract, the zeta potential decreases, and the

Table 6.5. Physical Properties of the Colloidal Fraction of a Soil with Different Degrees of Saturation with Sodium and Calcium (Mattson, 1928)

Degree of Saturation of Cation-Exchange Capacity with Indicated Cation, %		Volume per Gram of Material after Imbibition of Water, cm^3	Proportion of Material in Suspension, %	Electrical Migration per Second per Volt per Centimeter, μ
Sodium	Calcium			
0	100	1.9	2	1.7
5	95	2.0	3	1.9
10	90	2.0	2	1.7
20	80	2.1	13	2.9
30	70	2.3	53	3.0
40	60	3.2	88	3.3
50	50	5.1	97	3.5
75	25	6.5	99	3.5
100	0	7.1	99	3.5

particles tend to attract each other. In dilute suspensions of clay, the particles collect in floccules and settle rapidly if the suspension is not disturbed. In the presence of excess salt, even sodium-saturated clays are flocculated. If the salt content is reduced sufficiently by leaching or dilution, however, the floccules disperse.

These theoretical concepts may be made more tangible from consideration of the data by Bower and Goertzen (1955) in Table 6.6. The experiments were performed on a silty clay loam soil that was saturated with exchangeable calcium or sodium and then treated with known quantities of solutions of calcium chloride or sodium chloride in different concentrations to produce saturation of the soil with solution. A quantity of the solution was then removed under a matric suction of 0.5 bar and was analyzed for chloride. Because of negative adsorption, the concentration of chloride found by analysis was higher than the concentration of chloride in the solution added.

On the basis of the quantity of water present in the soil at the saturation percentage, the concentration of chloride in the solution added, and the concentration of chloride in the saturation extract, the authors calculated a value for "adsorbed water," which is the quantity of water from which chloride would have to be excluded completely to account for the difference in concentration of chloride between the solution added and the saturation extract. With the aid of data on the specific surface of the soil (164 m^2/g), they calculated the thickness of the layer of adsorbed water on the surface of the soil particles. Although theoretically there is no layer of adsorbed water from which chloride is excluded completely, these calculations provide conventional values that give some concept of the differences in thickness of the ionic atmospheres around calcium- and sodium-saturated particles of soil in the presence of different concentrations of salt. The conventional figures indicate that the ionic atmospheres were about three times as thick with sodium as with calcium in exchangeable form at low concentrations of salt but that at high concentrations of salt the thickness was much reduced and was about the same for the two ions. These conventional figures probably represent a considerable underestimate of the true thickness of the ionic atmospheres. According to theory, and also in agreement with other data by Bower and Goertzen, the concentration of exchangeable cations in the ionic atmospheres gradually decreases with distance from the surface of the particles outward into the solution. As an aid in providing a scale of distances, an oxygen ion is considered to have a diameter of about 2.7 Å.

The deflocculation or dispersion of clays produced so readily in the laboratory and illustrated in Table 6.5 may occur only to a small degree in soils in the field because these soils are not subjected to the forces used in the laboratory to separate the particles. Currently, there is some doubt that the foregoing theory as presented is an adequate explanation for the physical effect of sodium in natural soils. According to Edwards and Bremner's (1967) microaggregate theory mentioned in Chapter 1,

Table 6.6. Calculated Amount and Thickness of Layers of Adsorbed Water, Free of Chloride, in a Silty Clay Loam Soil Saturated with Exchangeable Calcium or Sodium in the Presence of Two Concentrations of Calcium or Sodium Chloride (Bower and Goertzen, 1955)

Exchangeable Cation	Salt Added	Chloride per Liter of Saturation Extract, m.e.	Adsorbed Water in Soil	
			Amount, %	Thickness, Å
Calcium	$CaCl_2$	12.6	9.6	5.9
	$CaCl_2$	1054.0	4.5	2.7
Sodium	NaCl	15.7	27.6	16.8
	NaCl	1073.0	5.2	3.2

the primary bond in structural units is clay-polyvalent cation-organic matter, the polyvalent cation being held electrostatically to cation-exchange sites in both clay and organic matter. Sodium ruptures the bond by exchanging for the polyvalent cation. As the proportion of sodium in the exchangeable cations increases, an increasing proportion of the bonds would be broken. Aggregates would then swell to an increasing degree, although they might not disperse without input of energy to break the remaining bonds. The pressure of overlying soil could conceivably cause the swelled aggregates to deform and occupy many of the larger pores that were present before swelling. Whatever may be the precise nature of the microscopic changes in soil associated with high exchangeable sodium and low salt, the conductivity of soil for water is decreased.

An example of the effect of sodium on water movement is shown in Fig. 6.18 for samples of a clay loam soil that had been treated in the laboratory to produce different degrees of sodium saturation under both acid and alkaline conditions. The hydraulic conductivity was reduced almost to zero at exchangeable-sodium-percentages above 20, based on the sum of exchangeable calcium, magnesium, potassium, and sodium. (This formulation omits exchangeable aluminum in acid soils, a procedure that the authors found to give improved correlations with the results obtained with alkaline soils.)

As an example from field conditions, Fireman and Reeve (1949) reported hydraulic conductivities of 1.9 and less than 0.01 cm/hour in samples from the surface horizon of a silty clay loam soil at two adjacent

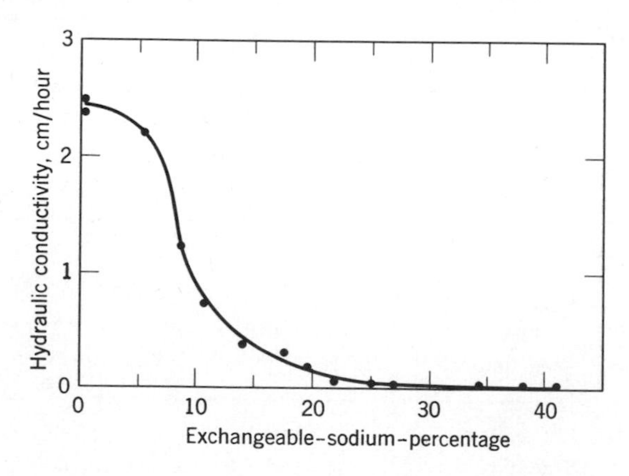

Figure 6.18. Hydraulic conductivity of a clay loam soil under acid and alkaline conditions versus exchangeable sodium as a percentage of the sum of exchangeable calcium, magnesium, potassium, and sodium. (Martin et al., 1964)

locations in Idaho. The exchangeable-sodium-percentages of the samples from the respective areas were 4 and 54. The low-sodium, permeable area was carrying a good crop; the high-sodium, impermeable area was bare. The bare area was clearly unsuitable for production of dry-land crops because of low hydraulic conductivity if for no other reason. A hydraulic conductivity of 0.01 cm/hour is equivalent to only 88 cm/year. Thus the soil would have to remain flooded for perhaps 6 months before enough water could be absorbed to produce a good crop. Once such an impermeable soil becomes wet, the aeration probably would be too poor for the production of dry-land crops because the soil would remain essentially saturated with water over a long period of time. And, finally, reclamation would be difficult or perhaps impracticable because of the difficulty of introducing calcium into the soil and removing sodium from it.

In some areas, rice is grown during reclamation of nonsaline-sodic soils by leaching, thus turning to advantage the low rate of movement of water through the soil. Figure 6.19 illustrates results of some of the work done by Overstreet and Schulz (1958) in this connection.

A number of experiments have been done in which plants have been grown on soils adjusted to different degrees of sodium saturation, with and without treatment with a synthetic organic aggregating agent to determine the effect of the deterioration of structure. Figure 6.20 shows the results of an experiment by Martin, Jones, and Ervin (1959) that illustrates the usual finding. If the treatment has prevented deterioration of soil structure, the vertical difference between the two lines results from direct and indirect influences of sodium on the physical properties of the soil; and the difference between a relative yield of 100 and the yield on the soil treated with the aggregating agent results from other effects. In this instance the physical and other effects produced about equal decreases in relative yield at an exchangeable-sodium-percentage of 13, whereas effects other than physical properties produced the greater decrease in yield at an exchangeable-sodium-percentage of 27.

The apparent physical effects are sometimes less and sometimes greater than those indicated by Fig. 6.20. Chang and Dregne (1955) conducted a similar experiment in which the physical effects appeared to be relatively minor in a clay loam soil cropped to cotton and alfalfa in the greenhouse. Allison (1952) conducted a field experiment with sweet corn in which the major effects of excess sodium appeared to be due to deterioration of structure of the fine sandy loam soil employed. The data in Table 6.7 show that the yield of corn was decreased by an increase in exchangeable-sodium-percentage from 3 to 29 in the absence of a soil-aggregating agent but not in its presence. These results

Figure 6.19. Appearance of plants before reclamation of a nonsaline-sodic, fine-sandy-loam soil in California and during the first two years of reclamation by flooding. Top: first year (before reclamation), cotton. Center: second year, rice. Bottom: third year, rice. Production of rice in California involves flooding of the soil for about five months. About two meters of water are required for this soil in the first season of rice production. In following seasons, the water requirement greatly increases as a result of increased permeability of the soil; after three seasons, the cost of water becomes an important economic consideration in the production of rice. (Overstreet and Schulz, 1958)

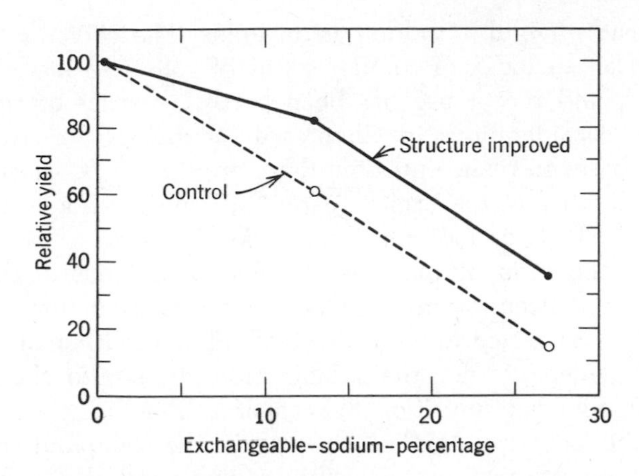

Figure 6.20. Relative yield of peach seedlings on a sandy loam soil adjusted to different exchangeable-sodium-percentages, with and without addition of vinyl acetate-maleic acid copolymer to improve the structure. The yield on the soil with an exchangeable-sodium-percentage less than 1 is given a relative value of 100 in each series. (Martin, Jones, and Ervin, 1959)

indicate the existence of major physical effects and the absence of substantial effects of other kinds. Inhibition of seedling emergence by the hard surface crust apparently was the most important of the physical effects. The stand was poor in the untreated soil at an exchangeable-sodium-percentage of 29; most of the plants that emerged came up through cracks in the crust.

Nutritional Effects. Perhaps the first question to be answered with regard to nutritional effects is the reason for choice of exchangeable

Table 6.7. Stand and Yield of Sweet Corn on a Sandy Loam Soil at Two Values of Exchangeable-Sodium-Percentage, with and without Application of 0.1% Calcium Carboxylate Polymer to Improve the Aggregation (Allison, 1952)

Exchangeable-Sodium-Percentage	Calcium Carboxylate Addition, %	Stand of Corn, %	Yield of Ear Corn per Plot, g
3	0	100	3700
3	0.1	100	3820
29	0	30	760
29	0.1	100	4050

sodium instead of soluble sodium as an index. The choice of exchangeable sodium as an index of sodium status of soils was made a number of years ago, and recent use has been partly a matter of tradition. In view of the considerations in Chapter 4 on the role of exchangeable and soluble bases in plant nutrition, there appears to be cause to reconsider the suitability of the exchangeable-sodium-percentage as an index of nutritional effects of sodium. If the ratio of uptake of ions by plants is related to the ratio of the ions in solution and not to the ratio of the activity or concentration of monovalent cations to the square root of the activity or concentration of divalent cations, neither the exchangeable-sodium-percentage nor the sodium-adsorption ratio should be the best obtainable index of nutritional effects of sodium.

Perhaps the next question of concern is the comparative value of the absolute quantity of exchangeable sodium and the exchangeable-sodium-percentage as an index of the nutritional status of soils with regard to sodium. The two indexes are of equal value for a given soil because the quantity of exchangeable sodium is equal to the product of the exchangeable-sodium-percentage and a constant factor (cation-exchange capacity/100). The critical test is the behavior of the two indexes with different soils. Figure 6.21 shows the results of an experiment by Bernstein and Pearson (1956) in which alfalfa was grown in large metal drums containing two different soils to which quantities of sodium bicarbonate had been added to vary the exchangeable sodium content. The

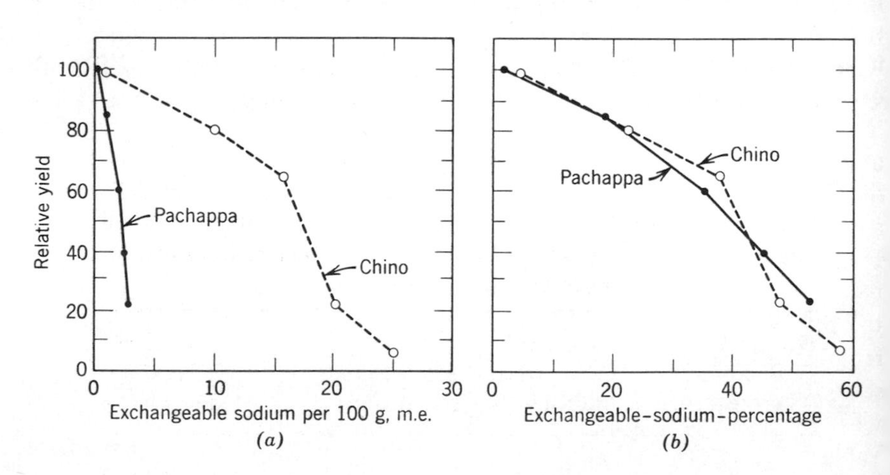

Figure 6.21. Relative yield of alfalfa versus (*a*) exchangeable sodium and (*b*) exchangeable-sodium-percentage on Pachappa and Chino soils treated with vinyl acetate-maleic acid copolymer to maintain good aggregation. The cation-exchange capacities of the two soils were 11 and 43 m.e./100 g. (Bernstein and Pearson, 1956)

yields of alfalfa are expressed on a relative basis with the yield at zero exchangeable sodium (extrapolated) taken as 100. Considerable divergence in plant responses between soils may be observed where relative yields are plotted against the exchangeable sodium but not where they are plotted against the exchangeable-sodium-percentage, thus indicating that the latter is the more suitable of the two indexes for comparing different soils.

Fireman and Wadleigh (1951) found from analyses of samples of many soils from western United States that the *p*H tends to increase with the exchangeable-sodium-percentage. Because the unfavorable effects of sodic soils on the growth of plants thus increase in severity with an increase in both exchangeable sodium and *p*H, it becomes important to determine whether the unfavorable effect is associated primarily with the sodium or the *p*H.

Several types of evidence indicate that excess exchangeable sodium has unfavorable effects that are not attributable to high hydroxyl-ion activity. Bower and Wadleigh (1949) grew different plants on ion-exchange resins and found that the yields decreased as the exchangeable-sodium-percentage increased, despite the fact that the *p*H value of all cultures was about 6.5. Martin and Bingham (1954) obtained yields of 18, 14, 12, and 3 g of avocado seedlings per culture of a loam soil with exchangeable-sodium-percentages of 1, 4, 7, and 14. The plants died where the exchangeable-sodium-percentage was 26. The respective *p*H values were 6.6, 6.8, 7.1, 7.3, and 7.7. Similar results were obtained by Martin, Bitters, and Ervin (1959) in an experiment with orange seedlings. Because of the low hydroxyl-ion activity in these experiments, it seems unlikely that the detrimental effect associated with increasing exchangeable sodium is attributable to hydroxyl ions.

Bower et al. (1954) obtained another type of evidence by making a statistical analysis of yields of sugar beets in relation to properties of the surface 30 cm of soil of field plots established on an area of a saline-sodic fine sandy loam in Washington. The yield of beets varied from 0.02 to 70 metric tons/hectare, the soil *p*H (measured on the saturated soil paste) from 7.8 to 9.6, the exchangeable sodium from 1.9 to 15.1 m.e/100 g, the specific conductivity of the saturation extract from 2.8 to 26.2 millimhos/cm, and the sodium-adsorption-ratio of the saturation extract from 8 to 407. The yield of sugar beets decreased with an increase in value of each of the soil properties taken individually. Where the soil properties were considered jointly, however, only the exchangeable sodium was of significant independent value in estimating the yield of beets. Thus the apparent effect of *p*H taken alone was attributable to its correlation with exchangeable sodium and not to an independent effect. The use of exchangeable sodium in this work is approximately equivalent to use of exchangeable-sodium-percentage because the cation-exchange capacity was fairly constant at 22 to 24 m.e./100 g. The value of exchangeable sodium in accounting for the yield of beets in this experiment is one form of evidence for the value of the exchangeable-sodium-percentage as an index of the nutritional status of soils with regard to sodium.

In the experiments just described, the unfavorable effects of exchangeable sodium on growth of plants do not appear to be attributable to a deterioration of soil structure and the secondary effects associated therewith. The manner in which detrimental nutritional effects of sodium are produced, however, is not well understood. The most specific theory is that excess sodium may induce deficiencies of other cations, particularly calcium and magnesium. Evidence supporting this theory is shown in Table 6.8. This table contains data on the yield and composition of alfalfa grown on a nonsaline-sodic soil with different treatments. The marked increases in yield and percentage content of calcium and magnesium from applications of the chlorides indicate that the alfalfa was deficient in both calcium and magnesium on this particular soil.

The action of sodium in inducing deficiencies of calcium and magnesium appears to be threefold. First, because sodium is comparatively loosely held in exchangeable form, the ions released to the soil solution in a fractional exchange are mostly sodium ions if the soil has a high exchangeable-sodium-percentage. Second, at the high *p*H values usually associated with excess exchangeable sodium in the absence of excess salts, the soil solution contains bicarbonate and carbonate ions that tend to precipitate calcium and magnesium as carbonates. The saturation extract of the soil used in the experiment described in Table 6.8, for example, contained 19.5 m.e. of bicarbonate and 8.8 of carbonate, but only 1.5 m.e. of calcium and 0.3 of magnesium per liter of solution. This concentration of calcium is well down in the deficiency range as judged from Fig. 5.17. The third action of sodium is exclusion of calcium and magnesium from absorption on a competitive basis because sodium is absorbed instead. Some evidence of a calcium-deficiency effect was obtained in the field experiment on sugar beets previously described. The yield of beets was found to increase significantly with the soluble-calcium-percentage in the cations in the saturation extract independently of the sodium effect, which was of greater significance.

Another theory relating to the detrimental effect of excess sodium is that plants are affected adversely by the excess sodium they absorb. This theory was discussed to some extent in connection with soil salinity in relation to plant growth. In the experiments by Ayers et al. (1951) on avocados (Fig. 6.15), the symptom associated with excess sodium

Table 6.8. The Yield and the Calcium and Magnesium Content of Alfalfa Grown on a Nonsaline-Sodic Soil with Different Treatments (Bower and Turk, 1946)

Soil Treatment	*p*H	Exchangeable-Sodium-Percentage	Yield of Dry Matter per Culture, g	Content in Dry Matter, %	
				Calcium	Magnesium
Control	9.6	46	0.1	0.6	0.3
Leached with water	9.8	45	0.1	—	—
$CaCl_2$, 12 m.e./100 g, followed by leaching with water	8.6	6	2.0	1.5	0.3
$CaCl_2$, 9 m.e./100 g, plus $MgCl_2$, 3 m.e./100 g, followed by leaching with water	8.7	8	3.7	1.5	0.5

was production of dead spots in the leaves. In investigations of sodium tolerance in various plants, Bower and Wadleigh (1949), Bernstein and Pearson (1956), and others have noted that some plant species accumulate sodium in their roots from high-sodium media but translocate little to the tops; whereas others translocate sodium readily to the tops and do not accumulate so much in the roots. For reasons as yet unknown, the tendency for sodium retention in the roots is associated with sensitivity to excess sodium.

Soil Alkalinity

Soil alkalinity may influence plant growth in ways other than those discussed in the preceding sections. High hydroxyl-ion concentrations no doubt have direct detrimental effects on plants. According to Olsen (1953), damage from hydroxyl ions occurs at *p*H 10.5 or higher. Jones (1961) obtained evidence for toxicity of aluminum to plants grown on the ash from pulverized coal at high *p*H values, which suggests that aluminum toxicity may occur at high *p*H values in nonsaline-sodic soils. In addition, the availability of various nutrients at times may be critical. Deficiencies of phosphorus, iron, and zinc are fairly common. All these elements have relatively low solubility under alkaline conditions, and the deficiency is alleviated by lowering the *p*H value. Boron and manganese also have relatively low solubility under alkaline conditions. Deficiencies of these elements, however, are not common in plants grown on naturally alkaline soils in arid regions but are more common in alkaline soils in humid regions. Nitrogen deficiency is common where irrigation is practiced on alkaline soils in dry regions because ordinarily the native supply of nitrogen is small. In this case, the alkaline conditions are not responsible for the deficiency but are merely associated with it. Both the alkaline conditions and the low native supply of nitrogen result from the limited rainfall under which the soils have developed.

Literature Cited

Abd-el-Malek, Y., and S. G. Rizk. (1963) Bacterial sulphate reduction and the development of alkalinity. II. Laboratory experiments with soils. III. Experiments under natural conditions in the Wadi Natrûn. *Jour. Appl. Bact.* **26**:14–19, 20–26

Aldrich, D. G., A. P. Vanselow, and G. R. Bradford. (1951) Lithium toxicity in citrus. *Soil Sci.* **71**:291–295.

Allison, L. E. (1952) Effect of synthetic polyelectrolytes on the structure of saline and alkali soils. *Soil Sci.* **73**:443–454.

Allison, L. E. (1964) Salinity in relation to irrigation. *Adv. Agron.* **16**:139–180

Ayers, A. D. (1950) Salt tolerance of avocado trees grown in culture solutions. *California Avocado Soc. Yearbook* **1950**:139–148.

Ayers, A. D., D. G. Aldrich, and J. J. Coony. (1951) Sodium and chloride injury of Fuerte avocado leaves. *California Avocado Soc. Yearbook* **1951**:174–178

Bernstein, L. (1961) Osmotic adjustment of plants to saline media. I. Steady state *Amer. Jour. Bot.* **48**:909–918.

Bernstein, L. (1962) Salt-affected soils and plants. *Arid Zone Res.* **18**:139–174

Bernstein, L. (1963) Osmotic adjustment of plants to saline media. II. Dynamic phase. *Amer. Jour. Bot.* **50**:360–370.

Bernstein, L., and H. E. Hayward. (1958) Physiology of salt tolerance. *Annual Rev. Plant Physiol.* **9**:25–46.

Bernstein, L., and G. A. Pearson. (1956) Influence of exchangeable sodium on the yield and chemical composition of plants: I. Green beans, garden beet clover, and alfalfa. *Soil Sci.* **82**:247–258.

Bingham, F. T., A. L. Page, and G. R. Bradford. (1964) Tolerance of plants to lithium. *Soil Sci.* **98**:4–8.

Bower, C. A. (1961) Prediction of the effects of irrigation waters on soils. *Arid Zone Res.* **14**:215–222.

Bower, C. A., and J. O. Goertzen. (1955) Negative adsorption of salts by soils *Soil Sci. Soc. Amer. Proc.* **19**:147–151.

Bower, C. A., W. G. Harper, C. D. Moodie, R. Overstreet, and L. A. Richards (1958) Report of the nomenclature committee appointed by the board of collaborators of the U.S. Salinity Laboratory. *Soil Sci. Soc. Amer. Proc.* **22**:27

Bower, C. A., C. D. Moodie, P. Orth, and F. B. Gschwend. (1954) Correlation of sugar beet yields with chemical properties of a saline-alkali soil. *Soil Sci.* **77**:443–451.

Bower, C. A., and L. M. Turk. (1946) Calcium and magnesium deficiencies alkali soils. *Jour. Amer. Soc. Agron.* **38**:723–727.

Bower, C. A., and C. H. Wadleigh. (1949) Growth and cationic accumulation by four species of plants as influenced by various levels of exchangeable sodium. *Soil Sci. Soc. Amer. Proc.* (1948) **13**:218–223.

Bower, C. A., and L. V. Wilcox. (1965) Soluble salts. *Agronomy* **9**:933–951.

Brown, J. W., C. H. Wadleigh, and H. E. Hayward. (1953) Foliar analysis of stone fruit and almond trees on saline substrates. *Proc. Amer. Soc. Hort. Sci.* **61**:49–55.

Campbell, R. B., C. A. Bower, and L. A. Richards. (1949) Change of electrical conductivity with temperature and the relation of osmotic pressure to electrical conductivity and ion concentration for soil extracts. *Soil Sci. Soc. Amer. Proc.* (1948) **13**:66–69.

Carter, D. L. (1962) A bibliography of publications in the field of saline and sodic soils (through 1961). *U.S. Dept. Agr., Agr. Res. Service, ARS 41–80.*

Chang, C. W., and H. E. Dregne. (1955) Effect of exchangeable sodium on soil properties and on growth and cation content of alfalfa and cotton. *Soil Sci. Soc. Amer. Proc.* **19**:29–35.

Danielson, R. E., and M. B. Russell. (1957) Ion absorption by corn roots as influenced by moisture and aeration. *Soil Sci. Soc. Amer. Proc.* **21**:3–6.

Eaton, F. M. (1935) Boron in soils and irrigation waters and its effect on plants, with particular reference to the San Joaquin Valley of California. *U.S. Dept. Agr. Tech. Bul. 448.*

Eaton, F. M. (1941) Water uptake and root growth as influenced by inequalities in the concentration of the substrate. *Plant Physiol.* **16**:545–564.

Eaton, F. M. (1942) Toxicity and accumulation of chloride and sulfate salts in plants. *Jour. Agr. Res.* **64**:357–399.

Eaton, F. M. (1950) Significance of carbonates in irrigation waters. *Soil Sci.* **69**:123–133.

Edwards, A. P., and J. M. Bremner. (1967) Microaggregates in soils. *Jour. Soil Sci.* **18**:64–73.

Ehlig, C. F. (1964) Salt tolerance of raspberry, boysenberry, and blackberry. *Proc. Amer. Soc. Hort. Sci.* **85**:318–324.

Eriksson, E. (1952) Cation-exchange equilibria on clay minerals. *Soil Sci.* **74**:103–113.

Fireman, M., and R. C. Reeve. (1949) Some characteristics of saline and alkali soils in Gem County, Idaho. *Soil Sci. Soc. Amer. Proc.* (1948) **13**:494–498.

Fireman, M., and C. H. Wadleigh. (1951) A statistical study of the relation between *p*H and the exchangeable-sodium-percentage of western soils. *Soil Sci.* **71**:273–285.

Gauch, H. G., and C. H. Wadleigh. (1944) Effects of high salt concentrations on growth of bean plants. *Bot. Gaz.* **105**:379–387.

Hayward, H. E., and L. Bernstein. (1958) Plant-growth relationships on salt-affected soils. *Bot. Rev.* **29**:584–635.

Hayward, H. E., and W. B. Spurr. (1944) Effects of isosmotic concentrations of inorganic and organic substrates on entry of water into corn roots. *Bot. Gaz.* **106**:131–139.

Hayward, H. E., and C. H. Wadleigh. (1949) Plant growth on saline and alkali soils. *Adv. Agron.* **1**:1–38.

Heald, W. R., C. D. Moodie, and R. W. Leamer. (1950) Leaching and pre-emergence irrigation for sugar beets on saline soils. *Washington Agr. Exp. Sta. Bul. 519.*

Indian Agricultural Research Institute. (1962) *Seminar on Salinity and Alkali Soil Problems.* Indian Agricultural Research Institute, New Delhi.

Jones, L. H. (1961) Aluminium uptake and toxicity in plants. *Plant and Soil* **13**:297–310.

Kelley, W. P. (1951) *Alkali Soils. Their Formation, Properties and Reclamation.* American Chemical Society Monograph No. 111. Reinhold Publishing Corp., New York.

Magistad, O. C., A. D. Ayers, C. H. Wadleigh, and H. G. Gauch. (1943) Effect of salt concentration, kind of salt, and climate on plant growth in sand cultures. *Plant Physiol.* **18**:151–166.

Martin, J. P., and F. T. Bingham. (1954) Effect of various exchangeable cation ratios in soils on growth and chemical composition of avocado seedlings. *Soil Sci.* **78**:349–360.

Martin, J. P., W. P. Bitters, and J. O. Ervin. (1959) Influence of exchangeable Na, and K and of excess lime on growth and chemical composition of trifoliate orange seedlings. *Proc. Amer. Soc. Hort. Sci.* **74**:308–312.

Martin, J. P., W. W. Jones, and J. O. Ervin. (1959) Influence of exchangeable K and Na in the soil on growth and chemical composition of Lovell peach seedlings and other crops. *Agron. Jour.* **51**:418–421.

Martin, J. P., S. J. Richards, and P. F. Pratt. (1964) Relationship of exchangeable Na percentage at different soil *p*H levels to hydraulic conductivity. *Soil Sci. Soc. Amer. Proc.* **28**:620–622.

Mattson, S. (1928) The influence of the exchangeable bases on the colloidal behavior of soil materials. *First Internat. Congr. Soil Sci. Proc. Papers, Comm.* **2**:185–198.

Moss, P. (1963) Some aspects of the cation status of soil moisture. Part I: The ratio law and soil moisture content. *Plant and Soil* **18**:99–113.

Olsen, C. (1953) The significance of concentration for the rate of ion absorption by higher plants in water culture. IV. The influence of hydrogen ion concentration. *Compt. Rend. Trav. Lab. Carlsberg, Sér. Chim.* **28**:488–498.

Overstreet, R., and R. K. Schulz. (1958) The effect of rice culture on a nonsaline sodic soil of the Fresno series. *Hilgardia* **27**:319–332.

Reeve, R. C., and C. A. Bower. (1960) Use of high-salt waters as a flocculant and source of divalent cations for reclaiming sodic soils. *Soil Sci.* **90**:139–144.

Reitemeier, R. F. (1946) Effect of moisture content on the dissolved and exchangeable ions of soils of arid regions. *Soil Sci.* **61**:195–214.

Slatyer, R. O. (1961) Effects of several osmotic substrates on the water relationships of tomato. *Australian Jour. Biol. Sci.* **14**:519–540.

Strogonov, B. P. (1964) *Physiological Basis of Salt Tolerance of Plants.* Israel Program for Scientific Translations, Jerusalem. (Published in the United States by Daniel Davey & Co., Inc., New York.)

U.S. Salinity Laboratory Staff. (1954) Diagnosis and improvement of saline and alkali soils. *U.S. Dept. Agr. Handbook 60.*

Wadleigh, C. H., and A. D. Ayers. (1945) Growth and biochemical composition of bean plants as conditioned by soil moisture tension and salt concentration. *Plant Physiol.* **20**:106–132.

Wadleigh, C. H., and M. Fireman. (1949) Salt distribution under furrow and basin irrigated cotton and its effect on water removal. *Soil Sci. Soc. Amer. Proc.* (1948) **13**:527–530.

Whittig, L. D., and P. Janitzky. (1963) Mechanisms of formation of sodium carbonate in soils. I. Manifestations of biological conversions. *Jour. Soil Sci.* **14**:322–333.

7 *Nitrogen*

According to geochemical theory (Stevenson, 1965b), the nitrogen now found in the atmosphere was present originally as ammonium compounds and nitrides in the solid matter of the earth. With development of heat in the earth, nitrogen was driven off into the atmosphere, where it existed mostly as ammonia. When the atmosphere became enriched with oxygen as a result of photosynthesis, the ammonia became oxidized to elemental nitrogen.

Although nearly 80% of atmospheric gas is nitrogen, nitrogen in the atmosphere is estimated at only 2% of the total nitrogen of the earth. Almost all the remainder is still present in rocks. The concentration of the original nitrogen in rocks, however, is extremely low.

In comparison with the original rocks, soils have been greatly enriched in nitrogen; nevertheless, the nitrogen present in soils is a negligible part of the total. The bulk of the soil nitrogen is in organic forms and presumably has been accumulated there from the elemental form in the atmosphere by fixation processes, the most important of which are thought to be of biological nature.

Circumstantial evidence indicates that most of the combined nitrogen absorbed from soil by plants under natural conditions has been accumulated in the soil from the atmosphere and is not the original nitrogen still present in the mineral matter. In agriculture, an important part of the nitrogen used by plants is sometimes supplied by nitrogen fertilizers. These are derived mainly from elemental nitrogen by industrial chemical fixation processes.

It is probably no exaggeration to state that growth of agricultural plants is limited more often by a deficiency of nitrogen than of any other nutrient. One reason for this is the relatively large requirement of plants for nitrogen. Viets (1965) calculated that plants contain more atoms of nitrogen than of any other element except hydrogen derived from soils. Although deficiency of water is more important than that of nitrogen, this deficiency is one of water itself for providing a suitable internal environment for plant processes and not a deficiency of the hydrogen plants derive from it for building organic substances.

The literature on nitrogen is extensive. Monographs dealing with subject matter related to this chapter include one on soil nitrogen edited by Bartholomew and Clark (1965), one on nitrogen in agriculture and nutrition by Balz et al. (1961), and one on nitrogen metabolism (which includes sections on soil nitrogen and nitrogen fixation but puts principal emphasis on nitrogen metabolism in higher plants) edited by Ruhland (1958). All these monographs contain articles by many authors, and reference will be made to some of these contributions in connection with individual topics throughout this chapter.

Content in Soils

The plowed layer of the majority of cultivated soils contains between 0.02 and 0.4% nitrogen by weight. How much is present in a particular case is determined largely by the general influence of climate and the type of vegetation controlled thereby, as these are modified by local influences of topography, parent material, and activities of man, and by the length of time these different factors have been in operation.

Effect of Climate and Vegetation

Climate plays a dominant part in determining the nitrogen content of soil through the influence of temperature and water supply on the activities of plants and microorganisms. The classical investigation of the relationship between climate and soil nitrogen content is that made by Jenny (1930). His results in Fig. 7.1 show a marked decrease in soil nitrogen content with an increase in annual temperature from 0°C in Canada to 20°C in southern United States. He considered (Jenny 1941, p. 213) that the effect of temperature on microbial activity was primarily responsible because, as indicated by the similarity of the yields of prairie hay over the area in question, there apparently was no pronounced effect of temperature on the quantity of organic material that was added to the soil each year prior to cultivation. In fact, his examination of the literature showed that some of the highest nitrogen percentages in mineral soils are found where the annual yield of vegetation is extremely low as a result of the low temperature.

Other factors remaining the same, the content of nitrogen in soil increases with the water supply. The vertical displacement of the nitrogen-temperature curves in Fig. 7.1 for the soils of the semihumid and humid grasslands is an indication of this relationship. The cause lies partly in the rate at which vegetation is produced and partly in the rate at which it is decomposed. Up to a limit, both the rate of production

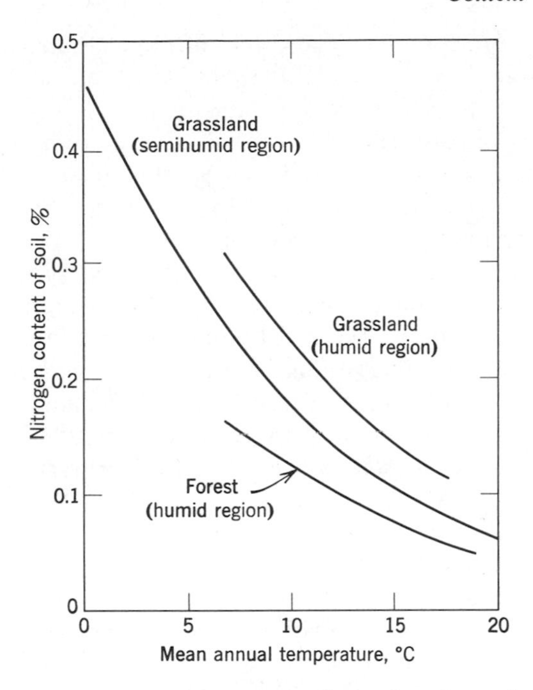

Figure 7.1. Soil nitrogen versus temperature relationships for soils developed under different conditions of climate and vegetation in the United States. (Jenny, 1930)

of vegetation and the content of nitrogen in soil increase with the water supply. Presumably, the increase in soil nitrogen in this range is due primarily to the rate of production of vegetation, with associated biological nitrogen fixation, and not to a decrease in rate of decomposition. Where the water supply exceeds that corresponding to the maximum rate of production of vegetation, however, the further increase that occurs in soil nitrogen is probably attributable to the effect of water supply in decreasing the rate of decomposition. Organic soils, which develop in shallow water, represent the extreme condition.

Jenny (1930) expressed the combined effects of temperature and water supply by the equation

$$N = 0.55e^{-0.08T}(1 - e^{-0.005H})$$

where N is the nitrogen percentage of the surface soil, e is the base of natural logarithms, T is the mean annual temperature in degrees centigrade, and H is the "humidity" factor. The humidity factor is the ratio of mean annual precipitation in millimeters to the absolute water-vapor saturation-deficit of the air in millimeters of mercury. It is a measure of effective precipitation. The values of the constants in the equation are those for grass vegetation on upland soils of sandy loam, loam, and silt loam texture. The equation states that, with any given humidity factor, the nitrogen content decreases with increase in annual temperature and approaches zero as a limit. At any given temperature, the nitrogen content increases with increase in humidity factor and approaches a limit fixed by the temperature.

The effects illustrated in Fig. 7.1 were observed as gradual changes over long distances. Similar effects may be found within relatively short distances where climate is modified by altitude or by direction of slope. Harradine and Jenny (1958) published a paper describing relationships between nitrogen content of soils and the rainfall and temperature at different altitudes in California. At a given altitude, the highest nitrogen contents were found in soils on north and northeast slopes and the lowest on southwest slopes.

The effect of climate is indirect in that it modifies the activities of higher plants and microorganisms that produce the nitrogen transformations. Although vegetation is related to climate, it is not controlled completely by climate and thus may be investigated as an independent factor in determining soil nitrogen. As illustrated by Fig. 7.1, for example, the nitrogen content at a particular temperature and under a given climatic regime is greater in soils developed under grass than under forest vegetation. Effects of a more local nature may be observed within a distance of a few kilometers [see, for example, the paper by White and Riecken (1955)] or within a few meters, as, for example, at the boundary between one type of natural vegetation and another where other factors were originally similar.

Effect of Topography

Marked variations in nitrogen content of soils occur within local areas in response to differences in topography. In effect, the local climate, primarily the effective precipitation, varies with slope. Steep slopes are relatively dry because of runoff. Consequently, little vegetation is produced, and little nitrogen is stored in the soil. The runoff, in turn, causes rapid erosion of the surface soil, which generally contains the highest percentage of nitrogen. On the other hand, accumulation in depressions of a part of the water from surrounding areas increases the effective precipitation, the production of vegetation, and the storage of nitrogen in the soil. These relationships were discussed and illustrated by Ellis (1938).

Where agricultural plants are dependent on soil nitrogen, the differences in nitrogen availability associated with topographic features often have significant effects. Data published by Engelstad et al. (1961) from experiments on corn on loess-derived soils in southwestern Iowa illustrate the effect of topography on both nitrogen availability and water availability. In a year with sufficient rainfall for corn on all slopes, the increase in yield from nitrogen fertilization became smaller with a decrease in slope, presumably because soil nitrogen supply increased with a decrease in slope. In a dry year, the increase in yield from nitrogen fertilization was small regardless of the slope, indicating that nitrogen was not an important limiting factor; but the yields increased with a decrease in slope, presumably because water availability increased with decreasing slope.

Effect of Mineral Components

The mineral portion of soil influences the environment for plants and microorganisms through the air and water relations and fertility of soils. Furthermore, the mineral portion of soil tends to combine with the

Table 7.1. Average Nitrogen Content of Upland Soils Differing in Texture in Northeastern Iowa (Walker and Brown, 1936)

Soil Texture	Nitrogen Content of Soil, %
Sand	0.027
Fine sand	0.042
Sandy loam	0.100
Loam	0.188
Silt loam	0.230

organic fraction, as will be mentioned subsequently in the section on stability of nitrogen in soils. Probably all these factors are involved in the influence of the mineral portion of the soil on the nitrogen content. Table 7.1 shows the magnitude of the mineral-fraction effects in the environment of northeastern Iowa. These data illustrate the usual tendency for soils with the finer texture to have the higher content of nitrogen.

Profile Distribution

Figure 7.2 shows the relationship of nitrogen content to depth in three soil profiles in Iowa, selected to illustrate the patterns associated

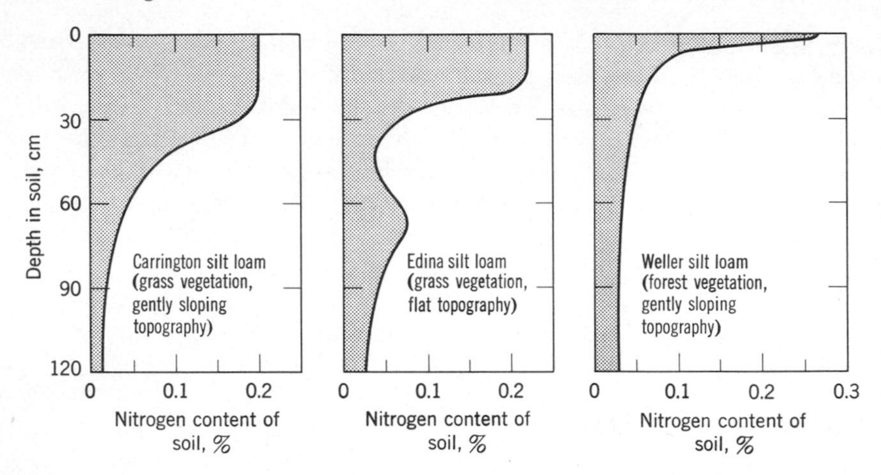

Figure 7.2. Vertical distribution of nitrogen in three soil profiles. (Pearson and Simonson, 1939)

with different conditions. Although the maximum nitrogen content occurs in the surface layer of all these soils, the distribution with depth differs among soils. Weller silt loam, developed under forest vegetation, has a greater relative concentration of nitrogen near the surface than do the other two soils, both developed under grass. This difference probably results from the addition of a greater proportion of the organic matter on the surface of the Weller profile. The irregular curve in the Edina silt loam profile is connected with occurrence of a gray A2 horizon and a B horizon high in clay. Figure 1.31 gives a picture and description of this same profile.

The depth of accumulation of nitrogen varies in accordance with the accumulation of organic matter. In soils of arid and semiarid regions where water penetration is shallow, the depth to which organic matter and nitrogen accumulate is also shallow. In most soils, however, the bulk of the organic matter and nitrogen is found in the surface 50 cm of the profile.

Forms in Soils

Elemental Nitrogen

Elemental nitrogen (N_2) is present in gaseous form in the soil atmosphere and in dissolved form in the soil water. In dry soils, it is present in adsorbed form on the solid surfaces. Elemental nitrogen is not ordinarily considered, however, because it is not of direct significance to plants that do not form symbiotic associations with nitrogen-fixing microorganisms and because it is present in ample supply for both symbiotic and nonsymbiotic nitrogen fixation.

Inorganic Combinations

In inorganic combined form, nitrogen occurs in soils as nitrous oxide (N_2O), nitric oxide (NO), nitrogen dioxide (NO_2), ammonia (NH_3), ammonium (NH_4^+), nitrite (NO_2^-), and nitrate (NO_3^-). The first four of these are gases and ordinarily are not present in concentrations great enough to be detected. The last three are ionic forms found in the soil solution. Nitrite and nitrate occur exclusively or almost exclusively as freely diffusible ions in the soil solution. Most of the ammonium occurs in exchangeable and nonexchangeable forms. Only a little occurs in ionic form in the soil solution. Ordinarily the ammonium in exchangeable and soil-solution forms and the nitrite and nitrate collectively constitute less than 2% of the total nitrogen in soils. These forms are of great qualitative importance, however, because they are the forms used by plants. Papers by Cheng and Bremner (1965) and Bremner (1965) may be consulted for information on analysis of soils for these constituents.

The fact that some soils have the property of retaining ammonium in nonexchangeable form against extraction with potassium chloride or other salts has been known for many years, but little consideration was given to the possibility that substantial amounts of nonexchangeable ammonium may exist in soils until Rodrigues (1954) found that treatment of soils with a mixture of sulfuric and hydrofluoric acids caused liberation of considerable quantities of ammonium from certain soils. Hydrofluoric acid dissolves silicate clay minerals which had been implicated in previous research as being responsible for retention of ammonium in nonexchangeable form.

Indications now are that nonexchangeable ammonium may occur in three general forms. The first is crystalline compounds formed from constituents in solution. Tamini et al. (1963) obtained evidence from X-ray diffraction for the formation of a crystalline ammonium, iron, aluminum phosphate (taranakite) in soils treated simultaneously with ammonium and soluble phosphate. The second is inclusion of ammonium in primary silicate minerals such as micas and feldspars, where it appears to occupy sites normally occupied by potassium (Adams and Stevenson, 1964). The third is interlayer positions in minerals such as vermiculite and illite in sites where potassium also may be held in nonexchangeable form. The third form is probably of greatest importance in soils containing substantial quantities of vermiculite and illite, and most of the research has been done on these types of minerals and on soils containing them.

In analysis of samples of a number of soil profiles from Illinois and elsewhere by a method that would probably include the third form and part of the second, but not the first, Stevenson and Dhariwal (1959) found quantities of nonexchangeable ammonium ranging from less than 0.05 to about 1.7 m.e./100 g of soil. The values sometimes decreased with depth in the soil but more often remained about the same or increased with depth. Nonexchangeable ammonium accounted for 4 to 8% of the total nitrogen present in samples of various surface soils. With the use of a similar method of analysis on soil profiles from Saskatchewan, Hinman (1964) found quantities of nonexchangeable ammonium ranging from 7% of the total nitrogen in the surface soil to as much as 58% of the total nitrogen at a depth of 91 to 122 cm. The total amount of nonexchangeable ammonium to this depth ranged from 2.9 to 5.2 metric tons/hectare.

Organic Combinations

The organic nitrogen added to soils in plant and animal residues is largely proteinaceous in nature. The microbial attack to which these materials are subjected in soils probably results in nearly complete disappearance of the original protein and its partial replacement by microbial protein, with the remainder of the nitrogen being changed to inorganic or elemental forms. To what extent the nitrogen present as microbial protein remains as such or is transformed to still other forms is not known. Proteins certainly exist in soil, but the only isolation of protein that appears to have been accomplished is the separation and crystallization of a mixture of substances having urease activity (Briggs and Segal, 1963). If the amounts of protein present exceed those occurring in microbial tissue and undecomposed plant and animal residues, they seem to be attached firmly to other substances in such a way that their isolation is prevented.

One indication that proteins are present is provided by the presence of a number of amino acids in substantial quantities in extracts obtained by heating soils with 6-normal hydrochloric acid (Bremner, 1965a, 1965c, 1966; Stevenson, 1965a). The amino acids released are of the same types that occur in proteins, plus a few additional ones. If proteins were present in soils, this hydrolysis procedure would cause their breakdown to amino acids. About 20 to 40% of the total nitrogen of surface soils can be accounted for by amino acids. In addition to these amino acids liberated on hydrolysis, soils contain relatively small quantities of free amino acids that may be removed to varying degrees by water, ethyl alcohol, ammonium acetate, and other extractants. These amino acids probably exist in part free in the soil and in part in microbial tissue. The amounts are negligible in terms of accounting for the nature of soil organic nitrogen, but they may be of significance to plants because plants can absorb at least some of them. Paul and Schmidt (1961) identified fifteen amino acids in ammonium acetate extracts of soil and found that the total amount of those identified was equivalent to about 0.5 μg of nitrogen per gram of untreated soil and 15 μg/g of soil that had been treated with glucose and nitrate and incubated 3 days.

A second form of evidence for the existence of protein in soils is the liberation of amino acids from soil organic matter by the enzyme papain. About one-third of the total amount of amino acids has been released from extracted soil organic matter upon digestion with this enzyme (Scharpenseel, 1962; Scharpenseel and Krausse, 1962). Papain is effective in splitting the peptide bonds that link amino acids in proteins.

Another theory is that the bound amino acids released upon acid hydrolysis are combined with aromatic ring substances (Kononova, 1961; Flaig, 1950; Swaby and Ladd, 1962). According to this theory, polyphenols present in plants and microorganisms are oxidized enzymatically to quinones; amino acids and quinones then condense to form large three-dimensional molecules of varying size, with amino acids of different kinds interspersed in random order. The first steps may be represented as

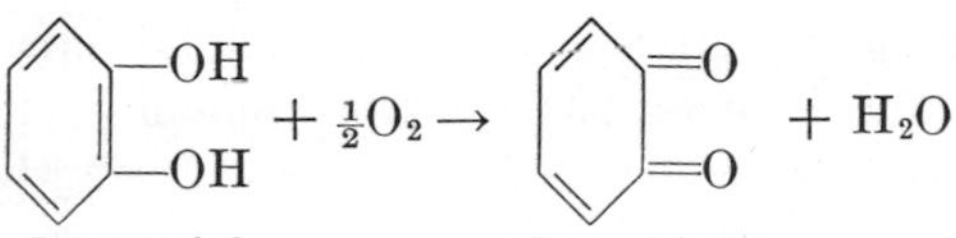

$$\text{Pyrocatechol} + \tfrac{1}{2}O_2 \rightarrow \text{o-Benzoquinone} + H_2O$$

Pyrocatechol o-Benzoquinone

and

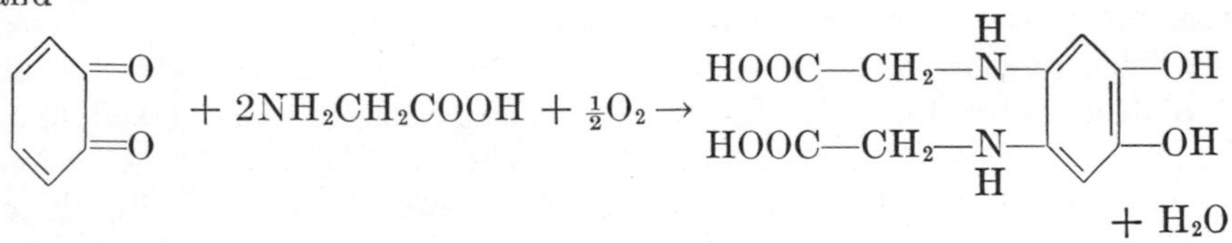

$$\text{o-Benzoquinone} + 2NH_2CH_2COOH + \tfrac{1}{2}O_2 \rightarrow (HOOC\text{—}CH_2\text{—}NH\text{—})_2C_6H_2(OH)_2 + H_2O$$

o-Benzoquinone Amino-acetic acid (glycine)

Further condensation then occurs through the =NH groups to form larger molecules. Swaby (1962a) suggested the structure

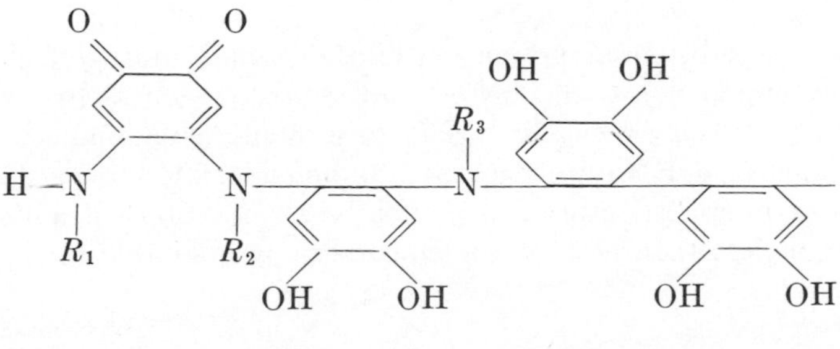

[Structure: quinone ring with two =O groups, bearing H—N—R_1 and N—R_2 groups; N—R_2 linked to a dihydroxyphenyl ring (OH, OH), which is linked to N—R_3, linked to a dihydroxyphenyl ring (OH, OH), linked to a further dihydroxyphenyl ring (OH, OH)]

where R_1, R_2, and R_3 represent different amino acids or other nitrogenous components. Substances of this type have been formed *in vitro* and in cultures of microorganisms, and the parent phenols and amino acids have been recovered by acid hydrolysis. Incidentally, this theory provides a source of the carboxyl groups that have been identified as the sites of much of the cation-exchange capacity of soil organic matter.

Still another theory is that some of the bound amino acids in soils are present as mucopeptides (complexes of amino acids and amino sugars) and teichoic acids (organic phosphorus polymers containing the amino acid alanine). These substances are important constituents of bacterial cell walls. This theory was discussed by Bremner (1965c).

In addition to amino acids, soil organic nitrogen contains amino sugars (hexoseamines), including glucosamine and galactosamine. Like the amino acids, the amino sugars are bound in the soil and are released by acid hydrolysis (Bremner, 1965a, 1965c, 1966; Stevenson, 1965). The nitrogen present in these forms has been found to comprise 5 to 10% of the total nitrogen in surface soils. Traces of other nitrogenous organic compounds have been identified in soil extracts (Bremner, 1965c, 1966).

The sum of 20 to 40% amino-acid nitrogen and 5 to 10% amino-sugar nitrogen amounts to 25 to 50% of the total nitrogen of surface soils accounted for as identifiable organic compounds. If 5% of the total nitrogen is taken as an average figure for nonexchangeable ammonium in surface soils, a substantial balance of 45 to 70% of the nitrogen evidently remains to be accounted for.

A number of investigations have shown that the content of organic nitrogen in soil may be increased considerably by reaction with ammonia. This process is of practical significance in connection with use of anhydrous ammonia as a fertilizer. The organic nitrogen of soil may be increased also by reaction with nitrite. The availability of the nitrogen to plants is reduced by such reactions. Literature dealing with these reactions was reviewed by Bremner (1965c), Mortland and Wolcott (1965), and Nõmmik (1965).

Stability in Soils

Although a substantial proportion of the organic nitrogen in soils is present as amino acids, the behavior of soil nitrogen is by no means analogous to that of either amino acids or proteins. The minute quantities of free amino acids in soils appear to be associated with metabolic activities of microorganisms and plants. Small additions of amino acids disappear in less than a week under conditions favorable for microbiological activity (Schmidt et al., 1957; Driel, 1961). Similarly, proteins are rapidly decomposed upon addition to soil (Kelley, 1914). In contrast, the nitrogen of soils resists microbial attack and is changed to inorganic forms at the rate of only 1 or 2% annually under temperate or cool climates.

The current theories to account for the stability of the soil organic nitrogen may be considered in three groups. The first group relates to the chemical nature of soil nitrogen. According to one of the theories discussed in the preceding section, polyphenols, amino acids, and other nitrogenous substances are condensed into large molecules of heterogeneous composition in soil organic matter. Swaby and Ladd (1962) advanced essentially the following reasons for supposing that this sort of structure would resist decomposition: (1) The close-linked, three-dimensional pattern would confine the action of enzymes to the outer surface. (2) The attachment of each unit to more than one other unit by covalent bonds that might be rather different would confer stability because enzymes tend to be specific, and a single enzyme hence might not be capable of splitting off a unit such as an amino acid. If more than one enzyme were needed, assemblage of the proper ones in the juxtaposition required to produce the splitting would occur infrequently for reasons of probability and the physical difficulty associated with the relatively large size of the enzyme molecules. The same reasons could be advanced for stability of complexes formed between polyphenols and proteins in a reaction analogous to that in the tanning of leather (Handley, 1954; Davies et al., 1964), with the exception that the peptide bonds would remain and presumably would be more labile than the bonds between amino acids and polyphenols postulated previously.

The second group of theories has to do with the physicochemical and physical condition of the nitrogen. As mentioned in connection with soil structure in Chapter 1, there is evidence for interaction of organic and mineral constituents of soils. Demolon and Barbier (1929) found that treatment of a mixed suspension of ammonium "humate" and ammonium clay with enough potassium chloride to flocculate the clay caused some of the organic matter to be carried down with the clay. The proportion of the organic matter carried down increased with decreasing *p*H of the medium. In the absence of clay, the organic matter remained in suspension or solution. More recently, a number of investigations (for example, Ensminger and Gieseking, 1939; McLaren and Peterson, 1961; Talibudeen, 1955; Armstrong and Chesters, 1964) have shown that proteins are adsorbed by clays. The proteins penetrate between layers of montmorillonite-type clays and cause wide spacings. Armstrong and Chesters (1964) reported spacings up to 64 Å. The cation-exchange capacity of the clay is reduced by the combination, and there is some evidence that amino groups of the proteins act as cations to satisfy the net negative charge on the clay (Ensminger and Gieseking, 1941). Breakdown of protein (Pinck et al., 1954) and plant residues (Allison et al., 1949) has been found to be retarded by the presence of clay. Moreover, enzymatic hydrolysis of protein has been found to be decreased to a greater extent in the presence of bentonite (montmorillonite), having high cation-exchange capacity, than in the presence of kaolinite, having low exchange capacity (Ensminger and Gieseking, 1942). Work of Jackman (1964) on accumulation of organic carbon and nitrogen in soils under field conditions in New Zealand indicates that the clay mineral "allophane" is relatively effective in stabilizing organic matter against decomposition.

Rovira and Greacen (1957) proposed the theory that occurrence of part of the soil organic matter in locations physically inaccessible to microorganisms stabilizes the organic matter against breakdown. They noted that up to 60% of the pore space in some soils of fine texture consists of pores less than 1 μ in diameter, a size too small to permit entry of microorganisms. In one of their experiments, they compressed samples of an air-dried and remoistened clay soil into an annulus between two concentric metal rings and then subjected the soil to a shearing force by a flat metal plate that was turned while pressing against the surface of the soil. The soil was then removed from the shearing device and was broken by hand into fragments less than 2 mm in diameter. Oxygen uptake was measured on a small sample at 30°C. The remainder of the sample was stored at the same temperature and sheared by the same method on the following day. The procedure was repeated on a third day. Oxygen uptake by a sample that had not been subjected to shearing was measured as a control. The results in Fig. 7.3 show that a marked increase in oxygen uptake occurred after each shearing treatment. Supplementary tests showed that (1) uptake of oxygen was inhibited completely where mercuric chloride was added, (2) incubation of the sheared and unsheared soil in pure oxygen instead of air increased oxygen uptake to only a small extent, and (3) uptake of oxygen during incubation of the soil after treatment with glucose was the same in sheared and unsheared soil. The experiments thus verify that the enhanced uptake of oxygen by the sheared soil was a biological effect and that the effect was due to an increase in available substrate and not to an increase in aeration associated with the shearing treatment.

The foregoing data by Rovira and Greacen (1957) provide no direct information on stability of the nitrogen, but other experiments indicate that exposure of new surfaces increases the susceptibility of soil nitro-

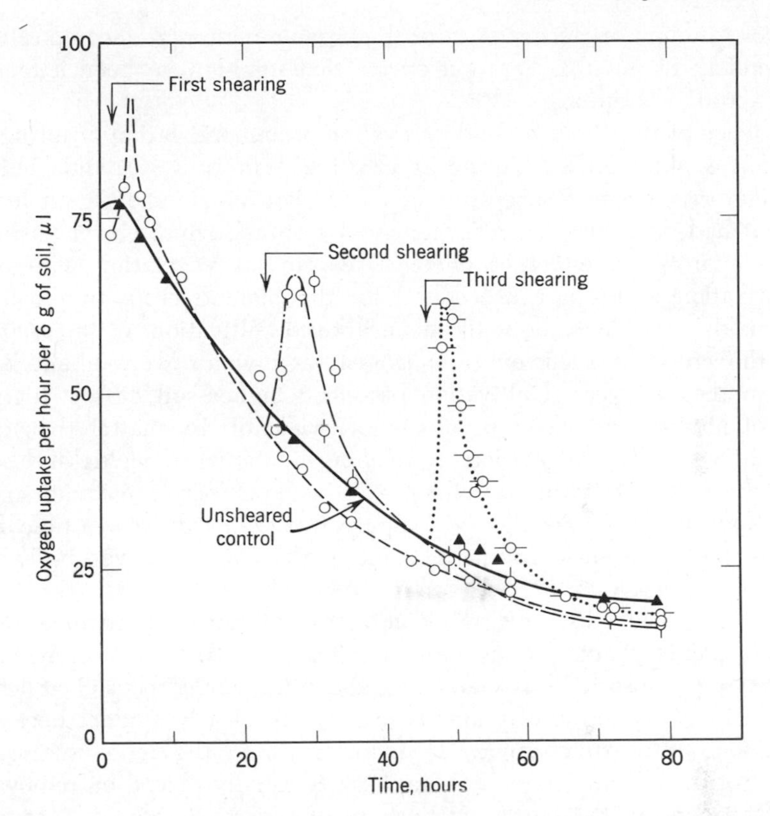

Figure 7.3. Uptake of oxygen by a clay soil during incubation in a moist condition at 30°C with and without successive shearing treatments designed to expose new surfaces. (Rovira and Greacen, 1957)

genous substances to attack by microorganisms. For example, Seifert (1963) found that the rate of production of nitrate was increased where soil aggregates greater than 2 mm in diameter were frozen before incubation. The effect of freezing increased with aggregate size. Similarly, when large aggregates were crushed without freezing before incubation, the rate of nitrate production was increased (Seifert, 1963a). Waring and Bremner (1964) found that the rate of change of soil nitrogen to extractable mineral form during incubation under both aerobic and anaerobic conditions was increased as a result of grinding air-dry soil to finer sizes before incubation. With 52 soils, decreasing the mesh size from less than 10 to less than 80 produced an increase of from 25 to 124% in the mineral nitrogen produced under anaerobic conditions. Use of anaerobic conditions in this investigation provides evidence for an effect of grinding on accessibility of the organic matter to decomposition independent of possible aeration effects that might have been a factor under aerobic conditions.

Evidence of the effect of field cultivation on mineralization of nitrogen in soil was obtained by During et al. (1963) in New Zealand. These investigators grew two successive crops of chou moellier on a silt loam soil that had previously been in perennial ryegrass, white clover pasture and compared the effect of chemical control of vegetation with use of cultivation to prepare a seedbed for the planted crop. In the first year, yields were the same with chemicals and cultivation. In the second year, the crop was deficient in nitrogen; and yields were about half those in the first year. Cultivation produced higher soil nitrate, higher yield of nitrogen in the crop, and higher yield of dry matter than did chemical control of vegetation. A relationship between yields and soil nitrate was found within treatments as well as between treatments, which indicates that the nitrogen supply was primarily responsible for the observed differences in yields. Some of the data are given in Table 7.2.

The third mechanism of stabilization of soil nitrogen involves the action of plants. Plants produce organic nitrogen in the soil from mineral nitrogen they absorb, thus stabilizing the nitrogen in the soil at least temporarily. This action of plants is usually not clearly evident because the mineral nitrogen changed to organic form in the roots is always less than the total absorbed, and emphasis is usually placed on removals of soil nitrogen in the harvested portions of plants. If, however, conditions should be such that the capacity of the plants to generate organic

Table 7.2. Nitrate-Nitrogen in Soil and Yields of Dry Matter and Nitrogen in Chou Moellier in the Second Year Following a Perennial Ryegrass, White Clover Pasture (During et al., 1963)

Cultivation[1]		Nitrate-Nitrogen per Gram of Soil on Oct. 25, μg	Crop Yield per Hectare, kg	
First Year	Second Year		Dry Matter	Nitrogen
None	None	16	3470	39
Cultivation	None	12	2770	30
Cultivation	Cultivation	36	4850	53

[1] Cultivation consisted of plowing, rolling, discing, levelling, and harrowing. Chemicals were applied to control the vegetation where no cultivation was used.

nitrogen in the soil exceeds the production of mineral nitrogen, application of fertilizer nitrogen may increase the organic nitrogen content of the soil. The stabilizing influence will then be clearly evident. The data in Table 7.3, obtained in a field experiment in South Africa, indicate a modest increase in content of soil nitrogen from continued fertilization of grassland with nitrogen. Similar numerical data were obtained by Theron (1965) in another experiment, also in South Africa, in which nitrogen fertilizer was applied four times annually over a period of

Table 7.3. Content of Total Nitrogen and Organic Carbon in a Soil in South Africa after Growth of Perennial Grasses (*Chloris gayana* and *Paspalum dilatatum*) for Seven Years with Annual Applications of Different Quantities of Nitrogen as Ammonium Sulfate (Haylett and Theron, 1955)

Nitrogen Applied per Hectare Annually, kg	Total Nitrogen Content of Soil, %	Organic Carbon Content of Soil, %
0	0.105	1.61
53	0.110	1.66
106	0.113	1.75
212	0.117	1.65

years to virgin veld. In this experiment, the botanical composition of the grasses was changed by the nitrogen fertilizer so that high-nitrogen plots carried mostly species different from those on low-nitrogen plots. Further discussion of stabilization of nitrogen in grassland will be found in the section on soil-plant interactions.

Mineralization and Immobilization

The principal source of nitrogen used by plants that do not fix nitrogen in symbiosis with microorganisms is the mineral forms of nitrogen in soils, not the organic forms that constitute the bulk of the soil nitrogen. In unfertilized soils, the mineral forms of nitrogen are derived almost entirely from decomposition of organic nitrogenous compounds. Because of the great importance of this process, its complexity, and the many factors that may affect it, much research has been done in an attempt to discover the underlying principles. Reviews of research on this subject include those by Harmsen and van Schreven (1955), Harmsen and Kolenbrander (1965), and van Schreven (1965).

Microbiological Processes

The various steps in the process of mineral nitrogen formation and transformation in aerobic soils follow the order

$$\text{Organic nitrogen} \xrightarrow[\text{(Slow)}]{\substack{\text{Nonspecialized}\\ \text{organisms}}} NH_4^+ \xrightarrow[\text{(Fast)}]{\substack{\text{Specialized}\\ \text{autotrophs}}} NO_2^- \xrightarrow[\text{(Very fast)}]{\substack{\text{Specialized}\\ \text{autotrophs}}} NO_3^-.$$

These are all microbiological processes. Generally, the rate-controlling step in the over-all process is the conversion of organic nitrogen to ammonium rather than the conversion of ammonium-nitrogen to nitrite and then to nitrate. Consequently, ammonium and nitrite usually do not accumulate in labile form but are converted rapidly to nitrate, which is the principal form of inorganic combined nitrogen absorbed by plants.

The organic-nitrogen → ammonium-nitrogen transformation does not require a specific kind of organism but is carried out by many different organisms and hence is said to be effected by nonspecialized organisms. There are indications that nitrite or nitrate can be produced in traces by a number of microorganisms. The ammonium → nitrite → nitrate transformations are said to be carried out by specialized organisms, however, because the principal oxidation does seem to be done by such organisms. Greatest importance is attached to *Nitrosomonas,* which oxidizes ammonium to nitrite, and to *Nitrobacter,* which oxidizes nitrite to nitrate. The term "autotroph" applied to these organisms signifies that they do not depend on oxidation of organic compounds for energy. They are capable of growing on inorganic substrates, using carbon dioxide as the source of carbon, and deriving the energy needed for reduction of carbon dioxide from the energy released in the oxidation of ammonium or nitrite, as the case may be. These processes require elemental oxygen and hence do not take place under anaerobic conditions. Under anaerobic conditions, the over-all process stops with the step of ammonium formation.

The population of *Nitrosomonas* and *Nitrobacter* varies with the supply of substrate. According to Alexander (1965), the numbers of these organisms per gram of unfertilized soil are rarely more than 100,000 and frequently less than 100, particularly if the soil is acid. If the concentration of substrate is increased, however, the numbers rise and may exceed one million per gram. This variation of population with substrate is one of the reasons for supposing that *Nitrosomonas* and *Nitrobacter* are predominant among microorganisms involved in ammonium oxidation.

The production of mineral nitrogen in soil is accompanied by the reverse process, in which nitrate and other mineral forms of nitrogen are converted back into organic forms. This reverse process is nonspecialized and is caused by incorporation of nitrogen in organic compounds synthesized in the course of metabolism of the microorganisms present. Hence, if the rate of mineral nitrogen production is calculated from the difference between the amounts of mineral nitrogen present in soil at two different times, the figure obtained actually represents the net result of two opposing processes and not the absolute rate of production of mineral nitrogen. Work with tracers has verified qualitatively that nitrogen mineralization and immobilization proceed concurrently. Attempts thus far to estimate quantitatively the absolute rates of each process from tracer data, however, have involved assumptions that are not in accord with known facts. The results, therefore, must be looked upon with some reservation.

Nomenclature

The term *ammonification* has been used for the first step of the microbiological process described in the preceding section, namely, transformation of organic nitrogen to ammonium. *Nitrification* is the term applied to oxidation of ammonium to nitrite and nitrate. Particularly in older literature, however, nitrification sometimes means formation of nitrite and nitrate from any other source of nitrogen. Another term sometimes used is *nitrate production,* which means formation of nitrate from any other source of nitrogen. If the sum of ammonium and nitrite remains relatively constant, the increase of nitrate-nitrogen over a period of time is about equal to the net increase in mineral nitrogen.

Two terms commonly used in current literature on nitrogen transformations in soils are *nitrogen mineralization* and *nitrogen immobilization.* Nitrogen mineralization means the change of organic nitrogen to mineral form without distinction among the several forms of mineral nitrogen as end-products. Nitrogen immobilization means the change of mineral nitrogen to organic form. Mineralization and immobilization are usually considered to be microbiological processes, but higher plants carry on a process analogous to immobilization. *Assimilation* is sometimes used in the same sense as immobilization. In older literature, *denitrification* sometimes means any loss of nitrite plus nitrate not caused by leaching or absorption by higher plants. In this sense, immobilization is one kind of denitrification. Currently, however, denitrification is used only to refer to processes by which gaseous forms of nitrogen are produced from nitrite and nitrate. In this sense, immobilization and denitrification are distinct processes. Although *nitrogen fixation* involves a change of nitrogen to organic form, the nitrogen is derived from elemental rather than so-called mineral or combined inorganic forms. Nitrogen fixation and immobilization are thus distinct processes. Ammonium fixation in nonexchangeable inorganic form is not considered as either immobilization or nitrogen fixation, and reaction of ammonium and nitrite with soil organic matter to produce nonexchangeable organic nitrogen may or may not be considered as immobilization according to whether one includes strictly chemical reactions or limits use of the term to reactions brought about biologically.

Factors Affecting Mineralization and Immobilization

Total Nitrogen. The total nitrogen content of soils provides a measure of the quantity of substrate undergoing decomposition. Experience has shown that the net quantity of nitrogen mineralized in samples of soil upon incubation under conditions of temperature and water content suitable for microbiological activity (say, 25°C and a matric suction of 0.33 bar) is approximately proportional to the total quantity of nitrogen. The total quantity of nitrogen or substrate is not the sole controlling factor, however, as exemplified by Allison and Sterling's (1949) finding that the coefficients of correlation of nitrogen mineralized during incubation of a group of soil samples with the total nitrogen content of the samples ranged from 0.59 to 0.84 in different phases of an investigation of the association of these two variables. The principal reason for the failure of such correlation coefficients to approach unity, indicating a perfect linear relationship, is probably that the nature of the substrate and other conditions in the different soils are not sufficiently similar. The analytical measurements can be made in a reasonably accurate and precise manner.

Substrate Composition. The subject of substrate composition will be considered first from the standpoint of the behavior of substances added to soils. This aspect has received most attention and is of much practical significance. Composition of soil organic matter will be considered later.

Organic materials added to soils are attacked enzymatically by microorganisms and are split into small molecular or ionic units, which, together with the organic and mineral constituents already present, constitute the nutrient medium from which microorganisms derive energy for metabolism and components for building cellular tissue. Mineralization and immobilization of nitrogen occur continuously in microbial metabolism, and the magnitude and direction of the net effect is greatly influenced by the nature and quantity of organic material added. Figure 7.4 illustrates schematically the effect of addition of two different kinds of substrates on the mineral-nitrogen, organic-nitrogen balance in soils.

Section (*a*) of the figure shows the situation in uncropped soils that remain in warm, moist condition without addition of organic material or loss of mineral nitrogen by leaching. The content of mineral nitrogen gradually increases because the net quantity of nitrogen present in microbial tissue remains approximately the same while a small portion of the large excess of soil organic matter undergoes gradual decomposition. Figure 7.4*b* shows how the course of events is changed by addition

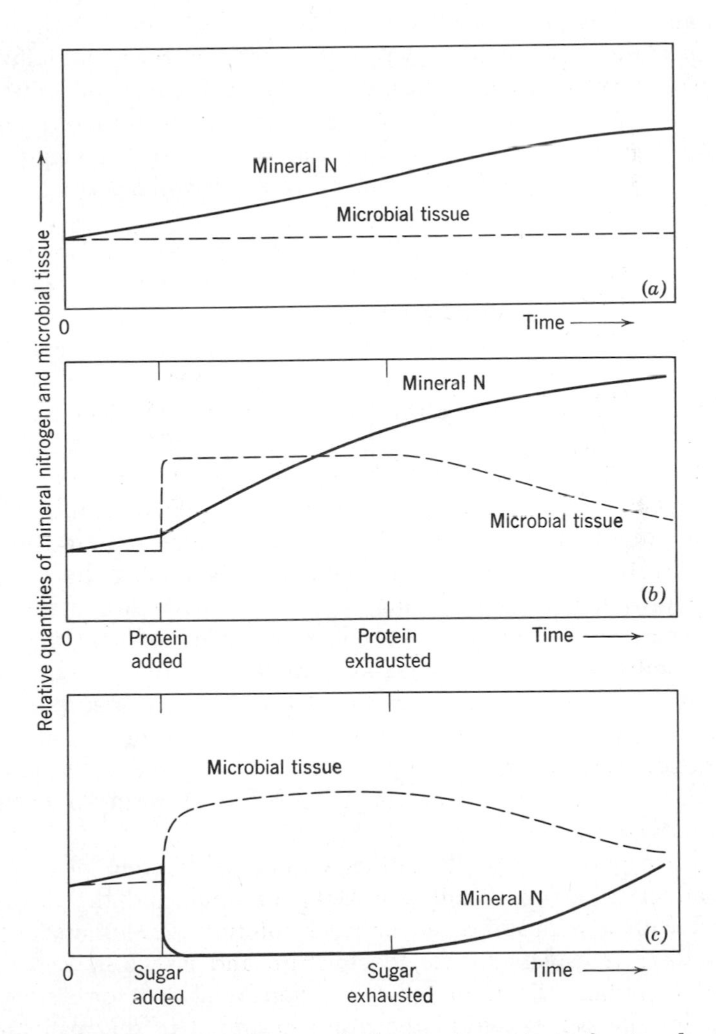

Figure 7.4. Schematic representation of changes in content of mineral nitrogen and microbial tissue with time in fallow soil. (*a*) No organic matter added. (*b*) Protein added. (*c*) Sugar added.

of proteinaceous material. The microbial population increases rapidly to a limiting value determined by the environment. The content of mineral nitrogen continues to increase beyond the time when the protein is exhausted because of liberation of mineral nitrogen from the soil and because of a decrease in the microbial population. Eventually, the mass of microbial tissue in the soil approaches its initial value, and the content of mineral nitrogen in the soil remains above that in case (*a*). In section (*c*) of Fig. 7.4, addition of carbonaceous material provides an energy supply deficient in nitrogen. The microbial population then increases, but at the expense of mineral nitrogen initially present and subsequently produced in the soil. The content of mineral nitrogen remains at a low level until the carbonaceous material is exhausted and the microbial population decreases in size. If the length of time is great enough, the content of mineral nitrogen will surpass that present initially.

Changes of the type described depend on the relative availability of carbonaceous and nitrogenous food material for the microorganisms. If the nitrogenous material is in excess, nitrogen is mineralized; if the carbonaceous material is in excess, nitrogen is immobilized. Although other mineral elements required by microorganisms have on occasion been found to affect mineralization and immobilization of nitrogen, the availability of these elements in most plant residues and soils is such that their addition does not greatly alter the behavior of the nitrogen.

The relative availability of carbon and nitrogen in organic materials for the growth of soil microorganisms is usually represented by the ratio of carbon to nitrogen. This method of expressing the composition has the advantage of emphasizing the fact that both carbon and nitrogen play a part in determining the effect of added organic materials on net mineralization or immobilization. Another method of expressing the composition of organic materials in relation to their effects on nitrogen mineralization and immobilization is simply the use of the nitrogen percentage. A comparison of these methods, as applied to the data in an experimental investigation of the availability of nitrogen of different crop residues to plants, is shown in Fig. 7.5.

In this experiment, 1-g quantities of eleven different plant materials were added to 200 g of soil plus 200 g of sand, and the mixture was treated with a minus-nitrogen nutrient solution. Pregerminated tomato plants were transplanted into the mixture and grown 4 weeks. At the end of this time, the tops and roots were analyzed for nitrogen. The values for the percentage of the nitrogen added in the plant materials that was recovered in the plants are plotted in Fig. 7.5 against both the nitrogen percentage and the ratio of carbon to nitrogen in the origi-

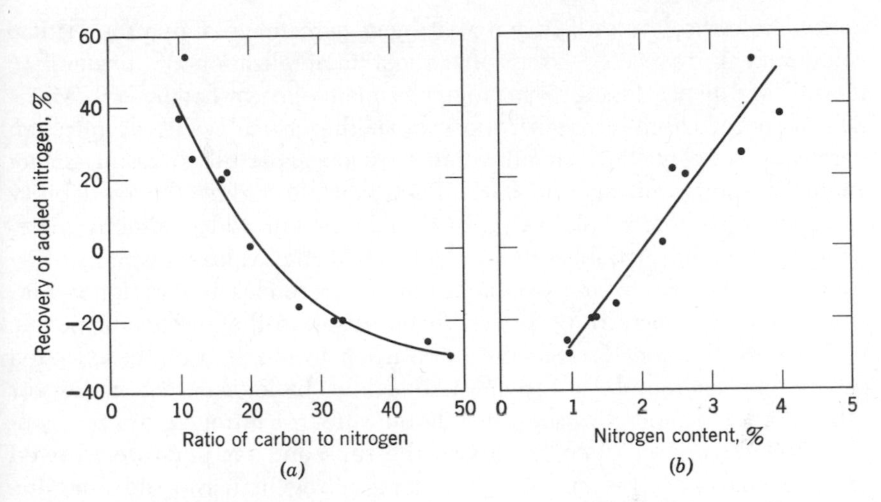

Figure 7.5. Percentage recovery of nitrogen by tomato plants in four weeks from eleven different plant residues added to soil versus (*a*) ratio of carbon to nitrogen in the original plant residues and (*b*) nitrogen percentage in the original plant residues. (Iritani and Arnold, 1960)

nal material. The percentage recovery of the added nitrogen in the tomato plants represents an index of the net percentage mineralization of the added nitrogen. The scatter of points about the line indicating the general trend is perhaps a little smaller where the composition effects are expressed as ratios of carbon to nitrogen than as nitrogen percentages, indicating that the former method of representation is somewhat the better in this instance. The evaluation obtained by the two methods usually gives similar comparative ratings for different materials because variations in ratios of carbon to nitrogen result primarily from variations in nitrogen percentages, as may be inferred from the fact that, in the eleven plant materials employed to obtain the data in Fig. 7.5, nitrogen percentages ranged from 0.94 to 4.01; but carbon percentages ranged only from 38.8 to 46.7

For practical purposes, considerable importance is attached to the composition value associated with zero recovery of added nitrogen. In the experiment in Fig. 7.5, the critical values are a carbon-to-nitrogen ratio of about 22 and a nitrogen percentage of about 2. Ratios of carbon to nitrogen below 22 are associated with positive recoveries and ratios above 22 with negative recoveries. Nitrogen percentages above 2 are associated with positive recoveries and percentages below 2 with negative recoveries. In other words, materials with carbon-to-nitrogen ratio below the critical value and with nitrogen percentage above the critical value may be expected to result in net mineralization of nitrogen in the soil and hence to supply nitrogen to plants grown on the soil. Materials with carbon-to-nitrogen ratios above the critical value or nitrogen percentages below the critical value may be expected to result in net immobilization of nitrogen in the soil and hence to reduce the availability of nitrogen to plants below that in the absence of added organic materials. Iritani and Arnold (1960) found that the values given in Fig. 7.5 for uptake of nitrogen by tomato plants were closely correlated with values for net mineralization of nitrogen in soil in the absence of plants.

The critical values for the ratio of carbon to nitrogen or the nitrogen percentage, as found by either chemical or biological techniques for estimating net mineralization or immobilization of nitrogen, are not constant but depend on such factors as the time and temperature allowed for decomposition, the supply of mineral nitrogen in the soil, and the quantity and composition of the organic material. In most instances, the critical ratio of carbon-to-nitrogen lies between 15 and 33; and the critical nitrogen percentage lies between 1.2 and 2.6. Harmsen and van Schreven (1955) reviewed the literature on this subject.

The nitrogen percentage and the ratio of carbon to nitrogen are to be regarded only as first approximations for evaluating the properties of organic materials that affect nitrogen mineralization and immobilization. Such incomplete characterizations could hardly be expected to provide a precise index of the behavior of diverse materials. There is some evidence that, with a given carbon-to-nitrogen ratio, nitrogen mineralization decreases with increasing content of lignin in the material (Peevy and Norman, 1948) and with decreasing content of water-soluble nitrogen (Iritani and Arnold, 1960). According to Bartholomew (1965), immobilization and subsequent mineralization of nitrogen in soil take place more rapidly upon addition of soluble carbohydrates than on addition of ordinary plant residues. Moreover, the mineralization or immobilization of nitrogen that occurs during decomposition of plant materials under favorable conditions takes place most rapidly during approximately the first month. Afterward, the change in organic nitrogen is slow, even though mineral nitrogen is present for immobilization and the ratio of carbon to nitrogen is in the range in which mineralization would be expected in fresh plant residues. All these observations suggest there are differences in susceptibility of organic materials to decomposition that are not taken into account by the ratio of carbon to nitrogen.

An observation reported by Bartholomew (1965) indicates that the nitrogen in its original form in plant materials is readily available to microorganisms. He treated plant materials with N^{15}-labeled nitrogen in mineral form and found that during decomposition of the plant materials by microorganisms the ratio of N^{15} to N^{14} in organic form increased rapidly and soon approached the ratio of N^{15} to N^{14} in mineral form. This process is not one of kinetic exchange, as with exchangeable bases, because the nitrogen in proteins is firmly attached by covalent bonding; rather, it is a process of splitting of amino groups from proteins by enzymes, mixing of the ammonium nitrogen thus produced with the tagged mineral nitrogen, and withdrawal of nitrogen from the mixture of mineralized nitrogen and tagged mineral nitrogen to form proteins in the tissue of microorganisms. The term biological interchange has been used to describe the process.

With regard to mineralization and immobilization in relation to the composition of soil organic matter, it may be helpful to recall the remarks introducing the subject of mineralization of nitrogen in organic materials added to soils. Such materials are the principal immediate source of soil nitrogen. From the nutrient medium consisting of the enzymatically digested organic material in the soil, microorganisms use selectively the components needed for building cellular tissue, thereby fitting the diverse materials added into the pattern of composition of microbial tissue. The carbon oxidized to supply energy, and the other constituents supplied by the organic materials in excess of the needs of the microorganisms, are eliminated in simple inorganic forms such as carbon dioxide, ammonium, and phosphate. In addition to the microbial tissue pattern, there is another chemical pattern in the more stable soil organic matter. The units fitted into this pattern may be derived from both the enzymatically digested organic materials added initially and the microbial cells produced thereon, as the latter die and serve as a secondary source of organic matter.

The two general chemical patterns of microbial tissue and stable soil organic matter in soil cannot be distinguished by direct analysis, and analyses that are made after addition of organic materials such as plant residues are complicated by the existence of an undetermined amount of residual, undecomposed material. To avoid complications from the undecomposed material, van Driel (1961) added small quantities of a number of different amino acids to soil and incubated the mixtures long enough for the amino acids to disappear. He took the difference between the quantities of carbon and nitrogen added and those mineralized as the quantities immobilized. A plot of his data for twelve amino acids in Fig. 7.6 verifies the existence of a chemical pattern in soil into which all the amino acids were fitted. Although equal quantities of nitrogen were added in all amino acids, the figures for nitrogen immobilized range from 2 to 55% of the total added. Where the amino acids

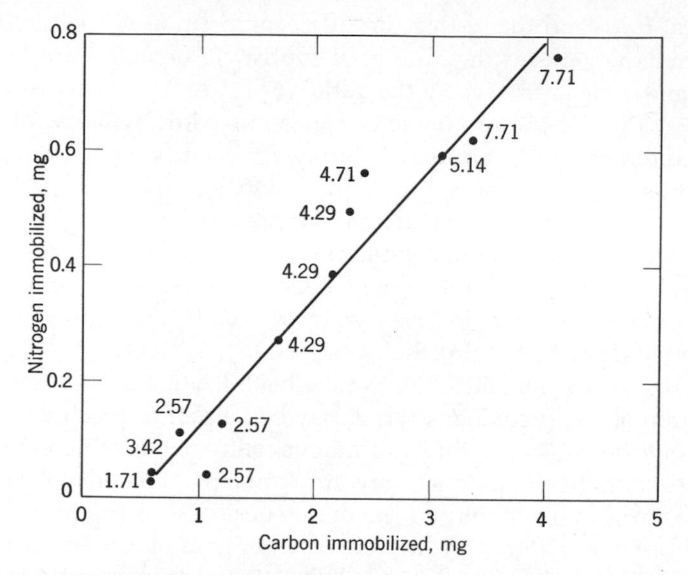

Figure 7.6. Nitrogen and carbon immobilized upon incubation of twelve different amino acids with soil in quantities to supply 1.4 mg of nitrogen per 20 g of soil. The number beside each point is the number of milligrams of carbon present per milligram of nitrogen in the original amino acid. (van Driel, 1961)

were low in carbon, the chemical pattern in the soil was accommodated by elimination of most of the nitrogen as ammonium. Where more carbon was present in the amino acids added, more carbon was immobilized; more nitrogen then was required. Because the incubation time was only 10 days, the relationship in Fig. 7.6 between the quantities of carbon and nitrogen immobilized probably reflects principally the selective utilization of substances for the building of microbial tissue and to only a secondary extent the production of stable soil organic matter.

An experiment illustrating the modification of added organic material to fit the chemical pattern of soil organic matter was performed by Simon and Barbier (1963). These authors incubated wheat straw with a subsoil for 5 years, with and without a supplemental addition of ammonium sulfate at the beginning of the incubation period. The subsoil was treated initially with 0.1% of a synthetic aggregating agent to improve the structure, after which the soil contained 1.34% carbon and 0.126% nitrogen. As shown in Fig. 7.7*c*, the initial ratios of carbon to nitrogen in the soil were rather different as a result of the nitrogen addition. With increasing time of incubation, the ratios decreased and became more similar; and, by the end of the 5-year period, they were almost the same. The convergence of the ratios of carbon to nitrogen is evidence for the tendency of organic materials of diverse composition to be altered to residual soil organic matter of similar composition. The data for the content of organic carbon and nitrogen in Figs. 7.7*a* and *b* are significant in another way in that they indicate the importance

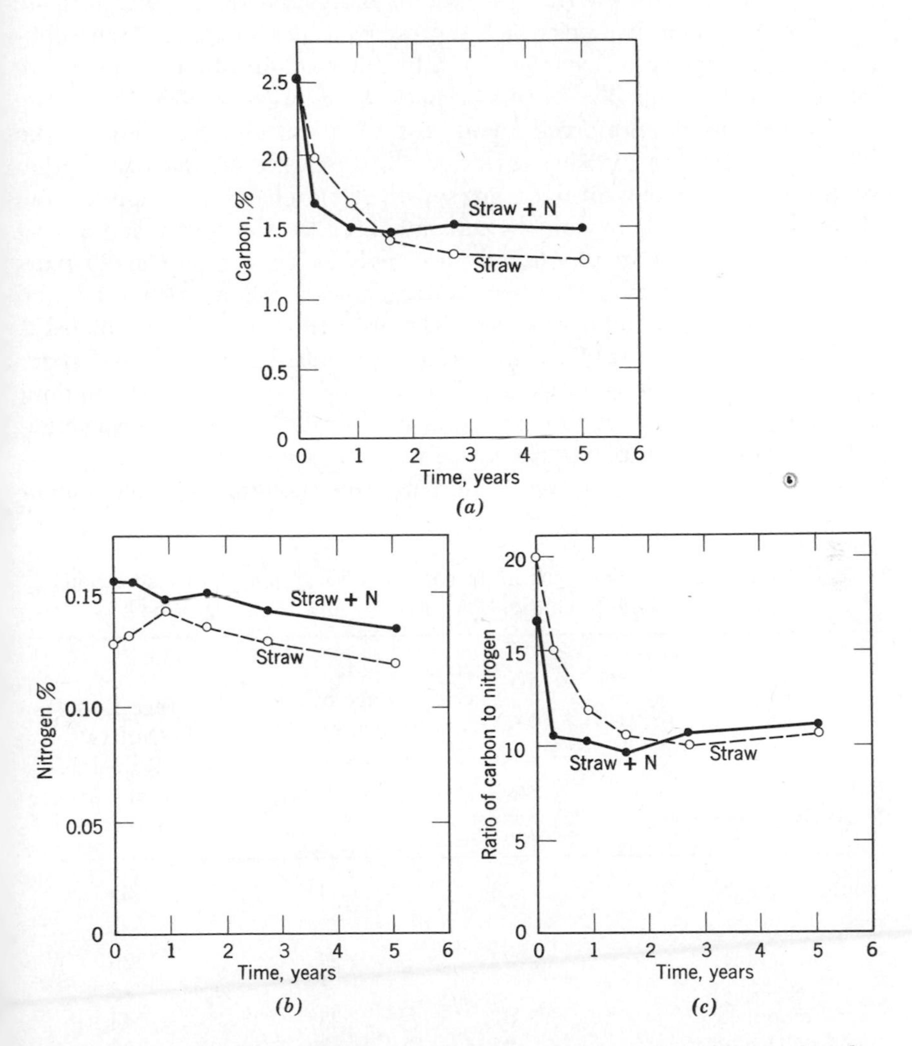

Figure 7.7. (*a*) Carbon content, (*b*) nitrogen content, and (*c*) ratio of carbon to nitrogen in a soil treated with wheat straw, with and without an initial application of ammonium sulfate, during incubation without cropping for a period of five years. (Simon and Barbier, 1963)

of nitrogen in forming stable organic matter in soil. Luecken et al. (1962) obtained similar evidence.

Keeney and Bremner (1964) investigated the question of composition of soil organic matter in relation to nitrogen mineralization by determining the proportions of different chemical forms of nitrogen in ten pairs of virgin and cultivated soils, mostly from Iowa. Their results, summarized in Table 7.4, show that the loss of individual fractions of organic nitrogen due to cultivation, which represents a measure of their susceptibility to mineralization, was not greatly different. In other words, there was a tendency for the chemical pattern of organic nitrogen in the virgin soils to be preserved upon loss of a considerable part of the nitrogen. [Bremner (1965c) reviewed the literature on the distribution of forms of organic nitrogen in soils.] Nonexchangeable ammonium showed a singular behavior, the amount in cultivated soils being almost identical with that in virgin soils. In analyses on a group of 25 pairs of virgin and cultivated soils from Saskatchewan, Hinman (1964) found that the content of total nitrogen decreased from an average of 0.23% in the virgin soils to 0.15% in the cultivated soils; but the nonexchangeable ammonium averaged 0.017% in both groups, thus verifying the finding of Keeney and Bremner. In connection with availability of nonexchangeable ammonium to plants, it may be mentioned that native nonexchangeable ammonium behaves differently from nonexchangeable ammo-

Table 7.4. Nitrogen Fractions in Ten Pairs of Virgin and Cultivated Soils and Losses Due to Cultivation (Keeney and Bremner, 1964)

Nitrogen Fraction	Average Content as Percentage of Total Nitrogen		Average Loss Due to Cultivation as Percentage of Total Nitrogen in Virgin Soils
	Virgin Soils	Cultivated Soils	
Nonhydrolyzable	25	24	39
Hydrolyzable			
Ammonium	22	25	29
Hexosamine	5	5	28
Amino acid	27	23	43
Unidentified	21	23	35
Nonexchangeable ammonium	5	7	0

nium produced in soil by addition of ammonium. Axley and Legg (1960) found that uptake of nitrogen by oats and corn from soils treated with urea or ammonium sulfate was little affected by the capacity of the soil to fix added ammonium in nonexchangeable form unless much potassium was present to block the release of the nonexchangeable ammonium to exchangeable form. Results leading to the same conclusion were obtained by Walsh and Murdock (1963). Presumably, the native nonexchangeable ammonium occupies the most stable positions, which are then not accessible to added ammonium.

Porter et al. (1964) used a somewhat different type of fractionation of the nitrogen of virgin and cultivated soils and found that, upon cultivation, there was a greater loss of nitrogen from a "nondistillable, acid-soluble" fraction than from two other fractions. Their nondistillable, acid-soluble fraction corresponds approximately to the sum of the amino acid and unidentified fractions in Table 7.4, and so their findings appear to be in general agreement with the table in respect of the behavior of these two fractions of soil nitrogen.

Another approach that has been used in an attempt to characterize the soil organic matter with regard to mineralization of nitrogen is measurement of the ratio of carbon to nitrogen. This use of the carbon-to-nitrogen ratio follows directly from the characterization of plant residues discussed previously. Figure 7.8 summarizes results of three investigations of the relationship between the ratio of carbon to nitrogen in soil organic matter and the mineralization of soil nitrogen during incubation in the laboratory. Although a downward trend of percentage mineralization of soil nitrogen with an increase in the ratio of carbon to nitrogen is evident in results of two of the three investigations, the scatter of points is so great that the ratio of carbon to nitrogen must be considered a poor index of the susceptibility of soil nitrogen to mineralization. Probably an important reason for the poor correlations is that fresh organic matter is added to soils continually; thus the organic matter is a mixture of a relatively stable part, which may have fairly similar composition from one soil to the next, with an unstable, more rapidly decomposing part that is more characteristic of the source of organic matter added. Small amounts of the unstable organic matter will not have much effect on the ratio of carbon to nitrogen in the soil organic matter as a whole; however, because they may supply most of the microbial substrate, they may have a disproportionately great influence on nitrogen mineralization.

A secondary reason for the limited value of the ratio of carbon to nitrogen in soils as an index of the susceptibility of the nitrogen to mineralization is that the measurement of total nitrogen includes nonexchangeable ammonium, which is an inorganic form of nitrogen, and

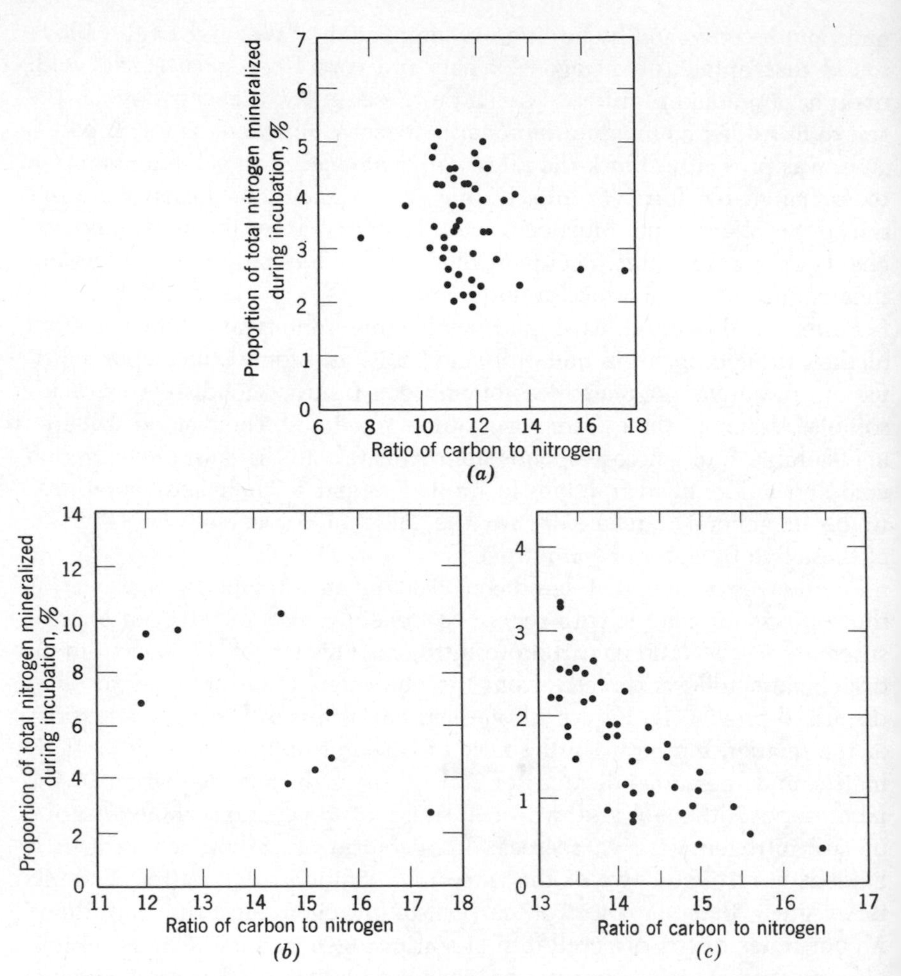

Figure 7.8. Percentage of total nitrogen mineralized during incubation versus ratio of carbon to nitrogen in soils in investigations by (*a*) Thompson et al. (1954), (*b*) Lopez and Galvez (1958), and (*c*) Richer and White (1946). In (*a*), different soils were incubated under aerobic conditions. In (*b*), different soils were incubated under anaerobic conditions. In (*c*), samples were from different areas of a single soil in a field experiment in which different quantities and kinds of organic materials and fertilizer nitrogen had been applied over a period of years. Incubation of samples was under aerobic conditions.

which (in native form) apparently contributes little to the quantities of ammonium and nitrate measured as mineralized nitrogen. Although nonexchangeable ammonium represents only a small percentage of the total nitrogen of surface soils, and hence would not have much influence on carbon-to-nitrogen ratios as an index of the susceptibility of nitrogen of surface soils to mineralization, the influence would be significant in subsoil samples containing a relatively high proportion of the total nitrogen as nonexchangeable ammonium.

The existence of nonexchangeable ammonium provides an explanation for the long-puzzling observation that the ratio of organic carbon to total nitrogen usually decreases from the surface downward in soil profiles. Investigators now are finding that this decrease is mostly due to nonexchangeable ammonium. Sometimes the ratios of organic carbon to organic nitrogen (total nitrogen minus mineral nitrogen, most of which is nonexchangeable ammonium) increase with depth.

With the exception of analyses for nonexchangeable ammonium, determination of quantities of different chemical forms of soil nitrogen has not proved particularly helpful in characterizing soil nitrogen as regards mineralization. The best chemical method now available is based on susceptibility of nitrogen-bearing substances in soil to hydrolysis and dissolution without regard to the chemical form.. This subject will be discussed further in the section on indexes of nitrogen availability.

Soil pH. Differences in soil *p*H value are sometimes considered to have an important influence on mineralization of nitrogen in soils. The principal experimental basis for this view is that nitrogen mineralization increases when calcium carbonate is added to acid soils in short-term laboratory experiments. As explained in Chapter 5, this effect appears to be only temporary. Nevertheless, it may be important as long as it lasts. The *p*H effect seems to be due to an increase in susceptibility of a portion of the soil organic nitrogen to mineralization. After this fraction has been mineralized, the *p*H effect diminishes or disappears. Evidence from a field experiment by Mulder (1950) in the Netherlands indicates that a period of 5 years may be required for the effect to disappear. The mechanism of the effect has not yet been explained.

The absence of a marked *p*H effect on change of organic nitrogen to ammonium after dissipation of the temporary effect of liming acid soils may be accounted for on the basis that this transformation is brought about by many different organisms. Hence the change in nature of the soil population with change of soil *p*H merely substitutes one group of organisms for a somewhat different group that does the same thing.

The transformations of ammonium to nitrite and of nitrite to nitrate,

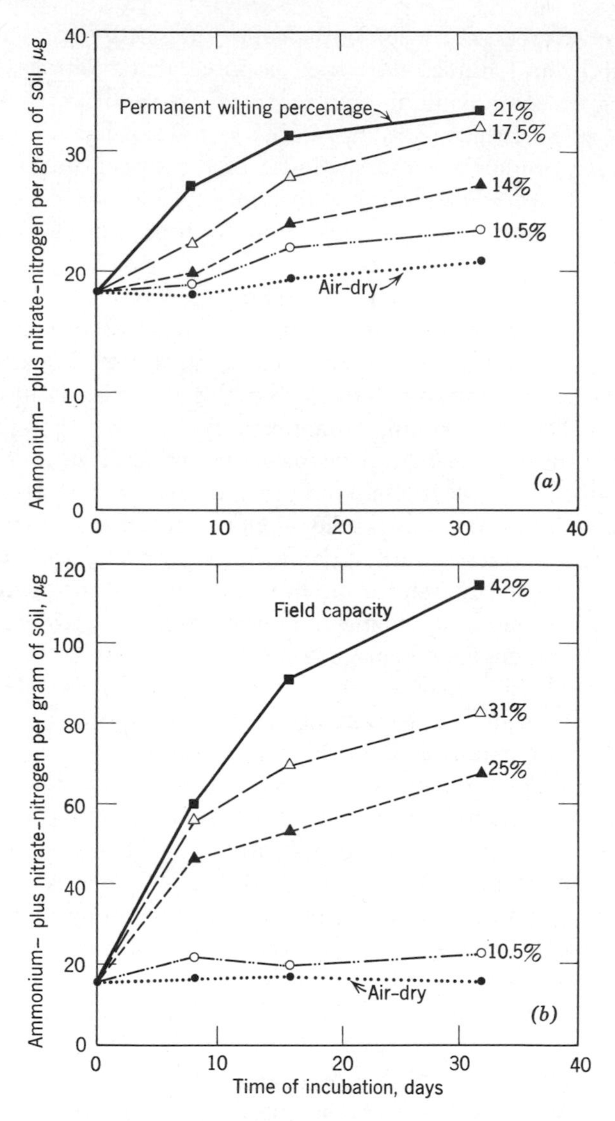

Figure 7.9. Ammonium- plus nitrate-nitrogen in a soil during incubation at 35°C in two experiments. The numbers at the right of individual lines are the water percentages in the soil during incubation. (Robinson, 1957)

however, are influenced by soil *p*H; and the effects may be of practical consequence. This subject will be discussed subsequently under the heading of factors affecting ammonium oxidation and nitrate reduction.

Water Supply. Figure 7.9 shows the sum of the ammonium- and nitrate-nitrogen found in a soil during incubation at different water contents at 35°C in two experiments. The results in Fig. 7.9*a* show that the mineral nitrogen increased steadily as the water content of the soil during incubation increased from that corresponding to the air-dry condition to 21%, which corresponds to the permanent wilting percentage. The results in Fig. 7.9*b* show that the mineral nitrogen increased further as the water content of the soil during incubation was increased to 42%, which corresponds to the field capacity.

Figure 7.10 shows a plot of the increase in ammonium-nitrogen produced under anaerobic conditions against the increase in the sum of ammonium-, nitrite-, and nitrate-nitrogen produced under aerobic conditions during incubation of samples of 39 soils for 2 weeks at 30°C. The anaerobic conditions were obtained by adding enough water to cover the soil to a depth of 5 to 6 cm and stoppering the tubes. The incubation under aerobic conditions was done by mixing the soil with an excess of sand and by moistening the mixture to approximately the

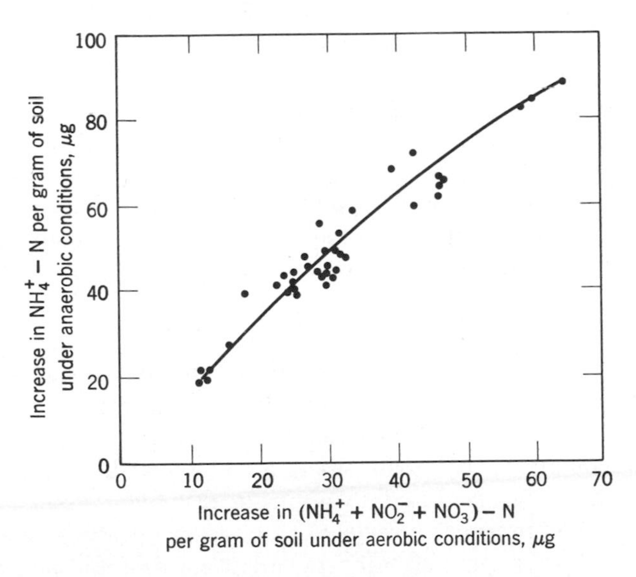

Figure 7.10. Ammonium-nitrogen produced under anaerobic conditions versus (ammonium + nitrite + nitrate)-nitrogen produced under aerobic conditions during incubation of samples of 39 soils for two weeks at 30°C. (Waring, 1963)

field capacity. The values obtained under anaerobic and aerobic conditions were closely related, as shown by the relatively small scatter of points. Moreover, all values for ammonium-nitrogen produced under anaerobic conditions exceeded the comparable values for the sum of ammonium-, nitrite-, and nitrate-nitrogen produced under aerobic conditions.

The results of these investigations suggest that mineralization of soil organic nitrogen occurs at a gradually increasing rate as the water content of soil is increased from the air-dry condition to complete submergence. These investigations, however, do not cover the range from field capacity to submergence. This range is the most difficult to study because of possible loss of nitrate by denitrification, a subject to be considered in the section on ammonium oxidation and nitrate reduction. Early work of Kelley (1914) on mineralization of nitrogen added to soil in the form of dried blood showed greater mineralization of nitrogen in unsaturated soils than in soils that were submerged.

The adaptation of the population of nitrogen-mineralizing organisms to different levels of water content of soil is evidently broader than that of plants. The fact that mineralization of nitrogen and growth of plants are not affected equally by the water content of soil has some significant implications in terms of plant behavior. As might be anticipated from the continuation of nitrogen mineralization in soils that have been dried below the permanent wilting percentage, soil dryness often restricts the yield of plants relatively more than the yield of nitrogen in plants. For example, in an experiment on oats in Utah, Greaves and Carter (1924) found that the yield of dry matter of oat grain without irrigation was 57% as great as that with irrigation, whereas the corresponding figure for yield of nitrogen in the grain was 73%.

In instances such as the foregoing, the rate of mineralization of nitrogen may be less limiting to growth of plants under dry conditions than under moist conditions. Where the content of mineral nitrogen in the soil is high under moist conditions, a considerable percentage accumulation of nitrogen in plants may occur if the soil dries. Much of the accumulation generally occurs as nitrate. The nitrate content of plants occasionally is so high under conditions of drought and high soil nitrate that it is thought to poison livestock. As much as 2.8% nitrate-nitrogen by weight has been reported to occur in dry plant tissue (Gilbert et al., 1946). A comprehensive review of literature on nitrate accumulation in plants and apparent poisoning of animals was published by Wright and Davison (1964), who suggested that forages containing more than 0.34 to 0.45% nitrate-nitrogen should be considered potentially toxic. In the Netherlands, the toxic condition is thought to be a magnesium deficiency induced in the animals by the high concentration of potassium that usually accompanies high nitrate; veterinarians are said to treat affected animals successfully by injecting a magnesium salt (private communication from C. T. de Wit).

Accumulation of nitrogen in plants under dry conditions is probably favored where the mineral nitrogen is distributed uniformly throughout the moist part of the soil from which water is removed or where it is concentrated in the portion of the soil from which the last water is removed before permanent wilting. In some instances, however, soil dryness limits the yield of nitrogen in plants relatively more than the yield of dry matter. The appropriate conditions are found where the subsoil supplies water but little nitrogen and the surface soil supplies nitrogen but little water. Most of the mineralization of soil nitrogen occurs in the surface part of the soil, as may be inferred from the data in Table 7.5. As the soil dries, upward movement of water by evaporation carries nitrate from the moist portion of the soil into the dry portion above. Gadet et al. (1961) applied nitrate-nitrogen to different layers of a soil profile that was originally moistened uniformly; then they determined the distribution of nitrate at intervals. No rain was allowed to fall on the soil during the test period. Regardless of the depth of application of the nitrate, the most pronounced movement was always upward. The greatest effects were noted with applications made near the surface of the soil. For example, where the nitrate was applied in the layer 5 to 10 cm in depth, sampling 54 days later showed that 72% of the

Table 7.5. Production of Nitrate during a Ten-Day Incubation of Samples Taken from Different Depths in a Fine Sandy Loam Soil (Ramig and Rhoades, 1963)

Depth of Sampling, cm	Total Nitrogen, %	Nitrate-Nitrogen Produced during Incubation	
		μg/g of Soil	% of Total Nitrogen
0–15	0.073	6.9	0.95
15–30	0.070	0.9	0.13
61	0.055	0.1	0.02
91	0.033	0.0	0.00
122	0.021	0.0	0.00
152	0.013	0.0	0.00
183	0.010	0.0	0.00

added nitrate was in the 0- to 2.5-cm layer, 26% was in the 2.5- to 5-cm layer, 2% was in the 5- to 10-cm layer (which originally contained 100%), and 1% was in the 10- to 15-cm layer. The nitrate present at a given moment in a given layer may thus in time be moved upward almost quantitatively into drier soil above. Fertilizer nitrogen also is applied to the surface part of the soil and sometimes is applied on the surface without mixing it with the soil.

Accordingly, the greatest concentration of mineral nitrogen is frequently in the upper part of the soil, and its uptake by plants is inhibited when this part of the soil becomes dry. If plants have not absorbed an excess of mineral nitrogen from the surface part of the soil before uptake is reduced by drought, and if the supply of subsoil water is adequate to permit continued growth, onset of drought may cause the plants to exhibit nitrogen deficiency instead of water deficiency. This behavior has been used as an argument for deep placement of nitrogen fertilizers in humid regions affected by summer droughts.

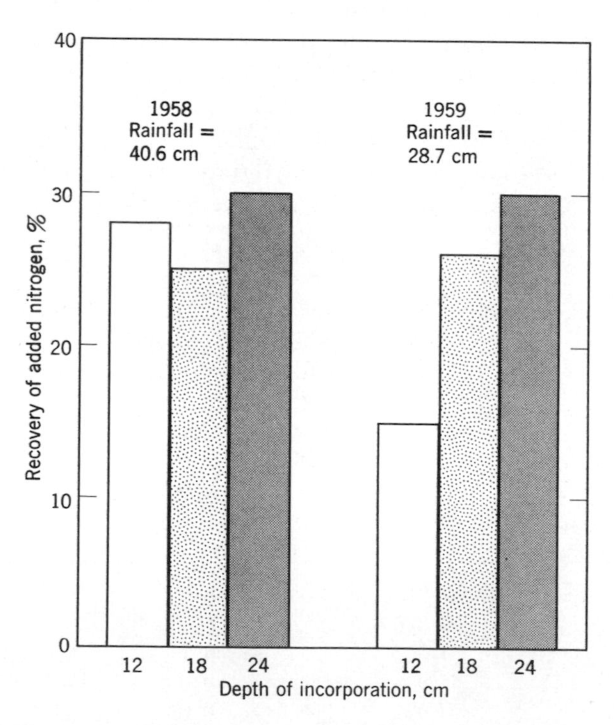

Figure 7.11. Percentage recovery in corn of the nitrogen applied in manure and plowed down to different depths in two years differing in rainfall. Rainfall data are for April to September only. (Koepke, 1962)

Koepke (1962) described an experiment in Germany in which manure was plowed down to three different depths in each of five successive seasons. The soil was a loamy sand in the surface 60 cm, below which the texture was a sandy loam. The subsoil was very poor in nutrients, but the *p*H value was 6.2. Various crops were grown in the year of application of the manure. In the year following the application, a small-grain crop was grown without further fertilization to test the residual effect. Figure 7.11 gives values for percentage recovery of added nitrogen in the corn grown in the year of application in 1958, when the rainfall from April to September was 40.6 cm, and in 1959, when the rainfall in the same period was 28.7 cm. Little effect of depth of application is evident in the more moist season, but a pronounced increase in recovery with increasing depth of application may be seen in the drier season. The small-grain crops grown to test the residual effect in a succession of four seasons responded to depth of application of manure in two comparatively moist and two comparatively dry seasons in the same way as did the corn.

Figure 7.12 shows the rainfall and soil nitrate in 1959. These values correspond to the nitrogen recovery data for 1959 in Fig. 7.11. No consis-

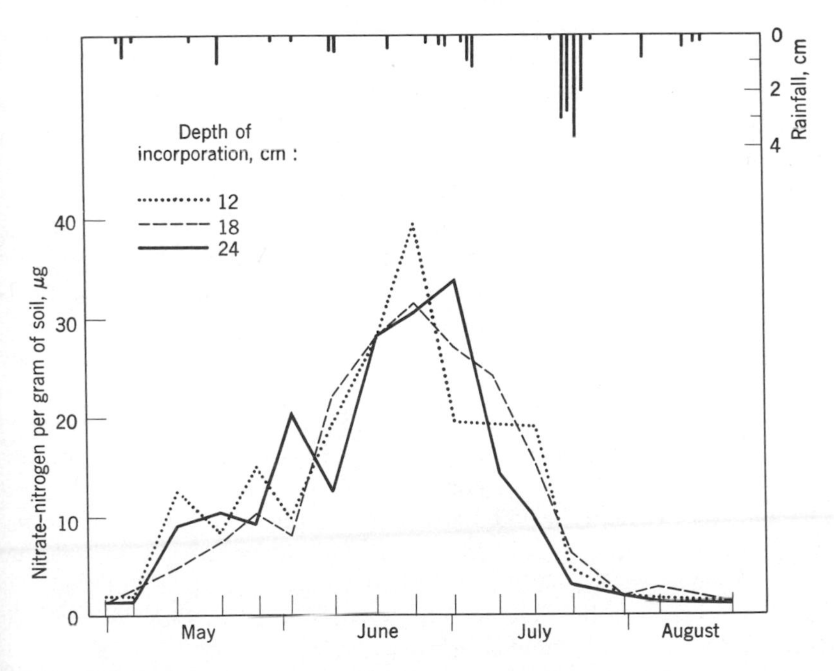

Figure 7.12. Nitrate-nitrogen in a loamy sand soil under corn with manure plowed down to different depths, together with distribution of rainfall. (Koepke, 1962)

tent effect of depth of application on soil nitrate may be seen during the early part of the season. The peak value corresponds to the shallow application. Beginning in the latter part of June, the soil nitrate dropped rapidly, primarily because of growth of the crop. This is when rapid absorption of nitrate began. Rainfall shown in this period was too small to produce any substantial leaching. After the light rainfall at the end of June, there was a period of several weeks with no rainfall. Although the initial drop in soil nitrate was most pronounced with the shallow incorporation, the nitrate level in the soil remained about the same during the period without rainfall. With the deeper incorporations of manure, however, soil nitrate continued to fall during the dry period. The difference between these treatments may be explained on the basis that much of the soil nitrate was derived from the nitrogen added in the manure (this was shown to be the case in 1958), and hence the location of the nitrate in the soil profile depended on the depth to which the manure was incorporated. During the dry period, when much of the nitrogen was being absorbed, the uptake of nitrate from the soil was limited to a greater extent where it was close to the surface in relatively dry soil than where it was deeper in more moist soil. Perhaps an additional factor of significance in accounting for the differences among depths of incorporation is that in the latter part of July a succession of rains totaling about 12 cm within a week may have caused enough downward movement of nitrate to prevent the crop from using much of the nitrate that had accumulated without absorption in the upper part of the soil.

The findings of Mitchell (1957) in New Zealand may be cited as a a second example. Mitchell noted that surface applications of nitrogen fertilizer to ryegrass-clover pastures had low efficiency during a dry season. With the onset of drought, white clover continued to grow vigorously for some time after the rate of growth of ryegrass had decreased. Because the clover did not have a deeper root system than the grass, he explained the observation on the basis that growth of clover was not dependent on absorption of combined nitrogen from the soil, whereas growth of grass was inhibited when the principal supply of nitrogen in the surface part of the soil was cut off by dryness.

Drying and Freezing. Related to water supply is the effect of drying of soil. Mineralization of nitrogen takes place more rapidly in soil that is incubated in a moist condition after air-drying than in soil that has been maintained continuously in a moist condition. This behavior is illustrated in Fig. 7.13, which represents the nitrate-nitrogen produced during laboratory incubation of samples taken at different dates during a given season from a silty clay loam soil in Nebraska.

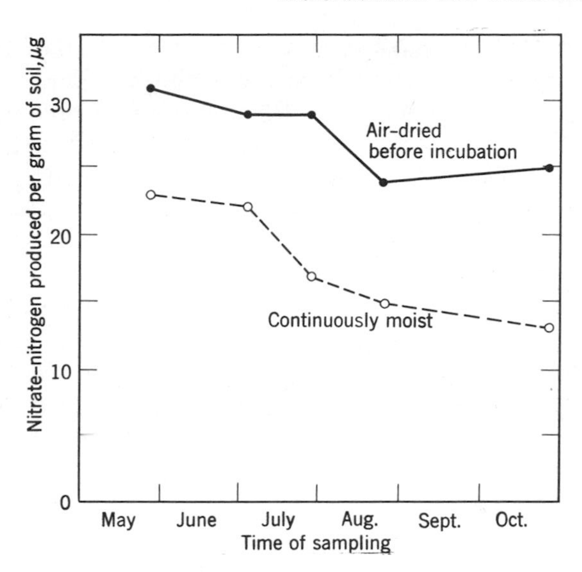

Figure 7.13. Nitrate-nitrogen produced during incubation of soil samples for eight weeks, with and without previous drying. The values are averages of eight field treatments. (Landrau, 1953)

The effect of drying on nitrogen mineralization in soil has been known for many years. At one time, the effect was accounted for by the "partial sterilization" theory. According to this theory, drying was supposed to kill certain soil organisms which, when active, inhibited the activity of other organisms that brought about nitrogen mineralization. Khalil (1929) reasoned that, if the partial sterilization theory were valid, one might expect that (1) the organisms remaining after drying would be more efficient than an equal number before drying, (2) the number of organisms would be decreased by drying and would remain at a relatively low level for several weeks, (3) the effect of drying would disappear if the dried soil were reinoculated with organisms from the moist soil before incubation, and (4) the effect of drying would disappear if, before drying, the soil were partially sterilized by some other method. Khalil examined each of these consequences of the theory and obtained negative results in each case, thus discrediting the theory. His opinion was that drying increases the susceptibility of some of the soil organic matter to decomposition.

Evidence supporting Khalil's view was obtained by Achromeiko (1928). Achromeiko measured the quantity of water-soluble organic matter in terms of milligrams of oxygen from permanganate required to oxidize the organic matter in a liter of 1-to-2 soil-to-water extract. He found that the oxygen equivalent was increased from 46 to 241 mg when a soil was dried three times by exposure to the sun. During subsequent incubation of the continuously moist and remoistened samples for 7 months, the oxygen equivalent of the extract dropped to 27 mg in the continuously moist samples and 36 mg in the remoistened samples. Additional evidence has been obtained in more recent investigations. Birch (1958) found that carbon dioxide production by soil during incubation under moist conditions was increased by previous drying of the soil at 100°C and that an increase occurred after each drying. After 43 oven-drying and moist-incubation cycles, the soil he employed had lost 38% of its carbon content. In subsequent work, Birch (1959) found that air-drying increased the organic nitrogen in water extracts. The organic nitrogen extracted increased from 7.5 μg/g of soil that had been maintained continuously moist for 4 weeks to 21.5 μg/g of soil that was air-dried for 35 days.

The large loss of organic carbon in some of Birch's work must have involved the relatively stable portion of the soil organic matter. One theory to account for it is that the swelling and shrinking associated with wetting and drying physically disrupt to some extent the otherwise resistant masses of stable organic matter, with exposure of new surfaces and release of some parts to the soil solution. This theory is a corollary of the theory proposed by Rovira and Greacen (1957) and discussed previously in the section on stability of soil nitrogen. Supporting evidence was obtained by Soulides and Allison (1961), who found that wetting and drying of soils decreased the water-stable aggregates. To what extent this theory will account for the results is as yet uncertain. Patrick and Wyatt (1964) found that a soil lost about 20% of its organic carbon and total nitrogen in five cycles of alternate submergence and drying where the soil in the dry part of each cycle contained about 18% water. The water content of this soil at the permanent wilting percentage was 12%. It seems unlikely that drying to this extent would be sufficient to cause the physical disruption suggested.

Soulides and Allison (1961) found that the sum of exchangeable and soluble ammonium in soils was increased by drying in the absence of a subsequent incubation period for microbial activity. Although the origin of this ammonium is not known, the fact that drying of soil may cause release of considerable nonexchangeable potassium (see Chapter 9 on potassium) suggests that the ammonium may have been released from nonexchangeable form. Finally, it may be pointed out that the discrediting of the partial sterilization theory as an explanation for the effect of drying on soil nitrogen mineralization does not mean that effects of drying on the microorganisms are of no consequence. Some of the active microbial tissue in moist soil is undoubtedly killed by air-drying or by more drastic drying treatments, and this tissue may then be expected to undergo autolysis and decomposition. As an indication of the effect of drying on microorganisms, Payne et al. (1956) and Takai and Harada (1959) reported that air-drying of soil increased the concentration of free amino acids in water extracts of soil.

The effect of drying on subsequent mineralization of soil nitrogen is probably of some significance in at least thin surface layers of many soils. In areas where long periods without rainfall are accompanied by high temperatures, application of irrigation water or onset of rains is accompanied by release of considerable mineral nitrogen. The importance of this effect in relation to plant growth in such areas was discussed by Birch (1958).

Freezing of soil has been found by a number of investigators to increase the mineralization of nitrogen during subsequent incubation. Most of the recent work has involved placing soil in freezing units at −10 to −20°C, although Mack (1963) used liquid nitrogen for freezing purposes. The effect of freezing may result from disruption of soil aggregates and exposure of new surfaces to attack by microorganisms in a manner similar to that suggested for drying; however, the mechanism does not appear to have been investigated. As compared to the pronounced effect of drying on subsequent mineralization of nitrogen, the effect of freezing seems to be moderate and sometimes absent. Results of laboratory work on the effect of freezing, together with citations to other work, may be found in a paper by Harding and Ross (1964).

Certain evidence from measurements on samples of soil taken at different times from the field is suggestive of an effect of freezing on subsequent mineralization of soil nitrogen. Richardson (1938) took samples of soil periodically during 2 years from a given area in England and determined the nitrogen mineralized during incubation under standard conditions in the laboratory with results as shown in Fig. 7.14. The nitrogen mineralized shows a cyclic behavior, being high in late winter and decreasing through the spring and summer. Landrau's (1953) results in Fig. 7.13 show a similar decrease in nitrogen mineralized in the laboratory in samples taken through the spring and summer in Nebraska. These results, of course, might be accounted for in various ways; the cause has not been determined. Eagle (1961) found that the nitrogen mineralized upon incubation of samples of soil taken at different times during the year in England decreased during the spring and summer. In his work, however, maximum mineralization occurred in samples taken at about the end of April, a finding that would not be expected if freezing were responsible for the peak.

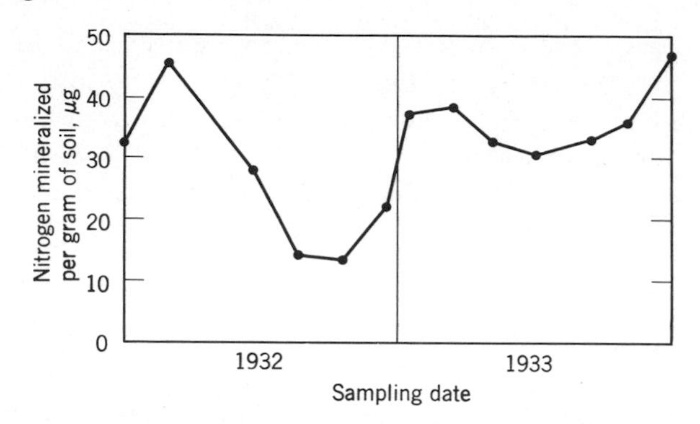

Figure 7.14. Nitrogen mineralized in samples of soil taken from grassland at different dates and incubated in the laboratory under standard conditions. (Richardson, 1938)

Temperature. Mineralization of nitrogen takes place slowly at temperatures near freezing, because of the restricted microbiological activity, and increases up to temperatures above those at which plants grow. Thompson (1950) found maximum accumulation of mineral nitrogen in soils after incubation at 60 to 70°C and lesser accumulations at lower or higher temperatures. He did not determine losses of nitrogen in gaseous forms, however, and so the maximum at 60 to 70°C may have been merely the consequence of a balance between mineralization and volatilization.

The rate at which plants require nitrogen also varies with temperature. Plants will absorb mineral nitrogen at low temperatures at which they make no appreciable growth. As the temperature increases, uptake of nitrogen and growth increase up to limiting values determined by the nature of the plants, the supply of nitrogen in the soil, and other environmental factors. Because the rates of nitrogen mineralization in soils and of nitrogen use by plants are different functions of temperature, the sufficiency of soil nitrogen for plant development may vary with the temperature. Many different results may be obtained according to the circumstances.

From experiments conducted over a period of several years in England, Blackman (1936) concluded that temperature was the factor that limited the growth of pasture herbage when the soil temperature was below 5.4°C at a depth of 10 cm. Added nitrogen was absorbed, but no growth occurred below this temperature. At temperatures of 5.4 to 8.3°C the herbage grew, but the rate of growth was limited by the supply of soil nitrogen; evidence for this is the fact that herbage production was much increased by application of nitrogen fertilizer. The greatest increase in early growth of herbage from nitrogen fertilization was obtained when the soil temperature rose slowly through this range. At higher temperatures, the rate of increase of herbage production with soil temperature in the absence of nitrogen fertilizer was accelerated; Blackman interpreted this behavior to be a result of increased mineralization of soil nitrogen. Blackman's work was mainly on pastures in which perennial ryegrass was the dominant species. In experiments on soft chess (*Bromus mollis*) at different elevations in California, Jones et al. (1963) found that the grass made little growth and responded little to nitrogen fertilization at soil temperatures below 9.2°C. The greatest increase in growth from nitrogen fertilization occurred between soil temperatures of 9.2 and 12.8°C. This grass evidently required a higher temperature for growth than did the ryegrass used in Blackman's experiments.

Quite a different result was obtained in unpublished work by Roger McHenry, J. W. Fitts, L. T. Alexander, and H. F. Rhoades on nitrogen fertilization of blue gramagrass pasture in western Nebraska. Application of nitrogen fertilizer in early spring produced a rank growth of wild lettuce but had little effect on the grass. In this instance, the growth of wild lettuce apparently was controlled by the nitrogen supply, whereas the growth of grass was controlled by the temperature. Blue grama is a "warm-season" grass that makes little early spring growth. Sexsmith and Pittman (1963) noted that germination of wild oats was increased by application of nitrogen fertilizers early in the spring, which indicates that in some instances the development of plants at low temperatures in the spring may be modified by an effect of nitrogen supply on seed dormancy.

Soil-plant Interactions. Mineral nitrogen is produced in soils as a by-product of microbial metabolism. Higher plants are not known to bring about mineralization of soil nitrogen. Instead, they are presumed to depend on nitrogen mineralized microbiologically. Accumulated evidence nevertheless indicates that net mineralization of nitrogen is not entirely independent of the growth of plants. For example, Goring and Clark (1949) maintained samples of soil in fallow condition in containers in the greenhouse and grew plants alongside in comparable samples of the same soil. At intervals, they analyzed the fallow and cropped soils for mineral nitrogen and analyzed the tops and roots of plants in the cropped soil for their content of nitrogen. They represented the sum of the nitrogen in the plants and in mineral form in the soil as the net nitrogen mineralized. A summary of their results in Fig. 7.15 shows that the amount of nitrogen mineralized in the fallow and cropped soil was about the same during the first 6 weeks, but after this time

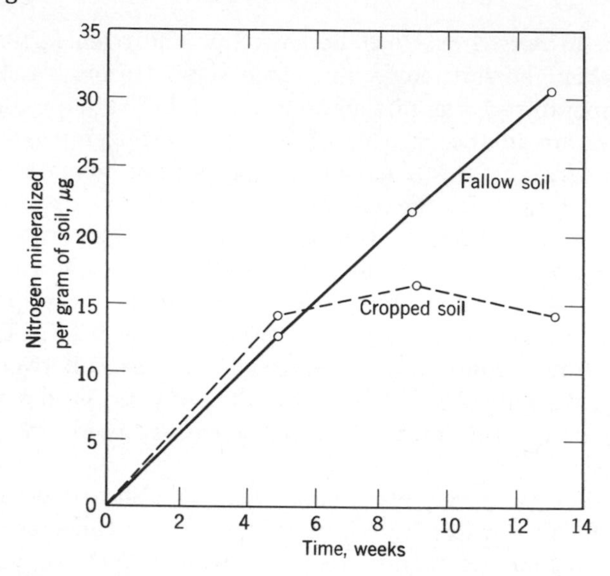

Figure 7.15. Mineralization of nitrogen in cropped and fallow soil over different lengths of time. (Goring and Clark, 1949)

more nitrogen was mineralized in the fallow soil than in the cropped soil. The presence of the crops thus decreased nitrogen mineralization.

Examination of the results obtained with the individual crops showed that nitrogen mineralization decreased with an increase in total root weight and with an increase in number of microorganisms around the plant roots. When samples of cropped and fallow soil were incubated in the laboratory (Clark, 1949) for 16 weeks after completion of the experiment described in Fig. 7.15, nitrogen mineralization in previously cropped soil exceeded the mineralization in previously fallowed soil. The increase in mineralization due to the prior presence of a crop was approximately equal to the decrease in mineralization due to the presence of the crop while the crop was growing. Because the increase in number of microorganisms around the roots probably resulted from organic materials supplied by the roots, and because growth of these organisms would immobilize mineral nitrogen, the results may be accounted for in part by the theory that the organic materials supplied by the plant roots induced microbial immobilization of nitrogen in the soil. Because plant roots cannot be separated quantitatively from soil, the results may be accounted for partly by the allied theory that some of the mineral nitrogen absorbed from the soil and converted to organic form in the roots was left behind in the soil when the roots were separated. The difference in total nitrogen mineralized between cropped and fallow soil appears to be much too large to be accounted for by incomplete recovery of living roots from the soil, which lends emphasis to the former theory.

Further evidence for the theory of immobilization of nitrogen by microorganisms acting on root exudates and sloughed root material was obtained by Legg and Allison (1961). They found that the percentage recovery of N^{15}-tagged nitrogen by oats in sand cultures during a 3-day period, after the oats had been grown with different levels of untagged nitrogen, increased with the level of untagged nitrogen added at the beginning of the growth period. They explained this observation on the basis that relatively little immobilization of tagged nitrogen by the microorganisms should occur where the needs of the microorganisms were already satisfied by the high levels of untagged nitrogen. From unpublished data involving use of N^{15} as a tracer, Bartholomew and Clark (1950) inferred that both total mineralization and total immobilization of nitrogen were greater in cropped soil than in fallow soil and hence that the smaller net mineralization in cropped soil was a consequence of a greater increase in immobilization than in mineralization.

There has been considerable speculation about soil-plant interactions in mineralization and immobilization of nitrogen in connection with the pronounced tendency of perennial grasses to become deficient in nitrogen, or "sod-bound." This behavior is exemplified in Fig. 7.16 by data of Haylett and Theron (1955) on the yield of grass with different annual applications of nitrogen fertilizer. Why perennial grasses should have a greater tendency than other common crops to behave in this way

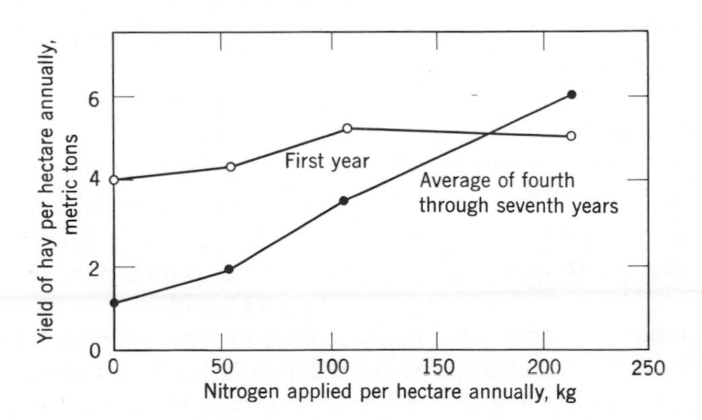

Figure 7.16. Yield of perennial grasses (*Chloris gayana* and *Paspalum dilatatum*) with different quantities of ammonium sulfate during a seven-year experiment in South Africa. (Haylett and Theron, 1955)

is probably not explainable by any single theory. The following may be suggested as elements of an explanation: (1) Cultivation ceases when the grassland is established; hence the effect of cultivation in enhancing the susceptibility of soil nitrogen to mineralization is eliminated. (2) Grass roots are fibrous in nature, and they permeate the soil throughout the year; hence they are in position to absorb mineral nitrogen rapidly. (3) Perennial grasses may produce a relatively large mass of roots; hence, according to the observations of Goring and Clark previously described, considerable nitrogen may be immobilized in the soil as a result of exuded and sloughed organic material. (4) Much of the nitrogen in the roots is carried over from one year to the next in living tissue and thus is not subject to mineralization. (5) The nitrogen in dead roots is mineralized slowly because of the high ratio of energy material to nitrogen in the roots and the deficiency of mineral nitrogen in the soil for decomposition. Harmsen and van Schreven (1955) and Harmsen and Kolenbrander (1965) reviewed this subject in detail.

A different type of relationship between plant growth and nitrogen mineralization is presumed to occur in roots infected with mycorrhizal fungi. The word *mycorrhiza* means fungus root. Mycorrhizae are of two general types. In the more spectacular type (Fig. 7.17*a*, Fig. 7.18), the small roots are completely encased in a sheath of fungal tissue, which commonly has a thickness of 20 to 40 microns and comprises 20 to 30% of the total root volume and 34 to 45% of the total dry weight (Harley, 1959, p. 27). Filaments of fungal tissue extend outward from the sheath into the soil and inward into the cortex or outer portion of the root proper. This type of mycorrhiza is common on forest trees in temperate regions. The less spectacular type (Fig. 7.17*b*) forms no sheath of fungal tissue and consists of filaments that are partly external and partly within or between the cortical cells of the root, along with other forms shown in the figure. This type of mycorrhiza occurs on many plants, including members of the grass family.

Development of mycorrhizae is favored by a high-carbohydrate, low-nitrogen condition in the plants (Richards, 1965). Development of mycorrhizae then improves the nitrogen nutrition of the plants. Table 7.6 provides evidence for the improvement of nitrogen nutrition. The greater relative increase in nitrogen content than in yield of dry matter with inoculation indicates that the increase in nitrogen content resulted at least in part from a specific effect of the mycorrhizae on nitrogen availability. There are different theories to account for the improved nitrogen supply to mycorrhizal plants on low-nitrogen soils. One is that the improved nitrogen supply is a consequence of the greater area of contact

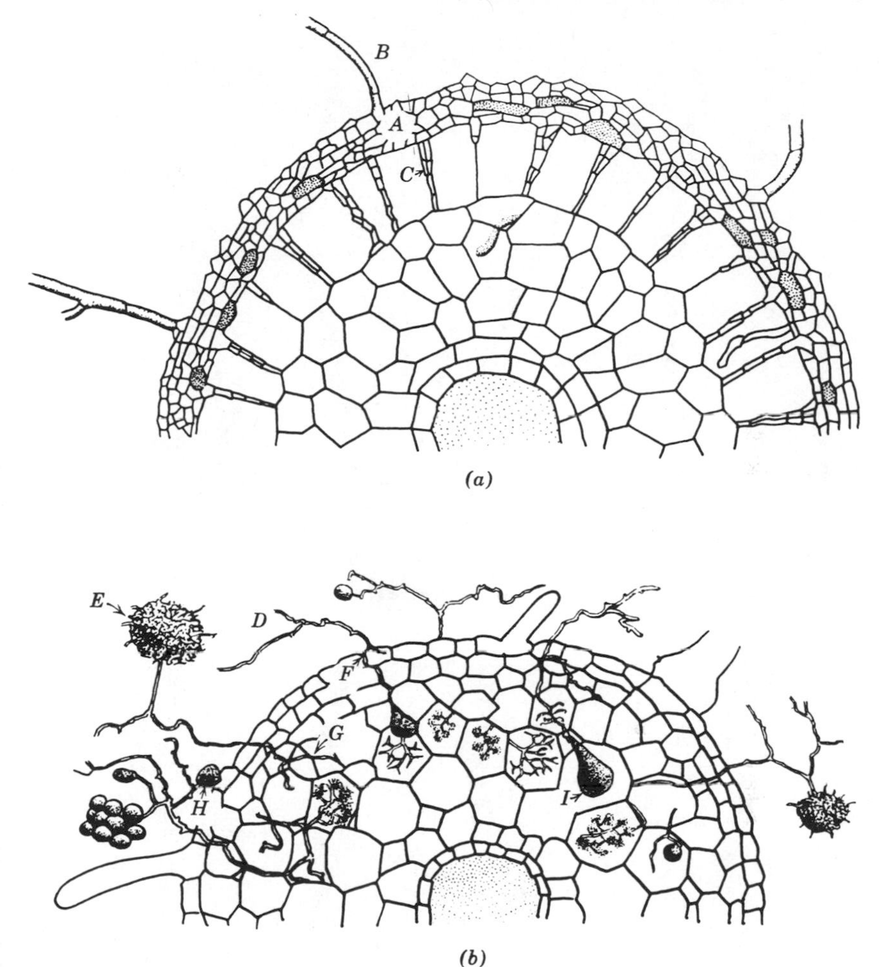

Figure 7.17. Cross section of roots infected with mycorrhizal fungi. (*a*) Ectotrophic mycorrhizal rootlet showing (*A*) sheath of fungal tissue, (*B*) external filament of fungal tissue, and (*C*) penetration of the fungus between the outer cells of the root proper. This type of mycorrhiza is common in forest trees of temperate regions. (*b*) Endotrophic mycorrhizal rootlet. Endotrophic mycorrhizae form no sheath of fungal tissue. Shown in the figure are a variety of fungal forms, including (*D*) external filament of fungal tissue, (*E*) fruiting body on external filament, (*F*) internal filament between host cells, (*G*) internal filament within host cell, (*H*) swollen vesicle or sac external to the root, and (*I*) swollen vesicle or sac within host cell. This type of mycorrhiza is widely distributed in many kinds of plants. (Harley, 1965)

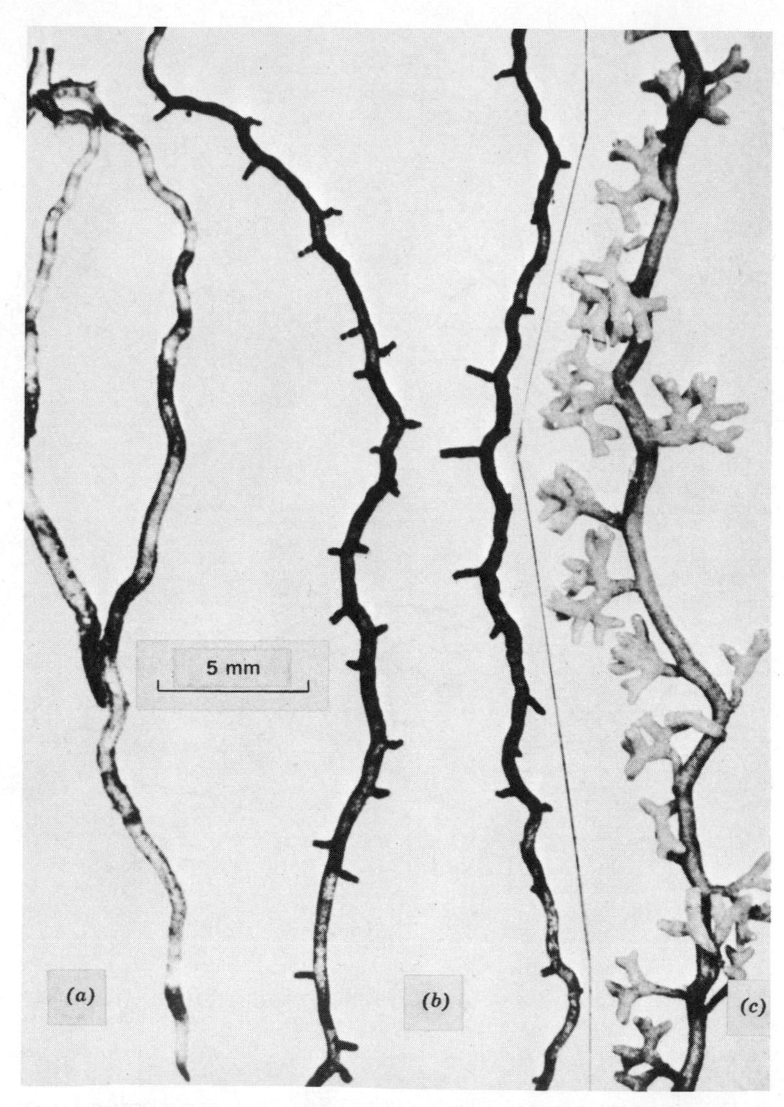

Figure 7.18. Development of roots of *Pinus* under different conditions. (*a*) Roots without mycorrhizae obtained without inoculation where the nitrogen supply was ample. (*b*) Roots without mycorrhizae obtained without inoculation where the nitrogen supply was deficient. (*c*) Roots with mycorrhizae obtained with inoculation where the nitrogen supply was deficient. Note the enhanced development of short roots compared with *b*. (Hatch, 1937)

between soil and fungus of mycorrhizal roots than between soil and the effective absorbing surface of nonmycorrhizal roots. This theory is reasonable on the basis of superficial examination of mycorrhizal and nonmycorrhizal roots (see Fig. 7.18 and other photographs in a bulletin by Hatch, 1937). According to a second theory, mycorrhizal fungi can utilize soil nitrogen that is unavailable to higher plants (usually considered to be organic nitrogen). This theory is reasonable in that soil microorganisms do mineralize soil nitrogen. Tests on mycorrhizal fungi, however, do not indicate that as a class they have a marked quality of hydrolyzing complex organic substances, although some have enzyme systems that are effective in hydrolyzing lignin. Critical experiments to test the effect of mycorrhizal fungi on soil organic nitrogen mineraliza-

Table 7.6. Yield and Nitrogen Content of Coniferous Tree Seedlings Grown on an Infertile Soil with and without Inoculation with Mycorrhizal Fungi (Finn, 1942)

	Yield per Plant, mg	
	Dry Matter	Nitrogen
Control (*A*)[1]	155	1.17
Inoculated (*B*)	223	2.05
B/*A*	1.44	1.75

[1] Some infection of noninoculated controls occurred.

tion have not been done. A third theory is that elemental nitrogen is fixed by the association of mycorrhizal fungi and higher plants. This theory is supported by certain experiments, some with N^{15} (Stevenson, 1959). Other results have been negative.

In all the theories it is implied that part of the nitrogen gained by the fungus is passed on to the plants. From the work of Melin and Nilsson (1952, 1953), it is known that combined nitrogen may be absorbed from solution by mycorrhizal fungi of pine roots and transferred in part to the host. The source of the extra nitrogen found in mycorrhizal plants on low-nitrogen soils now appears to be the primary question to be resolved from the standpoint of soil-plant relationships. The mycorrhizae no doubt immobilize in their tissue some nitrogen absorbed from the soil; however, classification of the subject matter of mycorrhizae under the heading of mineralization and immobilization may be less appropriate than the heading of symbiotic nitrogen fixation, depending on the outcome of investigations on the source of the extra nitrogen. For further information, the books by Kelley (1950) and Harley (1959) and the summary account by Harley (1965) may be consulted.

Ammonium Oxidation and Nitrate Reduction

Nitrogen that has been released as ammonium from organic forms in soil may remain as ammonium or may undergo further transformations to other forms. These transformations are sensitive to soil conditions. Factors affecting ammonium oxidation and nitrate reduction and some of the consequences for plants will be considered in this section.

Aeration

Transformation of ammonium to nitrite and nitrate requires oxygen. Competing for oxygen with the autotrophic ammonium-oxidizing organisms are numerous other organisms, for which oxidation of carbon compounds serves as the primary source of energy. Both oxidations occur simultaneously as long as a supply of elemental oxygen is present. In the absence of elemental oxygen, ammonium oxidation ceases. Certain organisms then are capable of using oxygen derived from nitrite, nitrate, or both in place of elemental oxygen. Removal of oxygen from nitrite and nitrate reduces these substances chemically. The products of the reduction are mostly gaseous forms of nitrogen, including nitrogen dioxide (NO_2) nitrous oxide (N_2O), and elemental nitrogen (N_2). Formation and loss of gaseous forms of nitrogen by biological reduction of nitrite and nitrate is known as denitrification. Only a little of the nitrogen originally present as nitrite or nitrate is reduced to ammonium. Literature on this subject was reviewed by Broadbent and Clark (1965).

The concentration of elemental oxygen required to support ammonium oxidation and to inhibit denitrification is comparatively low. In work with soil, Greenwood (1962) arrived at the figure of 3 micromoles of oxygen per liter of water as the concentration of elemental oxygen required to support ammonium oxidation at half its rate in an atmosphere in which oxygen is not limiting and as the concentration required to support denitrification at half its rate in an atmosphere of nitrogen. This figure may be compared with the concentration of 236 micromoles of oxygen per liter of water in equilibrium with air at 25°C (calculated from handbook data).

The foregoing work by Greenwood was carried out in such a way as to cause movement of solution through the soil aggregates and hence to reduce the importance of diffusion in supplying oxygen to microorganisms in interior locations. In further experiments, he measured the rates of denitrification and ammonium oxidation in water-saturated aggregates at different oxygen percentages in the external air under conditions in which transport of oxygen to internal locations was primarily by diffusion. He found in these experiments that the results depended on the size of the aggregates and could be accounted for by the theory that consumption of oxygen by the microorganisms caused the concentration of oxygen to be lower inside the aggregates than at the surface. According to this theory, soil can be a mosaic of aerobic and anaerobic environments under appropriate circumstances. The balance between ammonium oxidation and denitrification may then be seen to depend on factors such as the distance of diffusion of oxygen through soil water to interior sites away from the air-filled pores, the concentration of oxygen in the air-filled pores, and the concentrations of ammonium and nitrate in the aerobic and anaerobic locations. An additional significant factor is the populations of organisms that bring about the transformations. The populations tend to follow the concentration of substrate, and they are affected by other factors.

In view of the complex and dynamic nature of the processes that affect the balance between ammonium oxidation and denitrification, and also in view of deficiencies in methods of analysis, the considerable diversity of experimental findings that has been reported in the extensive literature on this subject is to be expected. The most general observations have been that loss of nitrogen from nitrate occurs in wet soils high in decomposable organic matter. The decomposable organic matter supports the population of microorganisms that uses the oxygen, and the water inhibits replenishment of oxygen from the atmosphere. These matters were discussed in Chapter 3.

One of the significant implications of the theory of associated aerobic and anaerobic environments is that such circumstances will promote continuous loss of nitrogen by denitrification. If the concentration of nitrate in the soil water is initially uniform, nitrate will first be lost from the environments that become anaerobic. Loss of nitrate from these environments will establish a concentration gradient, and nitrate produced in aerobic environments will then diffuse to the anaerobic environments, where it will be subject to loss. Although this concept is at least 50 years old, critical experiments in which existence of the process has been verified experimentally apparently have not been carried out. Various experiments have been performed, however, in which samples of soil have been subjected to alternating aerobic and anaerobic conditions; these provide clear evidence of loss of nitrogen. Figure 7.19 shows the

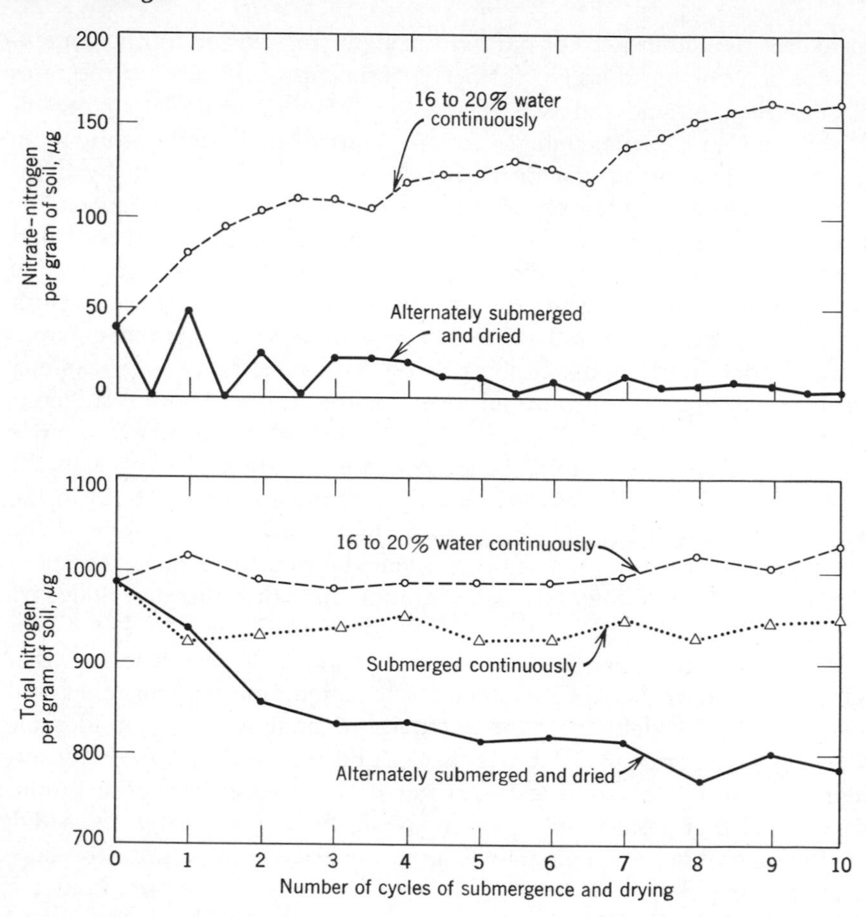

Figure 7.19. Nitrate- and total-nitrogen in a silt loam soil incubated at 35°C under three different conditions for 60 weeks: (1) incubated at a water content of 16 to 20% (water content at permanent wilting percentage = 12%); (2) submerged continuously; and (3) alternately submerged and dried. In the third treatment, the soil was submerged in the first half of each cycle, then dried by evaporation to a water content of about 18%, and maintained at that value for the last part of the cycle. Each cycle required five to six weeks. (Patrick and Wyatt, 1964)

results obtained by Patrick and Wyatt (1964). This experimental work was performed in connection with investigations of nitrogen transformations in soil used for rice culture.

The data in Fig. 7.20 may serve as a basis for speculation about the significance of concurrent ammonium oxidation and nitrate reduction in soil that is wet but not submerged, a situation commonly found in practice. This figure gives values for increases in ammonium- and nitrate-nitrogen and for the sum of these values in soil incubated with the water content adjusted to different matric suctions. In this experiment, air was passed continuously over 50-g samples of soil. The percentages of the soil volume occupied by air-filled pores were 0, 6, and 12 at matric suctions of 0, 0.05, and 0.3 bar. The figure shows a marked decrease in accumulation of nitrate at matric suctions below 0.5 bar (corresponding approximately to the field capacity) and a marked increase in ammonium between a suction of 0.05 bar and 0 bar. The accumulation of ammonium, however, does not balance the decrease in nitrate, so that the sum of ammonium and nitrate accumulated at a suction of 0 bar is less than that at 0.5 bar. The sum is at a minimum at 0.05 bar. Evidence discussed previously provides no indication of a minimum in nitrogen mineralization at a matric suction of 0.05 bar and suggests that mineralization at 0 bar is equal to or greater than that at 0.5 bar. The deficit of mineral nitrogen found in the soil at matric suctions below 0.5 bar therefore may represent loss of nitrate-nitrogen by denitrification. To establish whether the deficit does, in fact, represent loss by denitrification, however, would require additional information. Cooper and Smith (1963) found that the sequence of nitrogen

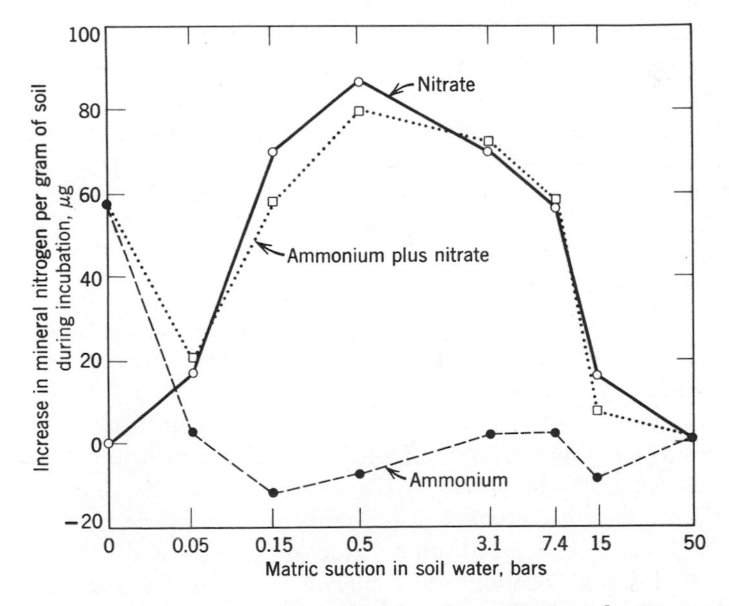

Figure 7.20. Increase in ammonium-, nitrate-, and ammonium- plus nitrate-nitrogen in a silt loam soil incubated for two weeks at 30°C with the water content adjusted to different matric suctions. (Miller and Johnson, 1964)

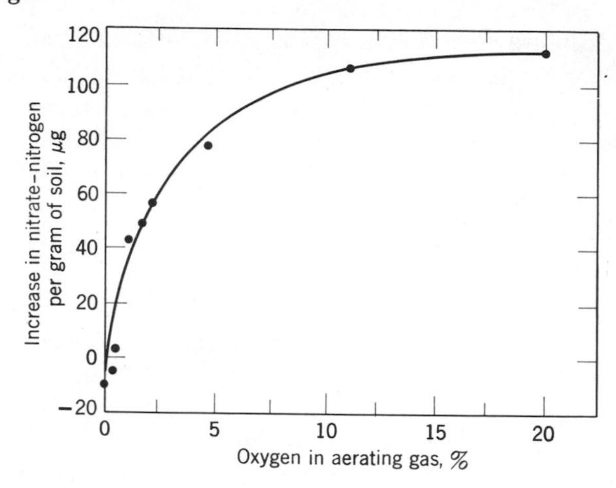

Figure 7.21. Increase in nitrate-nitrogen during incubation of a silt loam soil with ammonium sulfate for three weeks at 30°C under continuous aeration with air-nitrogen mixtures differing in oxygen percentage. (Amer, 1949)

forms during denitrification was nitrate, nitrite, nitrous oxide (gas), and nitrogen (gas).

Figure 7.21 provides an example of a different kind. The data in this figure were obtained by use of a silt loam soil adjusted to 60% of the water-holding capacity (approximately the field capacity). Mixtures of nitrogen and air to give different oxygen percentages were passed continuously through 25-g samples of the soil. Each sample received ammonium sulfate in a quantity equivalent to 200 μg of nitrogen per gram of soil. The quantities of nitrate- and nitrite-nitrogen were measured after an incubation period of 3 weeks. Nitrite-nitrogen was found only at the lower oxygen percentages. Because the quantities did not exceed 0.4 μg/g of soil, they were neglected. One may note from the figure that the maximum accumulation of nitrate-nitrogen occurred at an oxygen percentage approximating that in ordinary air and that accumulation at half the maximum rate occurred at a concentration of about 2% oxygen in the aeration stream. The 2% concentration of oxygen corresponds to about 20 micromoles/liter of water, which is approximately seven times higher than Greenwood's (1962) value of 3 micromoles/liter. The discrepancy presumably is due primarily to the need for a high oxygen concentration external to the aggregates to supply oxygen to internal locations by diffusion. Now, if ammonium oxidation and denitrification both proceed at half their maximum rate at about 3 micromoles of oxygen per liter, as found by Greenwood (1962), and if oxidation of ammonium is inhibited to some degree even where the concentration of oxygen in the air external to the soil aggregates approaches that in the atmosphere, one may infer that, conversely, nitrate reduction and loss by denitrification is not inhibited completely even where the concentration of oxygen in the air external to the soil aggregates approaches that in the atmosphere. This inference, of course, applies to the soil and the conditions associated with the data in Fig. 7.21.

Losses of nitrogen in gaseous forms by strictly chemical processes and the significance of the over-all losses will be discussed subsequently in the section on nitrogen balance.

Water Supply

In dry soils, as in wet soils, there is a tendency for accumulation of ammonium without oxidation to nitrate. Figure 7.22 gives results of one investigation of mineral nitrogen transformation in soil under relatively dry conditions. In this experiment, there was essentially no accumulation of nitrate where the water content of the soil was below the permanent wilting percentage. At the permanent wilting percentage, the ammonium content of the soil decreased with time, and nitrate accumulated in excess of the loss of ammonium, indicating increased nitrogen mineralization and rapid oxidation of ammonium. The principal explanation for the differences in ammonium oxidation under the dry soil conditions of this experiment is probably not a matter of oxygen supply but rather the broader collective adaptation of the many types of ammonium-producing organisms than of the more specialized ammonium-oxidizing organisms. Another factor in the behavior observed in Fig. 7.22 may be the population of nitrifying organisms. These organisms seem to be sensitive to drying, as suggested by occurrence of a lag period in ammonium oxidation when air-dried soils are moistened and incubated (Sabey et al., 1959). Accordingly, ammonium oxidation at a given water content in the dry range could conceivably be greater if the desired water content is obtained by drying a moist soil than by moistening a dry soil. A paper illustrating both the effect of water content of soil on ammonium and nitrate production and the lag in nitrate production in remoistened soils was published by Reichman et al. (1966).

Soil pH

The autotrophic ammonium-oxidizing organisms are sensitive to acidity, and successful culture of these organisms in the laboratory involves

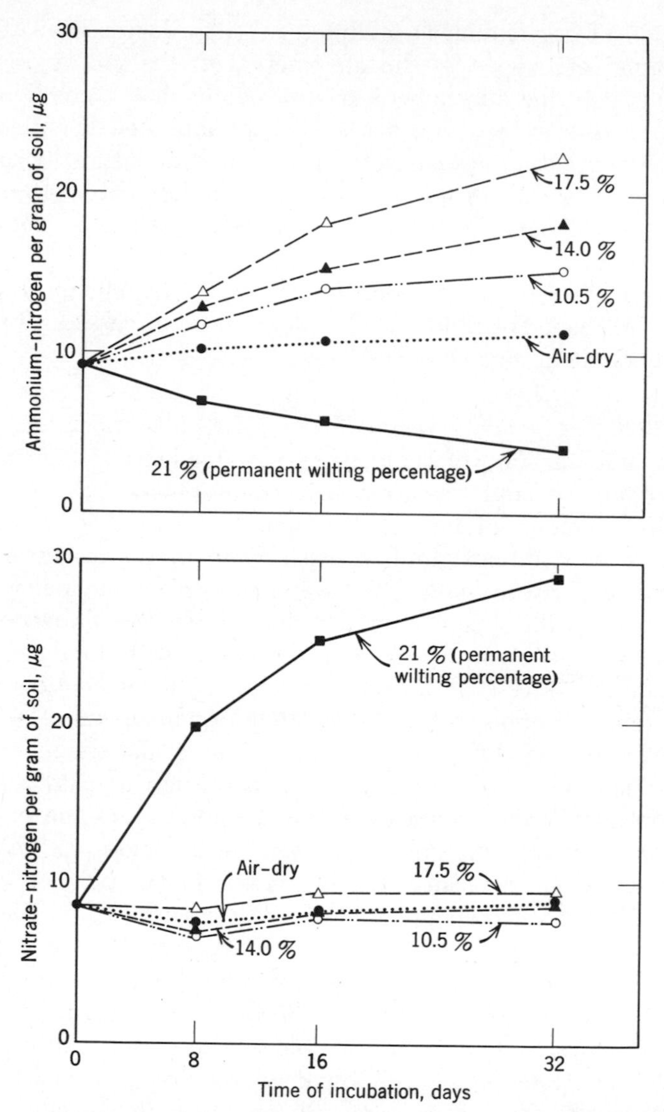

Figure 7.22. Content of ammonium- and nitrate-nitrogen in soil during incubation at 35°C at different water contents from air-dry to the permanent wilting percentage. Numbers in the graphs refer to water percentages in the soil. (Robinson, 1957)

use of buffered, mildly alkaline media, sometimes containing calcium carbonate. In soil, ammonium oxidation takes place at *p*H values below those at which the organisms isolated from soil do well in synthetic media. Nevertheless, there is evidence that ammonium oxidation is inhibited by low soil *p*H values (Fig. 7.23), that the population of autotrophic ammonium-oxidizing organisms is greater in neutral and alkaline soils than in acid soils (Wilson, 1928; Morrill, 1959), and that inoculation of limed acid soil with other soil that produces nitrate rapidly may increase the rate of nitrate production (Shih, quoted by Blue et al., 1964).

The autotrophic ammonium-oxidizing organisms are sensitive also to the combination of high concentrations of ammonium and high *p*H values developed in soil by anhydrous ammonia, urea, and other fertilizers that contain or produce ammonium hydroxide. Table 7.7 illustrates the inhibition of ammonium oxidation associated with application of increasing quantities of anhydrous ammonia to soil. Because of this inhibition, there is a tendency for oxidation to proceed most rapidly in peripheral locations around the fertilizer zones where ammonium concentrations and *p*H values are not excessive. As oxidation proceeds, nitric acid is produced, the *p*H values decrease, and the soil may become strongly acid. The central part of the fertilizer zone is the last to become acid. The behavior, of course, varies with the conditions. In calcareous

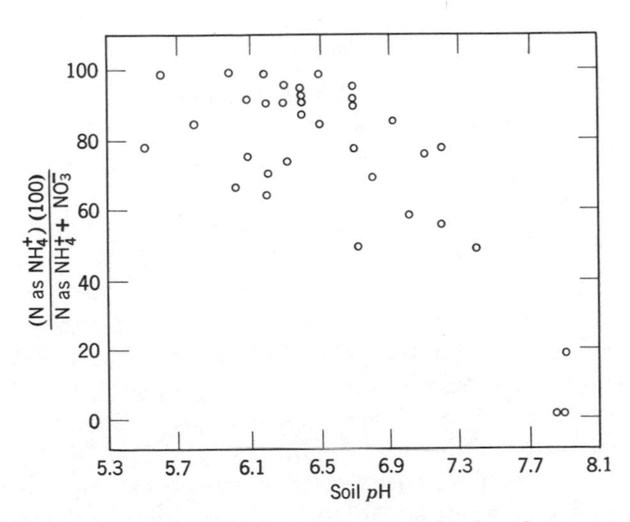

Figure 7.23. Increase in ammonium–nitrogen in air-dried and remoistened soils of different *p*H values as a percentage of the increase in ammonium– plus nitrate–nitrogen during an incubation period of seven days at 35°C. (Eagle, 1961)

soils, for example, high *p*H values result from applications of anhydrous ammonia or urea, but the nitric acid subsequently produced reacts with the carbonates and does not make the soil acid. Movement of water through the fertilizer band disperses the nitrate and enlarges the zone of fertilizer-affected soil. Papers dealing with transformations of nitrogen in soils treated with anhydrous ammonia in localized applications include those by McIntosh and Frederick (1958) and Nõmmik and Nilsson (1963).

Table 7.7. Nitrate Production and *p*H Values in a Fine Sandy Soil Treated with Different Quantities of Anhydrous Ammonia (Eno et al., 1955)

Ammonia-Nitrogen Applied per Gram of Soil, μg	Soil *p*H		Nitrate-Nitrogen per Gram of Soil after Two Weeks, μg
	Initial	After 2 Weeks	
0	6.5	7.1	24
173	8.2	5.8	139
302	8.5	5.6	164
402	8.6	7.2	64
773	9.0	8.8	0

Stojanovic and Alexander (1958) found that in a soil maintained at *p*H 7.7 most of the oxidized ammonium appeared as nitrate during 12 days at initial concentrations of ammonium-nitrogen up to 250 μg/ml but as nitrite at concentrations of 890 and 1780 μg of ammonium-nitrogen per milliliter. Nitrite accumulated only while ammonium remained in solution and then disappeared rapidly. Nitrite was oxidized rapidly to nitrate at *p*H 7.7 in the absence of ammonium, and the authors pointed out that other investigators had found this oxidation to occur at higher *p*H values as well if ammonium was absent. From this and other information, partly derived from the literature, they inferred that (1) high concentrations of ammonium are more inhibitory to organisms that oxidize nitrite to nitrate than to organisms that oxidize ammonium to nitrite and (2) the greater inhibition at high *p*H values than at low *p*H values may be a consequence of toxicity of ammonia present at high *p*H values.

Several instances of nitrite toxicity to plants from oxidation of ammonium under alkaline conditions have been recorded. Figure 7.24 shows the results of an experiment in which a marked accumulation of nitrite occurred during nitrification of the ammonium developed from an application of urea to a soil having an initial *p*H value of 7.4. The urea hydrolyzed rapidly, releasing ammonium in the soil. The peak concentration of ammonium-nitrogen was nearly 300 μg/g of soil, and the corresponding peak *p*H value was 8.9. Court et al. (1962) reported that volatilization of ammonia from the soil during the first week could be detected by smell. The concentration of ammonium fell after the first week and nitrite appeared, reaching a maximum concentration in the fourth week. The decline of nitrite after the fourth week was accompanied by an increase of nitrate. The soil *p*H value had dropped to about 6.2 in the fourth week and remained there during the fifth and sixth weeks. Occurrence of the greatest inhibition of plant growth during the time of high nitrite concentration suggests the importance of nitrite toxicity. The authors reported that many of the plants appeared to be dead at the end of tests made during the fourth and fifth weeks but that in the sixth week the plants in contact with the urea-treated soil

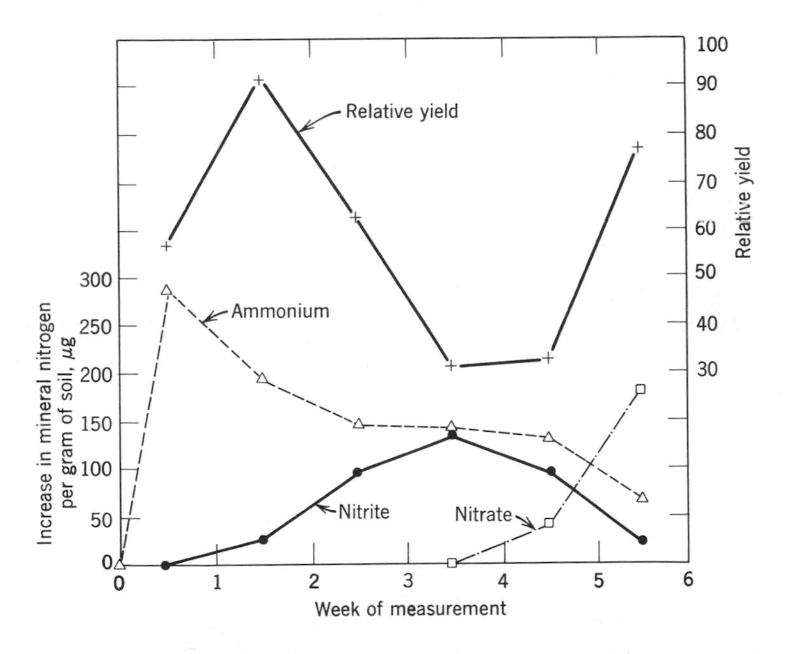

Figure 7.24. Increase in ammonium-, nitrite-, and nitrate-nitrogen in a sandy loam soil (initial *p*H 7.4) during incubation after addition of urea, and yield of corn seedlings as a percentage of the yield of controls without application of urea. The plant-response measurements were made by a technique in which pregerminated plants were in contact with the soil for only the week indicated. (Court et al., 1962)

Figure 7.25. Results of an investigation of nitrite toxicity in a calcareous loam soil (*p*H 7.7) from California. All comparable cultures received equal quantities of fertilizer phosphorus and equal quantities of fertilizer nitrogen as ammonium nitrate or as indicated. Applications of nitrogen were 125 or 188 μg/g of soil in the various experiments. The test crop was lettuce. (*a*) Stunted yellow plants with ammonium nitrate. (*b*) Vigorous green plants with potassium nitrate. (*c*) Effect of number of plants. At a stand of five plants per culture, all plants are stunted and yellow. These plants made little or no growth during the first month after emergence. At a stand of 100 and 200 plants, all plants are green. These plants showed no growth lag after emergence. (*d*) Effect of inhibition of nitrification by heating the soil before planting. Plants are stunted and yellow on the unheated soil but vigorous and green on the heated soil. Nitrite was found

seemed as healthy as the controls. From the figure, it may be noted that most of the nitrite had disappeared by the sixth week.

In another investigation involving a calcareous soil, Paul and Polle (1965) found that application of ammonium nitrate or ammonium sulfate produced stunted yellow lettuce plants. The condition of poor growth and yellow leaves persisted for about a month. Then the leaves became green and normal growth commenced. Application of nitrogen as potassium nitrate, however, resulted in green plants that grew vigorously from emergence. Figure 7.25*a,b* shows the contrast between plants treated with ammonium nitrate and potassium nitrate. The potassium supplied by the potassium nitrate apparently was unimportant in this comparison, because separate tests showed that the soil was well supplied with potassium. Nitrite toxicity was suspected on account of the temporary period of growth inhibition produced by ammonium salts but not by potassium nitrate, and additional experiments were undertaken to determine whether nitrite toxicity was involved. In one experiment, the number of plants per container was varied; this showed that the growth improved as the number of plants increased, as would be expected if a soluble inhibitor were being absorbed and diluted in greater quantities of plant tissue. Results of this test are shown in Fig. 7.25*c*.

Paul and Polle reasoned further that, if nitrite toxicity were the cause of the poor growth with ammonium sources of nitrogen, a treatment to inhibit nitrite formation should improve growth. Accordingly, they determined the response of the plants to prior heating of the soil and found that growth upon addition of ammonium-nitrogen was poor on the unheated soil and good on the heated soil (Fig. 7.25*d*). No nitrite was produced in the heated soil, but quantities up to 74 μg of nitrite-nitrogen had accumulated per gram of unheated soil after 46 days. (This experiment was done at lower temperatures than the others, which may account for the slowness of nitrite accumulation.) To test further the effect of treatments to inhibit microbial formation of nitrite, the authors added a nitrification inhibitor (see the following section on chemical inhibitors) with an ammonium source of nitrogen. Plants grown in the presence of the inhibitor were green and vigorous, whereas those grown on the ammonium-treated soil without the inhibitor

---

in the unheated soil but not in the heated soil. (*e*) Effect of inhibition of nitrification by application of 2-chloro-6-(trichloromethyl) pyridine with the fertilizer. Plants are stunted and yellow in the control but vigorous and green in the presence of the inhibitor. Nitrite was found in the control soil, but only a trace was found at one sampling date in the presence of the nitrification inhibitor. (Photographs courtesy of J. L. Paul)

were yellow and stunted (Figure 25*e*). In another experiment, the authors found that injecting a culture of *Nitrobacter* into the soil almost completely prevented the accumulation of nitrite and caused good growth of plants. These experiments on inhibition of nitrification, incidentally, provide evidence that the poor growth with ammonium sources of nitrogen was not attributable to the ammonium as such.

In still other tests, the growth of plants varied inversely with the content of nitrite-nitrogen found in the soil with the various treatments. Additions of nitrite-nitrogen to soil in concentrations of the order of those found in the ammonium-salt treatments inhibited the growth.

Although these results provide clear evidence for nitrite toxicity, it should be noted that additions of nitrite to the soil at planting did not affect the plants in the same way as nitrite developed from ammonium nitrate. Nitrite added at planting reduced germination and did not result consistently in development of the leaf-yellowing symptoms. The authors suggested in this connection that the timing might be important: the plants treated with nitrite in their experiments received all the nitrite at planting, whereas those grown with nitrite developed microbiologically were subjected to a gradually increasing concentration of nitrite.

Experiments on the response of plants to nitrite in culture solutions have shown that the toxicity of a given concentration of nitrite-nitrogen increases as the *p*H decreases. Bingham et al. (1954) estimated that the concentration of nitrite-nitrogen required to produce a 50% reduction in yield of barley, bean, and tomato was 40 to 50 μg/ml at *p*H 6, 5 to 7.5 μg/ml at *p*H 5, and 1 to 3 μg/ml at *p*H 4. The fact that the concentrations of nitrous acid-nitrogen (calculated with the aid of the dissociation constant of nitrous acid) associated with a 50% yield reduction were mostly in the range of 0.1 to 0.25 μg/ml, regardless of the *p*H, led them to suggest that the toxicity was due to nitrous acid.

The lower toxicity of nitrite to plants under alkaline conditions than under acid conditions is fortunate from the practical standpoint in view of the tendency for nitrite to accumulate in higher concentrations in alkaline soils than in acid soils. At *p*H 8, a concentration of 50 μg of nitrite-nitrogen per milliliter decreased the yield of tomato plants from 72 g to 55 g in the work of Bingham et al. (1954). Taking this as a yield reduction of significant magnitude, the authors suggested that concentrations of 50 μg or more of nitrite-nitrogen in the soil solution in the root zone in alkaline soils would probably be toxic to plants. In soil having a water content of 20% by weight, this concentration in the soil solution would correspond to 10 μg/g of soil.

Concentrations of nitrite-nitrogen as high as 140 μg/g of alkaline soil have been observed in the laboratory in soil treated with anhydrous ammonia (Duisberg and Buehrer, 1954). Chapman and Liebig (1952) measured concentrations as high as 90 μg/g of alkaline soil treated with large quantities of urea in the field. These values are the extremes. Most values reported in laboratory and field work are much lower. Nevertheless, occasional values above the suggested tolerance level of 10 μg/g of soil have been reported, and it seems likely that nitrite toxicity is sometimes significant under field conditions. Experimental investigation of nitrite toxicity incidental to application of ammonium-bearing or ammonium-producing fertilizers in the field, however, is difficult. The inhibition of nitrite oxidation due to fertilization is not general throughout the soil but exists only in the portions containing the high concentrations of ammonium. Nitrite is a freely diffusible ion and may be moved out of the zones of inhibition by diffusion or mass movement with water into parts of the soil where its oxidation proceeds more readily. Consequently, nitrite usually is not found in the entire root zone, its existence is only transitory, and it is accompanied by varying quantities of ammonium and nitrate, which have beneficial effects.

Temperature

If ammonium oxidation were inhibited sufficiently by low temperatures, ammonium-bearing fertilizers applied in late autumn could conceivably remain without oxidation and without appreciable loss by leaching during the winter and early spring, to be nitrified only when the soil becomes warm enough for crops to grow. Autumn application of nitrogen fertilizer would aid producers in distributing their manufacture and sales over a greater proportion of the year instead of having them concentrated in the winter and spring. Also, in some areas, it is more convenient to apply nitrogen in the autumn because of weather, soil conditions suitable for transport of machinery, and competition with other work.

The optimum temperature for ammonium oxidation is usually found to be in the range from 30 to 35°C. Lower temperatures do inhibit ammonium oxidation, as illustrated by data of Sabey et al. (1956) in Fig. 7.26; the inhibition is gradual, however, and it does not approach the "ideal" situation for most efficient autumn application of ammonium-bearing fertilizer. Frederick (1956) found that in some soils the rate of nitrification at temperatures of 0 to 2°C was great enough to oxidize completely in 2 months a quantity of 55 kg of ammonium-nitrogen per hectare. More recent work on this subject and references to the literature may be found in a paper by Anderson and Boswell (1964).

Chapman and Liebig (1952), Tyler et al. (1959), and Justice and

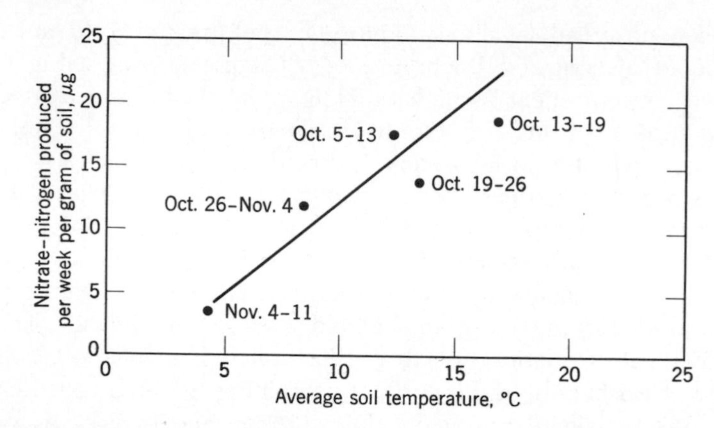

Figure 7.26. Rate of production of nitrate-nitrogen in ammonium-fertilized soil in the field in Iowa at different temperatures in the autumn. Each measurement was made on previously unfertilized areas of soil to which ammonium sulfate was added at the beginning of the test period to provide 112 and 224 kg of ammonium-nitrogen per hectare. (Sabey et al., 1956)

Smith (1962) all observed a tendency for persistence of nitrite at low temperatures in calcareous soils treated with ammonium. Justice and Smith (1962) found, in addition, that oxidation of ammonium was much slower and persistence of nitrite much longer in a calcareous soil incubated with ammonium sulfate at 35°C than at 25°C. These findings may all be attributable to combined inhibitory effects of temperature and ammonium on ammonium-oxidizing organisms at high *p*H values.

Chemical Inhibitors

The capability to oxidize ammonium rapidly and completely was once held to be a desirable soil quality, probably because of the impression that nitrate was the appropriate source of nitrogen for plants and ammonium was unsuitable. This stigma attached to ammonium as a source of nitrogen in plant nutrition has now largely disappeared. Currently, retention of mineral nitrogen in the form of ammonium is looked upon more favorably; in fact, the view is growing that ammonium oxidation may do more harm than good. Oxidation of ammonium increases the susceptibility of mineral nitrogen to loss by leaching and denitrification, it increases soil acidity, and it produces nitrite toxicity under some conditions. By prolonging the persistence of soil and fertilizer nitrogen in the form of ammonium, chemical inhibitors of ammonium oxidation could conceivably increase the efficiency of a given quantity of nitrogen in crop production. Absence of undesirable side-effects and low cost are, of course, other features desirable in such an inhibitor.

That ammonium oxidation in soil may be inhibited by certain chemicals has long been known. Alexander (1965) published a table listing many inhibitors and including citations to the literature describing their effects. Probably the most extensive research on any single inhibitor has been done on 2-chloro-6-(trichloromethyl) pyridine. Table 7.8 gives some of Goring's (1962) data showing that none of the nitrogen applied to soil as ammonium sulfate could be recovered as ammonium after incubation for 16 weeks in the absence of the inhibitor but that the major part of the nitrogen was still present as ammonium after addition of the inhibitor at concentrations of 1 μg or more per gram of soil where the incubation was carried out at 10 or 21°C. At 32°C the inhibition was less effective, and a higher concentration of inhibitor was required. Additional data not reproduced in the table show that the inhibition persisted almost unchanged at 24 weeks at the lower temperatures, but at 32°C the effect had disappeared at all concentrations of inhibitor employed. Loss of effectiveness with time appears to be caused by loss of the chemical by a combination of volatilization and degradation to 6-chloro-picolinic acid (Redemann et al., 1964).

The chemical 2-chloro-6-(trichloromethyl) pyridine appears to be remarkably specific for the autototrophic ammonium-oxidizing organisms.

Table 7.8. Ammonium-Nitrogen in a Sandy Loam Soil of *p*H 7.3 after Incubation for Sixteen Weeks at Different Temperatures in the Presence and Absence of Ammonium Sulfate and with Different Concentrations of 2-Chloro-6-(Trichloromethyl) Pyridine as a Nitrification Inhibitor (Goring, 1962)

Inhibitor Added per Gram of Soil, μg	Ammonium-Nitrogen per Gram of Soil, μg					
	Incubated at 10°C		Incubated at 21°C		Incubated at 32°C	
	Control	$(NH_4)_2SO_4$[1]	Control	$(NH_4)_2SO_4$[1]	Control	$(NH_4)_2SO_4$[1]
0	0	0	0	0	0	0
1	8	192	0	166	0	40
2	8	192	14	170	0	12
5	8	192	14	170	14	96
10	8	194	14	180	16	150

[1] Two-hundred micrograms of nitrogen per gram of soil.

Shattuck and Alexander (1963) found that the inhibitor affected both *Nitrosomonas,* which oxidizes ammonium to nitrite, and *Nitrobacter,* which oxidizes nitrite to nitrate; but the activity of the former organisms was inhibited much more than that of the latter. Production of nitrate by several strains of a nitrifying fungus (*Aspergillus flavus*) was not affected by concentrations of the inhibitor that effectively controlled ammonium oxidation by *Nitrosomonas.* Growth of various other heterotrophic and autotrophic organisms was unaffected by the inhibitor. The specificity of 2-chloro-6-(trichloromethyl) pyridine for reducing ammonium oxidation by *Nitrosomonas* and *Nitrobacter* and the marked reduction of ammonium oxidation that occurs in the presence of the inhibitor is one of the reasons for supposing that these bacteria are predominant among microorganisms capable of ammonium oxidation in soil.

Field experiments with cotton, sweet corn, and sugar beets under irrigation in California (Swezey and Turner, 1962) showed that both the yield of the crop and the nitrogen content were increased where 2-chloro-6-(trichloromethyl) pyridine was applied with anhydrous ammonia or ammonium hydroxide as sources of nitrogen. In these experiments, the nitrogen was applied as a side dressing while the crops were growing. Turner and Goring (1966) summarized research on this inhibitor.

Reddy (1962) and Soubies et al. (1962) found that dicyandiamide greatly delayed the oxidation of ammonium applied to soil as ammonium sulfate or urea. Soubies et al. (1962) found that field application of dicyandiamide (also called cyanoguanidine) with urea in late November in France inhibited ammonium oxidation completely or almost completely until the end of March, with the consequence that loss of nitrate by leaching during the winter months was greatly reduced. By the end of June, however, most of the ammonium had been oxidized. In the experiments on autumn application of urea to wheat, they found that yields were increased where dicyandiamide was added in the proportion of 11.5% of the nitrogen applied as urea. This proportion of dicyandiamide supplies an additional 7.7% nitrogen, which could have accounted for part of the extra yields. Reddy (1962) obtained no benefit from application of dicyandiamide with ammonium sulfate under field conditions in Georgia, where the application presumably was made in the spring. Toxic effects of the inhibitor on various crops were noted where increasing concentrations of dicyandiamide were applied with sodium nitrate. Apparently, dicyandiamide must be used with caution unless a period of detoxification preceeds planting.

Although performance of these inhibitors of ammonium oxidation seems to fall somewhat short of the ideal, the possibilities have been by no means exhausted. With further investigation, inhibitors with better properties may be developed. Hamamoto (1966) reported active research on this subject in Japan and investigation of the action of substances other than those mentioned here.

Indexes of Nitrogen Availability to Plants

Nitrogen present in organic forms associated with the soil solids appears to be unavailable to plants, but that present in soluble organic forms in the soil solution is, no doubt, available in part (Ulrich et al., 1964; Miller and Schmidt, 1965). Investigation of the significance of the soluble organic forms to plants is difficult because of their low concentration, the variety of forms present, and the rapidity with which they are converted to other forms. There is little doubt that the bulk of the chemically combined nitrogen susceptible to absorption by plants at a given moment is usually the sum of the exchangeable ammonium and the ammonium, nitrite, and nitrate in the soil solution. This concept is generally accepted.

Although substantial amounts of mineral nitrogen can be caused in the laboratory to accumulate in soils from the soil organic matter, such amounts are usually not found in the field. Nitrate, which is present as a freely diffusible anion in the soil solution, is removed periodically by downward movement of excess water. Moreover, depletion of nitrate from soil by plant roots occurs through distances of several centimeters. The mineral nitrogen found in soil therefore represents a balance between accumulation and removal. The balance fluctuates but usually is such that the amount of mineral nitrogen present (excluding nonexchangeable ammonium) is equivalent to only a small proportion of the total nitrogen plants absorb during a season or a year. Figure 7.27 illustrates the fluctuation of ammonium and nitrate in unfertilized soil under coffee trees as found by repeated sampling and analysis during a number of consecutive months. The causes of the fluctuations cannot be explained in detail, but it is evident that the downward trend of nitrate during the early months is associated with relatively high rainfall and rapid growth.

Where the adequacy of soil mineral nitrogen for plants is concerned, one is usually interested, as a first approximation, in the over-all effects during a given year or growing season. It is evident from Fig. 7.27 that if one were to use the content of ammonium and nitrate in the soil as an index of the availability of nitrogen for plants, the index value obtained would vary a great deal, depending on the day on which

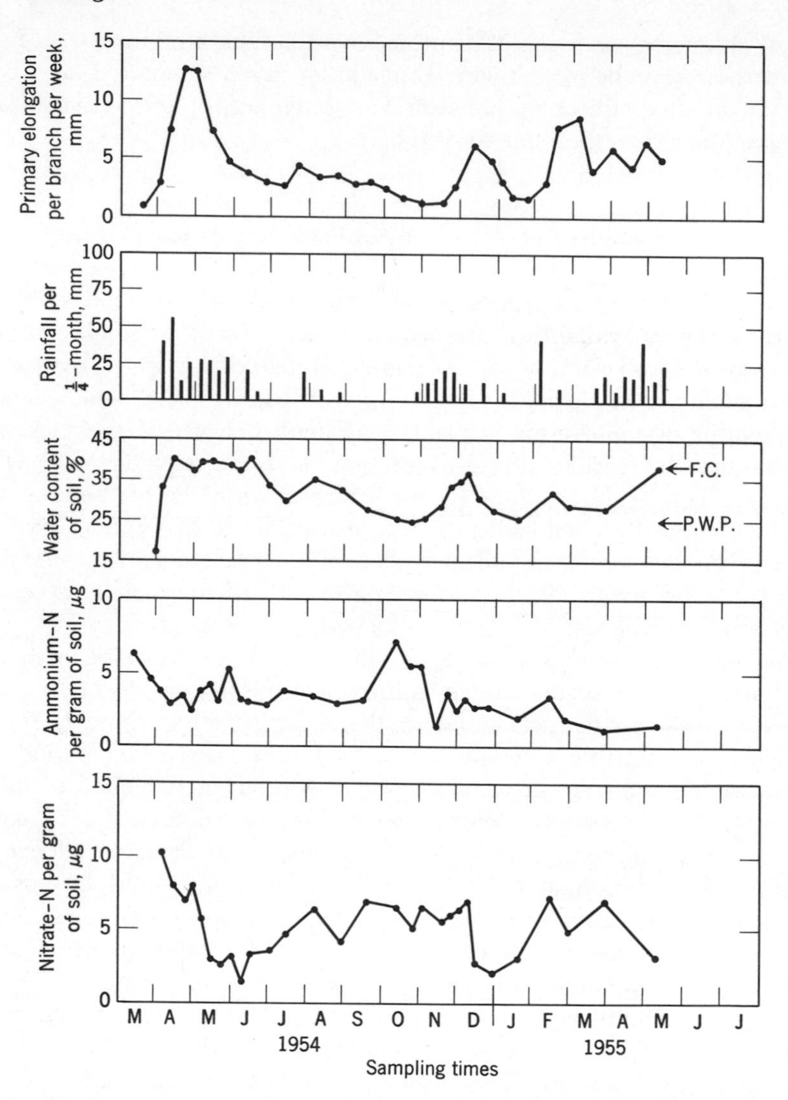

Figure 7.27. Ammonium and nitrate found in periodic samples of the surface 61 cm of soil under unfertilized, mature, unshaded coffee trees on a loam soil in Kenya, with associated data on tree growth, rainfall, and water content of the soil. F.C. = field capacity; P.W.P. = permanent wilting percentage. (Robinson, 1960)

the sample happened to be taken; and the value on the day of sampling would not provide much information about values to be expected on days other than those immediately preceding and following the date of sampling. On the other hand, if the soil were analyzed frequently enough, one could obtain integrated values for ammonium and nitrate over the period of interest. The integrated values would provide an assessment of the balance between accumulation and removal that would probably be related to the degree of sufficiency of the soil nitrogen for plants.

Integrated values are not obtained for practical purposes because of the large amount of work involved. Instead, various alternative means are employed in an attempt to obtain from analysis of a given sample an index of the mineral nitrogen a soil will supply to plants over a longer period of time. Bremner (1965b) reviewed the various methods that have been proposed.

Of the many methods, the only ones that appear to be used extensively in soil-testing work for field crops are measurement of the total content of organic matter and measurement of the nitrogen mineralized during incubation of a sample of soil in the laboratory. The content of organic matter provides an index of the total nitrogen or the total substrate from which mineral nitrogen is produced. The reason for measuring organic matter instead of total nitrogen is that an approximate value for organic matter can be obtained much more easily than can a value for total nitrogen (the method most commonly used was described by Allison, 1965). Measurement of the nitrogen mineralized during incubation of a sample of soil is a more time-consuming process; but most of the time is involved in incubation, which requires no attention if performed in appropriate ways. The incubation method is probably used more extensively than any other. It has the advantage of providing automatically an integration of the effects of amount and composition of substrate that affect the natural process in the field. The mineral nitrogen present at the time of sampling may or may not be taken into account. Sometimes only nitrate production is measured, but it appears preferable to measure the sum of ammonium, nitrite, and nitrate. Nitrite is usually present only in traces and hence is usually of no significance. The various forms may be determined individually if desired, or the sum of the three forms may be determined in a single analysis. Methods were described by Bremner (1965b).

Figure 7.28 gives an example of results obtained by use of organic matter, total nitrogen, and nitrogen mineralized during incubation as indexes of nitrogen uptake by rice plants grown on ten different soils in the greenhouse. The comparative scatter of the points representing

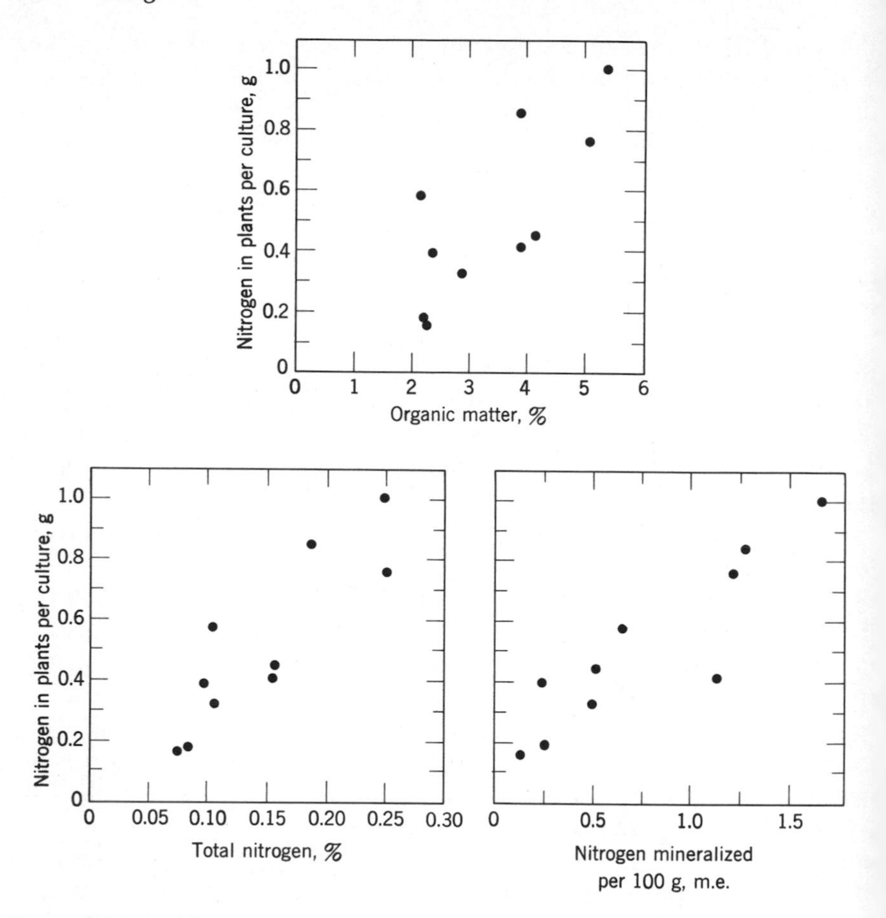

Figure 7.28. Yield of nitrogen in rice plants grown in the greenhouse on ten soils versus content of organic matter, total nitrogen, and mineralizable nitrogen in the soils. (Lopez and Galvez, 1958)

individual soils shows that the precision of the three indexes increased in the order named.

Chemical indexes of nitrogen availability based on measurement of a fraction of the organic nitrogen have generally given disappointingly poor results. The only one that has proved comparable in value to the incubation method is that of determining the total nitrogen in a boiling water extract of soil. The method was proposed originally by Livens (1959, 1959a) and was modified by Keeney and Bremner (1966).

Indexes of nitrogen availability obtained in any manner on single samples of soil are empirical in nature and cannot be expected to provide information on the many uncontrolled variations in conditions from one place to another in the field and from time to time at a given location. Some of these variations, but not all, can be taken into account in interpreting the laboratory measurements if suitable supplementary information is available.

Nitrogen Balance

In each annual cycle, some of the organic nitrogen in the soil is mineralized and some mineral nitrogen is immobilized. Some of the nitrogen is removed by plants, and some is returned in plant residues; some is lost to the atmosphere, and some is returned; some may be lost by leaching and some added by fertilization. Some may be lost be erosion or added by deposition.

A general conception of the significance of some of these processes in terms of quantities of nitrogen and annual rates of change may be derived from results of an experiment by Jansson (1963) in Sweden in which 500-mg quantities of N^{15}-labeled nitrogen in the form of sodium nitrate were added at the beginning of a 6-year experiment to 5.5-kg quantities of a sandy loam soil on which oats were grown outdoors each year. The containers allowed free drainage, and a pan was present beneath each to catch any drainage water during periods of high rainfall. The drainage water was poured back on the soil when watering was necessary, thus eliminating losses by leaching. Analyses of the crops and the soil each year then made it possible to determine how much of the labeled nitrogen had been recovered in the crops, how much was left in the soil, and how much had been lost by volatilization.

Figure 7.29 shows that most of the loss of labeled nitrogen to the atmosphere and most of the recovery in the oats occurred in the first year. Although a substantial quantity of labeled nitrogen remained in the soil, only a little of it was recovered in the oats after the first year. In years after the first, the rate of recovery of the labeled nitrogen in the oats averaged only 3.4% of the quantity of labeled nitrogen present in the soil at the time of planting. At this rate of removal, 20 years would be required for the recovery in crops of half of the labeled nitrogen residual in the soil after the first year. Another 20 years would be required for recovery of half of the remainder, and so on. Theoretically, an infinitely long time would be required to recover all the labeled nitrogen from the soil.

To judge from the recovery of over half of the initial application of labeled nitrate-nitrogen in the oats in the first year and from the low rate of recovery in succeeding years, most of the labeled nitrogen

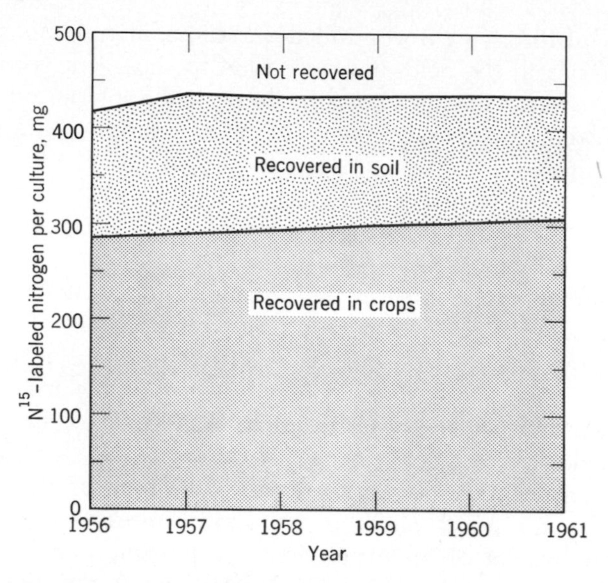

Figure 7.29. Balance sheet for 500 mg of N^{15}-labeled nitrogen added as sodium nitrate in 1956 to 5.5-kg quantities of a sandy loam soil that were cropped annually to oats. (Jansson, 1963)

present in the soil after the first year was not in the form of nitrate. No doubt most of it was present in organic forms as a result of immobilization by microorganisms and incorporation into organic forms in the roots. The recovery of added nitrogen in the oats after the first year then represents an index of the net mineralization of labeled nitrogen residual in the soil in organic forms.

Additional inferences about the length of time given atoms of nitrogen remain in soil may be made from radiocarbon dates of the organic carbon. In unpublished work, R. V. Ruhe obtained apparent ages of 410, 840, 1545, and 16,500 years for samples for organic carbon from the surface soil and successively greater depths in a soil profile developed in loess under grass vegetation in Iowa (the same soil type pictured in Fig. 1.31). The oldest sample was from organic matter in the A horizon of a soil buried beneath the modern profile at a depth 265 cm below the present surface. The word "apparent" is used in connection with these ages because the values given correspond to the ages if all the carbon had been incorporated from the atmosphere into the organic matter at the same time before the present and if all the carbon incorporated had been at isotopic equilibrium with the carbon dioxide in the atmosphere. These conditions are not fulfilled, and so the calculated ages must be regarded with reservation. Nevertheless, considerations that will not be detailed here support the view that the length of time since incorporation of most of the atmospheric carbon into organic carbon in this soil profile is measurable at least in centuries. By inference, the same is true of the organic nitrogen.

Although much of the organic nitrogen in soil may remain as such for a long time, mineralization of nitrogen from this source is of vital importance in supplying nitrogen for plants; and the rate of nitrogen mineralization in a given soil increases with the amount of nitrogen present in these relatively stable organic forms. It is therefore of concern whether the total nitrogen content is low or high and whether current management practices in a particular instance are causing a decrease or an increase. The ways in which nitrogen is lost and gained by soils and the significance of the various processes to plants will be considered in this section.

Losses

Crop Removal. The loss of nitrogen by crop removal from the harvested crop land of the United States was estimated by Lipman and Conybeare (1936) to be 28 kg/hectare for the year 1930. The figure of 28 kg refers to the nitrogen content of the portion of the crop that is harvested, not to the total. The total was probably about 40 to 45 kg. On the basis of Lipman and Conybeare's estimate of 3200 kg/hectare as the average nitrogen content of the plowed layer of soil on which these crops were grown, removals of 28 and 45 kg are seen to be equivalent to about 0.9 and 1.4% of the total nitrogen in the plowed layer.

Leaching. Lipman and Conybeare (1936) estimated the loss of nitrogen from crop land of the United States by leaching to be 26 kg/hectare annually. This figure is similar to their estimated removal of 28 kg of nitrogen per hectare in harvested crops in 1930.

Most of the mineral nitrogen lost from soil by leaching is in the form of nitrate. Collison and Mensching (1930) found that more than 99% of the nitrogen in the leachate from lysimeters in New York was present as nitrate. Less than 1% was present as ammonium, and only a trace was present as nitrite. The relatively low losses of ammonium- and nitrite-nitrogen may be attributed to the low initial concentration and, in the case of ammonium, to retention by the soil. Because nitrate usually constitutes the major portion of the equivalents of anions in solution in acid soils, most of the equivalents of cations lost by leaching are balanced by nitrate. Work of various investigators on this subject was reviewed by Raney (1960).

Loss of nitrate by leaching occurs more readily from soils of coarse texture than from those of fine texture (Morgan and Street, 1939) as a result of perhaps two basic causes. First, soils of coarse texture have the lower water-holding capacity. Consequently, 1 cm of drainage water is a higher percentage of the total water and contains a higher percentage of the total nitrate in soils of coarse texture than in those of fine texture. Moreover, more total drainage of water occurs from soils of coarse texture than from those of fine texture. Second, the proportion of the total pore space made up of small pores within aggregates is smaller in soils of coarse texture than in soils of fine texture. Hence, on the average, the distance nitrate must move by diffusion through small pores within aggregates to reach the large pores among aggregates in which the most rapid movement of water occurs is less in soils of coarse texture than in those of fine texture. Thus there is less detention of nitrate within aggregates during rapid movement of water through soils of coarse texture than through those of fine texture. According to literature reviewed by Harmsen and Kolenbrander (1965), downward displacement of nitrate upon entry of 10 cm of water into soil originally at the field capacity is about 45 cm in sandy soils, 30 cm in soils with 20 to 40% silt plus clay, and 20 cm in clay soils.

Loss of nitrate by leaching increases with nitrate concentration and loss of water. Loss is greater from bare soil than from soil with growing plants because of the absence of plants to absorb nitrate and water. In New York, Lyon et al. (1930) recorded an average annual loss of 56 cm of water through fallow soil and 40 cm through cropped soil. Corresponding losses of nitrogen in kilograms per hectare annually were 76 and 6. In Germany, Pfaff (1963) measured leaching losses of less than 10, 80, and 160 kg of nitrogen per hectare annually from soil under grass, grape vine, and bare fallow. The time at which the loss occurs depends on the water regime in the soil and the soil depth employed. In some situations the loss may not take place until the year after the nitrate is produced. Such a case was reported by Collison and Mensching (1930).

A number of investigations have been made in northern European countries on the nitrogen relations of crops as a function of rainfall during the preceding winter months. For example, Fig. 7.30 gives the results of an experiment conducted by van der Paauw (1962) in the Netherlands. This figure shows that the yield of rye decreased with increasing winter rainfall where no nitrogen fertilizer was applied but that there was no appreciable trend of yields with prior rainfall where a fairly heavy quantity of nitrogen fertilizer was applied—results that would be expected if the winter rainfall removed nitrate present in

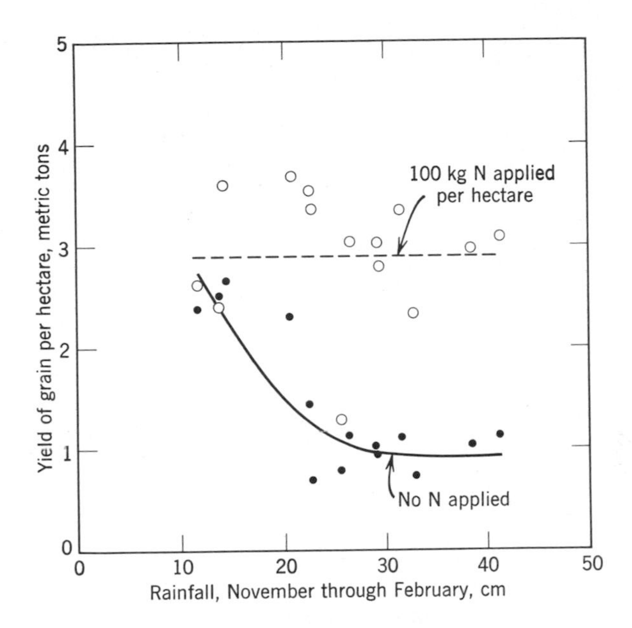

Figure 7.30. Yield of rye with and without nitrogen fertilization versus winter rainfall in the Netherlands during a period of 14 years. The experiment was on a sandy soil, high in organic matter, with a peat subsoil. (van der Paauw, 1962)

the soil during the autumn and if the fertilizer were merely substituting for the nitrate lost. In other work, van der Paauw (1959) found that preventing 28% of the total rainfall during the winter period from November through February from entering the soil produced an increase in subsequent crop yield equivalent to that obtained by a spring application of 33 kg of fertilizer nitrogen per hectare.

Mineral nitrogen supplied in fertilizer or produced in soil may remain for a long time without much loss if the soil remains fairly dry and if other conditions are suitable. In an environment in which drainage occurs, however, much of the value of nitrogenous fertilizer may be dissipated if the interval between application and absorption of nitrogen by the crop is too long. Figure 7.31 illustrates the course of downward movement of nitrate from a surface application of calcium nitrate to a sandy loam soil on which no plants were allowed to grow. Analyses were made after different periods of time during which the soil received the natural rainfall. The values obtained from analyses of the controls have been subtracted, so that the data in the figure are increases due to fertilization. It is evident from these data that significant movement of nitrate occurred with only moderate rainfall. The depth of movement

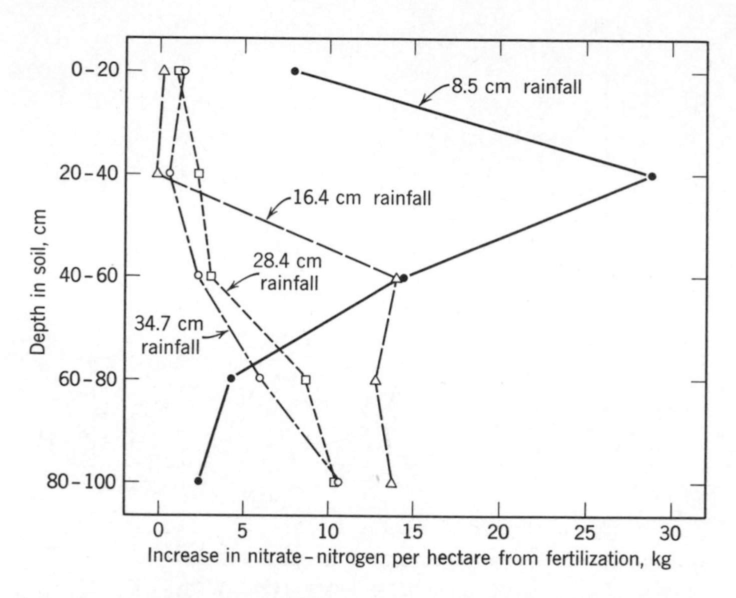

Figure 7.31. Increase in content of nitrate-nitrogen at different depths in a fallow sandy loam soil as a result of fertilization with calcium nitrate equivalent to 60 kg of nitrogen per hectare. The measurements recorded for cumulative totals of 8.5, 16.4, 28.4, and 34.7 cm of rainfall after fertilization were obtained 34, 66, 107, and 153 days after fertilization. The total recovery of added nitrogen at the respective times of measurement amounted to 58, 41, 25, and 21 kg/hectare. (Soubies et al., 1952)

at which the nitrate may be considered lost is rather arbitrary and, in practical situations, depends on the crop and the soil. In experiments in which crops were used to measure the effect, Pearson et al. (1961) found that the availability or effective quantity of nitrogenous fertilizer applied in the autumn for a spring crop was 13 to 69% as great as that of spring applications; the average was 49%. There was no appreciable difference among nitrogen sources, which indicates that the ammonium in ammonium-bearing or ammonium-producing sources was oxidized to nitrate. These experiments were conducted in southeastern United States in a region of relatively high winter rainfall where the soil is not frozen during the winter.

Experiments have been done also on the residual effect of nitrogen fertilization of a crop in one year on the supply of nitrogen for the crop grown the year following. In the Netherlands, van der Paauw (1963) found that the residual value of nitrogen fertilizer applied the year preceding the test crop averaged about 15% of the value of nitrogen applied directly to the test crop in years when the rainfall from November through February was 20 cm and decreased almost to zero with rainfall of 30 cm or more. Relatively high winter rainfall and winters in which the soil is mostly unfrozen are conducive to loss of nitrate by leaching. Higher values (17 to 63%) for residual nitrogen were reported by Dumenil et al. (1961) from Iowa, where winter rainfall is lower and the soil is frozen much of the winter.

Winter cover crops offer an effective means of reducing loss of nitrogen by leaching and erosion during the autumn, winter, and early spring in areas where the soil is not frozen during the winter. As indicated by the data from an experiment by Jones (1939) in Table 7.9, however, the benefit to the following crop from the presence of a nonlegume during the winter may be considerably less than the benefit anticipated from the nitrogen saved. No doubt, the reason for this is that much of the nitrogen is retained in organic form in residues of the cover crop. With a leguminous cover crop, the benefit may exceed the saving because of the nitrogen fixed.

Change to Gaseous Forms. Numerous experiments have demonstrated the existence of a negative nitrogen balance in soils (Allison, 1955, 1965). That is to say, the sum of the nitrogen removed from soil by plants, leaching, and erosion, and the nitrogen residual in the soil, is less than the sum of the nitrogen added and the nitrogen present initially in the soil. For example, Table 7.10 shows the nitrogen balance in an experiment in which plants were grown in containers of soil with different quantities of nitrogen fertilizer. The balance was positive (indicating some fixation) under conditions of marked deficiency of nitrogen for production of plants but negative where the supply was adequate. The nitrogen that cannot be accounted for in experiments of this sort is

Table 7.9. Nitrogen Lost by Leaching and Nitrogen Absorbed by a Summer Crop of Sudangrass from Different Soils with and without Winter Cover Crops in Alabama (Jones, 1939)

	Nitrogen Lost by Leaching per Hectare, kg		Nitrogen in Sudangrass per Hectare, kg	
Cover Crop during Winter	Norfolk Sandy Loam	Decatur Clay Loam	Norfolk Sandy Loam	Decatur Clay Loam
None	155	19	26	126
Nonlegume (oats)	36	0	36	64
Legume (vetch)	123	9	221	247

presumed to have been lost in gaseous forms. Loss of nitrogen in gaseous forms from soils is known to occur as a result of biological processes in which microorganisms participate and as a result of chemical processes that take place either in association with the biological transformations or independently of them. Both types of losses will be discussed in this section.

Table 7.10. Nitrogen Balance in a Soil Culture Experiment with Different Additions of Nitrogen Fertilizer (Pinck et al., 1948)

Nitrogen Added as Urea per Hectare, kg	Total Yield of Dry Matter per Culture in Five Crops, g	Nitrogen Gain or Loss per Hectare, kg
0	33	+56
224	91	+4
449	133	−21
898	182	−35
1346	180	−172

Biological reduction of nitrate and nitrite to gaseous forms of nitrogen is known as denitrification. Denitrification is a result of substitution of nitrate or nitrite for oxygen as a hydrogen acceptor in enzymatic oxidation reactions, and it takes place if oxygen is deficient. Many kinds of microorganisms have the capacity for denitrification. Gaseous products found by analysis include nitrogen dioxide (NO_2), nitrous oxide (N_2O), and nitrogen (N_2). Denitrification was considered to some extent under ammonium oxidation and nitrate reduction, and only the over-all significance will be considered here. Broadbent and Clark (1965) discussed the significance of two classes of biologically induced losses of gaseous forms of nitrogen from soils: losses that take place at a high rate in a short time and losses that take place at a low rate over a long time.

Losses occur rapidly and in a short time in warm soils saturated with water and containing much readily decomposable organic matter. Losses under these conditions are extensive if the soil initially contains much nitrate. These conditions are not characteristic of cultivated upland soils, but they may occur for short times during and after heavy rains or flood irrigation. In soils used for lowland (flooded) rice, the conditions of warmth and wetness frequently prevail, and readily decomposable organic matter may be moderately abundant. But because the possibility of loss is well recognized, nitrate fertilizers are not used, and loss by denitrification is confined mostly to soil nitrate present at the time of flooding.

In the opinion of Broadbent and Clark (1965), the rapid losses that take place in a short time are probably less significant, in total, than the slow losses from denitrification under anaerobic or nearly anaerobic conditions in isolated locations in presumably well aerated soil. They estimated the magnitude of the latter loss at 10 to 15% of the annual mineral nitrogen input and considered it almost inevitable in practical agriculture.

Broadbent and Clark attached little significance to loss of gaseous forms of nitrogen resulting from movement of nitrate from the surface into lower parts of the soil where the water content may be high. The basis for their view is that the subsoil ordinarily contains too little readily decomposable organic matter to support much microbial reduction of nitrate and that the downward movement usually occurs in winter when the temperature is low and microbial activity is limited. According to their view, loss of nitrate from downward movement in the soil is more likely to be brought about by passage of nitrate as such below the root zone than by denitrification.

These authors considered also the possibility of formulating new nitrogen fertilizers that would inhibit loss of nitrogen from denitrification. The major question appeared to them to be the economics of producing and using such fertilizers versus the cost of applying enough extra conventional nitrogen fertilizer to compensate for losses of nitrogen that may be expected as elemental nitrogen and gaseous oxides.

With regard to chemical losses of gaseous forms of nitrogen, attention has been focused on reactions of nitrite and ammonium. These will be considered in turn.

In an investigation of mineral nitrogen transformations in soils, Clark et al. (1960) observed losses ranging from 3 to 172 μg of mineral nitrogen per gram of soil during incubation of samples of 41 soils with additions of 400 μg of urea nitrogen per gram of soil. (Urea undergoes rapid enzymatic hydrolysis to ammonium carbonate in warm soil and hence may be classed as mineral nitrogen from the standpoint of the reactions the nitrogen undergoes subsequent to hydrolysis.) When samples of the same soils were incubated with nitrate-nitrogen, the quantities recovered ranged from 95 to 105% of those added, which indicates that the losses were not attributable to immobilization or denitrification. A check on volatilization of ammonia indicated that the losses were small and unrelated to the observed deficits. The greatest deficits were found in soils that contained nitrite at the end of the incubation period, which led to the implication of nitrite. In an additional series of incubations of the same soils after addition of 400 μg of nitrogen as potassium nitrite per gram of soil, the nitrogen deficits exceeded those found with urea as the source of nitrogen. This experiment confirmed the supposition that the loss of nitrogen was related to the presence of nitrite.

Nelson (1967) incubated sterile and nonsterile samples of 11 soils under moist but aerobic conditions for a week after an addition of 100 μg of nitrite-nitrogen per gram of soil and then determined the recovery of the added nitrogen as nitrite and nitrate. The total percentage recoveries, averaged over all soils, were 48 under sterile conditions (produced by prior heating of the soils in an autoclave) and 52 under nonsterile conditions. Loss of nitrite-nitrogen thus was almost as great in the absence of microbial activity as in its presence, which shows that the principal reaction or reactions responsible for the loss were not biological in nature.

Several chemical reactions have been implicated in nonbiological loss of nitrogen from nitrite. Experimental evidence concerning these reactions was reviewed by Allison (1963), Broadbent and Clark (1965), and Nelson (1967).

The most important reaction seems to be the spontaneous decomposition of nitrous acid,

$$2HNO_2 = NO + NO_2 + H_2O.$$

Nelson's (1967) examination of the nitrogenous gases evolved from sterile, nitrite-treated soil showed the presence of nitric oxide (NO), as indicated in the equation, but only when the atmosphere contained no oxygen. In the presence of low concentrations of oxygen, nitric oxide was oxidized rapidly to nitrogen dioxide (NO_2),

$$2NO + O_2 = NO_2.$$

Nitrogen dioxide is in equilibrium with nitrogen tetroxide (N_2O_4),

$$2NO_2 = N_2O_4,$$

and it dissolves in water, with which it reacts,

$$2NO_2 + H_2O = HNO_2 + HNO_3.$$

A cyclic process thus exists in which part of the nitrite that decomposes in the first reaction is regenerated in the last, along with some nitrate. Further information on some of these transformations was reviewed by Cheng and Bremner (1965).

Nelson (1967) examined the foregoing reactions under various conditions and found that loss of nitrite from buffered solutions, with formation of nitrate, took place rapidly at *p*H 4 and below, slowly at *p*H

5 to 7, and not at all at *p*H 8. The proportion of the nitrogen recovered as nitrate varied with the opportunity for loss of the gaseous nitrogen dioxide before the secondary reaction with water. In sterile soils, similarly, more nitrite disappeared and more nitrate was formed at low than at high *p*H values. Moreover, the proportions of added nitrite-nitrogen lost as nitrogen dioxide and recovered as nitrate varied with the opportunity for loss of the gas before occurrence of the secondary reaction with water. In soils, however, the quantities of nitrogen dioxide formed by decomposition of nitrite were much greater than in buffered solutions at the same *p*H values. Significant quantities of nitrogen dioxide were evolved from soils with *p*H values of 7 and above. Further experiments showed that addition of various soil minerals to a buffered solution containing nitrite at *p*H 5 did not increase the decomposition of nitrite. To explain these observations, Nelson surmised that the greater loss of nitrite from soils than from buffered solutions at the same *p*H occurred in the ionic atmospheres around soil particles, where the hydrogen-ion activity exceeds that indicated by a *p*H measurement on the soil by the glass-electrode method. (The high concentration of salts in the buffer solution would eliminate the local effect around the soil particles. See Chapter 5 for discussion of this phenomenon of hydrogen-ion activity.)

The second most important reaction in nonbiological decomposition of nitrite with loss of gaseous nitrogen from soil seems to be an interaction between nitrite and soil organic matter. Nelson's (1967) investigation of gaseous forms of nitrogen evolved from sterile soils treated with nitrite showed that, in addition to nitrogen dioxide, the gases included nitrogen(N_2) and nitrous oxide (N_2O). Nitrogen and nitrous oxide are not produced by spontaneous decomposition of nitrous acid, but they are formed by reduction of nitrous acid by organic matter. Nelson found that extracted soil organic matter, prepared lignins, and soils high in content of organic matter all promoted decomposition of nitrite. He related the reactivity of these substances to their content of phenolic groups. Phenol and other phenolic compounds of low molecular weight were found to react with nitrite under mildly acid conditions, yielding nitrous oxide and nitrogen as gaseous products.

Phenolic substances underwent another reaction with nitrite in which part of the nitrogen was retained in organic form. For example, in a buffered solution containing phenol and nitrite at *p*H 5, the following was the distribution of the original nitrite-nitrogen among various forms after 24 hours in one of Nelson's (1967) experiments: nitrite, 23%; nitrate, 4%; nitrogen dioxide, 4%; nitrogen plus nitrous oxide, 25%; and organic nitrogen, 44%. Production of nitrate and nitrogen dioxide in the reaction may be accounted for by spontaneous decomposition of some of the nitrite at the same time as the reaction with the phenol was taking place. Control analyses without phenol showed recovery of 81% of the original nitrite-nitrogen as nitrite, 3% as nitrate, and 15% as nitrogen dioxide. None was found in nitrogen, nitrous oxide, or organic forms. In an experiment with a soil (containing 9% organic carbon) that was kept in a buffer solution at *p*H 5 for 24 hours, the distribution of added N^{15}-labeled nitrite-nitrogen at the end of the experiment was as follows: nitrite, 65%; nitrate, 3%; nitrogen dioxide, 10%; nitrogen, 12%; nitrous oxide, 1%; and organic nitrogen, 10%. Controls without soil, with ignited soil, with quartz sand, and with a mixture of quartz sand and clay minerals found in soils yielded no nitrogen, nitrous oxide, or organic nitrogen, which verifies the connection between these products and soil organic matter. The fraction referred to here as organic nitrogen was called "fixed" nitrogen by Nelson because it was determined by a method that would not distinguish between organic nitrogen and nonexchangeable ammonium. The reaction of nitrite with organic matter does not seem to form ammonium, however, as indicated by the control analyses for the synthetic mixture of quartz sand and clays. This mixture contained vermiculite, a mineral that fixes ammonium in nonexchangeable form (see Chapter 9 for further description of the properties of this mineral in reactions involving ammonium or potassium), but no "organic" nitrogen was formed in it upon treatment with nitrite. Führ and Bremner (1964) investigated the properties of the N^{15}-labeled, nitrite-derived nitrogen that was fixed by soil organic matter and found that only about half of it was released by hydrolysis with 6-normal hydrochloric acid for 12 hours. Up to 80% of the labeled nitrogen released during hydrolysis was ammonium. Bremner and Führ (1966) found that most of the nitrogen released by acid hydrolysis of certain simple nitroso compounds ($R{-}N{=}O$, where R is an organic group) was present in the hydrolysate as ammonium, which suggests that a nitrosation reaction may be responsible for the fixation of nitrite-derived nitrogen in organic form by soil organic matter.

Two other reactions proposed to account for loss of nitrite-nitrogen from soils are decomposition of ammonium nitrite,

$$NH_4NO_2 = N_2 + 2H_2O,$$

and interaction of nitrous acid and aliphatic amino groups,

$$RNH_2 + HNO_2 = ROH + H_2O + N_2,$$

where R is an aliphatic organic substance.

Nelson (1967) examined the ammonium nitrite reaction and obtained no losses of gaseous nitrogen by this mechanism from relatively high concentrations of ammonium and nitrite in acid or alkaline solutions and moist soils. If alkaline soils containing high concentrations of ammonium and nitrite were air-dried, however, losses of N_2 from decomposition of ammonium nitrite were found. The loss due to air-drying was large in neutral sand, but in alkaline or acid soil was less significant than losses due to other mechanisms.

The interaction of nitrous acid and aliphatic amino groups with release of nitrogen gas is known as the Van Slyke reaction. It is used for quantitative determination of alpha-amino nitrogen in amino acids in the absence of certain other types of amino groups that react similarly. The reaction is carried out with a high concentration of nitrite in glacial acetic acid under an atmosphere of nitric oxide to inhibit spontaneous decomposition of nitrite. Nelson (1967) found that addition of various amino acids and other substances [including urea, $(NH_2)_2CO$] to a buffer solution containing nitrite at *p*H 5 had little or no effect on the rate of disappearance of nitrite from solution, which confirms the views of previous investigators that the reaction is of little or no significance under soil conditions because of the low concentration of nitrite and reactive amino groups and because of the high *p*H. The finding of no significant loss due to the presence of urea is worthy of particular comment because it had once been thought that significant loss of nitrogen might occur in this way from urea added as fertilizer. Nelson's findings verify previous work by Sabbe and Reed (1964) on the nitrite-urea reaction.

Broadbent and Clark (1965) pointed out two approaches that might be used to control the losses of gaseous forms of nitrogen associated with the presence of nitrite. One would be to inhibit ammonium oxidation and thus to avoid formation of nitrite; the other would be to cause rapid oxidation of nitrite, thus limiting the concentration of nitrite, which in turn would limit the reactions leading to loss. They considered the former approach impractical for general use and the latter unusable for the present because of insufficient information on conditions required.

Ammonium has been implicated in losses of gaseous forms of nitrogen from soils in the reaction with nitrite mentioned in preceding paragraphs and also in direct loss of ammonia by volatilization. Under alkaline conditions the ammonium ion is unstable and subject to decomposition to ammonia gas:

$$NH_4OH = NH_3 + H_2O.$$

Loss of ammonia may be strictly an inorganic process, as where anhydrous ammonia or ammonium hydroxide is added as fertilizer, or it may follow biological processes, as where urea is hydrolyzed in soil to ammonium carbonate. If enough urea is added, or if local concentrations around individual pellets are great enough, the soil becomes alkaline; and ammonia may be volatilized.

Figure 7.32 illustrates the effect of several factors that influence loss of ammonia by volatilization. First, a significant decrease in loss of ammonia from the application of urea may be seen to be associated with an increase in cation-exchange capacity of the soil. This behavior is probably attributable principally to the relationship between the cation-exchange capacity and the *p*H attained after hydrolysis of urea. The lower the cation-exchange capacity, the higher would be the soil *p*H after hydrolysis, the lower would be the stability of the ammonium ion, and the greater would be the volatilization of ammonia. Losses of nitrogen as ammonia from ammonium sulfate, with one exception, may be seen to be far lower than the losses from urea applied to the same soils. Ammonium sulfate is a neutral salt and does not produce

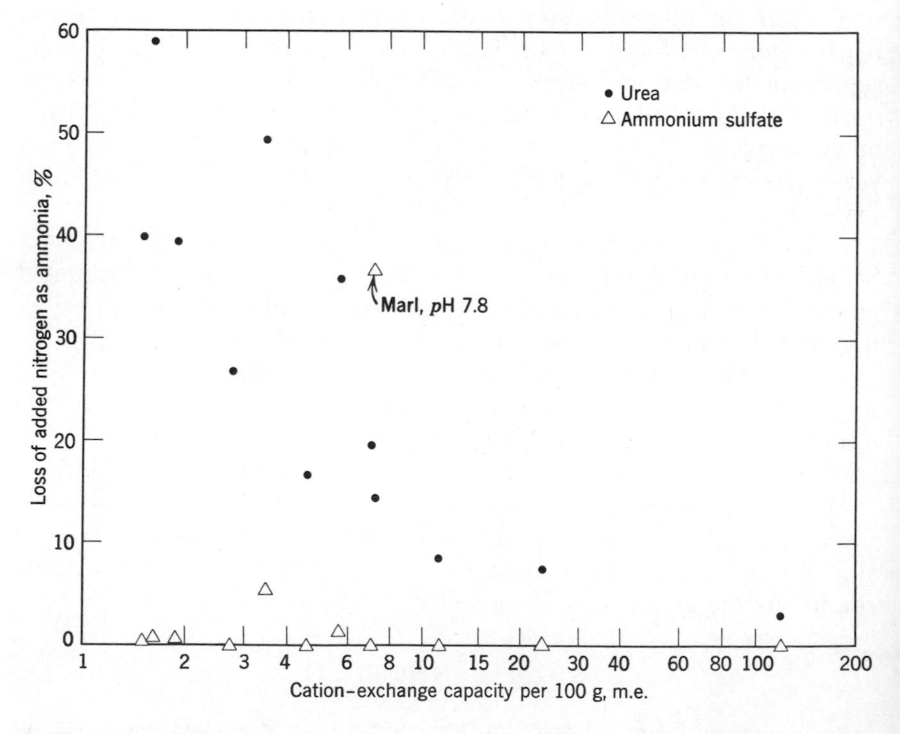

Figure 7.32. Volatilization of ammonia during the first week following surface application of urea and ammonium sulfate to bare, moist soils differing in cation-exchange capacity. The nitrogen addition was equivalent to 112 kg/hectare. (Volk, 1959)

high pH values upon addition to soils. All the soils in the group included in Fig. 7.32 were acid except for marl (calcareous), having a pH value of 7.8. This soil is marked in the figure and represents the exceptional instance in which a high loss of ammonia occurred from an application of ammonium sulfate. In this instance, the instability of the ammonium sulfate may be accounted for by the alkaline environment provided by the soil.

Losses of ammonia by volatilization are most pronounced if the source of ammonia is at the surface of the soil because of the limited opportunity for ammonia, once volatilized there, to be readsorbed by the soil at another location where lower pH values and lower ammonium concentrations would favor retention. Application of even anhydrous ammonia as fertilizer results in little loss if the application is made 10 cm or more below the soil surface with good covering. Loss of ammonia is promoted also by soil dryness, presumably because of limitation of the volume of solution available to retain it. High temperatures also favor ammonia volatilization. Water content and temperature, however, are of secondary importance in comparison with the exchange capacity and pH. Where surface applications are concerned, the opportunity for downward movement into the soil is of some importance. If the source of ammonia is urea, the rate of hydrolysis may be involved. For example, in cold soils, 6 weeks or more may be required for hydrolysis. During this time, a rain could wash much of the urea down into the soil, where loss on subsequent hydrolysis would be reduced. Dryness also inhibits hydrolysis of urea and hence inhibits volatilization of ammonia. Papers in which these matters are treated include those by Volk (1959), Chao and Kroontje (1964), Puh (1964), and Gasser (1964).

The findings in the foregoing paragraphs with regard to ammonia volatilization are based on laboratory and field experiments in which plant growth was not involved. Such experiments can be done relatively easily by completely enclosing the soil in a glass vessel or by covering the soil with a glass vessel that encloses a small volume of air. The atmospheric ammonia may then by trapped by diffusion into sulfuric acid or by passing a stream of ammonia-free air through or over the soil and then bubbling it through sulfuric acid as it leaves the enclosed volume. A more elaborate arrangement would be needed to make measurements while plants are growing under environmental conditions similar to those in the field, and such measurements apparently have not been made.

Experiments have been done, however, to determine the comparative value of nitrogenous fertilizers under conditions in which differences in volatilization of ammonia would be expected from the experiments

Table 7.11. Recovery of Nitrogen in the Above-Ground Parts of Corn Plants from Ammonium Sulfate and Urea Mixed with or Surface-Applied to a Fine Sandy Loam Soil at Two *p*H Values in a Greenhouse Experiment (Terman and Hunt, 1964)

	Recovery of Added Nitrogen, %			
	Unlimed soil, *p*H 5.2		Limed soil, *p*H 7.5	
Source of Nitrogen	Mixed	Surface	Mixed	Surface
Urea	76	30	72	18
Ammonium sulfate	62	67	77	80

on soils; these indicate that the general concepts derived from measurements on soils apply in the presence of plants. For example, Table 7.11 gives results from an experiment in which the recovery of nitrogen applied as ammonium sulfate and urea by both mixing with the soil and application on the surface was determined on a soil with and without liming. A marked difference in recovery of fertilizer nitrogen due to placement is evident with urea but not with ammonium sulfate, and the placement effect is more pronounced with the limed soil than with the unlimed soil.

In a field experiment on a loamy sand soil (*p*H 5.5), Jackson and Burton (1962) found that the yields of bermudagrass obtained from applications of 224 kg of nitrogen per hectare were 15.2 and 16.5 metric tons per hectare with urea applied on the surface and at a depth of 15 cm and were 17.9 and 16.3 metric tons with ammonium nitrate applied in the same ways. These results again indicate comparative inefficiency of urea applied to the soil surface. Half of the nitrogen in ammonium nitrate could be lost by volatilization as ammonia under the proper conditions, but ammonium nitrate does not make the soil alkaline as does urea. Jackson and Burton, incidentally, obtained evidence suggesting that some of the loss of ammonia from urea applied to the surface of established bermudagrass sod resulted from hydrolysis of the urea to ammonium carbonate by urease in the sod before the urea touched the soil.

Erosion. Precise information on loss of nitrogen by erosion is lacking except for small experimental areas that do not represent an unbiased sample of the landscape as a whole. On the basis of available data, however, Lipman and Conybeare (1936) estimated the annual loss of nitrogen by erosion from the crop land of the United States to be 27 kg/hectare. This figure corresponds approximately to the estimated removal of nitrogen in harvested crops.

In estimating the loss of nitrogen by erosion, Lipman and Conybeare assumed that the material lost had the same composition as the soil. This supposition represents only an approximation. Erosion is a selective process. Although the degree of selectivity varies, the percentage content of nitrogen and other nutrients in solids lost by erosion generally exceeds the corresponding percentage content in the soil from which they were eroded. Massey and Jackson (1952) found that, in 177 measurements of erosion made at different locations and times in Wisconsin, the percentage content of nitrogen in the eroded solids averaged 2.7 times greater than the percentage content of nitrogen in the original soil. The estimates of erosion losses made by Lipman and Conybeare thus appear to be conservative.

In consequence of the concentration of total and mineralizable

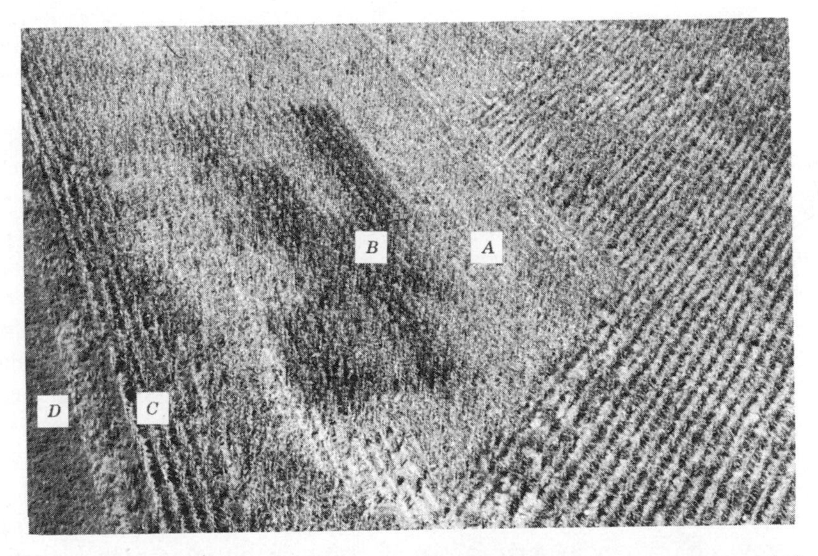

Figure 7.33. Aerial view of a field of nitrogen-deficient corn on an eroded silt loam soil in Iowa. Features marked in the figure include the following: (*A*) Light-colored, nitrogen-deficient plants in the main part of the field. (*B*) An irregular area with darker colored corn resulting from application of different quantities of nitrogen to individual replicated plots in a field experiment. Plots with corn of the darkest color received nitrogen in the quantity of 202 kg/hectare. (*C*) An area with darker colored corn along a fence line where relatively little erosion had occurred. (*D*) An adjacent field planted to grass-legume meadow. (Photograph courtesy of L. C. Dumenil)

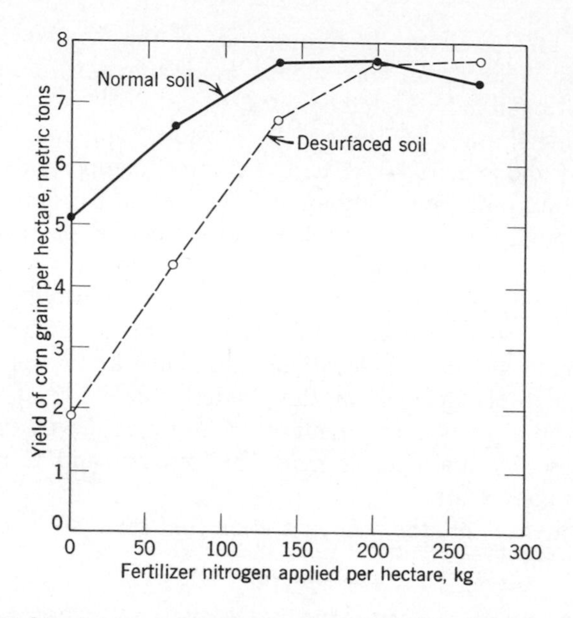

Figure 7.34. Yield of corn versus quantity of fertilizer nitrogen on a silt loam soil in its normal condition and after removal of the surface soil. (Engelstad and Shrader, 1961)

nitrogen in the upper part of the soil profile (Fig. 7.2 and Table 7.5), the soil nitrogen absorbed by plants originates principally in the upper part of the profile; thus loss of surface soil by erosion has significant consequences in terms of availability of nitrogen to plants. Figure 7.33 illustrates the nitrogen deficiency of corn associated with loss of soil by erosion. Figure 7.34 shows that a marked reduction in yield of corn was associated with artificial removal of the surface layer of a silt loam soil developed under grass vegetation in Iowa. Phosphorus and potassium fertilizers were applied to both the normal and desurfaced areas in an attempt to eliminate the effect of desurfacing on the response of the plants to the difference in supply of these two nutrients. Almost identical yields were obtained on the normal and desurfaced areas in the presence of a sufficient quantity of fertilizer nitrogen, which verifies that the difference in yield in the absence of nitrogen fertilizer was due primarily to the loss of soil nitrogen from the upper part of the profile in the desurfacing operation.

Gains

If there were no natural processes by which the content of combined nitrogen in soils could be increased at the expense of elemental nitrogen, crop production on unfertilized soils eventually would cease for lack of available nitrogen. Soils may gain combined nitrogen by four generally recognized processes. These are nonsymbiotic fixation, symbiotic fixation, addition in rainfall, and fertilization. Nonsymbiotic (sometimes called asymbiotic) fixation is the process by which certain free-living organisms change elemental nitrogen into organic combination. The term free-living means in this connection that the nitrogen-fixation process occurs when the organism is grown in pure culture, under suitable conditions, and is not dependent on a special association between the organism concerned and another kind of organism. Symbiotic fixation is the process by which elemental nitrogen is changed to organic forms by the symbiosis or association between two kinds of plants. The symbiosis used most widely to advantage in agriculture is that between leguminous plants and the bacteria found in nodules on their roots. In addition to these biological fixation processes, fixation of nitrogen at the soil surface by photochemical action has been claimed, although not substantiated.

Nonsymbiotic Fixation. The genera of nonsymbiotic nitrogen-fixing organisms may be classified as follows:

- Bacteria
  - Aerobic
    - *Azotobacter* (heterotrophic)
  - Anaerobic
    - *Clostridium* (heterotrophic)
    - *Aerobacter* (heterotrophic)
    - *Methanobacterium* (autotrophic)
    - *Rhodospirillum* (photosynthetic)
    - *Chromatium* (photosynthetic)
    - *Chlorobium* (photosynthetic)
    - *Rhodomicrobium* (photosynthetic)
- Blue-green algae (aerobic and photosynthetic)
  - *Nostoc*
  - *Calothrix*

This classification, modified from one given by Wilson (1958), includes only organisms that have been shown by use of N^{15}-labeled nitrogen gas to fix atmospheric nitrogen. The relatively high sensitivity of the tracer method makes the N^{15} technique useful for verifying fixation. Unequivocal demonstration of the property of nitrogen fixation by the conventional method of determining the increase in content of total combined nitrogen in the culture medium is difficult if the difference due to activity of the organism is small in relation to the experimental errors of measurement of the nitrogen content of the cultures.

Yocum (1960) and Jensen (1965) listed a number of other organisms for which they considered evidence for nonsymbiotic nitrogen fixation to be good. These include *Beijerinckia* and *Derxia,* which are aerobic, heterotrophic bacteria like *Azotobacter*, and a number of algae.

Attention of research workers has been centered on the bacteria *Clostridium* and *Azotobacter. Clostridium* was isolated by Winogradsky (1893). He found that it is capable of fixing nitrogen under only anaerobic conditions in pure culture but that it can act under aerobic conditions as well if certain other bacteria are present. Fixation of elemental nitrogen is inhibited by ammonium. *Clostridium* has been found distributed widely in soils throughout the world. The organism develops best at *p*H values near neutrality although it can be found in soils as acid as *p*H 5.

Azotobacter was isolated by Beijerinck (1901). *Azotobacter* fixes nitrogen under aerobic conditions in pure culture and under semi-anaerobic conditions in symbiosis with certain other bacteria. *Azotobacter* uses combined nitrogen if present in available form. Concentrations of ammonium or nitrate nitrogen as low as 5 μg/ml reduce fixation, although considerably higher concentrations are needed to inhibit fixation completely. *Azotobacter* is distributed widely but is more sensitive to soil conditions than is *Clostridium.* It is sensitive to phosphorus deficiency, and the common types are of limited occurrence in soils having *p*H values below 6. Certain species are more tolerant, however. A species of *Azotobacter* described by Starkey (1939) developed and fixed nitrogen in cultures as acid as *p*H 3.1. Jensen (1955) described another acid-tolerant isolate of *Azotobacter*.

To determine the fixation of nitrogen taking place nonsymbiotically in soil, Delwiche and Wijler (1956) placed variously treated samples of two soils from lawns in an enclosed glass chamber, in which the atmospheric nitrogen was labeled with N^{15}, and then analyzed the soil after an incubation period, with results as shown in Table 7.12. The first two treatments employed plugs of soil taken from lawns containing grass but no legumes. The glass chamber was illuminated during the incubation period, and conditions inside were such that the grass remained alive where the plugs of soil were placed upright. Where the plugs were inverted, the grass decayed. Fixation was negligible where the grass was living or where the soil was incubated after removal of the grass but was much greater where the grass was present and decaying and was far greater where the soil was incubated after addition of sucrose. These observations indicate the importance of organic matter, low in nitrogen, as an energy source for the nitrogen-fixing organisms.

Hassouna and Wareing (1964) described a situation in nature in which nonsymbiotic fixation of nitrogen seems to be significant and to be related to the supply of organic matter as an energy source. They found that ordinary crop plants suffered extreme nitrogen deficiency when grown on dune sand but that certain species of plants that colonize sand dunes under natural conditions grew well even though they were not legumes. In one experiment, the authors grew *Ammophila arenaria* from small pieces of rhizome in quantities of dune sand and found that the total nitrogen present in sand and plant increased from 230 mg at the beginning of the experiment to 419 mg at the end of the experiment 2 years later. In the same interval of time, the nitrogen content of control sand without a plant increased from 225 to 242 mg. They found large numbers of nitrogen-fixing bacteria including *Azotobacter* on and adjacent to root surfaces but relatively small numbers in the

Table 7.12. Nitrogen Fixed Nonsymbiotically by Soils under Different Conditions in the Laboratory from an Atmosphere Containing Elemental Nitrogen Labeled with N^{15} (Delwiche and Wijler, 1956)

Treatment	Nitrogen Fixed per Hectare Annually, kg[1]	
	Soil *A*, *p*H 7.8	Soil *B*, *p*H 6.2
Soil with grass living	0.1	0.5
Soil inverted, with grass decaying	131	40
Soil alone, grass removed	2.4	0.9
10 g of soil + 780 mg of sucrose	3650	—

[1] Calculated for a depth of 15 cm, assuming a bulk density of 1.2, from data obtained during an incubation period of 31 days at 21°C.

bulk soil. Their view was that root exudates provided the energy source for the nitrogen-fixing microorganisms. On death and decomposition of the organisms, the plants benefited from the fixed nitrogen. Although the relative gain in nitrogen in this experiment is impressive, the absolute gain due to the presence of the plants is relatively small, amounting to only 35 kg of nitrogen per hectare annually.

Hüser (1963) investigated with the aid of N^{15} the nonsymbiotic fixation of nitrogen in the surface organic horizons of soil from a beech forest. Although this situation is one in which the supply of organic matter was high, Hüser's results led to an estimate of fixation of only

1.5 kg of nitrogen per hectare annually. For organic matter to promote fixation, the availability of fixed nitrogen must be low; and perhaps it was not low enough in Hüser's experiment. Table 7.13 gives the results of another experiment by Delwiche and Wijler to determine the effect of nitrate on the fixation that occurred upon incubation of 100-g samples of soil after addition of 1 g of glucose as an energy source. These data illustrate the inhibitory effect of moderate and high concentrations of nitrate. Supplementary measurements showed that *Azotobacter* was present in the soil and that the numbers increased greatly during incubation, whether or not the growth was accompanied by much fixation.

Table 7.13. Nonsymbiotic Fixation of Elemental Nitrogen in Samples of a Soil (*p*H 7.8) Incubated 40 Days at 21°C with Addition of 1% Glucose and Various Quantities of Nitrate-Nitrogen in an Atmosphere Containing N^{15}-Labeled Nitrogen (Delwiche and Wijler, 1956)

Nitrate-Nitrogen per Gram of Soil, μg		Nitrogen Fixed per Gram of Soil, μg
Initial	Final	
23	3	19.6
93	11	7.0
162	26	0.5
302	153	0.4
582	415	0.4

Because of the variety of microorganisms known to fix nitrogen nonsymbiotically and the probability that this property is shared by other organisms not now known to be nitrogen fixers, it seems likely that most, if not all, soils contain nitrogen-fixing microorganisms. Nonsymbiotic nitrogen fixation no doubt occurs under field conditions; and, from what is known from laboratory studies of the factors involved, one may expect that the amount fixed should be greatest where the supply of available energy material is ample and mineral nitrogen is deficient. The occurrence of nonsymbiotic fixation under field conditions, however, has still not been verified satisfactorily by experimental measurements. The principal difficulty in estimating the amount of nonsymbiotic fixation under field conditions has been that such estimates have had to be made from the change in content of total nitrogen in the soil during a period of time. The change in total nitrogen is a net change, representing the algebraic sum of all gains and all losses. The net change is usually so small as to be nonsignificant compared with the experimental errors of measurement. Most estimates of nonsymbiotic fixation under field conditions are in the range from 0 to 60 kg of nitrogen per hectare annually. After reviewing the evidence pertaining to nonsymbiotic fixation in the field, Jensen (1965) reached the opinion that the process is of relatively little significance in agriculture in general, a notable exception being in growth of lowland rice. Jensen considered that the nitrogen supply of rice in Asiatic countries has probably been provided for many centuries by nonsymbiotic fixation by algae, which may grow profusely on the surface of the soil. Allison (1965) discussed the evidence for the importance of nonsymbiotic fixation in forests in tropical areas with high rainfall.

According to Cooper (1959), application of preparations of *Azotobacter* and other organisms to seeds of nonleguminous crop plants to increase crop yields is a standard practice in the Soviet Union. Preparations of *Azotobacter* are said to benefit 50 to 70% of the field crops to which they are applied and to produce increases in yield up to 20%. Although the value of such inoculation might be supposed to result from an enhancement of nitrogen fixation, with subsequent release of fixed nitrogen in forms available to plants, Samtsevich (1962) reported that various theories involving other effects, such as production of growth-promoting substances, have been proposed to account for the beneficial results.

In countries other than the Soviet Union, a number of experiments have shown no benefit from inoculation of seeds of nonlegumes with preparations of *Azotobacter;* however, some positive results have been obtained. For example, Sundaro Rao et al. (1963) obtained no significant increases in yield of wheat in India from inoculation with *Azotobacter.* In England, Brown et al. (1964) reported the results of 13 field experiments with various crops in which the effect of inoculation was tested in the presence and absence of nitrogen fertilizer. In these 26 comparisons they obtained increases in yield from inoculation ranging from -12 to $+15\%$. Two of the increases and one of the decreases in yield were statistically significant at the 10% level of probability. At this level of probability, one would expect, on the average, a statistically significant difference in one of ten experiments if the treatment had no real effect. Their 3 statistically significant results in 26 are thus what one would expect if the inoculation were without effect. The total number of field trials susceptible to study in this way is as yet too small to say with confidence whether the inoculation is valuable. In countries other than the Soviet Union, the relatively small and infrequent benefits from the treatment have as yet discouraged inoculation of seeds with *Azotobacter* in practical agriculture. Because of the low cost, however, the treatment would be profitable even if the increases in yield were relatively small; accordingly, a careful examination of the effects of inoculation is desirable on a wider scale than heretofore employed. A more extensive review of this subject was published by Jensen (1965).

Symbiotic Fixation. Discovery of the cause of the anomalous nitrogen nutrition of legumes was the objective of a number of unsuccessful researches in the nineteenth century. Hellriegel and Wilfarth (1891) finally connected this behavior with the root nodules present on leguminous plants. They demonstrated that the growth of nonlegumes depends upon the quantity of mineral nitrogen supplied, but that this is not necessarily true of legumes (Table 7.14). They found that leguminous plants gained considerable nitrogen and grew well if they developed root nodules when planted in a medium free of nitrogen except for the trace supplied in an inoculum of soil. On the other hand, where the soil inoculum was added but subsequently was sterilized, the roots of leguminous plants did not develop nodules. These plants made poor growth and did not gain nitrogen.

Hellriegel and Wilfarth theorized from these experiments that the bacteria in the root nodules assimilate elemental nitrogen from the air

Table 7.14. Yield of Oats and Peas and Nitrogen Balance in Quartz Sand Variously Inoculated or Fertilized with Nitrogen (Hellriegel and Wilfarth, 1891)

Nitrogen Added as Nitrate, mg	Inoculation	Dry Weight of Plants, g		Nitrogen in Plants Gained (+) or Lost (−), mg[1]	
		Oats	Peas	Oats	Peas
0	Not inoculated	0.6	0.8	−20	−25
0	Inoculated but sterilized	—	0.9	—	−23
0	Inoculated	0.7	16.4	−19	+386
112	Not inoculated	12.0	12.9	−49	+38
112	Inoculated	11.6	15.3	−47	+107

[1] Relative to nitrogen added in seed, medium, and nitrate.

and that the plants then use some of the nitrogenous compounds synthesized by the bacteria. Beijerinck (1888) isolated the root-nodule organism, which he called *Bacillus radicicola*. Later the genus name was changed to *Rhizobium*. Schloesing and Laurent (1892) found that the gain in nitrogen by the plants was balanced by a corresponding loss of nitrogen from the atmosphere, thus verifying Hellriegel and Wilfarth's theory about the source of the nitrogen. More recently, verification has been obtained of the long-standing theory that the root nodules are the site of the fixation process. Magee and Burris (1954) found with the aid of N^{15} that nodules will fix nitrogen for a short time after being cut from the plants. That the nodule bacteria are directly responsible for nitrogen fixation by the nodules, however, has never been established. On the contrary, there is some evidence suggesting that fixation is done by the host plant in a structure formed only in response to invasion of the root by the bacteria.

Most species of legumes that have been examined are known to bear root nodules. A few apparently do not form such nodules. A few nonlegumes have nodules resembling those of legumes and fix nitrogen symbiotically. Little is known about the organisms in the root nodules of nonlegumes, but they are not members of the genus *Rhizobium*.

As an effective nodule develops after initial infection of a root by rhizobia in the soil, conducting tissue is formed between the main conducting tissue of the host and parenchyma cells that surround the colony of rhizobia in the nodule; transfer of substances between the host and bacteria is thus facilitated. Individual nodules, once formed, generally do not remain effective throughout the remainder of the life of the host but degenerate in time; on perennial species, however, a given nodule may last for years.

Root nodules of legumes are of various sizes and shapes, and they differ in effectiveness in fixing nitrogen. According to Nutman (1965), the maximum rate of fixation per unit of nodular tissue is fairly constant, but below this the fixation may occur at any rate down to zero. The best visual criterion of effectiveness is probably the color. Effective nodules have pink or reddish centers, and ineffective nodules have greenish-white centers. In addition, there is a tendency for effective nodules to be relatively large in size, few in number, and concentrated on the main roots and for ineffective nodules to be relatively small in size, large in number, and scattered on the lateral or secondary roots. Both effective and ineffective nodules may occur on a single plant. The number per plant may range from none to hundreds or even thousands. Three different nodule patterns in a given species are illustrated in Fig. 7.35.

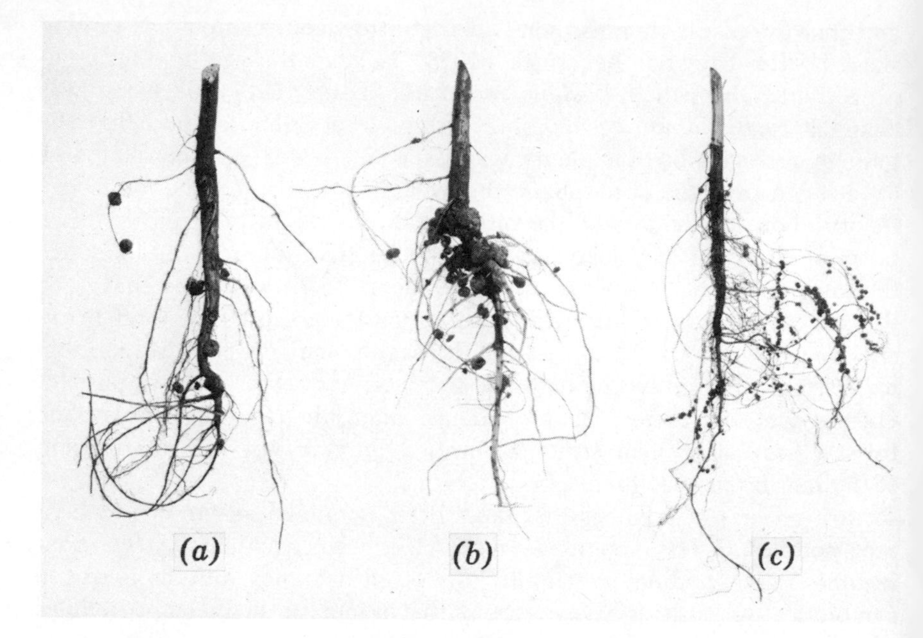

Figure 7.35. Three typical patterns of nodulation of soybean plants: (*a*) nodulation with effective rhizobia present in the soil; (*b*) nodulation with effective rhizobia supplied by inoculation of the seed; and (*c*) nodulation with ineffective rhizobia present in the soil. (Photograph courtesy of C. R. Weber)

Rhizobia live in soil independently of the host, where, as far as is known, they perform as heterotrophic bacteria and fix no nitrogen. Their competitive ability is strongly dependent on higher plants. In the presence of the host and other plants, numbers of rhizobia in the soil increase greatly, apparently because of critical growth factors supplied by root exudates. (The extent of specificity in enhancement of multiplication of specific strains of rhizobia in the presence of a host species or variety they will nodulate effectively is currently unknown.) Additional rhizobia may be added to the soil from root nodules on host plants. The free-living rhizobia in the soil provide inoculum if the host is planted subsequently. Rhizobia, however, do not necessarily persist in soil indefinitely in the absence of a host. Sometimes they disappear rapidly, depending on soil conditions. Consequently, the host plant might not be well inoculated if planted again after a few years. Moreover, a given soil may contain a mixture of species and strains, some effective and others ineffective.

In contrast to the practice of inoculating seeds of nonlegumes with

Azotobacter, inoculation of seeds of legumes with *Rhizobium* has long since passed the controversial stage and has become a widely accepted agricultural practice in many countries. Various commercial organizations are engaged in producing and distributing cultures of rhizobia for use with specific crops. The commercial culture is ordinarily applied to the seed just before planting.

The value of artificial inoculation may well be negligible in many practical instances where the host is grown frequently on the same field; however, the cost of inoculation is so small, and the possible loss of yield from inadequate inoculation or from inoculation with ineffective strains is so great, that artificial inoculation is usually looked upon as cheap insurance and is ordinarily practiced without specific tests of effectiveness in either field or laboratory. Direct isolation of rhizobia from soil is difficult and uncertain. The only practical way to determine the effectiveness of rhizobia in a soil is to grow each desired species of legume on the soil and to compare the resulting nodulation, growth, and nitrogen fixation with those obtained where a strain known from previous trials to be effective is used as artificial inoculum. Because this procedure is so expensive in comparison with inoculation, it is used only in research. A modification of this procedure may be used in testing commercial or other inocula. An extensive account of procedures and problems of artificial inoculation of legumes was published by Vincent (1965).

In the broad sense, symbiosis is the living together of two dissimilar organisms in more or less intimate association. In the more usual, narrow sense, the term refers to associations that are favorable to both the partners in the association. The association between rhizobia and legumes is generally considered to be a symbiosis in the narrow sense (although there is room for debate on this view under certain circumstances). A single bacterium, from which the colony in an individual nodule is almost invariably formed, finds in the root a suitable culture medium that permits it to multiply extensively. From the plant, the rhizobia in the root nodules receive all their requirements for nitrogen fixation and growth, including elemental nitrogen, mineral nutrients, and carbohydrates (and perhaps other growth factors). The elemental nitrogen is dissolved in the plant sap as an incidental constituent. Although the microorganisms in the nodules consume constituents that could otherwise be used by the plant for its own growth and metabolism, the fixed nitrogen returned to the plant from the nodules increases the degree of sufficiency of nitrogen in the plant; this in turn results in an increase in growth and photosynthesis, so that the net result is a gain.

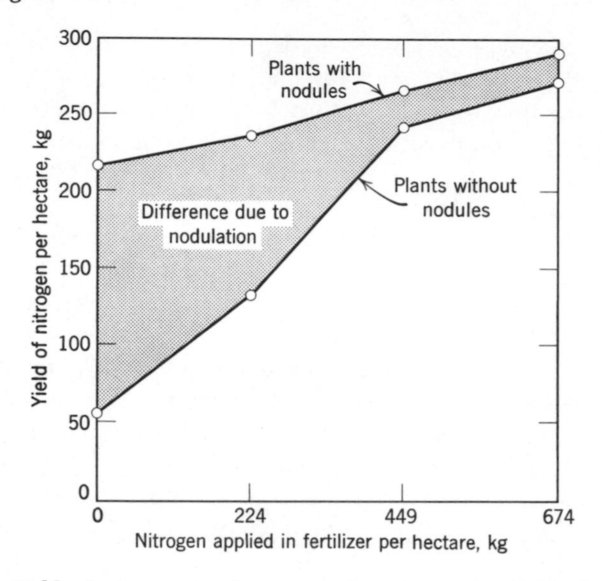

Figure 7.36. Yield of nitrogen in above-ground portions of nodulated and nonnodulated soybeans grown on soil treated with ground corn cobs to immobilize the soil nitrogen. (Weber, 1966)

Various aspects of the symbiotic relationship between rhizobia and legumes have been accounted for conceptually by the theory that the relative supplies of carbohydrate and nitrogen in the plants are a suitable expression of the balance between the rhizobia and the host. According to Wilson and Fred (1939), invasion of the host plant by rhizobia takes place in response to a high ratio of carbohydrate to nitrogen in the plant. If, however, the inoculation is delayed until definite nitrogen deficiency has set in, development of the nodules may be limited by the nitrogen supply. This relationship presumably accounts for the observation that application of small quantities of nitrogenous fertilizer sometimes results in improved stands of legumes.

If plants with well-developed and effective nodules are placed in darkness, photosynthesis stops. Soon afterward, nitrogen fixation stops, and the nodules begin to degenerate. According to the carbohydrate-nitrogen theory, the cessation of nitrogen fixation and the degeneration of the nodule result from a low carbohydrate condition in the plants. Under intermediate conditions, in which the supply of carbohydrates and other substances to the nodules is limited, nitrogen fixation proceeds, but at a low rate. The fixation of nitrogen thus is closely tied to the welfare of the plant, which, in turn, is affected by environmental conditions.

Another factor in the symbiosis that has been explained in terms of the carbohydrate-nitrogen theory is the decrease in fixation of atmospheric nitrogen that occurs as the nitrogen supply in the soil is increased. Figure 7.36 shows the data obtained in an experiment with nodulated and nonnodulated soybeans. The difference in yield of nitrogen due to nodulation, which is an estimate of fixation of atmospheric nitrogen, decreased from 159 kg/hectare in plants on control plots to 19 kg in plants on plots receiving 674 kg of fertilizer nitrogen per hectare. Figure 7.37 shows the relative amounts of nodules on the variously treated plants and the appearance of the nodulated roots. A decrease

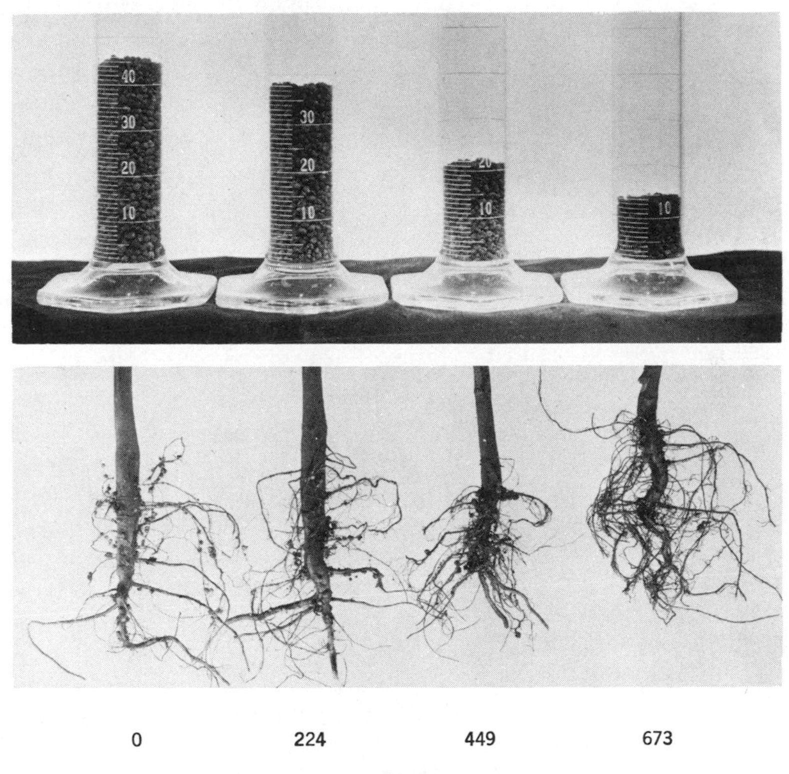

Figure 7.37. Nodulation of soybean plants grown on soil treated with different quantities of nitrogen fertilizer. Above: Relative volumes of nodules. Average weights of nodules in a cylinder of soil 9.5 cm in diameter and 7.6 cm in length in the zone of nodule production around four plants were 7.6, 5.2, 2.4, and 1.6 g in treatments from left to right. Below: Appearance of typical roots with nodules. (Weber, 1966a)

in nodulation as well as in nitrogen fixation was evidently associated with nitrogen fertilization. In the comparison shown in Fig. 7.36 an application of about 410 kg of fertilizer nitrogen per hectare was needed to produce a yield of nitrogen in plants without nodules equivalent to that obtained without fertilization in plants with nodules. Conditions in this comparison are abnormal in that the soil was treated with a large quantity of ground corn cobs to immobilize the native soil nitrogen. In another comparison in the same experiment on soil not treated in this way, fixation decreased from 84 kg/hectare on the control plots to 43 kg on plots receiving 168 kg of fertilizer nitrogen per hectare. In this comparison, an application of 168 kg of fertilizer nitrogen per hectare was not enough to produce a yield of nitrogen in nonnodulated plants equal to that in nodulated plants grown without nitrogen fertilizer. According to the carbohydrate-nitrogen theory, lowering the supply of carbohydrate in the plants by nitrogen fertilization should reduce the transfer of carbohydrate to the microorganisms in the root nodules and hence should inhibit nitrogen fixation.

The carbohydrate-nitrogen theory is useful as a general guide to performance of the symbiosis, as indicated by the preceding paragraphs. The phenomena in question are complex, however, and the theory provides little enlightenment on the mechanisms involved. Development of knowledge about mechanisms may be expected to lead to a number of more detailed and explicit theories. With regard to the effect of nitrogen fertilization on fixation, see, for example, the paper by Tanner and Anderson (1964).

A matter of great practical significance related to the ratio of carbohydrate to nitrogen is the transfer of fixed nitrogen from leguminous plants to nonleguminous plants growing in association with them in meadows, pastures, and other situations. The basic concept and a possible mechanism are indicated in Table 7.15. In this experiment, the oats grown with the peas contained more nitrogen where the peas were inoculated than where they were not. The sand in which the two plants were grown also contained more nitrogen where the peas were inoculated. Fixed nitrogen apparently was exuded from inoculated pea plants into the sand around the roots, from which it was removed by the oats.

The phenomenon of nitrogen release from the legumes and of transfer to associated nonlegumes was at one time a rather controversial subject until it was discovered that an important factor in the differences in results obtained by different investigators was the environmental conditions. Sometimes the phenomenon is important, as in Table 7.15, but other times it is not. Wilson and Wyss (1938) obtained no evidence for exudation of nitrogen from legumes grown in association with non-

Table 7.15. Yield and Nitrogen Content of Peas and Oats Grown in Association in Sand Culture, with and without Inoculation of the Peas with *Rhizobium,* and Apparent Exudation of Fixed Nitrogen (Virtanen et al., 1937)

	Dry weight, g		Nitrogen Content, mg			Apparent Proportion of Fixed Nitrogen Exuded, %
	Peas	Oats	Peas	Oats	Sand	
Peas not inoculated	0.6	0.4	12	3	9	—
Peas inoculated	2.9	1.4	45	16	26	48

legumes in containers under outdoor conditions in Wisconsin from early June to September. Where certain legumes were grown with reduced light, however, evidence was obtained for considerable exudation. They reported also that exudation occurred at a relatively low temperature. They explained their results in essentially the following manner. Where conditions are favorable for rapid photosynthesis, the carbohydrate produced is ample to meet the requirements of the rhizobia and to convert to protein the fixed nitrogen supplied by the rhizobia; exudation then does not occur. On the other hand, where photosynthesis is reduced to the proper degree, the rhizobia still receive enough carbohydrate to permit fixation; but there is not enough carbohydrate to permit rapid growth of the plants and full utilization of the fixed nitrogen to form protein. Fixed nitrogen then tends to accumulate, and some is exuded. They suggested that exudation of fixed nitrogen from leguminous plants grown in the field might be expected to occur in regions where weather during the growing season is cool and cloudy but not where it is hot and sunny.

The importance of exudation of fixed nitrogen from leguminous plants grown under field conditions has still not been determined directly because of the experimental difficulties involved. If a nonlegume contains more total nitrogen per hectare where it is associated with a legume than where it is grown alone, as in an instance reported by Johnstone-Wallace (1937), there is little doubt that a transfer of nitrogen from the legume to the nonlegume has occurred. If, on the other hand, the nonlegume contains less nitrogen per hectare where it is associated with the legume than where it is grown alone, as frequently is the case, it is possible that a transfer has occurred; but the evidence is not clear. Exudation of simple nitrogenous organic compounds, an important one of which is aspartic acid COOH—CH_2—CH(NH_2)—COOH, of course is not the only way in which transfer of nitrogen may occur. Sloughed nodules and root tissue may also contribute. A way to estimate the over-all transfer of nitrogen under field conditions despite the complications was worked out by Walker et al. (1954). Their procedure will be discussed in the last section of this chapter.

Leguminous plants fix nitrogen symbiotically, but they also absorb nitrogen from the soil. The proportions they derive from the two sources may vary widely according to the supply of nitrogen in the soil, the effectiveness of nodulation, and the environmental conditions. Evaluation of the amount of nitrogen fixed thus requires some way of partitioning the total nitrogen into soil and atmospheric sources. Two different methods have been used to determine the total amount of nitrogen fixed by legumes under field conditions: (1) analysis of the legume grown with and without nodules under otherwise comparable conditions; and (2) determination of the net increase in total amount of nitrogen in the soil-plant system over a period of time. The first method is unsatisfactory for use with many soils because rhizobia already present in the soil cause some inoculation, and the nitrogen thus fixed cannot be allowed for. Another limitation of this method is that, even where the control plants are not nodulated, the control and nodulated plants may not absorb equal amounts of nitrogen from the soil. Two other limitations are that the roots cannot be recovered quantitatively from soil, and the fixed nitrogen may have been exuded in part into the soil. A modification that has been used occasionally in circumstances where some nodulation occurs without artificial inoculation is to employ as the control a nonlegume. This modification has all the limitations just mentioned except the first. In the second method, the precision of the measurements is limited by the fact that the net increase one wishes to determine is small in comparison with the amount of nitrogen present initially. The net increase in a single year is usually so small that it is likely to be within the experimental errors of measurement. Use of the second method thus involves an experiment several years in length. Because of the loss of nitrogen by leaching in most such experiments, provision must be made for collecting the leachate for analysis. Even when due precautions are taken in such experiments, the data obtained represent only net changes. To attribute the net increase in nitrogen content of the soil-legume system to symbiotic fixation involves the questionable assumption that gains from nonsymbiotic fixation and losses from formation of gaseous products are the same in the soil bearing leguminous plants as in the control soil.

Because of the experimental difficulties, measurements of symbiotic fixation of nitrogen in the field are of limited accuracy. Nevertheless, the many measurements that have been made leave no doubt that legumes may fix significant amounts of nitrogen in the field. Giöbel (1926) made a summary of nitrogen fixation data reported in the literature and found a range from 56 to 323 kg/hectare annually. Many more recent experiments have been performed, and data from some of these were summarized by Allen and Allen (1958) and Vincent (1965). Fixation of as much as 670 kg/hectare by a single crop in a single season has been reported.

Another way of evaluating the contribution of legumes to the nitrogen supply in soil under field conditions is to determine the residual effect of the legumes on plants that follow them. The results are often spectacular, and many measurements have been made of the magnitude of the effects under different conditions. Only one example will be given. Stickler et al. (1959) seeded several legumes and mixtures of legumes with oats in the field and allowed the legumes to grow during a single season. In the autumn, before plowing the various plots, they measured the total nitrogen content of the legume tops and of roots excavated to a depth of 76 cm. In each of the following 2 years, corn was grown as a test crop. In the year the legumes were grown, comparable plots were planted to oats but without a legume. Corn was then grown on these plots during the following 2 years with different quantities of ammonium nitrate. Ammonium nitrate was applied only in the spring of the first year of corn. The average results obtained in the 2 years of corn are shown in Fig. 7.38. The yields obtained following all the legumes are substantially above the yield following oats with no nitrogen fertilization of the corn. With two of the legumes, the yields are about equal to the maximum obtained from fertilization with ammonium nitrate. The increases in yield from the other legumes are equivalent to those obtained with about 20 to 40 kg of nitrogen as ammonium nitrate per hectare. The nitrogen found in the legumes ranged from about 70 to 165 kg/hectare. Part of this nitrogen, of course, came from the soil. In connection with the beneficial effects of legumes on the supply of nitrogen for the crop that follows, one must keep in mind that most of the nitrogen is in the above-ground portions; hence, if the tops are harvested, the nitrogen added to the soil in the roots may be less than the soil nitrogen absorbed by the plants.

The brief coverage of symbiotic nitrogen fixation in this section has emphasized the contribution of the process to the nitrogen supply of soils and plants. Wilson (1940), Allen and Allen (1958), and Nutman

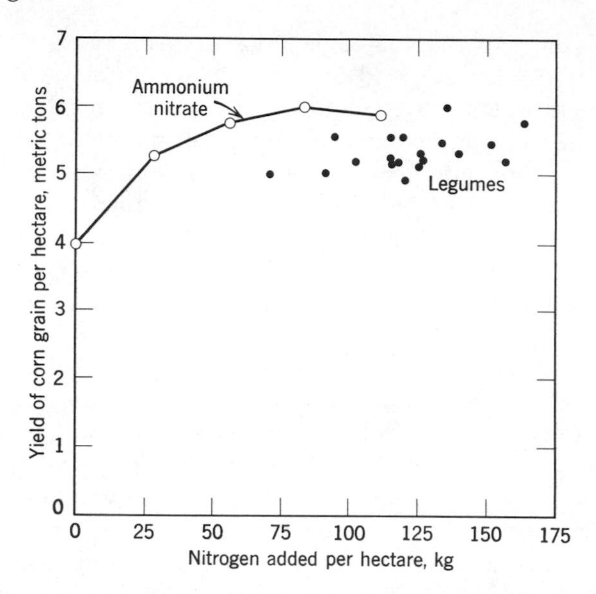

Figure 7.38. Average yield of corn grain in two years following oats alone or following a single season of various legume species and mixtures seeded with oats. Quantities of nitrogen added in the legumes are the quantities found by analysis of top growth and roots excavated to a depth of 76 cm. Quantities of ammonium nitrate are those added to corn grown following oats alone. The ammonium nitrate was added in the spring to the first crop of corn, and none was added to the second crop. The experiment was conducted on a silty clay loam soil in Iowa. (Stickler et al., 1959)

(1965) have published far more extensive accounts, parts of which are related to the aspect emphasized here. In addition, there has been a great deal of physiological and biochemical work; and a number of reviews of this subject matter are available. Although the biochemical and physiological aspects are accorded only scant attention here, they are significant from a long-range standpoint because they are directed toward basic understanding of the mechanisms and conceivably might eventually provide the basis for incorporating the characteristic of symbiotic nitrogen fixation into nonleguminous crop plants. The significance of this goal to practical agriculture needs no elaboration.

Combined Forms from the Atmosphere. Small amounts of nitrogen are brought down each year in the rainfall. The many available data were reviewed by Eriksson (1952), who found that the range reported was from 1 to 62 kg/hectare annually but that most values were below 10 kg.

Yaalon (1964) reported that the concentration of ammonium-nitrogen in rain water in Israel, where most soils are calcareous, increased with the soil temperature and was much higher in 1962–63 than in 1922–24. He pointed out that use of nitrogen fertilizers had increased greatly during the time between the two series of analyses and suggested that the ammonium nitrogen brought down in the rainfall had originally been volatilized from the soil.

Junge (1958) reported the results of a survey of ammonium and nitrate concentrations in rain water over the United States in the form of maps representing results obtained during different intervals of time. According to his interpretation of the results, the ammonium and nitrate in the rain water originate largely over land surfaces and not over the oceans. He related the low concentrations of ammonium in rainfall in southeastern United States to the fact that the acid soils there would tend to react with ammonia. The high values in certain areas of the western and north-central states he attributed to the combined effects of use of ammoniacal fertilizers and the presence of soils with high *p*H values that would permit some volatilization of ammonia. The pattern of nitrate concentrations was somewhat related to that of ammonium concentrations. Junge was able to exclude photochemical formation in the high atmosphere, lightning, and industrial activities as major sources; and he suggested instead that oxidation of ammonia in the atmosphere and release of oxides of nitrogen from soils are the two most likely sources. The nitrogen added to soils in combined forms from the atmosphere thus appears to represent for the most part a return of fixed nitrogen previously lost from soils and not an addition of nitrogen fixed in the atmosphere.

Accumulation of ammonium in soils from atmospheric sources is not necessarily limited to addition by rainfall. Ammonia introduced into air at low concentrations is adsorbed directly by clays (Brown and Bartholomew, 1963). Malo and Purvis (1964) obtained evidence that air-dry soil exposed to the atmosphere in New Jersey adsorbed ammonia from the air. The average concentration of atmospheric ammonia-nitrogen in their work was 0.047 mg/m^3, which is about three times the concentration generally taken to be "normal."

Slow Changes

If environmental conditions and soil treatments remain approximately the same over a long period of years, there is opportunity for the nitrogen cycle to approach a steady state wherein annual rates of gain and loss are equal and the total nitrogen content is constant. On the other hand, when conditions are modified, as by a change in management practices, the balance is disturbed. After the change, the rate of loss exceeds the rate of gain, or vice versa, as a new equilibrium is approached. Figure

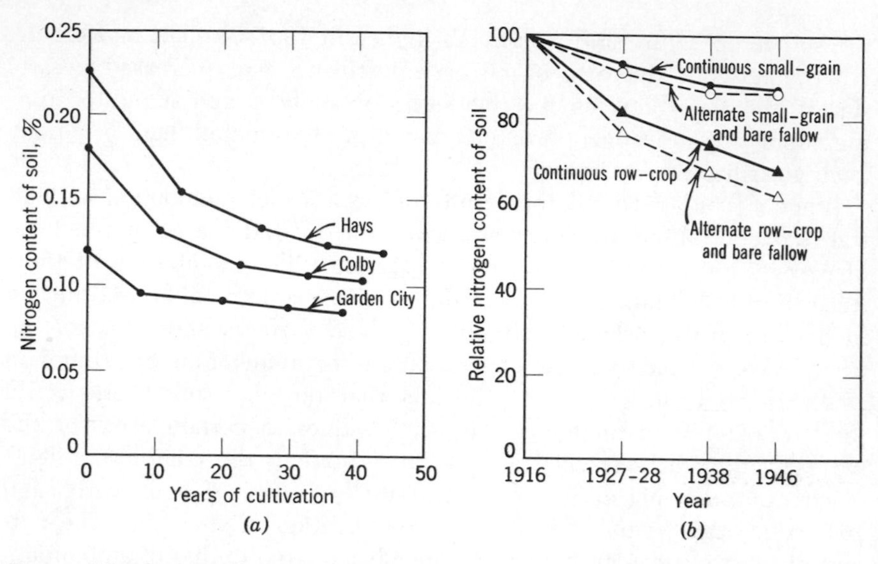

Figure 7.39. Nitrogen content of the surface 18 cm of soils developed under grass vegetation in western Kansas. (*a*) Nitrogen content of soils at three locations versus number of years since cultivation began. Results are average data from plots with various cropping systems. (*b*) Relative nitrogen content of soils at Hays with different cropping systems versus time. The nitrogen percentage in the soil in 1916 is given a relative value of 100 in each cropping system. Cultivation at this location began in 1903. (Hobbs and Brown, 1957)

7.39 represents an example of an excess of losses over gains when virgin soils developed under grass vegetation were brought into cultivation. Part *a* of the figure shows the changes that occurred at three locations, results at each location being averages of various cropping systems. Figure 7.39*b* shows the change in nitrogen content under different cropping systems at a given location. These results indicate the importance of cultivation in promoting losses of nitrogen. The trends indicate also that a stable content of nitrogen in the soil had not been attained at the last sampling.

Figure 7.40 shows the opposite trend upon establishment of grass-legume pastures in New Zealand. Figure 7.40*a* gives results for three soils, illustrating different types of behavior. The pastures on the sandy loam and the silt loam were established on areas that had been in old pasture. The pastures were plowed; and the land was cropped for 1 year, replowed, and then put back to pasture. The soils apparently had accumulated nitrogen almost to the equilibrium level under the old pasture. The pasture on the sand was established on an area that had been in gorse and bracken fern and then in temporary pastures. The nitrogen content of the soil apparently was much below the equilibrium content for the grass-legume vegetation at the time of establishment of the pasture. One additional significant point about these data is the unusual relationship between soil texture and the equilibrium nitrogen content. The two soils having the higher nitrogen content contain allophane, but the soil with the lowest nitrogen content does not. This mineral seems to stabilize organic matter in soils. Figure 7.40*b* shows the vertical distribution of nitrogen in a soil profile at the time of establishment of grass-legume pasture and 30 years later. These results indicate that accumulation of nitrogen in the soil under grass-legume pasture occurred from the surface downward. Presumably, the time required to attain the equilibrium content of nitrogen in individual layers of soil would increase with increasing depth of the layer in the soil.

Various investigators have expressed the slow changes in nitrogen content of soils with time in the form of mathematical equations. The equations are sometimes entirely empirical and sometimes based on assumptions about the contribution of different processes such as mineralization of organic nitrogen, resynthesis of organic nitrogen from mineralized nitrogen, and nitrogen fixation, each with adjustable parameters to ac-

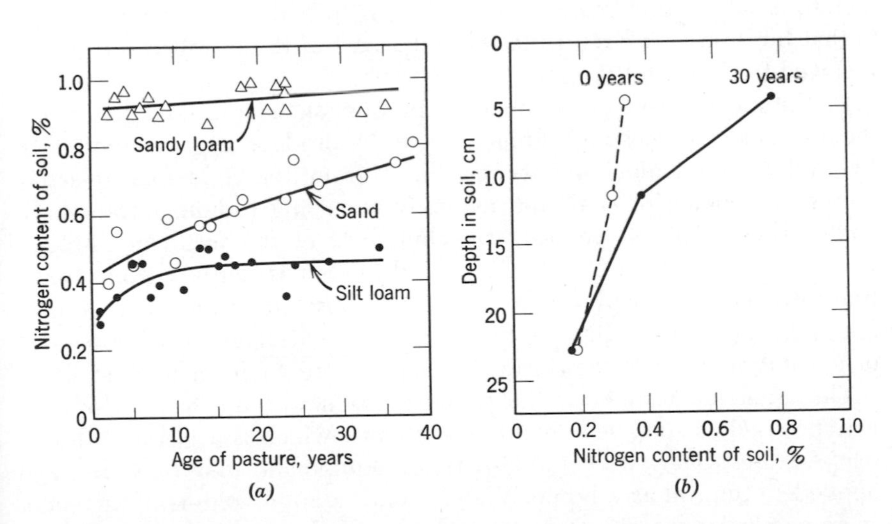

Figure 7.40. Nitrogen in soils under grass-legume pasture vegetation in New Zealand. (*a*) Nitrogen content of the surface 7.6 cm of soil versus time after establishment of permanent pasture on three soils. (*b*) Vertical distribution of nitrogen in a soil profile at the time of pasture establishment and 30 years later. (Jackman, 1964)

commodate variations from one soil to the next. Although the assumptions may not be entirely valid, the equations are valuable in focusing attention on factors involved in the changes and in estimating equilibrium values by extrapolating observed data. Some of the equations employed were reviewed by Stevenson (1965b). A paper including experimental observations illustrating the complexity of the relations involving soil development on a long-term basis under natural conditions was published by Dickson and Crocker (1953).

Nitrogen Fertilizers

Nitrogen fertilizers are mentioned incidentally in connection with other topics in this chapter because of their value in experimental work to investigate soil-plant relationships involving nitrogen. The purpose of this section is to convey a general concept of the nature, importance, and use of nitrogen fertilizers.

Animal manures are sometimes classed as nitrogen fertilizers, and important amounts of nitrogen are added to soils in this way. But because of their great weight per unit of nutrient content, animal manures are usually used on the farm on which they are produced. In a sense, therefore, addition of manures is analogous to addition of crop residues in that both processes return to the soil a part of the nitrogen previously removed from the soil by plants.

In the early years of commercial production of nitrogen fertilizers, the principal sources of nitrogen were by-product organic materials, by-product ammonium sulfate produced from the ammonia released during the coking of coal, and naturally occurring sodium nitrate from Chile. These sources are still used, but most of the fertilizer nitrogen now used (more than 90% in the United States) is derived from industrial nitrogen-fixation processes. By far the most important of these processes is the combination of nitrogen with hydrogen in the presence of a catalyst at high temperature and pressure to form ammonia.

The ammonia formed as the primary product of the fixation process may be used as such as anhydrous ammonia, which is a gas at ordinary temperatures, but, with suitable pressure equipment, may be stored and applied to the soil as a liquid. Water retains a large volume of ammonia as ammonium hydroxide, which can be handled without pressure equipment. Ammonium hydroxide is used as a fertilizer, frequently with ammonium nitrate, urea, or both in the same solution. The ammonia synthetically produced may be converted to other forms, of which urea and nitrate are the principal ones. Urea is the product of a reaction between ammonia and carbon dioxide. This extra step adds to the cost of production but results in a solid product that is easily handled and has a high content of nitrogen (46%). Ammonium nitrate may be made by first reacting ammonia with the oxygen of air at a high temperature in the presence of a catalyst to produce nitrogen dioxide, which reacts with water to form nitric acid (plus nitric oxide), and then by reacting the nitric acid with more ammonia. Ammonium nitrate, like urea, is a solid containing a high percentage of nitrogen (about 33% in the conditioned material sold for fertilizer purposes). Ammonium sulfate may be prepared by passing ammonia into sulfuric acid. The fertilizer-grade material contains about 21% nitrogen. There are many modifications in manufacturing processes of the materials described here. Other manufactured nitrogenous fertilizers are produced in smaller quantities. Further information on these subjects may be found in monographs edited by Sauchelli (1960, 1964) and in a review by Nelson (1965).

In the year 1962–63, the world output of nitrogen fertilizers was distributed among the various forms as follows (Food and Agriculture Organization of the United Nations, 1964): ammonium nitrate, 28%; ammonium sulfate, 21%; urea, 9%; calcium nitrate, 4%; sodium nitrate, 3%; calcium cyanamide, 2%; organic materials, 1%; other forms (including ammonium phosphate and solutions), 32%. In the same period, 39.8% of the fertilizer nitrogen was consumed in Europe, 31.1% in North and Central America, 15.0% in Asia, 8.7% in the Soviet Union, 3.3% in Africa, 1.7% in South America, and 0.4% in Oceania (Australia and New Zealand). In 1961–1962 the average consumption of fertilizer nitrogen in kilograms per hectare of arable land was 28.7 in Europe, 13.4 in North and Central America, 5.3 in Asia (excluding Mainland China and Korea), 3.8 in the Soviet Union, 1.5 in Africa, 1.3 in Oceania, and 8.3 in the world as a whole.

Nitrogen fertilizers are adapted to application in various ways, according to the nature of the material and the circumstances. Most of the nitrogen fertilizer applied is incorporated into the soil either directly by the distributor or indirectly by cultivation operations. Anhydrous ammonia and ammonium hydroxide are usually placed below the soil surface by special distributors because of volatility of the ammonia, but they may also be applied to and distributed in irrigation water. Much solid nitrogen fertilizer is applied to the surface of the soil, with dependence on water movement to carry it down into the root zone. Urea is absorbed effectively by foliage and may be sprayed on plants. Nitrogen fertilizer may be applied at the time of planting annual crops, or it may be applied in part or entirely at some later time during the season. With crops that grow over a long period of time, application of nitrogen fertilizer at intervals is almost always the preferred method.

Quantities of nitrogen applied vary widely. Occasionally, applications of the order of 5 or 10 kg/hectare are made if the nitrogen is present with other nutrients in mixed fertilizers. Applications made for the nitrogen effect are seldom less than 20 kg/hectare or more than 200 kg/hectare in a single year. The most appropriate application depends on the supply of nitrogen in the soil, the nature of the crop, the cost of application, the value of the crop, the environmental conditions, and other factors. Of the quantities applied as fertilizer, recoveries of the order of 40% are commonly obtained in the form of increases in the nitrogen content of the above-ground portion of plants in the year of application.

Function in Plants

Nitrogenous compounds make up a significant part of the total weight of plants. In a plant that contains 1.6% nitrogen, for example, about 10% of the plant weight is contributed by compounds containing nitrogen. Nitrogen occurs in both inorganic and organic forms in plants. The inorganic forms of combined nitrogen usually make up only a small proportion of the total. The nitrogen absorbed by plants from soils is taken up principally as nitrate, which is the only inorganic form that can and does accumulate in quantity in plants without injurious effects. Nitrate is assimilated by reduction to ammonium, followed by incorporation into organic forms. Ammonium and nitrite are usually found only in small quantities, if at all.

As regards both numbers of compounds and amounts of nitrogen involved, organic forms predominate. Most of the organic nitrogen of plants is present in proteins. Proteins are high-molecular-weight compounds made up of amino acids that are linked together through the amino ($—NH_2$) and carboxyl ($—COOH$) groups. According to Waldschmidt-Leitz (1958, pp. 277–314), twenty-one different amino acids form the building units of normal plant proteins. Others are present in special proteins and in free form in plants. With the exception of water, proteins are the main constituents of protoplasm. From the standpoint of the functioning of nitrogen, protoplasmic proteins are important because, acting together, they synthesize not only themselves but also the other organic compounds of plants.

The proteins of cells of the vegetative parts of plants are thought to be primarily functional in nature. Many of them are enzymes; others are nucleoproteins, some of which are present in chromosomes. Nitrogen is present in both the protein and nucleic acid portions of nucleoproteins. Proteins thus serve both as catalysts and directors of metabolism. Functional proteins are not stable substances; rather, they exist in a continuous state of flux, being both broken down and reformed. The amino acid links used for repair may be newly formed; or they may come from hydrolysis of some other protein, such as a reserve protein. Much of the protein in seeds, for example, may be classified as reserve protein because during germination it is hydrolyzed and not reformed and hence appears to serve only as a source of amino acids for the development of other proteins in the seedling.

In addition to its function in proteins, nitrogen plays a part in other processes. Nitrogen is a component of the chlorophyll pigments that give plants their green color. Because chlorophyll is necessary for the process of photosynthesis, nitrogen may be said to play an essential part in photosynthesis. Nitrogen is found in hormones, which are organic substances that exert important regulatory effects on metabolism when present in only minute quantities. Nitrogen is a component of the respiration-energy carrier, adenosine triphosphate, which will be discussed in Chapter 8.

Further reading on this subject will be found in books on plant physiology, including the one by Meyer et al. (1960).

Nitrogen Supply and Plant Behavior

Deficiency Symptoms

Under conditions of nitrogen deficiency, the leaves are small, the stems are thin and upright, and lateral shoots are few; hence the growth has a sparse appearance. The leaves usually have a pale, yellowish-green color in the early stages of growth, because of limited synthesis of chlorophyll, and may develop yellow, red, or purple colors from anthocyanin pigments as they grow older. Symptoms are most pronounced on older leaves.

Specific symptoms may differ somewhat from one kind of plant to another. In corn, the leaf tip becomes yellow; and the yellowing follows the midrib toward the base of the leaf. The edges of the leaf remain green for a longer period of time. With potato, on the other hand, the color of the margins of the lower leaflets fades to a yellow. In tomato and apple, the yellowish green of nitrogen-deficient foliage is associated with the production of purple anthocyanin pigments. Acute nitrogen deficiency in peach trees results in leaf colors varying from a yellowish green at the tip of long current-season twigs to a reddish yellow with numerous red and brown spots at the base of the twigs. Symptoms of nitrogen deficiency in citrus leaves vary with the time at which the deficiency occurs. If a nitrogen shortage occurs when growth begins, the young leaves are a light, yellowish green color; and the color of the veins is slightly lighter than that of the tissue between. If a nitrogen shortage occurs during the summer and autumn while the fruit is maturing, leaves develop a mottled green and yellow pattern. Under conditions of severe deficiency, the leaves ultimately become yellow and are shed.

The book by Wallace (1961) and the book edited by Sprague (1964) should be consulted for further information. These books include both reproductions of color photographs and descriptions of nitrogen deficiency symptoms in many plants.

The occurrence of deficiency symptoms is outward evidence of internal competition for nitrogen among the various parts of the plant, and the difference in intensity of the symptoms that may occur on different leaves on the same plant is evidence of the relative competitive ability of the leaves. Breakdown and resynthesis of nitrogen compounds in plants take place continuously, and the breakdown products may be moved from one part to another under conditions of internal competition; hence a leaf that appears green and normal at one time may later become yellow because of loss of nitrogen. Generally, old leaves are poorer competitors than young ones. Fruit is a strong competitor. As a quantitative example of a relationship between deficiency symptoms and yields, the yield of corn grain was found by Viets et al. (1954) to decrease 940 kg/hectare with each increase of one leaf per plant showing visual evidence of nitrogen deficiency at time of silking. Although no leaves showed nitrogen deficiency symptoms if the nitrogen percentage in certain index leaves was 2.2 or above, the nitrogen content corresponding to the maximum yield attainable by increasing the supply of nitrogen was estimated as 2.8%. In the range between 2.2 and 2.8%, the plants could be deficient in nitrogen, as far as production of grain was concerned, without providing visual evidence in the form of deficiency symptoms on the leaves.

Carbohydrate Utilization

As the supply of nitrogen to plants increases, the tendency is for the carbohydrate content to decrease. For example, Hasegawa et al. (1962) found that nitrogen fertilization decreased the starch content of leaf sheaths of rice plants at each of four stages of growth. Baumeister (1939) found that the nitrogen percentage in wheat grain increased and the starch percentage decreased with increasing application of nitro-

Table 7.16. Yield and Sugar Content of Sugar Beets with Different Applications of Nitrogen as Ammonium Nitrate (Walker et al., 1950)

Nitrogen Applied per Hectare, kg	Yield of Beets per Hectare, Metric Tons	Sucrose in Beets, %	Yield of Sucrose per Hectare, Metric Tons
0	46	18	8.2
90	54	17	9.0
180	59	16	9.7
269	61	15	9.0

gen. The greatest decrease in starch content occurred with quantities of nitrogen approaching and exceeding the application corresponding to the maximum yield of grain. This effect of nitrogen supply is explained on the basis that nitrogen promotes growth of additional tissue, in which the carbohydrates produced by photosynthesis are used.

The nitrogen-carbohydrate balance has many indirect effects on plant behavior, some of which are described in subsequent sections. It is of direct importance in the culture of sugar-producing crops. Gardner and Robertson (1942) and others have found that the sucrose percentage in sugar beets decreases with an increase in content of nitrate in the beets. Usually, the maximum yield of sugar is reached with less nitrogen than is the maximum yield of the plant as a whole. This behavior is illustrated in Table 7.16 by results of an experiment on nitrogen fertilization of sugar beets conducted in California. The depressing effect of nitrogen on the sugar percentage may not be found with low or moderate applications of nitrogen (Haddock, 1952; Ulrich, 1954) because of early depletion of added nitrogen from the soil and growth response to the extra nitrogen absorbed.

Succulence

If the nitrogen supply and other factors are favorable for growth, the tendency is for utilization of carbohydrates to form more protoplasm and more cells rather than for deposition of carbohydrate to thicken the cell walls. Cells produced under such conditions tend to be large and to have thin walls. Letham (1961) observed that fruit from apple trees fertilized with nitrogen contained larger cells than fruit from unfertilized trees or from trees fertilized with phosphorus, nitrogen and phosphorus, or nitrogen, phosphorus, and potassium. The incidence of internal breakdown of the fruit during storage increased with the average cell volume.

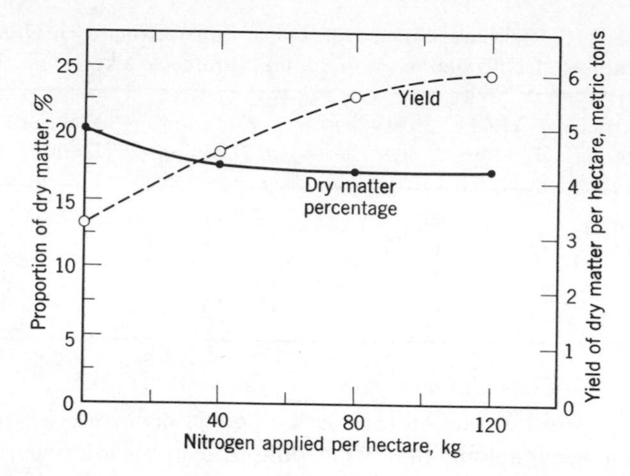

Figure 7.41. Dry matter percentage and yield of Sudangrass with different applications of nitrogen. (Siebert, 1939)

Because protoplasm is largely water, high-nitrogen plants under conditions favorable for growth contain a relatively high proportion of water and a low proportion of dry matter and are said to be succulent. Figure 7.41 shows the effect of nitrogen supply on yield and dry-matter content of Sudangrass.

The characteristic succulence of plants high in nitrogen is of special significance in the production of certain fiber crops. In hemp, for example, the fiber cells are larger and have thinner walls if the plants are grown with a high supply of nitrogen than with low nitrogen (Scheel, 1936). Such hemp fiber has low mechanical strength. Table 7.17 shows

Table 7.17. Yield and Fiber Content of Hemp Grown with Different Applications of Nitrogen Fertilizer (Unpublished data, Iowa Agricultural Experiment Station, 1944)

Nitrogen Applied per Hectare, kg	Yield of Green Plants per Hectare, Metric Tons	Yield of Fiber as Percentage of Green Weight	Yield of Fiber per Hectare, kg		
			Line	Tow	Total
0	23.1	4.0	780	152	932
56	29.4	3.8	833	292	1125
112	32.4	3.6	690	478	1168

the behavior of hemp with different applications of fertilizer nitrogen. The fiber percentage decreased with an increase in nitrogen supply; hence the yield of fiber was not increased to as great a degree with nitrogen fertilization as was the total yield. The main effect of the nitrogen supply on the fiber was an increase in the yield of the fraction classified as tow. Tow is the short, commercially undesirable fiber derived from fibers that are originally short and from long fibers that are broken in processing. Fiber quality thus deteriorated with an increase in supply of nitrogen in the experiment. Several investigations of the effect of nitrogen supply on cotton fiber have been made. Gulati and Ahmad (1947) reported a decrease in strength of fiber with an increase in supply of nitrogen to the plants. The more usual finding, however, is that such characteristics as length, diameter, and breaking strength of cotton fibers are affected only slightly by nitrogen supply (Sturkie, 1947; Hamilton et al., 1956).

Root Growth

Within the range of practical interest, an increase in the supply of nitrogen causes the growth of the above-ground portion of plants to increase relatively more than the growth of roots. Figure 7.42 shows the results of an experiment on oats. These data are given as an example because they cover the complete range from extreme deficiency to ex-

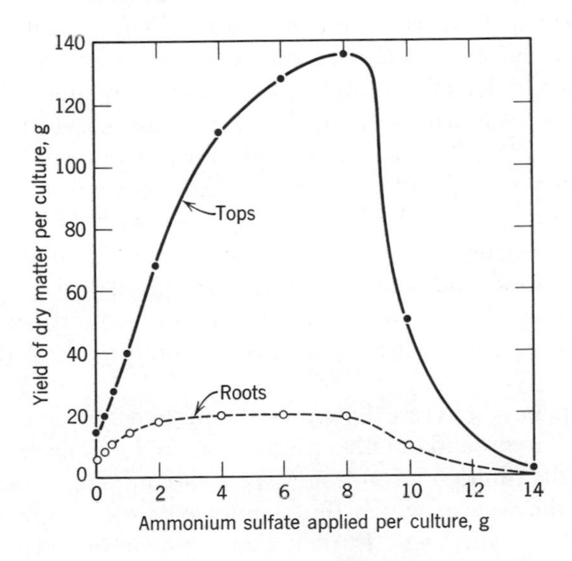

Figure 7.42. Yield of oat tops and roots in sand cultures with different applications of ammonium sulfate. (Meyer and Storck, 1927)

treme excess. Experiments done under field conditions usually cover only a limited portion of the range, but the results generally agree with those in Fig. 7.42. As an example of a field experiment covering a relatively broad range, Holt and Fisher (1960) found that the yield of bermudagrass herbage during one growing season was increased from 2.4 metric tons per hectare on control plots to 25.0 tons on plots receiving a total of 1.8 tons of nitrogen per hectare, whereas corresponding yields of roots plus rhizomes were 0.8 and 0.9 metric ton per hectare. Troughton (1957) reviewed the results of various experiments on this subject.

Irregular distribution of nitrogen in soil may influence the growth of roots in a manner precisely the opposite of what might be expected from the over-all effect described in the preceding paragraph. Wiersum (1958) found that production and growth of lateral roots in media deficient in nitrogen was greater in local areas relatively high in nitrate than in adjacent areas deficient in nitrogen. Duncan and Ohlrogge (1958) found that corn roots proliferated in low-nitrogen, low-phosphorus soil containing localized applications of nitrogen and phosphorus fertilizers together but not in soil receiving only nitrogen fertilizer (see also Fig. 7.43). That these observations are related to the supply of nutrients for growth is verified by the converse findings that the effects did not occur in the absence of deficiencies. Duncan and Ohlrogge (1958) observed root-proliferation effects only in the secondary roots of corn and not in the primary roots, which were produced before exhaustion of seed reserves. Similarly, Brouwer (1962) noted that the nitrogen supply in solution around a corn root had no effect on growth of the root if the remainder of the root system was adequately supplied with nitrogen but that it had a marked effect if the other roots were not well supplied with nitrogen. In a field experiment in which corn was grown on a loamy sand soil in Nebraska, Linscott et al. (1962) observed that roots were associated almost exclusively with old root channels, old roots, or decaying organic materials to the full depth of root penetration in unfertilized soil and in nitrogen-fertilized soil below a depth of about 1 meter. In the surface meter of nitrogen-fertilized soil, however, the roots were not restricted in this way but proliferated throughout the soil.

The discrepancy between the effect of nitrogen supply on the over-all production of roots and on the production in local areas in instances of irregular distribution of nitrogen is more apparent than real and is related to the way in which the character of root systems is affected by the supply of nitrogen. If plants are grown in nutrient solutions in which the concentration is uniform throughout, the tendency is for low-nitrogen roots to be long, thin, and sparsely branched and for high-

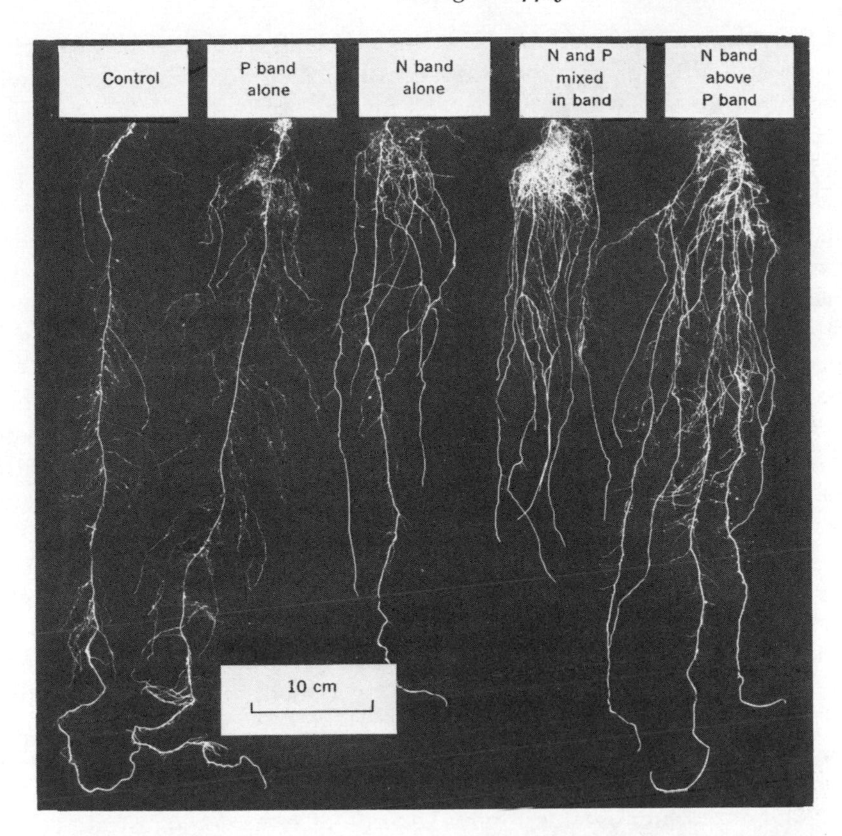

Figure 7.43. Growth of single roots of soybean plants in relation to placement of nitrogen and phosphorus salts. The plants were grown in expanded, unfertilized vermiculite, and one root of each was led laterally through a tube into a vertical, 1-liter, graduated cylinder of a silt loam soil. The picture shows root development in the soil. Localized placements of phosphorus and nitrogen consisted of 0.75 g of monobasic calcium phosphate monohydrate and 0.398 g of ammonium sulfate mixed with a 50-ml volume of soil in the upper portion of the cylinder. Note the proliferation of roots where the nitrogen and phosphorus were mixed in a single band. (Photograph courtesy of A. J. Ohlrogge; Wilkinson, 1961)

nitrogen roots to be short, thick, and well branched [see the literature reviewed by Wiersum (1958)]. Accordingly, the proliferation of a portion of the root system in local, high-nitrogen areas may be seen to be associated with the tendency of the entire root system to be short, thick, and well branched in uniform media high in nitrogen.

One basic theory may be proposed to account for both the general and local effects of nitrogen supply on root production (Brouwer, 1962).

If the principal limiting factor is nitrogen, plants contain excess carbohydrates in both tops and roots because use of carbohydrates for protein formation and growth is limited. Nitrogen absorbed by the roots then tends to react with carbohydrates in the roots. Growth of tops remains limited relative to growth of roots because much of the nitrogen absorbed is used in the roots. As the supply of nitrogen increases, more of the nitrogen reaches the tops and causes use of carbohydrates there for protein synthesis and growth. Consequently, less carbohydrate remains for translocation to roots, and growth of roots then is limited relative to growth of tops. Proliferation of roots in the vicinity of a locally high supply of nitrogen in an otherwise nitrogen-deficient soil may be accounted for by extension of the theory to operation among different parts of the root system as well as between tops and roots. If the two primary deficiencies should be nitrogen and some other nutrient, with the latter the more important, a moderate general or local increase in supply of nitrogen should have little effect on the extent or distribution of roots. But if the two nutrients are applied together, the effect should be more pronounced than if either is applied singly, on the basis of the theory of carbohydrate utilization.

In addition to accounting for the over-all effect of nitrogen on growth of tops versus roots, this theory is in agreement with the observation (Troughton, 1957) that effects of ammonium are more extreme than those of nitrate. Ammonium combines rapidly with organic acids in plants, whereas nitrate must first be reduced to ammonium before it is metabolized into organic forms. The rate of reduction of nitrate differs among plants, but in general it seems to be much slower than the rate at which ammonium nitrogen is changed to organic forms; hence there is opportunity for more translocation of nitrate than of ammonium. Considerable quantities of nitrate may be stored in plants if soil nitrate is high. Accordingly, in low-nitrogen plants, in which the supply of carbohydrates in the roots is ample, nitrogen absorbed from the soil may be metabolized more rapidly and used to a greater extent in root growth if the nitrogen is taken up as ammonium than if it is taken up as nitrate. In high-nitrogen plants, on the other hand, carbohydrates are kept at a lower level throughout if the nitrogen is absorbed as ammonium than if it is absorbed as nitrate, thus favoring greater growth of tops at the expense of roots. In other words, the effect of nitrate on growth tends to be more distributed than that of ammonium.

To account for the tendency of low-nitrogen roots to be long and sparsely branched and of high-nitrogen roots to be short and well-branched, another theory has been proposed, namely, that high nitrogen increases the concentration of auxins or plant hormones, which then causes inhibition of terminal growth of roots and production of laterals.

Torrey (1950) found that detached root tips of peas increased in length when placed in a suitable nutrient medium, but no lateral roots were produced. Elongation of detached root tips was much reduced in a similar medium modified by addition of 1 μg of indoleacetic acid (a plant auxin) per milliliter, but the roots then produced many laterals. Bosemark (1954) found that roots of wheat grown in a nutrient solution increased in length 27 mm in 6 days in the presence of 10^{-4}-molar sodium nitrate and only 3 mm in the presence of 10^{-2}-molar sodium nitrate. Addition of α-parachlorophenoxyisobutyric acid (an antiauxin) caused the growth to increase to 32 mm in 10^{-4}-molar sodium nitrate and to 21 mm in 10^{-2}-molar sodium nitrate. The antiauxin thus had little effect on the low-nitrogen roots but largely overcame the inhibition of elongation of high-nitrogen roots. He explained these results on the basis of work by Avery et al. (1936) showing that nitrogen increased the concentration of natural auxin in plants. Presumably, the high concentration of auxin inhibited root elongation, and the antiauxin counteracted in part the inhibition due to the natural auxin. More recent investigations have shown that the auxin relationships are relatively complex (Street, 1960); hence the theory that the nitrogen supply controls the character of the root system through its effect on auxin production may require elaboration or revision.

The increase in ratio of tops to roots with increase in nitrogen supply should not be allowed to obscure the fact that nitrogen is essential for the growth of roots and that it is an important fertilizer for root crops grown on soils having low nitrogen availability. An adequate nitrogen supply is necessary to obtain rapid elaboration of leaves in the early part of the growth cycle. Without this, the total capability of the plant for photosynthesis is diminished. At the proper time, the growth of leaf must be checked by a decrease in the content of nitrogen (or in some other way). Growth of leaf cannot be continued too long because, with a growing season of a given length, the time available for development of the storage organ decreases with increase in length of time the carbohydrate supply is used primarily for top growth. These relationships are exemplified by the data of Woodman and Paver (1944) on yield of turnips supplied with different levels of nitrogen at different times. Table 7.18 shows that the best root growth was obtained where the high nitrogen level was applied early rather than late. Results of a somewhat similar experiment with sugar beets were published by Loomis and Nevins (1963).

Fruiting

With reference to tomato, Kraus and Kraybill (1918) recognized that high-nitrogen plants are unfruitful or excessively vegetative even

Table 7.18. Yield of Turnip Tops and Roots with Nitrogen Applied in Different Concentrations at Different Times (Woodman and Paver, 1944)

Nitrogen per Milliliter of Culture Solution in Growth Period Indicated, μg			Yield of Dry Matter, g	
Early	Intermediate	Late	Tops	Roots
4	4	4	0.4	1.4
66	4	4	1.1	6.6
66	66	4	2.3	14.4
66	66	66	2.8	13.1
4	66	66	2.6	9.7
4	4	66	1.5	4.0

though conditions are favorable for carbohydrate synthesis. With some reduction in nitrogen supply, vegetative growth decreases and fruitfulness increases. With further reduction in nitrogen supply, both vegetation and fruitfulness are diminished. Stated another way, the maximum yield of fruit is obtained with a lower supply of nitrogen than is the maximum yield of vegetative parts; and the ratio of fruit to vegetative parts is at a maximum with a certain supply of nitrogen and decreases with either an increase or a decrease in nitrogen.

The behavior described by Kraus and Kraybill is characteristic of many plants. With the small-grain cereal crops, the maximum ratio of grain to straw usually occurs in the range of marked deficiency of nitrogen. Hence the ratio of grain to straw usually decreases with an increase in supply of nitrogen. Under extreme nitrogen deficiency, however, many plants are barren; application of nitrogen then increases the ratio of grain to straw (Hellriegel et al., 1898). Increases in ratio of grain to straw from nitrogen fertilization within the practical range of crop production may be obtained by delayed application of the fertilizer (Halliday, 1948). With corn, the maximum ratio of grain to vegetative parts seems to occur at a relatively high level of nitrogen. Corn is subject to barrenness if markedly deficient in nitrogen, and increasing the supply of nitrogen increases the relative yield of grain to a greater extent than that of leaves and stalk. The ratio of grain to leaves and stalk does not seem to decrease with increasing supply of nitrogen within the range of practical interest. Figure 7.44 gives an example.

In some instances, the effect of nitrogen supply on fruitfulness is associated with its effect on rate of plant development. Ryle (1963) found that with timothy planted outdoors in vermiculite in England at intervals from October 16 to the following June 25, plants receiving a nutrient solution high in nitrogen (150 μg/ml) usually required fewer days until emergence of the inflorescence than did plants receiving a solution low in nitrogen (15 μg/ml). All shoots were fertile if the date of planting was April 30 or earlier, whether the nitrogen supply was high or low. With later planting dates, the nitrogen supply became more critical in determining flowering. Inflorescences emerged on September 9 from plants seeded on June 25 and maintained with the high nitrogen supply, but none emerged at any time during the season from comparable plants with the low nitrogen supply. Analogous behavior has been observed in table beets. This plant does not normally produce seed in the first year it is grown. Mann (1951) observed that the proportion of the plants that produced seed the first year was increased from 3% with no supplemental nitrogen to 9% with 1.5 metric tons of ammonium sulfate per hectare.

Still another type of behavior was observed by Williams (1963) in work with apples. He observed that the self-fruitfulness of a particular variety of apples varied from orchard to orchard and from year to year and that self-fruitfulness was greatest in years of profuse blossom and with heavy applications of nitrogen. He was able to connect the self-fruitfulness primarily with the longevity of the embryo sacs and second-

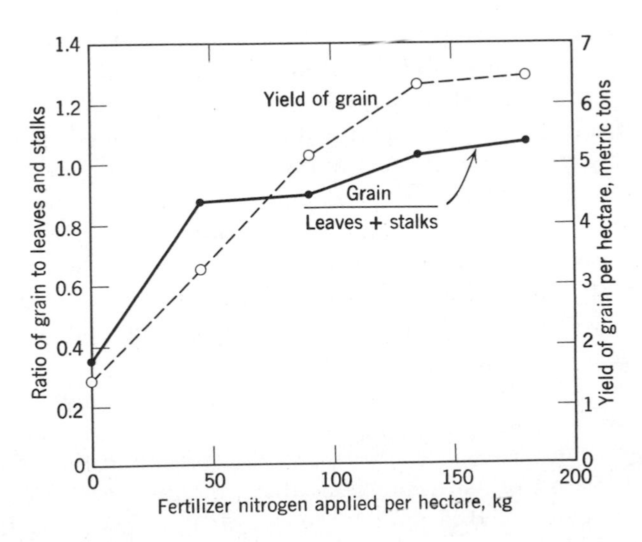

Figure 7.44. Yield of corn grain and ratio of grain to leaves and stalks with different applications of nitrogen fertilizer in North Carolina. (Krantz, 1949)

arily with the rate of growth of pollen tubes. Embryo sacs from low-nitrogen trees had a much shorter life than those from trees receiving optimum nitrogen during the preceding summer. Where the pollen was derived from the same variety as the stigma on which it fell, the growth of pollen tubes through the stigma and style occurred at a relatively slow, borderline rate; and the rate of growth was somewhat greater where the pollen was produced under high-nitrogen conditions than under low-nitrogen conditions. Under these circumstances, small differences in longevity of the embryo sacs and in rate of growth of pollen tubes may be relatively critical.

Williams (1963) noted further that the tendency of trees of the variety of apples he investigated to bear fruit on alternate years could be eliminated by summer application of nitrogen fertilizer. The effect of nitrogen supply on fruiting of apples depends on the variety and conditions, according to literature reviewed by Weeks and Southwick (1956), and may be accounted for by the theory that competition exists between fruit and fruit buds. Apple trees produce fruit buds during the summer of the year before the buds open and produce fruit. Formation of fruit begins before development of fruit buds for the next year and continues until after bud development is completed. Because both fruit and buds are dependent on the same supply of minerals and organic substances in the tree, development of fruit buds is inhibited by the presence of developing fruit in years of fruit production. In intervening years, when there is little or no fruit, the minerals and organic substances are available for development of fruit buds. According to this theory, increasing the supply of nitrogen should decrease the tendency of trees for biennial bearing to the extent that the critical factor in competition between fruit and fruit buds is the supply of nitrogen. But biennial bearing may persist irrespective of the supply of nitrogen if carbohydrate or some other factor is critical in development of buds. The use of hormone sprays to reduce fruit set on apple trees, a common practice in commercial production, may be related to the foregoing theory. With a reduction in number of apples produced per tree, competition is reduced. Fruit buds then develop concurrently with the apple crop, and the trees bear fruit each year.

Lodging

Lodging is the inelastic displacement of plants from the vertical position. Usually wind provides the force that produces the lateral displacement. The force applied to the base of the plant, however, is determined not only by wind pressure but also by the height and weight of the plant and the weight of rain that may be adhering to the plant.

The common observation that susceptibility of plants to lodging increases with the nitrogen supply may be attributed to the action of nitrogen in increasing the weight, height, and leaf area without at the same time increasing the strength of the basal parts sufficiently to counteract the increase in moment of inertia of the plant with respect to the base. Sometimes the resistance to breaking decreases.

For example, Fig. 7.45 shows results of an experiment on corn in which lodging increased markedly with density of stand and to a smaller extent with increase in nitrogen supply and in which the breaking load of the third internode decreased markedly with density of stand. (The breaking load is the weight required to break the third internode when the stalk was laid on supports 15.2 cm apart and the weight was applied midway between the supports.) The increase in lodging with nitrogen supply in this experiment may be attributed to the greater moment of inertia from the greater weight and somewhat greater height of the ear and probably also to greater weight of the stalk and leaves, although stalk and leaf weight was not recorded. The breaking load was almost unaffected by nitrogen supply (average breaking loads over the various densities of stand were 49, 51, 51, and 50 kg with application of 0, 112, 224, and 449 kg of nitrogen per hectare).

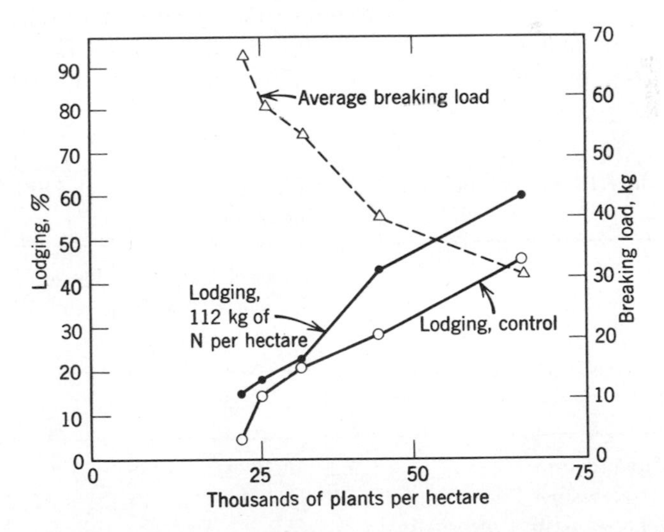

Figure 7.45. Lodging of plants and breaking load of third internode of corn with different applications of nitrogen fertilizer and different densities of stand. Lodging was similar with application of 112, 224, and 449 kg of nitrogen per hectare. Values for breaking load are average results over applications of 0, 112, 224, and 449 kg of nitrogen per hectare. (Nelson, 1958)

Modern varieties of hybrid corn usually produce only one stalk per plant regardless of the nitrogen supply. The number of stalks produced by a single plant of so-called small-grain crops, such as wheat and oats, however, increases with the supply of nitrogen. With such crops, therefore, the effect of nitrogen supply is similar to the combined effects of nitrogen fertilization and density of stand in the experiment on corn just described. Lodging is a more important problem with small grains than with corn, and a number of investigations have been made on the characteristics of small-grain crops in relation to lodging [see the paper by Mulder (1954) for a review of this research]. With small grains, the effect of nitrogen supply on lodging has been associated primarily with the diameter and wall thickness of the stalk. Table 7.19 gives results of measurements made on rice plants fertilized with different quantities of nitrogen. In this experiment, both the diameter and wall thickness of the stalks decreased with an increase in supply of nitrogen.

The tendency of cereal crops to lodge if well supplied with nitrogen limits the yield that could otherwise be produced by application of nitrogen fertilizers. Three different methods have been employed in an attempt to counteract the lodging: breeding of varieties with stiffer stalks, application of hormone-type chemicals to shorten the stalk, and delaying the application of nitrogen. Some success has been attained with all three methods. The first method is the simplest from the standpoint of the producer, but the character of stiffness of stalk has not

Table 7.19. Characteristics of Rice Plants Fertilized with Different Quantities of Nitrogen (Morita and Taya, 1957)

Plant Characteristic and Units of Measurement	Numerical Value of Characteristic with Indicated Application of Nitrogen per Hectare, kg				
	0	38	57	76	95
Length of plant, mm	955	1034	1045	1081	1086
Number of stems per hill	14.8	14.4	15.6	17.2	17.8
Measurements on third internode					
External diameter of straw, mm	3.82	3.55	3.44	3.39	3.36
Straw wall thickness, mm	1.80	1.59	1.48	1.44	1.44
Breaking load, g	156	142	136	124	118
Lodging, %	—	—	60	90	90

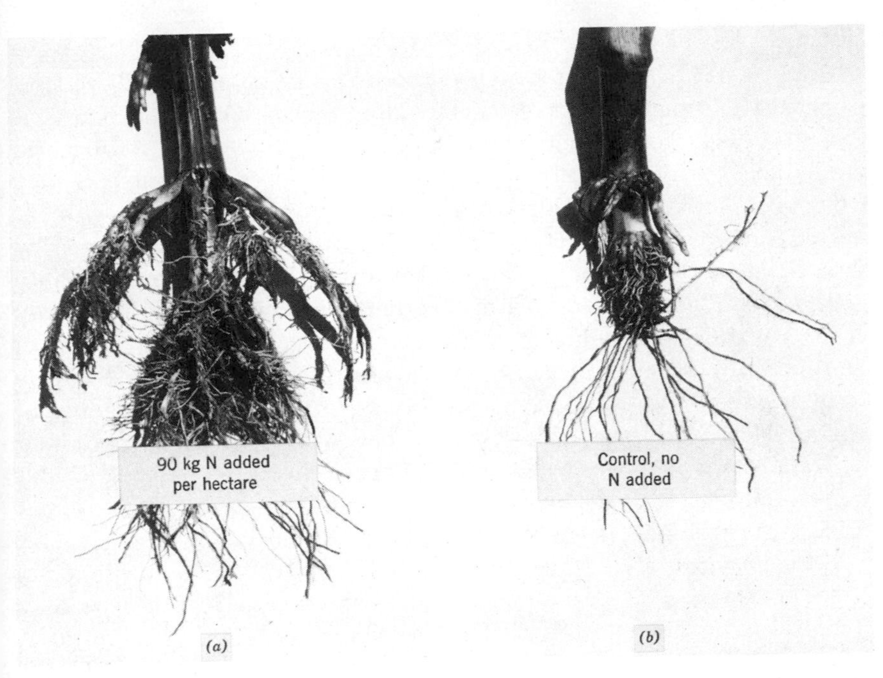

Figure 7.46. Growth of new roots on corn plants with original roots severed by root worms in an experiment in Nebraska. (*a*) Ninety kilograms of fertilizer nitrogen per hectare. (*b*) Control. Corresponding yields were 5.9 and 2.6 metric tons of grain per hectare, and percentages of plants lodged 30° or more on September 10 were 35 and 82. (Hill et al., 1948)

been developed to the desired degree. Hormone-type chemicals employed thus far are likely to decrease the yield of grain if they are effective in reducing lodging (Reichard and Schönbrunner, 1961; Linser and Kühn, 1962). The third method (Singh and Takahashi, 1962) may increase the efficiency of a limited application of nitrogen (Halliday, 1948); however, the absolute response that can be produced in this way is less than that attainable with an adequate supply of nitrogen throughout the growth period in the absence of lodging.

In some instances the stem remains intact, and lodging results from partial uprooting of the plant. Such lodging may occur, for example, with corn affected by root worms. Hill et al. (1948) found that lodging of corn associated with root-worm injury was decreased by application of nitrogen fertilizer. Their observations indicated that nitrogen fertilization did not affect the population of root worms but that it promoted rapid growth of new roots after losses due to root-worm damage (see Fig. 7.46).

Time of Maturity

Nitrogen fertilization often delays the maturity of plants. The consequences are sometimes of economic importance. Thus Stakman and Aamodt (1924) found that heavy applications of nitrogen fertilizer to wheat delayed the maturity 7 to 10 days in one experiment and 4 to 8 days in another. These delays in maturity were associated with increased lodging and infection of the plants with stem rust. In New York, Boynton (1948) noted that peaches attain their best market quality if they ripen in relatively warm weather. Fruit matures more slowly on trees fertilized heavily with nitrogen than on low-nitrogen trees. In the southern Hudson Valley, where peaches ripen late in August during warm weather, the delay in maturity is no particular disadvantage. In western New York near Lake Ontario, however, peaches do not ripen until late in September, when the weather is likely to be cool. Under these conditions, a delay in ripening may have important consequences because the best market quality may never be attained.

Nitrogen fertilization does not invariably delay the maturity date. In some instances it has no measurable effect, and in others it may cause the crop to mature earlier. In considering the effect of nitrogen on maturity, at least four factors must be taken into account: (1) the

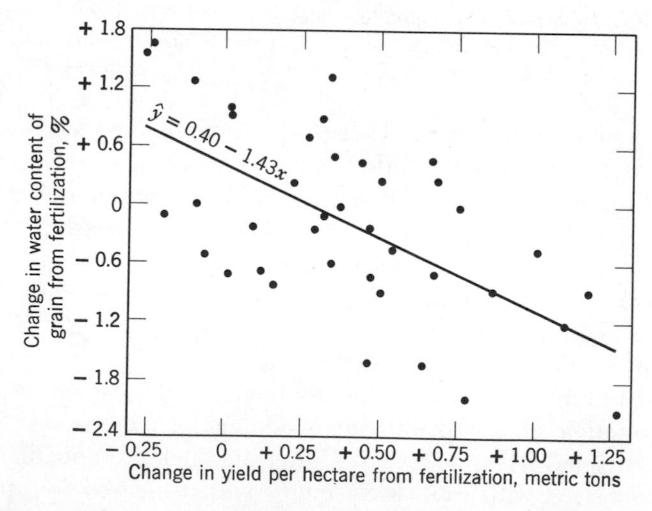

Figure 7.47. Change in water percentage in corn grain at harvest versus change in yield of corn from fertilization with 45 kg of nitrogen per hectare in experiments conducted in Iowa in 1944. (Unpublished data, Iowa Agricultural Experiment Station, 1944)

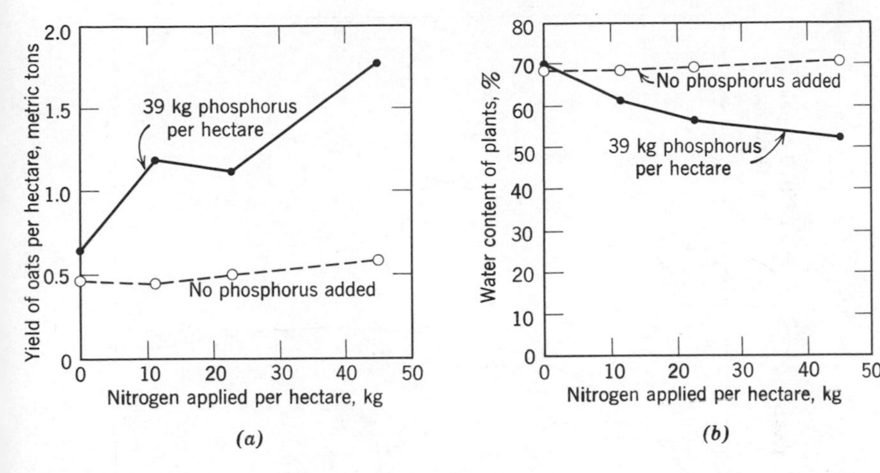

Figure 7.48. Response of oats to different quantities of nitrogen with and without phosphorus ferulization. (*a*) Yield of grain. (*b*) Water content of plants one week before harvest. (Unpublished data, Iowa Agricultural Experiment Station, 1947)

degree of nitrogen deficiency; (2) the quantity of nitrogen applied; (3) the time of nitrogen application; and (4) the nature of the crop.

The connection between the degree of nitrogen deficiency, the quantity of nitrogen applied, and the effect on maturity is illustrated in Figs. 7.47 and 7.48. In Fig. 7.47, the effect of nitrogen fertilizer on maturity is measured in terms of the change in water percentage of corn grain at harvest. Each point represents the result of a separate experiment in which the application of fertilizer nitrogen was equivalent to 45 kg/hectare. The greater the increase in yield from fertilization, the lower was the water content (i.e., the earlier was the maturity) of the nitrogen-fertilized corn relative to that of the unfertilized corn. In Fig. 7.48, the effect of nitrogen fertilization on maturity is measured in terms of the water content of oat plants a week before harvest. The results were obtained in a single experiment with oats in which different quantities of nitrogen fertilizer were applied at two levels of phosphorus fertilizer. Where no phosphorus was applied, nitrogen applications had little effect on either yield or water content of the plants. Where phosphorus was applied, the yield was increased considerably by nitrogen and the water content of the plants became progressively lower (i.e., the maturity was earlier) with an increase in the quantity of nitrogen applied. Similar results were obtained by Peterson and Ballard (1953) in an experiment with sweet corn in which different quantities of nitrogen fertilizer and irrigation water were applied and maturity was

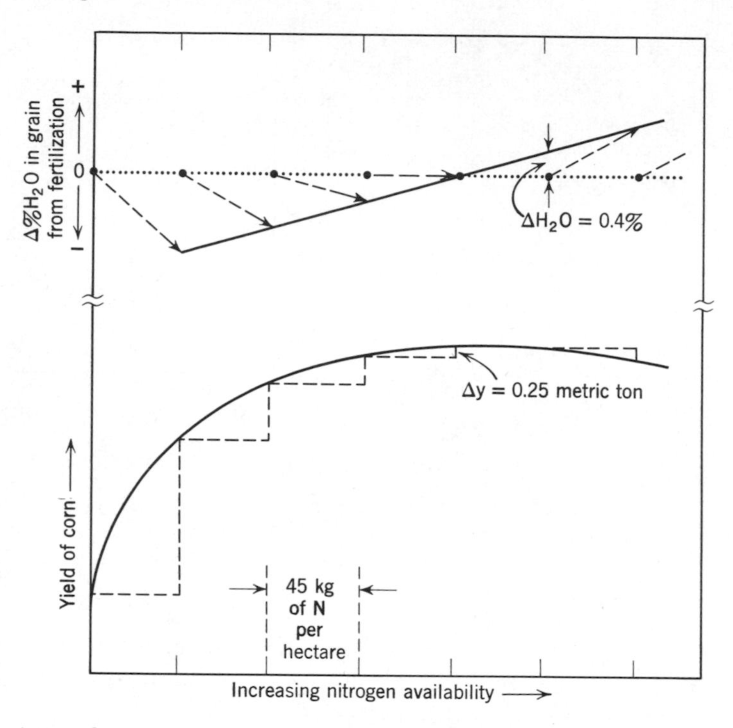

Figure 7.49. Schematic representation of change in maturity of corn with application of 45-kg increments of nitrogen to soils differing in nitrogen supply.

measured in terms of the date on which the corn was suitable for harvesting.

The foregoing observations may be combined to yield the schematic diagram in Fig. 7.49, which portrays the significance of the regression equation obtained in the experiments on corn in Fig. 7.47. Where the unfertilized corn was deficient in nitrogen and application of 45 kg of nitrogen per hectare produced a relatively large increase in yield, a relatively large decrease in the water content of the grain occurred. As the nitrogen supply in the unfertilized soil approached that conducive to the maximum yield, the increase in yield and the decrease in water content of the grain from fertilization became smaller. Where the soil nitrogen supply was such that a 0.25-ton increase in yield resulted from fertilization, there was no effect of fertilization on the water content of the grain. Where there was no increase in yield from fertilization, the water content of the grain was increased 0.4%. Thus the application of 45 kg of nitrogen per hectare resulted in earlier maturity where the corn was markedly deficient in nitrogen, it had no effect where the soil supply of nitrogen was slightly deficient, and it resulted in later maturity where the soil supply was ample.

The effect of the third factor, the time of application, is indicated by data in Table 7.20. Application of nitrogen to oats at the time of planting had essentially no effect on the water content of the plants 1 week before harvest. The water content of the plants gradually increased with delay in time of application to 5 weeks after emergence and then decreased with the application 7 weeks after emergence.

Observations indicate that the earlier maturation connected with nitrogen fertilization of plants initially deficient in nitrogen is associated with more rapid elaboration of plant parts. Thus the inflorescence may appear earlier on nitrogen-fertilized oats than on unfertilized, nitrogen-deficient oats. Silks and tassels may appear earlier on corn plants well supplied with nitrogen than on those deficient in nitrogen (Glover, 1953; Peterson and Ballard, 1953). Excessive nitrogen may delay maturity by diverting carbohydrate into vegetative growth and by decreasing the concentration of other nutrients in the vegetative tissue. Induced deficiencies of other nutrients may result in a decrease in rate of fruit development and maturation.

The behavior just described for such plants as corn and oats does not hold for certain other plants. Plants such as cotton and snap beans do not have distinct stages of fruit formation and fruit development; rather, these periods overlap. The supply of nitrogen determines in part how long the plant continues to produce new flowers. If the nitrogen supply is relatively high, additional flowers are produced after nitrogen-deficient plants have ceased flowering. While these flowers are being produced, development of the existing fruits is retarded, perhaps because

Table 7.20. Yield and Water Content of Oats with Nitrogen Fertilizer Applied at Different Times (Unpublished data, Iowa Agricultural Experiment Station, 1947)

Quantity of Nitrogen Applied per Hectare and Time of Application	Yield of Oat Grain per Hectare, Metric Tons	Water Content of Oat Plants 1 Week before Harvest, %
None	1.4	60
90 kg at planting	2.9	59
90 kg 1 week after emergence	3.1	61
90 kg 3 weeks after emergence	2.8	63
90 kg 5 weeks after emergence	2.6	65
90 kg 7 weeks after emergence	1.5	62

Table 7.21. Numbers of Flowers and Bolls Present on Unfertilized and Nitrogen-Fertilized Cotton Plants of Different Ages (Crowther, 1934)

Ammonium Sulfate Applied per Hectare, kg	Flowers per Plant		Bolls per Plant	
	133 Days	253 Days	133 Days	253 Days
0	10	13	0.8	8
641	17	23	0.5	15

of diversion of part of the carbohydrate to the new growth. Nitrogen fertilization thus delays the maturation of the crop. These relationships are illustrated by data of Crowther (1934) on nitrogen fertilization of cotton. Table 7.21 shows the effect of nitrogen fertilization on the numbers of flowers and bolls per plant on two different dates, and Fig. 7.50 shows the cumulative yield of seed cotton at different dates of picking. It is evident from Fig. 7.50 that a given percentage of the

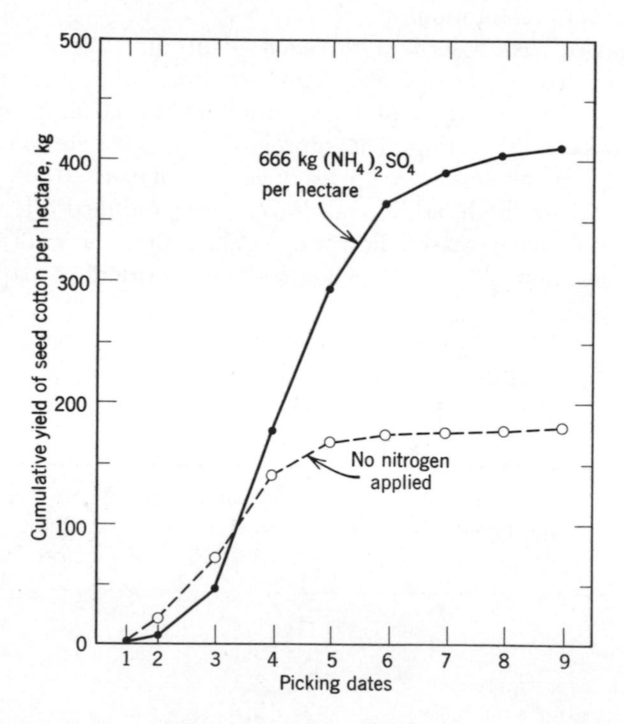

Figure 7.50. Cumulative yield of seed cotton at successive picking dates with different applications of ammonium sulfate. (Crowther, 1934)

total yield was obtained at an earlier date from unfertilized plants than from nitrogen-fertilized plants; that is, nitrogen fertilization delayed the maturation.

Winter-Hardiness

In an experiment on nitrogen fertilization of intermediate wheatgrass in Saskatchewan, Lawrence (1963) observed that survival of the grass in the spring amounted to 78, 42, and 6% where applications of ammonium nitrate in the preceding August had been equivalent to 0, 37, and 74 kg of nitrogen per hectare. These observations represent a rather extreme example of the effect of nitrogen in decreasing winter-hardiness, which is the usual response in instances where a measurable effect is observed. In Lawrence's experiment, a severe stress was placed on the plants by a combination of the following conditions: the application of nitrogen fertilizer late in the season, the occurrence of a drought before application of the fertilizer, and the rigor of Saskatchewan winters.

Experiments with results that do not fit the general pattern of nitrogen effects are reported occasionally. Higgins et al. (1943) found that injury to trunks of peach trees in Georgia at temperatures below freezing diminished with an increase in nitrogen fertilization. Injury at the heaviest application of nitrogen was only one-thirtieth as great as that without nitrogen fertilization. On the same trees, injury to flower buds was essentially independent of the application of nitrogen.

The inconsistency of the effects of nitrogen supply suggests that this factor only modifies in some way a more complex process. Levitt (1956, pp. 116–118) proposed a theory of winter-hardiness involving the location of ice crystals formed in tissue at low temperatures and the resistance of protoplasm to the dehydration associated with formation of ice crystals. In terms of this theory, winter-hardiness should tend to decrease with an increase in nitrogen supply because cell size increases with nitrogen supply and because small cell size is correlated with hardiness. Moreover, growing plants are not winter-hardy, and a high nitrogen supply tends to keep plants in a growing condition.

Practical experience dictates the limitation of autumn applications of nitrogen in instances in which winter-killing may occur, even though nitrogen applied at this time might otherwise be used effectively. Lawrence (1963) observed that winter-hardiness of intermediate wheatgrass in Saskatchewan was greater if nitrogen fertilizer was applied in the spring than in the autumn. Boynton (1948) reported that orchardists in New York avoid autumn applications of nitrogen fertilizers to apple trees because nitrogen applied at this time may increase the damage to the trunk from low temperatures during the winter. In instances such as these, the greater winter-hardiness of plants that receive nitrogen fertilizer in the spring is presumably the result of exhaustion of the effects of the applied nitrogen before the following winter.

Disease Incidence

Development of plant diseases often varies with the nitrogen supply. The response is not uniform, however, because of the variation in nature of plants and diseases and the environmental conditions that affect both. In some cases disease incidence increases with nitrogen supply, and in others the reverse is true.

Nitrogen supply may affect disease incidence in several ways. One of these is through influences on the environment that result in changes in initial infection. In an investigation of the effect of nitrogen on infection of bean plants by *Rhizoctonia solani,* a pathogenic soil-borne fungus, Davey and Papavizas (1960) made three significant observations: (1) Addition of ammonium nitrate or a urea-formaldehyde polymer to soil alone or after treatment of the soil with plant materials generally decreased the infection. (2) Suppression of the disease was greatest with addition of the plant material that was highest in nitrogen and most readily decomposable (soybean hay) and least with the plant material that was lowest in nitrogen and least readily decomposable (sawdust). (3) Colonization of buried buckwheat stems by the fungus was decreased by the same treatments of nitrogen fertilizers and plant materials that decreased infection of the bean plants. These observations indicate that the suppression of the disease resulting from the treatments was a consequence of competition between the disease organism and other organisms in the soil. The authors suggested that the competition effect might be merely a consequence of an increase in the concentration of carbon dioxide in the soil. In this connection they cited research by Durbin (1959) showing that strains of *Rhizoctonia solani* attacking the host at or near the ground line are especially sensitive to an excess of carbon dioxide. The strains of the pathogen encountered by Davey and Papavizas (1960) had the trait of attacking plants at the ground line. If sensitivity to carbon dioxide was indeed responsible for the effect of nitrogen, the connection of nitrogen with the environmental conditions influencing infection in this instance was definitely indirect.

Influences of nitrogen supply on infection of plants by airborne diseases are in some instances equally indirect. For example, infection by airborne diseases is usually greater at high than at low atmospheric relative humidity. By encouraging rank vegetative growth, an increase in nitrogen supply may raise the relative humidity around the plants and thereby increase infection by such diseases as mildews and rusts. The effect of nitrogen in this respect may be particularly important if an increase in nitrogen supply induces lodging because air circulation around lodged plants is much reduced. The effect of nitrogen supply on incidence of rice blast due to the fungus *Piricularia oryzae* is more direct. Suryanarayanan (1958) found that germination of spores of the fungus was induced by glutamine and that nitrogen fertilization increased the concentration of this nitrogenous compound in the guttation fluid on rice leaves. An increase in nitrogen supply thus created a more favorable external substrate for development of the vegetative form of the fungus. Still another example is the effect of urea sprays in reducing bean rust (Cosper and Schuster, 1953). Urea inhibits germination of the spores on agar plates and presumably has a similar effect on the surface of bean leaves.

The nitrogen supply may also alter the suitability of plant tissues as a medium for development of the disease, once entry has been accomplished. For example, Nightingale (1936) noted that when apple twigs were inoculated by introducing the fire-blight bacterium into the tissue, lesions developed on succulent twigs but not on hardened twigs. The same was true in succulent and hardened portions of the same twig. Twigs and leaves of high-nitrogen trees were more susceptible than those of low-nitrogen trees, the susceptibility being associated with the succulence. Juice from succulent, high-nitrogen twigs provided a relatively good culture medium, whereas that from hardened, low-nitrogen twigs did not. Growth of the organism was increased by addition of asparagine (a nitrogen source) to juice of hardened twigs and was decreased by addition of sugar to juice of succulent twigs. The effect of nitrogen supply on development of viruses is another instance of internal suitability. Viruses consist of protein and nucleic acid; and virus particles increase in numbers only within cells, where the presence of the virus redirects cell metabolism into production of additional virus particles. Multiplication of virus particles in plant cells has been found to increase with the supply of nitrogen and phosphorus (Pound and Weathers, 1953; Papasolomontos and Wilkinson, 1959), both of which are components of the virus structure. Because both viruses and normal plant cells contain proteins and nucleoproteins, the theory has been proposed that the two classes of end-products compete for the same building units.

In experiments on wheat, McNeal et al. (1958) found that the yield-response of plants to nitrogen fertilization was greater on both an absolute and relative basis if the plants were produced from healthy seed than from seed infected with barley-stripe virus (Fig. 7.51). In terms of the

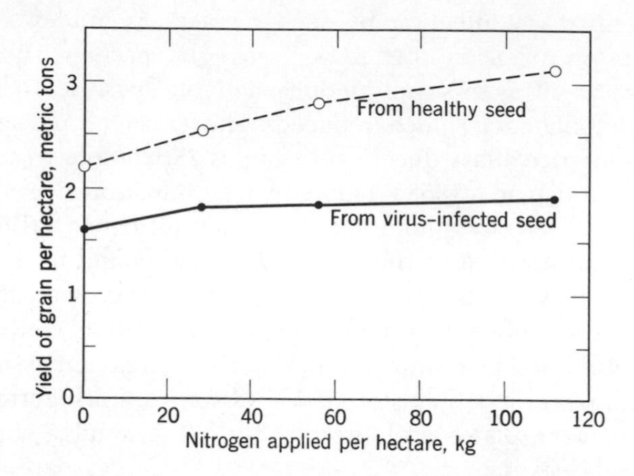

Figure 7.51. Average yields of eight varieties of wheat from virus-infected and healthy seed with different applications of ammonium nitrate. (McNeal et al., 1958)

competition theory, this observation may be interpreted to indicate that the proportion of the building units of proteins and nucleoproteins directed into virus production increased with the supply of nitrogen to the plants.

There is evidence also that the nitrogen supply may alter the concentration of plant substances that are toxic to certain microorganisms. According to McNew (1953), Parker-Rhodes found that the wheat plant produces a substance toxic to *Puccinia glumarum,* one of the fungi responsible for wheat rust, and that high-nitrogen plants produce more of the toxin than do low-nitrogen plants. Király (1964) observed, on the contrary, that nitrogen fertilization increased the susceptibility of wheat to infection by *Puccinia graminis,* another rust-producing fungus, and that the increase in susceptibility to the disease was associated with a marked decrease in concentration of phenolic substances in the leaves at the time of emergence of the inflorescence. Evidently no single, simple pattern exists.

In addition to influencing disease susceptibility in one or more of the ways mentioned in preceding paragraphs, the nitrogen supply influences the capability of plants to escape the effects of certain diseases by growth. The disease-escape effect is important with the so-called "take-all" of cereals. Take-all is a fungal disease that is carried over from one crop to the next in diseased crop residues in the soil. Roots of the new crop are infected when they contact this scattered inoculum. Garrett (1948) found that nitrogen fertilization increased the susceptibility of the host to the disease, as evidenced by the extent of damage to infected roots. At the same time, however, the total number of crown roots and the proportion remaining uninfected were increased by fertilization. The nitrogen-fertilized plants thus were enabled partially to escape the disease by virtue of the greater number of uninfected crown roots produced. A similar explanation may be given for the effect of nitrogen fertilization in reducing lodging of corn because of root-worm injury (see Fig. 7.46).

Plant Competition

In his review on competition among crop and pasture plants, Donald (1963) commented that, in growing what is considered to be a "healthy" field crop, man creates such intense competition among plants that individual plants produce much less than their potential yield if competition were absent. It is by such competition, which results from an induced deficiency of one or more environmental factors, that maximum yields per unit area are achieved.

The nitrogen supply of soils is one of the environmental factors for which important competition exists. The importance of competition for nitrogen is attributable to perhaps two causes. First, nitrogen deficiency is common. Second, soil nitrate exists in the soil water without a solid-phase reserve that dissolves in response to removal of nitrate; hence removal of nitrate from scattered locations induces inward diffusion from adjacent soil and reduces the nitrate concentration throughout. Consequently, plants grown in association draw nitrate from a common source. Removal of nitrate by one plant thus reduces the supply of nitrate for all the others. Plants in sufficient density can exhaust a soil of nitrate and keep the supply at a low level despite continuous nitrate production. An illustration of this phenomenon is shown in Fig. 7.52.

Some consequences of competition among plants of a given species for nitrogen are illustrated by Fig. 7.53, which shows yields of corn at different densities of stand and with different supplies of nitrogen. The important points to note in the present connection are that addition of nitrogen increased the yield at a given density of plant population and that the maximum yield obtainable by increasing the population was attained at a greater density as the nitrogen supply increased. The first observation verifies the importance of nitrogen as a limiting factor, and the second verifies the existence of competition for nitrogen. If no competition for nitrogen had existed, the population corresponding to the maximum yield should not have increased with the nitrogen supply.

Perhaps the most extensive research on competition among different

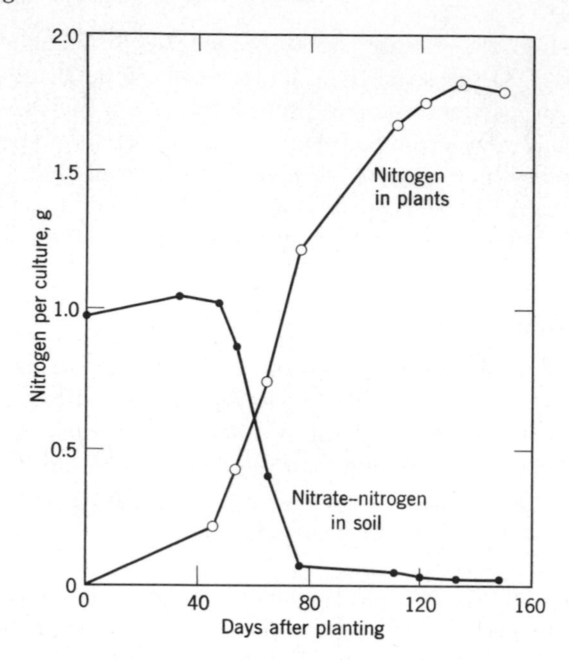

Figure 7.52. Nitrogen present as nitrate in a container of soil and nitrogen in barley plants grown on the soil. One gram of nitrogen per culture is approximately equivalent to 120 μg of nitrogen per gram of soil. (Bishop, 1930)

species has been done with grasses and legumes because these plants often are grown together in practice and because the relative dominance of the different species varies with the environmental conditions. The capability for fixing elemental nitrogen makes the legumes largely independent of mineral nitrogen in the soil; thus they will grow well alone on soils low or high in nitrogen if other factors are favorable. Grasses, however, depend on nitrogen from the soil and hence offer increasing competition to the associated legumes as the nitrogen availability in the soil increases. Experiments with tracer nitrogen (Walker et al., 1956) indicate that the grass components of grass-legume mixtures absorb almost all the mineral nitrogen in the soil and the legume components almost none. Where such is the case, one may infer that there is little competition between grasses and associated legumes for nitrogen because the nitrogen is derived from different sources. The competition thus must be primarily for growth factors other than nitrogen, such as water, light, and mineral nutrients, and the nature of the factor for which competition is most critical will vary with the circumstances.

Although it may be approximately true from one standpoint that grasses and legumes grown in association do have different sources of nitrogen, the bare statement in the preceding paragraph is an oversimplification in the sense that grasses may derive some nitrogen from the associated legume and, hence, indirectly, from symbiotic nitrogen fixation. The transfer of nitrogen from legumes to associated grasses was mentioned previously in the section on symbiotic fixation. In a theoretical analysis of the nitrogen nutrition of grasses and legumes grown in association, Walker et al. (1954) reasoned that the amount of nitrogen transferred from the legume to the grass must be nil in a pure stand of grass and that, if transference occurs in the presence of a legume, it should increase with the proportion of legume in the mixture. Because the yield of nitrogen in grass increases approximately linearly over a considerable range with the quantities supplied by nitrogen fertilizers, these investigators hypothesized that in this linear range the yield of nitrogen in the grass component of the vegetation G from each source in the soil is proportional to the quantity of nitrogen supplied by that source. As sources, they considered the nitrogen mineralized in the soil

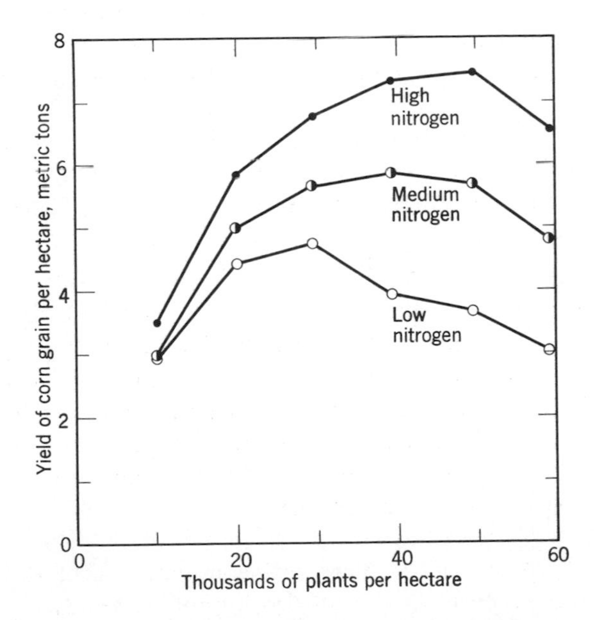

Figure 7.53. Yield of corn with different densities of stand and different levels of nitrogen. The high and low levels of nitrogen were obtained by growing the corn following different crop rotations. The intermediate level was obtained by adding fertilizer nitrogen at 78.5 kg/hectare to the soil used for the low level. (Lang et al., 1956)

S, the nitrogen added in the fertilizer F, and the nitrogen released into the soil by the legume component of the vegetation. The quantity of nitrogen released by the legumes cannot be measured directly but might be expected to be proportional to the yield of nitrogen in the legume component L. These considerations lead to the equation $G = aS + bL + cF$, where a, b, and c are parameters. If the legume absorbs nitrogen from the supply of mineral nitrogen in the soil, the values of the parameters a and c would be expected to vary with the proportion of legume in the mixture. As mentioned in the preceding paragraph, however, experiments with tracer nitrogen indicate that the legume component of grass-legume mixtures absorbs almost none of the mineral nitrogen in the soil. Therefore c should be substantially independent of the proportion of legume in the mixture, and the product aS should be substantially constant. If aS is replaced by k, a constant, the equation becomes

$$G = k + bL + cF.$$

Walker et al. (1954) fitted this equation to the data from an experiment conducted in New Zealand on a ryegrass-white clover meadow in which the variables were fertilizer nitrogen and frequency of cutting. Yields of nitrogen in the grass and in the clover were affected by both these variables. The parameters had the values $k = 40$, $b = 0.65$, and $c = 0.69$, where the units of G, L, and F are kilograms per hectare. Because the values of b and c were so nearly the same, the experimenters averaged the two and expressed the results by the simplified equation

$$G = 40 + 0.67(L + F).$$

The relatively close agreement between calculated and observed values shown in Fig. 7.54 supports the theoretical basis of the equation. The slope-constant 0.67 indicates that the ratio of the portion of the yield of nitrogen in the grass derived from the legume to the yield of nitrogen in the legume is 0.67. This value was found applicable to several other experiments. Peterson and Bendixen (1961) obtained a slope constant of 0.72 in an experiment in California.

The work of Walker and co-workers is of interest not only because of the light it sheds on the quantitative importance of the transfer of nitrogen from legume to grass but also because it yields further expression of the competitive relationships involved. For example, in the experiment described, cutting the meadow four times instead of twice reduced the dominance of the grass and increased the yield of nitrogen in the legume from 118 to 174 kg/hectare. At the same time, the yield of nitro-

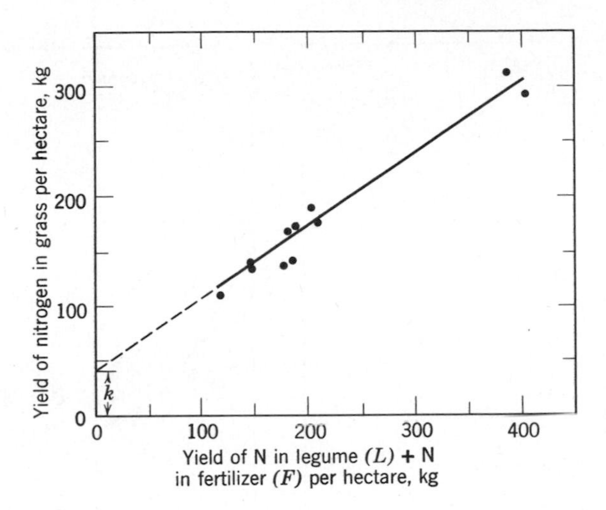

Figure 7.54. Yield of nitrogen in grass versus yield of nitrogen in clover plus nitrogen applied in fertilizer to a mixed planting of ryegrass and white clover. (Walker et al., 1954)

gen in the grass increased from 111 to 137 kg/hectare. Because the values of k and cF remain constant in this comparison, the equation evidently predicts an increase in nitrogen content of the grass G with an increase in nitrogen content of the legume L. As a second example from the same experiment, increasing the fertilizer nitrogen F by 75 kg/hectare caused the nitrogen in the grass G to increase by 53 kg/hectare and caused the nitrogen in the legume L to decrease by 51 kg/hectare, so that the net increase in yield of nitrogen from fertilization was only 2 kg/hectare. From the standpoint of nitrogen economy, fertilization thus had the effect of trading some fertilizer nitrogen for legume nitrogen. If L is added to both sides of the equation to obtain the total yield of nitrogen, one may perceive that if an increase of 53 in G and an increase of only 2 in $G + L$ are obtained with addition of fertilizer nitrogen, the value of L must be decreased by fertilization.

De Wit et al. (1966) investigated legume-grass competition in a different way. They grew *Glycine javanica* (a tropical legume) and *Panicum maximum* (a grass) on soil in a greenhouse in the Netherlands in a series of cultures containing, respectively, 8, 6, 4, 2, and 0 plants of the former species and 0, 1, 2, 3, and 4 plants of the latter species. The legume was grown with and without inoculation with rhizobia, and each series was repeated with no nitrogen added and with a total of nearly 10 g of fertilizer nitrogen per 7 kg of soil in each culture during

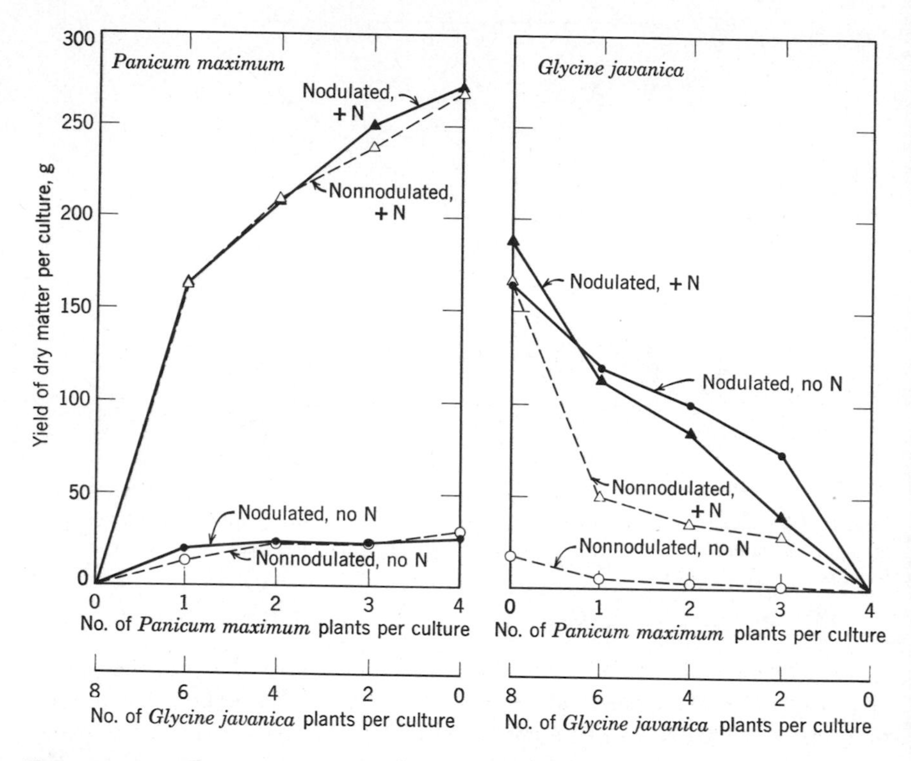

Figure 7.55. Yield of a grass (*Panicum maximum*) and a legume (*Glycine javanica*) grown in association, with variables of proportionate numbers of the two species, nitrogen fertilization, and nodulation of the legume. (de Wit et al., 1966)

the time required to obtain seven cuttings of the herbage. Because the soil did not contain the rhizobia needed to inoculate the legume, no nodulation occurred in the absence of inoculation.

In Fig. 7.55 the values for the total yield of dry matter in seven cuttings of herbage are plotted against the numbers of plants of the two species present per culture. From the results, it is evident that nodulation benefited the legume in both the presence and absence of fertilizer nitrogen and that this effect was greatest (in percentage) where the legume was in competition with grass. As may be inferred from the findings by Walker et al. (1956) mentioned previously, the greater benefit to the legume from nodulation where the legume was in competition with grass was a consequence of nitrogen deficiency in the nonnodulated legume resulting from the greater competitive ability of the grass in taking up mineral nitrogen from the soil.

Because the yield of the nodulated legume grown alone was about the same in the presence and absence of fertilizer nitrogen, the yield of nodulated legume presumably was independent of the nitrogen supply in the soil. Accordingly, the lower yield of the nodulated legume in competition with 2 or 3 grass plants in nitrogen-fertilized cultures than in comparable unfertilized cultures was probably due to competition for growth factors other than nitrogen. The grass grew taller than the legume in the presence of fertilizer nitrogen and thus offered competition for light, which may have been of some significance. Competition for light under the experimental conditions was probably less than would be the case in the field because a closed canopy in the field would permit less access of light to lower parts of the canopy than would be true in the greenhouse where the plants were separated by groups in individual culture vessels.

The pair of species investigated by de Wit et al. (1966) was well suited to associated growth in that the changes in relative yields of grass and nodulated legume components in successive cuttings did not lead to elimination of either component by the final harvest. In the absence of nitrogen fertilization and nodulation, however, competition from the grass completely eliminated the legume in the last three cuttings.

There was some evidence for transfer of nitrogen from legume to grass in this experiment; but the transfer was comparatively small, as indicated by the fact that the yields of grass were but little affected by nodulation of the legume grown in association. In this experiment, the transfer was considerably less than that found by Walker et al. (1954) and Peterson and Bendixen (1961). Data on yields of nitrogen corresponding to the yields of herbage in Fig. 7.55 may be found by consulting the original research report.

Literature Cited

Achromeiko, A. (1928) Der Einfluss des Pulverisierens und Trocknens des Bodens auf dessen Fruchtbarkeit. *Zeitschr. Pflanzenernähr. Düng. Bodenk.* **11A**:65–89.

Adams, R. S., Jr., and F. J. Stevenson. (1964) Ammonium sorption and release from rocks and minerals. *Soil Sci. Soc. Amer. Proc.* **28**:345–351.

Alexander, M. (1965) Nitrification. *Agronomy* **10**:307–343.

Allen, E. K., and O. N. Allen. (1958) Biological aspects of symbiotic nitrogen fixation. *In* W. Ruhland (Ed.) *Handbuch der Pflanzenphysiologie*. Band VIII. *Der Stickstoffumsatz*. Springer-Verlag, Berlin.

Allison, F. E. (1955) The enigma of soil nitrogen balance sheets. *Adv. Agron.* **7**:213–250.

Allison, F. E. (1963) Losses of gaseous nitrogen from soils by chemical mechanisms involving nitrous acid and nitrites. *Soil Sci.* **96**:404–409.

Allison, F. E. (1965) Evaluation of incoming and outgoing processes that affect soil nitrogen. *Agronomy* **10**:573–606.

Allison, F. E., M. S. Sherman, and L. A. Pinck. (1949) Maintenance of soil organic matter: I. Inorganic soil colloid as a factor in retention of carbon during formation of humus. *Soil Sci.* **68**:463–478.

Allison, F. E., and L. D. Sterling. (1949) Nitrate formation from soil organic matter in relation to total nitrogen and cropping practices. *Soil Sci.* **67**:239–252.

Allison, L. E. (1965) Organic carbon. *Agronomy* **9**:1367–1378.

Amer, F. M. (1949) Influence of oxygen percentage and moisture tension on nitrification in soils. M.S. Thesis, Iowa State Univ., Ames, Iowa.

Anderson, O. E., and F. C. Boswell. (1964) The influence of low temperature and various concentrations of ammonium nitrate on nitrification in acid soils. *Soil Sci. Soc. Amer. Proc.* **28**:525–529.

Armstrong, D. E., and G. Chesters. (1964) Properties of protein-bentonite complexes as influenced by equilibration conditions. *Soil Sci.* **98**:39–52.

Avery, G. S., Jr., P. R. Burkholder, and H. B. Creighton. (1936) Plant hormones and mineral nutrition. *Proc. Nat. Acad. Sci.* **22**:673–678.

Axley, J. H., and J. O. Legg. (1960) Ammonium fixation in soils and the influence of potassium on nitrogen availability from nitrate and ammonium sources. *Soil Sci.* **90**:151–156.

Balz, O., W. Busch, H. Hamacher, H. H. Herlemann, H. Krebs, W. Nicolaisen, G. Reinken, P. Rintelen, F. Scheffer, E. Schulze, H. Stotz, J. Tinbergen, T. Tjebbes, H. Ullrich, and W. Wittich. (1961) *Der Stickstoff. Seine Bedeutung für die Landwirtschaft und die Ernährung der Welt.* Fachverband Stickstoffindustrie, Düsseldorf.

Bartholomew, W. V. (1965) Mineralization and immobilization of nitrogen in the decomposition of plant and animal residues. *Agronomy* **10**:285–306.

Bartholomew, W. V., and F. E. Clark. (1950) Nitrogen transformations in soil in relation to the rhizosphere microflora. *Fourth Internat. Cong. Soil. Sci. (Amsterdam) Trans.* **2**:112–113.

Bartholomew, W. V., and F. E. Clark (Eds.) (1965) *Soil Nitrogen. Agronomy 10.* American Society of Agronomy, Madison.

Baumeister, W. (1939) Der Einfluss mineralischer Düngung auf den Ertrag und die Zusammensetzung des Kornes der Sommerweizen Pflanze. *Bodenk. u. Pflanzenernähr.* **12**:175–222.

Beijerinck, M. W. (1888) Die Bakterien der Papilionaceenknöllchen. *Bot. Zeitschr.* **46**:726–735, 742–750, 758–771, 782–790, 798–804.

Beijerinck, M. W. (1901) Ueber oligonitrophile Mikroben. *Centbl. Bakt.* (Abt. 2) **7**:561–582.

Bingham, F. T., H. D. Chapman, and A. L. Pugh. (1954) Solution-culture studies of nitrite toxicity to plants. *Soil Sci. Soc. Amer. Proc.* **18**:305–308.

Birch, H. F. (1958) The effect of soil drying on humus decomposition and nitrogen availability. *Plant and Soil* **10**:9–31.

Birch, H. F. (1958) The effect of soil drying on humus decomposition and nitrogen *Plant and Soil* **11**:262–286.

Bishop, L. R. (1930) The nitrogen content and "quality" of barley. *Jour. Inst. Brewing* **36**:352–364.

Blackman, G. E. (1936) The influence of temperature and available nitrogen supply on the growth of pasture in the spring. *Jour. Agr. Sci.* **26**:620–647.

Blue, W. G., C. F. Eno, N. Gammon, Jr., and D. F. Rothwell. (1964) Timing liming applications to obtain the maximum beneficial effect in clover-grass pasture establishment on virgin flatwoods soils. *Soil and Crop Sci. Soc. Florida Proc.* **24**:162–166.

Bosemark, N. O. (1954) The influence of nitrogen on root development. *Physiol. Plant.* **7**:497–502.

Boynton, D. (1948) Your fruit trees will tell you about their nitrogen needs. *Amer. Fert.* **109,** No. 6:7–8.

Bremner, J. M. (1965) Inorganic forms of nitrogen. *Agronomy* **9**:1179–1237.

Bremner, J. M. (1965a) Organic forms of nitrogen. *Agronomy* **9**:1238–1255.

Bremner, J. M. (1965b) Nitrogen availability indexes. *Agronomy* **9**:1324–1345.

Bremner, J. M. (1965c) Organic nitrogen in soils. *Agronomy* **10**:93–149.

Bremner, J. M. (1966) Nitrogenous compounds. *In* A. D. McLaren and G. H. Peterson (Eds.) *Soil Biochemistry.* Marcel Dekker, Inc., New York.

Bremner, J. M., and F. Führ. (1966) Tracer studies of the reaction of soil organic matter with nitrite. *In The Use of Isotopes in Soil Organic Matter Studies,* pp. 337–346. Pergamon Press, London.

Briggs, M. H., and L. Segal. (1963) Preparation and properties of a free soil enzyme. *Life Sci.* **2**:69–72.

Broadbent, F. E. (1956) Tracer investigations of plant residue decomposition in soils. *U.S. Atomic Energy Comm. Rept. No. TID-7512*:371–379.

Broadbent, F. E., and F. Clark. (1965) Denitrification. *Agronomy* **10**:344–359.

Broadbent, F. E., and A. G. Norman. (1947) Some factors affecting the availability of the organic nitrogen in soil—a preliminary report. *Soil Sci. Soc. Amer. Proc.* (1946) **11**:264–267.

Brouwer, R. (1962) Nutritive influences on the distribution of dry matter in the plant. *Netherlands Jour. Agr. Sci.* **10**:399–408.

Brown, J. M., and W. V. Bartholomew. (1963) Sorption of gaseous ammonia by clay minerals as influenced by sorbed aqueous vapor and exchangeable cations. *Soil Sci. Soc. Amer. Proc.* **27**:160–164.

Brown, M. E., S. K. Burlingham, and R. M. Jackson. (1964) Studies on *Azotobacter* species in soil. III. Effects of artificial inoculation on crop yields. *Plant and Soil* **20**:194–214.

Chao, T., and W. Kroontje. (1964) Relationships between ammonia volatilization, ammonia concentration and water evaporation. *Soil Sci. Soc. Amer. Proc.* **28**:393–395.

Chapman, H. D., and G. F. Liebig, Jr. (1952) Field and laboratory studies of nitrite accumulation in soils. *Soil Sci. Soc. Amer. Proc.* **16**:276–282.

Cheng, H. H., and J. M. Bremner. (1965) Gaseous forms of nitrogen. *Agronomy* **9**:1287–1323.

Clark, F. E. (1949) Soil microorganisms and plant roots. *Adv. Agron.* **1**:241–288.

Clark, F. E., W. E. Beard, and D. H. Smith. (1960) Dissimilar nitrifying capacities of soils in relation to losses of applied nitrogen. *Soil Sci. Soc. Amer. Proc.* **24**:50–54.

Collison, R. C., and J. E. Mensching. (1930) Lysimeter investigations: I. Nitrogen and water relations of crops in legume and non-legume rotations. *New York (Geneva) State Agr. Exp. Sta. Tech. Bul. 166.*

Cooper, G. S., and R. L. Smith. (1963) Sequence of products formed during denitrification in some diverse western soils. *Soil Sci. Soc. Amer. Proc.* **27**:659–662.

Cooper, R. (1959) Bacterial fertilizers in the Soviet Union. *Soils and Fertilizers* **22**:327–333.

Cosper, H., and M. L. Schuster. (1953) Effect of urea on the incidence of bean rust. *Agron. Jour.* **45**:74–75.

Court, M. N., R. C. Stephen, and J. S. Waid. (1962) Nitrite toxicity arising from the use of urea as a fertilizer. *Nature* **194**:1263–1265.

Crowther, F. (1934) Studies in growth analysis of the cotton plant under irrigation in the Sudan: I. The effect of different combinations of nitrogen applications and water-supply. *Ann. Bot.* **48**:877–913.

Davey, C. B., and G. C. Papavizas. (1960) Effect of dry mature plant materials and nitrogen on *Rhizoctonia solani* in soil. *Phytopathology* **50**:522–525.

Davies, R. I., C. B. Coulson, and D. A. Lewis. (1964) Polyphenols in plant, humus, and soil. III. Stabilization of gelatin by polyphenol tanning. *Jour. Soil Sci.* **15**:299–309.

Delwiche, C. C., and J. Wijler. (1956) Non-symbiotic nitrogen fixation in soil. *Plant and Soil* **7**:113–129.

Demolon, A., and G. Barbier. (1929) Conditions de formation et constitution du complexe argilo-humique des sols. *Compt. Rend. Acad. Sci.* **188**:654–656.

Dickson, B. A., and R. L. Crocker. (1953) A chronosequence of soils and vegetation near Mt. Shasta, California. II. The development of the forest floors and the carbon and nitrogen profiles of the soils. *Jour. Soil Sci.* **4**:142–154.

Donald, C. M. (1963) Competition among crop and pasture plants. *Adv. Agron.* **15**:1–118.

Driel, W. van. (1961) Studies on the conversions of amino acids in soil. *Acta Bot. Neerl.* **10**:209–247.

Duisberg, P. C., and T. F. Buehrer. (1954) Effect of ammonia and its oxidation products on rate of nitrification and plant growth. *Soil Sci.* **78**:37–49.

Dumenil, L., J. J. Hanway, and J. T. Pesek. (1961) Evaluation of residual fertilizer effects by chemical composition of plant parts. *Trans. 7th Internat. Congr. Soil Sci.* **3**:212–219.

Duncan, W. G., and A. J. Ohlrogge. (1958) Principles of nutrient uptake from fertilizer bands. II. Root development in the band. *Agron. Jour.* **50**:605–608.

Durbin, R. D. (1959) Factors affecting the vertical distribution of *Rhizoctonia solani* with special reference to CO_2 concentration. *Amer. Jour. Bot.* **46**:22–25.

During, C., G. S. Robinson, and M. W. Cross. (1963) A study of the effects of various intensities of cultivation on yields of chou moellier and on soil nitrate levels. *New Zealand Jour. Agr. Res.* **6**:293–302.

Eagle, D. J. (1961) Determination of the nitrogen status of soils in the West Midlands. *Jour. Sci. Food Agr.* 12:712–717.

Ellis, J. H. (1938) *The Soils of Manitoba.* Manitoba Econ. Survey Board, Winnipeg.

Engelstad, O. P., and W. D. Shrader. (1961) The effect of surface soil thickness on corn yields: II. As determined by an experiment using normal surface soil and artificially-exposed subsoil. *Soil Sci. Soc. Amer. Proc.* **25**:497–499.

Engelstad, O. P., W. D. Shrader, and L. C. Dumenil. (1961) The effect of surface soil thickness on corn yields: I. As determined by a series of field experiments in farmer-operated fields. *Soil Sci. Soc. Amer. Proc.* **25**:494–497.

Eno, C. F., W. G. Blue, and J. M. Good, Jr. (1955) The effect of anhydrous ammonia on nematodes, fungi, bacteria, and nitrification in some Florida soils. *Soil Sci. Soc. Amer. Proc.* **19**:55–58.

Ensminger, L. E., and J. E. Gieseking. (1939) The adsorption of proteins by montmorillonitic clays. *Soil Sci.* **48**:467–473.

Ensminger, L. E., and J. E. Gieseking. (1941) The adsorption of proteins by montmorillonitic clays and its effect on base-exchange capacity. *Soil Sci.* **51**:125–132.

Ensminger, L. E., and J. E. Gieseking. (1942) Resistance of clay-adsorbed proteins to proteolytic hydrolysis. *Soil Sci.* **53**:205–209.

Eriksson, E. (1952) Composition of atmospheric precipitation. I. Nitrogen compounds. *Tellus* **4**:215–232.

Finn, R. F. (1942) Mycorrhizal inoculation of soil of low fertility. *Black Rock Forest Papers* **1**:116–117.

Flaig, W. (1950) Zur Kenntnis der Huminsäuren. I Mitteilung. Zur chemischen Konstitution der Huminsäuren. *Zeitschr. Pflanzenernähr., Düng., Bodenk.* **51**:193–212.

Food and Agriculture Organization of the United Nations. (1964) *Fertilizers: an annual review of world production, consumption and trade.* 1963.

Frederick, L. R. (1956) The formation of nitrate from ammonium nitrogen in soils: I. Effect of temperature. *Soil Sci. Soc. Amer. Proc.* **20**:496–500.

Führ, F., and J. M. Bremner. (1964) Untersuchungen zur Fixierung des Nitrit-Stickstoffs durch die organische Masse des Bodens. *Landw. Forsch. Sonderheft* **18**:43–51.

Gadet, R., L. Soubiès, and F. Fourcassié. (1961) Remontée des nitrates dans les couches superficielles du sol en période estivale. *Compt. Rend. Acad. Agr. France* **47**:897–910.

Gardner, R., and D. W. Robertson. (1942) The nitrogen requirement of sugar beets. *Colorado Agr. Exp. Sta. Tech. Bul. 28.*

Garrett, S. D. (1948) Soil conditions and the take-all disease of wheat. IX. Interaction between host plant nutrition, disease escape, and disease resistance. *Ann. Appl. Biol.* **35**:14–17.

Gasser, J. K. R. (1964) Some factors affecting losses of ammonia from urea and ammonium sulphate applied to soils. *Jour. Soil Sci.* **15**:258–272.

Gilbert, C. S., H. F. Eppson, W. B. Bradley, and O. A. Beath. (1946) Nitrate accumulation in cultivated plants and weeds. *Wyoming Agr. Exp. Sta. Bul.* 277.

Giöbel, G. (1926) The relation of the soil nitrogen to nodule development and fixation of nitrogen by certain legumes. *New Jersey Agr. Exp. Sta. Bul. 436.*

Glover, J. (1953) The nutrition of maize in sand culture. I. The balance of nutrition with particular reference to the level of supply of nitrogen and phosphorus. *Jour. Agr. Sci.* **43**:154–159.

Goring, C. A. I. (1962) Control of nitrification by 2-chloro-6-(trichloromethyl) pyridine. *Soil Sci.* **93**:211–218.

Goring, C. A. I., and F. E. Clark. (1949) Influence of crop growth on mineralization of nitrogen in the soil. *Soil Sci. Soc. Amer. Proc.* (1948) **13**:261–266.

Greaves, J. E., and E. G. Carter. (1924) The influence of irrigation water on the composition of grains and the relationship to nutrition. *Jour. Biol. Chem.* **58**:531–541.

Greenwood, D. J. (1962) Nitrification and nitrate dissimilation in soil. II. Effect of oxygen concentration. *Plant and Soil* **17**:378–391.

Gulati, A. N., and N. Ahmad. (1947) Effect of fertilizers on fineness of cotton.

Proc. Conf. Cott. Grow. Prob. India **1946**:245–251. (*Soils and Fertilizers* **11**, Abstr. 1293. 1948.)

Haddock, J. L. (1952) The nitrogen requirement of sugar beets. *Proc. Amer. Soc. Sugar Beet Technol.* 7:159–165.

Halliday, D. J. (1948) Nitrogen for cereals. *Jealott's Hill Res. Sta. Bul. 6.*

Hamamoto, M. (1966) Isobutylidene diurea as a slow acting nitrogen fertiliser and the studies in this field in Japan. *Fertiliser Soc. Proc.* No. 90.

Hamilton, J., C. O. Stanberry, and W. M. Wootton. (1956) Cotton growth and production as affected by moisture, nitrogen, and plant spacing on the Yuma Mesa. *Soil Sci. Soc. Amer. Proc.* **20**:246–252.

Handley, W. R. C. (1954) Mull and mor formation in relation to forest soils. *Forestry Comm.* (London) *Bul. 23.*

Harding, D. E., and D. J. Ross. (1964) Some factors in low-temperature storage influencing the mineralisable-nitrogen of soils. *Jour. Sci. Food Agr.* **15**:829–834.

Harley, J. L. (1959) *The Biology of Mycorrhiza.* Leonard Hill (Books) Limited, London.

Harley, J. L. (1965) Mycorrhiza. *In* K. F. Baker and W. C. Snyder (Eds.) *Ecology of Soil-Borne Plant Pathogens. Prelude to Biological Control,* pp. 218–229. University of California Press, Berkeley.

Harmsen, G. W., and G. J. Kolenbrander. (1965) Soil inorganic nitrogen. *Agronomy* **10**:43–92.

Harmsen, G. W., and D. A. van Schreven. (1955) Mineralization of organic nitrogen in soil. *Adv. Agron.* **7**:299–398.

Harradine, F., and H. Jenny. (1958) Influence of parent material and climate on texture and nitrogen and carbon contents of virgin California soils. I. Texture and nitrogen contents of soils. *Soil Sci.* **85**:235–243.

Hasegawa, G., K.-I. Nishikawa, M. Kogasaki, and S. Suzuki. (1962) Influence of N, P and K fertilization on starch content in leaf sheath of rice plant. (Translated Title) *Sci. Rpts. Hyogo Univ. Agr., Ser. Agr. 5,* No. 2:74–76.

Hassouna, M. G., and P. F. Wareing. (1964) Possible role of rhizosphere bacteria in the nitrogen nutrition of *Ammophila arenaria. Nature* **202**:467–469.

Hatch, A. B. (1937) The physical basis of mycotrophy in *Pinus. Black Rock Forest Bul. 6.*

Haylett, D. G., and J. J. Theron. (1955) Studies on the fertilisation of a grass ley. *Union of South Africa, Dept. Agr. Sci. Bul. 351.*

Hellriegel, H., and H. Wilfarth. (1891) Recherches sur l'alimentation azotée des graminées et des légumineuses. *Ann. Sci. Agron.* (1890) **7**:84–175, 189–352.

Hellriegel, H., H. Wilfarth, H. Römer, and G. Wimmer. (1898) Vegetationsversuche über den Kalibedarf einiger Pflanzen. *Arb. deut. Landw. Gesell., Heft 34.*

Higgins, B. B., G. P. Walton, and J. J. Skinner. (1943) The effect of nitrogen fertilization on cold injury of peach trees. *Georgia Agr. Exp. Sta. Bul. 226.*

Hill, R. E., E. Hixson, and M. H. Muma. (1948) Corn rootworm control tests with benzene hexachloride, DDT, nitrogen fertilizers and crop rotations. *Jour. Econ. Entomol.* **41**:392–401.

Hinman, W. C. (1964) Fixed ammonium in some Saskatchewan soils. *Canadian Jour. Soil Sci.* **44**:151–157.

Hobbs, J. A., and P. L. Brown. (1957) Nitrogen and organic carbon changes in cultivated western Kansas soils. *Kansas Agr. Exp. Sta. Tech. Bul. 89.*

Holt, E. C., and F. L. Fisher. (1960) Root development of Coastal bermudagrass with high nitrogen fertilization. *Agron. Jour.* **52**:593–596.

Hüser, R. (1963) Probleme zur biologischen Luftstickstoffbindung in Waldböden. *Zeitschr. Pflanzenernähr., Düng., Bodenk.* **103**:220–226.

Iritani, W. M., and C. Y. Arnold. (1960) Nitrogen release of vegetable crop residues during incubation as related to their chemical composition. Soil Sci. **89**:74–82.

Jackman, R. H. (1964) Accumulation of organic matter in some New Zealand soils under permanent pasture. I. Patterns of change of organic carbon, nitrogen, sulphur, and phosphorus. II. Rates of mineralisation of organic matter and the supply of available nutrients. *New Zealand Jour. Agr. Res.* **7**:445–471, 472–479.

Jackson, J. E., and G. W. Burton. (1962) Influence of sod treatment and nitrogen placement on the utilization of urea nitrogen by Coastal bermudagrass. *Agron. Jour.* **54**:47–49.

Jansson, S. L. (1963) Balance sheet and residual effects of fertilizer nitrogen in a 6-year study with N^{15}. *Soil Sci.* **95**:31–37.

Jensen, H. L. (1965) Nonsymbiotic nitrogen fixation. *Agronomy* **10**:436–480.

Jenny, H. (1930) A study on the influence of climate upon the nitrogen and organic matter content of the soil. *Missouri Agr. Exp. Sta. Res. Bul. 152.*

Jenny, H. (1941) *Factors of Soil Formation.* McGraw-Hill Book Co., New York.

Johnstone-Wallace, D. B. (1937) The influence of grazing management and plant associations on the chemical composition of pasture plants. *Jour. Amer. Soc. Agron.* **29**:441–455.

Jones, M. B., C. M. McKell, and S. S. Winans. (1963) Effect of soil temperature and nitrogen fertilization on the growth of soft chess (*Bromus mollis*) at two elevations. *Agron. Jour.* **55**:44–46.

Jones, R. J. (1939) Nitrogen economy in different systems of soil and crop management. *Alabama Agr. Exp. Sta. Ann. Rpt.* **50**:13–15.

Junge, C. E. (1958) The distribution of ammonia and nitrate in rain water over the United States. *Trans. Amer. Geophys. Union* **39**:241–248.

Justice, J. K., and R. L. Smith. (1962) Nitrification of ammonium sulfate in a calcareous soil as influenced by combinations of moisture, temperature, and levels of added nitrogen. *Soil Sci. Soc. Amer. Proc.* **26**:246–250.

Keeney, D. R., and J. M. Bremner. (1964) Effect of cultivation on the nitrogen distribution in soils. *Soil Sci. Soc. Amer. Proc.* **28**:653–656.

Keeney, D. R., and J. M. Bremner. (1966) A chemical index of soil nitrogen availability. *Nature* **211**:892–893.

Kelley, A. P. (1950) *Mycotrophy in Plants.* Chronica Botanica Co., Waltham, Mass.

Kelley, W. P. (1914) Rice soils of Hawaii: their fertilization and management. *Hawaii Agr. Exp. Sta. Bul. 31.*

Khalil, F. (1929) The effect of drying on the microbiological processes in soils. *Zentbl. Bakt., Parasitenk. Infektionskr.* (II) **79**:93–107.

Király, Z. (1964) Effect of nitrogen fertilization on phenol metabolism and stem rust susceptibility of wheat. *Phytopath. Zeitschr.* **51**:252–261.

Koepke, V. (1962) Ausnutzung des Stickstoffs unterschiedlich gelagerter Stallmistarten auf leichtem Boden. *Albrecht-Thaer-Archiv* **6**:585–596.

Kononova, M. M. (1961) *Soil Organic Matter. Its Nature, Its Role in Soil Formation and in Soil Fertility.* Pergamon Press, New York.

Krantz, B. A. (1949) Fertilize corn for higher yields. *North Carolina Agr. Exp. Sta. Bul.* **366.**

Kraus, E. J., and H. R. Kraybill. (1918) Vegetation and reproduction with special reference to the tomato. *Oregon Agr. Exp. Sta. Bul. 149.*

Landrau, P., Jr. (1953) Influence of cropping and cultural practices on the seasonal trends in nitrification rates of soils used for growing corn in Nebraska. *Puerto Rico Univ. Agr. Exp. Sta. Tech. Paper 10.*

Lang, A. L., J. W. Pendleton, and G. H. Dungan. (1956) Influence of population and nitrogen levels on yield and protein and oil contents of nine corn hybrids. *Agron. Jour.* **48**:284–289.

Lawrence, T. (1963) The influence of fertilizer on the winter survival of intermediate wheatgrass following a long period of drought. *Jour. British Grassland Soc.* **18**:292–294.

Legg, J. O., and F. E. Allison. (1961) Role of rhizosphere microorganisms in the uptake of nitrogen by plants. *Trans. 7th Internat. Congr. Soil Sci.* **2**:545–550.

Letham, D. S. (1961) Influence of fertilizer treatment on apple fruit composition and physiology. I. Influence on cell size and cell number. *Australian Jour. Agr. Res.* **12**:600–611.

Levitt, J. (1956) The hardiness of plants. *Agronomy* **6.**

Linscott, D. L., R. L. Fox, and R. C. Lipps. (1962) Corn root distribution and moisture extraction in relation to nitrogen fertilization and soil properties. *Agron. Jour.* **54**:185–189.

Linser, H., and H. Kühn. (1962) Lagerungshemmende bzw. standfestigkeits-stärkende Düngemittel auf Basis von gibberellinsäureantagonistischen Stoffen der Gruppe CCC (Chlorcholinchlorid). *Zeitschr. Pflanzenernähr., Düng., Bodenk.* **96**:231–247.

Lipman, J. G., and A. B. Conybeare. (1936) Preliminary note on the inventory and balance sheet of plant nutrients in the United States. *New Jersey Agr. Exp. Sta. Bul. 607.*

Livens, J. (1959) Contribution a l'etude de l'azote mineralisable du sol. *Agricultura* (Louvain) **7**:27–44.

Livens, J. (1959a) Recherches sur l'azote ammoniacal et organique du sol soluble dans l'eau bouillante. *Agricultura* (Louvain) **7**:519–532.

Loomis, R. S., and D. J. Nevins. (1963) Interrupted nitrogen nutrition effects on growth, sucrose accumulation and foliar development of the sugar beet plant. *Jour. Amer. Soc. Sugar Beet Technol.* **12**:309–322.

Lopez, A. B., and N. L. Galvez. (1958) The mineralization of the organic matter of some Philippine soils under submerged conditions. *Philippine Agr.* **42**:281–291.

Lueken, H., W. L. Hutcheon, and E. A. Paul. (1962) The influence of nitrogen on the decomposition of crop residues in the soil. *Canadian Jour. Soil Sci.* **42**:276–288.

Lyon, T. L., J. A. Bizzell, B. D. Wilson, and E. W. Leland. (1930) Lysimeter experiments: III. Records for tanks 3 to 12 during the years 1910 to 1924 inclusive. *New York* (Cornell Univ.) *Agr. Exp. Sta. Mem. 134.*

Mack, A. R. (1963) Biological activity and mineralization of nitrogen in three soils as induced by freezing and drying. *Canadian Jour. Soil Sci.* **43**:316–324.

Magee, W. E., and R. H. Burris. (1954) Fixation of N_2^{15} by excised nodules. *Plant Physiol.* **29**:199–200.

Malo, B. A., and E. R. Purvis. (1964) Soil absorption of atmospheric ammonia. *Soil Sci.* **97**:242–247.

Mann, H. H. (1951) The effect of manures on the bolting of the beet plant. *Ann. Appl. Biol.* 38:435–443.

Massey, H. F., and M. L. Jackson. (1952) Selective erosion of soil fertility constituents. *Soil Sci. Soc. Amer. Proc.* **16**:353–356.

McIntosh, T. H., and L. R. Frederick. (1958) Distribution and nitrification of anhydrous ammonia in a Nicollet sandy clay loam. *Soil Sci. Soc. Amer. Proc.* **22**:402–405.

McLaren, A. D., and G. H. Peterson. (1961) Montmorillonite as a caliper for the size of protein molecules. *Nature* **192**:960–961.

McNeal, F. H., M. A. Berg, M. M. Afanasiev, and T. J. Army. (1958) The influence of barley stripe mosaic on yield and other plant characters of 8 spring wheat varieties grown at 4 nitrogen levels. *Agron. Jour.* **50**:103–105.

McNew, G. L. (1953) The effects of soil fertility. *In Plant Diseases*, United States Dept. Agr. Yearbook 1953, pp. 100–114.

Melin, E., and H. Nilsson. (1952) Transport of labelled nitrogen from an ammonium source to pine seedlings through mycorrhizal mycelium. *Svensk Bot. Tidskr.* **46**:281–285.

Melin, E., and H. Nilsson. (1953) Transfer of labelled nitrogen from glutamic acid to pine seedlings through the mycelium of *Boletus variegatus* (Sw.) Fr. *Nature* **171**:134.

Meyer, B. S., D. B. Anderson, and R. H. Böhning. (1960) *Introduction to Plant Physiology*. D. Van Nostrand Co., Inc., Princeton, New Jersey.

Meyer, R., and A. Storck. (1927) Ueber den Pflanzenertrag als Funktion der Stickstoffgabe und der Wachstumszeit bei Hafer. *Zeitschr. Pflanzenernähr. Düng. Bodenk.* **10A**:329–347.

Miller, R. D., and D. D. Johnson. (1964) The effect of soil moisture tension on carbon dioxide evolution, nitrification, and nitrogen mineralization. *Soil Sci. Soc. Amer. Proc.* **28**:644–647.

Miller, R. H., and E. L. Schmidt. (1965) Uptake and assimilation of amino acids supplied to the sterile soil:root environment of the bean plant (*Phaseolus vulgaris*). *Soil Sci.* **100**:323–330.

Mitchell, K. J. (1957) The influence of nitrogen and moisture supply on the growth of pastures during summer. *Empire Jour. Exptl. Agr.* **25**:69–78.

Moore, A. W. (1966) Non-symbiotic nitrogen fixation in soil and soil-plant systems. *Soils and Fertilizers* **29**:113–128.

Morgan, M. F., and O. E. Street. (1939) Seasonal water and nitrate leachings in relation to soil and source of fertilizer nitrogen. (A second report on Windsor lysimeter series "A"). *Connecticut* (New Haven) *Agr. Exp. Sta. Bul. 429.*

Morita, N., and N. Taya. (1957) Dynamical studies on the straw of paddy-rice with reference to lodging. Part I. The dynamical characteristics of straw of paddy-rice with various amount of nitrogen fertilizer and barnyard manure, and the relation between these characteristics and lodging. *Hirosaki Univ. Faculty Agr. Bul.* **3**:52–59.

Morrill, L. G. (1959) An explanation of the nitrification patterns observed when soils are perfused with ammonium sulfate. *Diss. Abstr.* **20**:3005.

Mortland, M. M., and A. R. Wolcott. (1965) Sorption of inorganic nitrogen compounds by soil materials. *Agronomy* **10**:150–197.

Mulder, E. G. (1950) Effect of liming an acid peat soil on microbial activity. *Trans. Fourth Internat. Congr. Soil Sci.* **2**:117–121.

Mulder, E. G. (1954) Effect of mineral nutrition on lodging of cereals. *Plant and Soil* **5**:246–306.

Nelson, C. E. (1958) Lodging of field corn as affected by cultivation, plant population, nitrogen fertilizer, and irrigation treatment at the Irrigation Experiment Station, Prosser, Washington. *U.S. Dept. Agr., Agr. Res. Service, Prod. Res. Rpt. 16.*

Nelson, D. W. (1967) Chemical transformations of nitrite in soils. Ph.D. Thesis, Iowa State University, Ames, Iowa.

Nelson, L. B. (1965) Advances in fertilizers. *Adv. Agron.* **17**:1–84.

Nightingale, A. A. (1936) Some chemical constituents of apple associated with susceptibility to fire-blight. *New Jersey Agr. Exp. Sta. Bul. 613.*

Nõmmik, H. (1965) Ammonium fixation and other reactions involving a nonenzymatic immobilization of mineral nitrogen in soil. *Agronomy* **10**:198–258.

Nõmmik, H., and K.-O. Nilsson. (1963) Nitrification and movement of anhydrous ammonia in soil. *Acta Agr. Scand.* **13**:205–219.

Nutman, P. S. (1965) Symbiotic nitrogen fixation. *Agronomy* **10**:360–383.

Paauw, F. van der. (1959) Stikstofbehoefte in afhankelijkheid van het weer in de voorafgaande winter. *Landbouwk. Tijdschr.* **71**:679–689.

Paauw, F. van der. (1962) Effect of winter rainfall on the amount of nitrogen available to crops. *Plant and Soil* **16**:361–380.

Paauw, F. van der. (1963) Residual effect of nitrogen fertilizer on succeeding crops in a moderate marine climate. *Plant and Soil* **19**:324–331.

Papasolomontos, A., and R. E. Wilkinson. (1959) The effect of nitrogen, phosphorus and potassium levels on growth and virus content of excised tobacco-mosaic-virus-infected tomato roots. *Phytopathology* **49**:229.

Patrick, W. H., Jr., and R. Wyatt. (1964) Soil nitrogen loss as a result of alternate submergence and drying. *Soil Sci. Soc. Amer. Proc.* **28**:647–653.

Paul, E. A., and E. L. Schmidt. (1961) Formation of free amino acids in rhizosphere and nonrhizosphere soil. *Soil Sci. Soc. Amer. Proc.* **25**:359–362.

Paul, J. L., and E. Polle. (1965) Nitrite accumulation related to lettuce growth in a slightly alkaline soil. *Soil Sci.* **100**:292–297.

Payne, T. M. B., J. W. Rouatt, and H. Katznelson. (1956) Detection of free amino acids in soil. *Soil Sci.* **82**:521–524.

Pearson, R. W., H. V. Jordan, O. L. Bennett, C. E. Scarsbrook, W. E. Adams, and A. W. White. (1961) Residual effects of fall- and spring-applied nitrogen fertilizers on crop yields in the Southeastern United States. *U.S. Dept. Agr. Tech. Bul. 1254.*

Pearson, R. W., and R. W. Simonson. (1939) Organic phosphorus in seven Iowa soil profiles: distribution and amounts as compared to organic carbon and nitrogen. *Soil Sci. Soc. Amer. Proc.* **4**:162–167.

Peevy, W. J., and A. G. Norman. (1948) Influence of composition of plant materials on properties of the decomposed residues. *Soil Sci.* **65**:209–226.

Peterson, H. B., and J. C. Ballard. (1953) Effect of fertilizer and moisture on the growth and yield of sweet corn. *Utah Agr. Exp. Sta. Bul. 360.*

Peterson, M. L., and L. E. Bendixen. (1961) Plant competition in relationship to nitrogen economy. *Agron. Jour.* **53**:45–49.

Pfaff, C. (1963) Das Verhalten des Stickstoffs im Boden nach langjährigen Lysimeterversuchen. I. Mitteilung. *Zeitschr. Acker- und Pflanzenbau* **117**:77–99.

Pinck, L. A., F. E. Allison, and V. L. Gaddy. (1948) The effect of green manure crops of varying carbon-nitrogen ratios upon nitrogen availability and soil organic matter content. *Jour. Amer. Soc. Agron.* **40**:237–248.

Pinck, L. A., R. S. Dyal, and F. E. Allison. (1954) Protein-montmorillonite complexes, their preparation and the effects of soil microorganisms on their decomposition. *Soil Sci.* **78**:109–118.

Porter, L. K., B. A. Stewart, and H. J. Haas. (1964) Effects of long-time cropping on hydrolyzable organic nitrogen fractions in some Great Plains soils. *Soil Sci. Soc. Amer. Proc.* **28**:368–370.

Pound, G. S., and L. G. Weathers. (1953) The relation of host nutrition to multiplication of turnip virus 1 in *Nicotiana glutinosa* and *N. multivalvis*. *Phytopathology* **43**:669–674.

Puh, Y. (1964) Effects of soil temperature, mineral type, cation ratio, and degree of base saturation on volatilization losses of ammonia. *Diss. Abstr.* **24**:4898–4899.

Ramig, R. E., and H. F. Rhoades. (1963) Interrelationships of soil moisture level at planting time and nitrogen fertilization on winter wheat production. *Agron. Jour.* **55**:123–127.

Raney, W. A. (1960) The dominant role of nitrogen in leaching losses from soils in humid regions. *Agron. Jour.* **52**:563–566.

Reddy, G. R. (1962) Effect of dicyandiamide on transformation, loss and plant uptake of fertilizer nitrogen from Georgia soils. *Diss. Abstr.* **23**:1844–1845.

Redemann, C. T., R. W. Meikle, and J. G. Widofsky. (1964) The loss of 2-chloro-6-(trichloromethyl)-pyridine from soil. *Jour. Agr. Food Chem.* **12**:207–209.

Reichard, T., and J. Schönbrunner. (1961) Vorläufiger Bericht über Versuche zur Verhinderung der Getreidelagerung. *Bodenkultur* **12A**:29–40.

Reichman, G. A., D. L. Grunes, and F. G. Viets, Jr. (1966) Effect of soil moisture on ammonification and nitrification in two Northern Plains soils. *Soil Sci. Soc. Amer. Proc.* **30**:363–366.

Richards, B. N. (1965) Mycorrhiza development of loblolly pine seedlings in relation to soil reaction and the supply of nitrate. *Plant and Soil* **22**:187–199.

Richardson, H. L. (1938) The nitrogen cycle in grassland soils: with especial reference to the Rothamsted Park grass experiment. *Jour. Agr. Sci.* **28**:73–121.

Richer, A. C., and J. W. White. (1946) Fate of applied green manure and inorganic nitrogen fertilizers and their residual effect upon the soil. *Pennsylvania Agr. Exp. Sta. Bul.* **483**:101–120.

Robinson, J. B. D. (1957) The critical relationship between soil moisture content in the region of wilting point and the mineralization of natural soil nitrogen. *Jour. Agr. Sci.* **49**:100–105.

Robinson, J. B. D. (1960) Nitrogen studies in a coffee soil. I. Seasonal trends of natural soil nitrate and ammonia in relation to crop growth, soil moisture and rainfall. *Jour. Agr. Sci.* **55**:333–338.

Rodrigues, G. (1954) Fixed ammonium in tropical soils. *Jour. Soil Sci.* **5**:264–274.

Rovira, A. D., and E. L. Greacen. (1957) The effect of aggregate disruption on the activity of microorganisms in the soil. *Australian Jour. Agr. Res.* **8**:659–673.

Ruhland, W. (Ed.) (1958) *Handbuch der Pflanzenphysiologie*. Band VIII. *Der Stickstoffumsatz*. Springer-Verlag, Berlin.

Ryle, G. J. A. (1963) Studies in the physiology of flowering of timothy (*Phleum pratense* L.). III. Effects of shoot age and nitrogen level on the timing of inflorescence production. *Ann. Bot.* (N. S.) **27**:453–465.

Sabbe, W. E., and L. W. Reed. (1964) Investigations concerning nitrogen loss through chemical reactions involving urea and nitrite. *Soil Sci. Soc. Amer. Proc.* **28**:478–481.

Sabey, B. R., W. V. Batholomew, R. Shaw, and J. Pesek. (1956) Influence of temperature on nitrification in soils. *Soil Sci. Soc. Amer. Proc.* **20**:357–360.

Sabey, B. R., L. R. Frederick, and W. V. Bartholomew. (1959) The formation of nitrate from ammonium nitrogen in soils: III. Influence of temperature and initial population of nitrifying organisms on the maximum rate and delay period. *Soil Sci. Soc. Amer. Proc.* **23**:462–465.

Samtsevich, S. A. (1962) Preparation, use and effectiveness of bacterial fertilizers in the Ukrainian SSR. Microbiology **31**:747–755; **31**:923–933.

Sauchelli, V. (Ed.) (1960) *Chemistry and Technology of Fertilizers.* American Chemical Society Monograph No. 148. Reinhold Publishing Corp., New York.

Sauchelli, V. (Ed.) (1964) *Fertilizer Nitrogen, Its Chemistry and Technology.* American Chemical Society Monograph No. 161. Reinhold Publishing Corp., New York.

Scharpenseel, H. W. (1962) Studies with C^{14}- and H^{3}-labelled humic acids as to the mode of linkage of their amino-acid fractions and with $S^{35}O_4^{--}$, S^{35}-methionine and S^{35}-cystine regarding the sulphur turnover in humic acid and soil. *In Radioisotopes in Soil-Plant Nutrition Studies,* pp. 115–133. International Atomic Energy Agency, Vienna.

Scharpenseel, H. W., and R. Krausse. (1962) Aminosäureuntersuchungen an verschiedenen organischen Sedimenten, besonders Grau- und Braunhuminsäurefraktionen verschiedener Bodentypen (einschliesslich C^{14}-markierter Huminsäuren). *Zeitschr. Pflanzenernähr., Düng., Bodenk.* **96**:11–34.

Scheel, R. (1936) Einfluss der Düngung auf Ertrag und Faserausbildung des Hanfes. *Ernährung der Pflanze* **32**:322–327.

Schloesing, T., fils, and E. Laurent. (1892) Recherches sur la fixation de l'azote libre par les plantes. *Ann. Inst. Pasteur* **6**:65–115.

Schmidt, E. L., H. D. Putnam, and E. A. Paul. (1957) Behavior of amino acids in soil. *Bact. Proc., Soc. Amer. Bact.* **57**:11.

Schreven, D. A. van. (1965) Quelques aspects microbiologiques du métabolisme de l'azote dans le sol. *Ann. Inst. Pasteur* **109,** Suppl. No. **3**:19–49.

Seifert, J. (1963) The influence of soil freezing on the intensity of nitrification. *Acta Univ. Carolinae Biol.,* pp. 265–270. (*Soils and Fertilizers* **27,** Abstr. 1523. 1964.)

Seifert, J. (1963a) The influence of crushing structural soil aggregates on nitrification intensity. *Acta Univ. Carolinae Biol.,* pp. 271–273. (*Soils and Fertilizers* **27,** Abstr. 1522. 1964.)

Sexsmith, J. J., and U. J. Pittman. (1963) Effect of nitrogen fertilizers on germination and stand of wild oats. *Weeds* **11**:99–101.

Shattuck, G. E., and M. Alexander. (1963) A differential inhibitor of nitrifying organisms. *Soil Sci. Soc. Amer. Proc.* **27**:600–601.

Siebert, H. (1939) Der Einfluss von steigenden Stickstoffgaben auf Ertrag und Güte einiger Zwischenfrüchte. *Landw. Jahrb.* **87**:112–158.

Simon, G., and G. Barbier. (1963) Influence favorable de l'azote sur la formation d'humus par la paille enfouie dans le sol. *Compt. Rend. Acad. Agr. France* **49**:1206–1209.

Singh, J. N., and J. Takahashi. (1962) Effect of varying dates of top-dressing of nitrogen on plant characters leading to tendency to lodging in rice. *Soil Sci. Plant Nutr.* **8**:169–176.

Soubies, L., R. Gadet, and M. Lenain. (1962) Possibilité de controler dans le sol la transformation de l'azote ammoniacal en azote nitrique. Applications pratiques dans l'utilisation des engrais azotés. *Compt. Rend. Acad. Agr. France* **48**:798–803.

Soubies, L., R. Gadet, and P. Maury. (1952) Migration hivernale de l'azote nitrique dans un sol limoneaux de la région Toulousaine. *Ann. Agron.* **3**:365–383.

Soulides, D. A., and F. E. Allison. (1961) Effect of drying and freezing soil on carbon dioxide production, available mineral nutrients, aggregation, and bacterial population. *Soil Sci.* **91**:291–298.

Sprague, H. B. (Ed.) (1964) *Hunger Signs in Crops.* Third edition. David McKay Co., New York.

Stakman, E. C., and O. S. Aamodt. (1924) The effect of fertilizers on the development of stem rust of wheat. *Jour. Agr. Res.* **27**:341–380.

Starkey, R. L. (1939) The influence of reaction upon the development of an acid-tolerant *Azotobacter. Proc. Third Comm. Internat. Soc. Soil Sci.*, pp. 142–150.

Stevenson, F. J. (1965) Amino sugars. *Agronomy* **9**:1429–1436.

Stevenson, F. J. (1965a) Amino acids. *Agronomy* **9**:1437–1451.

Stevenson, F. J. (1965b) Origin and distribution of nitrogen in soil. *Agronomy* **10**:1–42.

Stevenson, F. J., and A. P. S. Dhariwal. (1959) Distribution of fixed ammonium in soils. *Soil Sci. Soc. Amer. Proc.* **23**:121–125.

Stevenson, G. (1959) Fixation of nitrogen by non-nodulated seed plants. *Ann. Bot.* **23**:622–635.

Stickler, F. C., W. D. Shrader, and I. J. Johnson. (1959) Comparative value of legume and fertilizer nitrogen for corn production. *Agron. Jour.* **51**:157–160.

Stojanovic, B. J., and M. Alexander. (1958) Effect of inorganic nitrogen on nitrification. *Soil Sci.* **86**:208–215.

Street, H. E. (1960) Hormones and the control of root growth. *Nature* **188**:272–274.

Sturkie, D. G. (1947) Effects of some environmental factors on the seed and lint of cotton. *Alabama Agr. Exp. Sta. Bul. 263.*

Sundaro Rao, W. V. B., H. S. Mann, N. B. Paul, and S. P. Mathur. (1963) Bacterial inoculation experiments with special reference to *Azotobacter. Indian Jour. Agr. Sci.* **33**:279–290.

Suryanarayanan, S. (1958) Role of nitrogen in host susceptibility to *Piricularia oryzae* Cav. *Current Science* **27**:447–448.

Swaby, R. J. (1962) Discussion. *Trans. Joint Meeting Comm. IV & V, Internat. Soc. Soil Sci.*, New Zealand, 1962, p. 217.

Swaby, R. J., and J. N. Ladd. (1962) Chemical nature, microbial resistance, and origin of soil humus. *Trans. Joint Meeting Comm. IV & V, Internat. Soc. Soil Sci.*, New Zealand, 1962, p. 197–202.

Swezey, A. W., and G. O. Turner. (1962) Crop experiments on the effect of 2-chloro-6-(trichloromethyl) pyridine for the control of nitrification of ammonium and urea fertilizers. *Agron. Jour.* **54**:532–535.

Takai, Y., and I. Harada. (1959) Prolonged storage of paddy soil samples under several conditions and its effect on soil microorganisms. I. On the drainable paddy clay loam soil. *Soil and Plant Food* **5**:46.

Talibudeen, O. (1955) Complex formation between montmorillonoid clays and amino-acids and proteins. *Trans. Faraday Soc.* **51**:582–590.

Tamini, Y. N., Y. Kanehiro, and G. D. Sherman. (1963) Ammonium fixation in amorphous Hawaiian soils. *Soil Sci.* **95**:426–430.

Tanner, J. W., and I. C. Anderson. (1964) External effect of combined nitrogen on nodulation. *Plant Physiol.* **39**:1039–1043.

Terman, G. L., and C. M. Hunt. (1964) Volatilization losses of nitrogen from surface-applied fertilizers, as measured by crop response. *Soil Sci. Soc. Amer. Proc.* **28**:667–672.

Theron, J. J. (1965) The influence of fertilizers on the organic matter content of the soil under natural veld. *South African Jour. Agr. Sci.* **8**:525–534.

Thompson, L. M. (1950) The mineralization of organic phosphorus, nitrogen and carbon in virgin and cultivated soils. Ph.D. Thesis, Iowa State University, Ames, Iowa.

Thompson, L. M., C. A. Black, and J. A. Zoellner. (1954) Occurrence and mineralization of organic phosphorus in soils, with particular reference to associations with nitrogen, carbon, and *p*H. *Soil Sci.* **77**:185–196.

Torrey, J. G. (1950) The induction of lateral roots by indoleacetic acid and root decapitation. *Amer. Jour. Bot.* **37**:257–264.

Troughton, A. (1957) The underground organs of herbage grasses. *Commonwealth Bur. Pastures and Field Crops Bul. 44.*

Turner, G. O., and C. A. I. Goring. (1966) N-Serve . . . a status report. *Down to Earth* **22**, No. 2:19–25.

Tyler, K. B., F. E. Broadbent, and G. N. Hill. (1959) Low-temperature effects on nitrification in four California soils. *Soil Sci.* **87**:123–129.

Ulrich, A. (1954) Growth and development of sugar beet plants at two nitrogen levels in a controlled temperature greenhouse. *Proc. Amer. Soc. Sugar Beet Technol.* **8**, Part 2:325–338.

Ulrich, J. M., R. A. Luse, and A. D. McLaren. (1964) Growth of tomato plants in presence of proteins and amino acids. *Physiol. Plantarum* **17**:683–696.

Viets, F. G., Jr. (1965) The plant's need for and use of nitrogen. *Agronomy* **10**:503–549.

Viets, F. G., Jr., C. E. Nelson, and C. L. Crawford. (1954) The relationships among corn yields, leaf composition and fertilizers applied. *Soil Sci. Soc. Amer. Proc.* **18**:297–301.

Vincent, J. M. (1965) Environmental factors in the fixation of nitrogen by the legume. *Agronomy* **10**:384–435.

Virtanen, A. I., S. von Hausen, and T. Laine. (1937) Investigations on the root nodule bacteria of leguminous plants: XX. Excretion of nitrogen in associated cultures of legumes and non-legumes. *Jour. Agr. Sci.* **27**:584–610.

Volk, G. M. (1959) Volatile loss of ammonia following surface application of urea to turf or bare soils. *Agron. Jour.* **51**:746–749.

Waldschmidt-Leitz, E. (1958) Pflanzliche Eiweisskörper. *In* W. Ruhland (Ed.) *Handbuch der Pflanzenphysiologie.* Band VIII. *Der Stickstoffumsatz.* Springer-Verlag, Berlin.

Walker, A. C., L. R. Hac, A. Ulrich, and F. J. Hills. (1950) Nitrogen fertilization of sugar beets in the Woodland area of California. I. Effects upon glutamic acid content, sucrose concentration and yield. *Proc. Amer. Soc. Sugar Beet Technol.* **6**:362–371.

Walker, R. H., and P. E. Brown. (1936) The phosphorus, nitrogen and carbon content of Iowa soils. *In* P. E. Brown (1936) *Soils of Iowa.* Iowa Agr. Exp. Sta. Spec. Report 3.

Walker, T. W., A. F. R. Adams, and H. D. Orchiston. (1956) Fate of labeled nitrate and ammonium nitrogen when applied to grass and clover grown separately and together. *Soil Sci.* **81**:339–351.

Walker, T. W., H. D. Orchiston, and A. F. R. Adams. (1954) The nitrogen economy of grass legume associations. *Jour. British Grassland Soc.* **9**:249–274.

Wallace, T. (1961) *The Diagnosis of Mineral Deficiencies in Plants by Visual Symptoms.* Second ed. Chemical Publishing Co., Inc., New York.

Walsh, L. M., and J. T. Murdock. (1963) Recovery of fixed ammonium by corn in greenhouse studies. *Soil Sci. Soc. Amer. Proc.* **27**:200–204.

Waring, S. A. (1963) A rapid procedure for the estimation of nitrogen availability in soils. Ph.D. Thesis, University of Queensland, Brisbane, Queensland.

Waring, S. A., and J. M. Bremner. (1964) Effect of soil mesh-size on the estimation of mineralizable nitrogen in soils. *Nature* **202**:1141.

Weber, C. R. (1966) Nodulating and nonnodulating soybean isolines: I. Agronomic and chemical attributes. *Agron. Jour.* **58**:43–46.

Weber, C. R. (196Ca) Nodulating and nonnodulating soybean isolines: II. Response to applied nitrogen and modified soil conditions. *Agron. Jour.* **58**:46–49.

Weeks, W. D., and F. W. Southwick. (1956) The relation of nitrogen fertilization to annual production of McIntosh apples. *Proc. Amer. Soc. Hort. Sci.* **68**:27–31.

White, E. M., and F. F. Riecken. (1955) Brunizem-gray brown podzolic soil biosequences. *Soil Sci. Soc. Amer. Proc.* **19**:504–509.

Wiersum, L. K. (1958) Density of root branching as affected by substrate and separate ions. *Acta Botan. Neerland.* **7**:174–190.

Wilkinson, S. R. (1961) Influence of nitrogen and phosphorus fertilizers on the morpholoy and physiology of soybean roots. Ph.D. Thesis, Purdue University, Lafayette, Indiana.

Williams, R. R. (1963) The effect of nitrogen on the self-fruitfulness of certain varieties of cider apples. *Jour. Hort. Sci.* **38**:52–60.

Wilson, J. K. (1928) The number of ammonia-oxidizing organisms in soils. *Proc. Papers First Internat. Congr. Soil Sci., Comm.* **3**:14–22.

Wilson, P. W. (1940) *The Biochemistry of Symbiotic Nitrogen Fixation.* University of Wisconsin Press, Madison.

Wilson, P. W. (1958) Asymbiotic nitrogen fixation. *In* W. Ruhland (Ed.) *Handbuch der Pflanzenphysiologie.* VIII. *Der Stickstoffumsatz,* pp. 9–47. Springer-Verlag, Berlin.

Wilson, P. W., and E. B. Fred. (1939) The carbohydrate-nitrogen relation in legume symbiosis. *Jour. Amer. Soc. Agron.* **31**:497–502.

Wilson, P. W., and O. Wyss. (1938) Mixed cropping and the excretion of nitrogen by leguminous plants. *Soil Sci. Soc. Amer. Proc.* (1937) **2**:289–297.

Winogradsky, S. (1893) Sur l'assimilation de l'azote gazeux de l'atmosphére par les microbes. *Compt. Rend. Acad. Sci.* **116**:1385–1388.

Wit, C. T. de., P. G. Tow, and G. C. Ennik. (1966) Competition between legumes and grasses. *Versl. Landbouwk. Onderzoek.* 687.

Woodman, R. M., and H. Paver. (1944) The effect of time of application of inorganic nitrogen on the turnip. *Jour. Agr. Sci.* **34**:49–56.

Wright, M. J., and K. L. Davison. (1964) Nitrate accumulation in crops and nitrate poisoning in animals. *Adv. Agron.* **16**:197–247.

Yaalon, D. H. (1964) The concentration of ammonia and nitrate in rain water over Israel in relation to environmental factors. *Tellus* **16**:200–204.

Yocum, C. S. (1960) Nitrogen fixation. *Annual Rev. Plant Physiol.* **11**:25–36.

8 *Phosphorus*

Phosphorus is invariably classed as one of the macronutrients, but its content in plants is considerably less than that of nitrogen, potassium, and calcium. As a limiting factor, however, phosphorus is more important than calcium and probably more important than potassium. In terms of use in fertilizers the world over, phosphorus ranks a little lower than potassium; and the amounts of both are less than one-fourth the amount of nitrogen in comparisons made on a weight basis.

In contrast to certain inorganic combined forms of nitrogen, which are not highly stable in soils and are lost readily by volatilization and leaching, phosphorus is relatively stable in soils and is not lost readily in either way. The high stability (low solubility) of phosphorus in soils is the immediate cause of deficiencies of soil phosphorus for plants. If the solubility could be increased, the small quantities of phosphorus would soon be of first importance.

Content in Soils

The total content of phosphorus in soils is relatively low. Lipman and Conybeare (1936) obtained 0.062% as an average value for the content of phosphorus in the plowed layer of the crop land of the United States, a value considerably smaller than the corresponding figures of 0.14% for nitrogen and 0.83% for potassium. A generalized map of the phosphorus content of soils of the United States published by Parker et al. (1946) shows that soils of the Atlantic and Gulf Coastal Plain contain less than 0.017% phosphorus. This is the most extensive low-phosphorus area. Most soils contain between 0.022 and 0.083% phosphorus, but soils of a large area in the Northwest contain 0.087 to 0.13%. Wild (1958) published a paper on the distribution of phosphorus in soils of Australia and delineated large areas of soils containing less than 0.017% phosphorus. Soils of most of the remaining area for which data were available were in the range from 0.017 to 0.053% phosphorus. His average figure was 0.030%, which is about half of the corresponding value obtained by Lipman and Conybeare (1936) for soils of the United States. As might be expected from the lower phosphorus content of the soils, deficiency of phosphorus for plants is more acute in Australia than in the United States. Wild (1958) attributed the comparatively low phosphorus content of Australian soils to losses of phosphorus by leaching under current and earlier environmental conditions and not to a low content of phosphorus in the parent rock.

Phosphorus released in soluble form in soils from weathering of primary phosphorus-bearing minerals and additions of plant residues and fertilizers combines primarily with the clay fraction. As a result, the phosphorus percentage of the clay fraction usually exceeds that of the coarser particle-size fractions (Failyer et al., 1908). Moreover, the phosphorus percentage of the soil as a whole usually increases as the texture becomes finer, if other conditions are similar; this tendency is illustrated in Table 8.1. The selective enrichment of the clay fraction with phosphorus in the upper part of the profile during soil development is indicated by analyses of a soil under forest vegetation in Nigeria (Bates and Baker, 1960) showing that the phosphorus contents of the fine sand and clay fractions were 0.02 and 0.17% in the surface 5 cm of soil and 0.02 and 0.03% in the 51- to 76-cm layer of soil. Goel and Agarwal (1960) observed a similar trend in two "mature" soil profiles in India but not in an "immature" profile.

Table 8.1. Phosphorus Content of the Surface Seventeen Centimeters of Soils Derived from Glacial Deposits in Northeastern Iowa (Walker and Brown 1936)

Soil Texture	Phosphorus Content of Soil, %
Sand	0.040
Fine sand	0.037
Sandy loam	0.043
Fine sandy loam	0.045
Loam	0.057
Silt loam	0.064

In soil profiles developed on seemingly uniform parent materials, the minimum phosphorus percentage usually occurs in the lower A or upper B horizon (Winters and Simonson, 1951). An example is shown in Fig. 8.1. The minimum in the phosphorus percentage presumably results

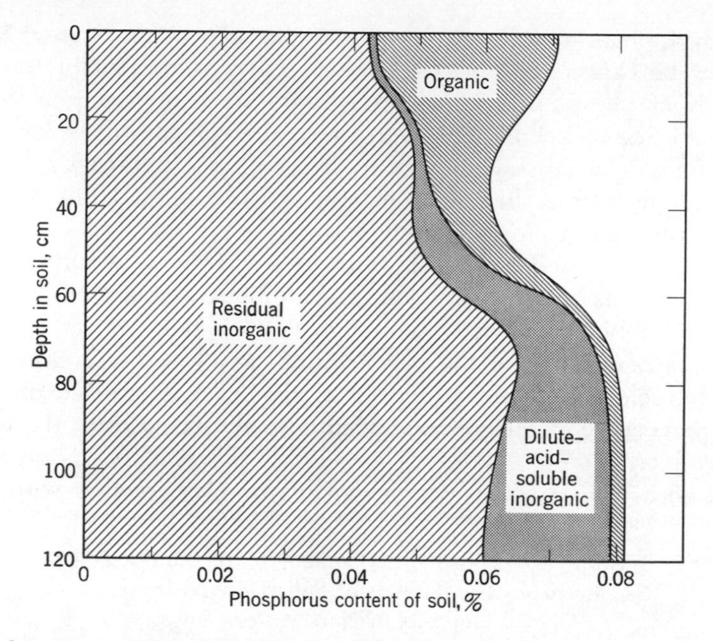

Figure 8.1. Vertical distribution of phosphorus in a soil developed on loess under grassland vegetation in Iowa. (Pearson and Simonson, 1939; Pearson et al., 1940)

from absorption of phosphorus by plants and loss of phosphorus by leaching. The higher content in the surface soil may be attributed to return of some of the phosphorus absorbed by plants and to retention of this phosphorus by the surface soil against rapid downward movement by leaching. If the original phosphorus content of the loessial parent material was uniform vertically, the upper part of the profile in Fig. 8.1 has lost a part of its original content of phosphorus. Both Runge (1963) and Fenton (1966) obtained evidence of accumulation of phosphorus at depths of about 1.5 to 2 meters in well-drained soils of Iowa developed on loess. These depths are greater than those generally included in the soil profiles.

Losses from Soils

Crop Removal

Lipman and Conybeare (1936) estimated that the average amount of phosphorus removed in the harvested portion of crops in the United States in 1930 was 4.3 kg/hectare. The total removal of phosphorus, including that in the crop residues returned to the soil, was perhaps

5 or 6 kg/hectare. The quantity 6 kg of phosphorus per hectare is equal to about 0.4% of the average content of phosphorus in the plowed layer of soil. It may be recalled from Chapter 7 that the total removal of nitrogen in harvested crops in 1930 averaged about 28 kg/hectare, a quantity equal to 0.9% of the total nitrogen in the plowed layer of soil. The removal of phosphorus in crops thus is lower than the removal of nitrogen, both in absolute terms and in relation to the amounts present in soils.

Leaching

The concentration of phosphorus in the soil solution is usually less than 0.1 μg/ml and rarely greater than 1 μg/ml. Consequently, the loss of phosphorus from soils by leaching is exceedingly low even if considerable drainage occurs. Lipman and Conybeare (1936) neglected the loss of phosphorus by leaching as being of no significance in their study of the annual balance of plant nutrients in soils of the United States. Because of the low solubility and limited movement of soil phosphorus, the deposition of phosphorus in soils in areas of man's early habitation is now of value as an indicator in archaeology (Dauncey, 1952; Jakob, 1955; Cook and Heizer, 1965).

The cumulative loss of soil phosphorus by leaching over a period of thousands of years, however, may attain significance. Wild (1961) estimated the loss of phosphorus from twelve soils derived from granite in Australia on the assumption that no zirconium was lost during soil development (zirconium is relatively resistant to loss by leaching) and obtained an average figure of 50% loss. All soil profiles except one showed a loss according to this method of calculation. The annual rainfall at the sites of the various profiles ranged from 46 to 483 cm. The one profile that showed no loss (but an apparent gain of 22%) occurred at the site with the lowest rainfall. Somewhat different estimates were obtained on the basis of other methods of calculation, but all methods indicated substantial losses. Another form of evidence is the decrease in content of phosphorus in soils with an increase in rainfall shown in Fig. 8.2. This decrease in total phosphorus content was associated with progressively greater deficiency of phosphorus for plants in field trials at the various sites.

The concentration of phosphorus in saturated solutions of most phosphate fertilizers is many thousands of times greater than that of phosphorus in soil solutions, and much fertilizer phosphorus could be lost by leaching if this level of solubility persisted after addition of the fertilizers to soils. Most soils, however, have the capability of reacting rapidly with soluble phosphates and reducing their solubility. Therefore,

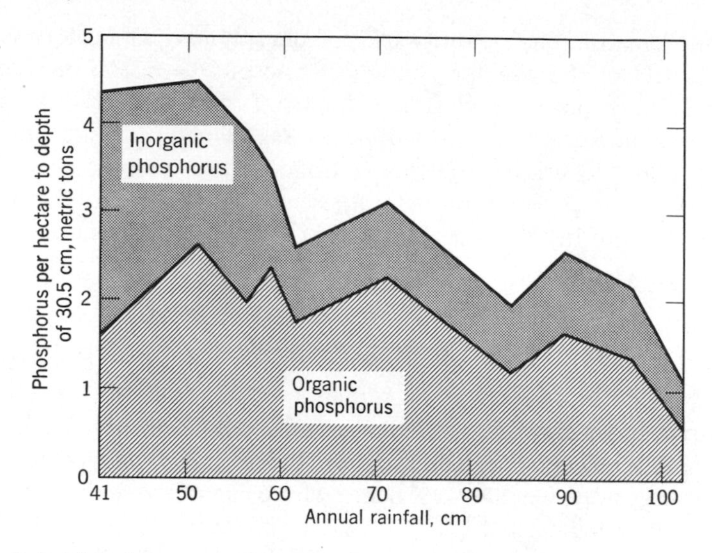

Figure 8.2. Phosphorus in surface 30.5 cm of soils developed on mica-schist in New Zealand versus annual rainfall. (Ludecke, 1962)

the phosphorus of fertilizers tends to remain near the point of application. For example, Fig. 8.3 shows that application of 135 kg of phosphorus as superphosphate per hectare on April 25 to a permanent pasture in Wisconsin markedly increased the dilute-acid-soluble phosphorus found in samples of soil taken October 15 but that the effect was confined

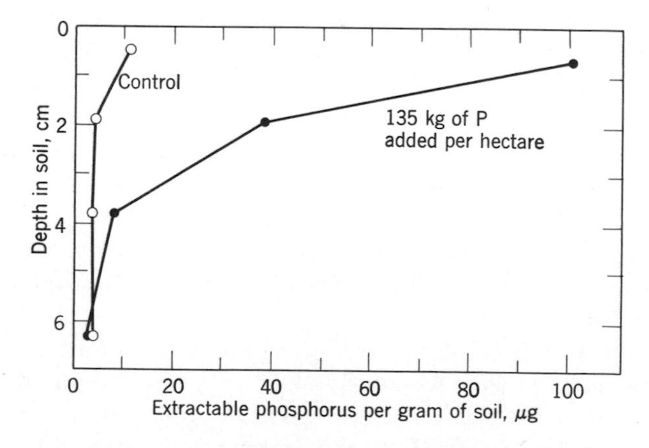

Figure 8.3. Vertical distribution of extractable phosphorus in control and superphosphate-fertilized silt loam under permanent pasture in Wisconsin. Phosphorus was extracted by a solution containing 0.045-normal ammonium sulfate and 0.002-normal sulfuric acid. (Midgley, 1931)

to approximately the surface 5 cm. In an experiment in Connecticut with a sandy loam soil, Morgan and Jacobson (1942) added as fertilizer the equivalent of 3372 kg of phosphorus per hectare in an 11-year period and measured an annual loss of only 0.1 kg of phosphorus per hectare through the 46-cm depth of soil used.

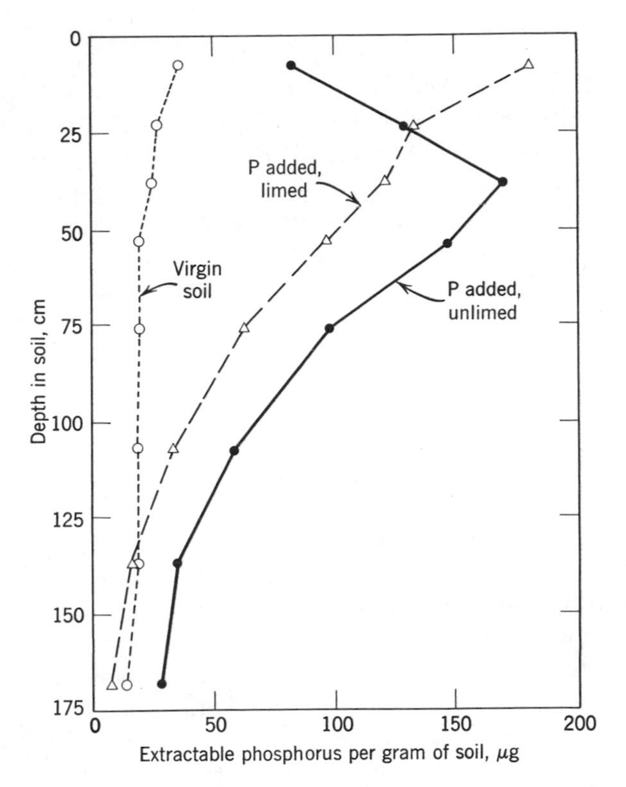

Figure 8.4. Vertical distribution of extractable phosphorus in a fine sandy soil in virgin condition and after a period of fertilization with superphosphate in quantities to supply a total of 2140 kg of phosphorus per hectare. Fertilized areas were planted to citrus. The limed soil was maintained at *p*H 5.5 to 6, but the *p*H of the unlimed soil decreased during the experiment to a value of 4 at the end. The extractant employed was a solution containing 0.03-normal ammonium fluoride and 0.025-normal hydrochloric acid. (Spencer, 1957)

Probably the only soils from which loss of fertilizer phosphorus by leaching is significant on a short-term basis are sands and peats that have little tendency to react with phosphorus. Spencer (1957) found considerable downward movement of fertilizer phosphorus in a sandy soil in Florida, as indicated by the data on extractable phosphorus in the soil profile (Fig. 8.4). Similar trends were observed in measurements of total phosphorus. In experiments conducted on sandy soils in western Australia, Ozanne (1962) observed movement of considerable fertilizer phosphorus in a single season to a depth exceeding 75 cm; and Ozanne et al. (1962) found that loss of fertilizer phosphorus by leaching from the surface 10 cm of soil under clover exceeded the uptake of phosphorus by the crop.

Volatilization

Under oxidizing conditions and at the *p*H values of soil, phosphorus is stable as orthophosphate and not subject to volatilization. Tsubota (1959) reported, however, that phosphorus was evolved in gaseous form from soil incubated anaerobically with additions of compost and dibasic ammonium phosphate. Losses amounting to 47 and 73 μg of phosphorus per gram of two different soils during incubation for 42 days at 24°C were recorded. The gaseous phosphorus was trapped and oxidized in nitric acid and presumably was evolved as phosphine (PH_3). In experiments on anaerobic liquid media, the same author identified phosphite (PO_3^{---}) and hypophosphite ($H_2PO_2^-$), which are reduction products of orthophosphate (PO_4^{---}).

Erosion

Lipman and Conybeare (1936) estimated the loss of phosphorus by erosion from the crop land of the United States at 11.9 kg/hectare annually. This figure is substantially higher than that for the removal of phosphorus in the harvested portion of crops. The values are not directly comparable, however, for two reasons.

First, the crop-producing potentiality of the phosphorus in the two forms is not the same. If the two forms were evaluated on the basis of their effectiveness in supplying plants with phosphorus, the phosphorus removed by plants would be found much more valuable, kilogram for kilogram, than that removed by erosion.

Second, the loss of phosphorus by both plant removal and erosion is compensated to some extent. Loss by plant removal is compensated in part by return to the soil of some of the phosphorus in forms of greater availability than the phosphorus in the soil initially. This point is discussed further in the section on availability to plants. Loss by erosion is compensated to some extent by the phosphorus in the new material brought into the profile from beneath. Availability of subsoil phosphorus to plants differs greatly among soils, as indicated in Table 8.2. Except for the lower A and upper B horizons, the supply of phosphorus was adequate for growth of sweet clover throughout the Fayette silt loam profile. In the Lindley silt loam profile, only the upper A

Table 8.2. Yield of Sweet Clover with and without Heavy Phosphate Fertilization on Samples Taken from Different Horizons of Two Soils from Minnesota (Rost, 1939)

	Relative Yield per Culture			
	Fayette Silt Loam		Lindley Silt Loam	
Soil horizon sampled	No Phosphorus Added	44 kg of P as Superphosphate per Hectare	No Phosphorus Added	44 kg of P as Superphosphate per Hectare
Upper A	100	108	100	105
Lower A	68	101	80	124
Upper B	54	88	36	129
Lower B	94	88	21	118
C1	83	78	55	109
C2	84	83	28	111

horizon had an adequate phosphorus supply. All other horizons were deficient. The comparatively high availability of phosphorus in the Fayette subsoil is exceptional. More commonly, the availability of subsoil phosphorus is similar to or lower than that in the Lindley silt loam.

Forms in Soils

Phosphorus in soils occurs almost exclusively as orthophosphate, in which a central phosphorus atom is surrounded by and bound to four oxygen atoms. The phosphates in soils are derivatives of phosphoric acid, H_3PO_4.

The phosphates in soils may be divided into two broad categories, inorganic and organic. In inorganic forms, from one to three of the hydrogen ions of phosphoric acid are replaced by metallic cations. In organic forms, one or perhaps more of the hydrogen ions of phosphoric acid is eliminated in an ester linkage; the remaining hydrogen ions are replaced in part or completely by metallic cations.

The relative proportions of the phosphorus in these two categories vary widely. Organic phosphorus tends to increase and decrease with the content of organic matter and hence is comparatively low in subsoils and high in surface soils. Analyses of surface soils have shown values for organic phosphorus ranging from as low as 0.3% of the total (Patel and Mehta, 1961) to as high as 95% of the total (Kaila, 1956).

Inorganic

Inorganic phosphates in soil may be classified according to their physical, mineralogical, or chemical nature or combinations of these, but there is no consensus on the matter. For present purposes, the subject matter will be organized in a manner that follows to some extent the evolution of inorganic phosphates in soils.

Apatite, which accounts for 95% or more of the phosphorus in igneous rocks (Rankama and Sahama, 1950, p. 585), occurs characteristically as small crystals of sand and silt size that are present in small amount and distributed rather generally throughout the rocks. The mineral has been identified in the sand and silt fraction of a number of soils. McCaughey and Fry (1913) identified apatite in eleven of twenty-five soils from different parts of the United States. Plummer (1915–1916) found apatite in four of nine soils investigated in North Carolina. Leahey (1934–1935) found fresh-appearing grains of apatite in the leached layer of a podzol soil. Apatite was found throughout each of three soil profiles examined in Minnesota by Arneman and McMiller (1955) and throughout each of three soil profiles examined in Nebraska by Shipp and Matelski (1960). Shipp and Matelski determined the content of both apatite and total phosphorus in the sand and coarse silt fractions. Their data, plotted in Fig. 8.5, show not only that the content of apatite increased with the content of phosphorus but also that if one calculates the content of fluorapatite [$Ca_{10}F_2(PO_4)_6$] corresponding to the content of phosphorus, the theoretical values much exceed the values found experimentally. (Fluorapatite is the most common and most stable form of apatite.) Indications are that in most of the samples less than half of the phosphorus could be accounted for as fluorapatite.

Weathering results in gradual disappearance of fluorapatite and formation of secondary phosphates. Secondary phosphate minerals of various kinds have been found in soils. In Australia (1956), phosphate minerals belonging to the gorceixite [$BaAl_3(PO_4)_2(OH)_5 \cdot H_2O$], florencite [$CeAl_3(PO_4)_2(OH)_6$] series were identified in soils from Queensland, New South Wales, and South Australia. These minerals were concentrated in the particle-size fraction 0.5 to 5 μ in diameter and in some instances accounted for more than 70% of the total phosphorus. Soils in these areas have undergone extensive weathering, and it seems reasonable that the gorceixite, florencite minerals now contain phosphorus that originally was present mostly as apatite.

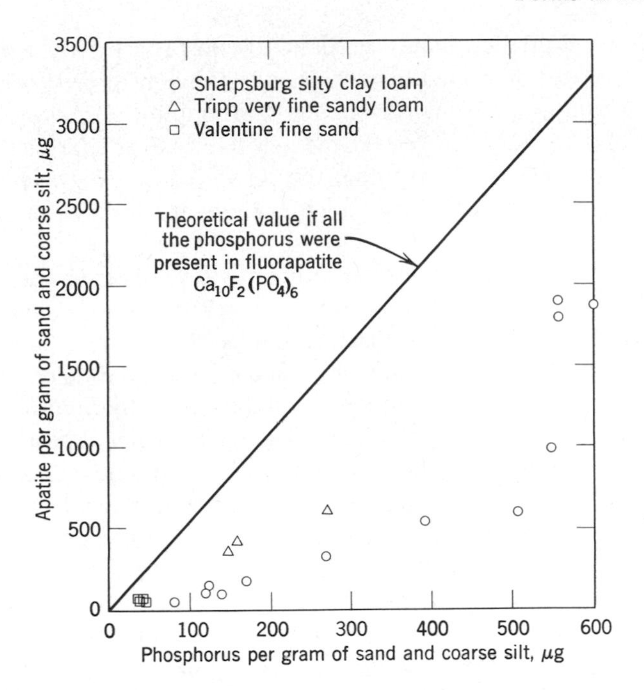

Figure 8.5. Apatite and total phosphorus in sand and coarse silt fractions of three soil profiles from Nebraska. (Shipp and Matelski, 1960)

Various other phosphatic minerals have been found in soils under more special conditions. The ferrous phosphate, vivianite [$Fe_3(PO_4)_2 \cdot 8H_2O$], has been identified under reducing conditions in buried alluvium (Dudley, 1890) and in peats (Okruszko, 1955; Bogdanović et al., 1963). A notable occurrence of vivianite is a layer up to 7.5 cm in thickness at a depth of about 75 cm in a peat bog at Labish Center, Oregon (J. E. Dawson, private communication). The comparatively large masses of vivianite described in the literature are not replacement forms that develop because of original existence of a high concentration of apatite or other phosphate mineral; rather, they represent a concentration of phosphorus and iron derived from a relatively large area by crystallization from iron- and phosphorus-laden waters that move past the developing crystalline mineral. Another phosphate mineral that has been identified in the sand, silt, and clay fractions of soil is wavellite [$Al_3(OH_3(PO_4)_2 \cdot 5H_2O$]. This mineral was identified in a soil from Florida developed on parent material containing 35 to 45% of this mineral, in effect a phosphate deposit (Dyal, 1953). Fieldes et al. (1960) and Schroo (1963) identified crandallite [$CaAl_3(PO_4)_2(OH)_6$] in soils developed on coral limestone on islands in the southern Pacific Ocean. These occurrences of wavellite and crandallite seem to be a consequence of accumulation of phosphate from the surrounding environment, which was probably sea water, followed by selective loss of more soluble constituents. There is also a theory that the accumulation of phosphate is derived from droppings of birds which congregate in great numbers on some islands.

For the most part, the phosphate minerals described in the preceding paragraphs were identified in the sand and silt fractions by petrographic microscopy. Where the phosphate minerals made up a large portion of the weight of the soil, they were sometimes identified also in the clay fraction by X-ray diffraction or differential thermal analysis. In the case of the gorceixite, florencite minerals, which constituted only a negligible proportion of the total soil weight, identification was made by X-ray diffraction following separation of the particle-size fraction in which the minerals were concentrated and following treatment of the separated size-fraction with hydrofluoric acid to dissolve the silicate minerals. Although the utility of these classical methods for identifying crystalline phosphates in soils has been by no means exhausted, the methods cannot be expected to yield a complete characterization of the inorganic phosphates in soil. An important reason for this is that much of the inorganic phosphorus, and in many soils most of it, occurs in the clay fraction, from which it cannot be separated by physical methods. The properties of the phosphatic minerals that might otherwise be measured by X-ray diffraction or thermal analysis are then masked by the great preponderance of nonphosphatic material.

The crystalline inorganic phosphates in soils occur only partly as discrete particles. Fry (1913) observed grains of apatite enclosed within grains of quartz in soils. Some of the crystalline phosphate may be present as a minor constituent of silicate minerals, where it substitutes for silicon. Work on mineralogical aspects of this subject was reviewed by Mason and Berggren (1941), who reported original data on phosphorus-bearing garnet and cited work on phosphorus-bearing zircon ($ZrSiO_4$) in which PO_4 groups substituted for as much as 25% of the normal complement of SiO_4 groups. Such phosphate groups presumably are incorporated into the structure of the mineral during crystallization and occur scattered throughout the structure. Experiments on the aluminosilicate clay mineral, kaolinite, indicate that some replacement of surface silicate groups by phosphate may occur subsequent to formation of the clay if the phosphate concentration is high in the external solution (Low and Black, 1950).

Three other methods have been used in an attempt to obtain information on the forms of inorganic phosphorus in soils. All are based on solubility criteria.

The first method is that of equilibrating various soils and known phosphate minerals individually with solutions of different *p*H, plotting the phosphorus concentration in solution against the *p*H after equilibration, and comparing the shape of the curves for soils with those for mineral standards. Calcium phosphates are more soluble in dilute hydrochloric acid than in dilute sodium hydroxide. Aluminum and ferric phosphates are more soluble in dilute sodium hydroxide than in dilute hydrochloric acid. Stelly and Pierre (1943) and others have noted suggestive similarities in work of this kind, as indicated in Fig. 8.6. The solubility of the phosphates is related to the *p*H of the original soil. Alkaline soils usually display a solubility-*p*H curve similar to that of apatite. Acid

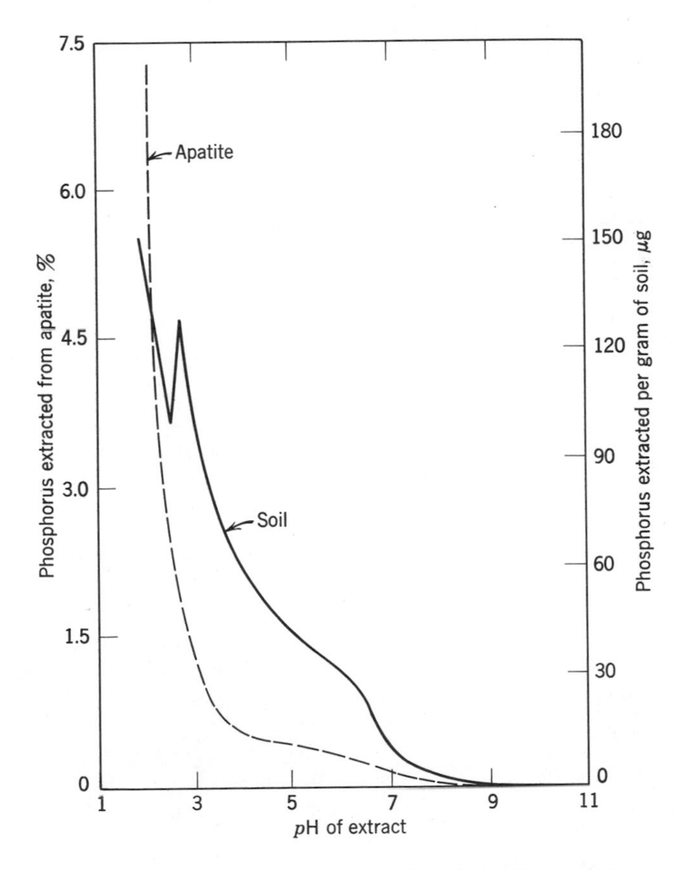

Figure 8.6. Phosphorus extracted at various *p*H values from apatite and the calcareous C horizon of a loam soil. (Stelly and Pierre, 1943)

soils usually have a solubility-*p*H curve similar to that of iron or aluminum phosphates. The initial reaction of the soil in Fig. 8.6 was *p*H 7.9.

The second method is that of extracting soil consecutively with different solutions, each designed to remove the phosphorus present in a known mineral form. A number of investigators have used an extraction sequence proposed by Chang and Jackson (1957). These authors outlined methods for extracting soil consecutively with normal ammonium chloride, half-normal ammonium fluoride, tenth-normal sodium hydroxide, half-normal sulfuric acid, three-tenths-molar sodium citrate plus sodium dithionite ($Na_2S_2O_4$), and half-normal ammonium fluoride. These extractants are supposed to remove water-soluble phosphate, aluminum phosphate, iron phosphate, calcium phosphate, occluded iron phosphate, and occluded aluminum phosphate, respectively. The term "occluded" is used to express the theory that the last two fractions are inside of iron oxides that are not dissolved by the first extractions. The combination of citrate and dithionite is used to remove iron oxides from soils without subjecting them to the strong acidity that would be necessary in the absence of a reducing agent (sodium dithionite).

By use of a modification of the Chang and Jackson method, Scheffer et al. (1960) analyzed different particle-size fractions of soils of Germany for phosphorus fractions and found a tendency for accumulation of iron and aluminum phosphates in the finer fractions and calcium phosphates in the coarser fractions (Table 8.3), as would be expected if apatite particles of sand and silt size were weathering to iron and aluminum phosphates associated with the finer size-fractions. In fractionations of

Table 8.3. Average Distribution of Forms of Phosphorus in Particle-Size Fractions of Nine Soils of Germany (Scheffer et al., 1960)

Particle Diameter, mm	Proportion of Total Phosphorus in Size-Fractions Present in Indicated Form, %		
	Iron and Aluminum Phosphates	Calcium Phosphates	Organic Phosphates
>0.2	13.0	60.0	23.9
0.06–0.2	13.7	44.6	22.7
0.02–0.06	27.9	29.6	12.5
0.006–0.02	31.3	12.3	42.9
0.002–0.006	29.6	18.7	34.7
<0.002	21.0	12.7	32.5

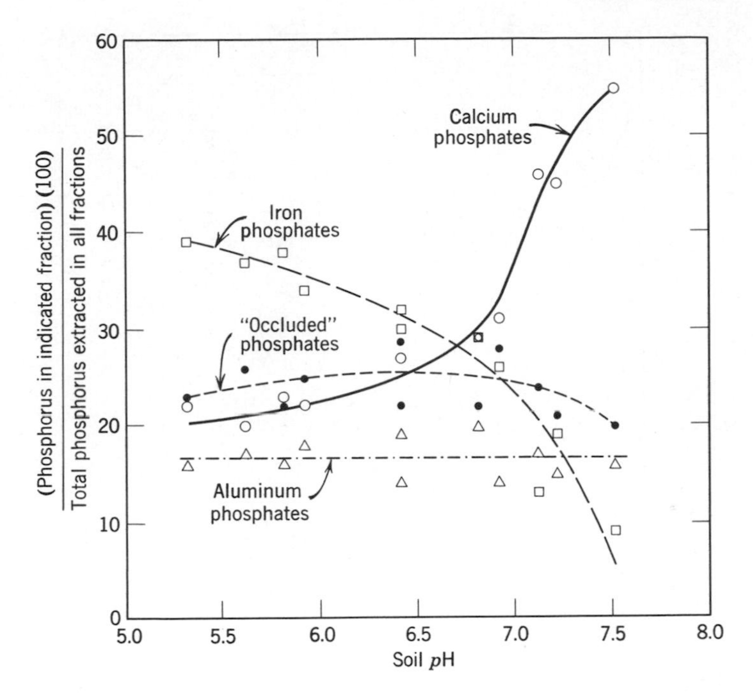

Figure 8.7. Distribution of extractable inorganic phosphorus among four forms in composite samples of loess-derived soils of Germany versus soil *p*H. (Schachtschabel and Heinemann, 1964)

phosphorus in a number of loess-derived soils in an area in Germany, Schachtschabel and Heinemann (1964) obtained the data in Fig. 8.7 relating the phosphorus fractions to soil *p*H. With an increase in soil *p*H, the calcium phosphate fraction increased, the iron phosphate fraction decreased, and the aluminum phosphate and occluded phosphate fractions remained about the same. The *p*H values of these soils provide an index of the weathering, but the differences are not large because of the uniformity of the parent material and the environmental conditions. Figure 8.8 gives a comparison of the distributions of phosphorus in three soils analyzed by Chang and Jackson (1958). These soils differ greatly in weathering, and a wide difference in phosphorus distribution may be observed. Calcium phosphates predominate in the slightly weathered soil, and occluded phosphates predominate in the strongly weathered soil.

Supplementary observations made in connection with the method of successive extractions provide additional information on the nature of the inorganic phosphorus in soils. Machold (1962, 1963) equilibrated

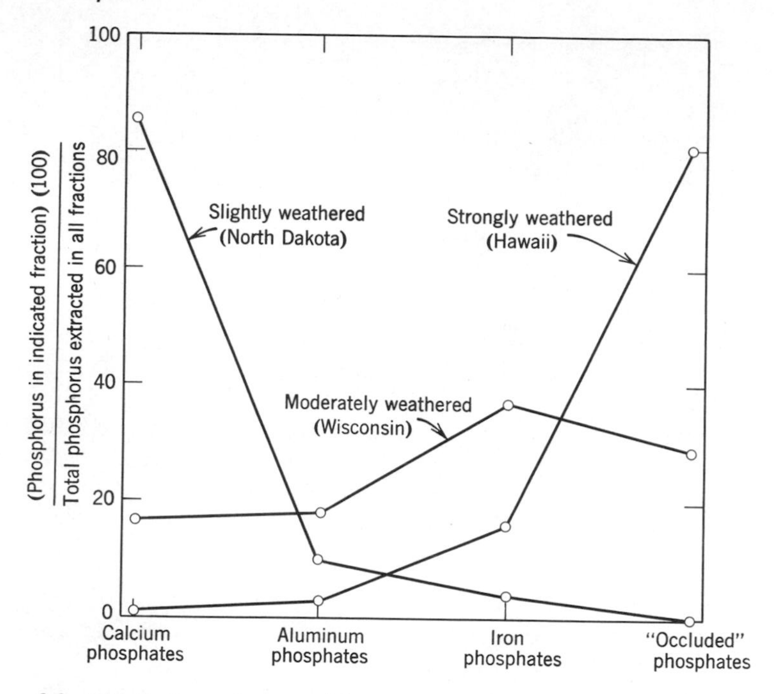

Figure 8.8. Distribution of extractable inorganic phosphorus among four forms in the A horizon of three soils representing different degrees of weathering. (Chang and Jackson, 1958)

samples of soils for 500 hours with a trace quantity of radioactive phosphorus as orthophosphate and then extracted the samples according to the scheme of Chang and Jackson and determined the radioactivity of the phosphorus in each fraction. Some of the results are given in Table 8.4. In preparation of this table, the comparative degree of labeling of the phosphorus in individual fractions was used to calculate conventional values for the proportion of the phosphorus in each fraction that had come to equilibrium with the radioactive phosphorus. Table 8.4 gives average data for seventeen soils on which measurements were reported for all five fractions. These values were calculated on the assumption that the water-soluble fraction of inorganic phosphorus equilibrated completely with the radioactive phosphorus. Indications are that the inorganic phosphorus present in the aluminum and iron phosphate fractions equilibrated with the radioactive phosphorus but that the phosphorus in the calcium phosphate and occluded phosphate fractions did so to only a limited degree.

The significance of these observations in terms of the nature of soil phosphorus is that radioactive phosphorus equilibrates to only a limited extent with crystalline phosphates. In parallel trials with strengite, a crystalline iron phosphate, Machold (1963) found that only 0.1% of the phosphorus was exchangeable with radioactive phosphorus. The corresponding figures for a natural apatite, a synthetic hydroxyapatite, and dibasic calcium phosphate dihydrate were 1, 2.4, and 20.7%. Although the last substance is more soluble than the phosphates in soils, the exchange was still small, presumably because most of the phosphorus remained in place inside the crystals during the 500-hour equilibration period and hence was not accessible for exchange with the phosphorus in the external solution. The data in Table 8.4 thus indicate that much of the extractable phosphorus in the soils either was not present in crystalline form or was present in crystalline units so small that crystallinity did not make the phosphorus inaccessible to exchange with phosphorus in solution. The extractable phosphorus in Table 8.4 includes the major part of the soil phosphorus. The remainder was no doubt mostly organic.

Table 8.4. Average Quantities of Extractable and Total Phosphorus in Seventeen Soils from Germany, and Conventional Percentage of Extracted Phosphorus Equilibrated with P^{32}-Labeled Orthophosphate in Five-Hundred Hours (Machold, 1962, 1963)

Extractant	Fraction	Phosphorus Extracted per Gram of Soil, μg	Conventional Percentage of Extracted Phosphorus Equilibrated with $P^{32}O_4$
Ammonium chloride (1 *N*)	Water-soluble phosphates	8	100
Ammonium fluoride (0.5 *N*)	Aluminum phosphates	112	116
Sodium hydroxide (0.1 *N*)	Iron phosphates	78	114
Sulfuric acid (0.5 *N*)	Calcium phosphates	153	17
Sodium citrate (0.3 *M*) + sodium dithionite	"Occluded" phosphates	80	5
Sum of phosphorus in five fractions extracted		431	—
Total phosphorus in soils		608	—

The comparative tendency of the phosphorus fractions to exchange with radioactive phosphorus in solution is of further significance in relation to availability of soil phosphorus to plants because quantities of phosphorus absorbed from soils by plants increase with the quantities of phosphorus that equilibrate with radioactive phosphorus added in solution. Machold (1962) found that the correlations between phosphorus absorbed from soils by rye seedlings and the quantities of soil phosphorus that equilibrated with P^{32} for 25, 50, 125, 250, and 500 hours were 0.85, 0.92, 0.78, 0.76, and 0.47, respectively. These observations may be interpreted to mean that the phosphorus less readily accessible to exchange with radioactive phosphorus added in solution was also less susceptible to uptake by plants.

The last method of investigating the forms of inorganic phosphorus in soils is that of comparing ion-activity products calculated for solutions in equilibrium with soils with the solubility-product constants for known phosphate minerals. If agreement is obtained, the phosphate in question is inferred to be present in the soil. To take hydroxyapatite as an example, the dissolution of the crystalline solid may be represented as

$$Ca_{10}(OH)_2(PO_4)_6 = 10Ca^{++} + 2OH^- + 6PO_4^{---}.$$

For sparingly soluble substances such as hydroxyapatite, the product of the activities of the ions formed in solution from dissolution of the solid (or the same ions from other sources), with the activity of each ion raised to the power equal to the number of ions of that species produced by dissolution of one molecule of the substance, gives a number known as the solubility-product. For hydroxyapatite, the solubility-product may be written

$$(a_{Ca^{++}})^{10}(a_{OH^-})^2(a_{PO_4^{---}})^6 = k.$$

Ideally, each sparingly soluble crystalline phosphate should have a solubility-product that has a constant value where equilibrium has been attained at a given temperature. As long as the solution is in equilibrium with the solid, the constant value should theoretically be maintained by suitable variation in activities of individual component ions in solution. Hence, if hydroxyapatite exists in soil, the extra calcium in solution derived from the exchangeable form should cause the activity of hydroxyl, phosphate, or both to be less than would be the case in the absence of the additional calcium. If, in addition to hydroxyapatite, the soil contains variscite [$Al(OH)_2H_2PO_4$], another sparingly soluble phosphate, and if this phosphate also is in equilibrium with the solution, the activities of ions in solution must satisfy the solubility-product of both minerals at the same time.

The solubility-product method has some complications. In addition to the matters of the *p*H-dependent equilibrium among the various phosphate ions ($H_2PO_4^-$, HPO_4^{--}, and PO_4^{---}) and the activity-coefficient corrections [an account of these subjects was given by Chaverri and Black (1966)], which are complex in themselves, the problem of equilibrium is troublesome. Shifts from one compound in the direction of another may be detected in pure systems in the laboratory in minutes (Rathje, 1960). Attainment of final equilibrium, however, is quite another matter; it is from the equilibrium condition that inferences about the presence of specific solid-phase phosphates are made. Sparingly soluble phosphates such as hydroxyapatite, which would be expected from their solubility-products to be stable in soils under certain conditions, are notoriously slow in approaching an equilibrium with the solution surrounding them. Larsen and Court (1961) found, for example, that 0.01-molar calcium chloride solutions equilibrated with samples of various surface soils from Great Britain were supersaturated with hydroxyapatite above about *p*H 6 and unsaturated with hydroxyapatite below this *p*H. According to their findings, hydroxyapatite should have been stable in soils above about *p*H 6 and should have precipitated from the supersaturated solutions until the ion-activity-product of the solutions became equal to the solubility-product constant for hydroxyapatite. Persistence of the state of supersaturation is evidence of the slowness with which equilibrium was attained.

Figure 8.9 gives an example of application of the solubility-product method to soils. The figure represents the results in a special way that permits a comparison of the composition of soil solutions with the composition of solutions that would be in equilibrium with several different crystalline calcium phosphates. A plot of $\frac{1}{2}p\mathrm{Ca} + p\mathrm{H_2PO_4}$ against $p\mathrm{H} - \frac{1}{2}p\mathrm{Ca}$ for such solutions, where the *p*'s refer to negative logarithms and the ion symbols to ion activities, yields a straight line with a characteristic location for each phosphate species. The figure shows that the values for samples of a number of soils of Great Britain, including some that had been fertilized in the field with superphosphate and others that had not, exhibited no particular tendency to follow the solubility line for any of the calcium phosphates. Analogous measurements by Weir and Soper (1963) on calcareous soils of Manitoba, some following phosphate fertilization, yielded points intermediate between the solubility lines for octacalcium phosphate and hydroxyapatite. Thus, although crystalline calcium phosphates may have been and probably were pres-

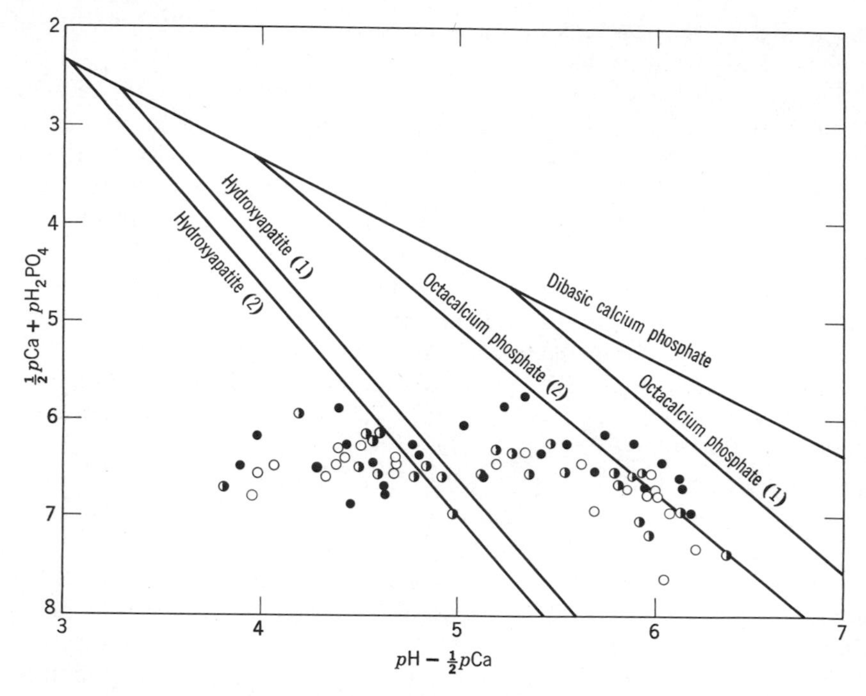

Figure 8.9. Plot of $\frac{1}{2}p\text{Ca} + p\text{H}_2\text{PO}_4$ against $p\text{H} - \frac{1}{2}p\text{Ca}$ for various calcium phosphates and soils. The numerals 1 and 2 following octacalcium phosphate and hydroxyapatite refer to apparent equilibrium values obtained by precipitation from supersaturated solutions and by dissolution of the solid in unsaturated solutions. Each point represents a separate sample of soil from Great Britain. Open circles are for samples with no added phosphorus. Half-shaded circles and solid circles are for samples with previous applications of 245 and 490 kg of phosphorus as superphosphate per hectare in the field. (Larsen and Court, 1961)

ent in some of the soils, the activities of ions in solution generally did not correspond to those in equilibrium with any of the calcium phosphate minerals shown.

The closest agreement obtained between ion-activity products in soil extracts and in solutions in equilibrium with phosphate minerals has been in the case of variscite, an aluminum phosphate. Lindsay et al. (1959) and Wright and Peech (1960) calculated the ion-activity-product $(a_{\text{Al}^{+++}})(a_{\text{OH}^-})^2(a_{\text{H}_2\text{PO}_4^-})$ for this mineral from analyses of extracts of various acid soils and obtained values close to $10^{-30.5}$, which they gave as the solubility product constant for variscite. Some of their data are summarized in Fig. 8.10 in a form analogous to that used in Fig. 8.9 for calcium phosphates.

Raupach (1963) argued, on the other hand, that this sort of agreement of the composition of soil extracts with the solubility line for variscite is not sufficiently good to establish the existence of variscite in the soil. In support of his view, one may note that the deviation of composition of soil extracts from the theoretical values for variscite (Fig. 8.10) is substantial if it is kept in mind that each scale division represents one-hundredfold variation in the vertical direction and tenfold variation in the horizontal direction. Raupach pointed out also that the concentration of aluminum in solution at a given pH value should decrease with an increase in concentration of phosphate if soil solutions are in equilibrium

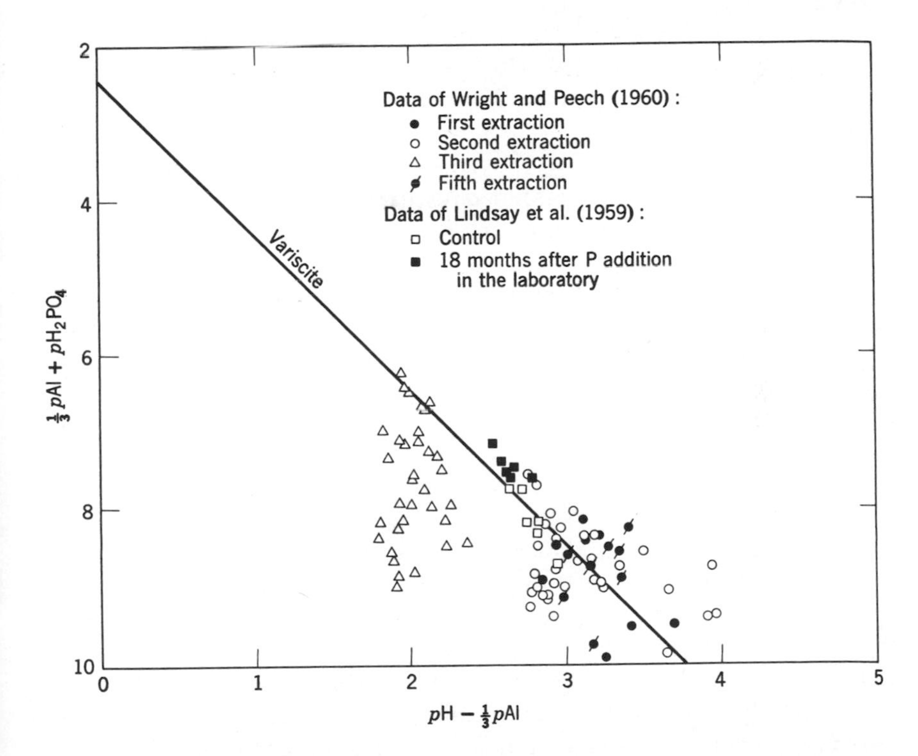

Figure 8.10. Plot of $\frac{1}{3}p\text{Al} + p\text{H}_2\text{PO}_4$ against $p\text{H} - \frac{1}{3}p\text{Al}$ for various soils, some with and others without previous phosphate fertilization in the field, and the line representing the solubility-product constant for variscite, $(a_{\text{Al}^{+++}})(a_{\text{OH}^-})^2(a_{\text{H}_2\text{PO}_4^-}) = 10^{-30.5}$. The open square symbols represent six soils without phosphate treatment, and the solid square symbols represent the same six soils 18 months after addition of 437 μg of phosphorus per gram of soil in the laboratory. All extractions except the third were made with 0.01-molar calcium chloride solution, and the third extraction was made with 0.01-molar calcium chloride solution and 0.001-molar hydrochloric acid.

with variscite. Plots of data from several investigations showed that the trend in this direction was in some instances not evident and in no instance as great as would have been expected had the supposed equilibrium existed.

From the standpoint of soil-plant relationships, the solubility-product method has the advantage that it deals with the solution phase, which is of direct significance in the nutrition of plants. The evidence, however, does not support the view that the phosphorus in solution in soils is at equilibrium with crystalline phosphate minerals. The results of the P^{32}-equilibration investigations mentioned previously support the view that the inorganic phosphorus absorbed from soils by plants comes from a substantial pool of phosphorus that is either not present in crystalline form or is present in crystals so small that their phosphorus equilibrates readily with P^{32}-labeled orthophosphate added in solution. The inorganic phosphorus available to plants is almost certainly included in the chemical fractions extracted according to the scheme proposed by Chang and Jackson (1957), and the part derived from noncrystalline or barely crystalline forms is no doubt combined with one or another of the cations after which their fractions are labeled but is dispersed in a matrix of nonphosphatic material.

Although the crystalline inorganic phosphates native to soils may be conceived to be of relatively little significance in supplying phosphorus to plants on a short-term basis, the same is not necessarily true of crystalline phosphates produced in soils upon addition of soluble phosphatic fertilizers. Research workers in the Tennessee Valley Authority have identified a host of crystalline phosphate compounds that may develop in soils around the site of phosphate fertilizer particles. These crystalline compounds develop in response to the high concentration of phosphate (and sometimes also the high acidity) produced in the soil around the particles. The great local excess of phosphate exhausts the supply of reactive divalent and trivalent cations in the soil solution and causes removal of exchangeable cations and dissolution of less soluble substances such as calcium carbonate, hydrous oxides of iron and aluminum, and even clays and other aluminosilicates. For example, the reaction between calcium carbonate and acid forms of fertilizer phosphorus, e.g., superphosphate, to form crystalline dibasic calcium phosphate may be represented as

$$Ca(H_2PO_4)_2 + CaCO_3 + 3H_2O = 2CaHPO_4 \cdot 2H_2O + CO_2\uparrow.$$

The reaction occurs rapidly, and its progress may be followed by observing the evolution of gas (Fig. 8.11) and the development of a white color in the soil from the newly formed, finely divided crystals of dibasic calcium phosphate (Fig. 8.12). The crystals may be identified by X-ray diffraction.

Crystalline phosphates formed in soils from fertilization are not necessarily stable under normal soil conditions. As the phosphate concentration in the soil solution decreases, because of diffusion of the soluble phosphate outward into the surrounding soil and reaction with the soil minerals, the crystalline phosphates formed upon fertilization tend to release soluble phosphate, at the same time either dissolving or undergoing alteration to a less soluble, more stable form.

The change with time in the location of points on the solubility diagram for calcium phosphates in one investigation is illustrated in Fig. 8.13, which gives measurements on water extracts of calcareous soil after treatment with monobasic calcium phosphate. In addition to the downward trend of points, indicating lower solubility, one may observe that there was no tendency for the composition of the solution to correspond to that for dibasic calcium phosphate, despite the probability that this compound was formed in the soil.

Lehr and Brown (1958) found that, after the time required for growth of two test crops in the greenhouse, octacalcium phosphate or apatite could be identified in fertilizer bands in alkaline soils to which monobasic

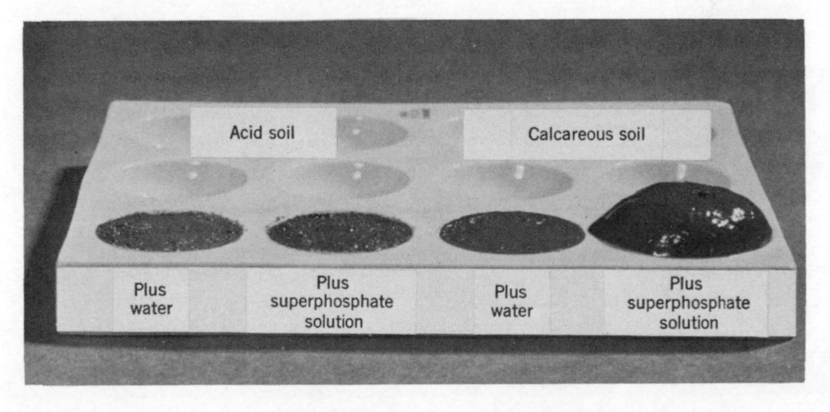

Figure 8.11. Evidence of interaction of superphosphate with soil. A concentrated solution derived from ordinary superphosphate decomposed part of the calcium carbonate in the calcareous soil, with release of carbon dioxide, which caused expansion of the soil. Initially, the surface of all samples of soil was flush with the top of the spot plate.

calcium phosphate had been added. Crystals of dibasic calcium phosphate added to acid soils were eroded, but those added to calcareous soils were largely altered to octacalcium phosphate. The order of the alteration is thought to be monobasic calcium phosphate, dibasic calcium phosphate, octacalcium phosphate, and hydroxyapatite, each of which

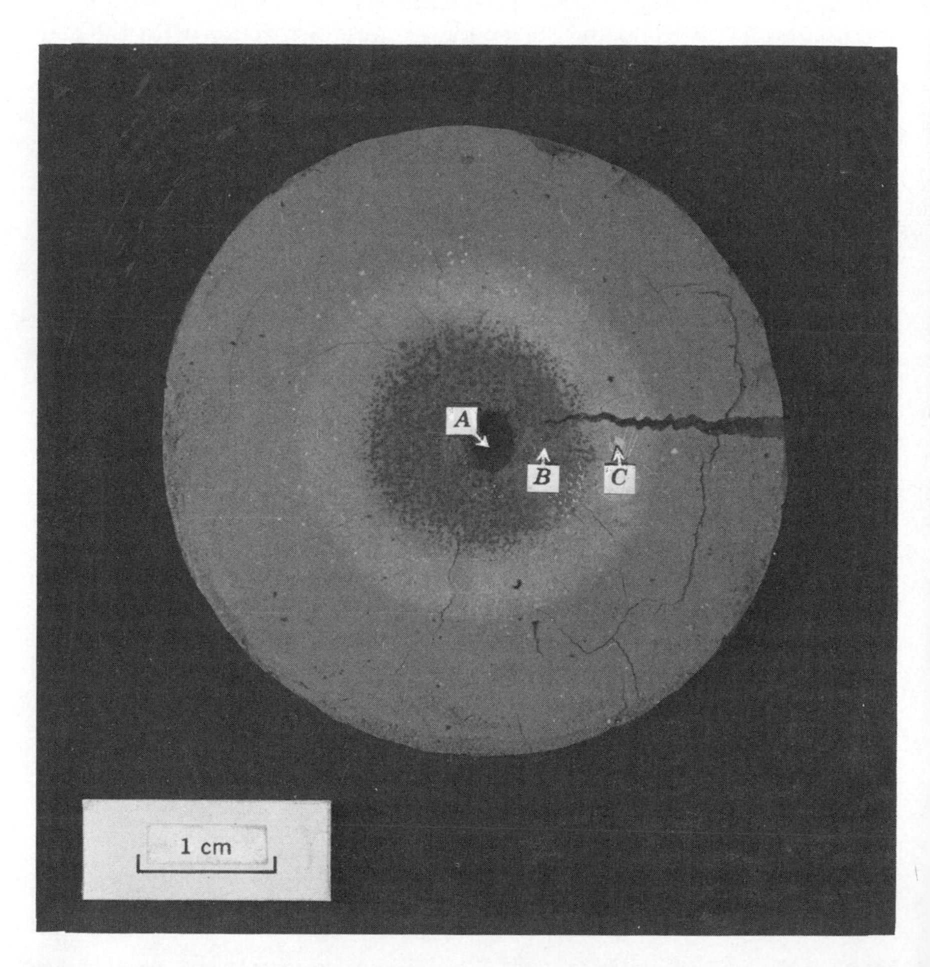

Figure 8.12. Photograph of surface of a calcareous silty clay soil three weeks after placement of 0.1 g of monobasic potassium phosphate in the opening *A* in the center. Dibasic magnesium phosphate trihydrate was identified in the dark-spotted area *B* outside the center opening, and dibasic calcium phosphate dihydrate was identified in the surrounding light-colored ring *C*. The soil was kept moist during the reaction period by addition of water from beneath. (Photograph courtesy of C. J. Racz)

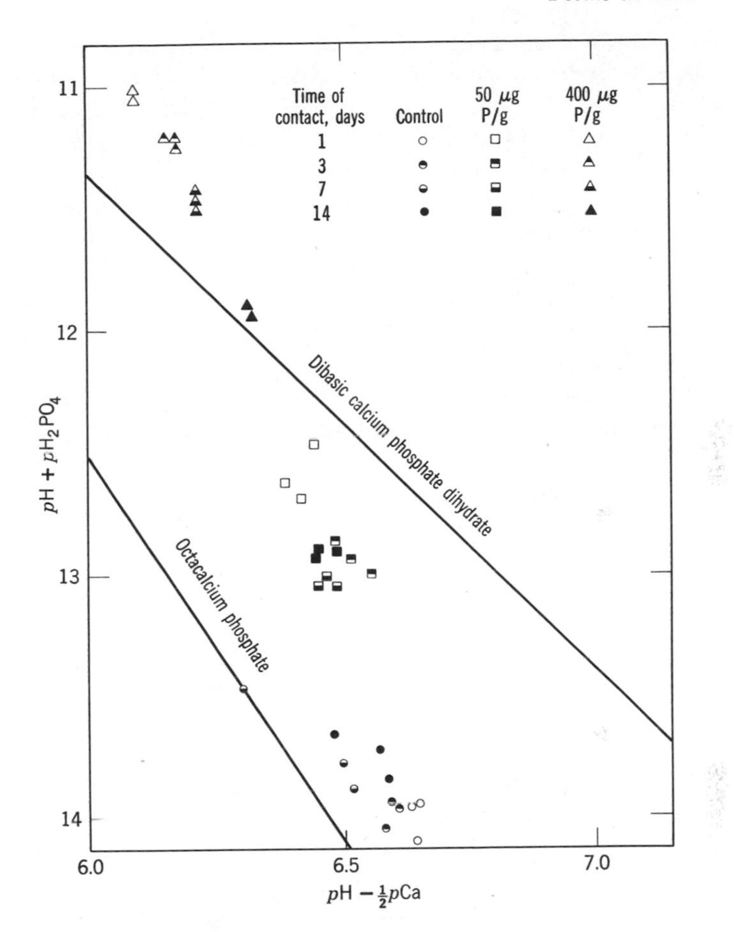

Figure 8.13. Plot of $pH + pH_2PO_4$ against $pH - \frac{1}{2}pCa$ for water extracts of calcareous soil various lengths of time after addition of different quantities of phosphorus as monobasic calcium phosphate. (Hagin and Hadas, 1962)

is a crystalline calcium phosphate. Equations for these reactions may be written as

$$Ca(H_2PO_4)_2 = CaHPO_4 + H_3PO_4,$$

$$4CaHPO_4 = Ca_4H(PO_4)_3 + H_3PO_4,$$

and

$$5Ca_4H(PO_4)_3 + 4H_2O = 2Ca_{10}(OH)_2(PO_4)_6 + 3H_3PO_4.$$

In each succeeding step, phosphoric acid is liberated; and the crystalline residue becomes more basic and less soluble. Lehr and Brown (1958) did not obtain definite evidence for the transformation from octacalcium phosphate to hydroxyapatite, but they did obtain evidence for an additional transformation of dibasic calcium phosphate dihydrate to the anhydrous form. This transformation is of some significance because the anhydrous form is less soluble than the dihydrate. In another investigation, Larsen et al. (1964) added granular dibasic calcium phosphate dihydrate to four acid and one slightly alkaline soil and after 26 months were able to recover 0 to 20% of the original addition in granules they could separate visually from the soil. X-ray examination showed only the original phosphate compound in the granules. After 37 months, no granules could be removed from any of the soils.

In work with iron phosphates that had been identified as reaction products of concentrated acidic phosphate solutions with soils, Taylor et al. (1963) found that additions of crystals of both calcium ferric phosphate [$H_4CaFe_2(PO_4)_4 \cdot 5H_2O$] and potassium ferric phosphate [$H_8KFe_3(PO_4)_6 \cdot 6H_2O$] increased the concentration of phosphorus in 0.01-molar calcium chloride extracts of an acid soil. During contact with the soil for 313 days, the calcium ferric phosphate underwent alteration to an amorphous solid, and the potassium ferric phosphate underwent alteration to a different crystalline substance that appeared to be strengite [$FePO_4 \cdot 2H_2O$ or $Fe(OH)_2H_2PO_4$].

Organic

The technical problem of identifying the organic phosphorus compounds of soils is facilitated by the fact that with care most of the organic phosphorus can be extracted from soil as organic phosphorus and not simply as inorganic orthophosphate. By suitable fractionation methods, the bulk of the organic phosphorus can be separated from the bulk of the nonphosphatic organic matter that has been extracted with it. Despite this substantial purification, however, the nature of the major part of the soil organic phosphorus still remains unknown. Barrow (1961) reviewed the literature on this subject.

Attempts have been made to determine in soils the content of various forms of organic phosphorus that are found in plants and microorganisms. Nucleic acids are major components of the organic phosphorus in plants and microorganisms, and Adams et al. (1954) investigated the presence of these substances in soil. The test used for nucleic acids was determination of the absorption of ultraviolet light in the wavelength range of 255 to 275 millimicrons by extracts of soils high in organic phosphorus. Nucleic acids absorb light in this wave band because of the nitrogen bases they contain. A test was obtained in this way for nucleic acid added to the soils, but there was no definite evidence for the presence of nucleic acid in the soils themselves. Anderson (1961) used a chemical method for determining the nitrogen bases of nucleic acids and found amounts of these bases that would account for 0.6 to 2.4% of the soil organic phosphorus. The proportions of the various bases were more characteristic of nucleic acids of microbial tissue than of plant tissue.

Phospholipids also are present in plants and microorganisms, and various investigators have examined soils for their presence. Phospholipids are fatty in nature and are conventionally extracted by use of fat solvents. Hance and Anderson (1963) found that extraction of soils with acetone, light petroleum, ethanol, benzene, methanol, and chloroform yielded values for lipid-type phosphorus accounting for 0.6 to 0.9% of the total soil organic phosphorus. Lower values were obtained by use of simpler extraction methods that had been used previously. Although the proportion of phospholipid phosphorus appears to be almost negligible, some evidence suggests that the true amounts may be greater than those found analytically. Goring and Bartholomew (1950) found that the presence of clay interfered with extraction of phospholipid phosphorus from microbial tissue. Use of solubility in fat solvents as a basis for designating a fraction of soil organic phosphorus as phospholipid is, of course, a questionable procedure in the absence of supplemental information on the nature of the phosphorus extracted. Accordingly, Hance and Anderson (1963a) hydrolyzed the extracts and examined the hydrolysates for glycerophosphate, choline, ethanolamine, serine, and inositol, which are constituents of phospholipids. The first three substances were identified, and glycerophosphate accounted for 52 to 86% of the total phosphorus in the lipid fraction. These findings verify that phospholipid phosphorus was present in the conventional lipid fraction and made up the major part of it.

The third group of organic phosphorus compounds found in soils is the inositol phosphates. Inositol is a saturated, cyclic, six-carbon alcohol with an alcoholic group on each carbon. The inositol phosphates are esters of inositol and phosphoric acid. Many different inositol phosphates are possible because there are several different isomers of inositol and because each isomer conceivably may exist with from one to six esterified phosphate groups per molecule. Four isomers of inositol have been discovered in inositol phosphates of soils (Cosgrove, 1963; Cosgrove and Tate, 1963), but only one of these has been found in the inositol phosphates of plants, an observation suggesting that the inositol phosphates of soils are not necessarily residual from plants but that they may be produced in soils by microorganisms. Further evidence of microbial synthesis of inositol phosphates in soils is provided by the finding that more inositol phosphate can be extracted from soil material after incubation with organic and inorganic nutrients than before (Caldwell and Black, 1958; Cosgrove, 1964). Anderson and Hance (1963) obtained evidence for the presence of soil inositol phosphates in a more complex substance, which was mainly carbohydrate but contained also some amino acids and some additional phosphorus of unidentified nature. Cosgrove (1963) estimated that 12 to 16% of the organic phosphorus of three soils was present as inositol phosphates containing six phosphate groups per molecule. These values compare with earlier estimates of 17% inositol hexaphosphate phosphorus, on the average, in the organic phosphorus of 49 soils (Caldwell and Black, 1958a). To judge from the work of Smith and Clark (1951), the total inositol phosphate phosphorus is at least twice the quantity present in forms with six phosphate groups per molecule.

If, then, the identified organic phosphorus is summed, one has about 2% of the total present in nucleic acids, 1% in phospholipids, and 35% in inositol phosphates, making a total of 38% accounted for. The nature of the remainder is unknown.

A number of investigations have been made of the relative amounts of organic carbon, total nitrogen, and organic phosphorus in soils. Although the findings show that the amounts of all constituents are correlated, the ratios among constituents are by no means constant among soils. Barrow (1961) summarized the results of many analyses in the form of relative weights of organic carbon, total nitrogen, and organic phosphorus, the value for nitrogen being taken in each case as 10. Average values for organic carbon ranged from 71 to 229 and those for organic phosphorus from 0.15 to 3.05 in different investigations. Soil organic matter thus seems to be more variable in respect of its ratios of carbon or nitrogen to phosphorus than its ratio of carbon to nitrogen. High ratios of organic carbon to organic phosphorus seem to occur in peats and poorly drained mineral soils and at high *p*H values in well drained mineral soils. Part of the variation is no doubt a consequence of methods of analysis. Methods for organic phosphorus are not so well developed or precise as those for organic carbon and nitrogen. To what extent the variation in ratios is associated with the nature and state of combination of the organic phosphorus compounds remains to be determined.

Availability to Plants

The forms of phosphorus that occur in soil parent materials, before growth of plants and formation of soil, are generally of low availability to plants. Each generation of plants converts soil phosphorus in quantities of the order of a few kilograms per hectare into plant phosphorus.

Perhaps one-third to two-thirds of the phosphorus in plants is inorganic, and the remainder is organic. Return of the plant residues to the soil then adds a small quantity of inorganic phosphorus, which has far greater solubility than the phosphorus in its original form in the soil material. The soluble phosphorus returned to the soil by plant residues does not revert quickly to stable crystalline forms but remains for a long time in forms of higher solubility. Hence the same slowness of approach to equilibrium that hinders identification of crystalline phosphates in soils from the composition of aqueous extracts is of benefit to plants in that it permits their use of energy from the sun to produce in soil a metastable condition of relatively high phosphate solubility that increases the availability of soil phosphorus to succeeding generations of plants.

The foregoing processes are inferred mainly from comparisons of soils with parent materials. Other indirect experimental evidence is provided from experiments by Larsen et al. (1965), who added superphosphate in quantities equivalent to 0, 245, and 490 kg of nonradioactive phosphorus per hectare to 24 sites in Britain and then followed the decline with time in the phosphorus subject to exchange with radioactive orthophosphate using a technique in which growing plants were used to measure the extent of isotopic dilution of the added $P^{32}O_4$. On 19 soils in which the time required for the increase in labile phosphorus from fertilization to decline to half its original value could be estimated satisfactorily, the "half-lives" were from 1 to 6 years, with one exception (the only peat soil tested), in which the half-life was estimated at 56 years.

Probably all forms of phosphorus in soils are of some significance in supplying phosphorus for plants on a long-term basis, as during soil development, but none of the forms that have been identified are known to be of significance in the short-term relationships that are of importance in determining the current availability of soil phosphorus. In investigations of availability of soil phosphorus to plants, the greatest value has been derived from empirical studies of the behavior of soil phosphorus. Because plants apparently do not take up the original phosphorus compounds as such, one may justify the empirical approach on the basis that the behavior of the phosphorus is of primary significance as far as absorption by plants is concerned, the forms being of importance only as they affect the behavior.

Source Used by Plants

The prevailing view is that most, if not all, of the phosphorus absorbed by plants from the soil pre-exists in the soil solution as inorganic orthophosphate. Several lines of evidence support this view:

1. No indigenous form of phosphorus other than orthophosphate has been found in soils (although traces of other forms no doubt occur there in the tissue of microorganisms).

2. Much of the phosphorus is present as inorganic orthophosphate.

3. Plants are not known to absorb phosphorus directly from the solid phase, and there is evidence that they do not absorb soil organic phosphorus present in the solution phase (Pierre and Parker, 1927).

4. The soil solution contains inorganic orthophosphate, and plants can absorb it almost quantitatively. Pierre and Parker (1927) found that corn would reduce the inorganic orthophosphate content of displaced soil solution to less than 0.007 μg of phosphorus per milliliter, and Olsen (1950) found that rye would decrease the concentration of inorganic orthophosphate in culture solution to 0.003 μg of phosphorus per milliliter. Uptake of phosphorus from even more dilute solutions has been reported, but at such low concentrations experimental difficulties become so great that it is difficult to tell whether net uptake of phosphorus is taking place.

Kinetics of Phosphate Transfer from Soil to Plant

Fried et al. (1957) represented the transfer of phosphate from soil to the interior of plants as a series of steps, which they expressed as equations:

$$P_{\text{soil}} + X_{\text{solution}} = P_{\text{solution}} + X_{\text{soil}},$$

$$P_{\text{solution}} + R = RP,$$

and

$$RP = P_{\text{inside plant}} + R'$$

where X is some anion in solution such as silicate, hydroxyl, or carbonate resulting in release of phosphate to the solution; R is the metabolically produced carrier in the root; and R' is the regenerated carrier, which may be the same as R. The general concept of the process is the same as that described in Chapter 4 for uptake of cations by plants. The first step represents release of phosphate from the soil; the second, combination of phosphate with a carrier site; and the third, release of phosphate inside the plant from the carrier-phosphate combination.

Attention will be directed here to the processes that occur at and outside of the surface at which active absorption of phosphate occurs by root cells. The absorption process itself will be dismissed with reference to the theory of Jackson et al. (1962) that phosphate uptake is effected by mitochondria (small organized bodies within the cytoplasm), which are "in direct communication" with the solution external to the cells. The carrier *R* and phosphate are joined by a high-energy bond produced by respiratory energy. Weigl (1963) found that adenosine triphosphate was the compound containing the most radioactive phosphorus after corn roots had been placed for 10 seconds in a solution containing radioactive inorganic orthophosphate. Because adenosine triphosphate is a high-energy phosphate, this finding verifies the theory of Jackson et al. and suggests that adenosine diphosphate in activated form is *R*. Further discussion of these compounds will be found in the subsequent section on the function of phosphorus in plants.

In examining the various steps in the process leading to absorption, it is convenient to proceed from the plant root back into the soil. Figure 8.14 shows the decrease in concentration of phosphorus with time in a vigorously aerated culture solution containing rye plants. Throughout

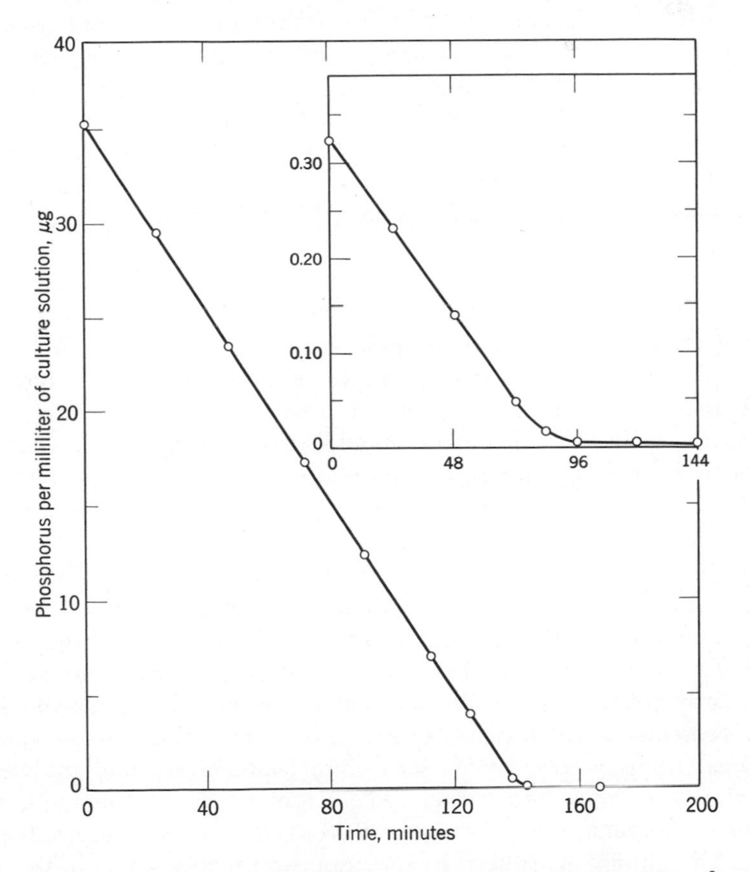

Figure 8.14. Concentration of phosphorus as orthophosphate in a vigorously aerated culture solution versus time in the presence of rye plants. (Olsen, 1950)

a wide range of concentrations from about 35 to 0.03 μg of phosphorus per milliliter, the concentration decreased linearly with time. In terms of the equations proposed by Fried et al., the linear range, in which the rate of phosphorus uptake is not affected by concentration, is interpreted as the range in which the concentration of P_{solution} is high enough to ensure that all the phosphate-specific carrier sites R are converted to the RP form. The rate of phosphorus uptake is then controlled by the third reaction, the release of phosphorus inside the plant from the carrier sites. The decrease in rate of uptake of phosphorus at external concentrations below about 0.03 μg/ml then is attributed to a decrease in the rate of formation of P_{inside} in the third reaction because of a decrease in concentration of RP. The decrease in concentration of RP in turn is attributed to a concentration of P_{solution} too low to convert all the phosphate-specific carrier sites R to RP in the second reaction.

In considering the meaning of these results in relation to subsequent discussion, it will be helpful to keep in mind that the entire root system was bathed in the solution and that the solution was intentionally agitated vigorously with the objective of keeping it constantly stirred and moving past the root, thereby eliminating the development of a concentration gradient in the solution around individual roots as a result of absorption of phosphorus by the roots. Although one cannot say whether or not the stirring of the solution was sufficient to accomplish this objective at concentrations of phosphorus below 0.03 μg/ml, the linear decrease of phosphorus concentration with time at concentrations above 0.03 μg/ml is evidence that at these higher concentrations absorption of phosphorus by the roots was not rapid enough in relation to movement of the solution to produce a depleted zone of low concentration at the root surface. Several other experimenters have obtained higher values for the critical concentration of phosphorus, but it is not clear to what extent the higher values resulted from partial depletion of phosphorus from solution around the roots.

Concentrations of inorganic phosphorus in solution in soils are low. Figure 8.15 gives a frequency distribution of concentrations of inorganic phosphorus in saturation extracts of many soils from midwestern United States. Concentrations obtained in these saturation extracts are probably not greatly different from the concentrations in solution at the lower water contents at which plants grow. In earlier analyses of soil solutions displaced from samples of 21 soils from southeastern and midwestern United States, Pierre and Parker (1927) found that the average concentration of inorganic phosphorus was 0.03 μg/ml. Concentrations of phosphorus in solution in soils therefore appear to be mostly of the same order of magnitude as Olsen's (1950) critical concentration of 0.03 μg/ml

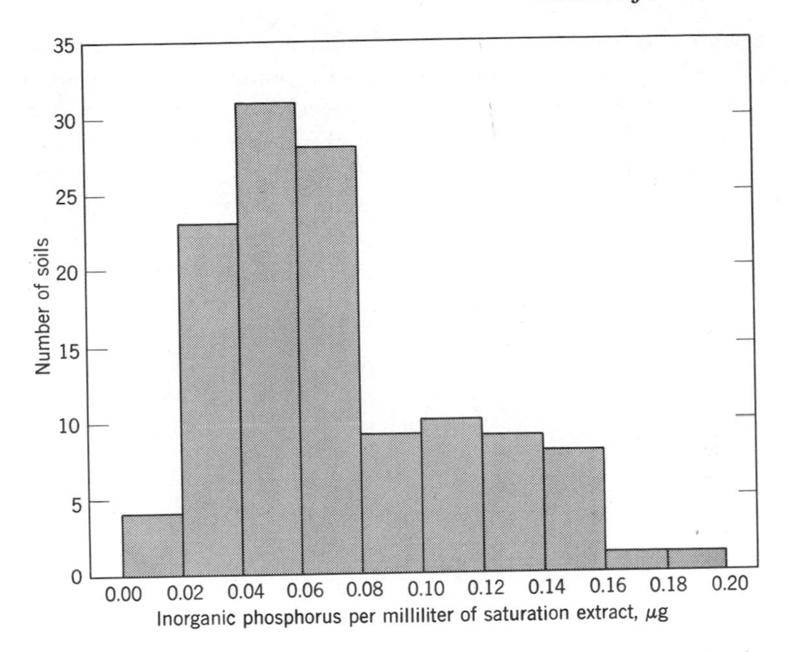

Figure 8.15. Frequency distribution of concentrations of inorganic phosphorus in the saturation extracts of samples of cultivated soils from the midwestern region of the United States. Not shown in the figure are 18 soils in which the concentrations were in the range from 0.2 to 1.2 μg of phosphorus per milliliter of saturation extract. Many of the soils had previously received phosphate fertilizers. (Barber et al., 1962)

in Fig. 8.14. Some soils are appreciably above and others appreciably below this value. Most, if not all, the high values in Fig. 8.15 are probably a consequence of recent treatment with phosphorus-bearing fertilizer.

If plants are grown on solutions displaced from soils, the inorganic phosphorus is rapidly absorbed because so little is present. Some idea of the rate may be obtained from the inset portion of Fig. 8.14 which shows that in Olsen's experiment about 25 minutes were required for depletion of the phosphorus concentration from 0.03 μg/ml to a much lower value that remained constant with time. The uptake of, say, 10 to 20 kg of phosphorus per hectare by an average crop during a single growing season requires hundreds of times as much phosphorus as can be found in solution at one time within the depth of rooting in soil with no more than 0.03 μg of phosphorus present per milliliter of soil solution. Accordingly, the first reaction, release of phosphorus from the soil to the solution, is seen to be of great importance.

Although there is no argument about the essentiality of continued release of phosphorus from soil solids to the solution, closer examination of the processes of uptake and release is needed to clarify the importance of release as a rate-limiting factor in uptake. According to the analysis of the last two steps in the phosphate uptake process, the concentration of phosphorus in solution controls the concentration of the carrier-phosphate complex and, through this, the rate of uptake. The significance of the first step in the process, release from the soil, may thus be examined in terms of its effect on the concentration of phosphorus in solution.

The net rate of release of phosphorus from the soil solids is zero when the solution has reached equilibrium with the solids. Accordingly, the rate of absorption of phosphorus by plants is infinitely greater than the net rate of release from soil at equilibrium. Absorption of phosphorus by plants thus must reduce the concentration of phosphorus in solution. The question then is at what reduction from the equilibrium concentration the rate of uptake by plants and the net rate of release from soil are equal. This question cannot be answered satisfactorily because of complications that will be discussed presently. Nevertheless, data are available that will illustrate the concept and provide a basis for discussion.

Figure 8.16 shows the rates of dissolution of inorganic phosphorus and the concentrations of phosphorus in solution when 5-g quantities of two soils moistened with 2 ml of water were subsequently shaken with 35 ml of water for different lengths of time before analysis. Shown also are numerical values for average rates of uptake of phosphorus by plants during growth on the soils in a separate experiment in which quantities of the soils were cropped intensively to sorghum. The average rate of uptake of phosphorus by plants from the high-phosphorus soil (0.039 μg/g of soil per hour) was equal to the net rate of dissolution of soil phosphorus at approximately 2 hours. At this time the concentration of phosphorus in solution was 0.42 μg/ml, a value not much below the maximum (obtained in 4 hours and not shown in the figure). Uptake of phosphorus from the low-phosphorus soil at the rate of 0.016 μg/g of soil per hour again corresponded approximately to the net rate of dissolution of soil phosphorus at 2 hours; and the concentration of phosphorus in solution at this time was 0.077 μg/ml, the maximum value obtained. To judge from these calculations, the rate of uptake of phosphorus by plants was so slow in comparison with the net rate of dissolution that dissolution of phosphorus from the soil solids was sufficient to maintain the concentration of phosphorus in solution almost at the maximum or equilibrium value.

The analysis in the preceding paragraph treats the soil-plant relationships as if the soil were a stirred nutrient solution in which rapid mass movement maintains the concentration of the solution uniform throughout. The circumstances thus are quite unrealistic. Under natural conditions, the roots are extended slowly through the soil, and only a little of the soil is contacted directly. Water, which serves as the pathway for movement, ordinarily occupies less than one-third of the soil volume and does not cover the roots completely (although the cell walls no doubt remain saturated with water). The principal mass movement of water significant in phosphorus transport is the movement to the roots of the water absorbed by the plants and lost in transpiration. Even if plants absorbed all the phosphorus moved to the roots in this way, however, the amount of phosphorus supplied would satisfy only a minor part of the needs because of the extremely low concentration of phosphorus in soil solutions.

The situation thus is more critical than would appear from a compari-

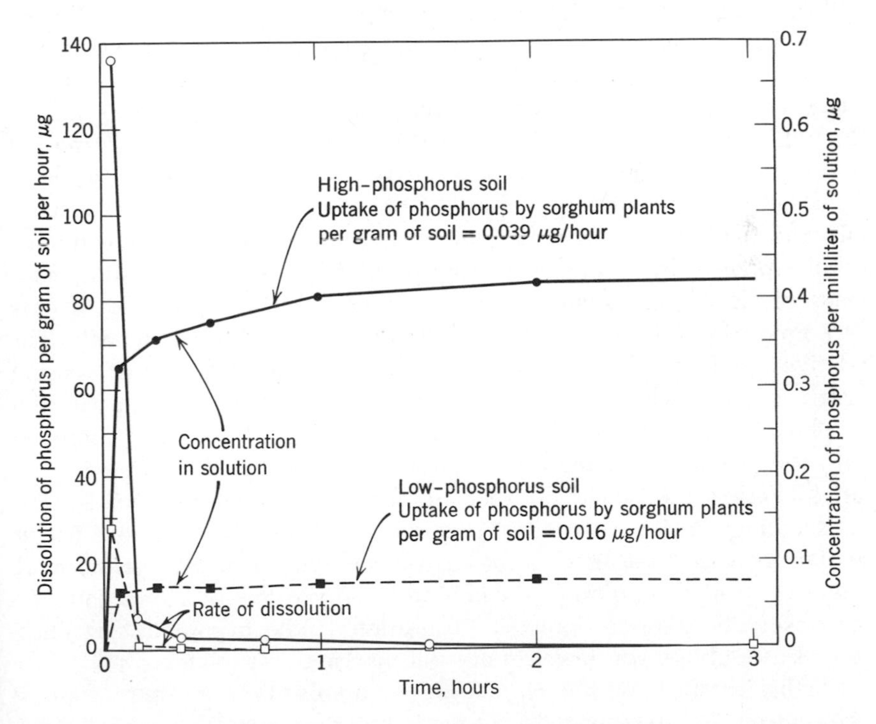

Figure 8.16. Concentration of inorganic phosphorus in solution and rate of dissolution of inorganic phosphorus from two soils suspended in water for different lengths of time together with rate of uptake of phosphorus by sorghum grown on the soils in the greenhouse. (Moser, 1956; Thompson, 1958)

son of Figs. 8.14 and 8.15 and from the analysis made in connection with Fig. 8.16. With the soil-plant relationships as they are, the phosphorus removed by roots from the adjacent soil solution must be replaced by phosphorus dissolved from the soil. As removal proceeds, the concentration of phosphorus in solution adjacent to the roots must decrease because of depletion of part of the total supply of labile phosphorus (even if the phosphorus in solution is continuously almost at equilibrium with that in the soil solids); and diffusion from more distant soil must then replace part of the phosphorus removed.

No way has been found to make direct measurements of the phosphorus concentration in solution adjacent to plant roots in soil. The concentration of phosphorus in 0.01-molar calcium chloride extracts of bulk samples of soil, however, was found by Mattingly (1958) to decrease as a result of growth of plants; this suggests that a decrease in concentration in the solution adjacent to the roots did occur as a result of uptake of phosphorus by the plants.

Where high concentrations of fertilizer phosphorus were placed locally in the soil in the absence of plants, Heslep and Black (1954) found that movement of the phosphorus through the soil increased with the water content of the soil, which verifies that the phosphorus moved through the water. Olsen et al. (1961) investigated the significance of water content of soil in the process of transfer of phosphorus to plants from soil having a uniform supply of phosphorus throughout. To maintain the matric suction in the soil water fairly constant while uptake of phosphorus by plants was occurring, they used germinating corn seeds in closed but aerated containers of soil for a test period of only 24 hours. Figure 8.17 gives the results of their work on four soils. The relative uptake of phosphorus from these soils increased with decreasing matric suction, which is in agreement with the view that the transfer of phosphorus from soil to root occurs through water and increases with the proportion of the cross-sectional area occupied by water. Similarly, Wesley (1965) found that uptake of phosphorus from soil by oat seedlings during a short test period at approximately constant matric suction increased with decreasing matric suction. This experimental work under controlled conditions provides an improved theoretical explanation for observations under practical conditions indicating that soil phosphorus availability was lower in dry soil than in moist soil.

Further analysis of the circumstances involved in transfer of phosphorus from soil to plant root was made by Olsen and Watanabe (1963), who pointed out that soils of coarse texture would contain lower volume percentages of water than would soils of fine texture at the same matric suction and hence that coarseness of texture should reduce the rate

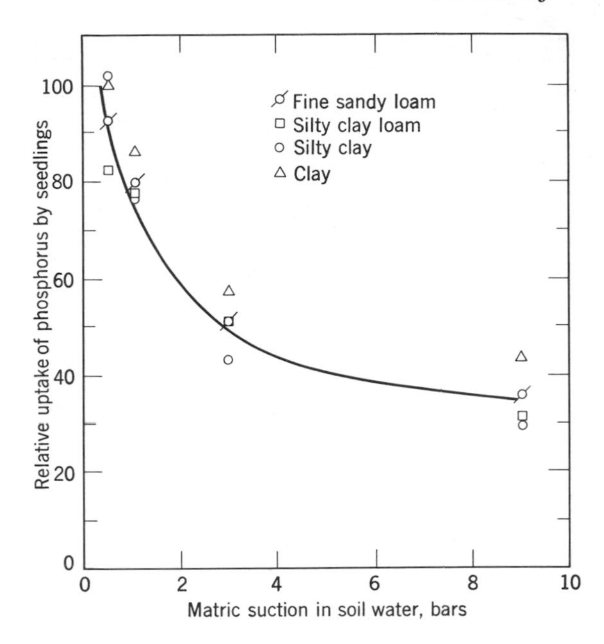

Figure 8.17. Relative uptake of phosphorus from soils by germinating corn seedlings versus matric suction in soil water during a 24-hour period in closed but aerated containers. Uptake at a matric suction of 0.33 bar is given a value of 100 in each soil. Each value shown is an average of results obtained with three levels of soil phosphorus. (Olsen et al., 1961)

of diffusion of phosphorus through the soil to the root. Moreover, at a given concentration of phosphorus in the soil solution, the quantity of labile phosphorus in the soil solids is less in soils of coarse than of fine texture (see Fig. 8.18). The importance of the labile phosphorus (which they defined as the phosphorus that would equilibrate readily with radioactive orthophosphate added in solution) is presumably that, being nearly in equilibrium with the phosphorus in solution, it serves as the primary source from which phosphorus is released upon absorption of phosphorus from the soil solution by plants. The diffusion effect related to the water content of the soil and the phosphorus-buffering effect related to the labile phosphorus in the soil act in the same direction to produce more effective transfer of phosphorus from soil solids to plant roots in soil of fine texture than of coarse texture. Removal of a given quantity of soil phosphorus by unit area of root surface should produce less decrease in phosphorus concentration at the root surface in soil of fine texture than of coarse texture because of the greater

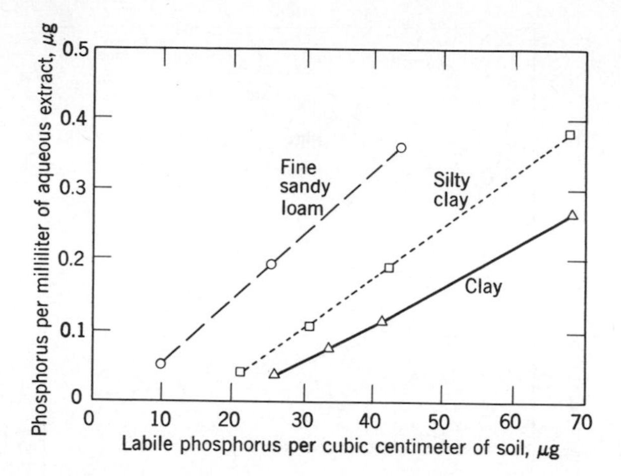

Figure 8.18. Concentration of phosphorus in an extract of one part of soil with ten parts of water by weight versus labile phosphorus found in three soils by the method of isotopic dilution of radioactive orthophosphate added in solution. The different levels of phosphorus were obtained by equilibrating the soils with concentrated superphosphate. (Olsen and Watanabe, 1963)

rate at which phosphorus may be moved to the root by diffusion from soil some distance away and because of the greater capacity of the soil adjacent to the root surface to supply phosphorus by dissolution and hence to resist a decrease in concentration. The authors obtained verification of the validity of their analysis by experiments on uptake of phosphorus by plants from three soils differing in texture, each adjusted to different levels of phosphorus by additions of soluble phosphorus and equilibration prior to the experiment. Two experiments were done, one with corn seedlings in which the uptake period was only 24 hours and the other with barley in which the uptake period was 60 days. The results, given in Fig. 8.19, show that uptake of phosphorus increased with fineness of texture in the 24-hour experiment as well as the 60-day experiment.

Stated another way, the results of the experiments in Fig. 8.19 indicate that a given concentration of phosphorus in solution does not have the same meaning to plants in different soils. Greater uptake is associated with a given concentration in soil of fine texture than of coarse texture.

Comparison of the results in Fig. 8.19 with those in Fig. 8.14 suggests that the critical concentration of phosphorus associated with the maximum rate of uptake is far greater in soil than in culture solution. The discrepancy is too great to be accounted for by the reduced cross-sectional area of water available for diffusion unless there is some discontinuity in this effect at the soil-root boundary. Local depletion of phosphorus at the root surface in soil also does not appear sufficient to account for the discrepancy, as indicated by calculations by the authors. One other factor of possible significance is interionic effects. According to a theory discussed in Chapter 4 on exchangeable bases, the product of the activity of any cation and of any anion in the ion atmosphere around soil particles, each activity being raised to a power equal to the reciprocal of the valence of the ion, is equal to the corresponding activity product in the bulk solution. For example,

$$(a_{Ca_i^{++}})^{1/2}(a_{H_2PO_{4i}^-}) = (a_{Ca_b^{++}})^{1/2}(a_{H_2PO_{4b}^-}),$$

where the subscripts i and b refer to the solution in the ionic atmospheres and the bulk solution. If there is some mechanism that keeps the activities constant in the ion atmospheres, the product of the ion activities in the bulk solution should remain constant despite differences in dilution and in concentration of electrolytes. In the case of an anion such as chloride, there is no mechanism to keep the activity constant in the ion atmospheres; but in the case of phosphate the labile phosphorus in the soil may serve the purpose. In any event, work of Larsen and Widdowson (1964) and Clark and Peech (1960) suggests that, under some conditions or in certain soils, the product $(a_{Ca^{++}})^{1/2}(a_{H_2PO_4^-})$ measured in solution is essentially constant at different ratios of 0.01-molar calcium chloride to soil and in the presence of different concentrations of salts. From these observations and certain other analyses on aqueous extracts of soils (e.g., Burd, 1948), one may suppose that the increase in concentration of calcium in solution that occurs at successively lower ratios of solution to soil is accompanied by a decrease in concentration of phosphorus. The phosphorus concentration values in Fig. 8.19 were measured in an aqueous extract of ten parts of water to one of soil by weight. The normal soil solution corresponds to only 1 to 3% of this amount of water. Hence the phosphorus concentration in the normal soil solution was probably somewhat less than that in the aqueous extracts, which would have the effect of decreasing the discrepancy between Figs. 8.14 and 8.19.

Up to this point, the discussion in this section has been presented from the point of view that the relation between inorganic soil phosphates and plant roots is a passive one in which roots absorb phosphate ions from the soil solution, and, in response, phosphate ions are released from the soil solids to the solution. By absorbing some substances and exuding others, plants may modify in a more active manner the ionic environment that affects phosphorus solubility.

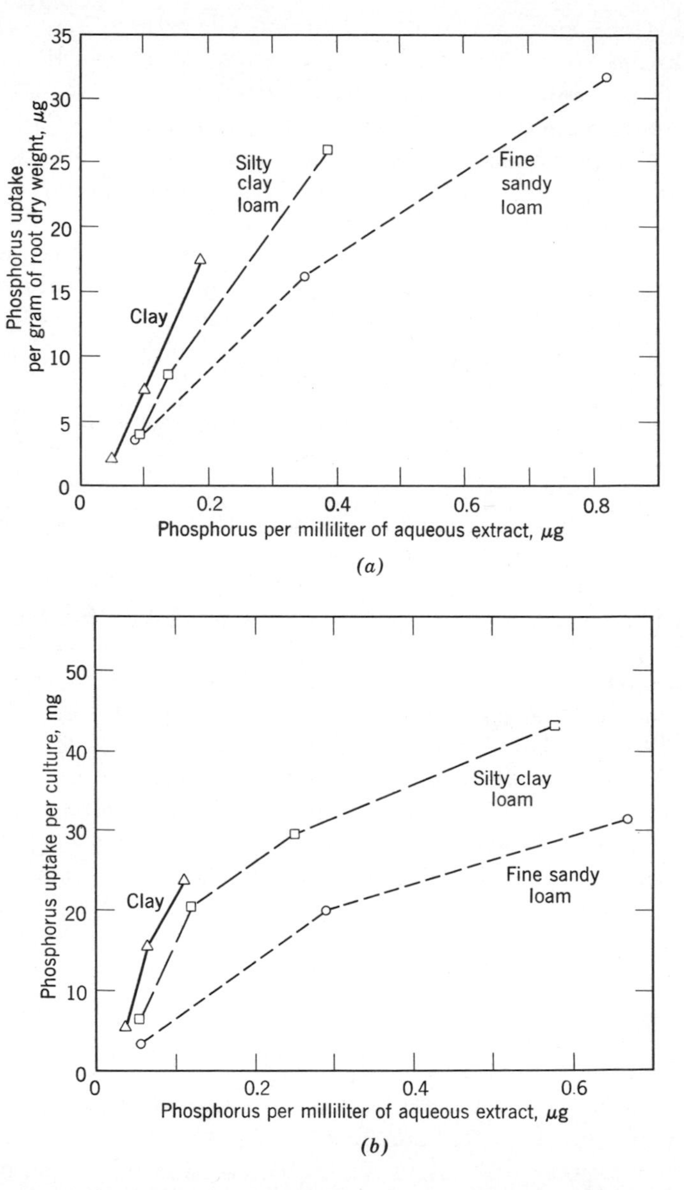

Figure 8.19. Uptake of phosphorus (*a*) by corn seedlings in 1 day and (*b*) by barley plants in 60 days from three soils versus concentration of phosphorus in an extract of one part of soil with ten parts of water by weight. The water content of each soil was maintained at a matric suction of 0.3 bar in the experiment with corn seedlings and was adjusted periodically to this value by watering in the experiment with barley. The soils are the same as those employed to obtain the data in Fig. 8.18. (Olsen and Watanabe, 1963)

The origin of the concept that plants actively influence the external ionic environment that modifies phosphorus solubility in soil is probably the classical experiments demonstrating the etching of polished marble (calcium carbonate) in contact with roots, presumably as a result of the action of hydrogen ions in carbonic acid exuded by the roots. In experimental work on calcium phosphates, Gerretsen (1948) observed no marked or consistent effect of roots on *p*H but did note greater uptake of phosphorus by plants from sand culture containing dibasic calcium phosphate (or other calcium phosphates with lower solubility) in the presence of microorganisms than in their absence. When he coated glass plates with a thin layer of agar containing precipitated dibasic calcium phosphate and placed them at an angle in the bottom of sand cultures, he found that clear zones due to dissolution of the calcium phosphate developed at random locations under the small roots of rape plants in nonsterile cultures but not in sterile cultures; with oats as test crop, the results were similar except that only a few solubilization spots could be found. By use of a microelectrode for determining *p*H values, he found that the clear zone around a colony of phosphate-dissolving bacteria on nutrient agar containing precipitated dibasic calcium phosphate was more acid (*p*H 4.5–4.8) than the surrounding unaffected portion of the agar (*p*H 5.5). Gerretsen interpreted these observations to mean that the roots themselves did not have much effect in dissolving the calcium phosphate but that the principal effect was due to colonies of microorganisms developing on organic materials exuded by the roots, greater exudation occurring from roots of rape than of oats. Gerretsen's views thus represent a substantial modification of the original carbon dioxide theory. In further work, Louw and Webley (1959) found that greater numbers of calcium phosphate-dissolving bacteria were isolated from cropped soil under oats than from uncropped soil. Most of these bacteria released lactic acid to the medium when supplied with glucose (Louw and Webley, 1959a). A few released 2-ketogluconic acid (Duff and Webley, 1959), which was relatively effective in dissolving calcium phosphates. The question now to be resolved is the extent to which the organisms effective in dissolving calcium phosphates in sand cultures and synthetic bacteriological media act in this capacity in soil where competition from other organisms is greater and buffering properties inhibit acidification.

An active effect of plants on another constituent of the ionic environment was the basis of Truog's (1916) theory that the difference in availability to various plants of the phosphorus of phosphate rock is caused by differences in calcium absorption by the plants. (Most phosphate rock is an impure form of apatite, a calcium phosphate found in soils;

hence the theory may be considered in reference to absorption of phosphorus by plants from soils containing apatite and other calcium phosphates.) Some plants absorb more calcium than others. Because the phosphorus concentration in solution should increase as the calcium concentration decreases, on the basis of the solubility-product principle, the concentration of phosphorus in solution should be greatest in the presence of plants that absorb the most calcium. In accord with the theory, sweet clover, for example, absorbs much calcium and grows relatively well with phosphate rock as a phosphorus source. Corn absorbs little calcium and grows relatively poorly with phosphate rock as a source of phosphorus.

The emphasis on absorption of calcium by plants in the Truog theory is incidental as far as the solubility-product principle is concerned. Uptake of calcium by plants is simply a way by which the activity of calcium in solution is reduced, and it has little or no bearing on uptake of phosphorus by plants. Leggett et al. (1965) found that the uptake of phosphorus by detached roots of barley increased with calcium concentration up to about 0.001-molar (20 μg/ml), which is in the deficiency range, but then was unaffected by a fivefold increase to 0.005-molar. Wild (1964) found that the yield and phosphorus content of plants were related to the phosphorus concentration in the nutrient solutions on which the plants were grown, and the calcium concentrations had no effect over a fivefold range. Hence, although the phosphate concentration may be influenced by the calcium concentration, and although the "phosphate potential" or ion-activity product, $(a_{Ca^{++}})^{1/2}(a_{H_2PO_4^-})$, which figures prominently in recent literature, may provide a convenient way of characterizing this relationship for individual soils, one must bear in mind that plants respond to the phosphorus and not to the ion-activity product. A given ion-activity product could represent, for example, 100 units of calcium and 1 of phosphorus or 1 unit of calcium and 100 of phosphorus.

Truog's (1916) theory of the active effect of plants in increasing phosphorus solubility by removing calcium from the soil solution involves uptake of calcium by the plant as a whole. Drake and Steckel (1955), on the other hand, proposed that the calcium is inactivated by the cation-exchange sites of the roots. According to their theory, availability of the phosphorus of phosphate rock should be greater to plants having roots with a high cation-exchange capacity than with a low cation-exchange capacity. Fox and Kacar (1964) tested the cation-exchange-capacity theory in an experiment in which seven plant species were grown in a calcareous soil with and without heavy phosphorus fertilization. The yield of the controls was then divided by the yield of the treated cultures to obtain a number reflecting the degree of sufficiency of the soil phosphorus for each species. This yield ratio was found to increase with the cation-exchange capacity of the roots, as would be expected from the theory. The range was from a yield ratio of 0.38 with wheat (cation-exchange capacity of roots 9 m.e./100 g) to 0.67 with sainfoin (cation-exchange capacity of roots 77 m.e./100 g).

At the same time as a high cation-exchange capacity of roots should theoretically increase the solubility of calcium phosphates by binding some of the calcium, it should theoretically cause exclusion of anions from the ion atmospheres around the roots and hence should reduce phosphate uptake. Elgabaly (1962) used this theory to explain his findings that chloride uptake by several plant species decreased with an increase in cation-exchange capacity of the roots. The relative importance of these two opposing effects on phosphate uptake is not known.

Another active but indirect effect of plants on inactivation of calcium in the ionic environment was postulated by Duff and Webley (1959), who attributed the special value of 2-ketogluconic acid for dissolving calcium phosphates to its tendency to form a soluble calcium salt having a low degree of dissociation. Plants presumably bring about this effect by exuding organic substances from the roots, which then are used by microorganisms, the microorganisms in turn producing the 2-ketogluconic acid.

Inorganic Phosphorus Evaluation

An important practical problem in soil-testing work is obtaining an index of the availability or effective quantity of phosphorus in soils. Many soils are deficient in phosphorus for production of economic crops, and consequently there is a need for evaluation of the phosphorus status of soils and interpretation of the results in terms of quantities of fertilizer phosphorus needed. This section will be concerned only with the problem of evaluation, which will be considered in relation to the subject matter covered previously on transfer of phosphorus from soil to plant.

In practical work in soil testing, the usual, if not invariable, method is that of treating a sample of soil with some extractant and measuring the quantity of phosphorus extracted. Such methods have in common the objective of separating soil inorganic phosphorus into two or more fractions on the basis of reactivity under specified conditions. Only the reactive fraction is measured. In theoretical discussions of phosphorus availability or phosphorus evaluation, on the other hand, frequent reference is made to the "intensity" and "capacity" or "quantity" aspects of soil phosphorus supply [see, for example, a paper by Mattingly (1965)]. The intensity aspect is usually identified with the concentration of phosphorus in solution and the quantity aspect with some quantity of labile phosphorus that includes the phosphorus in solution and a portion of the phosphorus in the soil solids. In terms of the carrier theory of ion uptake, the concentration of phosphorus in the external solution determines the proportion of the phosphorus-specific carrier sites present in the carrier-phosphate form and hence affects the rate of uptake of phosphorus by plants. Ideally, the quantity factor determines the quantity of phosphorus that can be released from soil between specified equilibrium concentrations.

In a group of similar soils, the intensity and quantity aspects of phosphorus supply are closely related; hence, in principle, either type of evaluation might be used equally well. In a group of dissimilar soils, however, the intensity and quantity aspects usually are not closely related; and it then becomes necessary to reach some decision as to which type of measurement is appropriate for the purpose at hand. Figure 8.20 shows the results of an experiment in which concentrations of phosphorus in aqueous 0.01-molar calcium chloride extracts may be used

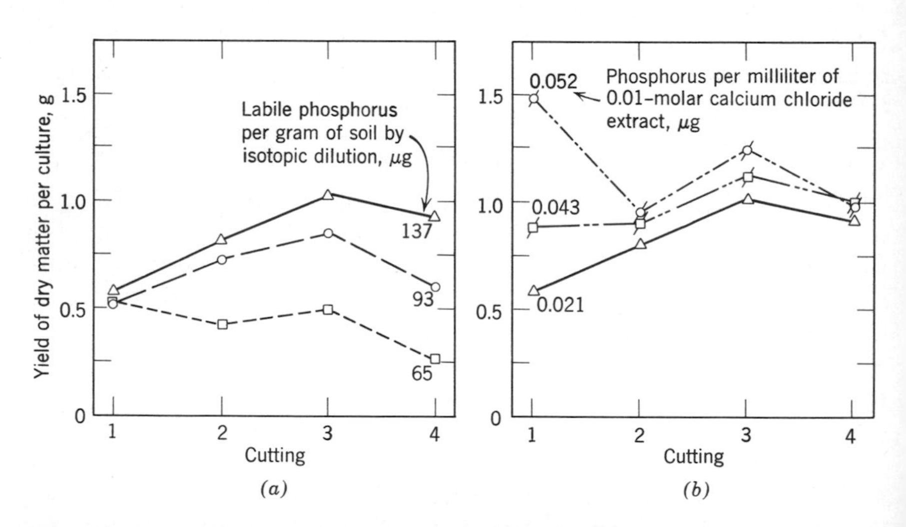

Figure 8.20. Yields of four successive cuttings of ryegrass obtained after 41, 64, 90, and 113 days on five calcareous soils classified into two groups according to phosphorus status. (*a*) Three soils with an initial phosphorus concentration of 0.022 to 0.025 μg of phosphorus per milliliter of aqueous 0.01-molar calcium chloride extract but with different initial values for labile phosphorus by the method of isotopic dilution. Values of the latter are shown in the figure. (*b*) Three soils with initial values for labile phosphorus equal to 137 to 139 μg/g but with different initial phosphorus concentrations in the calcium chloride extracts. Values of the latter are shown in the figure. (Mattingly et al., 1960)

as an index of the intensity factor and in which the quantities of labile phosphorus found by the method of isotopic dilution may be used as an index of the quantity factor of phosphorus supply. The figure shows yields of ryegrass in four successive cuttings in the greenhouse under conditions that were uniform for all the soils tested. Figure 8.20*a* shows yields on three soils that gave essentially the same concentration of phosphorus in 0.01-molar calcium chloride extracts but contained different quantities of labile phosphorus. Yields of the first cutting on the different soils were similar, but yields of succeeding cuttings diverged more and more and increased with the quantity of labile phosphorus. Part *b* of the figure shows yields on three soils that gave different concentrations of phosphorus in 0.01-molar calcium chloride extracts but contained essentially equal quantities of labile phosphorus. The increase in yield with the concentration of phosphorus in the 0.01-molar calcium chloride extracts was relatively large in the first cutting but small in succeeding cuttings. In this experiment, the significance of the intensity factor evidently decreased and that of the quantity factor increased with an increase in the amount of growth or the amount of phosphorus removed from the soil by plants. The results of this experiment provide some justification for the usual emphasis on the quantity factor in practical testing of diverse soils. Further evidence of similar nature was published by Mattingly et al. (1963). The explanation for the relatively poor showing of the calcium chloride extracts as phosphorus uptake continued was, of course, that the concentration changed, and it changed to different extents in different soils. No data on changes in concentration of phosphorus in the calcium chloride extracts were included in the papers just cited, but such data were published by Mattingly (1958) from similar experiments.

The most satisfactory laboratory method for providing an index of uptake of soil phosphorus by plants would evidently be one for which the relationship between yield of phosphorus in the plants and the soil phosphorus measured by the method is the same for all soils under similar conditions, as when plants are grown on quantities of the different soils in a greenhouse. Indications are, however, that no such method has been developed or in fact is likely to be developed.

The general method used to remove from soil enough phosphorus to obtain an index of the quantity factor in soil phosphorus supply, without simultaneously employing a large volume of solution in which the concentration of phosphorus would be too low for convenience in measurement, has been to treat samples of soil with a solution containing a chemical reagent that increases the solubility of the phosphorus. The extractants most commonly employed are probably the 0.025-normal hy-

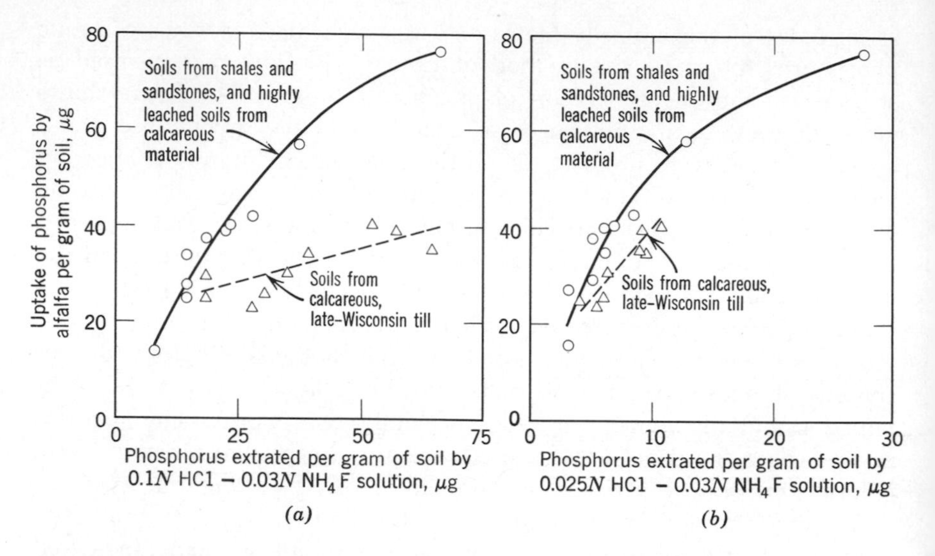

Figure 8.21. Phosphorus removed from two groups of Ohio soils by alfalfa versus (*a*) phosphorus extracted by 0.1*N* HCl – 0.03*N* NH_4F solution and (*b*) phosphorus extracted by 0.025*N* HCl – 0.03*N* NH_4F solution. (Thompson and Pratt, 1954)

drochloric acid, 0.03-normal ammonium fluoride extractant proposed by Bray and Kurtz (1945), and the 0.5-molar sodium bicarbonate extractant proposed by Olsen et al. (1954). Most such extractants are capable of providing a good index of uptake of soil phosphorus by plants if the soils concerned are similar enough. In a group of dissimilar soils, however, the significance of a given figure for phosphorus obtained in the laboratory usually depends on the nature of the soil. The situation is illustrated by Fig. 8.21, in which the phosphorus absorbed by alfalfa from two groups of soils in the greenhouse is plotted against the phosphorus extracted from the soils by two different reagents. Ideally, a single curve should represent the results for the two groups of soils, but with both extracting reagents the data for the two groups of soils evidently follow different trends.

From the standpoint of action of the extracting solution, the failure of a given value for extractable soil phosphorus to have the same significance to plants grown on different soils may be attributed to two general causes. First, the proportionate quantities of phosphorus extracted from different sources in the soil by the solution are not the same as those extracted by plants. In Fig. 8.21, for example, the divergence between the two groups of soils was much greater with the 0.1*N* acid extractant (*a*) than with the 0.025*N* acid extractant (*b*), presumably because the soils derived from calcareous, late-Wisconsin till contained much more acid-soluble phosphorus of low availability to plants than did the soils from shales and sandstones. The 0.1*N* acid extractant removed more of this phosphorus than did the 0.025*N* acid extractant (note the difference in scale). In some instances, part of the phosphorus dissolved by the reagent is reprecipitated in the soil before the phosphorus measurement is made. The sharp dip between *p*H 2 and 3 in the solubility versus *p*H curve for the soil in Fig. 8.6 probably can be accounted for in this way.

A second reason for the failure of a given value for extractable phosphorus to have the same significance in different soils is that the extractant may not act with the same intensity on the phosphorus of different soils. Extracting reagents are not specific for soil phosphorus, and they interact with various soil constituents other than phosphorus. Consequently, the intensity of action of the reagent on the phosphorus depends on the extent to which the reagent interacts with other constituents in the soil, which will depend on the nature of the soil. An example is given in Fig. 8.22. This figure shows a plot of the ratio of the phosphorus extracted from each of a group of acid soils from California

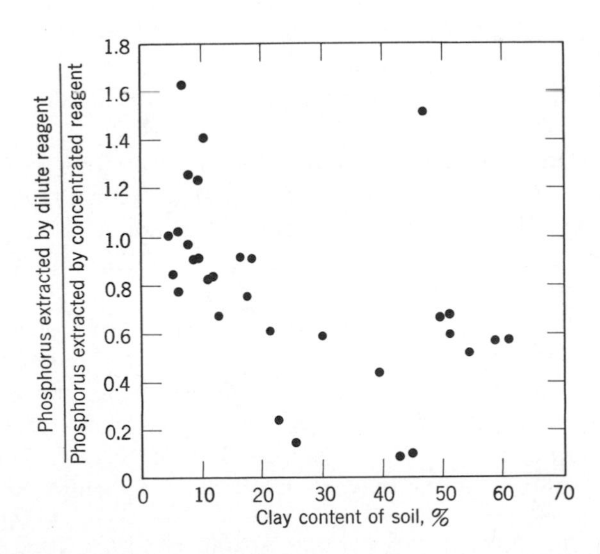

Figure 8.22. Ratio of phosphorus extracted by a dilute reagent (0.03-normal ammonium fluoride in 0.025-normal hydrochloric acid) to phosphorus extracted by a concentrated reagent (1-normal ammonium chloride followed by 0.5-normal ammonium fluoride) versus clay content of a group of acid soils from California. (Pratt and Garber, 1964)

by a dilute reagent (0.03-normal ammonium fluoride in 0.025-normal hydrochloric acid) to the phosphorus extracted by a concentrated reagent (1-normal ammonium chloride followed by 0.5-normal ammonium fluoride) against the clay content of the soil. The dilute reagent is popular for soil testing purposes, and the concentrated one is used to estimate the sum of water-soluble and aluminum phosphate in soils. As a result of differences in concentration of reagents and in ratio of solution to soil, one-hundred and nineteen times as much ammonium fluoride was used per gram of soil in the second reagent as in the first. Therefore depletion of the fluoride in the extractant by reaction with soil constituents other than phosphorus would be expected to have a greater effect on phosphorus extracted by the dilute reagent than by the concentrated reagent. The downward trend of the ratio in Fig. 8.22 with an increase in clay content of the soil (clay content provides an index of reactive substances in the soils) represents the behavior to be expected from exhaustion of the dilute reagent.

Acids are often used as extractants for soil phosphorus. If the acid is dilute and unbuffered, the final *p*H of the solution will increase with the buffering capacity of the soil against acid. Consequently, the intensity of action of the reagent on the soil will decrease with an increase in the buffering capacity of the soil against acid. If the extractant is strongly buffered, the final *p*H of the solution will be about the same as the initial *p*H. The intensity of action of the reagent on the soil will then increase with the difference between the initial *p*H of the soil and the *p*H of the reagent.

Because of the problems of interaction of extracting reagents with soils, attention has been directed to development of methods that overcome such objections. Two that involve extraction of soil phosphorus are treatment of soil with a large excess of water and treatment of soil with a strong-base-type anion-exchange resin. Extraction with water gives good results but is technically troublesome because of the difficulty of determining the extremely low concentrations of phosphorus in solution and the difficulty of removing soil particles from the extract. The colorimetric methods employed for measuring phosphorus in soil extracts involve development of a blue color in a strongly acid solution, and release of phosphorus by the reagents from a few suspended soil particles could lead to relatively large errors. Strong-base-type anion-exchange resins have little effect on soil *p*H and remove phosphorus from solution by an ion-exchange process. The resins usually are initially in the chloride form, so that chloride is the ion released to the solution. Anions other than phosphate are taken up by the resin; but the resin is added in large excess, so that soil anions other than phosphate do not interfere.

After the equilibration process, the resin is separated from the soil and treated with an electrolyte to replace the phosphate ions, which are then determined in the solution. The resin is usually added in the form of small beads, and separation is done by sieving. Hence the soil must contain no particles as large as the resin beads. In a different technique the resin is added in the form of a sheet or membrane, which facilitates separation from the soil. Although the anion-exchange-resin method is capable of good results, its practical use is discouraged by the tedious nature of the operations required.

Another method that has been widely investigated is determination of the labile phosphorus by isotopic dilution. This method does not involve extraction. Although reference to this method has been made on several occasions in this chapter, the process was not explained. Usually a sample of soil is first equilibrated with water. Then a trace of orthophosphate tagged with P^{32} is added. The tagged orthophosphate ions intermingle with the orthophosphate ions in solution and undergo exchange with the orthophosphate in the soil solids. Thus the radioactivity of the soil solids increases with time, and that of the solution decreases. In contrast to the analogous process described for calcium in Chapter 4, however, an apparent equilibrium is not reached quickly. Although the initial exchange occurs rapidly, the process continues at reduced rates for days. The distinction between exchangeable and nonexchangeable is thus by no means clear; rather, there is a gradation from phosphate ions that are readily exchangeable to those that have low exchangeability. The usual procedure is to analyze the solution for inorganic phosphorus and radioactivity at an arbitrary time and to calculate a value for the apparent amount of soil phosphorus that has equilibrated with the added radioactive orthophosphate on the assumption that the ratio of radioactive to nonradioactive inorganic phosphorus in solution is equal to the ratio of radioactive to nonradioactive inorganic phosphorus in a portion of the solid-phase inorganic phosphorus that has equilibrated completely with the radioactive phosphorus. This method is an adaptation of the general method of isotopic dilution, and values obtained by its use are frequently described as measurements of labile phosphorus by isotopic dilution or, more simply, as measurements of labile phosphorus.

Values of labile phosphorus usually provide a good index of the capacity of soils to supply phosphorus to plants. If widely different soils are concerned, however, the quantities of labile phosphorus found may reflect inadequately the relative uptake of phosphorus by plants, presumably because rather different concentrations of phosphorus may be maintained in solution by equal quantities of labile phosphorus (see Fig.

8.18 for an example). From the standpoint of practical use, the method has the disadvantages of involving tedious operations, difficulty in measurement of low concentrations of phosphorus, and special precautions in handling the radioactive phosphorus. A review on the isotopic dilution method was published by Machold (1964).

In none of the methods is any overt allowance made for the effect of soil texture noted in an earlier section; nevertheless, some allowance for this effect should be made automatically. With a given amount of extractable or labile phosphorus in soils of coarse and fine texture, the higher concentration of phosphorus would generally be expected in solution in those of coarse texture, and this would tend to increase the rate of uptake of phosphorus by plants on these soils; however, the lower water content would tend to reduce the rates of transfer and uptake. The compensating effect of these two factors may explain why soil texture is not recognized as an important factor in interpreting the results of soil analyses for phosphorus.

Another factor for which no allowance is made in measurements of extractable or labile soil phosphorus is the *p*H effect. Even with the anion-exchange-resin method, the results of which are not biased by interaction between extractant and nonphosphatic constituents of soils, van Diest et al. (1960) obtained a different relationship between yield of phosphorus in plants and phosphorus extracted from alkaline soils than from acid soils. They found they could make allowance for this *p*H effect by an adaptation of a theory about uptake of phosphorus by plants as $H_2PO_4^-$ and HPO_4^{--} that will be discussed subsequently in the section on effect of soil *p*H. In other words, their allowance for the *p*H effect was based on the variation in plant response to a given concentration of phosphorus that depends on the *p*H.

Organic Phosphorus

The organic phosphorus of soil is a part of the organic matter and tends to follow the pattern of accumulation and loss of organic matter as a whole, although there is no evidence for a fixed ratio of gains or losses of organic phosphorus to those of other organic constituents. Figure 8.23 illustrates an instance in which organic phosphorus increased with the content of organic carbon, nitrogen, and sulfur during a period of years following establishment of a permanent pasture.

Almost all the carbon and nitrogen and a part of the sulfur in soil in organic form are derived from atmospheric sources. The organic phosphorus, however, is derived almost entirely from inorganic phosphorus present in the soil and supplied in the fertilizer. Because plants are not known to absorb organic phosphorus compounds from the natural

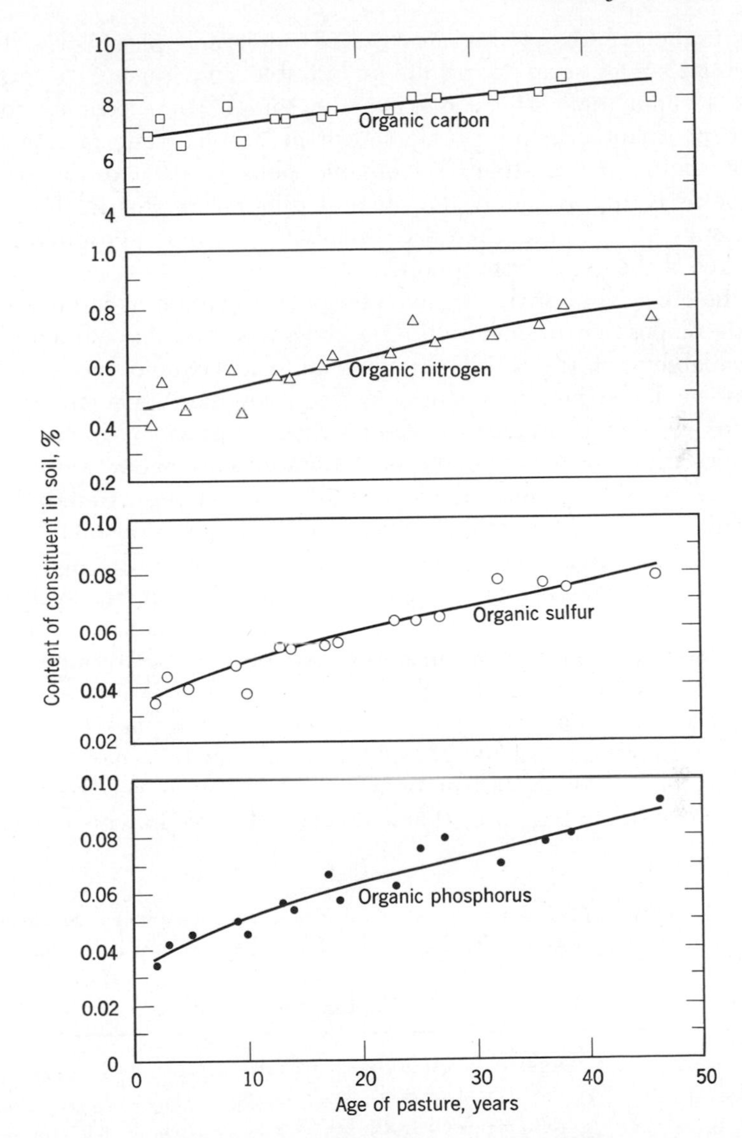

Figure 8.23. Content of organic carbon, nitrogen, sulfur, and phosphorus in surface 7.6 cm of a sandy soil different lengths of time after establishment of a permanent pasture in New Zealand. (Jackman, 1964)

supply in the soil, the gradual increase of soil organic phosphorus shown in Fig. 8.23 may seem to be an undesirable consequence of organic matter accumulation. The same may be said of the accumulation of organic phosphorus in the profile shown in Fig. 8.1. The fact that the sum of the dilute-acid-soluble inorganic phosphorus and the organic phosphorus is approximately the same at different depths in the profile would suggest that the dilute-acid-soluble inorganic phosphorus was the source of the organic phosphorus.

On the other hand, if the organic phosphorus content of soil decreases, from decomposition or mineralization processes, one may suppose that the availability of the soil phosphorus to plants would increase. One form of evidence for an increase in soil phosphorus availability as a result of organic phosphorus mineralization is provided by results of laboratory experiments in which incubation of soils under warm, moist conditions suitable for microbiological activity has been found to produce a decrease in total organic phosphorus and an approximately equal increase in extractable inorganic phosphorus. Because the mineralized phosphorus is presumably released in the soil solution, its availability to plants should be equivalent to that of phosphate fertilizer introduced into the soil solution at the same rate and distributed throughout the soil in the same manner. Although phosphate fertilizer cannot be applied in this way, some concept of the possible effect of the phosphorus fertilization resulting from organic phosphorus mineralization may be derived from data on the quantities of organic phosphorus mineralized. Table 8.5 summarizes data from a field experiment in Ghana, in which measure-

Table 8.5. Organic Phosphorus Content of Soil Under Tropical Conditions in Ghana at the Beginning and End of a Three-Year Period Following Felling the Forest Cover, and Average Values for Temperature and Water Content (Cunningham, 1963)

Exposure of Soil during Three-Year Period	Organic Phosphorus per Gram of Soil (surface 5 cm), μg		Average Daily Soil Temperature at Depth of 7.6 cm, °C[1]	Average Weekly Content of Water in Soil to Depth of 15.2 cm, %
	Initially	After 3 Years		
Shaded	334	271	27	33
Half-exposure	340	190	32	28
Full-exposure	353	176	38	21

[1] The average minimum temperature was 24°C in all exposures.

ments of organic carbon, nitrogen, and phosphorus in soil were made immediately after felling the forest cover and 3 years later. During the 3-year interval, no vegetation was allowed to grow; and the soil was either shaded, half-exposed to the sun, or fully exposed to the sun. Mineralization of organic phosphorus increased with exposure and temperature of the soil, half of the organic phosphorus being lost with full exposure to the sun. (Losses of organic carbon and nitrogen were similarly rapid.) On an area basis, the losses of organic phosphorus given in the table range from 38 to 106 kg of phosphorus per hectare, assuming a bulk density of 1.2. Expressed another way, the release of inorganic phosphorus per hectare from organic phosphorus mineralization was equivalent to the phosphorus contained in 435 to 1220 kg of ordinary superphosphate fertilizer (8.74% phosphorus). These figures are for the surface 5 cm of soil. Additional mineralization occurred at greater depths. The phosphorus fertilization provided by organic phosphorus mineralization in this case was evidently great.

The experiment just described was done under tropical conditions, and the high temperatures promoted far more rapid mineralization of organic phosphorus than occurs in temperate and colder environments. In contrast to these findings reported from Ghana, Mattingly and Williams (1962) found 440 μg of organic phosphorus still present per gram in a soil that had been buried since Roman times in Great Britain. The original organic phosphorus content is not known, but 970 μg of organic phosphorus were present per gram of modern surface soil at the site.

Figure 8.24 summarizes data obtained in analyses of soils from eleven experiment stations in the Great Plains region of the United States. One may see from the figure that cropping without manuring caused a slight increase in inorganic phosphorus and a marked decrease in organic phosphorus. The loss of phosphorus from the soil, presumably by cropping, thus was approximately equal to the loss of organic phosphorus. Manuring the cropped plots produced a marked increase in inorganic phosphorus but only a slight increase in organic phosphorus. Apparently the organic phosphorus added in the manure was mineralized almost completely.

That organic phosphorus mineralization contributes to the supply of labile inorganic phosphorus in soils seems most credible in instances in which the net content of organic phosphorus decreases. Nevertheless, it is reasonable to suppose that the process of mineralization is in operation even in instances in which the content of organic phosphorus increases over a period of years, as in Fig. 8.23, but that in such instances the gain of organic phosphorus by synthesis from inorganic phosphorus

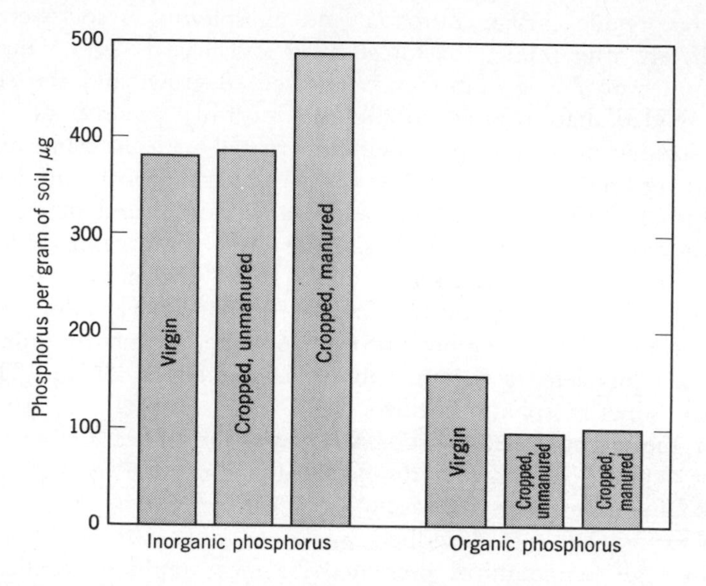

Figure 8.24. Inorganic and organic phosphorus in virgin and cropped soils at eleven experiment stations in the Great Plains region of the United States. The period of cropping ranged from 30 to 48 years at the different locations. (Haas et al., 1961)

derived from the soil exceeds the gain of inorganic phosphorus by mineralization. Plants might conceivably benefit from the increase in mineralization accompanying a gradual gain in organic phosphorus in soil, but whether they do remains to be determined.

Another form of evidence with regard to the significance of soil organic phosphorus in the phosphorus nutrition of plants is provided by analogy with organic nitrogen. The significance of organic nitrogen mineralization in the nitrogen nutrition of plants is no longer regarded as a controversial subject. Mineralization of organic phosphorus and organic nitrogen are similar processes. Moreover, the relative amounts of phosphorus and nitrogen mineralized from soil organic matter approximate the relative amounts of the two substances absorbed from soil by plants. Therefore it might be argued that the importance of organic phosphorus mineralization in the phosphorus nutrition of plants is equal to the importance of nitrogen mineralization in nitrogen nutrition.

The line of reasoning just described does not make allowance for the difference in behavior of the mineral forms of these two elements. The mineralized nitrogen is mostly nitrate, which is freely diffusible in soil and is readily absorbed from soil by plants. Recovery of 50% of nitrogen added as nitrate in the above-ground portion of a single crop is not uncommon. Phosphate, on the other hand, interacts extensively with the inorganic portion of soil, in consequence of which it is not freely diffusible and not readily absorbed. When superphosphate is mixed thoroughly with soil, the recovery of the added phosphorus in the crop seldom exceeds 20%. The susceptibility of mineralized organic phosphorus to absorption thus is probably considerably less than that of mineralized organic nitrogen.

Attention may be drawn also to a related aspect of the behavior of phosphorus and nitrogen that is of importance in considering the comparative significance of phosphorus and nitrogen mineralization. Because the soil supply of mineralized forms of nitrogen subject to absorption by plants (nitrate, nitrite, and exchangeable ammonium) at any given time is usually small relative to the total amounts of mineral nitrogen added by mineralization and absorbed by plants during a season, the nitrogen mineralized during a given season is evidently of importance to the current crop. With phosphorus, on the other hand, the soil supply of inorganic forms subject to absorption is usually much greater than the amounts added by mineralization and absorbed by plants. Consequently, one may argue that the phosphorus mineralized in a given season merely mixes with the much larger amount of inorganic phosphorus already present and is of little value to the current crop. According to this view, the process of organic phosphorus mineralization would attain substantial significance in phosphorus nutrition only over a period of years during which cumulative mineralization would have provided a substantial part, say, one-fourth, of the inorganic phosphorus from which the phosphorus used by plants is withdrawn.

Although organic phosphorus mineralization in previous years may have an important effect on the current availability of the inorganic phosphorus to plants, whether a given atom of soil phosphorus present currently in inorganic form was at one time present in organic form is of no significance from the standpoint of methods of evaluating the current phosphorus status of soils. The contribution of previously mineralized organic phosphorus to the current phosphorus status will be included automatically in indexes of phosphorus uptake by plants obtained by all the usual methods, which depend on measurements of inorganic phosphorus.

Several investigators have attempted to determine whether the current mineralization of organic phosphorus is significant in the phosphorus nutrition of plants or whether the contribution of organic phosphorus mineralization can be evaluated adequately by measurements of extractable inorganic phosphorus present at the time a crop is planted. Because of the difficulties of direct experimental testing under natural conditions, indirect methods have been used. The findings have not been consistent one way or the other. For example, in an investigation involving 92 field experiments in Norway, Semb and Uhlen (1954) found by statistical methods that at any given level of extractable inorganic phosphorus in the soil, as determined by laboratory tests, the response of the crop to phosphorus fertilization still decreased significantly with an increase in the total content of organic phosphorus in soils having *p*H values of 5.5 and above; below *p*H 5.5 the trend was not significant. On the other hand, Susuki et al. (1963) did experimental work on a similar pattern on samples of ten soils in Michigan and found that at any given level of inorganic phosphorus the uptake of phosphorus by plants grown on the soils in the greenhouse did not increase significantly with the content of total organic phosphorus.

The concept leading to use of total organic phosphorus in the work described in the preceding paragraph was probably that total organic phosphorus might provide an index of organic phosphorus mineralization in the soil while crops were growing, thereby evaluating indirectly an effect not taken into account by the usual measurements on the soil inorganic phosphorus. Organic phosphorus mineralization can be estimated more directly, but with relatively high experimental error; nevertheless, van Diest and Black (1959) found that uptake of phosphorus by plants from soils in the greenhouse was more closely related to estimates of organic phosphorus mineralized during incubation of the soils in the laboratory than to total organic phosphorus.

The significant positive relationships that have been found between uptake of phosphorus from soils by plants and the total quantities of organic phosphorus mineralized could conceivably arise in different ways. One theory is that the plants benefit from the microbial mineralization of organic phosphorus that takes place throughout the soil. This theory follows reasonably from evidence that mineralization of organic phosphorus does occur in this way and, by analogy, from evidence on the behavior of soil nitrogen. As mentioned previously, however, the nitrate-nitrogen that results from the processes of organic nitrogen mineralization and ammonium oxidation moves freely in soil and can be absorbed almost quantitatively by plants. The inorganic phosphorus produced from organic phosphorus mineralization, on the other hand, is subject to equilibration with soil solids adjacent to the site of release; hence it cannot be expected to move as readily through the soil or to be absorbed as extensively by plants, on a short-term basis, as is nitrate. Thus there is some doubt as to whether the modest quantities of inorganic phosphorus that seem to be produced by microbial mineralization throughout the soil are great enough to account for the degree of importance in phosphorus nutrition of plants that some of the experimental work indicates for organic phosphorus.

A second theory is that mineralization of organic phosphorus occurs more rapidly in soil adjacent to plant roots than in the bulk soil and that plants benefit from this enhanced local mineralization as well as from the mineralization in the bulk soil. This theory is supported by evidence that enzymes capable of releasing inorganic phosphorus from certain organic phosphorus compounds and from soil organic phosphorus extracted by hydrochloric acid and sodium hydroxide are present on root surfaces and in the mucilaginous coatings on roots [see, e.g., Bower (1949), Estermann and McLaren (1961), and Fig. 8.25]. Also, Hayashi and Takijima (1955) and Sekhon (1962) found a greater decrease in the total organic phosphorus content of cropped soil than of fallow soil. However, the supposed action of plant roots in enhancing the mineralization of organic phosphorus, as it occurs in the soil, has not yet been demonstrated. Rovira (1956) tested the effect of exudate from pea roots but obtained negative results.

In the Soviet Union, extensive use is made of "Phosphobacterin," a preparation containing spores of *Bacillus megatherium,* for inoculating crop seeds. This practice is justified by the increases in yield obtained. According to Cooper (1959), increases in crop yields from Phosphobacterin are obtained in about half the instances in which it is applied; the increases average about 10%. Benefits from this practice are usually attributed to action of the bacteria in mineralizing soil organic phosphorus and thereby providing more phosphorus for plants. Special emphasis is placed on the adaptation of the organisms to act in the soil immediately adjacent to plant roots. In tests of the preparations in field experiments outside the Soviet Union (Fiedler and Jahn-Deesbach, 1956; Smith et al., 1962), no significant increases in yield or phosphorus content of plants have been obtained from inoculation of the seed with Phosphobacterin. The value of the preparation for general use is thus a controversial subject.

Effect of Soil pH

Soil *p*H influences the availability of phosphorus to plants in several ways. Some of the effects are the result of chemical and biological processes that take place in soil in the absence of plants. These affect the supply of labile phosphorus and the concentration of phosphorus in the soil solution. Others involve the behavior of plants. These will be considered in turn.

The effect of *p*H on the behavior of aluminum and iron phosphates and associated hydrous oxides of iron and aluminum is an ill-defined

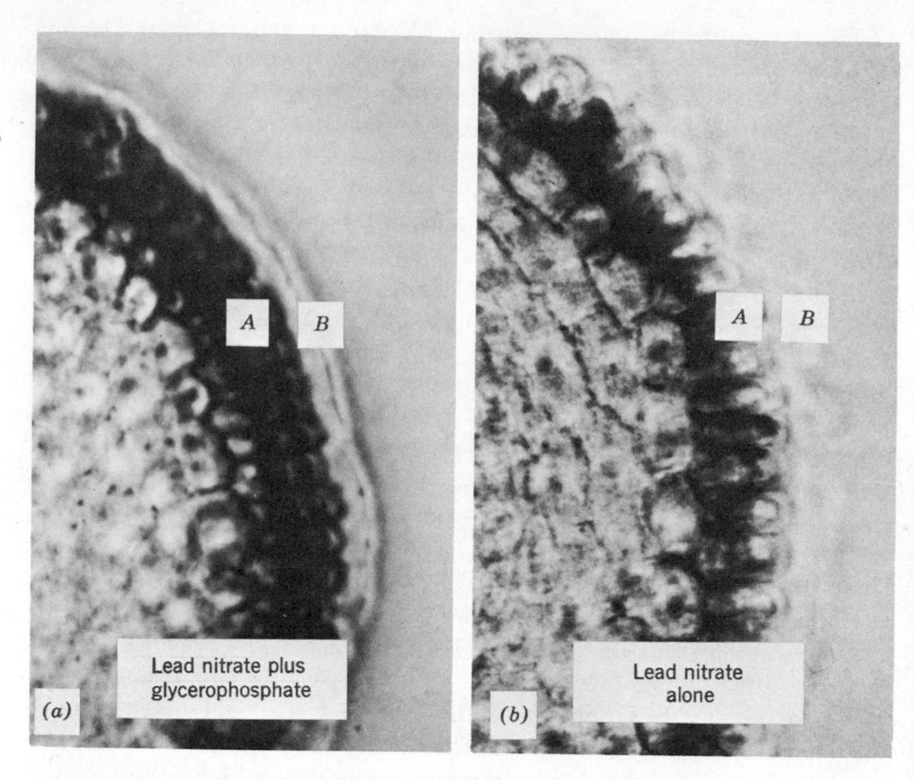

Figure 8.25. Photomicrographs of stained cross sections of barley roots showing results of a test for phosphatase activity. Phosphatase activity is indicated by the greater staining of walls of epidermal cells (*A*) and the external mucilaginous layer (*B*) of (*a*) than (*b*). Staining is due to lead sulfide, produced by addition of ammonium sulfide to the sections cut and mounted after the whole roots had been kept for two hours in a buffered solution of lead nitrate. The solution was washed off before the sections were cut. The root in (*a*) was exposed to a solution containing glycerophosphate as well as lead nitrate; hence phosphate released by action of phosphatase enzyme was precipitated in place as lead phosphate, which was subsequently converted to lead sulfide. In the absence of the extra phosphate derived from hydrolysis of the glycerophosphate, most of the lead was washed from the outer portion of the root in (*b*) before the section was made. (Estermann and McLaren, 1961)

chemical effect or series of effects that may be represented in an oversimplified way by the conventional equation

$$AlPO_4 + 3OH^- = Al(OH)_3 + PO_4^{---},$$

which emphasizes the role of hydroxyl-ion activity. With an increase in *p*H or hydroxyl-ion activity, aluminum and iron phosphates release phosphate in soluble form; and the aluminum and iron remain in insoluble form as the hydroxides. Conversely, with a decrease in *p*H or hydroxyl-ion activity, or with an increase in phosphate-ion activity, the tendency of aluminum and iron hydroxide to react with phosphate to form aluminum and iron phosphate increases. The reaction is highly concentration-dependent within the range of *p*H values and phosphate concentrations common in soils. This reaction is sometimes referred to as an anion exchange, which implies that soils have an anion-exchange capacity analogous to the cation-exchange capacity. Replacement of hydroxyl by phosphate does bear some similarity to the replacement of hydrogen in *p*H-dependent cation-exchange positions by other cations in that hydroxyl ions are not subject to replacement by phosphate at high *p*H values, whereas they may be replaced at low *p*H values. Nevertheless, the exchange between hydroxyl and phosphate depends on properties specific to the reaction of these two ions with aluminum and iron. Other common soil anions exchange with phosphate and hydroxyl either to only a small extent (sulfate) or practically not at all (chloride, nitrate). Description of this behavior in quantitative terms that are concrete and yet meaningful in terms of phosphorus availability is a complex matter because of such factors as differences in amount and reactivity of the hydrous oxides of iron and aluminum, the tendency of clays to decompose and release iron and aluminum at low *p*H values, and the slowness with which equilibrium conditions are approached.

When acid soils become anaerobic, the *p*H tends to increase. The increase in *p*H may be expected to affect aluminum phosphates in the manner just described. Reduction of ferric iron, however, releases in more soluble form the phosphorus combined originally as ferric phosphate; at the same time, the reactivity of the soil toward phosphate is decreased because of disappearance of hydrous ferric oxides. When aerobic conditions return, the ferrous iron is reoxidized to hydrous ferric oxides, which, being freshly formed, have great reactivity toward phosphorus. The phosphorus solubility then decreases. Because of these transformations, various aspects of which were discussed previously in the chapters on soil aeration (Chapter 3) and soil acidity (Chapter 5), acid soils that have low phosphorus availability for the usual externally aerated plants under aerobic conditions become nearly neutral and have much higher phosphorus availability for internally aerated plants under anaerobic conditions.

Another effect of *p*H has to do with calcium phosphates. This effect may be visualized in terms of the following conventional equation written for hydroxyapatite:

$$Ca_{10}(OH)_2(PO_4)_6 + 20H^+ = 10Ca^{++} + 2HOH + 6H_3PO_4,$$

which emphasizes the importance of both hydrogen- and hydroxyl-ion activity. Calcium phosphates dissolve as the *p*H decreases because of the tendency of hydrogen ions to associate with phosphate ions. With hydroxyapatite, which contains hydroxyl ions, the tendency of hydrogen ions to associate with hydroxyl ions to form water is another factor in the dissolution with a decrease in *p*H.

Still another effect of *p*H has to do with mineralization of organic phosphorus. Results of investigations reviewed by Black and Goring (1953) and more recent work by Halstead et al. (1963) and Kaila (1965) indicate that the usual effect is an increase in mineralization with an increase in soil *p*H. The effect of soil *p*H on organic phosphorus mineralization may be accounted for by the theory that raising the *p*H reduces the sorption of organic phosphorus compounds by hydrous oxides and hence increases their solubility and susceptibility to mineralization. In contrast to the temporary increase in carbon and nitrogen mineralization that occurs when the pH of acid soils is increased by treatment with calcium carbonate, the increase in rate of organic phosphorus mineralization appears to be permanent (Thompson et al., 1954).

The foregoing effects of *p*H have to do with processes that go on in soil independently of plants. Two additional effects involve both soil and plants. One is the relationship between aluminum toxicity and phosphorus nutrition is discussed in Chapter 5. There is evidence that excess aluminum in solution in strongly acid soils induces phosphorus deficiency in plants. High concentrations of aluminum in the soil solution are associated with low concentrations of phosphate. Moreover, aluminum enters the roots and seemingly inactivates some of the phosphorus absorbed by the roots. As the *p*H is increased, the aluminum is precipitated in the soil; and the presumed inactivation of phosphorus by aluminum in the plant no longer occurs. At the same time, alleviation of aluminum toxicity reduces the inhibition of root extension. An increase in phosphorus availability should then follow from the greater extent of the root system and the greater area of root-soil interface.

A second effect of *p*H involving both soil and plant has to do with the ionic form of orthophosphate. Orthophosphate may exist in solution as H_3PO_4, $H_2PO_4^-$, HPO_4^{--}, and PO_4^{---}. H_3PO_4 predominates in strongly acid solutions and PO_4^{---} in strongly alkaline solutions, but the proportions of these two forms are negligible within the *p*H range from 5 to 9. According to calculations by Olsen (1953), the percentages of the total inorganic orthophosphate present in solution as $H_2PO_4^-$ range from 99.3 at *p*H 5 to 50 at *p*H 7.2 and to 1.5 at *p*H 9. Corresponding figures for HPO_4^{--} are 0.6 at *p*H 5, 50 at *p*H 7.2, and 98.4 at *p*H 9. Hagen and Hopkins (1955) proposed the theory that plants absorb phosphorus as

$H_2PO_4^-$ and HPO_4^{--}, the two ions being attached to distinct carrier sites on the roots. Uptake of both ions is inhibited in a competitive manner by hydroxyl. The authors showed how the theory would account for their data on uptake of phosphorus by roots from solutions at different *p*H values and phosphorus concentrations. Although the details of the verification will not be given here, it is significant to note that uptake of phosphorus by plants generally decreased with an increase in *p*H, being less than one-fourth as great at *p*H 7.7 as at *p*H 5, even though the total concentration of phosphorus remained constant. Because the phosphorus in solution is present principally as $H_2PO_4^-$ at *p*H 5 and as HPO_4^{--} at *p*H 7.7, the *p*H effect indicates that $H_2PO_4^-$ was taken up more readily than HPO_4^{--}.

Soil *p*H evidently modifies several conditions that influence the availability of phosphorus to plants. The effect of *p*H on the systems mentioned is not consistently in either direction; and, because of differences in proportions and amounts of the components in different soils, the availability of soil phosphorus to plants is not determined by the *p*H. Nevertheless, experience has shown that, broadly speaking, there is a general effect of *p*H. Figure 8.26 summarizes the results of an extensive investigation of the relationship between soil *p*H and response of lettuce to phosphate fertilization of soils of California. In this work, samples were designated as low in phosphorus availability if the ratio of the yield of lettuce on the soil without phosphate fertilization in the greenhouse to the yield with heavy phosphate fertilization was less than 0.2. Nitrogen and potassium were added to all samples. Clearly, the frequency of occurrence of samples having low phosphorus availability was least in the approximately neutral group and greatest in the most acid and most alkaline groups.

The over-all relationship shown in Fig. 8.26 between soil *p*H and the proportion of soils markedly deficient in phosphorus for growth of plants was obtained with soils that were sampled in the field and brought to the greenhouse for testing without treatments to change the *p*H. If the *p*H of a given soil is altered, the short-term effect on the phosphorus status is usually in accordance with expectations from Fig. 8.26; however, there are exceptions.

The most frequent exception noted experimentally has been a decrease in phosphorus availability when acid soils are treated with calcium carbonate. In these instances, small applications of calcium carbonate are often beneficial. Only the larger applications are detrimental, and the detrimental effect does not seem to be permanent. For example, Pierre and Browning (1935) found that addition of calcium carbonate to eight of nine strongly acid soils of West Virginia resulted in lower yields

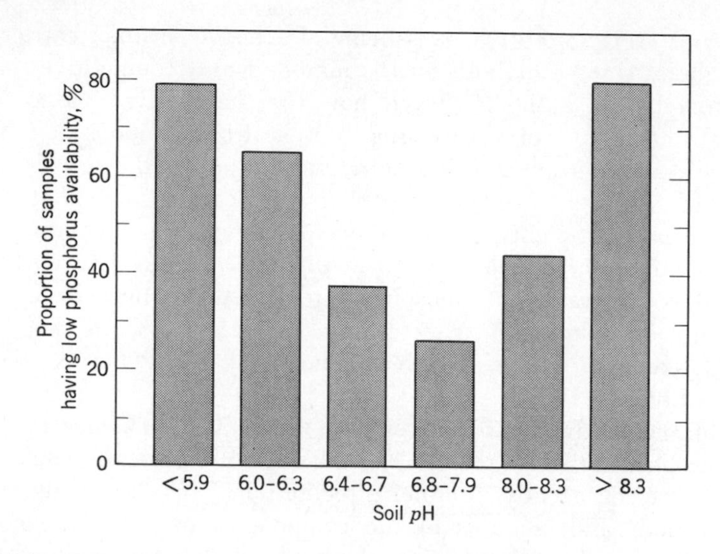

Figure 8.26. Percentage of samples having low phosphorus availability in groups classified according to *p*H in a survey of 448 samples of soils of California. (Jenny et al., 1950)

of alfalfa in the first year after the treatment where the *p*H was adjusted to 7.5 than to 6.5. The depression in yield was less marked in the second year after the treatment. In that year, five of the soils produced a higher yield at *p*H 7.2 than at *p*H 6.5. Corn similarly was affected adversely, yielding less on soil limed to *p*H 7 than to *p*H 5.8. That an induced phosphorus deficiency was at least partly responsible for the unfavorable effect of the higher *p*H value is indicated by the fact that the phosphorus percentage in the plants was lower and the increase in yield from phosphate fertilization was greater at *p*H 7 than at *p*H 5.8. Similar observations were reported by Lawton and Davis (1956), Sorteberg (1963), and Sorteberg and Dev (1964). The soils in which phosphorus deficiency is induced by application of calcium carbonate seem invariably to be relatively deficient in phosphorus in their initial acid condition. The cause of the induced deficiency has not been determined. From scattered experimental evidence, however, the usual effect of moderate additions of calcium carbonate to acid soils appears to be that of increasing the concentration of phosphorus in solution after a long time of reaction but of decreasing it after only a short time of reaction. Presumably, the initial effect is altered by gradual release of phosphorus from hydrous oxides of aluminum and iron and from organic sources, the time required for reversal depending on the rate of release, the *p*H to which the soil has been adjusted, and other factors.

Another exception has to do with acidification of soils that are originally approximately neutral and, hence, from Fig. 8.26, presumably in the optimum *p*H range with regard to phosphorus availability. Limited research suggests that moderate acidification sometimes produces an increase in phosphorus availability. This effect, which probably results primarily from dissolution of calcium phosphates, would be expected to decrease with time as a result of equilibration of the soluble phosphate with the hydrous oxides of aluminum and iron.

Fertilizers

Phosphorus-bearing fertilizers may be placed in three groups: natural organics, natural mineral phosphates, and processed mineral phosphates. Animal manures contain of the order of 0.1 to 0.4% phosphorus, partly in organic form and partly in inorganic form, and may be classed as low-grade phosphorus-bearing fertilizers. Manures are usually used on the farm on which they are produced. The phosphorus-bearing fertilizers employed in commercial operations are almost exclusively natural phosphates and processed phosphates, mostly the latter. Their phosphorus content is much higher than that of animal manures.

Phosphorus-bearing fertilizers in commercial production are derived almost entirely from natural deposits in which phosphorus is present in relatively high concentration. The principal types of phosphate deposits were classified by McKelvey et al. (1953) as igneous apatites, marine phosphorites, residual phosphorites, river pebble phosphorites, phosphatized rock, and guano. They listed apatite as one of the principal phosphatic minerals and sometimes as the only principal phosphatic mineral in each type of deposit. Phosphate deposits are widely distributed. Known deposits are present in largest quantity in North Africa, followed by the United States and the Soviet Union, in that order. Except for apatite and guano (and perhaps phosphatized rock), the classes of phosphate deposits designated by McKelvey et al. are called phosphorite or phosphate rock by geologists. In agriculture, the term "rock phosphate" is common.

Phosphate deposits are characteristically variable internally and sometimes can be segregated into high-grade and low-grade separates by washing and screening or by flotation. High-grade material is desired for economy in processing and shipping.

The high-grade natural-phosphate material may be finely ground and

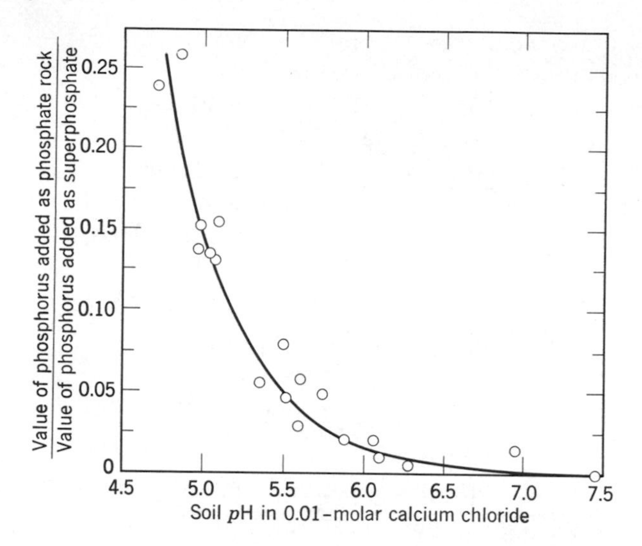

Figure 8.27. Ratio of value of phosphorus added as phosphate rock to that added as superphosphate versus soil *p*H. Values were derived from measurements of uptake of phosphorus by a test crop of sorghum from soil in the greenhouse under conditions of phosphorus deficiency and were calculated as follows: (yield of phosphorus in plants on soil treated with phosphate rock minus yield of phosphorus in plants on control soil) (quantity of phosphorus applied to soil treated with superphosphate)/(yield of phosphorus in plants on soil treated with superphosphate minus yield of phosphorus in plants on control soil) (quantity of phosphorus applied to soil treated with phosphate rock). Each point represents a different soil. (Peaslee et al., 1962)

applied directly to soil as a fertilizer. Because it is almost invariably an apatite, the solubility is exceedingly low; and, consequently, the effectiveness also may be exceedingly low. Results of field experiments on different soils indicate that the effectiveness of phosphate rock in supplying phosphorus for plants is less in neutral and alkaline soils than in acid soils. This general observation was verified in a greenhouse experiment in which the value of phosphate rock was tested on a number of soils under uniform conditions (Fig. 8.27). The availability of the phosphorus of phosphate rock to plants in different soils thus follows the general trend of increasing solubility of apatite phosphates with a decrease in *p*H, from which one may reasonably infer that an important factor in determining the value of the applied phosphorus to plants is the degree of dissolution of the phosphate rock in the soil. The degree of dissolution may vary greatly with the nature of the soil. For example, Mattingly and Talibudeen (1961) reported that labile soil phosphorus by the method of isotopic dilution was scarcely affected by application of phosphate rock to calcareous soil. In strongly acid soil, however, the increase in labile phosphorus from application of phosphate rock approached that from application of superphosphate.

As would be expected from the fact that apatite can be identified in the sand and silt fractions of many soils, phosphate rock does not quickly lose its identity upon addition to soil but may be detected by the characteristic X-ray diffraction pattern of the apatite it contains (Moschler et al., 1957). In contrast to soluble, processed, mineral phosphates, which decrease in value with time after application, phosphate rock may increase in value with time. The increase with time seems to occur in acid soils. Figure 8.28 gives an example showing the comparative value of superphosphate and phosphate rock for barley on an acid soil in Great Britain over a period of 5 years. In the comparison shown here, both fertilizers were added in quantities to supply 147 kg of phosphorus per hectare. Because the application of superphosphate is excessive, the value of this fertilizer did not decrease as rapidly in successive years as would be the case with a smaller application. Phosphate rock is usually added in relatively large quantities with the objective of substantially increasing the phosphorus content of the soil for a long residual effect. In the five successive crops in the experiment shown in Fig. 8.28, the total increase in phosphorus content of the fertilized crops

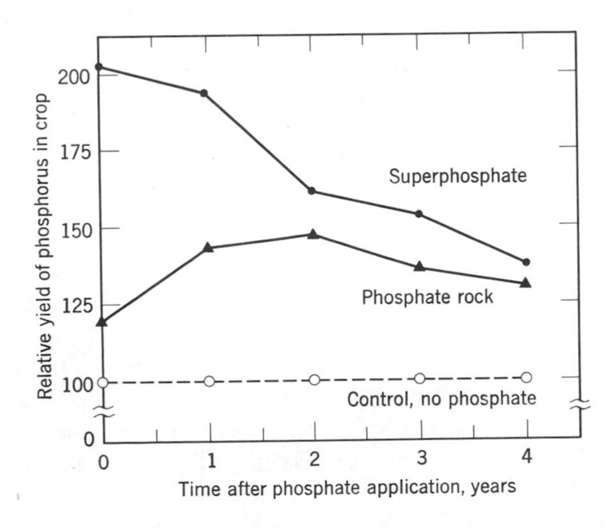

Figure 8.28. Relative yield of phosphorus in barley after a single application of superphosphate and phosphate rock equivalent to 147 kg of phosphorus per hectare to an acid soil in Great Britain. (Mattingly and Widdowson, 1963)

over the control crops amounted to only 12 kg/hectare in the case of superphosphate and 7 kg in the case of phosphate rock.

According to statistics prepared by the Food and Agriculture Organization of the United Nations (1965), world use of phosphate rock for direct application in 1963–64 consumed 0.63 million metric tons of phosphorus. The corresponding figure for processed mineral phosphate fertilizers was 5.3 million metric tons of phosphorus. The bulk of the phosphorus thus was used in the form of processed mineral phosphate fertilizers.

Of the processed mineral phosphate fertilizers, superphosphate (known also as single, ordinary, or normal superphosphate) is used in greatest quantities. In the 1963–64 fertilizer year, ordinary superphosphate accounted for 47% of the phosphorus in processed mineral phosphate fertilizers (Food and Agriculture Organization of the United Nations, 1965). Ordinary superphosphate is prepared by reacting phosphate rock with enough sulfuric acid to decompose the phosphate rock almost completely. The product is then mainly a mixture of monobasic calcium phosphate and calcium sulfate, with smaller amounts of iron and aluminum phosphates, unreacted phosphate rock, and other substances. The phosphorus content of the product varies with the composition of the original phosphate rock and other factors. Substances now classified as ordinary superphosphate contain from 7 to 9% phosphorus according to Sauchelli (1960, p. 129).

If a great enough excess of sulfuric acid is added to phosphate rock, the liquid phase contains phosphoric acid, and the solid phase contains calcium sulfate and other undissolved substances. If the solids are discarded and the phosphoric acid is reacted with more phosphate rock in proper proportions, the product is mainly monobasic calcium phosphate, together with iron and aluminum phosphates, a little unreacted phosphate rock, and other substances. The product is called double, triple, treble, or concentrated superphosphate; and the phosphorus content of fertilizers classified in this way is 19 to 21% (Sauchelli, 1960, p. 129). In 1963–64, concentrated superphosphate accounted for 14% of the phosphorus consumed in processed mineral phosphate fertilizers (Food and Agriculture Organization of the United Nations, 1965). World production of this fertilizer is mostly in the United States. Much phosphoric acid is produced also by an electric furnace process in which phosphate rock is reduced by coke. Elemental phosphorus is volatilized and subsequently reacted with oxygen to form phosphorus pentoxide, which is reacted with water to form phosphoric acid.

Basic slag is a by-product phosphate fertilizer produced in manufacture of steel from iron ores high in phosphorus. The phosphorus in basic slag is chiefly tetracalcium phosphate ($Ca_4P_2O_9$) together with some calcium silicophosphate. Basic slag accounted for 13% of the phosphorus in manufactured phosphate fertilizers in 1963–64 (Food and Agriculture Organization of the United Nations, 1965). World production of basic slag is concentrated almost exclusively in Europe.

In addition to these three types of processed mineral phosphate fertilizers, several others are in use. Collectively, these fertilizers accounted for 26% of the phosphorus in manufactured phosphate fertilizers in 1963–64 according to FAO statistics. Most of those in extensive use contain nitrogen. Ammonium phosphates are obtained by treatment of phosphoric acid with ammonia. Ammoniated superphosphates are superphosphates that have been treated with ammonia. Nitric phosphates are fertilizers in which nitric acid is used in place of sulfuric acid in treatment of phosphate rock and in which the product of the reaction is treated with ammonia. In addition, there are a number of fertilizers in which the apatite structure in phosphate rock is destroyed and the availability of the phosphorus increased by heat treatment, usually with addition of some other substance such as silica, sodium carbonate, or phosphorus pentoxide. Some of the fertilizers prepared by heat treatment are condensed phosphates or polyphosphates. These are derivatives of phosphoric acid and its salts in which individual molecules are linked into chains or rings by elimination of one molecule of water for each phosphate-to-phosphate bond. These condensed phosphates gradually hydrolyze to orthophosphates in the presence of water.

Most phosphate fertilizers do not dissolve completely in even a relatively large volume of water and are used as solids. Solid phosphate fertilizers are sometimes finely ground and mixed with the soil but frequently are applied in concentrated bands or in granules. Forms of low solubility such as phosphate rock give best results if finely ground and mixed with the soil. Fine grinding and mixing hasten the release of phosphorus from fertilizer particles to soil. Although the same general principle applies to highly soluble phosphate fertilizers such as superphosphate, the limiting factor with these is not the transfer of phosphorus from fertilizer to soil because the highly soluble forms are transferred within a few days. Rather, the more critical factor is the availability of the phosphorus that has reacted with the soil. The availability generally is greater where applications have been made in bands or granules than where the fertilizer has been finely ground and mixed with the soil; in calcareous soils, however, the reverse is often true.

Phosphate fertilizers of low solubility are relatively ineffective if applied to the surface of the soil and left there because little of the phosphorus they contain is washed down into the soil by rain. Where soluble phosphate fertilizers are applied to the surface of the soil, the phosphorus enters the soil and reacts mostly within a thin layer at the surface, which makes surface application a sort of localized application analogous to banding. Although fertilizer phosphorus should be absorbed efficiently from a band of soil at the surface if the soil were moist continuously, the surface portion of soil is often dry. Dryness inhibits uptake of phosphorus from soil by plants, as mentioned previously, and hence should decrease the uptake of phosphorus from the surface layer.

Occasionally, phosphate fertilizers are applied in solution. This practice is confined mostly if not entirely to phosphoric acid and ammonium phosphates, which leave no appreciable solid residue containing phosphorus. Phosphoric acid and ammonium phosphates are extremely soluble. The solutions may be applied in highly concentrated form in ways similar to those used for solid fertilizers, in dilute form to supply water and nutrients in transplanting tomatoes or other plants to the field, or in extremely dilute form by injection into irrigation water.

Phosphate fertilizers are ordinarily applied to soils at or just before the time of planting the crop to be fertilized. One reason for this is that the availability to plants of the phosphorus of the relatively soluble fertilizers such as superphosphate is greatest shortly after application and decreases with time.

Phosphorus-bearing fertilizers are applied most commonly in quantities to supply 8 to 20 kg of phosphorus per hectare. Where the fertilizer is applied with the seed or locally near the seed, however, as little as 4 or 5 kg of phosphorus may be used per hectare. Even small quantities such as these may have a marked effect on early growth of the crop. Sometimes small localized applications are supplemented by larger quantities added in other ways. Applications of relatively soluble phosphate fertilizers seldom exceed the equivalent of 80 kg of phosphorus per hectare. Applications of phosphorus as phosphate rock are occasionally greater but are not made annually.

Quantities of the order of 80 kg of phosphorus per hectare are far in excess of the quantities of phosphorus removed annually by cropping and leaching. Even the average applications of 8 to 20 kg/hectare exceed the annual removal by cropping and leaching in most instances. If the phosphorus is added in soluble forms, such as superphosphate, much of the residual phosphorus that remains in the soil is changed to forms that do not contribute directly to the supply of phosphorus for plants. For example, Mattingly (1958a) found that, of the increase in content of soil phosphorus resulting from fertilization of three soils with superphosphate for about 100 years in Great Britain, about two-thirds was nonlabile according to the method of isotopic dilution in vegetation

Table 8.6. Residual Value of Phosphorus Accumulated from Annual Applications of Superphosphate for a Century to a Soil in England (Mattingly, 1958a)

Soil Treatment	Phosphorus per Gram of Soil, μg		Increase in Labile Phosphorus as Percentage of Increase in Total Phosphorus
	Total	Labile[1]	
None	520	34	—
Superphosphate	1293	308	35

[1] Phosphorus found in soil by the method of isotopic dilution upon addition of P^{32}-labeled orthophosphate, growing plants on a quantity of the soil in the greenhouse, and analyzing the plants for total and radioactive phosphorus.

tests; and one-third was in the labile fraction from which the phosphorus used by plants was withdrawn. Table 8.6 gives the data obtained in vegetation tests on one soil. The increase in availability of phosphorus from accumulated residues of previous fertilizer applications may be too small to be significant if the quantities applied have been small. Where heavy applications have been made, however, the increase may be great enough to reduce substantially the current need for fertilization.

The greatest use of phosphorus for fertilizer purposes is in Europe and North America. Expressed in terms of millions of metric tons of phosphorus, the total use of all phosphorus-bearing fertilizers in the 1963–64 fertilizer year was 2.55 in Europe, 1.58 in North and Central America, 0.58 in the Soviet Union, 0.49 in Asia, 0.45 in Oceania, and 0.15 in Africa according to statistics published by the Food and Agriculture Organization of the United Nations (1965). (The figure for Africa is for 1962–63, and the figure for phosphate rock included in the total for North and Central America likewise is for 1962–63.) The ranking shown here for phosphorus use is similar to that given in Chapter 7 for nitrogen use.

Literature on the subjects of phosphorus fertilizers and phosphorus fertilization is extensive. Monographs dealing with these subjects include one on superphosphate by the United States Department of Agriculture and the Tennessee Valley Authority (1964), one on chemistry and technology of fertilizers edited by Sauchelli (1960), one on phosphates in agriculture by Sauchelli (1965), one on fertilizer technology and usage edited by McVickar et al. (1963), and one on efficient use of fertilizers edited by Ignatieff and Page (1958). A brief but comprehensive review on phosphate fertilizers was published by Nelson (1965).

Function in Plants

Unlike the nitrogen of nitrate and the sulfur of sulfate, which are reduced in the plant, the phosphorus absorbed from soils remains in the oxidized state and occurs in both organic and inorganic forms as the central atom of the phosphate group. Inorganic phosphorus is usually determined on an extract obtained by treating plant tissue with an acid such as trichloracetic acid that will remove the orthophosphate present in solution or in water-soluble form as well as the orthophosphate present in sparingly soluble forms. There are many organic forms, and these may be classified as storage and structural compounds and as compounds of intermediate metabolism.

The group of storage and structural compounds includes phytin, phospholipids, and nucleic acids. Phosphorus in seeds is stored principally as phytin, the calcium-magnesium salt of inositol hexaphosphoric acid. This compound is hydrolyzed enzymatically during germination; and the phosphate is thereby changed to the inorganic form, from which it may be used for other purposes by the developing seedling. Phospholipids are esters of glycerol or inositol with phosphoric acid, fatty acids, and sometimes other substances. The phosphoric acid may be present as a salt, or it may be esterified further with a nitrogen base. Phospholipids apparently act as storage material in seeds, and in growing plants they are involved in metabolism. They are thought to play an important part in selective permeability and ion transport. Nucleic acids are compounds of high molecular weight, and they are composed of units called nucleotides. Nucleotides contain one molecule each of a purine or pyrimidine base, ribose or desoxyribose sugar, and phosphoric acid and are linked by the phosphoric acid groups. Nucleic acids form the genes of plants, which determine hereditary qualities; and their role as genetic material is to determine the nature of each of the cellular proteins synthesized. Most of the cellular proteins are enzymes that catalyze individual metabolic reactions. Phosphorus occurs in certain of these enzymes, although it does not appear to be involved directly in the catalysis.

In plant metabolism, phosphorus plays a direct role as a carrier of energy. This role is made possible by the fact that the phosphate in several organic linkages may be split off by hydrolysis with a relatively high yield of energy. Phosphate groups that have this property are called high-energy phosphate. The most important carrier of high-energy phosphate is adenosine triphosphate, in which three phosphoric acid groups are joined linearly with elimination of a molecule of water on addition of each phosphate group to the organic portion of the molecule. The high-energy bonds are those joining the phosphate groups and not the bond joining the first phosphate to the organic portion of the molecule. (The ester linkages of phosphate groups in phytin, phospholipids, and nucleic acids are not high-energy bonds.) Although there are various kinds of high-energy bonds besides that of phosphorus, the phosphorus high-energy bond has the unique property of stability in water. By means of enzymes, the high-energy phosphate can be transferred from a source such as adenosine triphosphate to another compound, such as glucose, without dissipation of the energy in transit. Once the high-energy phosphate combines with the phosphate acceptor, however, the energy is transferred largely to the remainder of the molecule, and the phosphate becomes the low-energy type. The extra energy that has been transferred to the accepting compound makes that compound more reactive. For example, in oxidation of glucose to carbon dioxide and water, the first step is addition of phosphate from adenosine triphosphate. As the oxidation proceeds, with the aid of various enzymes, much of the energy liberated is recaptured as newly formed adenosine triphosphate. Complete oxidation of one molecule of glucose to carbon dioxide and water is thought to require use of two molecules of adenosine triphosphate and to involve production of forty, making a net production of thirty-eight. The energy thus captured in high-energy phosphate can then be used in synthesis of other substances requiring input of energy. For example, the energy needed to form starch from glucose is supplied by transfer of a high-energy phosphate group to each molecule of glucose. The phosphate is eliminated as inorganic phosphate when the glucose molecules are linked.

Finally, phosphorus plays a part in photosynthesis. The initial reaction in which light energy is trapped involves the splitting of water, in the presence of inorganic phosphate, adenosine diphosphate, and the coenzyme nicotinamide adenine dinucleotide phosphate, with production of the reduced form of nicotinamide adenine dinucleotide phosphate, adenosine triphosphate, and molecular oxygen. The reduced form of nicotinamide adenine dinucleotide phosphate then provides the energy for the hypothetical reaction $RH + CO_2 = RCOOH$, by which the carbon dioxide is fixed. R in this reaction is a five-carbon sugar containing two phosphate groups, and the intermediate product RCOOH splits to two molecules of a three-carbon compound, phosphoglyceric acid. At the same time, there is another reaction in which some of the light energy trapped by the splitting of water is transferred to adenosine diphosphate and inorganic phosphate, causing their association to form adenosine triphosphate. Energy present in the adenosine triphosphate then is used in synthesis of progressively more complex organic compounds from the initial product of carbon fixation.

Thus, from the standpoint of energy, synthesis of plant tissue may be considered to be a process in which light energy captured in water is transferred to organic compounds and used for reduction of carbon dioxide, followed by elaboration of the simple initial products of carbon dioxide reduction into more complex compounds of higher energy content at the expense of part of the energy derived from the initial photosynthetic reaction. Phosphorus plays an essential part as energy carrier throughout.

Although the functions of organic forms of phosphorus have been emphasized here, the auxiliary function of the inorganic phosphorus in the plant must not be overlooked. The inorganic phosphorus represents both the source of phosphate ions that enter the organic part of the various metabolic cycles and the repository of these ions as they leave. A given phosphate ion probably does not remain for a long time in either organic or inorganic form but changes frequently from one form to the other. If the concentration of phosphate ions in inorganic form is too low, the rate of the metabolic processes will be limited by the availability of phosphate ions for binding in organic form. The concentration of inorganic phosphorus in plants is, of course, influenced by the supply in the soil, and measurements of inorganic phosphorus in plant tissue are sometimes made to provide an index of the supply of phosphorus in the soil. Inorganic phosphorus is preferred for this purpose because it varies more with supply than does organic phosphorus.

The function of phosphorus in plants has been discussed here in only general outline, with omission of details concerning specific compounds, reactions, and enzymes. For more detailed coverage, current books on biochemistry and plant physiology should be consulted.

Phosphorus Supply and Plant Behavior

Deficiency Symptoms

In general, symptoms of phosphorus deficiency are not particularly pronounced or specific. Phosphorus deficiency thus may be difficult to diagnose from visual examination of plants. Chemical analysis of the plants or trials with phosphate fertilizer may be required.

Although extreme phosphorus deficiency may result in some yellowing of leaves, the more common symptom is a dark green or bluish-green color, which may be coupled with tints of bronze or purple. The purple coloration, due to anthocyanin pigments, is the most striking and most frequently mentioned symptom of phosphorus deficiency. As a diagnostic criterion, however, the purple coloration unfortunately is of little value for plants in general because of its low degree of specificity. Similar tints sometimes are produced by a deficiency of nitrogen or by other conditions. Moreover, some kinds of plants produce the color whether phosphorus is deficient or not. Others do not produce it even when phosphorus is deficient. The complete range of behavior is found in different varieties of corn.

Phosphorus deficiency produces certain effects that are similar to effects of nitrogen deficiency. With deficiencies of both elements, the stems of plants are thin, the leaves are small, and lateral growth is limited; defoliation is premature, beginning first with the lower leaves; blossoming is reduced, and the opening of buds in the spring is delayed. The similarity of plant response to phosphorus and nitrogen in these respects may result from two similarities in the two nutrients with regard to behavior in plants. First, of all the mineral nutrients, phosphorus and nitrogen play the most fundamental and all-pervading roles in metabolism. A deficiency of either element impedes metabolism in general with little unbalance. Second, under conditions of deficiency, both phosphorus and nitrogen are withdrawn from older tissue and translocated to meristematic tissue, where metabolism is more rapid.

Further discussion of phosphorus deficiency symptoms and citations to the literature will be found in the book by Wallace (1961) and in the book edited by Sprague (1964). Both these sources contain reproductions of photographs in color showing the comparative appearance of plants deficient in phosphorus and well-supplied with phosphorus.

Root Growth

Phosphorus sometimes is said to stimulate root growth, the implication being that phosphorus has some special effect on the growth of roots that it does not have on the above-ground portion of the plant. In examining this matter, it is important to consider what is meant by "root."

If "root" designates the subterranean storage tissue of root crops, the phosphorus supply does have a special effect. If a root crop is deficient in phosphorus, phosphorus fertilization usually increases the yield of roots relatively more than that of the above-ground parts. Table 8.7 gives an example from a field experiment with mangolds in England. This behavior may be accounted for by a theory patterned after the nitrogen-carbohydrate theory elaborated in Chapter 7. Translocation of carbohydrate to the roots is limited as long as leaf growth continues. The maximum leaf weight is attained at a later date by phosphorus-deficient plants than by phosphorus-fertilized plants. Consequently, translocation of carbohydrate to the storage tissue proceeds for a longer time in phosphorus-fertilized plants than in phosphorus-deficient plants.

Table 8.7. Average Yield of Leaves and Storage Roots of Mangolds with and without Application of Phosphate Fertilizer from 1904 to 1940 at the Rothamsted Experimental Station (Watson and Russell, 1943)

	Yield per Hectare, metric tons	
Plant Part	No Phosphate Added	Phosphate Fertilized
Leaves	5.8	5.8
Storage roots	16.3	20.4

On the other hand, if "root" refers to the absorbing roots, phosphorus does not seem to have any special "stimulating" effect. In fact, treatment of phosphorus-deficient plants with phosphate fertilizer ordinarily increases the yield of the above-ground parts to a greater extent than the absorbing roots. This effect is illustrated in Fig. 8.29, which gives

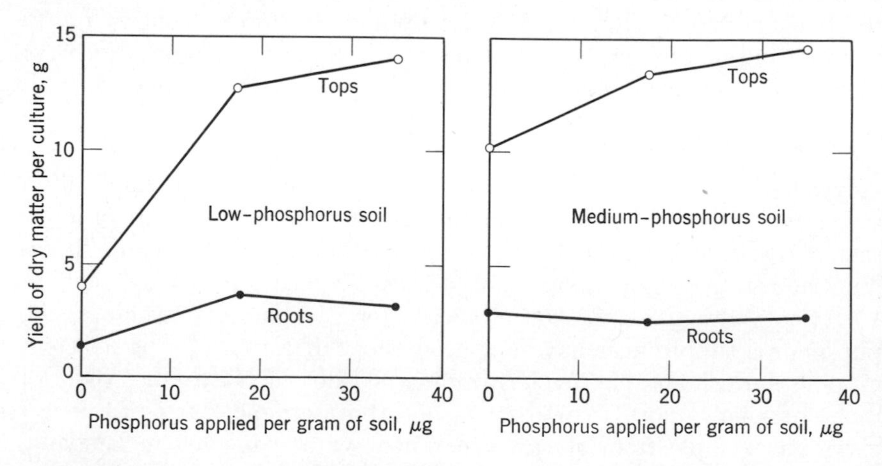

Figure 8.29. Yield of tops and roots of barley on fine sandy loam soils with different applications of phosphorus as concentrated superphosphate. (Power et al., 1963)

the response of the tops and roots of barley to phosphorus fertilization on two fine sandy loam soils differing in supply of phosphorus. One may note that the relative increase in yield from phosphorus fertilization was greater with tops than roots; also, the yield of tops continued to increase with phosphorus fertilization above the level at which there was no increase in yield of roots. An analogous response to nitrogen is shown in Fig. 7.42. The explanation given for the effect of nitrogen may be adapted to account also for the effect of phosphorus in Fig. 8.29. If conditions are made favorable for use of carbohydrate in growth, as when the supply of phosphorus is ample, the proportion of the carbohydrate translocated to the root decreases.

Further consideration of the nature of the two types of crops may aid in clarifying the difference in behavior described in the preceding two paragraphs. The storage root of mangold is not physiologically analogous to the root system of barley. In contrast to the storage root, the absorbing roots of mangold develop along with the above-ground parts and probably are affected by phosphorus supply in the same manner as the root system of barley. Work by Houghland (1947) on potatoes indicates that the effect of phosphorus supply on the relative growth of the absorbing roots (tubers not included) and above-ground parts follows the pattern indicated in Fig. 8.29 for barley; that is, phosphate fertilization increases the growth of the above-ground parts more than the growth of roots. In respect of disposition of carbohydrate produced by photosynthesis, the storage root of mangold and other root crops such as sugar beet is more similar to the grain of barley than to the absorbing roots. Development of both storage root and grain occurs in the latter part of the growing season after the vegetative parts and absorbing roots are well developed, and both serve as a repository for carbohydrate produced.

Development and Maturity

If soil phosphorus availability is high, young plants absorb phosphorus rapidly. By the time they have accumulated 25% of the seasonal total of dry matter, they may have absorbed as much as 50% of the seasonal total of phosphorus.

An ample supply of phosphorus promotes rapid development. For example, Glover (1953) found that corn reached the stages of tasseling and silking at an earlier date where the supply of phosphorus was ample than where it was deficient. Yamashita and Goto (1963) found that differentiation of the first flower cluster by tomato plants took place at progressively earlier dates as the supply of phosphorus increased. In perennials grown at different phosphorus levels, the difference in time required to reach a comparable stage of development is sometimes measurable in years. Specht (1963) found that some species of native heath vegetation on a phosphorus-deficient, sandy soil in Australia flowered 2 years earlier if the soil was treated with superphosphate.

Because of the difference in rates of development, plants deficient in phosphorus mature late compared with plants amply supplied with phosphorus. This principle is illustrated in Table 8.8 by the increase in percentage of the total yield of snap beans obtained in the early crop accompanying an increase in the phosphorus supply. As indicated by work of Crider (1927) and Franck (1931) and the data in Table

Table 8.8. Yield and Earliness of Maturity of Snap Beans with Different Applications of Fertilizer Phosphorus to a Sandy Soil in Alabama (Ware, 1938)

Phosphorus Applied per Hectare in Fertilizer, kg	Yield of Snap Beans per Hectare, hampers	Yield of Early Beans as Percentage of Total Yield
0	20	25
20	227	40
39	400	47
59	388	48

8.8, however, the date of maturity is advanced little or none by phosphorus fertilization unless the yield is increased. In experiments on potatoes, Dainty et al. (1959) found that the rate of development of the tubers was actually delayed by applications of superphosphate in excess of those that produced the maximum final yield. With regard to development and maturity, the response of plants to phosphorus supply thus bears considerable similarity to their response to nitrogen supply, as described in Chapter 7.

The effect of phosphorus fertilization on the time of maturity may be of practical importance with crops that utilize the full growing season, as is true of corn in the northern part of the United States. With other crops in particular environments, earliness of maturity may make possible the harvesting of the crop under relatively favorable weather conditions. Robertson (1927) commented on this matter in connection with experiments on phosphate fertilization of oats in northern Ireland, where the weather hazard is too much rain He noted that oats on phosphate-fertilized plots ripened 2 weeks earlier than oats on the control plots. In Oklahoma, Eck and Stewart (1959) found that the increase in yield of wheat from phosphorus fertilization became greater with an increase in temperature during the ripening period. Here the principal weather hazard during ripening of wheat is high temperatures, and earliness of maturity associated with an ample supply of phosphorus may aid the crop to escape the adverse effect of the heat. In other instances, the hazard may be of a different sort. With cotton, for example, frequent reference is made to the value of the early maturity associated with phosphorus fertilization as a means of escaping damage from the boll weevil.

Response throughout the Season

The relative response of crops to phosphorus fertilization usually is greatest early in the season and decreases gradually as maturity is approached, from which one may infer that the need for additional phosphorus is greater in the early part of the growth cycle than in the latter part. The usual behavior, illustrated in Table 8.9, probably results from interaction of several factors.

First, the supply of phosphorus seems to be more important early in the growth cycle than late. If plants have an adequate supply of phosphorus, the proportion of the seasonal total of phosphorus taken up during early growth characteristically exceeds the proportion of the seasonal total of dry matter produced. For example, in the experiment described in Table 8.9, 28% of the phosphorus absorbed by the fertilized plants by July 8 had been taken up by June 2; but only 16% of the

Table 8.9. Total Yield of Dry Matter and Yield of Phosphorus of Oats at Different Dates on Field Plots Differing with Regard to Fertilization with Superphosphate in Iowa (Unpublished data, Iowa Agricultural Experiment Station and U.S. Department of Agriculture, 1949)

	Yield of Dry Matter per Hectare			Yield of Phosphorus per Hectare		
Date of Sampling	No P Added, kg	39 kg P per Hectare, kg	Increase from P, %	No P Added, kg	39 kg P per Hectare, kg	Increase from P, %
June 2	570	760	33	1.6	2.5	56
June 22	3000	3400	13	6.1	8.1	33
July 8	4200	4700	12	7.3	9.0	23

yield of dry matter on July 8 had been accumulated by June 2. During the latter part of the growth cycle, much of the phosphorus that accumulates in the fruit is transferred from the previously absorbed supply contained in the vegetative parts and is not taken up from the soil during fruit development. If the supply of phosphorus is adequate to produce the maximum growth of vegetative parts, little additional phosphorus is needed during fruit production, at least with some plants (Gericke, 1925; Brenchley, 1929).

Second, the availability of fertilizer phosphorus tends to decrease relative to availability of soil phosphorus as the season progresses. This effect is probably due to several factors, including reaction of the fertilizer with the soil, preferential removal of water from the upper part of the soil where the fertilizer is located, and extension of roots into soil that does not contain fertilizer phosphorus.

Third, field observations indicate that phosphorus deficiency is more pronounced at low temperatures than at high temperatures. Accordingly, one may expect that in situations such as the one described in Table 8.9, in which the temperature increased through the season, the decrease in response to fertilization with time might be in part an effect of temperature. Because comparisons of temperature effects in the field usually

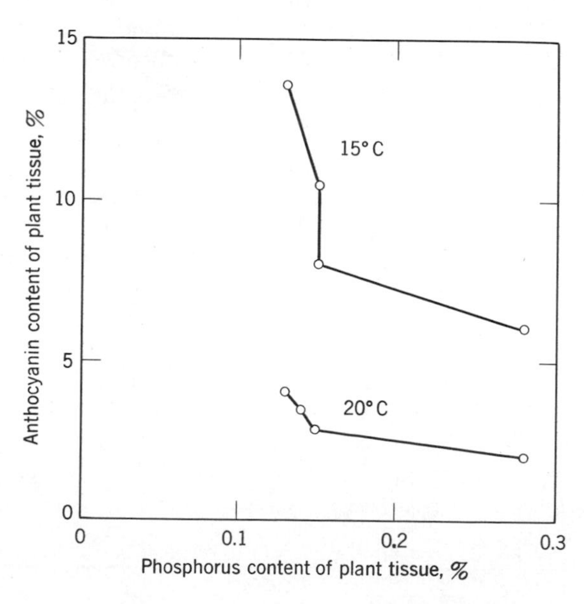

Figure 8.30. Anthocyanin percentage versus phosphorus percentage in corn plants grown in sand cultures with different phosphorus concentrations at two temperatures. Air temperatures were the same for all treatments. (Knoll et al., 1964)

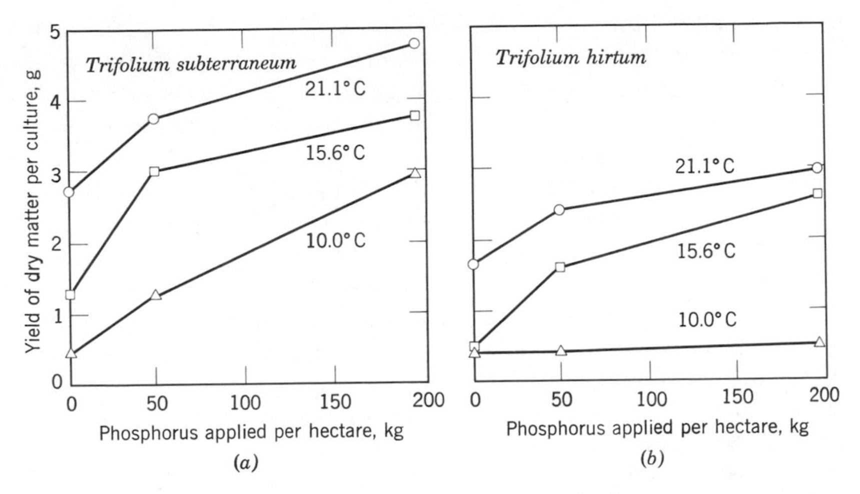

Figure 8.31. Yield of two species of clover on a silty clay loam soil with different applications of phosphoric acid at three temperatures. (McKell et al., 1962)

involve different years, locations, or both, however, they are not necessarily valid. Data in Fig. 8.30 obtained by Knoll et al. (1964) in experimental work conducted under controlled conditions suggest that the significance of the temperature effect in the field is likely to be over estimated if it is based on the purple coloration associated with phosphorus deficiency because production of anthocyanin pigments responsible for the color is promoted by low root temperature regardless of the phosphorus supply. Nevertheless, numerous controlled experiments have verified that phosphorus deficiency does tend to increase at low soil temperatures. The verification takes the form shown in Fig. 8.31*a*, in which the relative increase in yield of *Trifolium subterraneum* from fertilization may be seen to diminish with an increase in temperature. Several different factors, including increases in translocation of phosphorus from roots to tops, root growth (and hence area of root-soil contact), rate of phosphorus uptake, mineralization of soil organic phosphorus, and rate of reaction of fertilizer phosphorus with soil at higher temperatures, have been proposed to account for this effect. In view of the variety of factors and their dependence on the circumstances, the temperature effect may be expected to be rather variable, as indeed the results of experiments indicate. The data for *Trifolium hirtum* in Fig. 8.31*b* have been included in recognition of the complexity of the matter. These data were obtained in the same experiment as those for *Trifolium subterraneum*. Results obtained with *T. hirtum* at the higher two temperatures agree with the low-temperature, phosphorus-deficiency trend; however, at the lowest temperature, where the general trend would indicate the greatest response, the relative increase in yield from phosphorus fertilization was actually the least. In this case, the explanation is presumably that the lowest temperature was too low for appreciable growth of *T. hirtum,* with or without phosphorus fertilization. Numerous citations to literature on this subject, together with original experimental data, may be found in papers by Case et al. (1964) and Power et al. (1964,1964a).

The pattern of response of plants to phosphorus fertilization shown in Table 8.9 could conceivably have different implications concerning the most suitable time of application of phosphate fertilizers. For example, if the most important cause of the gradual decrease in relative response with time is loss of potency of the fertilizer because of reaction with the soil, delayed or repeated applications of phosphate fertilizer would appear useful. On the other hand, if the most important cause of the decrease in response is an increase in availability of soil phosphorus or a decrease in need by the plants with time, early applications of phosphate fertilizer would appear most useful. Early application seems to be the rule in practice, and the few field experimental data (e.g., Jones and Warren, 1954) available support this procedure. The practice of phosphorus fertilization thus differs from that of nitrogen fertilization, in which delayed applications are frequently made and in which the supposed benefit of delayed applications under some circumstances is supported by experimental data.

The comparative seasonal response of plants to phosphorus and nitrogen is sufficiently different to justify special comment. The decrease in response with time during the season, illustrated in Table 8.9 by results of an experiment in which phosphorus fertilizer was applied at planting, may occur also where nitrogen fertilizer is applied at planting. In other instances, the greatest relative increase in yield from nitrogen fertilization at planting may occur at a relatively late date, and there may be little or no increase in yield when the plants are young. A decrease in response to nitrogen fertilization with time during the season may be expected if the soil does not supply enough nitrogen to meet plant needs at the beginning of the season and if the nitrogen supplied by the fertilizer is considerably less than the total supplied by the soil throughout the season. Greater relative response late in the season than early may be expected if the soil supplies enough nitrogen at the beginning of growth but becomes deficient later.

A full explanation of these matters would be lengthy. In brief, however, it may be pointed out that the difference in behavior of nitrogen and phosphorus in soil appears to form an important part of the explanation. Soil may provide enough nitrogen for early growth of a crop by carryover from fallow or previous fertilization or cropping, and the crop may later become deficient in nitrogen because of exhaustion of the initial supply coupled with a low current supply. The supply of soil phosphorus, however, is much more stable and, under practical conditions, cannot be exhausted during a single season. Where delayed applications of fertilizer are concerned, the value of nitrogen for the current crop may greatly exceed the value of phosphorus. Delayed applications of fertilizer usually are placed on or near the surface of the soil for convenience or prevention of undue damage to roots, and these surface locations are likely to become relatively dry. Nitrogen added as nitrate may move downward into the root zone with a single rain, whereas phosphorus remains near the site of its original placement.

Disease Incidence

The most frequent reference to effects of phosphorus supply on incidence of plant diseases is in relation to those caused by fungi. Infection of barley by mildew on field plots in England was found by Last (1962) to decrease in the presence of phosphorus. Phosphorus counteracted to a large degree the effect of nitrogen in increasing the infection. The results are summarized in Fig. 8.32. In this experiment, nitrogen

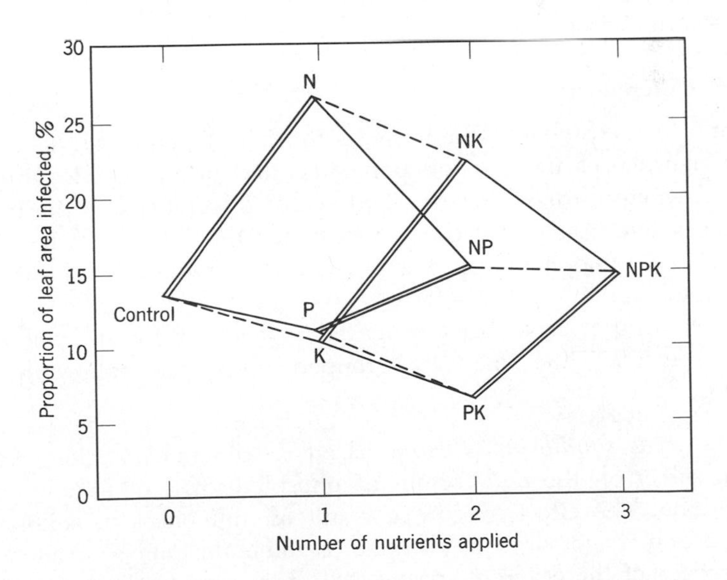

Figure 8.32. Infection of barley with powdery mildew on field plots in England with nitrogen, phosphorus, and potassium fertilizers applied singly and in all combinations. (Last, 1962)

and phosphorus were definitely deficient, but potassium was only slightly deficient, as indicated by average increases of 0.9, 0.8, and 0.2 metric tons of grain per hectare with applications of the three nutrients.

Similarly, the prevalence of fungal root-rots is greater in plants with a deficient supply of phosphorus than in those with an ample supply. This trend is noted particularly with plants in the seedling stage. The manner in which the effect of phosphorus is brought about, however, has received little attention. In some instances, the effect of phosphorus appears to be exerted not so much through a change in the inherent susceptibility of a particular root to infection as through an increase in the rate of growth and production of new roots. In such cases an adequate supply of phosphorus may be said to facilitate escape from the diseases rather than to increase resistance to them. The work of Vanterpool (1935) on pythium root-rot of wheat may be mentioned in this connection.

The effect of phosphorus supply in some instances is just the opposite of that described for fungal root-rots; that is, the susceptibility to the disease is greater where the supply of phosphorus is ample than where it is deficient. Several such instances were noted by McNew (1953), one of these being the virus diseases. Although it is not certain that the observed effect is traceable entirely to the phosphorus content of the plants, phosphorus is a constituent of the nucleoprotein molecules that constitute virus diseases.

Species Differences

Plant species probably differ more in their reaction to the supply of soil phosphorus than to soil potassium and nitrogen if legumes are excepted where nitrogen is concerned. Table 8.10 illustrates the marked differences found among certain crop plants that were grown individually in the same field experiment with and without application of superphosphate.

Various theories have been proposed to account for the differences among species. These may be grouped under the following three headings:

1. The *ionic-equilibrium theory,* based on differences among species in their effect on the equilibrium of phosphate and other ions in the soil solution. The effect may be a result of differences in acidity, calcium-ion activity, or organic phosphorus mineralization associated with the presence of the different species. All these aspects of the ionic-equilibrium theory have been discussed previously.
2. The *root-character theory,* based on differences among species with

regard to phosphorus-absorption characteristics and extent of the root system. This theory has not been discussed previously. Noggle and Fried (1960) investigated the phosphorus-absorption characteristics of three plant species by determining the uptake of phosphorus from solution on a short-term basis by detached roots of seedling plants. Some of their results are given in Fig. 8.33, which shows that the order of quanti-

Table 8.10. Ratio of Yield of Different Crops Obtained without Phosphate Fertilization to the Yield Obtained with 11.2 Metric Tons of Concentrated Superphosphate per Hectare on a Clay Loam Soil in California (Lilleland et al., 1942)

Crop	Yield of Control / Yield with Superphosphate Added
Squash	0.03
Cucumber	0.07
Corn	0.21
Wheat	0.38
Oats	0.41
Alfalfa	0.57
Wax beans	0.63
Almond	1.00[1]

[1] Ratio of cross-sectional areas of trunk.

ties of phosphorus absorbed during the test period was millet > barley > alfalfa. From the figure, one may note that the extrapolated uptake at zero time is in the same order; this suggests that the differences in rate of active uptake were associated with differences in number of phosphate-specific carrier sites per gram of root. Generally, one would expect the concentration of carrier sites per gram of roots to increase with a decrease in diameter of the roots because of the increase in surface area. In this experiment, however, there was no definite trend in this direction. Barley roots had the largest diameter, and millet and alfalfa roots were smaller and similar. Apparently there were real differences in character of the root surfaces. Because an increase in extent of a given kind of root provides both an increase in number of uptake sites and a positioning of these sites at additional locations, the root-extent part of the theory is self-evident in principle. In practice, however, this part of the theory has been difficult to subject to experi-

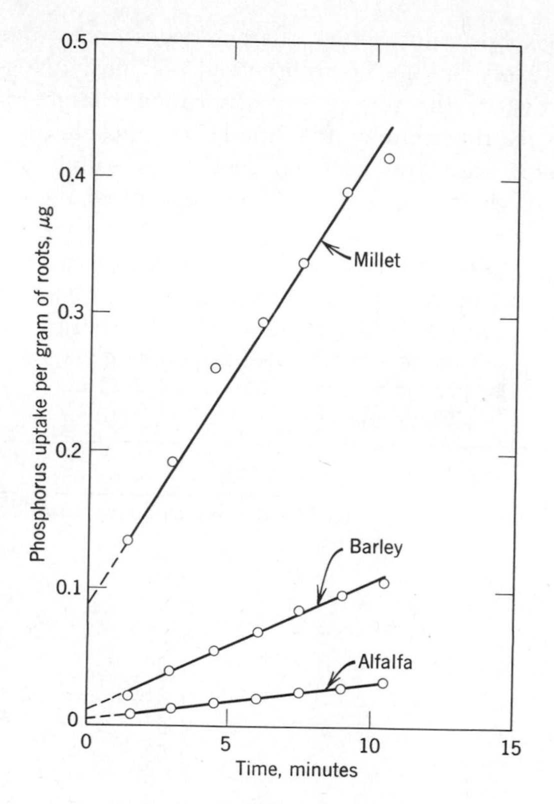

Figure 8.33. Short-term uptake of phosphorus by detached roots of millet, barley, and alfalfa at 30°C from a solution containing 15.5 μg of phosphorus as P^{32}-labeled orthophosphate per milliliter. (Noggle and Fried, 1960)

mental test in explaining species differences because the character of a given root varies from one portion to another, because methods of measuring root surface area have not been perfected, and because the significance of root extent is probably modified by the character of the plant to which the root is attached.

3. *The phosphorus-requirement theory,* based on differences among species in rate of phosphorus uptake. This theory is an expression of the common observation that slow-growing species, low in phosphorus content, are better adapted to growth under conditions of phosphorus deficiency than are fast-growing species high in content of phosphorus. In particular, perennial species that conserve a large part of the absorbed phosphorus in living tissue from one growing season to the next are highly efficient. In the experiment described in Table 8.10, for example,

almond trees needed no additional phosphorus to make good growth on a soil on which various annual plants were clearly in need of additional phosphorus.

Apparently a variety of causes may exist for the differences in degree to which the yield of different plant species can be increased by phosphorus fertilization if grown on a given soil. If any two species were to be chosen at random, they might well differ in respect of several and perhaps all the factors involved. Because concepts of the nature of soil-plant relationships involving phosphorus are still in the formative stage, and because the conditions thought to be significant cannot be measured quantitatively, essentially no progress has been made in assessing the relative importance of individual factors in particular cases. In practical work, such as testing soils for phosphorus supply and making inferences regarding fertilizer needs, therefore, knowledge is needed about the behavior of each crop.

Plant Competition

One of the consequences of differences among species in their response to different levels of supply of soil phosphorus is variation in relative competitive ability with the supply of phosphorus. Marked differences in botanical composition of mixed plantings and of natural vegetation thus may accompany variations in the phosphorus supply. Illustrative data have been published by Rossiter (1947), Specht (1963), and others. Usually the proportion of legumes in the total produce of mixed plantings of forage crops increases with an increase in the supply of phosphorus, which means that the competitive ability of the legumes is poorer under conditions of low than high soil phosphorus availability.

Relatively little research has been done on the soil aspects of plant competition in relation to phosphorus behavior, but some insight into the matter may be provided by discussion of results of an experiment carried out by Chambers and Holm (1965). These investigators grew a single snap bean test plant in the center of a field plot 1.2 meters square with one to four pigweed (*Amaranthus retroflexus*) plants grown around it in a circle with a radius of 15.3 cm. A solution containing P^{32}-labeled orthophosphate was injected in the soil in a circular pattern at five locations as follows in relation to the central test plant: (1) radius 7.6 cm, depth 7.6 cm, (2) radius 15.3 cm, depth 15.3 cm, (3) radius 15.3 cm, depth 30.5 cm, (4) radius 30.5 cm, depth 15.3 cm, and (5) radius 30.5 cm, depth 30.5 cm. Values for relative dry weight and relative radioactivity of the bean plants harvested 42 days after planting are given in Fig. 8.34. Evidently the uptake of radioactive phosphorus

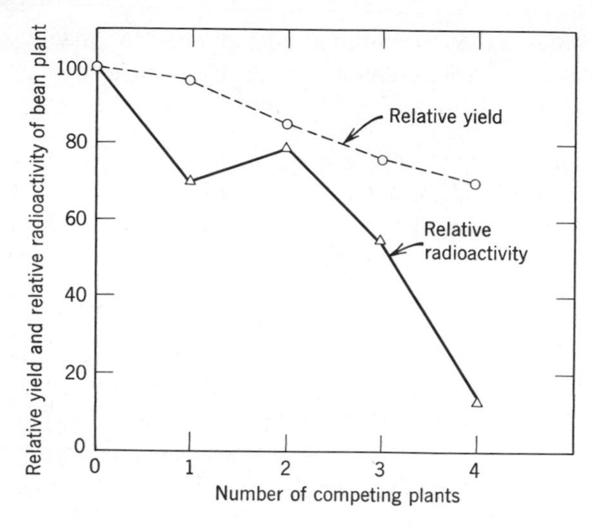

Figure 8.34 Relative yield and relative radioactivity of a single bean plant with zero to four competing pigweed (*Amaranthus retroflexus*) plants located around it in a circle with a radius of 15.3 cm. The radioactivity was derived from a solution of P^{32}-labeled orthophosphate injected into the soil in a pattern described in the text. (Chambers and Holm, 1965)

by the single central test plant was reduced to a much greater extent by competition from pigweed plants than was the yield of dry matter. From this behavior, one may conclude that competition for the radioactive phosphorus supplied as fertilizer was more severe than the competition for other growth factors leading to the production of plant tissue by the bean plant.

The question now to be considered is the manner in which the presence of adjacent plants could produce such an effect. One possibility is a direct competition for phosphorus in which uptake of phosphorus by one plant reduces the uptake by another plant by reducing the supply in the soil. This interpretation follows from the discussion of competition in Chapter 7, in which the point was made that plants may nearly exhaust the nitrate from the soil on which they grow, thus causing direct competition among associated plants for nitrate.

Phosphate behaves so differently from nitrate in soil, however, that there is reason to question the validity of the obvious interpretation given in the preceding paragraph for the experimental data shown in Fig. 8.34. If such competition were to account for the observations, plants should exhaust added fertilizer phosphorus from soil as they do nitrate; but plants rarely recover more than 20% of the added phosphorus in a single season.

Another type of experiment provides evidence that the behavior of phosphate is by no means analogous to that of nitrate where direct competition among roots is concerned. Tepe and Leidenfrost (1958) surrounded cylinders of exchange resins with layers of water-saturated soil differing in thickness and then determined the quantities of various ions taken up by the resin in 24 hours. The results in Fig. 8.35 indicate that uptake of ions by the resin did not affect the concentration of phosphate in the soil at a distance of 0.3 cm or more but that it affected the concentration of nitrate to a distance at least as great as 1.5 cm. The difference between phosphate and nitrate where plants are concerned is probably underestimated by this sort of experiment because the ion transport was accomplished by diffusion without the mass movement that would occur from uptake of water by plant roots. In any event, the evidence indicates that movement of phosphate through soil to roots occurs only through a thin layer of soil, whereas movement of nitrate occurs through a much greater distance. The implication of this behavior in terms of direct competition for ions among roots of different plants is that the roots must be close together before they

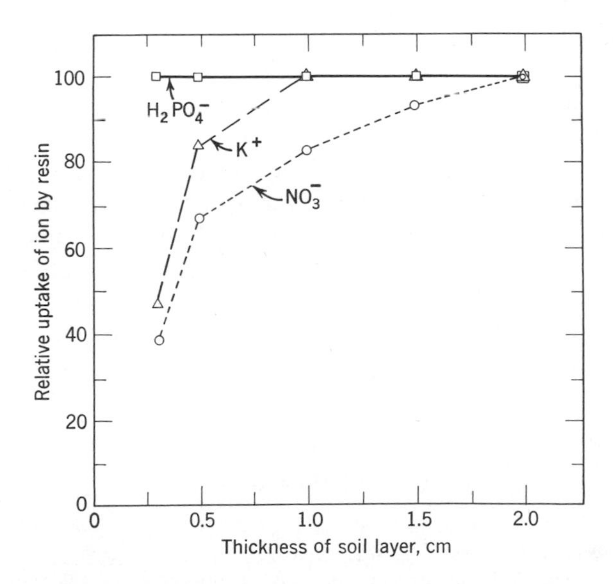

Figure 8.35. Uptake of ions by a cylinder of a mixed bed of cation- and anion-exchange resin from surrounding layers of a loam soil differing in thickness during a period of 24 hours. (Tepe and Leidenfrost, 1958)

will compete directly for phosphate ions but that they will compete directly for nitrate over distances of several centimeters.

These considerations about the behavior of phosphorus and nitrogen suggest that the concept of direct competition among roots for phosphorus was not responsible for the observations recorded in Fig. 8.34. An alternative and seemingly more reasonable explanation is that the competition for radioactive phosphorus was for the most part a consequence of exclusion of roots of the central test plant from part of the soil containing the radioactive phosphorus. The exact locations of the circles of radioactive phosphorus were given in connection with the description of the experiment to show that four of the five circular zones in which radioactive phosphorus was applied were located directly below or outside of the circle of competing plants; roots of competing plants thus should have colonized these zones first, thereby excluding to a considerable extent the roots of the test plant.

Because roots must be so close together to compete directly for soil phosphorus, it seems reasonable that the variation in botanical composition associated with differences in phosphorus supply in soils is not primarily a consequence of direct competition for phosphorus. It seems more likely that the order of cause and effect may be the following: (1) Some species are much more sensitive than others to a low supply of soil phosphorus. When grown on soil low in phosphorus, deficiency-sensitive species therefore will form a lower percentage of the total weight of vegetation than they do at high phosphorus levels even with a constant stand of all species and in absence of competition for growth factors other than phosphorus. (2) As a result of their relatively poor growth on soil low in phosphorus, the deficiency-sensitive species will lose part of their competitive ability for growth factors other than phosphorus. The deficiency-tolerant species, not so affected, may then eliminate some or all of the plants of the deficiency-sensitive species by competition for growth factors other than phosphorus, so that both the percentage of the total number of plants and the percentage of the total weight of the vegetation shift in the direction of an increase in the deficiency-tolerant species.

Literature Cited

Adams, A. P., W. V. Bartholomew, and F. E. Clark. (1954) Measurement of nucleic acid components in soil. *Soil Sci. Soc. Amer. Proc.* **18**:40–46.

Anderson, G. (1961) Estimation of purines and pyrimidines in soil humic acid. *Soil Sci.* **91**:156–161.

Anderson, G., and R. J. Hance. (1963) Investigation of an organic phosphorus component of fulvic acid. *Plant and Soil* **19**:296–303.

Arneman, H. F., and P. R. McMiller. (1955) The physical and mineralogical properties of related Minnesota prairie soils. *Soil Sci. Soc. Amer. Proc.* **19**:348–351.

Australia. (1956) *Commonwealth Scientific and Industrial Research Organization Ann. Rept.* **8**:18–19.

Barber, S. A., J. M. Walker, and E. H. Vasey. (1962) Principles of ion movement through the soil to the plant root. *Internat. Soc. Soil Sci., Trans. Joint Meeting Comm. IV & V* (New Zealand, 1962):121–124.

Barrow, N. J. (1961) Phosphorus in soil organic matter. *Soils and Fertilizers* **24**:169–173.

Bates, J. A. R., and T. C. N. Baker. (1960) Studies on a Nigerian forest soil. II. The distribution of phosphorus in the profile and in various soil fractions. *Jour. Soil Sci.* **11**:257–265.

Black, C. A., and C. A. I. Goring. (1953) Organic phosphorus in soils. *Agronomy* **4**:123–152.

Bogdanović, M., L. Nikolić, and S. Stojanović. (1963) Vivijanit iz tresetišta kod horgoša. *Zeml. Bilj.* **12**:77–81.

Bower, C. A. (1949) Studies on the forms and availability of soil organic phosphorus. *Iowa Agr. Exp. Sta. Res. Bul. 362.*

Bray, R. H., and L. T. Kurtz. (1945) Determination of total, organic, and available forms of phosphorus in soils. *Soil Sci.* **59**:39–45.

Brenchley, W. E. (1929) The phosphate requirement of barley at different periods of growth. *Ann. Bot.* **43**:89–110.

Burd, J. S. (1948) Chemistry of the phosphate ion in soil systems. *Soil Sci.* **65**:227–247.

Caldwell, A. G., and C. A. Black. (1958) Inositol hexaphosphate: II. Synthesis by soil microorganisms. *Soil Sci. Soc. Amer. Proc.* **22**:293–296.

Caldwell, A. G., and C. A. Black. (1958a) Inositol hexaphosphate: III. Content in soils. *Soil Sci. Soc. Amer. Proc.* **22**:296–298.

Case, V. W., N. C. Brady, and D. J. Lathwell. (1964) The influence of soil temperature and phosphorus fertilizers of different water-solubilities on the yield and phosphorus uptake by oats. *Soil Sci. Soc. Amer. Proc.* **28**:409–412.

Chambers, E. E., and L. G. Holm. (1965) Phosphorus uptake as influenced by associated plants. Weeds **13**: 312–314.

Chang, S. C., and M. L. Jackson. (1957) Fractionation of soil phosphorus. *Soil Sci.* **84**:133–144.

Chang, S. C., and M. L. Jackson. (1958) Soil phosphorus fractions in some representative soils. *Jour. Soil Sci.* **9**:109–119.

Chaverri, J. G., and C. A. Black. (1966) Theory of the solubility of phosphate rock. *Iowa State Jour. Sci.* **41**:77–95.

Clark, J. S., and M. Peech. (1960) Influence of neutral salts on the phosphate ion concentration in soil solution. *Soil Sci. Soc. Amer. Proc.* **24**:346–348.

Cook, S. F., and R. F. Heizer. (1965) Studies on the chemical analysis of archaeological sites. *Univ. Calif. Publ. Anthropol. 2.*

Cooper, R. (1959) Bacterial fertilizers in the Soviet Union. *Soils and Fertilizers* **22**:327–333.

Cosgrove, D. J. (1963) The chemical nature of soil organic phosphorus. I. Inositol phosphates. *Australian Jour. Soil. Res.* **1**:203–214.

Cosgrove, D. J. (1964) An examination of some possible sources of soil inositol phosphates. *Plant and Soil* **21**:137–141.

Cosgrove, D. J., and M. E. Tate. (1963) Occurrence of *neo*-inositol hexaphosphate in soil. *Nature* **200**:568–569.

Crider, F. J. (1927) Effect of phosphorus in the form of acid phosphate upon maturity and yield of lettuce. *Arizona Agr. Exp. Sta. Bul. 121.*

Cunningham, R. K. (1963) The effect of clearing a tropical forest soil. *Jour. Soil Sci.* **14**:334–345.

Dainty, J., R. D. Verma, and K. Simpson. (1959) Studies on the uptake of phosphorus from ^{32}P-labelled superphosphate by crops. II.-Potatoes. *Jour. Sci. Food Agr.* **10**:108–114.

Dauncey, K. D. M. (1952) Phosphate content of soils on archaeological sites. *Advanc. Sci.* **9**:33–36.

Diest, A. van, and C. A. Black. (1959) Soil organic phosphorus and plant growth. II. Organic phosphorus mineralized during incubation. *Soil Sci.* **87**:145–154.

Diest, A. van, H. W. Jespersen, R. F. White, and C. A. Black. (1960) Test of two methods for measuring a labile fraction of inorganic phosphorus in soils. *Soil Sci. Soc. Amer. Proc.* **24**:498–502.

Drake, M., and J. E. Steckel. (1955) Solubilization of soil and rock phosphate as related to root cation exchange capacity. *Soil Sci. Soc. Amer. Proc.* **19**:449–450.

Dudley, W. L. (1890) A curious occurrence of vivianite. *Amer. Jour. Sci.,* Ser. 3, **40**:120–121.

Duff, R. B., and D. M. Webley. (1959) 2-Ketogluconic acid as a natural chelator produced by soil bacteria. *Chem. and Ind.* **1959**:1376–1377.

Dyal, R. S. (1953) Mica leptyls and wavellite content of clay fraction from Gainesville loamy fine sand of Florida. *Soil Sci. Soc. Amer. Proc.* **17**:55–58.

Eck, H. V., and B. A. Stewart. (1959) Response of winter wheat to phosphate as affected by soil and climatic factors. *Agron. Jour.* **51**:193–195.

Elgabaly, M. M. (1962) On the mechanism of anion uptake by plant roots: 2. Effect of the cation-exchange capacity of plant roots on Cl^- uptake. *Soil Sci.* **93**:350–352.

Estermann, E. F., and A. D. McLaren. (1961) Contribution of rhizoplane organisms to the total capacity of plants to utilize organic nutrients. *Plant and Soil* **15**:243–260.

Failyer, G. H., J. G. Smith, and H. R. Wade. (1908) The mineral composition of soil particles. *U.S. Dept. Agr., Bur. Soils Bul. 54.*

Fenton, T. E. (1966) Soils, weathering zones, and landscapes in the upland loess of Tama and Grundy Counties, Iowa. Ph.D. Thesis, Iowa State University, Ames.

Fiedler, H. J., and W. Jahn-Deesbach. (1956) Über die Einwirkung von Mikroorganismen auf die organischen Phosphorverbindungen des Bodens. *Deut. Landw.* **7**:602–605.

Fieldes, M., G. Bealing, G. G. Claridge, N. Wells, and N. H. Taylor. (1960) Mineralogy and radioactivity of Niue Island soils. *New Zealand Jour. Sci.* **3**:658–675.

Food and Agriculture Organization of the United Nations. (1965) Fertilizers, an annual review of world production, consumption and trade, 1964. Food and Agriculture Organization of the United Nations, Rome.

Fox, R. L., and B. Kacar. (1964) Phosphorus mobilization in a calcareous soil in relation to surface properties of roots and cation uptake. *Plant and Soil* **20**:319–330.

Franck, O. (1931) Gödsling, mognadstid och kärnkvalitet. *Nordisk Jordbrugsforskning* **13**:282–290.

Fried, M., C. E. Hagen, J. F. Saiz del Rio, and J. E. Leggett. (1957) Kinetics of phosphate uptake in the soil-plant system. *Soil Sci.* **84**:427–437.

Fry, W. H. (1913) The condition of soil phosphoric acid insoluble in hydrochloric acid. *Jour. Ind. Eng. Chem.* **5**:664–665.

Gericke, W. F. (1925) Salt requirements of wheat at different growth phases. *Bot. Gaz.* **80**:410–425.

Gerretsen, F. C. (1948) The influence of microorganisms on the phosphate intake by the plant. *Plant and Soil* **1**:51–81.

Glover, J. (1953) The nutrition of maize in sand culture. I. The balance of nutrition with particular reference to the level of supply of nitrogen and phosphorus. *Jour. Agr. Sci.* **43**:154–159.

Goel, K. N., and R. R. Agarwal. (1960) Total and organic phosphorus in different size-fractions in genetically related soils of Kanpur in the Indian Gangetic alluvium. *Indian Jour. Soil Sci.* **8**:17–22.

Goring, C. A. I., and W. V. Bartholomew. (1950) Microbial products and soil organic matter: II. The effect of clay on the decomposition and separation of the phosphorus compounds in microorganisms. *Soil Sci. Soc. Amer. Proc.* (1949) **14**:152–156.

Haas, H. J., D. L. Grunes, and G. A. Reichman. (1961) Phosphorus changes in Great Plains soils as influenced by cropping and manure applications. *Soil Sci. Soc. Amer. Proc.* **25**:214–218.

Hagen, C. E., and H. T. Hopkins. (1955) Ionic species in orthophosphate absorption by barley roots. *Plant Physiol.* **30**:193–199.

Hagin, J., and A. Hadas. (1962) Solubility of calcium phosphate in calcareous soils. *Nature* **193**:1211–1212.

Halstead, R. L., J. M. Lapensee, and K. C. Ivarson. (1963) Mineralization of soil organic phosphorus with particular reference to the effect of lime. *Canadian Jour. Soil Sci.* **43**:97–106.

Hance, R. J., and G. Anderson. (1963) Extraction and estimation of soil phospholipids. *Soil Sci.* **96**:94–98.

Hance, R. J., and G. Anderson. (1963a) Identification of hydrolysis products of soil phospholipids. *Soil Sci.* **96**:157–161.

Hayashi, T., and Y. Takijima. (1955) Studies on utilization of soil organic phosphorus to crop plants (Part 3). On reduction of organic phosphorus content in soils caused by cropping. *Jour. Sci. Soil and Manure,* Japan **26**:215–218.

Heslep, J. M., and C. A. Black. (1954) Diffusion of fertilizer phosphorus in soils. *Soil Sci.* **78**:389–401.

Houghland, G. V. C. (1947) Minimum phosphate requirement of potato plants grown in solution cultures. *Jour. Agr. Res.* **75**:1–18.

Ignatieff, V., and H. J. Page. (1958) *Efficient Use of Fertilizers.* FAO Agricultural Studies No. 43.

Jackman, R. H. (1964) Accumulation of organic matter in some New Zealand soils under permanent pasture. I. Patterns of change of organic carbon, nitrogen, sulphur, and phosphorus. *New Zealand Jour. Agr. Res.* **7**:445–471.

Jackson, P. C., S. B. Hendricks, and B. M. Vasta. (1962) Phosphorylation by barley root mitochondria & phosphate absorption by barley roots. *Plant Physiol.* **37**:8–17.

Jakob, H. (1955) Die Bedeutung der Phosphatmethode für die Urgeschichte und

Bodenforschung. *In* W. Rothmaler and W. Padberg (Eds.) *Beiträge zur Frühgeschichte der Landwirtschaft.* **II**:67–85. *Wissenschaftliche Abhandlungen* No. 15, **2**:67–85. Deutsche Akademie der Landwirtschaftswissenschaften zu Berlin.

Jenny, H., J. Vlamis, and W. E. Martin. (1950) Greenhouse assay of fertility of California soils. *Hilgardia* **20**:1–8.

Jones, L. G., and G. F. Warren. (1954) The efficiency of various methods of application of phosphorus for tomatoes. *Proc. Amer. Soc. Hort. Sci.* **63**:309–319.

Kaila, A. (1956) Phosphorus in various depths of some virgin peat lands. *Maat. Aikak.* **28**:90–104.

Kaila, A. (1965) Effect of liming on the mobilization of soil phosphorus. *Maat. Aikak.* **37**:243–254.

Knoll, H. A., D. J. Lathwell, and N. C. Brady. (1964) The influence of root zone temperature on the growth and contents of phosphorus and anthocyanin of corn. *Soil Sci. Soc. Amer. Proc.* **28**:400–403.

Larsen, S., and M. N. Court. (1961) Soil phosphate solubility. *Nature* **189**:164–165.

Larsen, S., D. Gunary, and J. R. Devine. (1964) Stability of granular dicalcium phosphate dihydrate in soil. *Nature* **204**:1114.

Larsen, S., D. Gunary, and C. D. Sutton. (1965) The rate of immobilization of applied phosphate in relation to soil properties. *Jour. Soil Sci.* **16**:141–148.

Larsen, S., and A. E. Widdowson. (1964) Effect of soil/solution ratio on determining the chemical potentials of phosphate ions in soil solutions. *Nature* **203**:942.

Last, F. T. (1962) Effects of nutrition on the incidence of barley powdery mildew. *Plant Path.* **11**:133–135.

Lawton, K., and J. F. Davis. (1956) The effect of liming on the utilization of soil and fertilizer phosphorus by several crops grown on acid organic soils. *Soil Sci. Soc. Amer. Proc.* **20**:522–526.

Leahey, A. (1934–1935) Mineralogical and chemical studies on some of the inorganic phosphorus compounds in the soil. *Scientific Agr.* **15**:704–712.

Leggett, J. E., R. A. Galloway, and H. G. Gauch. (1965) Calcium activation of orthophosphate absorption by barley roots. *Plant Physiol.* **40**:897–902.

Lehr, J. R., and W. E. Brown. (1958) Calcium phosphate fertilizers: II. A petrographic study of their alteration in soils. *Soil Sci. Soc. Amer. Proc.* **22**:29–32.

Lilleland, O., J. G. Brown, and J. P. Conrad. (1942) The phosphate nutrition of fruit trees. III. Comparison of fruit tree and field crop responses on a phosphate deficient soil. *Proc. Amer. Soc. Hort. Sci.* **40**:1–7.

Lindsay, W. L., M. Peech, and J. S. Clark. (1959) Solubility criteria for the existence of variscite in soils. *Soil Sci. Soc. Amer. Proc.* **23**:357–360.

Lipman, J. G., and A. B. Conybeare. (1936) Preliminary note on the inventory and balance sheet of plant nutrients in the United States. *New Jersey Agr. Exp. Sta. Bul. 607.*

Louw, H. A., and D. M. Webley. (1959) The bacteriology of the root region of the oat plant grown under controlled pot culture conditions. *Jour. Appl. Bact.* **22**:216–226.

Louw, H. A., and D. M. Webley. (1959a) A study of soil bacteria dissolving certain mineral phosphate fertilizers and related compounds. *Jour. Appl. Bact.* **22**:227–233.

Low, P. F., and C. A. Black. (1950) Reactions of phosphate with kaolinite. *Soil Sci.* **70**:272–290.

Ludecke, T. E. (1962) Formulation of a rational fertiliser programme in tussock country. *Proc. New Zealand Grassland Assoc.* **24**:29–41.

Machold, O. (1962) Die Pflanzenaufnehmbarkeit des "labilen" Phosphats. *Zeitschr. Pflanzenernähr., Düng., Bodenk.* **98**:99–113.

Machold, O. (1963) Über die Bindungsform des "labilen" Phosphats im Boden. *Zeitschr. Pflanzenernähr., Düng., Bodenk.* **103**:132–138.

Machold, O. (1964) Die Verwendung der Isotopenverdünnungsanalyse zur Untersuchung des Phosphathaushaltes im Boden. *Phosphorsäure* **24**:300–318.

Mason, B., and T. Berggren. (1941) A phosphate-bearing spessartite garnet from Wodgina, Western Australia. *Geologiska Föreningens,* Stockholm, Förhandlingar **63**:413–418.

Mattingly, G. E. G. (1958) Phosphate concentrations in soil in relation to crop growth. *Rothamsted Exptl. Sta. Rpt.* **1957**:59–60.

Mattingly, G. E. G. (1958a) Phosphate residues in Rothamsted soils. *Rothamsted Exptl. Sta. Rpt.* **1957**:61.

Mattingly, G. E. G. (1965) The influence of intensity and capacity factors on the availability of soil phosphate. *Min. Agr., Fisheries and Food* (London) *Tech. Bul.* **13**:1–9.

Mattingly, G. E. G., M. Kuskizaki, and B. M. Close. (1960) Rates of growth and phosphorus uptake by ryegrass on calcareous soils. *Rothamsted Exptl. Sta. Rpt.* **1959**:44–45.

Mattingly, G. E. G., R. D. Russell, and B. M. Jephcott. (1963) Experiments on cumulative dressings of fertilisers on calcareous soils in southwest England. II. Phosphorus uptake by ryegrass in the greenhouse. *Jour. Sci. Food Agr.* **14**:629–637.

Mattingly, G. E. G., and O. Talibudeen. (1961) Isotopic exchange of phosphates in soil. Review of experimental techniques and results at Rothamsted 1952–60. *Rothamsted Exptl. Sta. Rpt.* **1960**:246–265.

Mattingly, G. E. G., and F. V. Widdowson. (1963) Residual value of superphosphate and rock phosphate on an acid soil. I. Yields and phosphorus uptakes in the field. *Jour. Agr. Sci.* **60**:399–407.

Mattingly, G. E. G., and R. J. B. Williams. (1962) A note on the chemical analysis of a soil buried since Roman times. *Jour. Soil Sci.* **13**:254–258.

McCaughey, W. J., and W. H. Fry. (1913) The microscopic determination of soil-forming minerals. *U.S. Dept. Agr., Bur. Soils Bul. 91.*

McKell, C. M., A. W. Wilson, and W. A. Williams. (1962) Effect of temperature on phosphorus utilization by native and introduced legumes. *Agron. Jour.* **54**:109–113.

McKelvey, V. E., J. B. Cathcart, Z. S. Altschuler, R. W. Swanson, and K. L. Buck. (1953) Domestic phosphate deposits. *Agronomy* **4**:347–376.

McNew, G. L. (1953) The effects of soil fertility. *In Plant Diseases, the Yearbook of Agriculture,* 1953, pp. 100–114. United States Department of Agriculture, Washington, D.C.

McVickar, M. H., G. L. Bridger, and L. B. Nelson (Eds.) (1963) *Fertilizer Technology and Usage.* Soil Science Society of America, Madison, Wisconsin.

Midgley, A. R. (1931) The movement and fixation of phosphates in relation to permanent pasture fertilization. *Jour. Amer. Soc. Agron.* **23**:788–799.

Morgan, M. F., and H. G. M. Jacobson. (1942) Soil and crop interrelations of various nitrogenous fertilizers. Windsor lysimeter series B. *Connecticut* (New Haven) *Agr. Exp. Sta. Bul. 458.*

Moschler, W. W., R. D. Krebs, and S. S. Obenshain. (1957) Availability of residual phosphorus from long-time rock phosphate and superphosphate applications to Groseclose silt loam. *Soil Sci. Soc. Amer. Proc.* **21**:293–295.

Moser, U. S. (1956) Correlation of soil and plant measurements of phosphorus availability. Ph.D. Thesis, Iowa State University, Ames.

Nelson, L. B. (1965) Advances in fertilizers. *Adv. Agron.* **17**:1–84.

Noggle, J. C., and M. Fried. (1960) A kinetic analysis of phosphate absorption by excised roots of millet, barley, and alfalfa. *Soil Sci. Soc. Amer. Proc.* **24**:33–35.

Okruszko, H. (1955) Torfowiska na terenie zlewni rzeki Omulwi. *Roczn. Nauk. Rol.* **71A**:407–441.

Olsen, C. (1950) The significance of concentration for the rate of ion absorption by higher plants in water culture. *Compt. Rend. Lab. Carlsberg, Sér. Chim.* **27**:291–306.

Olsen, S. R. (1953) Inorganic phosphorus in alkaline and calcareous soils. *Agronomy* **4**:89–122.

Olsen, S. R., C. V. Cole, F. S. Watanabe, and L. A. Dean. (1954) Estimation of available phosphorus in soils by extraction with sodium bicarbonate. *U.S. Dept. Agr. Cir. 939.*

Olsen, S. R., and F. S. Watanabe. (1963) Diffusion of phosphorus as related to soil texture and plant uptake. *Soil Sci. Soc. Amer. Proc.* **27**:648–653.

Olsen, S. R., F. S. Watanabe, and R. E. Danielson. (1961) Phosphorus absorption by corn roots as affected by moisture and phosphorus concentration. *Soil Sci. Soc. Amer. Proc.* **25**:289–294.

Ozanne, P. G. (1962) Some nutritional problems characteristic of sandy soils. *Internat. Soc. Soil Sci., Trans. Joint Meeting Comm. IV & V* (New Zealand): 139–143.

Ozanne, P. G., D. J. Kirton, and T. C. Shaw. (1961) The loss of phosphorus from sandy soils. *Australian Jour. Agr. Res.* **12**:409–423.

Parker, F. W., J. R. Adams, K. G. Clark, K. D. Jacob, and A. L. Mehring. (1946) Fertilizers and lime in the United States: resources, production, marketing, and use. *U.S. Dept. Agr. Misc. Publ. 586.*

Patel, J. M., and B. V. Mehta. (1961) Soil phosphorus fractionation studies. *Soil Sci. Soc. Amer. Proc.* **25**:190–192.

Pearson, R. W., and R. W. Simonson. (1939) Organic phosphorus in seven Iowa soil profiles: distribution and amounts as compared to organic carbon and nitrogen. *Soil Sci. Soc. Amer. Proc.* **4**:162–167.

Pearson, R. W., R. Spry, and W. H. Pierre. (1940) The vertical distribution of total and dilute acid-soluble phosphorus in twelve Iowa soil profiles. *Jour. Amer. Soc. Agron.* **32**:683–696.

Peaslee, D. E., C. A. Anderson, G. R. Burns, and C. A. Black. (1962) Estimation of relative value of phosphate rock and superphosphate to plants on different soils. *Soil Sci. Soc. Amer. Proc.* **26**:566–570.

Pierre, W. H., and G. M. Browning. (1935) The temporary injurious effect of excessive liming of acid soils and its relation to the phosphate nutrition of plants. *Jour. Amer. Soc. Agron.* **27**:742–759.

Pierre, W. H., and F. W. Parker. (1927) Soil phosphorus studies: II. The concentration of organic and inorganic phosphorus in the soil solution and soil extracts and the availability of the organic phosphorus to plants. *Soil Sci.* **24**:119–128.

Plummer, J. K. (1915–1916) Petrography of some North Carolina soils and its relation to their fertilizer requirements. *Jour. Agr. Res.* **5**:569–582.

Power, J. F., D. L. Grunes, G. A. Reichman, and W. O. Willis. (1964) Soil temperature and phosphorus effects upon nutrient absorption by barley. *Agron. Jour.* **56**:355–359.

Power, J. F., D. L. Grunes, G. A. Reichman, and W. O. Willis. (1964a) Soil temperature effects on phosphorus availability. *Agron. Jour.* **56**:545–548.

Power, J. F., D. L. Grunes, W. O. Willis, and G. A. Reichman. (1963) Soil temperature and phosphorus effects upon barley growth. *Agron. Jour.* **55**:389–392.

Pratt, P. F., and M. J. Garber. (1964) Correlations of phosphorus availability by chemical tests with inorganic phosphorus fractions. *Soil Sci. Soc. Amer. Proc.* **28**:23–26.

Rankama, K., and T. G. Sahama. (1950) *Geochemistry.* University of Chicago Press, Chicago, Ill.

Rathje, W. (1960) Zur gegenseitigen Umwandlung von calciumgebundener Phosphorsäure und aluminium- und eisen (3)-gebundener Phosphorsäure im Boden. *Plant and Soil* **13**:159–165.

Raupach, M. (1963) Solubility of simple aluminium compounds expected in soils. III. Aluminium ions in soil solutions and aluminium phosphates in soils. *Australian Jour. Soil Sci.* **1**:46–54.

Robertson, G. S. (1927) Experiments in Northern Ireland with various types of phosphatic fertilizers. *Northern Ireland Min. Agr. Jour.* **1**:7–36.

Rossiter, R. C. (1947) The effect of potassium on the growth of subterranean clover and other pasture plants on Crawley sand. *Australian Jour. Council Sci. Indust. Res.* **20**:389–401.

Rost, C. O. (1939) The relative productivity of some humid subsoils. *Soil Sci. Soc. Amer. Proc.* **4**:281–287.

Rovira, A. D. (1956) Plant root excretions in relation to the rhizosphere effect. III. The effect of root exudate on the numbers and activity of micro-organisms in soil. *Plant and Soil* **7**:209–217.

Runge, E. C. A. (1963) Influence of natural drainage on iron and phosphorus distribution in some Iowa soils. Ph.D. Thesis, Iowa State University, Ames.

Sauchelli, V. (Ed.) (1960) *Chemistry and Technology of Fertilizers.* Reinhold Publ. Corp., New York.

Sauchelli, V. (1965) *Phosphates in Agriculture.* Reinhold Publ. Co., New York.

Schactschabel, P., and G. Heinemann. (1964) Beziehungen zwischen P-Bindungsart und pH-Wert bei Lössboden. *Zeitschr. Pflanzenernähr., Düng., Bodenk.* **105**: 1–13.

Scheffer, F., A. Kloke, and K. Hempler. (1960) Die Phosphatformen im Boden und ihre Verteilung auf die Korngrössenfraktionen. *Zeitschr. Pflanzenernähr., Düng., Bodenk.* **91**:240–252.

Schroo, H. (1963) A study of highly phosphatic soils in a karst region of the humid tropics. *Netherlands Jour. Agr. Sci.* **11**:209–231.

Sekhon, G. S. (1962) Soil organic phosphorus and the phosphorus nutrition of plants. Ph.D. Thesis, Iowa State University, Ames.

Semb, G., and G. Uhlen. (1954) A comparison of different analytical methods for the determination of potassium and phosphorus in soil based on field experiments. *Acta Agric. Scand.* **5**:44–68.

Shipp, R. F., and R. P. Matelski. (1960) A microscopic determination of apatite and a study of phosphorus in some Nebraska soil profiles. *Soil Sci. Soc. Amer. Proc.* **24**:450–452.

Smith, D. H., and F. E. Clark. (1951) Anion-exchange chromatography of inositol phosphates from soil. *Soil Sci.* **72**:353–360.

Smith, J. H., F. E. Allison, and D. A. Soulides. (1962) Phosphobacterin as a soil inoculant. Laboratory, greenhouse, and field evaluation. *U.S. Dept. Agr. Tech. Bul. 1263.*

Sorteberg, A. (1963) Noen sider ved nitrogen- og fosforhusholdningen i lite omlaget myrjord den første tid etter oppdyrkingen. *Forskn. Fors. Landbr.* **14**:395–420.

Sorteberg, A., and G. Dev. (1964) Effect of liming peat soils on the availability of applied phosphate to plants. *Acta Agric. Scand.* **14**:307–314.

Specht, R. L. (1963) Dark Island heath (Ninety-Mile Plain, South Australia). VII. The effect of fertilizers on composition and growth, 1950–60. *Australian Jour. Bot.* **11**:67–94.

Spencer, W. F. (1957) Distribution and availability of phosphates added to a Lakeland fine sand. *Soil Sci. Soc. Amer. Proc.* **21**:141–144.

Sprague, H. B. (Ed.) (1964) *Hunger Signs in Crops.* 3rd ed. McKay Co., New York.

Stelly, M., and W. H. Pierre. (1943) Forms of inorganic phosphorus in the C horizons of some Iowa soils. *Soil Sci. Soc. Amer. Proc.* (1942)**7**:139–147.

Susuki, A., K. Lawton, and E. C. Doll. (1963) Phosphorus uptake and soil tests as related to forms of phosphorus in some Michigan soils. *Soil Sci. Soc. Amer. Proc.* **27**:401–403.

Taylor, A. W., E. L. Gurney, and J. R. Lehr. (1963) Decay of phosphate fertilizer reaction products in an acid soil. *Soil Sci. Soc. Amer. Proc.* **27**:145–148.

Tepe, W., and E. Leidenfrost. (1958) Ein Vergleich zwischen pflanzenphysiologischen, kinetischen und statischen Bodenuntersuchungswerten. I. Mitteilung. Die Kinetik der Bodenionen, gemessen mit Ionenaustauschern. *Landw. Forsch.* **11**:217–229.

Thompson, E. J. (1958) Evaluation of different laboratory indexes of soil phosphorus availability. M. S. Thesis, Iowa State University, Ames.

Thompson, L. F., and P. F. Pratt. (1954) Solubility of phosphorus in chemical extractants as indexes to available phosphorus in Ohio soils. *Soil Sci. Soc. Amer. Proc.* **18**:467–470.

Thompson, L. M., C. A. Black, and J. A. Zoellner. (1954) Occurrence and mineralization of organic phosphorus in soils, with particular reference to associations with nitrogen, carbon, and *p*H. *Soil Sci.* **77**:185–196.

Truog, E. (1916) The utilization of phosphates by agricultural crops, including a new theory regarding the feeding power of plants. *Wisconsin Agr. Exp. Sta. Res. Bul. 41.*

Tsubota, G. (1959) Phosphate reduction in the paddy field I. *Soil and Plant Food* **5**:10–15.

United States Department of Agriculture and Tennessee Valley Authority. (1964) *Superphosphate: Its History, Chemistry, and Manufacture.* Agricultural Research Service, U.S. Department of Agriculture, Washington, D.C.

Vanterpool, T. C. (1935) Studies on browning root rot of cereals. III. Phosphorus-nitrogen relations of infested fields. IV. Effect of fertilizer amendments. V. Preliminary plant analyses. *Canadian Jour. Res.* **13C**:220–250.

Walker, R. H., and P. E. Brown. (1936) The phosphorus, nitrogen and carbon content of Iowa soils. In P. E. Brown. *Soils of Iowa.* Iowa Agr. Exp. Sta. Spec. Rpt. 3.

Wallace, T. (1961) *The Diagnosis of Mineral Deficiencies in Plants by Visual Symptoms; A Colour Atlas and Guide.* 2nd ed., enlarged. Chemical Publ. Co., New York.

Ware, L. M. (1938) Influence of the major fertilizer elements on the earliness and yield of snap beans. *Proc. Amer. Soc. Hort. Sci.* (1937) **35**:699–703.

Watson, D. J., and E. J. Russell. (1943) The Rothamsted experiments on mangolds, 1872–1940. Part II. Effect of manures on the growth of the plant. Empire *Jour. Exptl. Agr.* **11**:65–77.

Weigl, J. (1963) Die Bedeutung der energiereichen Phosphate bei der Ionenaufnahme durch Wurzeln. *Planta* **60**:307–321.

Weir, C. C., and R. J. Soper. (1963) Solubility studies of phosphorus in some calcareous Manitoba soils. *Jour. Soil Sci.* **14**:256–261.

Wesley, D. E. (1965) Phosphorus uptake by oat seedlings from phosphorus fertilizer applied to a subsoil as affected by soil moisture. *Diss. Abstr.* **26**:606.

Wild, A. (1958) The phosphate content of Australian soils. *Australian Jour. Agr. Res.* **9**:193–204.

Wild, A. (1961) A pedological study of phosphorus in 12 soils derived from granite. *Australian Jour. Agr. Res.* **12**:286–299.

Wild, A. (1964) Soluble phosphate in soil and uptake by plants. *Nature* **203**:326–327.

Winters, E., and R. W. Simonson. (1951) The subsoil. *Adv. Agron.* **3**:1–92.

Wright, B. C., and M. Peech. (1960) Characterization of phosphate reaction products in acid soils by the application of solubility criteria. *Soil Sci.* **90**:32–43.

Yamashita, S., and Y. Goto. (1963) Effects of mineral nutrients on the flower-buds differentiation of crops. (Part 4) Influence of phosphorus on the first flower cluster differentiation of tomato plant. *Soil Sci. Plant Nutr.* **9**:202.

9 *Potassium*

Potassium is one of the macronutrients and usually is present in plants in quantities larger than those of any other mineral nutrient derived from soil with the exception of hydrogen and nitrogen. Some plant tissues accumulate much potassium. Tobacco leaves, for example, may contain as much as 8% potassium on the dry-weight basis and may show symptoms of potassium deficiency if the content falls much below 3%.

Potassium is a relatively abundant and widely distributed constituent of the surface rocks of the earth, making up an estimated 2.6% of the earth's crust by weight. Russell (1954) pointed out that, although potassium is removed from soil continually, most soils continue to supply potassium for a long time. Consequently, relatively few soils are so low in potassium that cropping is entirely dependent on fertilization. In contrast, there are large areas of soils on which no worthwhile crop can be grown without phosphorus fertilization and others on which only low yields are obtained without nitrogen fertilization.

The best general source of information on potassium in soils and plants is a series of symposia published by the International Potash Institute. The first "Potassium Symposium" was published in 1954, and additional volumes with the same title have been published periodically since.

Content in Soils

The content of potassium in mineral soils is usually much greater than that of nitrogen or phosphorus. According to estimates by Lipman and Conybeare (1936), the average total potassium content of the plowed layer of the crop land of the United States is 0.83%, which is 5.8 times as great as the nitrogen content and 13.4 times as great as the phosphorus content. A generalized map of the potassium content of the surface soils of the United States (Parker et al., 1946) shows that soils of the Coastal Plain from southern Virginia to the Mississippi Delta usually contain less than 0.3% potassium. In the south-central part of the United States is a large area of soils containing from 0.3 to 1%

potassium. The northeast quarter of the United States has a few soils with as little potassium as the southeast quarter, but most contain 1 to 2.5% potassium. Most soils of the western half of the United States contain 1.7 to 2.5% potassium, but in a strip along the northern part of the West Coast the content is 1 to 1.7%.

The potassium content of soils of the various regions is related to parent material and the degree of weathering. Soils of the Coastal Plain area, for example, have developed on transported materials that were strongly weathered and low in potassium at the time of deposition. The low potassium content of soils of the Pacific Northwest, on the other hand, may be attributed primarily to weathering in place and not to an initial dearth of potassium in the parent material.

With increasing depth in the soil, the potassium percentage usually remains about the same or increases in soils of initially uniform parent material. The deficit in the upper part of the profile may be attributed primarily to a combination of chemical weathering and leaching. Where strong weathering has reduced the potassium content of the entire profile to a low level, however, the surface soil sometimes has a higher potassium percentage than the subsoil. The action of plants in transporting potassium to the surface probably is responsible.

The potassium percentage varies with particle size in a less consistent manner than do nitrogen and phosphorus percentages. Failyer et al. (1908) analyzed separates of 27 soils from different parts of the United States and found that the potassium percentage in the clay fraction exceeded that of the silt fraction in 22 soils and that the potassium percentage in the silt fraction exceeded that in the sand fraction in 24 soils. The tendency, therefore, was for an increase in potassium percentage with a decrease in particle size. Salmon (1964) found from analysis of particle-size fractions of 30 samples of soil from Rhodesia that the highest potassium percentage occurred in the sand fraction (2 to 0.02 mm diameter) of 17 samples and decreased with decreasing particle size. The highest potassium percentage was in the silt fraction (0.02 to 0.002 mm diameter) in 11 samples and in the coarse clay fraction (0.002 to 0.0005 mm diameter) in 2. Thus Salmon's (1964) results differ from and are less consistent than those obtained by Failyer et al. (1908). An explanation for the various types of behavior is found in the potassium-bearing minerals present. The discussion of these minerals and some of their properties in the section on mineralogical forms of potassium in soils will provide some basis for understanding why the potassium percentage varies in such an inconsistent manner with particle size.

Losses from Soils

Crop Removal

Barber and Humbert (1963) published a table of values of potassium removal from soils by the principal world crops. Their figures are for relatively high yields and range from 45 kg/hectare in spinach, coffee, and tea up to 520 kg/hectare in sugarcane. Lipman and Conybeare (1936) estimated that in 1930 the average removal of potassium from the soils of the harvested crop land of the United States amounted to 19.5 kg/hectare. This figure refers only to the potassium in the harvested portion of the crops. The total removal of potassium was probably of the order of 35 to 45 kg/hectare. Forty kilograms per hectare is equivalent to about 0.21% of the average content of potassium in the plowed layer of soil. The total removal of potassium is much higher than removal of phosphorus and about equal to the removal of nitrogen. As a percentage of the total amount present in the soil, however, potassium is removed less rapidly than either phosphorus or nitrogen.

Leaching

Table 9.1 shows losses of potassium, nitrogen, and phosphorus by leaching from a fallow, sandy soil in South Carolina. In this experiment, the losses of potassium in both absolute amount and in percentage of the additions in the fertilizer were below the corresponding figures for nitrogen and above those for phosphorus, which is the usual behavior. Nevertheless, one must recognize that considerable variation in absolute and relative losses of the different nutrients occurs under different conditions. In the experiment in Table 9.1, the soil was sandy and low in fertility, no crop was present, and the loss of water by drainage was large. All these conditions are of significance in interpreting the results, and their relevance will be discussed in the following paragraphs.

With regard to cropping, loss of water, and soil fertility effects, pertinent information is provided by another treatment that was part of the experimental work reported in Table 9.1. In this treatment, fertilization was the same as shown for the fertilized soil in Table 9.1; but crotolaria, a legume, was grown each year. Drainage decreased to 44 cm annually; and losses of nitrogen, potassium, and phosphorus decreased to 24, 3, and 0.1 kg/hectare annually. The crop removed both water and nutrients and hence decreased both the amount of drainage water and the concentration of nutrients in it. The much greater decrease in loss of potassium than of nitrogen may be explained on the basis that the crop took up both nitrogen and potassium from the soil but, being a

Table 9.1. Annual Loss of Nitrogen, Phosphorus, and Potassium in Drainage Water from 109-Centimeter Columns of Fertilized and Unfertilized Fallow Sandy Soil in South Carolina (Allison et al., 1959)

	Annual Loss			
	Nitrogen, kg/hectare	Potassium, kg/hectare	Phosphorus, kg/hectare	Water, cm
Fertilized[1]	40	20	0.2	68
Unfertilized	30	13	0.1	67
Proportion of added nutrient lost in drainage water, %	56	47	1	—

[1] Annual additions per hectare were 18 kg of nitrogen, 15 kg of potassium, and 16 kg of phosphorus.

legume, added to the soil some nitrogen, a part of which subsequently appeared in the drainage water. The infertility of the soil is evident in the ratio of nitrogen to potassium in the drainage water from the fallow soil (Table 9.1). In soils of high fertility, the ratio of loss of nitrogen to potassium usually is much greater than the values shown in the table. For example, in analyses of drainage water from a 102-cm column of an unfertilized, fallow, highly fertile silt loam soil in Illinois, Stauffer (1942) recorded an annual drainage loss of 16 cm of water containing the equivalent of 86 kg of nitrogen and 1.5 kg of potassium per hectare.

The sandy nature of the soil used in the experiment described in Table 9.1 has another effect on potassium loss that requires explanation. To be lost in drainage water, potassium must be present in solution. But potassium moving downward through soil in solution equilibrates continuously with the exchangeable cations in the soil. A given potassium ion will spend most of the time in exchangeable form and only a little of the time in solution as a freely diffusible cation associated with a freely diffusible anion. Accordingly, the downward passage of potassium is much delayed. The extent of the delay may be perceived to increase with the cation-exchange capacity of the soil and with the ease of replacement of the exchangeable cations. The sandy soil used in this experiment had a low cation-exchange capacity. Much of the exchange capacity was probably occupied by aluminum, which would not be displaced readily by potassium. Hence loss of added potassium by leaching would take place readily.

Volk and Bell (1945) published the results of an experiment that illustrate the operation of principles of ion exchange in leaching of potassium and other ions from a 122-cm column of a sandy soil in Florida. At the beginning of the experiment, the soil was saturated with tap water and allowed to stand until drainage ceased. Then it received, at the surface, an application of salts containing calcium, ammonium, and potassium cations and nitrate, sulfate, and phosphate anions. The drainage water obtained from the soil as a result of natural rainfall then was analyzed at intervals, and the concentrations of the ions were plotted against the accumulated amount of drainage water.

The results in Fig. 9.1 show that concentrations of all ions were relatively low in the first drainage water, but then nitrate and cations (mostly calcium) emerged in definite peak concentrations. These peaks are due to the fertilizer applied at the beginning of the experiment. Although no doubt the nitrate ions emerging in the nitrate peak came mostly from the fertilizer and appeared early because they moved

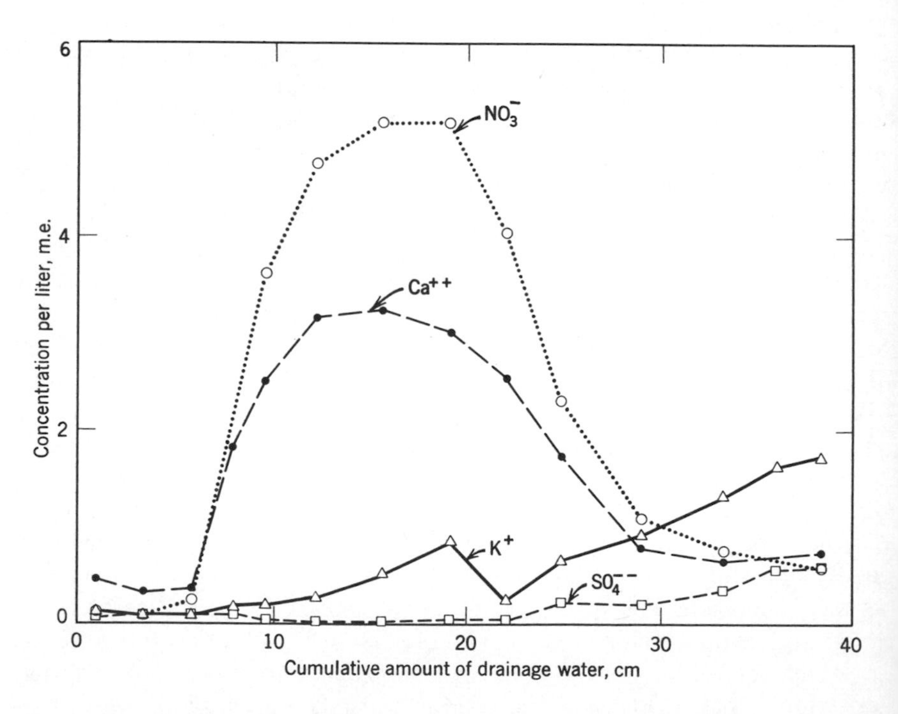

Figure 9.1. Concentration of ions in increments of drainage water obtained in the field in Florida from a column of a sandy soil that was fertilized at the surface after initial saturation with tap water. The fertilizer contained all ions for which analyses are shown. (Volk and Bell, 1945)

through the soil without attachment, the peak concentration of cations that appeared with the nitrate peak resulted from the electrical neutrality requirement for cations to accompany the nitrate. The cations emerging in the cation peak with the nitrate no doubt came mostly from the exchangeable form in the lower part of the column of soil, the fertilizer cations added to the surface of the soil being retained for the most part in exchangeable form in the upper part of the soil. Thus, in this first cation peak, the total concentration of cations was determined mostly by the concentration of nitrate, and the proportions of the cations were determined mostly by the proportions and ease of exchange of the exchangeable cations in the lower part of the soil column. At the end of the experiment, the concentration of potassium in the drainage water was increasing toward what would have been a peak concentration of potassium ions derived from the fertilizer. The concentration of sulfate in the effluent likewise was increasing at the end of the experiment toward a peak concentration of sulfate ions derived from the fertilizer. Emergence of sulfate was delayed because, unlike nitrate, sulfate is adsorbed to some extent by soil.

Liming of acid soils may significantly decrease the loss of potassium by leaching. For example, Shaw and Robinson (1960) found that leaching of potassium from a loam soil in Tennessee was decreased from 58 kg/hectare in the untreated control soil (*p*H 4.5) to 9 kg/hectare from the same soil after a heavy application of dolomitic limestone. These results were obtained in the first year after liming. The effect of liming on potassium loss may be explained on the basis of cation-exchange properties. Liming of acid soils has two effects on exchange properties that retard downward movement of potassium through soil. First, it causes replacement of exchangeable aluminum by calcium. Potassium in dilute solutions displaces exchangeable calcium more readily than exchangeable aluminum. Second, liming increases the cation-exchange capacity and hence increases retention of potassium in exchangeable form by a mass-action effect. In the experiment by Shaw and Robinson (1960), the reduction in leaching loss of potassium from liming was far greater in the first year than in succeeding years, as would be expected if the effect of liming were attributable to an alteration in the cation-exchange properties just mentioned. If the effect of liming had been primarily on release of potassium from nonexchangeable to exchangeable form, the effect would be expected to persist over a period of years.

A review paper on loss of potassium by leaching and on downward movement of applied potassium in soils was published by Munson and Nelson (1963).

Erosion

Considerable potassium may be lost from soils by erosion. Lipman and Conybeare (1936) estimated the annual loss of potassium by erosion from the crop land of the United States at 158 kg/hectare. The selective removal of finer particles during erosion depletes selectively the portion of the soil that is most important in supplying potassium to plants. The same is true of nitrogen, as pointed out previously in Chapter 7. Losses of potassium in this way are seldom taken as seriously as those of nitrogen, however, because of the difference in distribution of these two elements in soil profiles. Nitrogen is concentrated near the surface, which results in a decrease in nitrogen availability with increasing erosion. Potassium, on the other hand, is present throughout the profile. Loss of surface potassium by erosion, which may cause plants to develop in a surface soil having lower potassium availability, is compensated in part by addition of a new source of potassium in the deeper soil horizons that are brought nearer the surface.

Forms in Soils

Mineralogical Forms

Most of the potassium in soils is present in minerals classified as feldspars and micas. The most important of these are orthoclase and microcline feldspar, biotite and muscovite mica, and micaceous clays, known as illite.

The feldspars occur almost exclusively in the sand and silt fractions of soils but are found occasionally in the coarse clay. The feldspars may be of either primary or secondary origin but, if secondary, are thought to be inherited from the parent material and not formed during soil development.

Biotite and muscovite mica also occur mainly in the silt and sand fractions and may be of either primary or secondary origin. The forms of these minerals found in soils have been altered from the original, having lost some potassium and gained some water (Denison et al., 1929). Secondary biotite and muscovite are thought not to form under conditions of soils. Illite, the principal potassium-bearing mineral in the clay fraction of soils, is of secondary origin. There is some doubt among mineralogists whether the illite in soil clays is developed there. Evidence indicates that illite in soils is, in general, undergoing degradation. Addition of potassium fertilizer, however, may favor formation of illite, as will be explained further subsequently. Likewise, minerals similar to muscovite and biotite may form from degraded micas where potassium fertilization is practiced.

Mineralogical analyses of bottom sediments in the oceans indicate that illite is formed there by alteration of other minerals, particularly degraded illite (low in potassium) and kaolinite (Grim et al., 1949). Formation of illite in sediments on the ocean bottom provides an explanation for the low content of potassium in relation to sodium in ocean water. According to data summarized by Rankama and Sahama (1950), the ratio of potassium to sodium by weight is 0.92 in igneous rocks, 0.37 in river and lake water, and 0.04 in ocean water. Mackenzie (1955) reviewed the subject of potassium in clay minerals.

The occurrence of various potassium-bearing minerals in soils may be illustrated by Schachtschabel's (1937, 1937a) detailed mineralogical analyses of seven soils of Germany. The samples he analyzed contained an average of 2.3% potassium. On the average, the samples consisted of 27% by weight of potassium-bearing minerals (15.5% potassium feldspar and 11.3% micaceous minerals), 62% of nonpotassium-bearing minerals and large stones, and 5% of unidentified minerals. Of the feldspar, 14.5% was in the sand and silt, and 1% was in the clay. Of the micaceous material, 2.7% was in the sand and silt, and 8.6% was in the clay. The clay fraction of these soils averaged 40% micaceous material, which probably would be classified now as illite. Similar analyses of the silt and clay fractions of several soils of the United States were published by Reitemeier et al. (1951).

Chemical Forms

From the chemical standpoint, soil potassium often is divided into three categories: nonexchangeable, exchangeable, and water-soluble. Sometimes the nonexchangeable fraction is divided into two fractions on the basis of the potassium extracted by an acid, a cation-exchange resin, or a potassium-precipitating reagent.

In most soils, the great bulk of the potassium is nonexchangeable. As shown in Table 4.4, an average of 99.6% of the total potassium in samples of twenty soils of New Jersey was nonexchangeable. The remaining 0.4% included both the exchangeable and water-soluble forms. The water-soluble potassium usually constitutes only a small part of the sum of the water-soluble and exchangeable forms. Anderson et al. (1942) found that the exchangeable potassium content of surface samples of six soils from the humid and subhumid region of the United States ranged from 66 to 341 μg/g of dry soil (0.17 to 0.87 m.e./100 g).

If calculated on the same basis, the potassium content of the displaced soil solutions of the same soils ranged from 1 to 8 μg/g (0.003 to 0.02 m.e./100 g).

In most experimental work, soil-solution potassium is not measured independently but is included with exchangeable potassium. The two fractions are extracted simultaneously by the salt solutions used to remove exchangeable potassium for analysis, and this is one reason for the common practice. Another reason is that the soil-solution potassium is usually so low in relation to exchangeable potassium and in relation to uptake of potassium by plants that the portion in the soil solution is quantitatively of little significance. In humid regions, the content of exchangeable-plus-soluble potassium in soils usually is less than 200 μg/g and rarely is more than 600 μg/g. In arid regions the amount may be higher.

Although exchangeable potassium may be defined in the same terms used to define exchangeable bases, as in the chapter on that subject, research workers at present are coming to observe two special conventions in determining exchangeable potassium in soils. Both are of significance with regard to the relationship between exchangeable potassium and uptake of potassium from soils by plants.

The first convention is that an ammonium salt is used to displace the potassium. The reasons for this convention are that ammonium salts (particularly ammonium acetate) are commonly used to displace exchangeable bases and that the distinction between exchangeable and nonexchangeable forms of potassium is much more definite with ammonium salts than with other common salts and is less dependent on the technique. The results obtained by use of different displacing agents may be rather different, as shown in Fig. 9.2. One may note from the figure that release of potassium from the soil soon stopped where ammonium acetate was used but that it continued for a longer time and never ceased, within the limits of the experiment, where other cations were used. Ammonium acetate displaced less potassium than any of the other salts except calcium acetate. The low results obtained with calcium acetate and the similarity of the extraction pattern obtained with calcium and ammonium acetates may be attributed to the potassium present as an impurity in the calcium acetate solution. The authors found that all the soils they tested except the one shown in Fig. 9.2 took up potassium from the calcium acetate solution. Because of the potassium impurity, they found more release of potassium from the soils with a dilute calcium acetate solution (0.01 to 0.1 normal) than with a more concentrated solution (0.5 to 1 normal). In explanation of these results, it may be mentioned here that ammonium and potassium behave

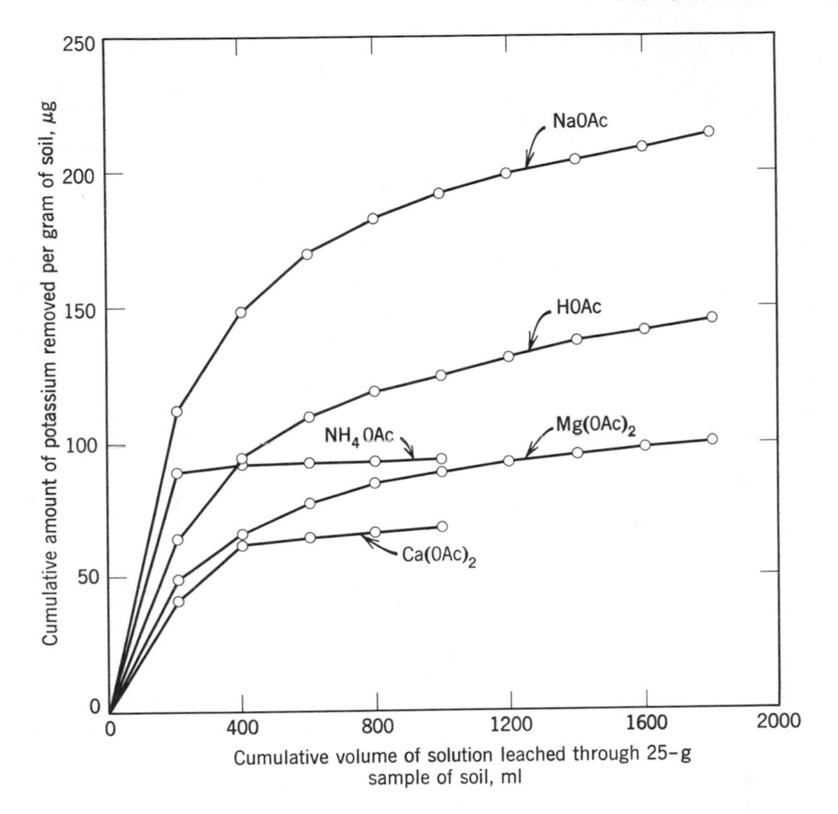

Figure 9.2. Cumulative amounts of potassium removed from a soil from New York by repeated leaching with 0.2-normal acetic acid and 0.5-normal solutions of various acetates. (Merwin and Peech, 1951)

similarly and that both are thought to prevent for the most part the release of potassium from interlayer positions in micaceous minerals. The singular behavior of potassium and ammonium in cation-exchange reactions involving micaceous minerals will be considered further in subsequent sections.

The second convention is that exchangeable potassium is being determined on undried soils. Use of air-dried samples is technically convenient, and earlier analyses were made on samples that were air-dried or occasionally on samples that had been oven-dried.

Many soils yield a different value for exchangeable potassium if the displacing solution is added to a field-moist sample than to an air-dried sample. The tendency is for higher values on air-dried samples than on field-moist samples, although the reverse may occur. Decreases due

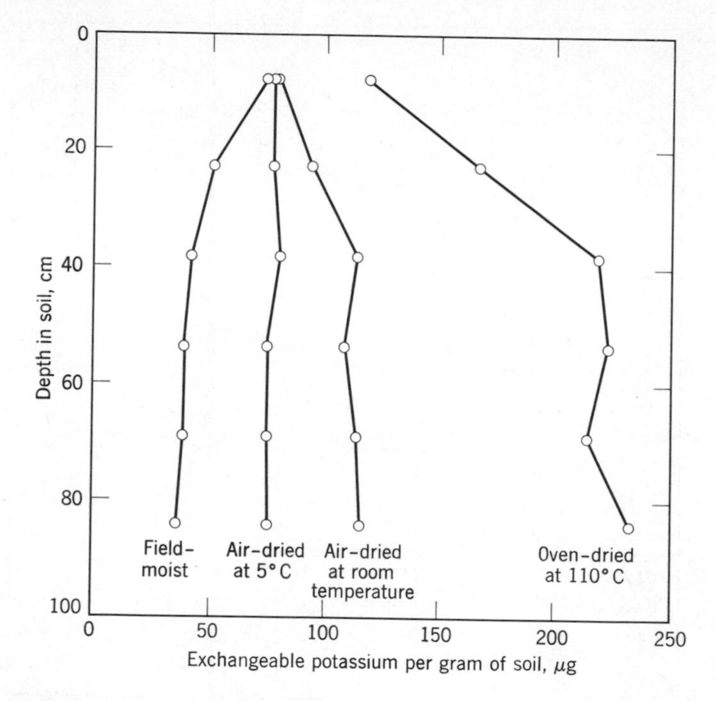

Figure 9.3. Exchangeable potassium at different depths in a silt loam soil in Indiana as determined on samples that were field-moist or dried in different ways before analysis. (Hanway et al., 1961)

to drying before analysis are usually encountered in samples high in exchangeable potassium. Subsoils usually are low in exchangeable potassium if analyses are made on field-moist samples. Analyses of subsoils made on dried samples are almost always higher than those on undried samples. Figure 9.3 shows results of analyses made on field-moist and variously dried samples of a silt loam soil from Indiana in which the effect of drying was pronounced.

Luebs et al. (1956) obtained a higher correlation between uptake of soil potassium by plants and exchangeable potassium in a group of undried soils in the greenhouse where the latter measurements were made on samples that had not been dried than on samples that had been air-dried before analysis. Their finding was subsequently confirmed by Hanway et al. (1961, 1962) in a large number of field experiments in north-central United States. Avoidance of drying samples of soil before analysis therefore seems advisable if analyses for exchangeable potassium are to represent the exchangeable potassium existing in soil in the field or if they are to be used as an index of availability to plants of the potassium in soils that have not been dried.

Availability to Plants

Uptake from Solution

Epstein et al. (1963) determined the rate of potassium removal by barley roots from solutions having different concentrations of potassium and explained their observations on the basis that two kinds of carrier sites govern binding and uptake of potassium. One kind of site operates at one-half the maximum velocity at a potassium concentration of about 0.7 μg/ml, and the other operates at one-half the maximum velocity at a potassium concentration of about 625 μg/ml.

Williams (1961) grew barley plants in culture solutions with low concentrations of potassium and found that the plants rapidly removed all the detectable potassium from solution (a concentration of 0.001 μg/ml was said to be detectable) if the potassium supply was not renewed. Where special arrangements were made to replace the potassium continuously as it was removed by plants, however, Williams found that the plants grew well with a concentration of only 0.01 μg/ml. Comparison of Williams' results with those obtained by Epstein et al. (1963) indicates that the carrier sites specific for potassium may be operating far below their maximum velocity and still provide enough potassium for plants.

The concentration of potassium in the soil solution varies, but most analyses give values considerably higher than Williams' 0.01 μg/ml. Figure 9.4 gives a frequency distribution of concentrations of potassium in water extracts obtained from samples of many soils after they had been equilibrated with water in an amount to saturate them. The concentrations fell more frequently in the range from 2 to 6 μg/ml than in any other similar range, but the concentrations in almost half the soils exceeded 6 μg/ml. These analytical values no doubt underestimate the concentrations of potassium that would have been present in solution at lower contents of water in the range at which most plants grow. At the same time, the concentrations are probably somewhat high in comparison with those that would have been found at saturation in the field because the process of drying soil in air usually increases the salt content and potassium content of the soil solution and because the process of saturation in the field is usually accompanied by downward movement of salts originally present in the surface layer. In reviews of literature on concentrations of potassium in displaced soil solutions, the range was given as 8 to 391 μg/ml by Fried and Shapiro (1961) and as 3 to 156 μg/ml by Barber (1962).

Although the potassium concentrations mentioned in the preceding paragraph are relatively high in comparison with Williams' value of 0.01 μg/ml, these concentrations are not necessarily maintained when plants are present. In Table 9.2 is an example giving measurements on soil solutions displaced at different times during the growing season of corn. The values are expressed in terms of milligram equivalents per liter to show the stoichiometric relations among the various ions. Because of the relatively large amount of potassium required by corn plants and the low concentration of potassium in solution in comparison with the other ions, potassium was removed selectively. For comparison with the numerical values in the preceding paragraphs, the concentrations of potassium may be changed to the same units used previously, giving values of 3.7 , 2.2, and 0.2 μg/ml of soil solution on June 26, July 12, and August 5, respectively. The value 0.2 is the average figure obtained from one field with a concentration of 0.79 μg of potassium per milliliter and four fields with a concentration of 0.00 μg/ml. Reduction of the concentration of potassium in the soil solution by roots provides an explanation for the fact that potassium deficiency occurs more

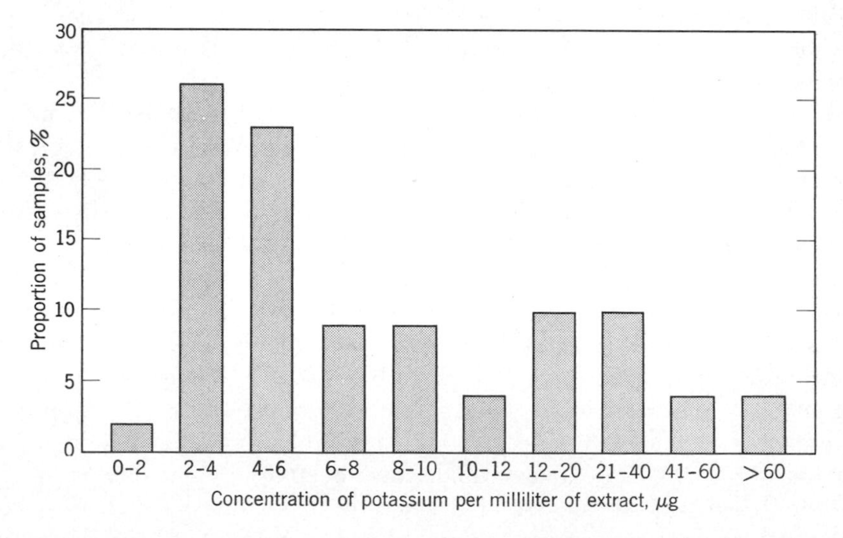

Figure 9.4. Frequency distribution of concentrations of potassium in aqueous extracts of samples of 142 cultivated soils of midwestern United States. The extracts were obtained after addition of enough water to the air-dry samples to saturate the soils with water. (Barber et al., 1962)

Table 9.2. Average Concentrations of Potassium and Other Ions in the Solutions Displaced from Five Soils under Corn[1] at Three Sampling Dates during the Growing Season in Iowa (Kelly, 1941)

Date of Sampling	Concentration of Indicated Ion per Liter of Displaced Soil Solution, m.e.			
	Potassium	Calcium	Magnesium	Nitrate
June 26	0.095	13.6	4.4	14.2
July 12	0.056	20.3	5.3	18.9
August 5	0.004	8.4	2.3	8.1

[1] Samples were taken from small open areas where four hills of corn had been removed, but the areas were small enough to permit roots of adjacent plants to enter and absorb nutrients.

frequently than would be expected from a comparison of Williams' value of 0.01 μg/liter with the concentrations of potassium found in displaced solutions and other water extracts of soils.

Loss from Exchangeable and Nonexchangeable Forms

In this section, emphasis will be placed on relationships between exchangeable and nonexchangeable forms in potassium nutrition of plants. Reference should be made to Chapter 4 for information on mechanisms by which plants may take up potassium ions present originally in exchangeable form.

The potassium content of soil solutions is usually small compared with the amounts of potassium absorbed by crops. Hence renewal during the season is needed if plants are to derive their potassium from the soil solution. Because cations in the soil solution remain continuously almost at equilibrium with those in exchangeable form, selective removal of potassium by plants from the soil solution should produce selective removal of potassium from exchangeable form. If, in fact, plants derive potassium from the exchangeable form, one should expect to find that removal of potassium from soil by plants would reduce the exchangeable potassium. Moreover, if exchangeable potassium is the only source other than the soil solution from which plants derive potassium, the reduction in exchangeable-plus-soluble potassium due to plant growth should equal the potassium removed by the plants.

Figure 9.5 shows the results of a field experiment in Hawaii in which eight crops of grass were grown in continuous cropping over a period

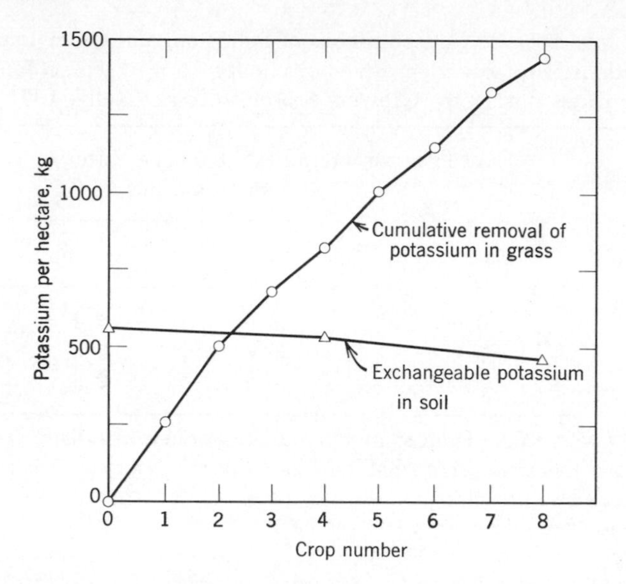

Figure 9.5. Cumulative removal of potassium in eight crops of napiergrass and exchangeable potassium in the surface 92 cm of a clay soil in Hawaii. Three or four crops were harvested per year. The soil was fertilized with nitrogen and phosphorus but received no potassium. (Ayres et al., 1947)

of somewhat more than 2 years, with measurements of potassium removed in the grass and of exchangeable-plus-soluble potassium in the surface 92 cm of soil. The existence of a decrease in exchangeable-plus-soluble potassium with cropping is in agreement with the theory that plants derive potassium from exchangeable form; however, the fact that the decrease was relatively small in comparison with the removal of potassium by plants provides evidence that most of the potassium harvested in the eight crops was not present in exchangeable form in the surface 92 cm of soil at the beginning of the experiment. Presumably, the potassium in the eight crops in excess of the decrease in exchangeable-plus-soluble potassium in the soil came mostly from the nonexchangeable form. Although there is little information on exchangeable soil potassium and on potassium removed by crops grown in the field over a period of years, the situation shown in Fig. 9.5 must exist rather generally if the potassium balance is determined over a long enough span of years. The exchangeable potassium in the soil within the root zone of ordinary crop plants is seldom sufficient for more than 10 or 15 years if the produce is removed; yet, plants continue to grow, and exchangeable potassium is not exhausted over much longer periods even if no appreciable addition of potassium is made. (Over a short time, however, the potassium removed by crops is sometimes balanced by an almost equivalent reduction in exchangeable potassium in the soil. An experiment with such results will be discussed subsequently, and a theoretical explanation for the differing findings will be developed.)

If, indeed, plants may take up potassium that was present in nonexchangeable form before their growth, an appreciation of the nature of the process is evidently important to an understanding of the potassium nutrition of plants growing in soils. Because weathering of potassium minerals in soils no doubt proceeds continuously, one may suppose that release of potassium by chemical breakdown of minerals such as potassium feldspars might provide the potassium that plants derive from nonexchangeable form. If this supposition is correct, one should expect the content of exchangeable-plus-soluble potassium to increase substantially if soils are kept moist and warm so that weathering will proceed, and if no plants are grown to remove the potassium.

Bray and DeTurk (1939) conducted an experiment that provides information on this question. They divided samples of each of ten Illinois soils into two parts, dried one part, and sealed the other part in moist condition. After incubation for 5 years, the moist samples were dried. Then exchangeable potassium in both sets of samples was determined. The results in Fig. 9.6 show that exchangeable potassium did increase

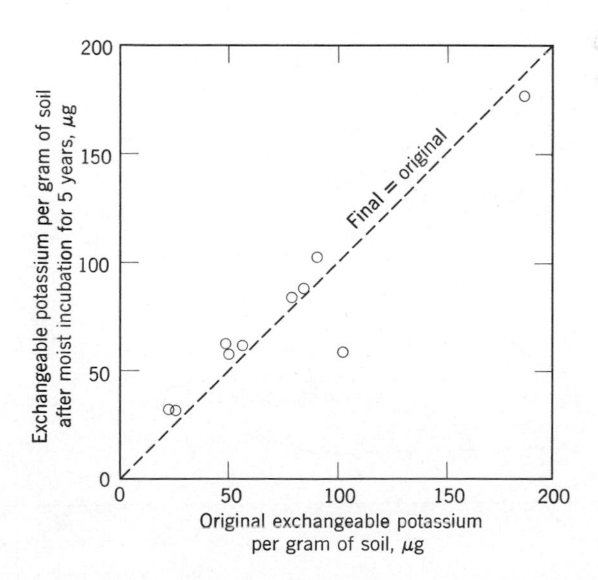

Figure 9.6. Exchangeable potassium in ten Illinois soils upon drying the samples after incubation in a moist condition for 5 years versus "original" exchangeable potassium determined on comparable samples that had been dried initially and stored in a dry condition for 5 years. (Bray and DeTurk, 1939)

in most of the samples during the 5-year period but that the increases were relatively small and by no means sufficient to account for the marked use of potassium from nonexchangeable form observed by Ayres et al. (1947) in the experiment mentioned previously. Addition of soluble potassium in the form of fertilizer, however, does increase the content of exchangeable-plus-solution potassium. For example, Reitemeier et al. (1951) found that samples of a loam soil from a long-term field experiment in Maine had contents of exchangeable-plus-soluble potassium ranging from 35 μg/g in the control receiving no potassium fertilizer to 456 μg/g in a sample receiving both potassium fertilizer and manure. The indications are therefore that exchangeable potassium would accumulate in quantity if soluble potassium were actually being added by weathering of minerals that contain nonexchangeable potassium. The smallness of the accumulation of exchangeable potassium in the soils in the experiment given in Fig. 9.6 thus is evidence that soluble potassium is not being added by weathering in quantities that would account for the uptake by plants from nonexchangeable form.

Additional information on the behavior of exchangeable and nonexchangeable potassium during cropping is provided in Fig. 9.7, which shows the cumulative loss of these two forms of potassium from a soil and the cumulative uptake of potassium by corn plants in intensive

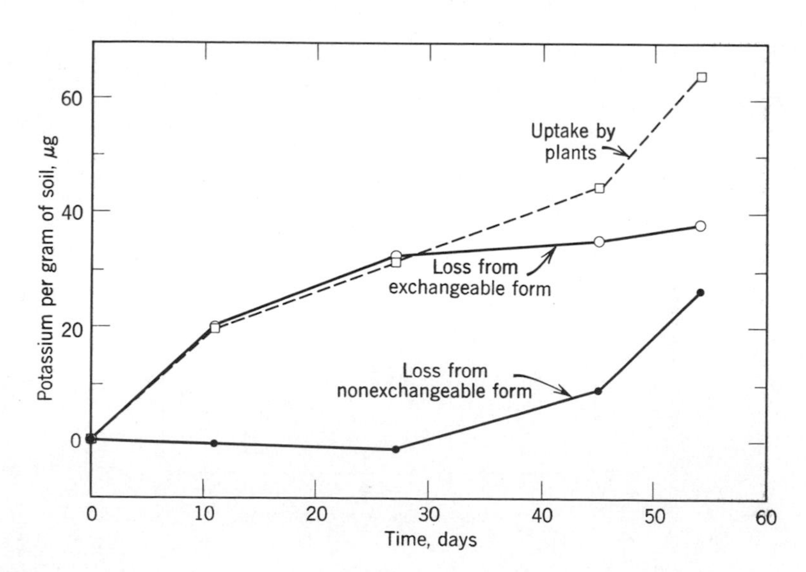

Figure 9.7. Uptake of potassium from 1 kg of a loam soil by six corn plants in the greenhouse and loss of potassium from exchangeable and nonexchangeable forms in the soil. (Grimes, 1966)

culture in the greenhouse. The figure shows that initially the potassium taken up by the plants could be accounted for almost entirely by a loss of exchangeable potassium. Toward the end of the cropping period, however, most of the potassium taken up by the plants could be accounted for by loss of initially nonexchangeable potassium from the soil. Although the tendency for greater loss of exchangeable potassium initially and greater loss of nonexchangeable potassium later is commonly observed, there is much variation from differences among soils and differences in intensity of cropping. In general, the more intensive the cropping and the lower the exchangeable potassium, the sooner loss of nonexchangeable potassium occurs and the greater is loss from this form relative to that from the exchangeable form. Results of an experiment on intensity of cropping are shown in Table 9.3. Evidently the smaller the quantity

Table 9.3. Uptake of Potassium by Fifteen Corn Plants During a Fifteen-Day Cropping Period from Different Quantities of a Silt Loam Subsoil in a Layer in a Sand Culture (Verma, 1963)

	Potassium per gram of soil, μg		
Quantity of soil per culture, g	Removed by Plants	Lost from Exchangeable Form	Lost from Nonexchangeable Form
100	100	−4	104
60	159	0	159
30	165	−9	174
5	415	−6	421

of soil supplied per plant, the greater was the loss of potassium from nonexchangeable form. An additional important point to note about this experiment is that an undried subsoil was used, and it lost no exchangeable potassium during cropping. This subsoil was low in exchangeable potassium at the beginning of the cropping (39 μg/g of soil), as are most subsoils that have not been dried before analysis.

Another significant experiment was conducted by Verma (1963). He subjected three soils to intensive cropping in the greenhouse to reduce the level of exchangeable potassium and to cause loss of much nonexchangeable potassium. He then removed the roots of the plants, allowed the soils to incubate in a moist condition, and analyzed the soils for exchangeable potassium at intervals, with results as shown in Fig. 9.8. The most important point to note in this figure is that the exchangeable

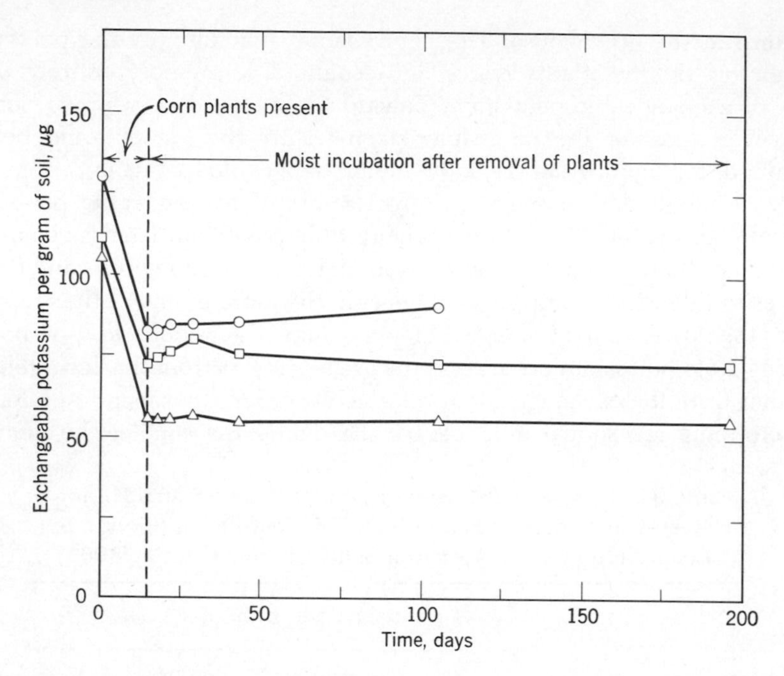

Figure 9.8. Exchangeable potassium in 60-g samples of three Iowa soils during an initial period of 15 days in which 15 corn plants were absorbing potassium followed by a much longer period of incubation of the moist soil after removal of the plants. The initial cropping was sufficiently intensive to cause release of nonexchangeable potassium equivalent to several times the observed decrease in exchangeable potassium. (Verma, 1963)

potassium remained almost constant in each soil during moist incubation after cropping. Now from the additional fact that from 78 to 91% of the potassium absorbed by the corn plants in this experiment was accounted for by loss of nonexchangeable potassium, one may infer that loss of potassium from nonexchangeable form took place rapidly while the plants were growing but almost ceased when the plants were removed. Presumably, the plants reduced the concentration of potassium in solution below some critical value at which release was inhibited; and, when they were removed, the concentration rapidly increased to the critical level and then prevented further release from nonexchangeable form. This mechanism will be considered further in the section on release from mineral forms.

The same general behavior seems to prevail under field conditions except that the removal of potassium by crops is much slower because

Table 9.4. Loss of Exchangeable and Nonexchangeable Potassium from Soil under Legume-Grass Meadow During a Three-Year Period in Scotland (Hemingway, 1963)

Fertilizers Added	Loss of Potassium from Soil per Hectare, kg[1]	
	Exchangeable	Nonexchangeable
None	179	36
Nitrogen	202	64
Potassium	11	−31
Potassium + nitrogen	191	1

[1] Potassium added in fertilizer was assumed to be added to the exchangeable form.

the cropping is much less intensive than in the greenhouse experiments just discussed. Table 9.4 gives the loss of exchangeable and nonexchangeable potassium from soil during a 3-year field experiment in which all cuttings of herbage were removed from a legume-grass meadow. The results indicate that loss of potassium was mainly from exchangeable form. Addition of nitrogen fertilizer increased the loss from both exchangeable and nonexchangeable forms. Nitrogen increased the growth of herbage, thereby increasing the intensity of cropping and presumably decreasing the concentration of potassium in solution. Addition of potassium without nitrogen apparently caused an increase in nonexchangeable potassium, a process known as potassium fixation. This process will be discussed in a subequent section.

Figure 9.9 shows in some detail the results from two of the treatments in the field experiment reported in summary in Table 9.4. Exchangeable potassium in the soil, determined after each cutting of herbage, is plotted against the cumulative amount of potassium removed in the herbage. With the potassium-fertilized plots, which received potassium chloride in February before each growing season, the first value for exchangeable potassium in each season was well above the last value the preceding autumn despite removal of potassium in the first cutting of herbage. Transfer of potassium from nonexchangeable to exchangeable form during the winter would have produced the same sort of discontinuity in results with the plots receiving no potassium fertilizer if sufficient transfer had occurred. The results provide evidence of a small release of potassium between the first year and the second but not between the second and the third.

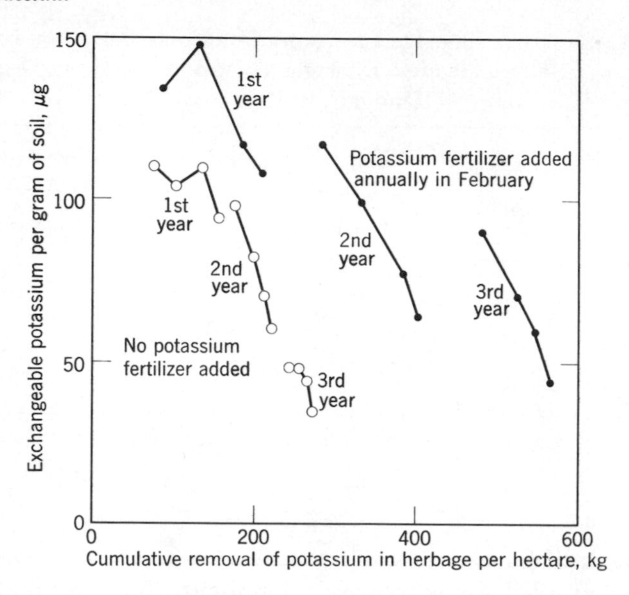

Figure 9.9. Exchangeable potassium in samples of soil taken after each cutting of legume-grass meadow versus cumulative removal of potassium in herbage, with and without potassium fertilization, in three successive years in a field experiment in Scotland. (Hemingway, 1963)

Further information on potassium behavior under field conditions was obtained by Grimes (1966). He found, like Hemingway (1963), that the exchangeable potassium in soils may decrease considerably during a single growing season in the presence of a crop. His data from three field experiments on silt loam soils indicated that the decrease in exchangeable soil potassium per gram of soil from spring to autumn in the presence of a crop of corn was as much as 50 μg if the exchangeable potassium was relatively high (125 μg) but was zero if the exchangeable potassium was relatively low (20 μg or less). In his experiments, he found higher values for exchangeable potassium in the spring than in the preceding autumn in soil under corn but not in soil under perennial legume-grass meadows. Moreover, the exchangeable potassium in soil under corn in the spring decreased with an increase in number of cuttings of hay removed the preceding year from the legume-grass meadow. For example, values of 65, 44, and 31 μg of exchangeable potassium per gram of the surface 15 cm of soil were recorded following a meadow of red clover and timothy in which 0, 1, and 2 cuttings of hay were removed. He interpreted his findings in terms of removal of potassium from the soil by the growing crops and return of potassium to the soil in the crop residues. In the case of perennial legumes and grasses in the meadow plantings, removal of the herbage would prevent return of this potassium; and carry-over of live roots from one year to the next would prevent return of much of the root potassium to the soil. In the case of corn, however, most of the potassium is present in the vegetative parts and roots and not in the grain; and return of vegetative parts and roots to the soil between growing seasons would have an effect similar to that of potassium fertilization. Grimes verified that potassium diffuses from dead plant residues under moist conditions and that plants absorb it.

These experiments lead to the following general concept of the cycling of potassium between soil and plants. Plants draw potassium from the soil solution and the exchangeable form, which continuously approaches an equilibrium with the soil solution. If the supply of potassium in these forms is high, they may sustain the entire loss occasioned by growth of plants. If the soil solution and cation-exchange sites are depleted sufficiently of potassium, potassium is released rapidly from nonexchangeable form; but the release almost stops as soon as the plants stop removing potassium. Once the potassium has been depleted to a level at which release from nonexchangeable forms occurs throughout the soil, the content of exchangeable potassium remains approximately constant despite much additional removal of potassium. As a result of recurrent additions of potassium in the form of plant residues, manures, or fertilizers, soils usually contain potassium in excess of the level needed to inhibit release from nonexchangeable form. Most of the potassium of plants is in the vegetative parts and underground parts; and, if these are removed, soil potassium is depleted rapidly. Removal of only selected parts (such as grain) that contain little potassium causes little depletion.

Subsoils are almost always low in exchangeable potassium. For example, in analyses of profiles of 48 mineral soils in the north-central region of the United States, Hanway et al. (1962) found an average content of exchangeable potassium amounting to 102, 52, 42, 38, 35, and 35 $\mu g/g$ of soil in successive 15-cm layers from the surface downward. Subsoils probably contain enough exchangeable and soluble potassium to stop further transfer from nonexchangeable to exchangeable form but possibly little more in most instances. The excess of exchangeable potassium over the level that stops release from nonexchangeable form probably occurs almost entirely in the upper part of the soil to which is added the potassium supplied by plant residues, manures, and fertilizers.

This concept is a generalization based on limited experimental evidence derived from work on soils containing micaceous minerals, the behavior of which will be considered in more detail in the section on release from mineral forms. Soils that do not contain micaceous minerals will probably behave differently with regard to release of potassium from nonexchangeable form.

Release from Particle-Size Fractions

Determination of the source of the potassium released from nonexchangeable form in soils has been the subject of a number of investigations. Several kinds of evidence indicate that the finer mineral fractions are of major importance.

McEwen and Matthews (1958) used the technique of correlating the content of individual particle-size fractions of soils of Ontario with the release of potassium from nonexchangeable form during continuous percolation of water saturated with carbon dioxide through the samples. They found that release of potassium increased with and was closely related to the clay content of the soils.

Merwin and Peech (1951) investigated the release from different particle-size fractions of four soils of New York in another way. They separated the soils into fractions of sand, silt, and clay size and then recombined the fractions in various ways. The recombined particle-size fractions were initially calcium saturated and were then incubated in a moist condition to allow potassium release to proceed. At the end of the incubation, the samples were leached to remove the exchangeable potassium that had accumulated; and the potassium removed was then determined. The results, summarized in Table 9.5, again indicate the importance of the finer fractions. Although the greatest release was from

Table 9.5. Relative Quantities of Nonexchangeable Potassium Released to Exchangeable Form from Different Particle-Size Fractions of Four Soils of New York During Incubation Following Removal of Exchangeable Potassium and Saturation with Calcium (Merwin and Peech, 1951)

	Relative Quantities of Nonexchangeable Potassium Released to Exchangeable Form from Indicated Fraction		
Basis of Calculation	Sand	Silt	Clay
Equal weights of all fractions	3	19	100
Relative weights of fractions found in the soils	13	62	100

the clay fraction, release from particles of silt size appears important where the relative release from the different particle-size fractions is calculated on the basis of the proportion in which the size-fractions occurred in the soils.

Still another technique was used by Arnold (1960a). He grew ryegrass on a soil of Great Britain in a greenhouse for 18 months and determined the uptake of potassium by the ryegrass from nonexchangeable form. He separated the soil into different particle-size fractions before and after cropping and analyzed the fractions for total potassium. He found that about 40% of the potassium removed by the crop could be accounted for by loss of potassium from the fraction less than 0.1 μ in diameter, and another 40% could be accounted for by loss of potassium from the 0.1- to 0.3-μ fraction. These two fractions together contained 25% of the total potassium in the soil. The remaining 20% of the potassium removed by the crop could be accounted for by loss of potassium from the fractions coarser than 0.3 μ in diameter, which contained about 75% of the total soil potassium.

Release from Mineral Forms

Although the behavior of potassium in soils can be determined only by studying soils, much has been learned from studying potassium-bearing minerals that contributes understanding to the behavior of potassium in soils. A special advantage in work with minerals is that the findings may be related to a particular structure and composition if pure mineral species are used. The two principal mineralogical sources of nonexchangeable potassium in soils are feldspars and micas, and consideration will be limited here to these two classes of minerals.

Potassium in feldspars is imbedded in a crystalline aluminosilicate structure (the chemical formula for orthoclase, for example, may be written as $KAlSi_3O_8$) in which each potassium atom is surrounded by and bonded to oxygen atoms. The bonds in the aluminosilicate part of the crystal structure extend completely around each potassium atom. Hence each potassium atom is isolated from each other potassium atom; and there is no access of the outside medium to the potassium, and vice versa, except by disruption of the aluminosilicate network.

Several observations indicate that release of potassium from feldspars is inhibited by development of a protective residual layer on the surface of the particles in response to mild chemical weathering of the type occurring in soils.

1. If potassium feldspars are broken into small pieces and then subjected to chemical weathering in the laboratory, a small portion of the

mineral dissolves. But the constituents do not appear in the solution in the same proportion as they occur in the mineral. Less aluminum and silicon are dissolved than would be expected from the potassium released. For example, Correns and von Engelhardt (1938) leached a quantity of finely ground adularia feldspar with distilled water in a reflux apparatus and then measured the constituents removed. The particles of feldspar were less than 2 μ in diameter; and the action of water was continued for 43 days, although the refluxing action (passage of water through the sample) took place for only 218 hours. This treatment removed from the feldspar 12.2% of the potassium, sodium, and calcium but only 1.4% of the silicon and 1.2% of the aluminum. The feldspar residue therefore contained relatively more aluminum and silicon and less potassium, sodium, and calcium than the original adularia.

2. The residue from treatment of finely ground feldspar with water, dilute acids, or salt solutions will retain in exchangeable form cations other than the ones originally present. For example, Nash and Marshall (1957) treated finely ground samples of seven different varieties of feldspar (0.5 to 1 μ in diameter) with normal ammonium chloride solution and then removed the excess ammonium chloride by washing the samples with alcohol. Subsequent treatment of the samples with normal potassium, magnesium, and strontium chlorides released from 0.4 to 3.2 m.e. of ammonium per 100 g of feldspar.

3. Verma (1963) found that release of potassium from nonexchangeable form in orthoclase feldspar under the influence of growing plants was rapid at first and then occurred more slowly (see Fig. 9.10), as would be expected from the theory of resistant surface layers. Release continued at a relatively high rate from certain micaceous minerals.

4. Feldspars occur mostly in the sand and silt fractions of soil and are absent from the clay fraction or occur there only in traces. Hence stability is associated with relatively large particle diameter, as would be expected from the theory of resistant residual surface layers.

Potassium in micas is imbedded in a crystalline aluminosilicate structure, but the potassium atoms lie in planes between molecular aluminosilicate networks. Although each potassium atom is surrounded by and bonded to oxygen atoms, two separate sets of bonds are involved, one set in the molecular aluminosilicate layer on one side of the potassium and the other in the aluminosilicate layer on the other side of the potassium. The molecular aluminosilicate layers are bonded together through the potassium ions (see Fig. 4.4 for a schematic diagram of the molecular structure of muscovite illustrating this bonding). No other electrostatic bonds cross the plane containing the potassium atoms.

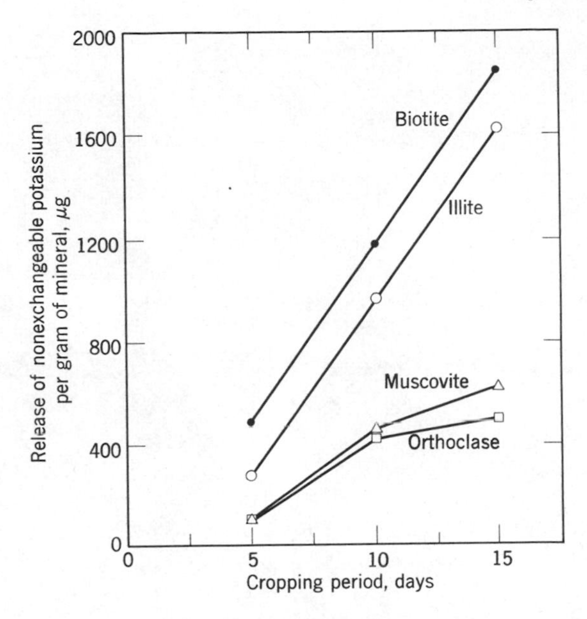

Figure 9.10. Release of nonexchangeable potassium from different minerals during intensive cropping of mixtures of the minerals with quartz sand. Particles of biotite, muscovite, and orthoclase were less than 50 μ in diameter, and those of illite were less than 20 μ in diameter. All minerals were added in quantities to supply the same total amount of potassium per culture. (Verma, 1963)

The potassium atoms are isolated from each other and from the external medium as long as the aluminosilicate layers remain drawn together by the intervening potassium atoms. The potassium plane is the weakest part of the crystal structure, however, as evidenced by the ease with which sheets of mica may be split physically into thinner sheets. This splitting occurs along the potassium planes and causes exposure of potassium atoms on the cleaved surfaces. Potassium atoms thus exposed are exchangeable.

Under chemical weathering in the soil, as in physical disintegration of mica crystals, the weakest part of the structure is the potassium plane. Access of all the potassium ions in a given plane to the external solution is provided if the enclosing silicate layers are pried apart. In the original mica structure, water molecules do not occur with the potassium between the molecular aluminosilicate layers. The distance from the surface of one molecular layer to the corresponding surface of the next layer in mica is about 10 Å. Although water molecules are adsorbed strongly to the surfaces of mica particles, potassium atoms hold the molecular

Figure 9.11. Degradation of micas in the laboratory in the presence of sodium tetraphenylboron, which continuously removes potassium from solution by precipitation as it is released from the minerals. All the particles shown were approximately 2.5 cm square and 1 mm in thickness before treatment with solution. (*a*) Edge view of five species of mica after treatment for 3 days. Muscovite has undergone no visible alteration. As a result of loss of potassium, the other species have expanded in a vertical direction perpendicular to the original molecular planes of potassium in the minerals. (*b*) Enlarged edge view of expanded lepidomelane after treatment for 3 days. The particle with an initial thickness of 1 mm has expanded along the original potassium planes to a thickness of 17 mm. (*c*) Surface view of biotite and muscovite after treatment for 18 months. These photographs, taken perpendicular

aluminosilicate layers together strongly enough to keep water molecules from entering and spreading the layers apart, which would make the potassium exchangeable. In time, however, as potassium atoms adjacent to the edges of the crystals are lost to the outer solution, water molecules do enter between the layers and spread the layers apart. This process has been investigated in the laboratory by observing the changes that occur when pieces of mica are placed in solutions of sodium tetraphenylboron. The reaction that occurs may be represented by

$$\text{K-mica} + \text{Na-tetraphenylboron} = \text{Na-mica} + \text{K-tetraphenylboron}.$$

Potassium tetraphenylboron is practically insoluble in an excess of sodium tetraphenylboron, and so the reaction proceeds to the right. Figure 9.11 provides graphic illustration of the rapidity with which certain micas may be degraded by use of sodium tetraphenylboron, which acts only on the potassium and does not attack the aluminosilicate structure. All that is needed to cause rapid release of potassium is a way to remove the potassium from solution and another cation to maintain electrical neutrality in the depleted residue.

To put the results of these laboratory observations in perspective in terms of the behavior of micas in soils, it seems reasonable that soil micas should release potassium rapidly provided the external concentration of potassium is low enough. Now, potassium-bearing micas are known to be present in many soils, and evidence indicates that they have been there, subject to chemical weathering, for thousands of years. From the laboratory findings just mentioned, the potassium should have been lost long ago if conditions were sufficiently favorable for potassium removal. Although some potassium has been lost, the conditions for potassium removal have evidently been far less effective than those employed in the laboratory.

Release of potassium from nonexchangeable, interlayer form in micas is primarily a cation-exchange reaction. Accordingly, if one represents

---

to the potassium planes, show how peripheral depletion of potassium from the edges inward has changed the appearance of the outer part of the specimens relative to the center. Note the greater penetration of the peripheral weathered zones in biotite than in muscovite. The light-colored areas across the central portion of the specimens result from removal of potassium between molecular planes that did not extend to the outer edges. (*d*) Enlarged edge view of specimen of biotite cut through the center after treatment for 18 months. Note the expansion that has occurred at the edges where potassium has been removed. Some expansion has occurred also in the central portion because of removal of potassium from a few planes through most or all of the specimen. (Photographs courtesy of A. D. Scott) the process in terms of interaction with soil solutions, an equation such as

$$\text{K-mica} + (\text{Ca, Mg, Na, Al, H, NH}_4)\text{-anion} = (\text{Ca, Mg, Na, Al, H, NH}_4)\text{-mica} + \text{K-anion}$$

may be written, where "anion" refers collectively to the freely diffusible anions in the soil solution. At equilibrium, the mica will have taken up some of each cation originally in solution and will have released some potassium to the solution. Application of the law of mass action in a qualitative way indicates that the total quantity of potassium released from a given quantity of mica at equilibrium should increase with the concentration of cations other than potassium in solution, with an increase in volume of solution, and with a decrease in concentration of potassium in solution before the reaction. Exchangeable cations enter the picture in that the potassium and other cations in the soil solution are substantially at equilibrium with those in exchangeable form. Ammonium is of special significance in that it behaves much like potassium in this particular reaction, as will be explained further in the next section. High concentrations of ammonium, like those of potassium, tend to keep the reaction from proceeding to the right.

Scott and Smith (1966) conducted an experiment that provides evidence on the concentrations of potassium needed to inhibit the reaction. In this experiment, finely divided micaceous minerals were equilibrated with a 1-normal sodium chloride solution. With each mineral species, the ratio of total potassium added in the mineral to total sodium in solution was varied; but in all instances the excess of sodium ion was great enough to cause practically quantitative release of all the potassium subject to exchange. Hence the potassium in solution represented practically all the potassium released from nonexchangeable form. Interpolation in Fig. 9.12 shows that the concentration of potassium in solution in micrograms per milliliter associated with a replacement of 5% of the potassium by sodium was about 30 with phlogopite, 15 with biotite, 7 with illite, and 0.3 with muscovite. These concentrations are in the same range as the concentrations of potassium in saturation extracts of soils in Fig. 9.4, which suggests that the concentrations of potassium in many soil solutions are high enough to keep most of the interlayer potassium in micaceous minerals present as such. Further information is needed, however, because saturation extracts are not soil solutions and because the quantitative significance of values measured in 1-normal sodium chloride solution in terms of soil solutions is not known.

Other evidence indicates that, at least qualitatively, the principles discussed in preceding paragraphs apply to the release of potassium

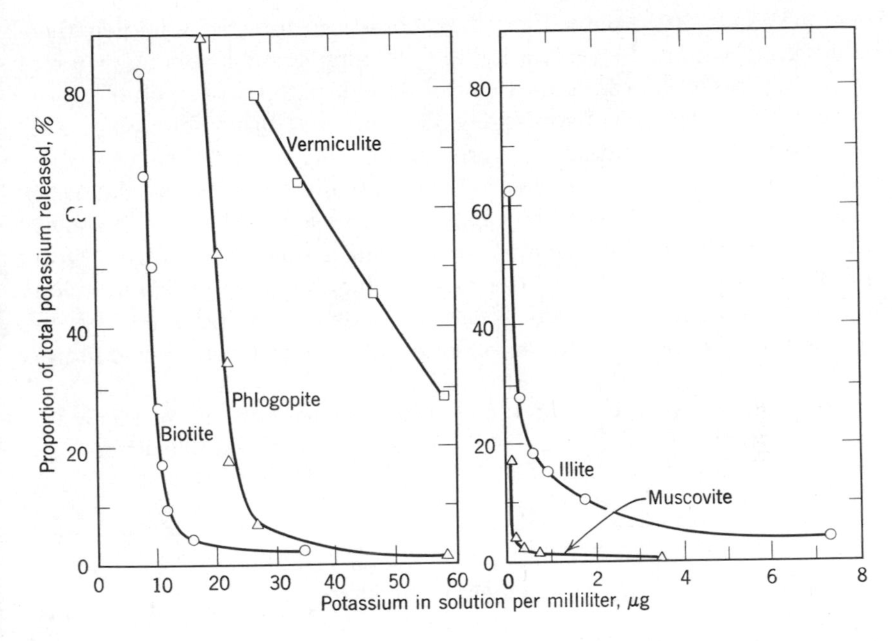

Figure 9.12. Release of potassium from different micaceous minerals suspended in 1-normal sodium chloride solution versus concentration of potassium in solution at equilibrium. In addition to sodium chloride, the solution in which the illite and muscovite were suspended contained 0.01-normal ethylene diamine tetraacetate. Illite particles were less than 2 μ in effective diameter; particles of all other minerals were 10 to 20 μ in effective diameter. Note the difference in horizontal scale between the two graphs. (Scott and Smith, 1966)

from micaceous minerals in the presence of plants. One may note, for example, that the order of decreasing release of potassium from micaceous minerals in sodium chloride solutions (biotite > illite > muscovite) in Fig. 9.12 is the same as the order of decreasing uptake of potassium from the minerals by plants in Fig. 9.10.

Mortland et al. (1956) made a detailed examination of the changes that took place in biotite when four successive crops of wheat were grown on mixtures of quartz sand and finely ground biotite with nutrients other than potassium added artificially. The biotite was added in different quantities to permit different degrees of depletion, and some of the material residual from the biotite was recovered from the sand for analysis after growth of the crops. Table 9.6 shows that cropping reduced the potassium percentage and increased the cation-exchange capacity of the residue from the biotite. The increase in cation-exchange capacity may be attributed to removal of potassium by plants, with consequent entry of cations other than potassium to satisfy the negative charges on the mineral layers and with spreading of the layers by water.

X-ray analysis showed that the diffraction pattern of biotite given by the original material had largely disappeared during cropping and was replaced by the pattern for vermiculite. The distance from the surface of one molecular layer to the corresponding surface of the next layer is about 10 Å in biotite but 14 Å in vermiculite. Figure 9.13 shows clearly a gradual change from a strong diffraction peak at 10Å with no diffraction peak at 14 Å in the original biotite to a weak diffraction peak at 10 Å with a diffraction peak of moderate intensity at 14 Å in the material residual from the application of 25 g of biotite per culture.

Table 9.6. Potassium Content and Cation-Exchange Capacity of Finely Ground Biotite and of the Material Residual from the Biotite after Growth of Four Crops of Wheat in Biotite-Quartz Cultures (Mortland et al., 1956)

Biotite Added per Culture, g	Total Potassium Content of Residue from Biotite, %	Cation-Exchange Capacity per 100 g of Residue from Biotite, m.e.
25	2.4	54
50	3.6	38
100	4.5	30
Original biotite	5.8	14

The results of the X-ray analysis thus provide additional understanding of the nature of the alteration that took place in the presence of plants. Vermiculite has a high cation-exchange capacity, the major part of which is attributable to interlayer sites (the ones that held the potassium in the original biotite). Confirmation that the product was vermiculite and not chlorite, another mineral with a 14-Å diffraction peak, was obtained by subjecting to X-ray analysis the material residual from the biotite after growth of plants and after saturating the exchange sites with potassium and heating the material to 110°C. When this was done, the 14-Å diffraction peak was replaced by one at 10 Å, indicating contraction of the layers and trapping of added potassium in the sites between the layers. Chlorite does not contract upon treatment in this way.

The foregoing experimental work was done under artificial conditions suitable for testing the principle of plant-induced release of potassium from micaceous minerals. Detection of analogous changes in potassium-

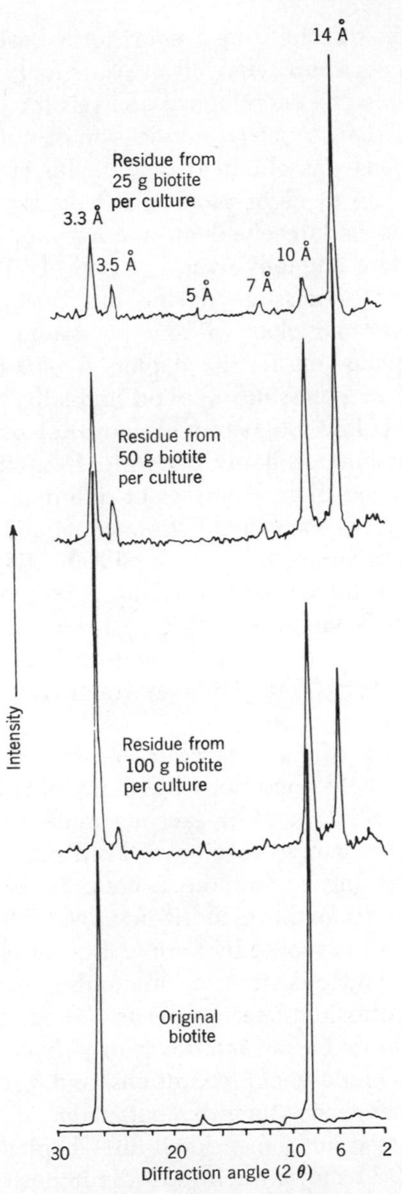

Figure 9.13. Intensity of X-ray diffraction at different angles from finely ground biotite and from the residues of finely ground biotite after growth of four consecutive crops of wheat on mixtures of quartz and biotite with nutrients other than potassium supplied artificially. The intensities of diffraction from the original biotite should be doubled to make them comparable with intensities from the other samples. Note particularly the gradual change in relative intensities of diffraction corresponding to 10 and 14Å in proceeding from the original biotite to the residue from biotite supplied at 25 g/culture. (Mortland et al., 1956)

bearing minerals in the field on a short-term basis is difficult because the quantities of potassium removed are so small, comparatively speaking, and the methods of mineralogical analysis for detecting the changes are not highly sensitive. A partial verification of operation of the process under field conditions was obtained by Scheffer et al. (1960) in investigating changes in soil of eight plots in a field experiment in Germany. Four of these plots had received an average application of 12.7 kg of potassium per hectare annually over a period of 77 years, and the other four had received no potassium during this time. The cation-exchange capacity of the soil from plots without potassium exceeded that of soil from comparable potassium-fertilized plots by 0.9 to 2.4 m.e./100 g. Because the quantity of potassium applied annually to the fertilized plots was probably well below the average removal of potassium in the crops, a net loss of potassium probably occurred from the fertilized plots as well as the unfertilized plots. Analyses of different particle-size fractions showed that, although the silt fraction contained about 25% feldspar and 35% micaceous minerals, the only significant difference in total potassium between the fertilized and unfertilized plots was in the clay fraction. Similarly, Van Ruymbeke (1963) reported that the total potassium content decreased and the cation-exchange capacity increased with age in soils of Belgian sea polders. The clay fraction of these soils contained about 50% illite.

Clearer verification of the alteration of micaceous minerals by loss of potassium under field conditions has been obtained in investigations of the vertical distribution of micaceous minerals in soils. These soil profile studies reflect changes that have taken place during soil development. The usual finding is that the content of 10-Å micas of silt and sand size decreases from the C horizon upward to the A horizon and that the reverse trend is shown by vermiculite- or chlorite-type minerals. These vermiculite- and chlorite-type minerals presumably are products of weathering of potassium-bearing micas present originally throughout the soil profile. Chlorite has an interlayer of polymeric hydroxyaluminum ions that develops under acid conditions and prevents occupation of the interlayer exchange positions by potassium and other cations (this process was discussed in some detail in Chapter 5 on soil acidity). For example, Fig. 9.14 shows tracings of the intensity of X-ray diffraction corresponding to different layer spacings in three particle-size fractions from each of four depths in a silt loam soil developed from a muscovite schist in Virginia. The figure shows that the height of the 14-Å peak relative to that of the 10-Å peak increased not only in proceeding from the greatest depth toward the surface but also in proceeding from particle-size fractions of relatively large size to those of small size. Similar

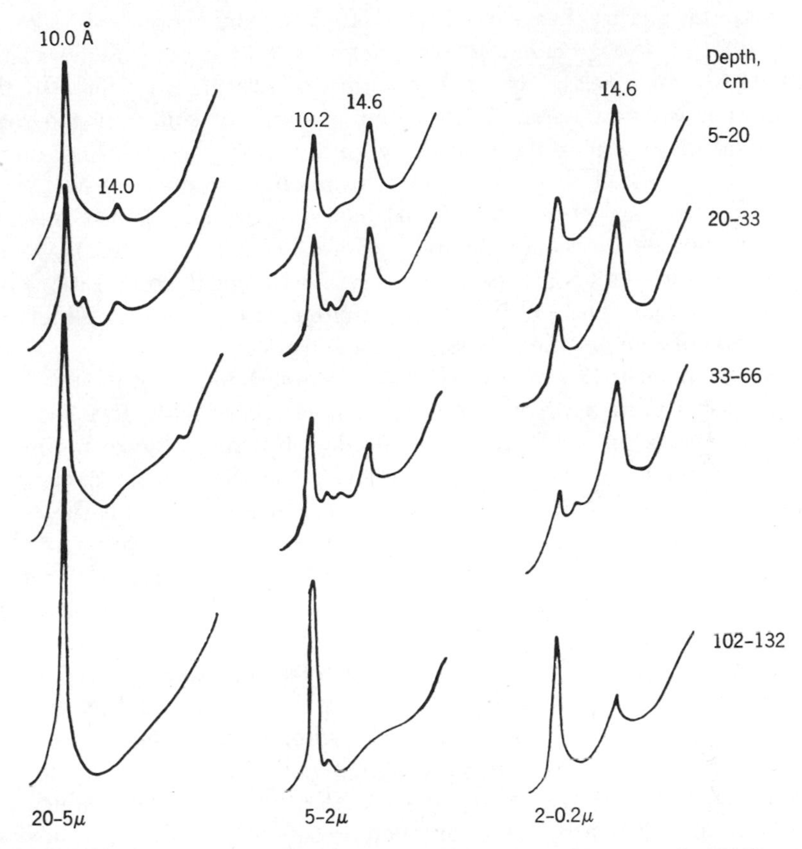

Figure 9.14. Smoothed X-ray diffractometer tracings from three particle-size fractions from four depths in a silt loam soil developed from muscovite schist in Virginia. The numbers associated with peak diffraction intensities indicate the distance from the surface of one molecular layer to the corresponding surface in the next layer. (Rich, 1958)

trends were observed in samples from two other profiles of the same soil type, but the tracings of X-ray diffraction intensities from each profile showed some characteristic features that differed from the tracings from the other profiles.

On the basis of the foregoing considerations relative to release of potassium from feldspars and micas in soils, one might expect the supply of potassium for plants to be related to the content of these minerals in soils. Some rather broad associations between potassium-bearing minerals in soils and the supply of potassium to plants in the field were reported in literature summarized by Arnold (1960). For the most part, however, this matter has been investigated in connection with experiments done in the greenhouse, where the technique of growing plants continuously on a relatively small quantity of soil greatly increases the rate of removal of potassium from unit quantity of soil over the rates of removal that occur in the field. In work of this kind, several investigators (e.g., Schachtschabel, 1937, 1937a; Cummings, quoted by Mortland, 1961; and Cook and Hutcheson, 1960) have noted a correlation between potassium uptake by plants and illite content of soils. Others, however, have found no significant association of mineralogical composition and potassium uptake. The publication by Reitmeier et al. (1951) includes a discussion of some of the problems in such work.

The fact that little success has been attained in attempts to relate the uptake of potassium by plants from nonexchangeable forms to the potassium-bearing minerals in soils is no doubt attributable to the methods of mineralogical analysis employed because there is no reasonable doubt that the potassium in question resides in potassium-bearing minerals. The X-ray method, on which principal reliance is often placed, is more suited to qualitative identification than quantitative determination; and it does not seem to be sensitive to some of the critical variations in the nature of the minerals that affect the release of potassium.

Because of the limited value of mineralogical methods for estimating the quantities of potassium-bearing minerals in the clay fraction of soils, chemical analysis is sometimes used to estimate the illite content, the usual assumptions being that all the potassium in the clay fraction is present as illite and that illite contains 6% potassium. The values for potassium are ordinarily used as such in investigations of potassium supply for plants. Arnold (1960a) used the total-potassium method in work on a group of 19 soils in Great Britain. He grew ryegrass on the soils for 18 months in the greenhouse, determined the uptake of potassium by the ryegrass and the changes in potassium fractions in the soil, and plotted the results in a form similar to that shown in Fig. 9.15. This figure gives a plot of potassium percentage in the fine clay fraction against the percentage content of fine clay in the soil. The data points representing individual soils are coded into three groups according to the amount of originally nonexchangeable potassium taken up by the ryegrass in the 18 months of intensive cropping. With the exception of results for two soils, the data may be seen to fall into three groups. A given uptake of potassium could be obtained from a soil in which the potassium percentage in the fine clay was high and the percentage content of fine clay was low or in soils in which the potassium percentage in the fine clay was low and the percentage content of fine clay was high.

One of the reasons for reproducing Fig. 9.15 here is that the author

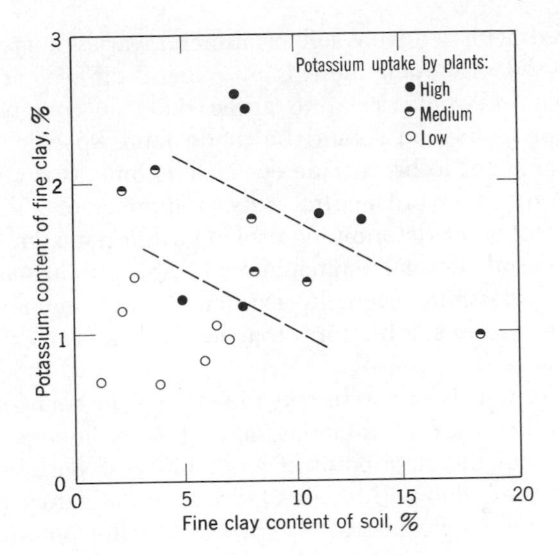

Figure 9.15. Potassium percentage in fine clay (<0.1 μ in effective diameter) versus fine clay content of 19 soils of Britain, with values classified according to relative uptake of potassium from nonexchangeable forms in 18 months of intensive cropping by perennial ryegrass grown on the soils in the greenhouse. Potassium uptakes in micrograms per gram of soil are classified as follows: >500 = high, 100–500 = medium, and <100 = low. (Arnold, 1960)

provided certain additional information to explain the behavior of the two exceptional soils in which the potassium uptake was high but in which analyses of the fine clay caused the soils to be classified with others that supplied little potassium to the ryegrass. The silt fraction of one of the exceptional soils was found to release considerable potassium when incubated with a hydrogen-saturated cation-exchange resin. Much of the potassium taken up from this soil by the ryegrass thus may have come from the silt fraction. The other exceptional soil was from an experimental field that had been heavily fertilized with potassium. Presumably, (1) the nonexchangeable potassium that developed in the fine clay of this soil from fertilization was released to plants more readily than an equivalent amount of nonexchangeable potassium in the other soils, or (2) the potassium was present mostly in particle-size fractions other than fine clay.

Fixation

Potassium fixation is a term applied to indicate a decrease in extractability of potassium as a result of interaction of originally soluble or extractable potassium with a soil or mineral. The magnitude of the fixation associated with a given soil or mineral varies a great deal with the method used. Although there is no general agreement on methods, most investigators would seem to agree that the foregoing definition of the concept of fixation should be made more specific to the extent that potassium is not to be considered fixed as long as it remains readily exchangeable by cations of neutral salts. A number of investigators employ the convention of determining exchangeable potassium by displacement with neutral, normal ammonium acetate and hence define fixed potassium as potassium originally extractable with ammonium acetate or originally added in soluble form that has become nonextractable with this reagent.

Potassium fixation is of interest in mineralogy in connection with the structure and behavior of minerals, and it is of interest in soil-plant relationships from the standpoint of availability of potassium to plants. In this section, attention will be directed first to the nature of the process and then to the significance of the process in the potassium nutrition of plants.

A great deal of research has been done on potassium fixation; and the status of knowledge on the subject is set forth in papers by Schuffelen and van der Marel (1955), Agarwal (1960), and Weaver (1958). The first two papers are reviews, and the third is primarily interpretive.

Nature of Potassium Fixation. Potassium is fixed in various minerals, which may be classified into three groups: (1) those that fix potassium by ion-exchange in sites that hold potassium in the unweathered mineral (examples are vermiculites and surface-weathered potassium feldspar); (2) those that fix potassium by ion-exchange in sites that do not necessarily hold potassium in the unweathered mineral (examples are certain zeolites); and (3) those that are newly synthesized (examples are complex phosphates of iron, aluminum, or both, containing also potassium, ammonium, or both).

The potassium-fixing properties of various minerals in these three classes have been discovered by special experiments with fairly pure minerals or in other ways. As yet, however, it is not possible to determine potassium fixation on a soil and allocate the fixation quantitatively among various minerals.

Schuffelen and van der Marel (1955) summarized the information in the literature on potassium fixation by various minerals and included original data on fixation by many minerals collected from different parts of the world. Of all the minerals they tested, the ones that fixed the most potassium per 100 g of mineral were vermiculites and zeolites. Zeolites are hydrous aluminosilicates with small channels in which the potassium presumably may be trapped. Zeolites do not seem to be common in soils, and little attention is accorded these minerals in research on potassium fixation. Vermiculite, as mentioned previously, is a layer-type mineral similar to micas except that (1) interlayer potassium has been replaced by other cations that remain exchangeable and (2) the molecular aluminosilicate layers are spread farther apart than they are in micas. Various other layer-type clay minerals with expanded layers fix smaller quantities of potassium, and they may fix substantial quantities under special conditions. Most of the research on potassium fixation has been done with vermiculite and other expanded clay minerals.

As was mentioned in the preceding section, potassium ions lie in a plane between the molecular aluminosilicate layers of micas. They are imbedded in openings in the surface layers of oxygen atoms in adjacent layers and are bonded electrostatically to the oxygens in both layers. The fit of the potassium ions is such that they spread the layers slightly as compared to the spacings found by X-ray diffraction if the micaceous minerals contain no interlayer potassium and are fully dehydrated. If micaceous minerals are depleted of potassium without loss of integrity of the aluminosilicate layers, other cations enter and balance the electrostatic charges that were originally satisfied by potassium; and water molecules enter and spread the molecular layers apart, producing an expanded mineral known as vermiculite. The water molecules are bonded strongly to the inner surfaces, as evidenced by the fact that they are not driven off even when the mineral is heated at 110°C. If the mineral is heated at 750°C, the water is driven off; the molecular layers then collapse to produce a spacing slightly less than that observed when potassium ions are present. The cations that permit such collapse (calcium, magnesium, sodium) are smaller than potassium and presumably fit readily into the openings in the oxygen layers with space to spare. (Incidentally, the gross expansion of vermiculite produced by heat treatment is thought to result from mechanical breakage of some of the layers by the steam produced when the interlayer water is volatilized and not from uniform spreading of the molecular layers. X-ray diffraction shows the molecular layers to be collapsed.)

If the cations that have replaced potassium in formation of vermiculite from mica are subsequently replaced by potassium, the molecular layers are drawn together; and the potassium is again entrapped in nonexchangeable form. This is the general theory of potassium fixation by expanded minerals. The process is simple enough in general outline, but the fixation that occurs seems to depend a great deal on the precise nature of the mineral and on the conditions.

With regard to the nature of the mineral, several factors are thought to be important. First, it should be mentioned that vermiculite and the other expanded clays that fix potassium all have the same basic molecular structure. All have the surface layers of oxygen atoms with the openings into which potassium can fit, and all have a negative charge on the molecular aluminosilicate layers that is balanced by cations located outside the layers; however, the minerals differ in charge per unit area of the molecular layers. The electrostatic charge per unit area is commonly called the "charge-density" and is expressed in units of milligram equivalents per 100 g, like cation-exchange capacity. In fact, if the balancing cations are exchangeable, the charge-density is equivalent to the cation-exchange capacity. According to the theory that electrostatic charges on the aluminosilicate layers cause bonding of adjacent layers through intervening potassium ions, the strength of bonding between adjacent layers should increase with the number of charges per unit area. In agreement with this concept, micas and vermiculites, which have a relatively high charge-density, hold potassium in nonexchangeable form to a much greater extent than does montmorillonite, which has a lower charge density.

Also involved in the bonding between adjacent layers and hence in the fixation of potassium is thought to be a charge-strength factor associated with the origin of the negative charges. Three general concepts are involved in this idea. First, the negative charges on the aluminosilicate layers result from a deficiency of cations in these layers (the principal cations are silicon, aluminum, and iron) in relation to oxygens and hydroxyls. Second, an individual negative charge may result from a cation deficiency in the outer, tetrahedral sheet of the aluminosilicate layer or from a cation deficiency in the inner, octahedral sheet. Third, the strength of the bond at the surface of the layer, where it is balanced by potassium, will be greater if the charge originates in the outer, tetrahedral sheet than in the inner, octahedral sheet. (Reference should be made to Chapter 4 for schematic structures of layer-silicate minerals and for further discussion to clarify these concepts.)

Three accessory conditions have been found to be of importance with regard to fixation of potassium by layer-silicate minerals: the nature of the balancing cations between layers, the particle size, and the degree of drying and heating. These will be discussed in turn.

If the balancing cations are polymeric hydroxyaluminum ions (see Chapter 5 for a discussion of these ions), fixation of potassium will be reduced or eliminated because the hydroxyaluminum ions keep the molecular aluminosilicate layers permanently expanded. On the other hand, if the balancing cations have the same charge and about the same size as potassium, they will tend to be nonexchangeable themselves and for this reason will inhibit fixation of potassium. Effects of ionic

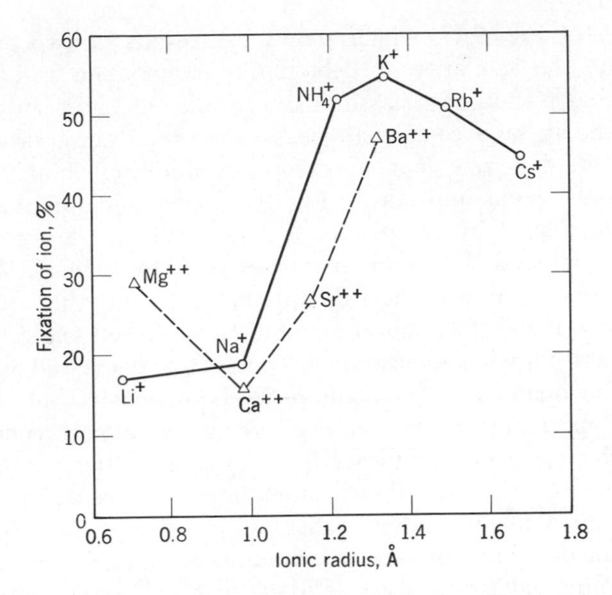

Figure 9.16. Percentage fixation of different cations by a soil clay versus ionic radius. Percentage fixation is defined as the decrease in the quantity of the cation released from the clay to soluble form caused by drying the clay at 100°C before treatment with hydrochloric acid equivalent to the cation-exchange capacity expressed as a percentage of the quantity of the cation released in the same way from the undried clay saturated with the cation in question. This method of expressing the results tends to separate the fixation effect from differences in replaceability of the individual cations by the hydrogen ion. (Page and Baver, 1939)

size and charge are illustrated in Fig. 9.16, which shows the fixation of various monovalent and divalent cations by clay separated from a soil in Ohio. The monovalent ions seem to form a regular series according to ionic size with peak fixation at the potassium size and decreasing fixation of smaller or larger cations, but the divalent cations do not follow this pattern. As indicated in the figure, the size of the ammonium ion is similar to that of the potassium ion, and the fixation of ammonium by the soil clay used was similar to the fixation of potassium.

Considerable research has been done on ammonium fixation not only because experiments involving ammonium alone or ammonium with potassium may provide information about potassium behavior but also because of the possible significance of ammonium fixation in connection with nitrogen behavior in soils. Ammonium fixation in the latter context was discussed at some length in Chapter 7.

At this point, consideration of ammonium fixation will be limited to introduction of Fig. 9.17, which shows the results of two experiments that indicate the similarity of behavior of ammonium and potassium. In Fig. 9.17*a*, fixation of potassium by samples of eight soils is similar to but, on the average, somewhat greater than the fixation of ammonium determined in the same manner except by addition of an ammonium salt instead of a potassium salt. In Fig. 9.17*b*, the amounts of ammonium and potassium fixed by samples of a single soil are shown where the ammonium was added first in quantities of 0 to 15 m.e./100 g of soil and the potassium was added subsequently in the quantity of 2.5 m.e./100 g of soil. Fixation of ammonium increased with the quantity of ammonium added, and fixation of potassium decreased with an increase in the quantity of ammonium fixed previously; but the sum of ammonium fixed and potassium fixed remained almost constant. The results of this experiment indicate that ammonium and potassium were fixed by the same sites and that if ammonium was fixed first, potassium merely occupied the remaining potassium-fixing sites that were not already occupied by ammonium. In an additional experiment, not shown in the figure, potassium was added first, in quantities of 0 to 15 m.e./100 g; and ammonium was added second, in quantities of 10 m.e./100 g. Prior potassium fixation then reduced ammonium fixation, but the sum of the two increased somewhat with the quantity of potassium added. At a given concentration of fixable cations, there was greater fixation of potassium than of ammonium, in accordance with the observations shown in Fig. 9.17*a* and in Fig. 9.16.

A second accessory condition that affects the fixation of potassium by expanded layer-silicate minerals is the particle size. This factor has been investigated with vermiculite, which may be obtained in particles of the order of a centimeter or more in diameter. Barshad (1954) found that sodium was far more effective than potassium or ammonium in replacing other exchangeable bases from vermiculite particles of large (gravel) size but that in particles small enough to pass a sieve with openings 0.1 mm in diameter the effect was practically absent. He explained these results by the theory that addition of potassium or ammonium promoted contraction of the molecular layers of the vermiculite at the edges of the particles where the exchange began and that the cations in the more remote inner locations were then trapped and nonexchangeable because of the restricted movement of cations in and out through the peripheral contracted zones. He noted the same trapping effect in an experiment on a subsoil that contained coarse, weathered biotite and had a high capacity to fix potassium and ammonium. He saturated the soil with barium, removed the excess barium salt with methyl alcohol, and then leached separate portions with neutral, 1-nor-

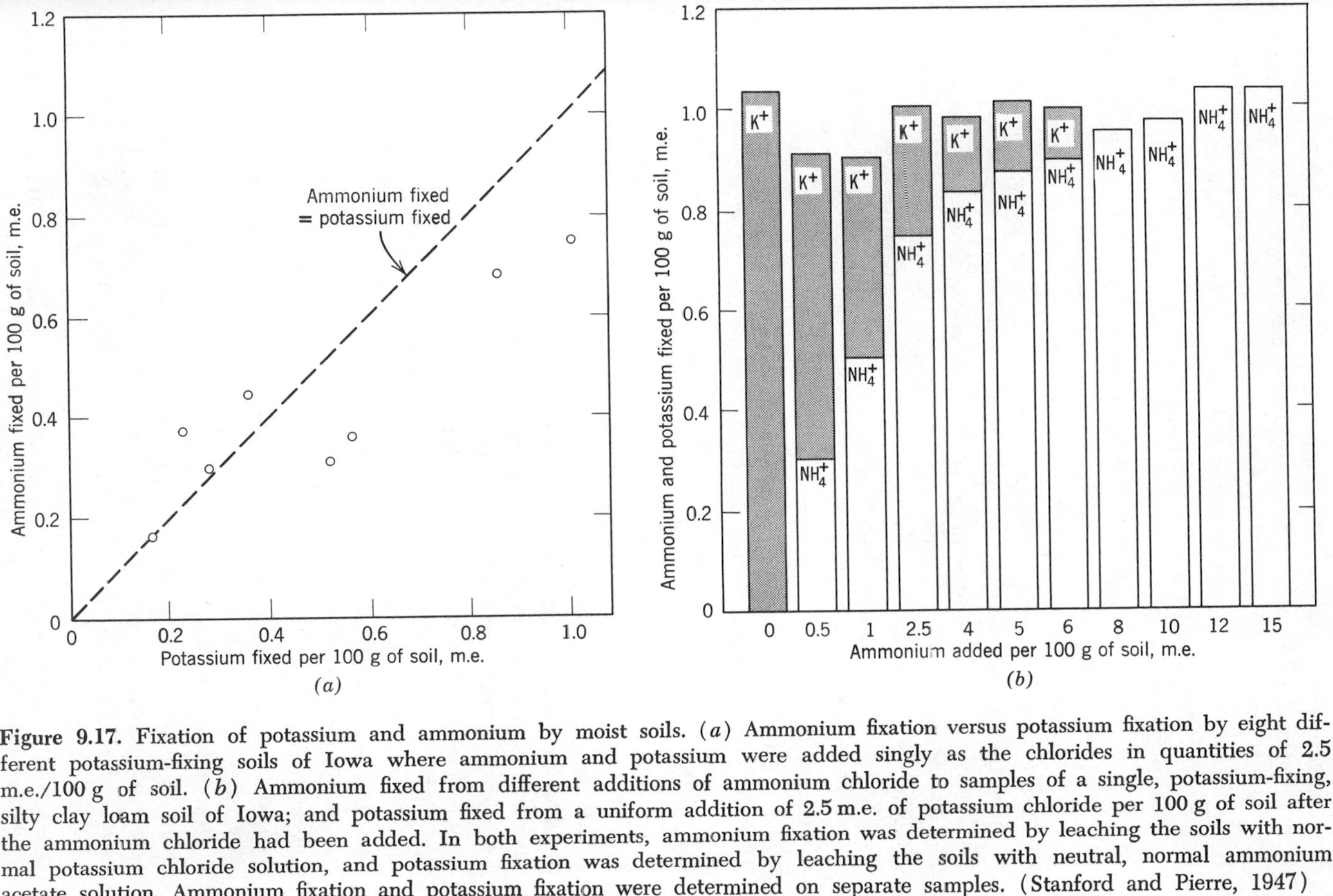

Figure 9.17. Fixation of potassium and ammonium by moist soils. (*a*) Ammonium fixation versus potassium fixation by eight different potassium-fixing soils of Iowa where ammonium and potassium were added singly as the chlorides in quantities of 2.5 m.e./100 g of soil. (*b*) Ammonium fixed from different additions of ammonium chloride to samples of a single, potassium-fixing, silty clay loam soil of Iowa; and potassium fixed from a uniform addition of 2.5 m.e. of potassium chloride per 100 g of soil after the ammonium chloride had been added. In both experiments, ammonium fixation was determined by leaching the soils with normal potassium chloride solution, and potassium fixation was determined by leaching the soils with neutral, normal ammonium acetate solution. Ammonium fixation and potassium fixation were determined on separate samples. (Stanford and Pierre, 1947)

mal ammonium acetate, with neutral, 1-normal sodium chloride, and with 1-normal hydrochloric acid until the filtrate was free of barium. The latter two reagents replaced 31 m.e. of barium per 100 g of soil, but the ammonium acetate replaced only 25 m.e. The difference of 6 m.e. represents the trapped barium.

A third accessory condition that influences the fixation of potassium by expanded layer-silicate minerals is the degree of drying and heating to which the material is subjected before the fixation is measured. In general, drying and heating markedly increase the fixation. The effects of drying and heating of layer silicates on potassium fixation are explained theoretically on the basis of loss of adsorbed water from interlayer positions and closure of the lattice layers about the potassium ions. Vermiculite, with a spacing of about 14 Å from the surface of one aluminosilicate layer to the corresponding surface of adjacent layers, contains adsorbed water between the molecular layers. If potassium ions are exchanged for the other balancing cations without drying the sample in the process, the spacing is decreased, often to about 12 Å, which must mean that the increase in strength of bonding between adjacent layers from introduction of potassium has caused some of the adsorbed water to be squeezed out. The potassium is not completely fixed when the mineral remains in this hydrated condition. If the material is air-dried at room temperature or at higher temperature, however, the spacing between the layers decreases; and more of the potassium becomes nonexchangeable. Existence of adsorbed water between the layers of micaceous minerals of high charge-density after saturation with potassium seems to be a metastable condition because the illites that presumably develop when potassium-depleted micaceous minerals take up potassium from sea water have the 10-Å spacing characteristic of muscovite and biotite. Montmorillonite, which has a relatively low charge-density, can be caused to fix potassium by saturating the exchange positions with potassium followed by oven-drying the mineral. Layer spacings of samples treated in this way are greater than those found in vermiculites after similar treatment, and subsequent release of potassium and expansion of the layers occur more readily than in vermiculites.

In research on potassium fixation in soils, considerable importance is attached to the distinction between fixation on drying and fixation occurring under moist conditions because soils remain for the most part in a moist condition as regards potassium fixation even though at times they may not contain enough water to support higher plants. Hence the fixation that occurs under moist conditions in the laboratory is inferred to be a phenomenon that would occur also in the field if an excess of potassium were added. If one disregards fixation of potassium in trace quantities, experiments indicate that some soils fix potassium under moist conditions and others do not. Perhaps most soils would fix some potassium under moist conditions with long exposure to a great enough excess of potassium. Many more soils show potassium fixation if they are dried and heated after treatment with excess potassium than if they are maintained in a moist condition.

One curious finding that has intrigued a number of investigators is that, although soils usually fix added potassium if they are dried or dried and heated, these same treatments often cause release of potassium from nonexchangeable form if the soil is dried without addition of potassium. A widely accepted theory to account for this behavior is that the content of exchangeable and soluble potassium in the potassium-treated soil is above an equilibrium condition in relation to nonexchangeable potassium, whereas the untreated soil is below it, and that drying and heating promote attainment of equilibrium.

The theory just mentioned seems now to be inadequate to account for experimental observations that have been made on potassium fixation and release. Luebs et al. (1956) found that the release of potassium from soils on drying took place when the soils they investigated approached the air-dry state. In laboratory experiments with samples of soil maintained in contact with air containing water vapor at different relative humidities, they found that the exchangeable potassium would remain relatively high at low humidities and low at high humidities. If the relative humidity was increased, the exchangeable potassium would decrease and vice versa. These observations verify the concept of an equilibrium between exchangeable and nonexchangeable potassium because the exchangeable potassium first approached a certain value at a given relative humidity and then remained approximately constant for a long time. But because the value for exchangeable potassium depended on the relative humidity, there was evidently no single equilibrium level.

The same authors made further analyses for exchangeable potassium in soils in the field to see if the values they obtained would vary from time to time with the water content of the soil. As shown in Fig. 9.18, the values obtained for exchangeable potassium were by no means constant and tended to vary inversely with the water content of the soil, as had been found in the laboratory work. This effect was largely confined to the upper part of the soil; the range of values for exchangeable potassium in micrograms per gram of soil was 64 to 141 in the surface 1.3 cm, 57 to 87 in the second 1.3 cm, and 52 to 65 in the next 2.5 cm. The lower layers did not become as dry as the surface layer.

The effect of drying on exchangeable potassium may be of some sig-

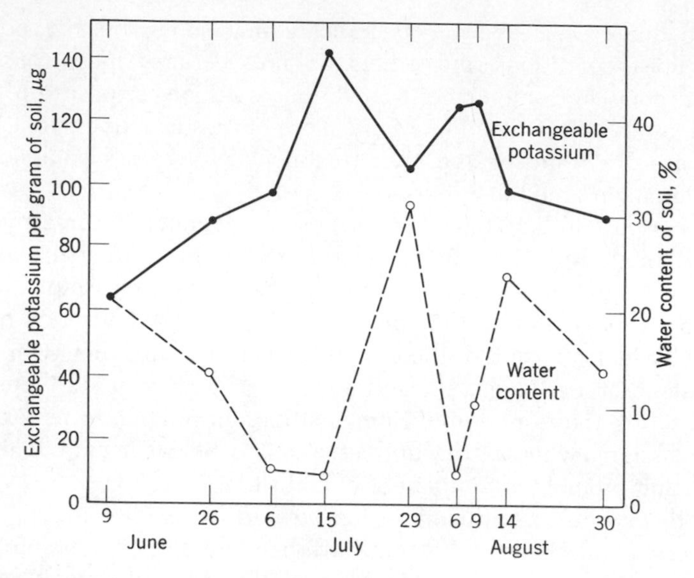

Figure 9.18. Content of exchangeable potassium and water in the surface 1.3 cm of a loam soil sampled in the field at different dates during a summer season in Iowa. (Luebs et al., 1956)

nificance to plants under field conditions if plant roots intermittently absorb potassium from the surface layer. As indicated by Fig. 9.19, the increase in exchangeable potassium produced when a soil was dried before growth of corn plants was associated with a substantial increase in uptake of potassium by the plants as compared with soil that was not dried. In the instance shown in the figure, the effect of drying was equivalent to the increase in potassium uptake obtained from addition of 36.5 μg of potassium as potassium chloride per gram of undried soil. In analogous trials with ten other soils of Iowa, the increases in uptake of potassium resulting from drying the soil before growth of plants were found to be equivalent to 8 to 60 μg/g. The potassium status of soil thus seems to be affected long enough by drying to influence the uptake of potassium by plants grown after the soil has been remoistened.

Development of knowledge of potassium behavior in relation to mineralogy has resulted in a new theory to the effect that potassium fixation is due to potassium-fixing minerals and release is due to potassium-bearing minerals and that both processes may occur in a given soil under appropriate conditions. Figure 9.20 shows the results of an experiment that verify this theory. In this experiment, 0.5-normal sodium acetate was used to displace exchangeable potassium from illite alone, vermiculite alone, and a mixture of the two minerals with and without addition of potassium. Illite (a potassium-bearing mineral) released potassium from nonexchangeable form when dried at 110°C in either the presence or absence of added potassium. Vermiculite (a potassium-fixing mineral), on the other hand, maintained a low, approximately constant value of native exchangeable potassium despite the drying and heating treatment but fixed most of the added potassium. A mixture of 90% illite with 10% vermiculite behaved like soils in that it released potassium on drying of the control samples but fixed potassium if potassium was added. As part of the same investigation, DeMumbrum and Hoover (1958) found that drying two soil clays and heating them at 110°C after addition of potassium caused release of potassium from nonexchangeable form from one clay containing illite and montmorillonite and fixation of potassium by the other clay containing illite, montmorillonite, and vermiculite.

As yet, the cause of release of potassium from nonexchangeable form that occurs when soils are dried is uncertain. A. D. Scott (unpublished) developed the theory that, although the X-ray diffraction pattern of micaceous minerals indicates a decrease of interlayer spacing when the

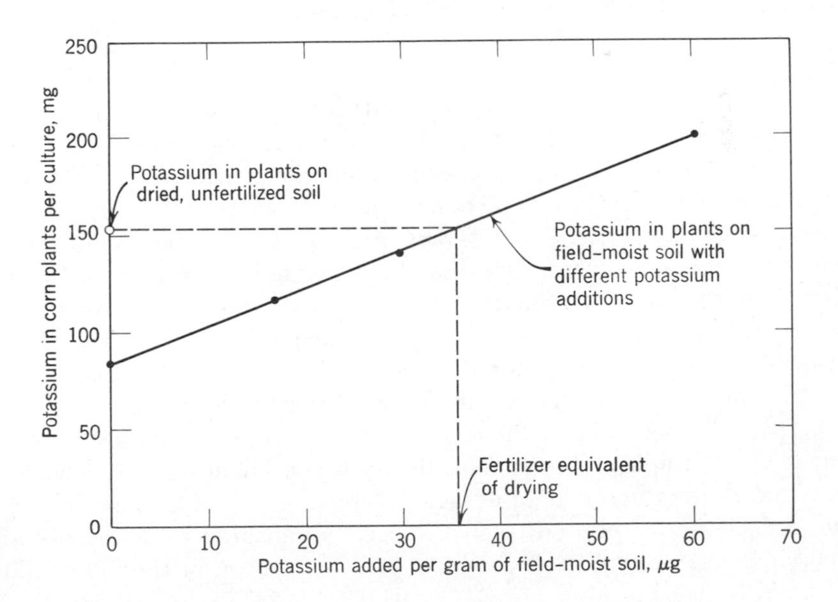

Figure 9.19. Biological assay of effect of drying on supply of native potassium for corn plants in a loam soil in greenhouse cultures. Drying the soil before cropping increased the exchangeable potassium, as determined by the neutral, normal, ammonium acetate method, from 47 to 85 μg/g. (Luebs et al., 1956)

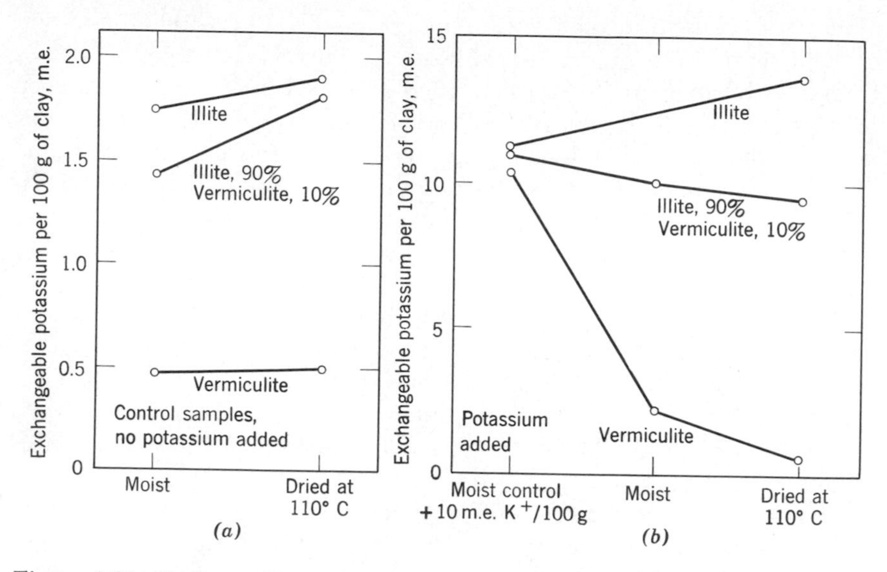

Figure 9.20. Exchangeable potassium in illite, vermiculite, and a mixture of the two with different treatments. (*a*) Control samples, no potassium added. (*b*) Samples after addition of 10 m.e. of potassium per 100 g. Values for "moist control + 10 m.e. K^+ per 100 g" are calculated. (DeMumbrum and Hoover, 1958)

expanded, weathered minerals are dried, certain of the interlayer spacings are increased by curling of the mineral layers when the mineral dries enough to cause loss of some of the interlayer water. The originally nonexchangeable potassium exposed on these more open layers then becomes exchangeable. The X-rays presumably do not show these widened interlayer spaces because there are only a few of them and their spacing varies. Diffraction of X-rays with enough intensity to be registered requires repetition of a given spacing. Figure 9.21 represents evidence on a macro basis for the theory. The figure shows that drying actually increased the over-all thickness of a particle of mica that had been partially degraded at the edges by immersion in a solution containing sodium tetraphenylboron. This theory is consonant with the observation that heating vermiculite at a high temperature causes loss of water and a decrease of the interlayer distance, as indicated by X-ray diffraction, at the same time as it produces gross expansion of the vermiculite particles.

The theory that the potassium fixation in soils under moist conditions occurs mostly in vermiculite-type minerals implies that this capacity to fix potassium is a consequence of prior removal of potassium (because vermiculite is thought to be derived from potassium-bearing micas). Thus, with greater prior removal of potassium, the fixation capacity should be increased. Welte and Niederbudde (1963) verified this inference in an experiment on samples of soil from a field plot in Germany that fixed no potassium under moist conditions, according to the procedure they used, in consequence of prior heavy fertilization with potassium. The soil was mixed with quartz sand in different proportions, with the total weight remaining constant at 11.5 kg per culture. The cultures were planted to ryegrass, and four cuttings were taken. Potassium analyses on the crop and soil and potassium fixation determinations on the soil then yielded the results shown in Fig. 9.22. Evidently cropping did generate capacity for potassium fixation in this soil, and the potassium-fixing capacity produced was about equal to the loss of potassium from nonexchangeable form during growth of the crop.

In consequence of the increase in fixing capacity of soil around the roots, subsequent equilibration may be expected in which potassium released from other locations is reintroduced into some of the newly formed exchange sites from which originally nonexchangeable potassium was removed. Detection of such adjustments in soils under natural conditions is not readily accomplished; but existence of the process in princi-

Figure 9.21. Effect of drying on the thickness of a flake of biotite mica that had been weathered 18 months in a solution of sodium tetraphenylboron. The weathered flake was cut in half; the section on the left was air-dried, and the section on the right was kept moist. The original outside edges were placed together, and the photograph was taken perpendicular to the newly cut edge. (Photograph courtesy of A. D. Scott)

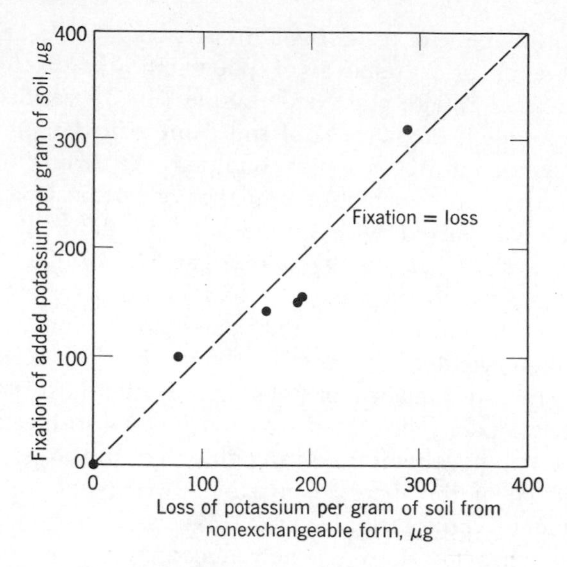

Figure 9.22. Fixation of added potassium under moist conditions, determined after growth of ryegrass (four cuttings), versus loss of potassium from nonexchangeable form in the soil during growth of the crop. The different experimental values were obtained with a single soil at different cropping intensities produced by varying the proportions of soil and sand in the culture vessels. The values of zero fixation and zero loss of potassium from nonexchangeable form were obtained on uncropped soil. (Welte and Niederbudde, 1963)

ple was verified by Mortland (1961), who suspended equal quantities of biotite mica and sodium-saturated soil clays in a large volume of 0.02-normal sodium chloride solution for 21 days and then determined the total potassium content of the clays after the mica had been removed by sieving. He found (Fig. 9.23) that the potassium percentage increased in all clays in the presence of biotite and that the lower the original potassium percentage in the clay the greater was the increase. The increases in potassium content of the clays were so great in relation to the concentration of potassium in the sodium chloride solution (7 to 9 μg/ml) that most of the increase must have been nonexchangeable potassium. X-ray diffraction analysis of the residue from the mica at the end of the equilibration period indicated the presence of vermiculite.

With regard to association of potassium fixation with other soil characteristics, three properties have been found significant: clay content, *p*H, and percentage potassium saturation. These will be discussed in turn.

Kaila (1965) found that potassium fixation by samples of many soils of Finland increased with the clay content. Schachtschabel and Köster

(1960) found that potassium fixation under moist conditions by marsh soils of Germany increased with the clay content but that, as shown in Fig. 9.24*a*, the fixation at a given clay content tended to be relatively low in the salt-water-marsh samples, intermediate in the lake-marsh samples, and high in the river-marsh samples. The clay content of soil may be considered a capacity or quantity factor in potassium fixation. Most of the fixation takes place in this fraction. Nevertheless, as indicated in Fig. 9.24 and as would be expected from the theory of potassium fixation, potassium fixation per gram of clay may differ enough from one soil to another to make clay content as such of limited value as an index of potassium fixation.

The effect of soil *p*H on potassium fixation has been investigated principally in connection with experiments on application of calcium carbonate. In some instances liming has increased potassium fixation, and in others it has had little effect. Kaila (1965) examined the *p*H effect in measurements of potassium fixation by a large number of soil samples from Finland and found a significant increase in fixation with an increase in soil *p*H.

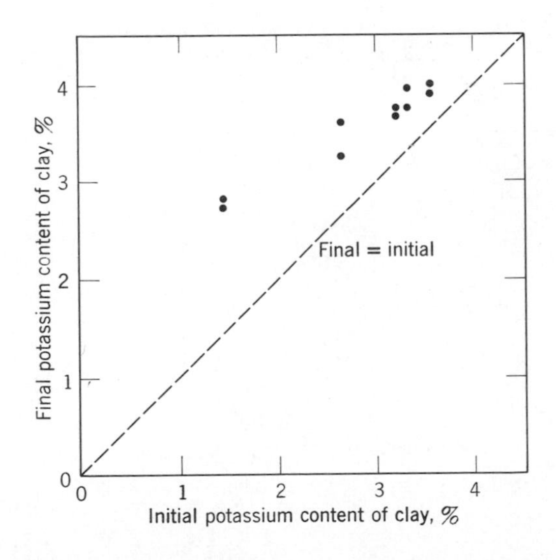

Figure 9.23. Potassium content of the clay from five soils of Michigan after contact for 21 days with an equal weight of biotite in 0.02-normal sodium chloride solution versus original potassium content of the clays. At the end of the reaction period, the biotite was removed by sieving before analysis of the clay for potassium. The two data points for each soil represent results obtained with different particle-size fractions of biotite. The greater increase in potassium content of four of the clays during equilibration occurred with the finer particle-size fraction. (Mortland, 1961)

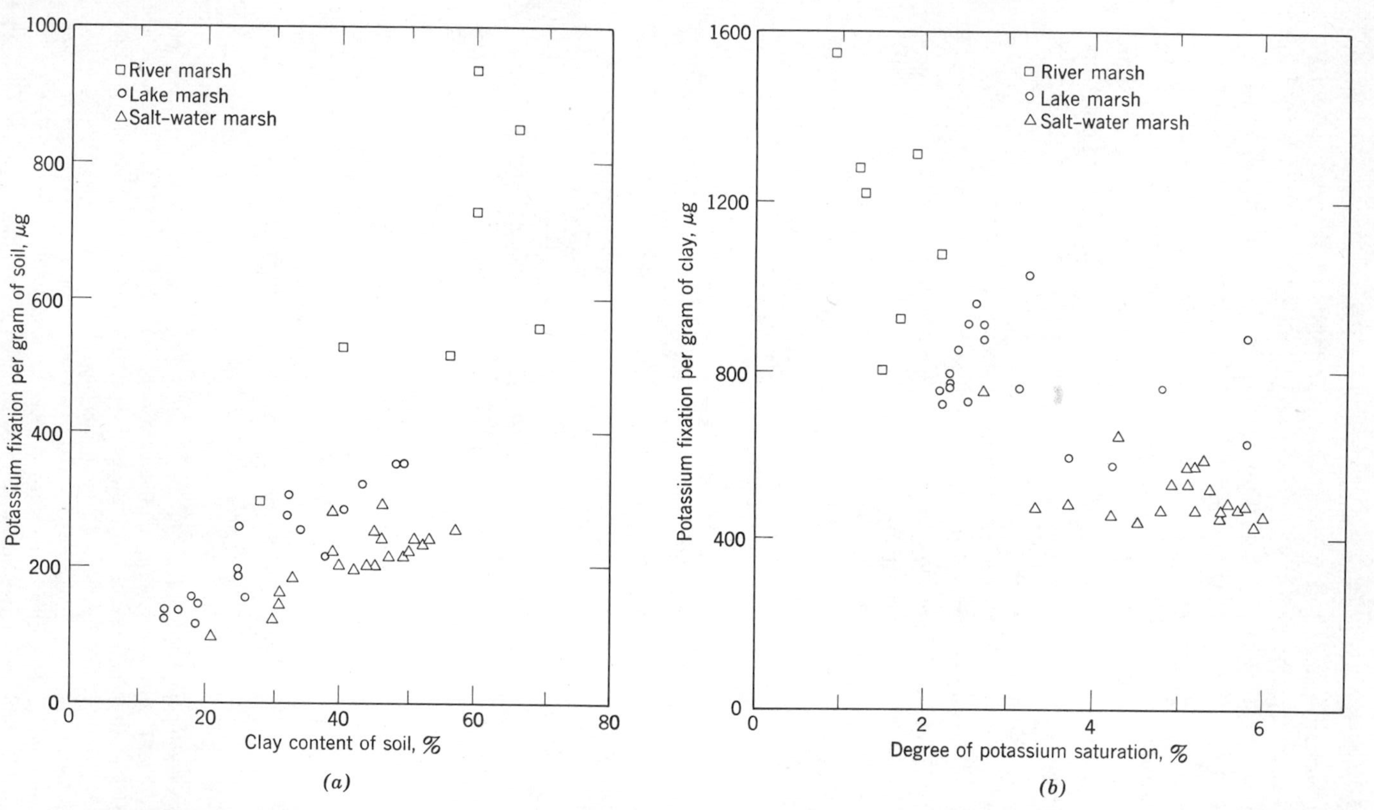

Figure 9.24. Potassium fixation under moist conditions by soils from three kinds of marshes in Germany. (*a*) Potassium fixation versus clay content of the soils. (*b*) Potassium fixation versus percentage saturation of the cation-exchange capacity of the soils with potassium. (Schachtschabel and Köster, 1960)

Page and Ganje (1964) found that fixation of potassium under moist conditions by soils containing vermiculite-type minerals was little influenced by *p*H in the range from 3 to 7 as long as the concentration of aluminum remained low. High concentrations of nonexchangeable interlayer aluminum resulting from leaching the soils with dilute acid prevented fixation of potassium at *p*H 3. Fixation occurred again when the acid-treated soils were adjusted to *p*H 7. Existence of a blocking effect of aluminum may be inferred also from earlier work by Barshad (1954) on fixation of ammonium by vermiculite. He found that fixation of ammonium at *p*H 7 was reduced if the vermiculite was treated with acid before fixation and increased if the vermiculite was treated with alkali before fixation. Treatment of acid-washed vermiculite with alkali restored the fixation capacity. Interlayer aluminum should have developed in the vermiculite as a result of the acid pretreatment and should have been removed by the alkali pretreatment.

The foregoing observations may be combined with concepts about vermiculite and soil acidity already discussed to provide a general theory regarding the *p*H effect: Concurrent with development and leaching of soils, loss of potassium from micas with formation of vermiculite may be expected. With development of acidity, polymeric hydroxyaluminum ions accumulate between the molecular layers of vermiculite, eliminating the sites they occupy as part of the cation-exchange capacity and as part of the potassium-fixing capacity. Accordingly, an increase in potassium fixation may or may not be associated with development of soil acidity, depending on the presence or absence of vermiculite-type minerals and the degree of blocking of interlayer potassium-fixation sites. Conversely, when acid soils are limed, an increase in potassium fixation may or may not result, depending on the presence or absence of vermiculite-type minerals and the degree of unblocking of interlayer potassium-fixing sites as the interlayer aluminum gradually disappears. The association of fixation with *p*H thus seems to vary with the circumstances, and so *p*H alone is not a suitable index of potassium fixation.

The association of potassium fixation with the percentage potassium saturation of control samples of soil was investigated by Chaminade (1936). In an experiment on samples of 20 soils of France, he found that percentage saturation of the cation-exchange capacity with potassium was a better index of potassium fixation under moist conditions than was the content of exchangeable potassium or the cation-exchange capacity. He concluded that soils having a degree of potassium saturation exceeding about 4.5% fixed little or no potassium, and those with potassium saturation below about 4% were definite potassium fixers.

Schachtschabel and Köster (1960) likewise noted that fixation of potassium under moist conditions by marsh soils of Germany decreased with an increase in percentage saturation of the cation-exchange capacity with potassium. Their data in Fig. 9.24*b* show that values for the three classes of marsh soils tend to fall in different ranges but that all the results may be expressed approximately by a single curve. The seeming existence of a single relationship in Fig. 9.24*b* for the same soils that gave somewhat different relationships in part *a* of the figure suggests that percentage potassium saturation is a more valuable index of the potassium-fixing tendency of soils than is clay content. Percentage potassium saturation is related to the proportion of potassium in the soluble cations, and the latter has application in the mass-law concept of potassium fixation and release discussed briefly in the preceding section. A similar plot of data by van der Marel and Venekamp (1955) on samples of 14 soils derived from river sediments and 13 soils derived from marine sediments in the Netherlands shows that potassium fixation decreased with an increase in percentage saturation of the cation-exchange capacity with potassium, with no indication of existence of separate relationships for the two groups of soils.

An additional experiment of significance in connection with the effect of percentage potassium saturation was conducted by Chaminade (1936). A field plot on low-potassium soil was fertilized each year with potassium sulfate equivalent to 208 kg of potassium per hectare, and a sample of soil was taken in late autumn of each year for measurement of exchangeable potassium and percentage potassium saturation. The results in Fig. 9.25 show that the annual increase in exchangeable potassium from the treatment was relatively small until the percentage potassium saturation reached a value of about 4%, and then it increased rapidly. Chaminade attributed the small increases in exchangeable potassium during the earlier years to fixation of the added potassium in nonexchangeable form and the larger increase at the end to lesser fixation at the higher percentage potassium saturation.

In summary, fixation of potassium in nonexchangeable form in soils is now thought to be a property of many minerals, of which expanded layer-silicates in the clay fraction are by far the most important. In these minerals, the potassium is thought to be held electrostatically on the outside of the molecular aluminosilicate layers as a balancing cation and to remain exchangeable if present on outer surfaces but to be subject to confinement in nonexchangeable form in closely fitting voids between the layers. If the expanded layer-silicate has a sufficiently high cation-exchange capacity (vermiculite), partial closure of the layers with loss of part of the interlayer water and fixation of much of the interlayer potassium occurs in an aqueous environment; and full closure to the

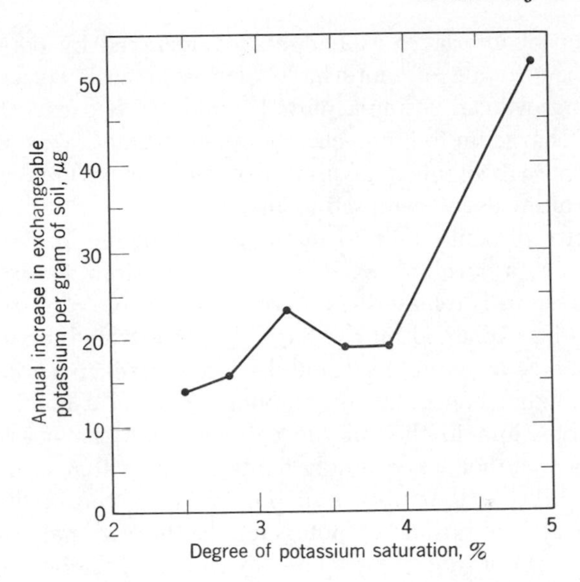

Figure 9.25. Annual increase in exchangeable potassium in soil samples taken each November from a field plot in France, to which was added potassium sulfate equivalent to 208 kg of potassium per hectare annually, versus percentage saturation of the cation-exchange capacity with potassium. Values from left to right in the graph were obtained in successive years. (Chaminade, 1936)

spacing of micas, with loss of most or all of the remaining interlayer water and fixation of more potassium, occur if the mineral is dried and heated. If the expanded layer silicate has a lower cation-exchange capacity (montmorillonite or related minerals), fixation of potassium occurs mostly when the mineral is dried and heated; and re-expansion and release of potassium occur comparatively readily.

The potentials to release potassium and to fix potassium coexist in soil. Release of potassium from nonexchangeable, interlayer form occurs rapidly if the external concentration of potassium is low enough and if the concentration of other cations is high enough. The cation-exchange positions thus produced are potential potassium-fixing sites as long as they are not blocked. Blockage may occur from the presence of ammonium or polymeric hydroxyaluminum ions. Uptake of potassium in nonexchangeable form by vermiculite-type minerals occurs even at potassium concentrations of the order of a few micrograms per milliliter, and vermiculite can probably be largely resaturated with nonexchangeable potassium at such concentrations. Higher concentrations may be expected to cause somewhat more fixation by increasing the proportion of interlayer exchange sites occupied by potassium, with consequent contraction of more molecular layers. All layers do not behave the same even in a single particle, perhaps because of differences in density of charge on the molecular aluminosilicate layers. Ocean water contains about 380 μg of potassium per milliliter, which accounts for regeneration of micas in ocean sediments.

The capacity of some soils to maintain for months or years, without loss by fixation, a level of exchangeable potassium in excess of that at which potassium is released from nonexchangeable form by micaceous minerals may be explained on the basis that the strongly fixing, vermiculite-type minerals are largely saturated with nonexchangeable potassium, and a much higher concentration of potassium and a much higher proportion of potassium in the soluble cations would be needed to cause much fixation by other less strongly fixing minerals that may be present. Another possibility is that uptake of potassium by vermiculite-type minerals at high concentrations of potassium in the external solution is inhibited by closure of molecular layers around the periphery of the particles before complete exchange has occurred.

Soil-Plant Relationships in Potassium Fixation. To conclude the subject of potassium fixation, an attempt will be made to place in perspective the biological significance of the process. First, some experiments described in a paper by Chaminade (1936) on potassium fixation will be considered. These provide different forms of evidence and at the same time illustrate techniques by which the biological significance of potassium fixation may be investigated.

In the first experiment, Chaminade added 830 μg of potassium as potassium chloride per gram of soil to a sample of potassium-fixing soil and incubated this sample along with an untreated control sample for 2 months. During this time, fixation of potassium in nonexchangeable form amounted to 294 μg/g of soil. At the end of the incubation, portions of both samples were leached to different degrees with calcium chloride to obtain three different levels of exchangeable potassium, with one series containing 294 μg of fixed potassium per gram of soil and the other series containing no fixed potassium. All samples were then subjected to intensive cropping by rye seedlings, and the seedlings were analyzed for potassium. The results in Fig. 9.26 show not only that the potassium-treated soil contained more exchangeable potassium than the control soil but also that the quantity of exchangeable potassium needed to obtain a given uptake of potassium was less where plants were grown on the potassium-treated soil than on the control. Presumably, the plants were making use of the potassium that had been fixed from the potassium added during the 2-month incubation period. If the difference in quantity of

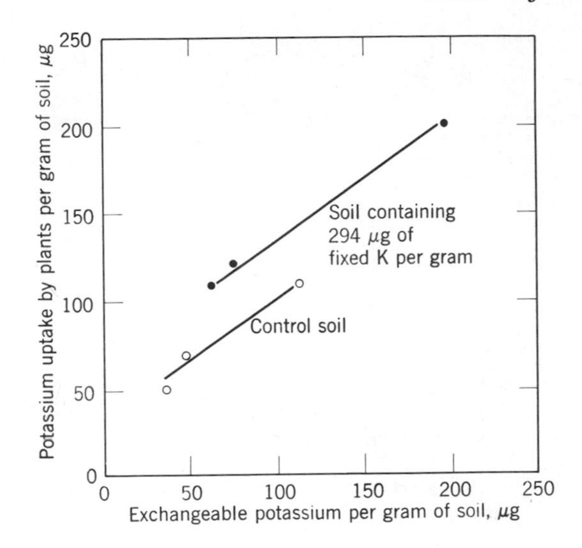

Figure 9.26. Uptake of potassium by rye seedlings from equal quantities of control soil and soil containing fixed potassium versus exchangeable potassium in the two samples. (Chaminade, 1936)

exchangeable potassium required to obtain equal uptake of potassium from the control and potassium-treated samples did in fact result from use of some of the fixed potassium, one may estimate the relative value of the exchangeable and fixed potassium in the following way: The horizontal distance between the two lines in the figure is equivalent to about 50 μg of exchangeable potassium per gram of soil. Dividing 50 by 294, the number of micrograms of fixed potassium per gram of soil, gives 0.17. The value 0.17 is sometimes called the availability-coefficient ratio because it gives the ratio of the biological effectiveness of one unit of the nutrient in one form to that of one unit of the nutrient in another form taken as the standard. In this instance, exchangeable potassium is taken as the standard, and one unit of fixed potassium had a biological value 0.17 as great as that of one unit of exchangeable potassium.

In the second experiment, Chaminade employed a potassium-fixing soil and added different quantities of potassium as potassium chloride to one series of samples in the autumn and maintained the samples in a moist condition during the winter. To another series of samples, which had been kept moist during the winter but with no potassium treatment, he added similar quantities of potassium in the spring. All containers were then planted immediately to barley, and the barley

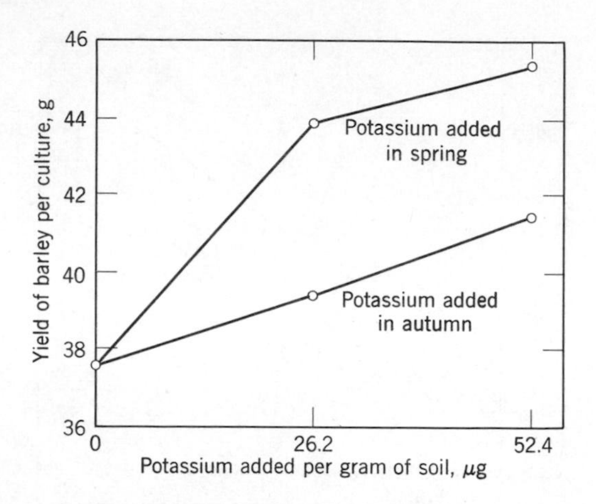

Figure 9.27. Yield of barley on soil with different additions of potassium as potassium chloride in the spring just before planting or in the preceding autumn. (Chaminade, 1936)

was grown to maturity. The results of this experiment are shown in Fig. 9.27. In this experiment, application of the equivalent of 52.4 μg of potassium per gram of soil in the autumn produced the same yield of barley as about 16 μg applied in the spring, which means that the availability coefficient of the added potassium as a whole was a little less than one-third as great if the addition was made in the autumn as if it was made in the spring. The decrease in availability due to autumn application may be said to result from fixation because the soil samples were not subjected to cropping or leaching during the incubation period.

The experiment just described was done with a soil that was known to fix added potassium. In another experiment on eight soils, Chaminade found that incubation for a year before cropping did not decrease the value of added potassium to plants unless the soil fixed potassium in nonexchangeable form.

Potassium fixation may be evidenced in several ways in the behavior of plants in the field. Many crop plants are deficient in potassium on potassium-fixing soils and do not respond in the usual way to potassium fertilization. If potassium fertilizer is applied to soils that produce potassium-deficient crops but fix little or no potassium, a plot of crop yield against potassium applied gives a curve that rises rapidly with small applications of potassium (of the order of 20 to 60 kg/hectare) and slowly with larger applications. If, on the other hand, potassium fertilizer is applied to potassium-fixing soils that produce crops similarly deficient in potassium, the yield does not increase so rapidly per unit of potassium added. For example, Page et al. (1963) reported that correction of potassium deficiency for cotton required quantities of fertilizer potassium in excess of 170 kg/hectare on potassium-fixing soils in the San Joaquin Valley in California in contrast to quantities of the order of one-third as much under usual circumstances of potassium deficiency. If potassium fixation is moderate, the response curve may have the same general shape as it does if there is little or no fixation except that the quantities required to reach the maximum yield are greater. With strong potassium fixation, the response curve may have a different shape, rising slowly with small applications and more rapidly with larger applications and finally reaching a maximum with still larger applications. Figure 9.28 shows the lower part of such a response curve. The potassium content of the leaf blades of the cotton plants 90 days after planting was only 0.9% of the dry matter with the heaviest application of potassium, which verifies that the plants were not high in potassium. This experiment was done in the greenhouse with one of the soils of the San Joaquin Valley that produced potassium-deficient cotton in the field and required heavy applications of potassium fertilizer. The soil in question contained vermiculite, and laboratory work on samples of soil from the various cultures in the greenhouse test showed that the cation-exchange capacity was reduced from 3.2 to 2.5 m.e./100 g upon application of the largest quantity of potassium employed. This reduction in cation-exchange capacity, calculated as potassium fixation, is equivalent to transfer of 65% of the added potassium to nonexchangeable form. Probably a response

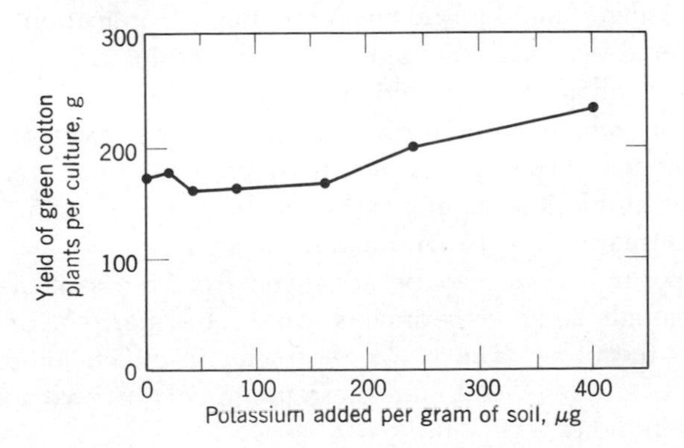

Figure 9.28. Yield of cotton plants with different applications of potassium as potassium sulfate to a potassium-fixing, fine sandy loam soil of California. (Page et al., 1963)

curve similar to the one shown in Fig. 9.28 would have been obtained in the field by Van Dijk (quoted by Middelburg, 1955) if a wide enough range of quantities of potassium had been applied. Van Dijk found that cassava grown on a clay soil in Java did not respond to potassium until potassium fertilization had been practiced for four consecutive years.

In addition to the gross effects evidenced by plant yields, there may be internal effects that give additional information on the effect of potassium fixation. Perhaps the most sensitive of these is the potassium content of the plant, to which may be attributed the yield effects more commonly determined. Breemhaar (1956) found that the potassium percentage in leaves of potato plants increased much more with a given application of potassium to a field in which the soil had little tendency to fix potassium than it did with an equal application of potassium to another field in which the soil fixed potassium strongly.

The experiments just discussed provide evidence for the view that potassium fixation has the detrimental effect of decreasing the availability coefficient of potassium added to soil. That is to say, if equal quantities of added potassium are present initially in nonexchangeable and exchangeable form, the former will be less effective than the latter because plants use potassium preferentially from the exchangeable form.

An opposing view has also been advanced to the effect that potassium fixation is a beneficial process because it provides a means by which soils may hold relatively large quantities of potassium in forms available to plants. Evidence for this position is shown in Table 9.7, which gives results obtained in greenhouse tests on samples of soil from three field plots on a potassium-fixing soil in Maine. One plot had received no potassium fertilizer, one had received an annual application of potassium in fertilizer, and one had received the fertilizer plus a heavy application of manure, which also contained potassium. The amount of potassium lost from nonexchangeable form during cropping evidently increased with the amount of potassium added in previous field treatments and with the accumulation of nonexchangeable potassium released from the soil upon treatment with 1-normal nitric acid.

These opposing views may be advanced from the same factual background, although some experiments would be more favorable to one and some to the other. The views represent, in effect, judgments about what is and is not beneficial; and these judgments involve considerations not inherent in the experimental data.

The view that potassium fixation is detrimental may be supported on two bases: (1) Over a period of years following potassium application, fixation will decrease the uptake of potassium by plants per kilo-

Table 9.7. Loss of Potassium from Exchangeable and Nonexchangeable Forms During Greenhouse Cropping of Soil from Field Plots in Maine That Had Accumulated Different Amounts of Potassium as a Result of Previous Field Treatments. (Reitemeier et al., 1951)

Potassium Applied per Hectare Annually in the Field, kg	Potassium per Gram of Soil, μg: Present in Soil at Planting: Exchangeable	Present in Soil at Planting: Nonexchangeable Released by 1-*N* HNO_3	Loss from Soil During Growth of Ladino Clover (15 Cuttings): Exchangeable	Loss from Soil During Growth of Ladino Clover (15 Cuttings): Nonexchangeable
0	36	146	5	49
74	92	201	59	66
74[1]	457	327	420	234

[1] Plus potassium added in 2.2 metric tons of manure per hectare.

gram added. Accordingly, more potassium will be required, at greater total cost, to eliminate the deficiency of potassium. (2) Although the potassium that has been fixed in nonexchangeable form may eventually be used, its principal use will occur after the exchangeable potassium has been reduced to a level low enough that failure to add more fertilizer potassium may be uneconomic. Moreover, return in crop value from money expended for the fixed potassium will be delayed; this also has the effect of decreasing the total return because the earning power of money invested in fixed potassium is zero as long as the potassium remains fixed.

The alternative view that potassium fixation is beneficial may be supported by arguments such as the following: (1) Potassium fixation retards loss of potassium by leaching and hence increases the efficiency of use of fertilizer. (2) Plants take up potassium much in excess of their needs if enough is present in exchangeable form. Potassium fixation reduces this excessive uptake and hence increases the efficiency of use of the fertilizer.

There are several ways in which the efficiency of potassium use on potassium-fixing soils may be improved: (1) changing the method of application of the fertilizer from the usual one of mixing the fertilizer with the soil to (*a*) localized application, (*b*) spray application to above-ground plant parts, or (*c*) injection into the plants; and (2) changing the cropping pattern from crops that require a high level of exchangeable potassium to those that grow well with a low level of exchangeable potassium. With regard to method of application, (*a*) seems to be employed most commonly, (*b*) to a small extent, and (*c*) rarely if ever. Adjustments in cropping pattern are widely employed as farmers observe how different crops behave on their fields. In general, however, the decision to use any of these practices is probably rarely based on explicit knowledge of potassium fixation by particular soils because too little information is available.

The significance of potassium fixation to plants growing in the field has received little attention except by Dutch scientists. Schuffelen and van der Marel (1955), who reviewed the work on this subject, were not convinced that the phenomenon has much effect on plant yields. In discussing Dutch field experiments, they noted that soils that fix potassium strongly produce normal yields of sugar beets, mangolds, oats, barley, and wheat if applications of 250 to 330 kg of potassium are made per hectare. Potatoes are more sensitive to potassium deficiency. In this connection, they cited experimental findings on two fields, one on soil that fixed much added potassium in nonexchangeable form and the other on soil that fixed little potassium. Annual applications of 241 kg of potassium per hectare were needed for 8 to 10 years to increase potato yields on the former soil to the level of those on the latter.

In interpreting the views of Schuffelen and van der Marel (1955) in terms of conditions elsewhere, it may be worth bearing in mind that applications of fertilizer per hectare in the Netherlands are among the highest in the world. Quantities of 100 to 150 kg of potassium per hectare, which they classed as normal annual applications for clay soils developed on marine and river sediments, would be considered high applications in most parts of the world. Accordingly, in an area where common annual applications are of the order of one-fourth of this range, potassium fixation might be considered more important.

Water Content and Aeration of Soil

Of the accessory factors that may affect the availability of soil potassium to plants, the water content of the soil has probably received most attention. Figure 9.29 shows the results of a field experiment in the Netherlands in which potatoes were grown with and without an application of potassium fertilizer in each of a number of consecutive years. The aspect of the results that seems to characterize the relation between water supply and availability of soil potassium is the gradually increasing response to potassium fertilization as the number of rainless days in-

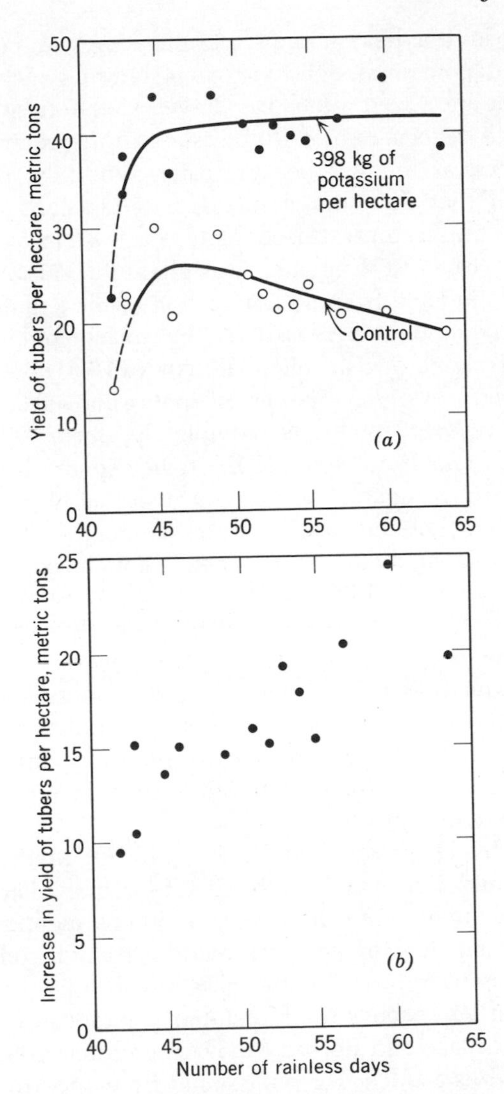

Figure 9.29. Response of potatoes to potassium fertilization versus number of rainless days in May, June, and July in 14 consecutive years in a field experiment in the Netherlands. (*a*) Yield of tubers with and without potassium fertilization. (*b*) Increase in yield from potassium fertilization. (van der Paauw, 1958)

creased. Because the heavy annual applications of potassium should have eliminated potassium deficiency as a limiting factor in growth of potatoes on the fertilized plots in all years, the trend of the results signifies that the deficiency of soil potassium in the control plots became more pronounced as the soil became drier. Similarly, Bruns (1935) in Germany found that the ratio of yield of field beans on control plots to the yield on potassium-fertilized plots was 0.37 in 5 moist years and 0.22 in 6 dry years. In England, Russell and Watson (1938) found that the increase in yield from an application of the sulfates of potassium, sodium, and magnesium increased as the spring and summer rainfall decreased. In Ireland, Walsh and Cullinan (1945) observed that crops which developed severe symptoms of potassium deficiency during a drought would recover, sometimes completely, during a subsequent wet period. Richards and Wadleigh (1952) reviewed results of field experiments by several investigators showing an increase in potassium percentage in plants with an increase in water content of the soil, and van der Paauw (1958) made a similar observation in connection with the experiment on potatoes in Fig. 9.29.

Although the effects just mentioned may be obvious, the manner in which they arise is not so obvious. Various theories may be proposed to account for the field behavior. In the field experiments, comparisons were usually made between years or between different periods within years; thus one cannot be certain that the effects are attributable strictly to the water content of the soil because such factors as temperature and sunshine were uncontrolled.

Experiments by Danielson and Russell (1957), Mederski and Stackhouse (1961), and Place and Barber (1964), in which water content was altered without varying the other conditions mentioned, have verified that potassium does move more readily through soil to plant roots if the water content approaches the one-third-bar percentage (field capacity) than if it approaches the fifteen-bar percentage (permanent wilting percentage). Place and Barber (1964) found that uptake of rubidium (sometimes used as a tracer for potassium) by plants from soil at different water contents increased linearly with the rate of diffusion of rubidium through the soil. From these experiments is derived the theory that potassium availability increases with water content of soil because both the proportion of the soil volume effective for diffusion of potassium and the volume of water that can carry potassium to roots by mass movement increase with the proportion of the soil volume occupied by water.

Another theory may be proposed on the basis of observations reported by Schuffelen (1954). He found that the molar ratio of potassium to magnesium in solution in equilibrium with a sandy soil increased from 0.5 at 15% water to 1.7 at 60% water. In a clay soil, the molar ratio increased from 0.2 at 11% water to 0.5 at 28% water. From these observations is derived the theory that an ample supply of water in soil increases the proportion of potassium in the soluble cations and hence increases the availability of potassium relative to calcium and magnesium. Schuffelen advanced the data on potassium and magnesium in explanation of the observation that magnesium deficiencies in plants in the Netherlands are more pronounced in wet springs than in dry springs. This observation is connected with potassium behavior in that magnesium deficiencies are frequently induced when soils low in magnesium are fertilized with potassium.

Finally, a third theory, based on the vertical distribution of water and exchangeable potassium, may be mentioned. If the water supply is ample throughout the soil, the availability of potassium to plants will generally be greatest in the surface layer because the highest concentration of exchangeable potassium and the highest concentration of roots are usually found there. During rainless periods, the soil dries from the surface downward; hence the availability of potassium in the most valuable portion of the soil is reduced, and plants become largely dependent on potassium in the lower horizons, where the availability is lower.

Considerations thus far have been limited to what seems to be the middle of the range of water content. It is unreasonable to expect, for example, that the absolute increase in yield from potassium fertilization observed by van der Paauw (1958) in Fig. 9.29 should continue to become greater as the number of rainless days increased. When conditions are dry enough, the absolute increase in yield from potassium fertilization should decrease. Bruns (1935) in Germany found that the absolute increase in yield from fertilization of field beans with potassium was less in 6 dry years than in 5 moist years; but the ratio of yield on the control plots to the yield on the fertilized plots was nevertheless lower in the dry years than in the moist years, signifying that availability of the potassium was reduced by soil dryness. Whether or not the ratio of yield on control plots to yield on potassium-fertilized plots would continue to decrease as soil becomes increasingly dry seems doubtful. Data from work by Barber (1961) in Indiana, shown in Fig. 9.30, suggest that it does not; however, the number of observations is insufficient to provide clear evidence one way or the other.

The middle part of the range in Fig. 9.30, where the points are connected by solid lines, may be interpreted as the portion of the range of water supply in which potassium availability increases with an in-

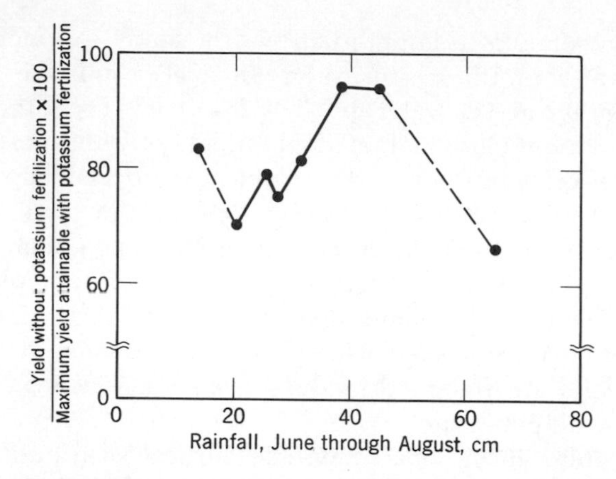

Figure 9.30. Yield of corn without potassium fertilization as percentage of maximum yield attainable with potassium fertilization in 8 years differing in summer rainfall in Indiana. (Barber, 1961)

crease in water content of the soil. The opposite trend with high rainfall, denoted by a broken line, may be interpreted as a decrease in availability of soil potassium with an increase in water content of the soil. Barber accounted for this trend on the basis that a high content of water in the soil resulted in poor aeration, and the poor aeration in turn decreased the availability of soil potassium to plants. Although the trend in question is a consequence of results of only one year that do not agree with the others, and hence cannot be considered significant, the results and Barber's interpretation are in agreement with other work done under controlled conditions that indicates the importance of aeration in uptake and retention of potassium by plants.

For a general review on this subject, a paper by Wallace (1958) may be consulted. A brief review of the aeration effect was published by Barber and Humbert (1963).

Laboratory Indexes of Availability

Considerable research has been done on laboratory indexes of availability of soil potassium to plants, but as yet no single measurement on the soil has proved to be consistently superior under varying circumstances. The situation is complex and involves both plant aspects and soil aspects. The plant aspects will be considered first because they provide background for the soil aspects.

In terms of the carrier theory, discussed in Chapter 4, potassium ions are taken up by roots at potassium-specific sites with little competition from other cations except at high concentrations of the latter. Inside the plant, however, a form of competition does develop because of the tendency of a given kind of plant to have a fairly constant sum of cation equivalents independently of the proportions of cations supplied externally. Accordingly, if the internal content of potassium is low, the internal content of calcium and magnesium tends to be high and vice versa. In long-term absorption, therefore, different cations may have mutual effects on uptake that are different from the short-term effects observed in experiments with detached roots. Hence, although the concentrations or activities of individual ions are emphasized in the carrier theory, which has to do with uptake only, the concentrations or activities of cations other than potassium are of some significance in potassium nutrition on a long-term basis.

When analyses of the external medium are made in an attempt to obtain an index of potassium availability to plants, the principal problems may be placed in two groups: (1) evaluation of the significance of potassium and other ions in solution and (2) evaluation of the significance of the changes in composition of the solution that accompany ion uptake by plants. The changes during growth of a single crop may be considerable, where potassium is concerned, because the crop ordinarily absorbs far more potassium than is present at one time in the soil solution and in some instances absorbs the equivalent of one-half or more of the exchangeable potassium. The extent of release of potassium from nonexchangeable form becomes important where a succession of several crops is concerned and sometimes is significant even for a single crop. For the most part, experimental work on indexes of potassium availability has dealt with a single type of measurement or with a comparison of several types of measurements as they relate individually to uptake of potassium, yield response to potassium, or some other biological measurement of potassium supply.

The matter of evaluating the significance of other ions as well as potassium in solution is usually ignored, but a little work has been done on it. In a greenhouse experiment on organic soils from eastern Canada, MacLean et al. (1964) found that the percentage saturation of the cation-exchange capacity with potassium gave a poorer correlation with the increase in uptake of potassium from fertilization than did the exchangeable potassium or the water-soluble potassium. Perhaps one reason for the poor showing of percentage potassium saturation in this work is the fact that soil *p*H values ranged from 4.2 to 7.0, whereas the exchange capacity was determined at pH 7 and hence would have involved much pH-dependent exchange capacity in the more acid of these organic soils. The best performance of percentage potassium saturation as an index of potassium availability would be expected where the cation-exchange capacity is determined at the pH of the soil. In another greenhouse experiment on mineral soils from Ontario, MacLean (1961) obtained a higher correlation of his biological response values with exchangeable potassium than with the percentage saturation of the cation-exchange capacity with potassium; the lowest correlation was with water-soluble potassium. In a group of field experiments in France, Strasman et al. (1958) found that the highest correlations between the relative increase in yield of potatoes from potassium fertilization per kilogram of potassium per hectare with soil measurements were obtained with the ratio of exchangeable potassium to the sum of metallic exchangeable cations and with the ratio of exchangeable potassium to exchangeable calcium. The correlation of plant response with percentage saturation of the cation-exchange capacity with potassium was lower, the correlation with exchangeable potassium was still lower, and the correlation with the concentration of potassium in the soil solution was essentially zero. Strasman et al. also made measurements of cation-exchange capacity at pH 7.

Because the cations in the soil solution continuously approach an equilibrium with the exchangeable cations, the percentage saturation of the cation-exchange capacity with potassium is related to the percentage of potassium in the cations in the soil solution. A closely related index is $p\text{K} - \frac{1}{2}p(\text{Ca} + \text{Mg})$, which is the negative logarithm of the activity ratio $(a_{\text{K}^+})/(a_{\text{Ca}^{++}} + a_{\text{Mg}^{++}})^{1/2}$ of ions in the solution. For acid soils the index $p\text{K} - \frac{1}{2}p(\text{Ca} + \text{Mg}) - \frac{1}{3}cp\text{Al}$ has been suggested, where c is a constant. A number of authors have used the first of the two indexes, and Tinker (1964) found that the response of oil palm to potassium fertilizer in field experiments in Nigeria was closely related to values obtained for this index in analyses on the soils. Moss (1964) found that the activity ratio in soils was closely related to the corresponding concentration ratio in the bark of cacao trees in field experiments. Neither of these investigators compared the activity-ratio method with any other. MacLean (1961) found that the efficiency of the activity ratio in estimating the relative uptake of potassium by oats in a greenhouse experiment with samples of different soils from Canada was about equal to the efficiency of water-soluble potassium. Exchangeable potassium and percentage saturation of the cation-exchange capacity with potassium gave higher correlations with the biological measurements.

One of the troublesome problems with activity ratios is the extent to which they may change as plants absorb potassium. MacLean (1961) found that the activity ratio before cropping in the experiment just mentioned was from 0.8 to 109 times as great as the activity ratio after cropping in the eleven soils tested. Most values were in the range from 2 to 9. The experimental problem is then to estimate to what extent the activity ratio will change in different soils and the significance of the change to plants. Even though plants may remove potassium selectively, the tendency is for concentrations of all cations to rise and fall together. Consequently, changes in the activity ratio with cropping are less pronounced than changes in water-soluble potassium. Exchangeable potassium and percentage saturation are affected to equal degrees by cropping (if the cation-exchange capacity remains constant) and are more stable than the activity ratio or the concentration of potassium in the soil solution. MacLean (1961a) published data showing water-soluble and exchangeable potassium and percentage potassium saturation in a group of soils before and after cropping.

The concentration of potassium in water-soluble form is not invariably an inferior index of the supply of soil potassium for plants, as indicated by results of a greenhouse experiment by Hood et al. (1956). These investigators found that the plant-response measurements they used on individual cuttings of ladino clover after the first were more highly correlated with the potassium content of water extracts of soils of New York than with the exchangeable potassium.

If the cropping causes loss of considerable nonexchangeable potassium, some measure of release from nonexchangeable form may be expected to prove beneficial. Several techniques have been used to provide such information. DeTurk et al. (1943) boiled a sample of soil in 1-normal nitric acid, Hunter and Pratt (1957) treated soil with various concentrations of sulfuric acid, and Garman (1957) and Gardner (1960) percolated dilute hydrochloric acid through soil. Ayers et al. (1947) used prolonged electrodialysis. Pratt (1951) incubated soil with a quantity of hydrogen-saturated cation-exchange resin and then determined the potassium content of the resin. Schulte and Corey (1965) treated soil with sodium tetraphenylboron and then determined the potassium extracted. Data of this kind have been related to the biological measurements in two general ways. One is to relate the biological measurement (usually the total potassium uptake by several crops or several cuttings of a perennial crop) to the total potassium extracted. The other is to separate the soil potassium into several fractions, such as exchangeable potassium, potassium released from nonexchangeable form by the method employed, and nonexchangeable potassium not released by the method employed, and to use a statistical method to obtain a single empirical equation relating the biological measurement to the quantities of all the fractions. The usual finding is that inclusion of a measure of release of potassium from nonexchangeable form gives a better correlation with the biological measurement than does exchangeable potassium alone.

In practical soil-testing work, extraction methods yielding something approximating exchangeable potassium are used almost exclusively. These methods have the advantage of comparative ease of execution in routine work. Although such methods may be criticized on the basis that they do not take full advantage of current information on potassium, extensive field work would be needed to verify the superiority of alternative methods that may seem theoretically to offer some improvement. One instance in which an improvement was made over the traditional measurement of exchangeable potassium for field application was reported by Semb and Uhlen (1954) in Norway. These investigators found that they could estimate the response of crops to potassium fertilization in 2-year experiments in the field more precisely on the basis of measurements of the total amount of potassium extracted by boiling soil with 1-normal nitric acid than on the basis of exchangeable potassium.

Two additional observations that have been made in connection with use of exchangeable potassium as an index of potassium availability in field experiments may be mentioned as specific examples of general concepts explained in previous chapters. The first is that a given content of exchangeable potassium in the soil seems to be of greater value to plants on soils of coarse texture than of fine texture (Hanway et al., 1962; Levy, 1964). This observation may be correlated with the difference in cation-exchange capacity and percentage potassium saturation of the soils and hence with the proportions in which potassium and other cations are released in a fractional exchange. Replacement of a given number of milligram equivalents of exchangeable bases will cause release of more potassium ions from sandy soils than from clayey soils with an equal content of exchangeable potassium. The second observation is that the usual index of availability provided by analysis of the plowed layer of soil is not necessarily the best. Titus and Boynton (1953) and Pratt et al. (1957) found that with tree crops a better index of potassium availability was provided by samples taken from the B horizon than from the surface horizon. Hanway et al. (1961, 1962) found that inclusion of values for exchangeable potassium in soil layers below the surface 15 cm provided better indexes of availability of soil potassium to alfalfa and corn than did analyses of the surface 15 cm only.

Fertilizers and Fertilization

Potassium Sources

Potassium is widely distributed, and large deposits of potential value for fertilizer production are found in many countries. For present purposes, potassium-bearing minerals may be considered to consist of one group with moderate to high solubility in water and another group with extremely low solubility in water. The low-solubility group includes the feldspars, micas, and certain other silicate minerals not mentioned previously in this chapter. Although some of these minerals contain considerable potassium (orthoclase feldspar contains 13.9% potassium and leucite contains 17.8%), the minerals must be treated to increase the solubility of the potassium for fertilizer purposes; and the cost of the treatment makes their use uneconomical in comparison with the soluble minerals.

The potassium sources in use for production of fertilizers are soluble salts. By far the most important of these in both extent and use are residual from evaporation of bodies of water. Those residual from earlier geologic time are covered by other materials and are generally mined at depths exceeding 200 meters. Recent deposits are exposed at the surface. According to estimates in a paper by MacDonald (1960), the countries with the largest known reserves of soluble potassium salts may be listed in order of decreasing reserves as follows: Germany, Canada, Israel-Jordan, Soviet Union, United States, France, and Spain. A number of other countries also are known to have deposits. In North America, large buried deposits of ancient origin are found in New Mexico, Texas, eastern Utah, and Saskatchewan. Surface deposits of more recent geologic time are being worked at Searles Lake in California and at the Bonneville salt flats in western Utah. Most of the production is from the New Mexico deposits. An account of these and other potassium sources was published by Reed (1953), and papers including information on the recently developed Saskatchewan deposits were published by MacDonald (1960) and Kapusta and Wendt (1963).

Fertilizer Processing

The potassium-bearing salt deposits contain a number of different forms of potassium. A table listing the minerals and their composition was published by Reed (1953) and MacDonald (1960). The principal ore in commercial use is sylvinite, which is a mixture of crystals of sylvite (potassium chloride) and halite (sodium chloride). Although the underground salt deposits are often thick, the sylvinite occurs in thin layers. These are mined mechanically, and the solid material is brought to the surface for processing. The potassium chloride is separated from the sodium chloride by flotation (a physical separation of the two solids) or by fractional crystallization of the potassium chloride from solution. These processes were described by Harley (1953), MacDonald (1960), and Kapusta and Wendt (1963). The product usually contains about 50% potassium (pure potassium chloride contains 52.4% potassium).

Potassium sulfate is produced commercially in the United States from langbeinite ($K_2SO_4 \cdot 2MgSO_4$) by treatment with potassium chloride followed by crystallization, from burkeite ($Na_2CO_3 \cdot 2Na_2SO_4$) by treatment with potassium chloride followed by crystallization, from potassium chloride and sulfur, and from potassium chloride and sulfuric acid. The processes were described by Harley (1953), MacDonald (1960), and Kapusta and Wendt (1963). Fertilizer-grade potassium sulfate usually contains 41 to 44% potassium (pure potassium sulfate contains 44.9% potassium). Potassium sulfate is marketed also as the mineral langbeinite which, in the New Mexico deposits, is mixed with halite and sylvite. The latter two minerals dissolve more rapidly than the langbeinite and are largely removed by fractional dissolution. The product contains about 18% potassium.

Fertilizer Statistics

According to statistics collected by the Food and Agriculture Organization of the United Nations (1965), 88% of the potassium in fertilizers manufactured in 1962–63 in France, Germany, Spain, and the United States was in the form of potassium chloride, 8% was in the form of potassium sulfate, and 4% was in other forms. In the United States, 93% was in the form of potassium chloride, 6% was in the form of potassium sulfate, and 1% was in other forms.

The same FAO source shows that, in 1962–63, 53% of the potassium used in fertilizer was consumed in Europe, 27% in North and Central America, 9% in the Soviet Union, 8% in Asia, and 1% each in Africa, Oceania, and South America. In the world as a whole (excluding mainland China and North Korea), an average of 2 kg of potassium was applied in fertilizer per hectare to agricultural land. The applications were greatest in Europe (17 kg) and least in Africa (0.1 kg). The figure for North and Central America was 3 kg. Individual countries with heavy use of fertilizer potassium included Belgium (94 kg), Barbados (82 kg), Germany (66 kg), and Japan (64 kg).

Fertilizers and Their Effects

A wide variety of potassium salts can be made and could be applied as fertilizer if desired. From the standpoint of benefit to plants and efficiency in transportation, the anion of the potassium salt should preferably contain a plant nutrient. Nutrients that occur in anion form include borate, molybdate, chloride, sulfate, nitrate, and phosphate. The first two are immediately ruled out on the basis of toxicity if the salt is to be used in quantities needed as a source of potassium. Some of the first potassium fertilizer produced from deposits in the United States had disastrous effects on crops because of toxicity of borate present as an impurity.

The need for chlorine as a nutrient is extremely small, and occurrence of chlorine deficiency in plants under practical conditions seems unlikely. Hence, from the standpoint of supplying something useful, the chloride salt appears to be a poor choice. The reason most of the potassium fertilizer is marketed as the chloride salt is, of course, that this form is generally the most economical to produce and transport. Although sulfur is a plant nutrient, the quantities of sulfur supplied as calcium sulfate in ordinary superphosphate meet the needs in many areas that would otherwise be deficient. The gradual replacement of ordinary superphosphate by concentrated superphosphate may be expected to increase the need for another source of sulfur such as potassium sulfate.

The chloride ion is not adsorbed to a significant degree by soils and is absorbed readily by plants, in which it is thought to occur mostly as an ion in solution, like potassium. A high content of chloride in plants increases the succulence, increases the thickness of the cuticle, and decreases the size of the epidermal cells. Sulfate is adsorbed to some extent by soils, is absorbed by plants less readily than is chloride, and in plants occurs in part in organic combination in proteins and amino acids and in part as sulfate ion in solution.

The comparative value of potassium chloride and potassium sulfate as fertilizers has been investigated in many experiments. Occasionally the chloride salt seems preferable to the sulfate, as in work by Kramer (1963) on strawberries. He noted that the yield of a variety that ripened during a dry period was nearly 30% higher where potassium was applied as the chloride than where it was applied as the sulfate. With two other varieties that ripened following rains, the yields where potassium was applied as the chloride differed by less than 10% from the yields obtained with the sulfate, one being above and the other below the yield with potassium sulfate. With cereals, there is usually no significant difference in results obtained with the two forms of potassium. For the most part, where recognizable differences do occur, crop yields and quality are better with the sulfate salt than with the chloride. Nevertheless, the fact that almost all the potassium used as fertilizer is applied as the chloride indicates that the benefits from the sulfate are not considered important enough to justify the higher cost of potassium sulfate. With two crops, tobacco and potatoes, however, the sulfate is rather commonly used. Tobacco leaf has poor fire-holding quality if it contains much chloride, and the yield and starch content of potato tubers are greater with the sulfate than with the chloride. A review of effects of chloride and sulfate was published by Burghardt (1962).

Potassium nitrate and potassium phosphate are superior for general use to all the other salts mentioned because both these substances supply two macronutrients that are frequently applied in large quantities in fertilizers. The disadvantage to use of potassium nitrate and potassium phosphate is economics. At present, their preparation is based on processing as a raw material the potassium chloride that is salable in its own right as a finished product in fertilizers.

Fertilizer Application

According to Ignatieff and Page (1958), potassium fertilizers should usually be applied at or just before the time of planting of annual crops; and the fertilizer should be in localized bands near the plants if the quantities applied are small (less than, say, 40 kg of potassium per hectare). This generalization is in agreement with de Wit's (1953) theory of fertilizer application, which predicts the greatest benefit from localized placement where small applications of fertilizer are made. The greatest benefit from localized applications near the plants should be obtained on soils that will fix much potassium in nonexchangeable form. In experiments on potassium fertilization of cereals on potassium-fixing soils in the Netherlands, Prummel (1957) found that band applications had an efficiency averaging 3.65 times as great as broadcast applications. These are among the highest efficiencies that have been recorded for localized applications of potassium fertilizer. On similar soils, however, efficiencies of localized and broadcast applications were about equal in experiments on potatoes and beets, which indicates that the nature of the crop is a significant factor. As yet, insufficient research has been done to correlate the relative efficiencies of localized and broadcast applications of potassium fertilizers with the potassium-fixing properties of soils.

Localized placement of potassium chloride or potassium sulfate produces a high local concentration of soluble salts. Hence, if the fertilizer is placed too near the seed, severe damage to the seedlings and loss of stand may result because of insufficient time for reduction of the salt concentration to a safe level before the seed starts to germinate. Applications with, directly above, or directly below the seed are especially unsuitable. Most of the movement of water is in a vertical direction, and so potassium salts placed above or below the seed may be moved into the seed zone with movement of water. Placement 5 cm or so to the side of the seed is preferable because then the seed zone will not be affected, and root development can occur in soil that does not contain excess salt. Within a short time, movement of the potassium salt into the surrounding soil will have reduced the salt concentration to a safe level. The marked local increase in potassium supply that has occurred may then increase the growth of roots in the fertilizer-affected soil. An illustration of this effect may be found in a paper by Cooke (1954).

For crops with a long growing period and for sands with low cation-exchange capacity, application of part of the potassium as a delayed side dressing may prove beneficial. For perennial crops, small annual applications are preferable to large, infrequent applications. In tropical regions, perennial crops of high value may receive potassium fertilizer several times annually.

The recovery of applied potassium in the first annual crop following application is not uncommonly 50% or more. Potassium thus ranks along with nitrogen and considerably above phosphorus with regard to efficiency of use. This difference in efficiency is probably one reason why the ratio of potassium applied to potassium in the crop is characteristically lower than the ratio of phosphorus applied to phosphorus in the crop. The most suitable applications of potassium vary greatly with the conditions, and much information on practices in different parts of the world was summarized by Ignatieff and Page (1958). Generally, in relation to world practice, annual applications up to 40 kg of potassium per hectare might be considered low, applications of 40 to 100 kg moderate, and applications exceeding 100 kg high.

Among crop plants, considerable difference exists in sensitivity to potassium deficiency. Crops may be compared in this respect in different ways. One way is to determine how much the yield of the various species or the content of potassium increases upon application of a given quantity of potassium fertilizer under comparable conditions. Results are preferably expressed on a relative basis. Figure 9.31 gives an example of differences among certain crop plants based on results of field experiments in England and Wales. In this example, the data are averages obtained from numerous experiments, but the crops were grown indi-

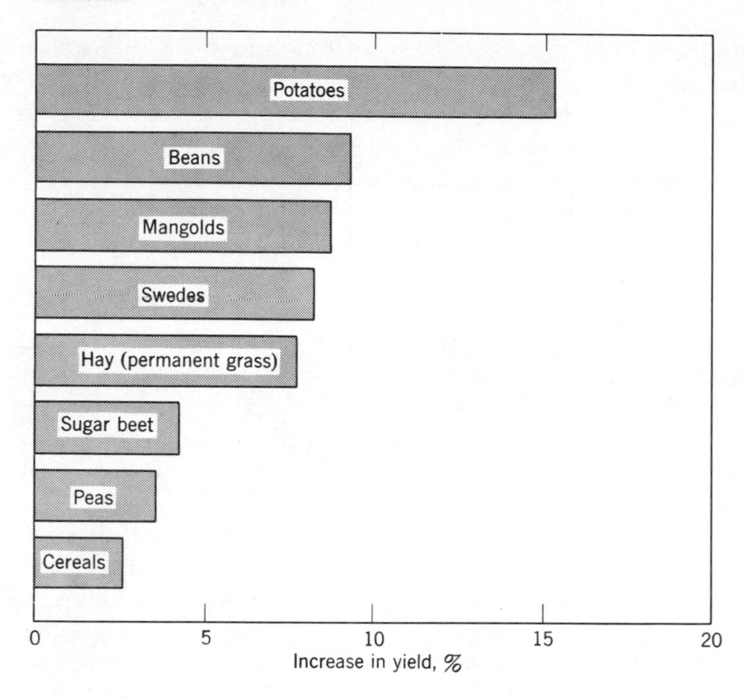

Figure 9.31. Percentage increase of yield of different crops from fertilization with 52 kg of potassium per hectare in experiments in England and Wales. (Boyd, 1956)

vidually and hence not necessarily under entirely comparable conditions. For this reason, the ratings should be qualified by representing them as the responses of the crops for the conditions under which the crops were grown.

Another rating may be obtained by growing the desired crops under comparable conditions and determining the yield of the controls without added potassium and the maximum yields attainable by application of potassium. The yields are then recalculated on a relative basis with the maximum yield of each crop taken as 100 or some other arbitrary figure. The lower the yield of the control on this relative basis, the lower is the degree of sufficiency of soil potassium or the greater is the sensitivity of the crop to potassium deficiency. Figure 9.32 shows the results of a comparison of this sort among four crops. In this case, the four crops were grown in a single field experiment in the Netherlands. One may note that wheat showed little sensitivity to potassium deficiency, in agreement with the entry for "cereals" in Fig. 9.31, but the position of beans and potatoes was reversed. This discrepancy illustrates the fact that the ratings one obtains are not entirely independent

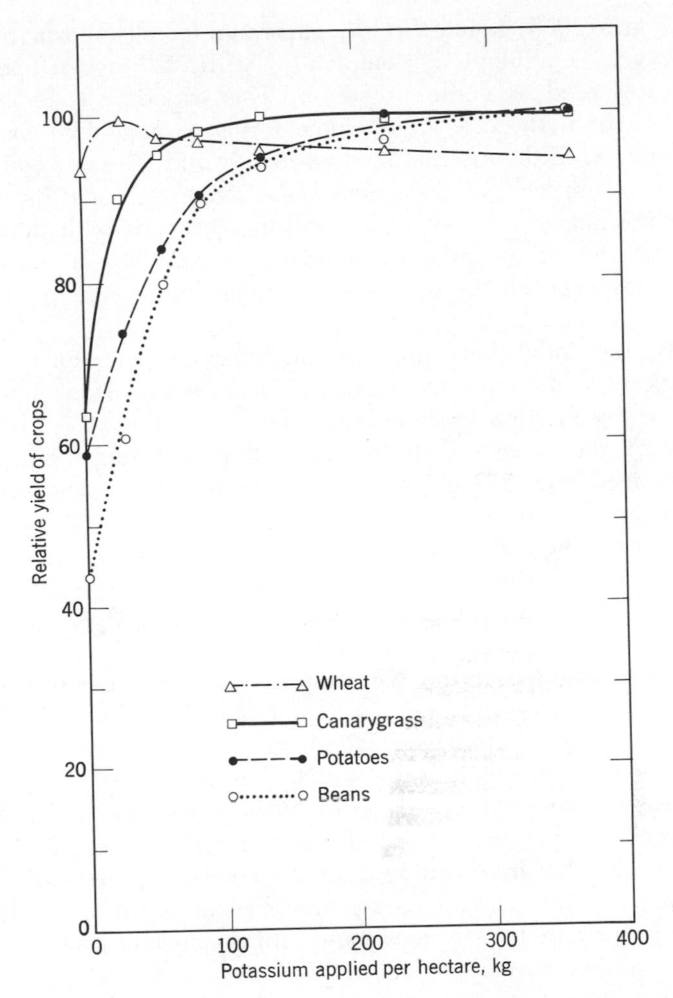

Figure 9.32. Relative yields of different crops with different annual applications of potassium in fertilizer in field experiments in the Netherlands. The maximum yield obtained from each crop on application of potassium fertilizer is given a value of 100. (van der Paauw, 1958)

of experimental conditions and hence may vary somewhat from one situation to another.

Function in Plants

Potassium, unlike nitrogen and phosphorus, does not seem to be present in nonionic combination with organic substances in plants. Scott and Hayward (1953) found that the potassium in cells of a marine alga, *Ulva lactuca,* equilibrated completely with K^{42}-tagged potassium in sea water used as culture medium. This experiment demonstrates the absence from the cells of nonionic forms of potassium that failed to equilibrate with the external medium. Eddy and Hinshelwood (1951) found that resting cells of *Bacterium lactis aerogenes* could be depleted of potassium almost completely by washing the cells with dilute acid. Removal of the potassium did not affect the viability of the cells, as would be expected had potassium been functioning in a structural capacity.

Nevertheless, indications are that the potassium in plants does not occur entirely in the aqueous cell sap. Olsen (1948, 1948a) found that the sap expressed from fresh beech leaves, after the leaves had been frozen to kill the tissue and destroy the differential permeability of the cells, contained only 78% of the total potassium, due allowance being made for the portion of the sap not extracted. More of the potassium was extracted as he added increasing amounts of water as an extractant, but 2% of the potassium was still unextracted where 400 ml of water were added to 10 g of previously frozen leaf tissue. Upon addition of 50 ml of 0.1-normal calcium chloride solution, however, he obtained quantitative extraction of the potassium. These experiments were performed on leaves sampled early in September. Tests on samples taken late in October just before leaf fall, when much of the nitrogen had migrated from the leaves into the wood, showed that all the potassium was present in the cell sap. Accordingly, he proposed the theory that the potassium not present in the cell sap during the summer is adsorbed by proteins and that this binding disappears with hydrolysis of the proteins before leaf fall. Olsen's theory that the potassium in plants is adsorbed in part by proteins is consonant with the current view that potassium functions in enzymatic reactions.

Two other lines of evidence substantiate the view that potassium plays a part in metabolic processes. First, the potassium content of plants increases with the rate of metabolism. Work of Hoagland and Broyer (1936) and Luttkus and Bötticher (1939) may be cited in this connection. Second, potassium is known to be required in three types of enzymatic reactions. Potassium is essential in certain enzymatically catalyzed reactions involving adenosine phosphates (Lardy, 1951; Black, 1951; Kachmar and Boyer, 1952; Stadtman, 1952). Adenosine phosphates are the energy carriers in metabolic processes in both plants and animals, as mentioned in Chapter 8 on phosphorus. In carbohydrate metabolism, the reactions requiring potassium are steps in the process by which energy is obtained from sugar. Potassium is necessary also in some of the syntheses. Webster and Varner (1954) found that potassium is required in the coupling of certain amino acids to form peptides, thus suggesting that potassium is essential in protein synthesis. Further work has verified the importance of potassium in protein synthesis (Lublin, 1964). Finally, work by Latzko (1959) indicates that potassium activates one of the first reactions in photosynthesis in which part of the light energy is captured in a reaction between adenosine diphosphate and inorganic phosphate to form adenosine triphosphate.

The action of potassium in these enzymatic processes has been verified in simple systems involving substrate, enzyme, and different concentrations of salts of potassium or other metals. The catalysis is thought to require the presence of potassium in ionic combination with the enzymes, the potassium being subject to exchange with other potassium ions in the medium or with cations other than potassium. Hence a high internal concentration of potassium and a high proportion of potassium in the total internal cations should increase the rate of transformation of the substrate under otherwise equal conditions.

Much biochemical and general physiological research has been done in relation to potassium supply to plants, and there is evidence for definite malfunction of metabolism under conditions of potassium deficiency. The complexity of metabolic processes and the inability to isolate potassium-bearing organic compounds, however, makes it difficult to interpret such experiments in terms of the manner in which potassium functions. Some of the observations made in this work will be mentioned in succeeding sections.

Sodium-Potassium Relations

Sodium is of importance in plant nutrition not only because sodium is required by at least a few plants (Allen and Arnon, 1955; Brownell and Wood, 1957) and is detrimental in excess, as noted in Chapter 6, but also because of its relation to potassium. Sodium and potassium are the two principal monovalent metallic cations in plants, and an increase in one generally brings about a decrease in the other. Sometimes the change in proportions of sodium and potassium in plants is brought about with relatively little alteration of the total equivalent concentration of metallic cations in the tissue (see, for example, a paper by van Itallie, 1938). Practical interest in sodium-potassium relations stems principally from a potassium-sparing action sometimes observed, which suggests the possibility of using sodium salts to replace a part or all of the more costly potassium salts used as fertilizer. In Great Britain, for example, Tinker (1965) found in 42 field experiments that either potassium chloride or sodium chloride, applied singly, produced a profitable increase in yield of sugar from sugar beets. In the presence of sodium chloride, however, the increase in yield from potassium chloride was comparatively small and no longer profitable.

Plant species behave rather differently with regard to uptake of sodium. Figure 9.33 shows the relative content of sodium and potassium in 21 species of plants grown together in the same nutrient solution. Initially, the molar concentrations of sodium and potassium in the solution were equal; yet 18 of the plants contained less sodium than potassium, and 6 contained less than 10% as many equivalents of sodium as potassium. Only 3 contained more equivalents of sodium than of potassium.

Plant species differ also with regard to yield response to sodium. (Effects of excess sodium were considered in Chapter 6. Emphasis here is on yield response to lesser concentrations.) Three effects are illustrated in Fig. 9.34. Results in Fig. 9.34*a* are from an experiment in which a moderate supply of sodium had no appreciable effect on plant yields. The sweet-orange seedlings used in this experiment responded markedly to an increase in supply of potassium but were essentially indifferent to sodium within the range used. Although the data are not given in the figure, analyses were made for sodium and potassium in the citrus leaves; and the findings showed concentrations of the order of 0.01% sodium and 0.2 to 1.4% potassium. Substitution of sodium for potassium in the leaves was thus quantitatively almost nonexistent. Sodium was taken up in quantity by the roots, but little was translocated to the tops. At least some other characteristically low-sodium plants behave similarly.

Figure 9.34*b* illustrates a second type of behavior in which sodium increased the yield where potassium was deficient but not where potassium was in plentiful supply. This type of response is generally designated as a partial substitution of sodium for potassium under conditions of potassium deficiency. With species showing such yield response, visual symptoms of potassium deficiency seem less severe on plants treated with sodium salts than on untreated control plants. The leaves of the sugar beets in the experiment from which the data in the figure were derived contained of the order of 1 to 3% sodium (far more than the sweet-orange seedlings in part *a* of the figure) and 2 to 6% potassium. A definite tendency was found for the sodium content to decrease as the potassium content increased.

Figure 9.34*c* illustrates a third type of response to sodium. The design of the experiment shown is not like the others in having two constant

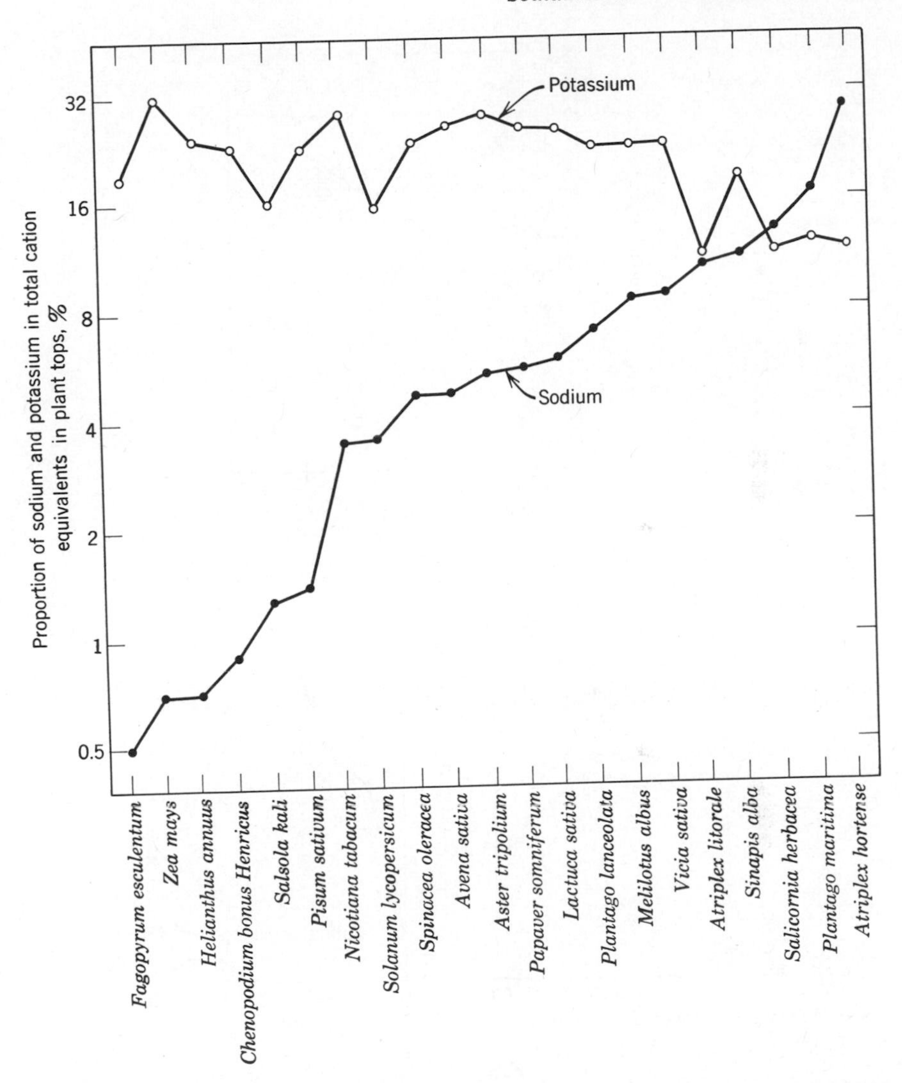

Figure 9.33. Sodium and potassium expressed as percentage of total cation equivalents in 21 plants grown together in the same culture solution in which sodium and potassium were initially present in concentrations of 2 m.e. each per liter of solution. Note the logarithmic scale. (Collander, 1941)

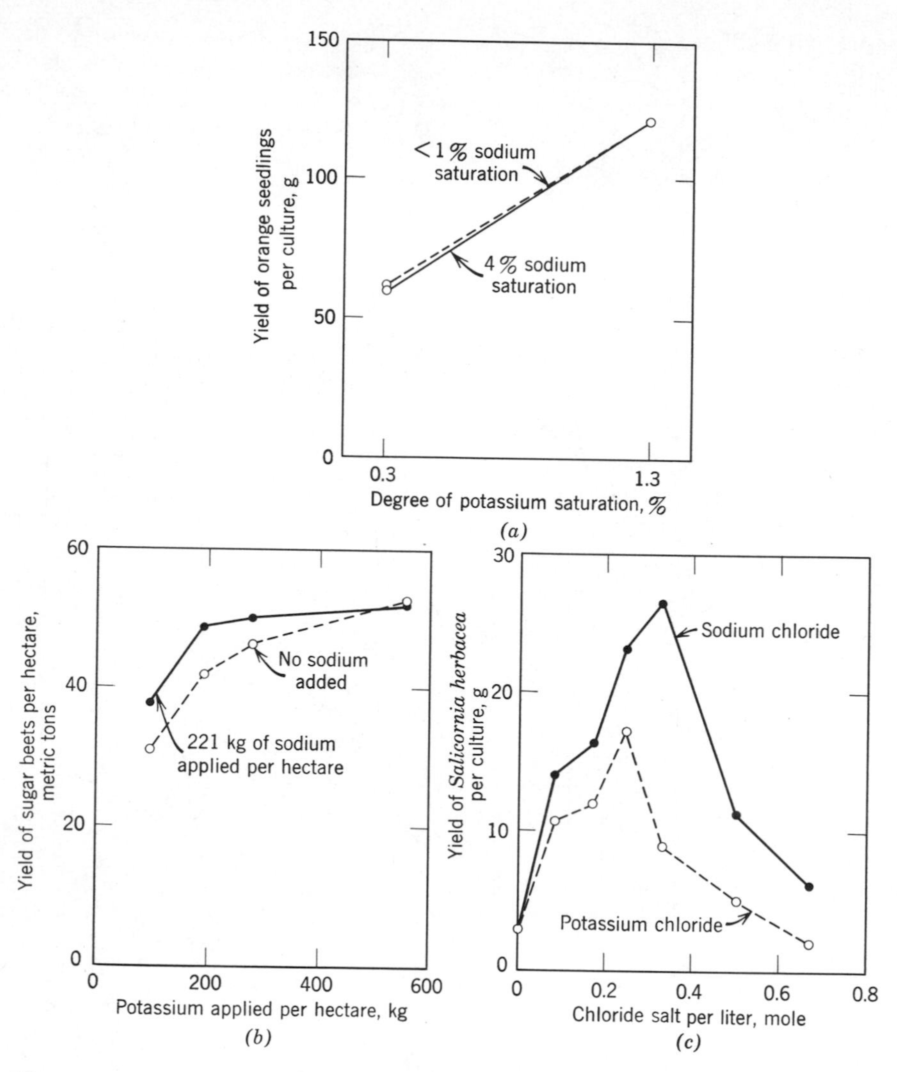

Figure 9.34. Response of different plants to sodium and potassium. (*a*) Yield of sweet orange seedlings at two degrees each of sodium and potassium saturation of the cation-exchange capacity of a loam soil of California. (Page and Martin, 1964) (*b*) Yield of sugar beets in Michigan with different applications of potassium to a muck soil with and without a supplemental application of sodium as sodium chloride. (Shepherd et al., 1959) (*c*) Yield of *Salicornia herbacea* in the Netherlands on 5 kg of soil treated with 2 liters of a nutrient solution containing different concentrations of sodium chloride or potassium chloride. The basal solution applied to all cultures contained, in addition, 0.0086 gram mole of potassium per liter and nutrients other than potassium. (van Eijk, 1939)

applications of sodium with each of several applications of potassium; rather, increasing quantities of sodium chloride and potassium chloride were added singly to soil that received a basal treatment of macronutrients including some potassium. The yield was increased by application of sodium even without addition of any potassium beyond that in the soil and the basal nutrient solution, and consistently higher yields were obtained with sodium chloride than with chemically equivalent quantities of potassium chloride. The fact that the maximum yield attained by use of sodium chloride exceeded the maximum yield attained by use of potassium chloride provides evidence that sodium had a favorable effect on yield independently of potassium. In a plant like the one used in this experiment, sodium would probably act as a partial substitute for potassium under conditions of potassium deficiency, but the experimental data in the figure do not permit one to decide whether or not this partial substitution effect occurs. As the name suggests, the plant used in the experiment grows naturally in saline environments and takes up much sodium and chlorine. *Salicornia herbacea* is one of the three species that contained more sodium than potassium in Collander's (1941) experiment (Fig. 9.33).

The range of plant responses to sodium shown in Fig. 9.34 is associated with the tendencies of the plants to take up sodium. The more readily the plants took up sodium, the greater was the beneficial effect. Most agricultural plants seem to be of the types illustrated by sweet orange and sugar beet in Fig. 9.34. According to a description by Brownell and Wood (1957), however, an important forage plant in arid regions of Australia (bladder saltbush—*Atriplex vesicaria*) is of the high-sodium type illustrated by *Salicornia herbacea* in Figs. 9.33 and 9.34. In the United States, another high-sodium plant, halogeton (*Halogeton glomeratus*), is important because it is widely distributed in western grazing areas and is toxic to livestock on account of the high concentrations of oxalate it contains (Williams, 1960). Classifications of crop plants with regard to response to sodium were published by Harmer and Benne (1945) and Lehr (1953).

The general subject of sodium-potassium relations under conditions of potassium deficiency was reviewed briefly by Baumeister (1958). An analysis of sodium-potassium relations where sodium salts are present in the salinity range was published by Heimann (1958).

Sodium-potassium relations have received far more attention than lithium-, cesium-, and rubidium-potassium relations because of the greater importance of sodium under natural conditions and the low cost of sodium salts. None of these alkali cations have been found to replace potassium completely for growth of higher plants. Nevertheless, as far as absorption is concerned, plants do not make much distinction between rubidium and potassium; and rubidium is sometimes used as a tracer for potassium because of its more convenient radioactive isotope. With regard to lower plants, Kellner (1955) described a mutant of the alga *Ankistrodesmus braunii* that would grow without potassium if rubidium was present.

Potassium Supply and Plant Behavior

Deficiency Symptoms

The two most commonly observed symptoms of potassium deficiency are (1) shortening of stems, producing a squat or bushy appearance in plants such as cereals that normally produce an extended stem or a rosette appearance in plants such as beet that normally grow from a crown without an extended stem and (2) browning and death of leaf tissue, commonly occurring at leaf margins (popularly known as "leaf scorch") or tips (known as "tip burn") but in some plants as spots in interveinal areas. Leaves may be narrow and crinkled with a tendency for the margin to curl toward the upper or lower surface.

Browning and death of the leaf tissue is usually preceded by other less marked symptoms, including chlorosis and tinting with bluish-green, purple, or orange colors, visible macroscopically, and by the shrinking, collapse, and browning of individual cells, visible microscopically. The necrotic or dead spots that characterize potassium deficiency are groups of cells that have collapsed. Bussler (1964) made an extensive investigation of the comparative microscopic morphology of tissues of potassium-deficient and normal plants.

Individual species have distinctive symptoms that are helpful in visual diagnosis. Descriptions of potassium-deficiency symptoms for a number of crops and reproductions of numerous colored photographs of deficiency symptoms may be found in a book by Wallace (1961) and in a book edited by Sprague (1964).

The occurrence of potassium deficiency symptoms is not a sign of incipient deficiency but rather an indication that potassium is a seriously limiting factor. Potassium fertilization thus may be of distinct benefit even though no foliar symptoms of deficiency appear. On the other hand, severe potassium deficiency symptoms early in the season (which indicate a real deficiency for growth at that time) may later disappear, with the consequence that the response of the harvested crop to potassium fertilization is relatively small. Interpretation of the symptoms in terms of remedial treatment thus may not be entirely straightforward. Walsh (1954) described some of the difficulties related to use of deficiency symptoms as a guide to fertilization of crops with potassium in Ireland.

Analysis of the plant or some index tissue may be helpful in refining the diagnosis of potassium deficiency beyond that permitted by visual symptoms of deficiency. In the case of sugarcane, for example, Schmehl and Humbert (1964) gave 1.5% potassium in leaf sheaths three to six (measured from the top down) as the upper limit at which deficiency symptoms occurred but noted that increases in yield from potassium fertilization could be obtained until the potassium content of this tissue reached 2.25%.

Development of symptoms of potassium deficiency is related to the internal behavior of potassium in plants. Retention of potassium in plant cells seems to require continuous expenditure of metabolic energy. For example, Luttkus and Bötticher (1939) observed that plant roots alternately lost potassium to a nutrient solution when maintained for a day or more in darkness and gained potassium from the solution when maintained a day or more in light. By use of radioactive potassium, Mengel (1964) found that potassium moved out of roots as well as into roots during net uptake and that net loss of potassium to the surrounding solution occurred upon addition of sodium azide, an inhibitor of metabolic reactions. Rathje (1961) found that the potassium concentration in the leaf sap in the afternoon was relatively high in the interveinal tissue and low in the veins of leaves of sugar beet and potato grown in the field in Germany. At night the concentration decreased in the interveinal tissue and increased in the veins. (Sodium, incidentally, manifested precisely the opposite behavior.) Kidson (1943) ashed an intact apple leaf showing marginal browning due to potassium deficiency and then sprayed the ash skeleton with a solution of sodium cobaltinitrite, which produces a bright yellow-orange precipitate with potassium. The result was a sharp differentiation between the original marginal brown area and the inner green area. The inner area yielded a much more dense precipitate than the outer area. The ash skeleton of a green control leaf from a tree that had been fertilized with potassium developed a comparatively uniform precipitate of potassium sodium cobaltinitrite over the entire surface. Potassium thus is not inactivated in cells but, under some circumstances, may move from one place to another.

If analyses are made of leaves at different positions on a single plant, definite differences in potassium content are sometimes found. As might be expected from the mobility of potassium and its function in metabolism, there is a tendency for potassium to occur in higher concentration in younger leaves than in older leaves. An illustration may be found in Table 9.8, which contains values for the potassium content in lower,

Table 9.8. Deficiency Symptoms and Potassium Content of Leaves of Tobacco with Different Applications of Potassium as Potassium Sulfate (Bowling and Brown, 1947)

Potassium Applied in Fertilizer per Hectare, kg	Symptoms of Potassium Deficiency	Potassium Content of Dry Matter, %		
		Lower Leaves	Middle Leaves	Upper Leaves
0	Very severe	0.6	0.6	1.3
22	Severe	1.0	1.6	2.2
45	Moderate	2.4	2.5	2.8
112	None	5.3	4.4	3.1
157	None	6.2	5.4	4.7

middle, and upper leaves of field-grown tobacco and includes observations on the severity of symptoms of potassium deficiency on the plants as a whole. The highest potassium percentage evidently occurred in the upper leaves where deficiency symptoms were present and in the lower leaves where deficiency symptoms were absent. Liebhardt and Murdock (1965) found potassium to behave similarly in corn. Appling and Giddens (1954) observed that, where soil potassium was deficient for cotton, potassium accumulation was greatest in the upper, immature leaves; and sodium accumulation was greatest in the lower leaves.

The higher concentration of potassium in the younger vegetative tissue may be produced in part at the expense of the older tissue. Greenway and Pitman (1965) found, for example, that 51 of the 129 μg of potassium entering per day into the youngest leaf of barley in the three-leaf stage came from sources within the plant. At the same time the oldest leaf was losing 63 μg of potassium per day. The selective efflux of potassium from the older leaves provides a partial explanation for the tendency of symptoms of potassium deficiency to appear soonest and to be most severe on the older leaves. Withdrawal of potassium may hasten senescence of the older leaves, as suggested by the finding that the rate of photosynthesis in older leaves declined more rapidly where the supply of potassium was low than where it was high (Alten et al., 1938).

Although translocation of potassium from old to new growth is no doubt a link in the chain of events that leads to development of visual symptoms of potassium deficiency in the old growth, the loss of potassium does not appear to be the immediate cause of the symptoms. There is evidence for action of specific toxic organic substances in this regard. When potassium is deficient, protein synthesis is limited and breakdown products accumulate. The latter include various amino acids that are normal components of proteins as well as other nitrogenous substances that are presumably derived from the primary components. Richards and Coleman (1952) discovered that barley plants deficient in potassium contained putrescine ($NH_2—CH_2—CH_2—CH_2—CH_2—NH_2$), a substance present only in traces in plants well supplied with potassium. The content increased with the severity of the visual symptoms of potassium deficiency, reaching maximum concentrations estimated at 0.15 to 0.2%. Coleman and Richards (1956) then added putrescine to plants in various ways and found that symptoms characteristic of potassium deficiency were thus produced even in plants high in potassium. Addition of potassium to deficient plants resulted in a reduction of their putrescine content. According to Richards (1954), low-potassium barley that received sodium, rubidium, or lithium developed lower concentrations of putrescine and had less severe visual symptoms than did barley without these other alkali metals. All this evidence supports the view that leaf tissue of barley is sensitive to putrescine and dies if the concentration is great enough.

Smith (1963) tested 13 plant species, including barley, and found putrescine in all, whether potassium was deficient or not. Also present was agmatine, a chemically related substance. In 11 species, potassium deficiency increased the concentration of putrescine, and in 6 species potassium deficiency increased the concentration of agmatine. Neither substance accumulated in potassium-deficient groundsel (*Senecio vulgaris*) or flax. Coleman and Richards (1956) had previously found that potassium-deficient flax accumulates arginine. The evidence now available indicates that arginine is formed first and that putrescine and agmatine are formed from it. According to work by Smith (1963), the accumulation of putrescine and agmatine results from action of an enzyme that removes a carboxyl group from arginine. His work showed that activity of this enzyme was greater in potassium-deficient barley than in barley well supplied with potassium. Presumably, therefore, the failure of potassium-deficient groundsel and flax to accumulate putrescine and agmatine is related to a low activity of the enzyme that acts on arginine. In tobacco, Takahashi and Yoshida (1960) found a marked accumulation of putrescine in plants deficient in phosphorus as well as in those deficient in potassium.

No doubt much remains to be learned about the biochemical processes leading to development of potassium-deficiency symptoms. For the present objective, however, the details of the processes are perhaps not of as much significance as the general concept that the visual symptoms are, in some plants, and perhaps in all, a consequence of toxic substances produced normally but present in much greater concentrations in plants deficient in potassium than in those well supplied with potassium.

Water Relations

The literature records many observations and statements to the effect that potassium fertilization "improves" the water relations of plants. Perhaps the most consistently reported observation is an increase in the water content and turgidity of plants with an increase in supply of potassium. Figure 9.35 gives an example, containing averages of many measurements on field-grown sugarcane in Hawaii. Humbert (1958) noted that leaves of potassium-deficient sugarcane have relatively low turgidity, and the tips commonly become frayed from blowing in the wind. Similarly, Anderson et al. (1929, 1930) in Connecticut noted that tobacco plants grown on field plots without potassium fertilization wilted sooner and more severely on hot days than did plants on comparable plots fertilized with potassium. Although the plants that did not receive potassium fertilizer were reduced in size, they showed no acute symptoms of potassium deficiency. Eckstein et al. (1937, pp. 38–39) reviewed similar observations by other investigators.

Turgidity effects such as those just mentioned could conceivably result from a relatively high water content in high-potassium plants indepen-

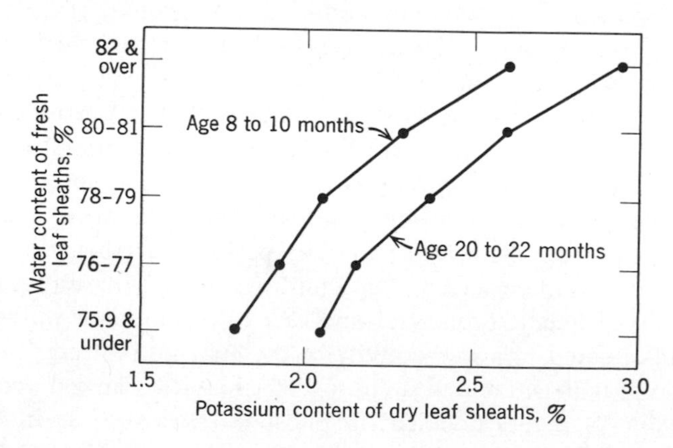

Figure 9.35. Water content versus potassium content of leaf sheaths of sugarcane in two age groups in Hawaii. Values given are averages of numerous measurements. (Humbert, 1958)

dently of loss of water by transpiration or from a relatively low loss of water from plants high in potassium. Evidence exists for both possibilities, but observations are not consistent, particularly those related to potassium supply and loss of water by transpiration. Experiments by Arland and Zwicker (1959) support both possibilities. These investigators found that the yield of expressed sap of potassium-deficient oat plants was much lower from the lower leaves than from the upper leaves even where the plants were well watered and the samples were taken early in the morning before the daytime transpiration could have had much effect. When the plants were detached from the roots and exposed to evaporation under uniform conditions, loss of water as a percentage of initial plant weight decreased with an increase in the prior supply of potassium to thc plants. Thus, despite a greater initial content of water in the high-potassium plants, loss of water by transpiration took place more readily from low-potassium plants. Mann (1924) obtained conflicting evidence with regard to the second possibility. He determined the loss of water from leaves detached from gooseberry plants grown in sand cultures supplied with a minus-potassium solution and with a complete nutrient solution. He found that the rate of loss was about the same if he used green leaves before the development of symptoms of potassium deficiency. After the initial symptoms had appeared, however, the leaves from potassium-deficient plants lost water more rapidly than those from potassium-sufficient plants. Experiments on shoots cut from apple trees and placed in water indicated that the rate of transpiration of leaves from potassium-deficient plants was above that of leaves from potassium-sufficient plants in the sunshine but below when the sun was covered by clouds.

Experiments with different cations and different kinds of plants indicate that high water content and turgidity are associated with high uptake of alkali-metal cations, particularly sodium (van Eijk, 1939; Richards, 1956; Williams, 1960); osmotic effects of the ions may be responsible. With regard to the effect of potassium supply on loss of water by transpiration from individual leaves or plants with equal exposure, the conflicting observations have resulted in various theories. Literature on this subject was reviewed by Biebl (1958).

Some of the observations of beneficial effects of potassium on the water economy of plants seem now to be merely an indirect consequence of the greater growth of the potassium-treated plants and not a specific potassium effect. As noted in Chapter 2, the amount of water lost by evapotranspiration per unit weight of plant decreases with an increase in production of plant tissue, whether the latter results from an extra supply of potassium or nitrogen or in some other way.

Carbohydrate and Nitrogen Metabolism

As noted in the section on function, potassium is essential for the action of enzymes that catalyze certain reactions in both carbohydrate metabolism and nitrogen metabolism. When plants become deficient in potassium, soluble forms of nitrogen usually accumulate in the tissues; and there is evidence that at least two of these compounds are toxic. Hsiao (1964) noted, however, that withholding of potassium reduced the rate of growth of corn leaves before free amino acids accumulated and before the protein content per leaf decreased; also, addition of potassium to severely deficient plants did not result in a reduction of amino acids until the plants had shown a growth response. These findings suggest that the primary cause of growth reduction in potassium deficiency is neither reduction of protein synthesis nor toxicity of accumulated soluble nitrogen compounds.

Although photosynthesis is reduced and respiration is increased by potassium deficiency, the usual finding is an accumulation of starch and sugars in plants with the onset of potassium deficiency. Probably the accumulation occurs because the limitation of use of carbohydrates in producing new proteins and new tissue is more severe than the limitation of net carbohydrate production. If the deficiency is severe, or if a moderate deficiency persists for a long time, potassium-deficient plants usually have a lower concentration of carbohydrate than controls with adequate potassium, presumably reflecting finally the effects of reduced photosynthesis and increased respiration. The matter is complex, and results may vary in consequence of interaction of effects of potassium supply with those of many other factors that influence carbohydrate behavior in plants. Hewitt (1963) reviewed the work on this subject.

Total accumulation of carbohydrates by potassium-deficient plants is limited not only by the decreased rate of photosynthesis and the increased rate of respiration but also by two other secondary factors: First, growth of plants is an exponential or compounding process in which the new growth per unit time depends on the initial growth; hence a decrease of a few per cent in efficiency of unit area of leaf may amount in time to a large difference in total product. Second, the lower leaves of potassium-deficient plants die prematurely, thus reducing the effective leaf area during the latter part of the growth period when the carbohydrate is being used mainly in fruit production. As a specific instance of this effect, Turner (1944) found in experiments on soils deficient in potassium for production of cotton that the loss of leaves by the time of the first picking amounted to 65, 53, and 31% on plots receiving potassium in quantities equivalent to 22, 45, and 90 kg/hectare.

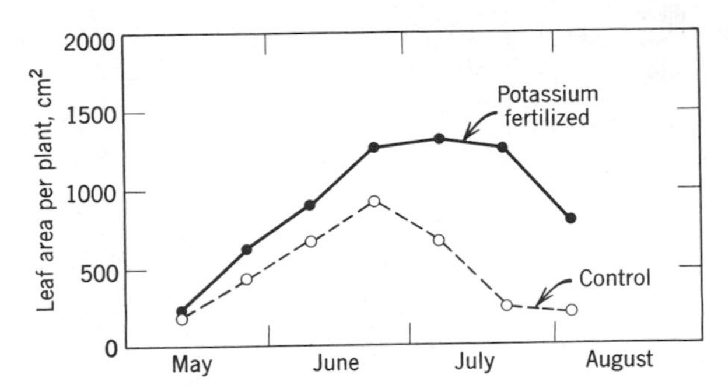

Figure 9.36. Total area of leaves during growth of control and potassium-fertilized potato plants. (Watson, 1956)

The premature loss of leaves from potassium-deficient potato is illustrated in Fig. 9.36.

A number of the effects of potassium deficiency on plant behavior seem to stem from a deficiency of carbohydrates. A direct practical consequence of potassium-carbohydrate relations is found in sugar crops. Application of potassium fertilizer to soils low in potassium may increase both the sucrose percentage in sugarcane and sugar beet and the total yield of cane or beet root (Samuels and Landrau, 1954; Harmer and Benne, 1941). Brummer's (1959) summary of many field experiments on potassium fertilization of sugar beet in Finland showed, however, that the maximum sugar percentage in the beets was obtained with approximately the application of potassium needed to produce the maximum yield of beets. Greater applications of potassium lowered the sugar percentage.

The two most frequently mentioned effects of potassium deficiency on crop performance, weak straw of cereals and poor fiber in fiber crops, also are associated with the low-carbohydrate condition of the plants. Anatomical changes in stems of barley grown on soil with different applications of potassium may be visualized from the data in Table 9.9. The stems of potassium-deficient plants evidently had a relatively low proportion of sclerenchyma (which is the woody, supporting tissue) and were composed of cells with relatively thin walls. Both these effects are conducive to lower rigidity of stems of potassium-deficient than of potassium-sufficient plants. Figure 9.37 shows differences in lodging of corn in an experiment with different applications of potassium. Baumeister (1958) reviewed German work on the relation between potassium supply and strength of straw, and Kono and Takahashi (1962, 1964) published two papers on the effect of potassium supply on the

Table 9.9. Characteristics of the Straw of Barley Fertilized with Different Quantities of Potassium in Germany (Acker, 1932)

Potassium Applied per Hectare, kg	Sclerenchyma in Wall of Straw, %	Cell Wall Thickness, μ		
		Epidermis Outer Wall	Parenchyma	Sclerenchyma
0	8.7	2.9	2.0	3.2
41	10.1	5.4	2.3	5.5
162	12.7	6.6	4.2	6.8

breaking strength of rice straw in Japan. The 1964 paper contains a review of the literature. From their research, Kono and Takahashi (1964) inferred that the beneficial effect of potassium on resistance of rice straw to lodging resulted from the relatively high turgidity and high content of cellulose plus hemicellulose associated with a high content of potassium in the stems.

Observations on straw characteristics and lodging in relation to potassium supply have caused to develop a popular impression that weak plants and lodging are symptomatic of potassium deficiency and hence are to be treated by application of potassium fertilizer. Although experimental evidence is not as extensive as desired, the popular impression seems to be a considerable oversimplification. Lodging may be affected by potassium supply but is not determined by it. From the physiological standpoint, the stiffness of straw desirable in cereals seems to be a consequence of turgidity and carbohydrate accumulation. High-potassium plants may be high in carbohydrates, but whether they are or not depends on other conditions. One of the important conditions is the nitrogen status of the plant. As the potassium supply increases from a level of severe deficiency, the nitrogen percentage in plants commonly decreases because of dilution of the nitrogen in the extra growth associated with potassium fertilization (see, for example, Table 9.10). Plants high in potassium may be low enough in nitrogen to permit carbohydrate accumulation. On the other hand, if the supply of both nitrogen and potassium is high, use of carbohydrates in new growth will be favored; and the plants may then not have the qualities of stiffness popularly associated with high-potassium plants. Further discussion of this subject, with examples, may be found in a classical bulletin by Nightingale et al. (1930). In any event, there seems to be no evidence that application of potassium fertilizer in quantities exceeding those needed to produce the maximum yield are effective in reducing lodging. Figure 9.37 suggests that minimum lodging was obtained with about the same quantity of potassium required to produce the maximum yield of corn. A similar finding was reported by Walker and Parks (1966) and Parks and Walker (1966).

Fiber cells are usually sclerenchyma cells, and they tend to respond to potassium supply as indicated for sclerenchyma tissue in Table 9.9. Anatomical investigations by Tobler (1929) and Alten and Goeze (1936) on various plants cultivated for fiber showed that fiber cells from potassium-deficient plants had thinner walls than those from plants with adequate potassium.

Although cotton fibers are not classified as sclerenchyma, they respond in a similar manner to variations in supply of potassium for growth of the plant. Nelson (1949) made two different measurements on cotton fibers that reflect the effect of potassium supply on fiber characteristics. With applications of 0, 25, 50, and 75 kg of potassium per hectare for cotton in field experiments in North Carolina, he obtained values of

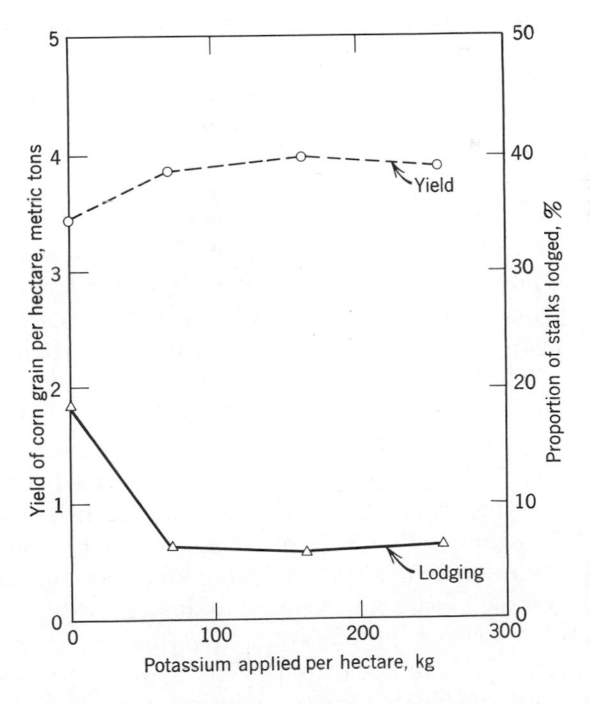

Figure 9.37. Average yield and lodging of four varieties of corn with different applications of potassium to a loam soil in Tennessee. Values for lodging are sums of lodging from root and stalk breakage. (Boswell and Parks, 1957)

Table 9.10. Characteristics of Wheat Grown in Sand Cultures with Different Applications of Potassium (Baumeister, 1939)

Potassium Added per Culture, g	Weight per Grain, mg	Yield per Culture, g		$\frac{G}{G+S}$	Starch Content of Grain, %	Nitrogen Content of Grain, %
		Grain (G)	Straw (S)			
0.25	16	8	29	0.22	45	3.4
0.50	28	21	42	0.33	52	2.7
1.00	30	21	47	0.31	54	2.6
2.00	32	22	50	0.31	55	2.7

3.3, 2.9, 2.8, and 2.8 cm^2 of fiber surface area per milligram of fiber weight (a measure of fineness) and values of 1.5, 1.7, 1.7, and 1.8 μg of fiber weight per centimeter of fiber length (an index of diameter and wall thickness of fiber).

In the cotton industry, a common empirical test of fiber quality is made by the so-called Micronaire instrument, which measures the rate of flow of air through a given weight of cotton fiber that has been compressed into a given volume. With such a technique, the rate of air flow would be expected to decrease with an increase in the proportion of the fiber volume occupied by the lumen (inner void) because the latter occupies space but does not contribute appreciably to either weight or air flow. The Micronaire scale is calibrated to correspond to the micrograms of fiber weight per 2.5 cm of fiber length of American cottons, and readings obtained are known simply as Micronaire values. Bennett et al. (1965) applied this test to cotton fiber produced in experiments with different applications of potassium (Fig. 9.38) and found that the lowest Micronaire value was obtained where the plants were grown with a marked deficiency of potassium, intermediate values were obtained in the range where there was little or no deficiency of potassium, and the highest value was obtained where the deficiency was moderate. These results would be interpreted to mean that the tendency was for thinnest cell walls in relation to lumen width in fiber from plants markedly deficient in potassium and for the thickest walls in relation to lumen width in fiber from plants with a moderate deficiency of potassium. Similarly, fiber strength was greatest in the range of moderate to slight deficiency of potassium. The test they used was a determination of the force required to break unit weight of a group of fibers cut to a given length. This sort of test has application to the use of fiber in making cloth; however, because the number of fibers used per test decreases with an increase in fiber weight per unit length, the test does not give the same sort of rating that would be obtained from determination of the force needed to break a single fiber.

Most research on the effect of potassium supply on quality of cotton fiber has understandably included control plants that were severely deficient in potassium, and the usual finding has been that application of potassium produced an increase in the proportion of fibers that had thick walls (frequently designated as mature fibers). The data in Fig. 9.38 do not entirely agree with the usual finding in that maximum fiber strength and Micronaire value were obtained in cotton that was somewhat deficient in potassium. The results of the experiment in question have been included because of their probable connection with nitrogen and carbohydrate metabolism. As an interpretation of the results, one may suggest that maximum thickness of fiber walls was not associated with maximum yield for two reasons. First, the plants may not have been high in carbohydrate because conditions for growth were made unusually favorable by irrigation and application of nutrients other than potassium (including nitrogen equivalent to 404 kg/hectare), and unusually high yields were obtained. Second, the high yields associated with the

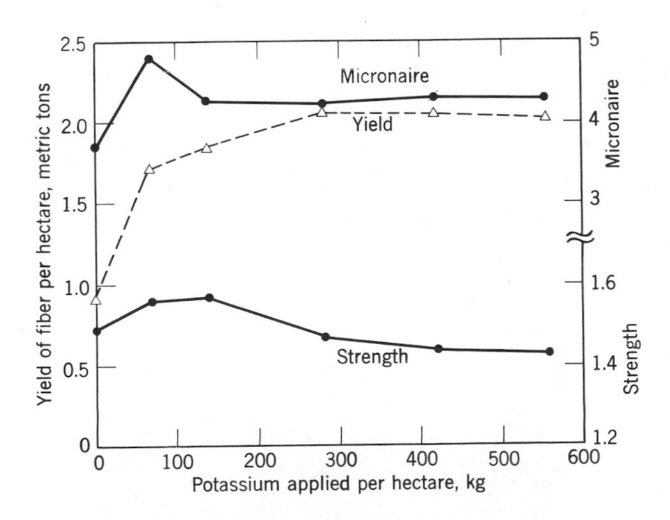

Figure 9.38. Yield, strength, and Micronaire of cotton fiber from plants grown with different applications of potassium to a sandy loam soil in Alabama. Strength is the load in kilograms required per milligram of fiber to break a specimen of fibers 1.5 cm in length. The meaning of Micronaire is explained in the text. (Bennett et al., 1965)

highest applications of potassium were obtained in part from bolls developed from flowers produced after the plants lower in potassium had stopped flowering (see Fig. 9.42, introduced later in another connection). Bolls produced late on the high-potassium plants thus may have been immature, with a high proportion of thin-walled fibers.

With regard to distribution of yield between grain and straw in grain crops, potassium supply has a distinct effect. Crops with a marked deficiency of potassium have a relatively low proportion of grain. The low-carbohydrate status of the plants is no doubt one reason for this behavior. Baumeister's (1939) data in Table 9.10 show the effect of potassium supply on the proportion of grain in the total yield as well as the starch content of the grain. The figures for starch are indicative of the degree of filling of the grain.

With regard to development of individual grains in cereal crops, there

Figure 9.39. Ears of corn from plants grown on low-potassium soil in Iowa without potassium fertilization and with 90 kg of potassium per hectare annually.

is a distinction between the effect of potassium on the one hand and nitrogen and phosphorus on the other. Grain size usually is not much reduced by deficiencies of nitrogen or phosphorus; an individual grain either is produced or it is not. With a deficiency of potassium, however, more grains are initiated than are filled, so that the grain is light in weight. An example of this potassium effect on wheat is shown in Table 9.10. With barley, Hellriegel et al. (1898) found that the average weight of one grain ranged from 5 mg, under conditions of extreme potassium deficiency, to 34 mg, with the greatest quantity of potassium they applied. In corn, the size of individual grains becomes smaller and the filling becomes poorer in proceeding from the basal to the distal portion of the ear. The ears are said to be "chaffy," and they tend to taper to a point (Fig. 9.39). This behavior seems to result from a limited supply of carbohydrate, and perhaps other substances, which in turn is associated with loss of the lower leaves of the plants. Ears with similar appearance can be produced on plants well supplied with potassium if the leaves are removed artificially after the cob has been produced but before the grain has filled.

Much attention has been devoted to potassium in relation to development of roots. Plants low in potassium are frequently high in nitrogen and low in carbohydrates, with the result that production of roots is limited. In the case of corn, "root lodging" or inelastic displacement of the stalks from a vertical orientation as a result of breakage of roots has been found by several investigators to be associated with low-potassium, high-nitrogen conditions. Table 9.11 illustrates not only the lodging from root breakage associated with application of potassium and other

Table 9.11. Development of Brace Roots of Corn and Lodging from Root Breakage on a Silt Loam Soil with Different Fertilizer Treatments in Wisconsin (Liebhardt and Murdock, 1965; and unpublished data)

Nutrients Added in Fertilizer per Hectare, kg			Yield of Grain per Hectare, metric tons	Proportion of Stalks Lodged, %	Proportion of Plants with Brace Roots, %	Area of Soil Surface Enclosed by Brace Roots per Plant with Brace Roots, cm^2
N	P	K				
0	0	0	4.5	2	53	32
180	0	0	4.3	24	42	45
180	0	149	4.7	3	79	136
180	79	0	4.3	50	42	65
180	79	149	7.4	10	85	298

nutrients but also one aspect of root development. In this instance, the lodging was evidently induced by application of nitrogen, made worse by application of phosphorus, and alleviated by application of potassium. The behavior of the brace roots presumably had much to do with the favorable effect of potassium. Corn plants develop roots from the lower nodes of the stem, and those roots that emerge from nodes above the ground are commonly called brace roots. Application of potassium increased both the percentage of the plants bearing brace roots and the surface area of soil included within the span of the brace roots.

In connection with their work on lodging and brace-root development in Table 9.11, Liebhardt and Murdock (1965) investigated the structure of the stalks and brace roots and found considerable breakdown of parenchymatous tissue where nitrogen and phosphorus were applied without potassium but not where potassium was added. These authors found no evidence of obstruction of the conducting tissue by disease organisms or otherwise. Hoffer and Carr (1923) and Hoffer and Trost (1923) found, however, that iron and aluminum compounds accumulated in nodes of corn plants deficient in potassium. According to Hoffer and Carr (1923), large numbers of the conducting vessels in the affected areas were clogged, which hindered translocation. Porter (1927) found that the nodal tissue eventually disintegrated and became infected with various organisms. The accumulation of iron and aluminum in the nodes of potassium-deficient corn plants deserves mention in another connection as well. Potassium seems to have a regulating effect on mobility of iron in plants. Application of potassium has been found to increase the solubility of iron and to alleviate iron deficiency (Bolle-Jones, 1955).

Special significance is attributed to potassium in production of root crops, particularly potatoes, which are relatively sensitive to potassium deficiency. Table 9.12 shows that yield and starch content of tubers continued to increase with quantities of potassium in excess of those needed to produce the maximum yield of absorbing roots. In contrast to cereal crops, in which the grain contains relatively little potassium and in which most of the potassium taken up remains in the vegetative parts, a good crop of potato tubers contains about twice as much potassium as do the tops. The relatively high requirement of potassium for tuber development in relation to the requirement for development of the vegetative parts of the potato is probably one reason why high levels of soil potassium are needed to prevent development of potassium deficiency in potatoes.

Miscellaneous Quality Factors

Of all the plant nutrients, potassium is mentioned most often in relation to crop quality. The low starch content of potassium-deficient pota-

Table 9.12. Starch Content of Potato Tubers and Yield of Different Parts of Potato Plants Fertilized with Different Quantities of Potassium (Wilfarth and Wimmer, 1902)

Potassium Applied per Culture, g	Yield of Dry Matter per Culture, g			Proportion of Total Dry Matter, %			Starch Content of Tuber Dry Matter, %
	Tubers	Roots	Tops	Tubers	Roots	Tops	
0	8	1	41	16	2	82	53
0.23	29	3	51	35	4	61	62
0.70	70	2	49	58	2	40	67
1.56	109	3	53	66	2	32	68
3.90	125	2	49	71	1	28	66

toes is illustrated in Table 9.12. Potassium-deficient potatoes also tend to be hollow in the center and to become dark in color after cooking. Potassium-deficient apples do not color well. Potassium deficiency results in premature dropping and poor coloration of tomato fruit (see Fig. 9.40) and in relatively poor keeping quality of various fruit and vegetable crops. With tobacco, the aroma and fire-holding quality are impaired. With barley, the brewing quality is impaired. With peas, the seed coats are toughened. With cabbage, the heads are not solid, and the color and flavor of sauerkraut are poor. Potassium deficiency results in a decrease in acidity of tomatoes and citrus fruits and in a decrease in oil content of oil seeds such as soybeans and tung. Effects of potassium

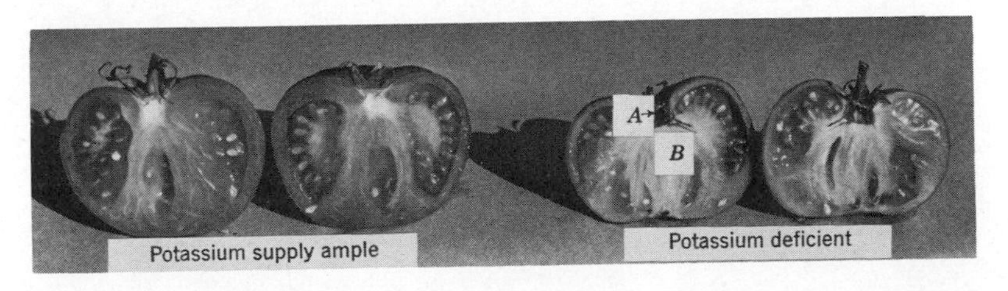

Figure 9.40. Quality effects of potassium supply on tomatoes. Note the decayed area *A* at the point of attachment of potassium-deficient fruit to the stem. Fruit from potassium-deficient plants drops prematurely, which reduces both size of fruit and development of the red color. The pigment in normal fruit is more concentrated in the outer portion than in the center. With the over-all decrease in pigment content of the potassium-deficient fruit, the core *B* is abnormally light in color. (Photograph courtesy of G. E. Wilcox and the American Potash Institute)

on carbohydrate and nitrogen metabolism may be suspected in most of these phenomena, if not all, and have been verified in a few instances. Further information on crop quality factors in relation to potassium supply may be found in publications by Eckstein et al. (1937), Lawton and Cook (1954), and Sprague (1964).

Time of Maturity

Frank (1931) recorded the observations shown in Fig. 9.41 for oats grown in Norway on soils differing in supply of potassium. The maturity evidently was delayed where potassium was deficient. Alten and Gottwick (1939) found that the length of the growing period of soybeans was 181 days under conditions of potassium deficiency and 157 days where potassium fertilizer was applied. Yields of soybeans were 562 and 967 kg/hectare, respectively. Lagatu and Maume (1932) noted that grapes developed less rapidly and matured less completely where potassium was deficient than where the supply was ample. Unpublished data of L. C. Dumenil at the Iowa Agricultural Experiment Station indicate that potassium deficiency delays the maturity of corn.

In contrast to the foregoing observations, which represent the type reported most frequently, deficiency of potassium sometimes causes plants to mature earlier; and this is true of cotton. The response of cotton is well shown in Fig. 9.42. At the first two pickings, the potassium-fertilized plants yielded less than the controls. At later pickings, the potassium-fertilized plants outyielded the controls; and they contin-

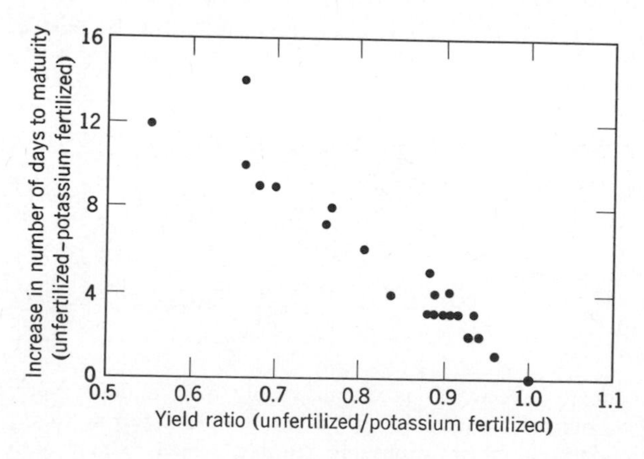

Figure 9.41. Increase in number of days required for oats to reach maturity without potassium fertilization versus ratio of yield of unfertilized to yield of potassium-fertilized oats grown on different soils. (Franck, 1931)

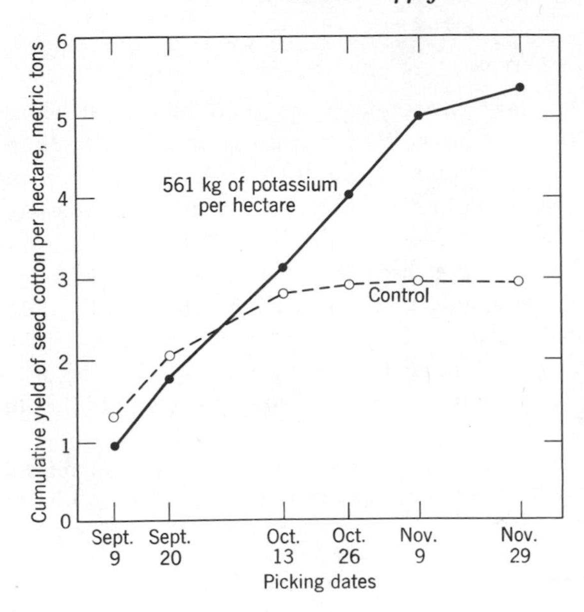

Figure 9.42. Cumulative yield of seed cotton at different dates on control and potassium-fertilized plots on a sandy loam soil in Alabama. (Bennett et al., 1965)

ued to produce cotton for a month after the controls had ceased production. Two related phenomena contribute to the "early maturity" of potassium-deficient cotton plants. First, the flowering stage of these plants is cut short by potassium deficiency, so that the process of boll maturation begins while the potassium-sufficient plants continue to produce new flowers and bolls. Second, the leaves of the potassium-deficient plants fall early, so that carbohydrate production and translocation come to an early halt. (In the experiment from which Fig. 9.42 was derived, the plants on the control plots had lost all their leaves by September 1.) Although the bolls become dry, they are not necessarily mature in the sense of being fully developed. Some bolls fail to open. The behavior of potassium-deficient cotton is thus partly a matter of early maturity and partly a matter of premature death. Figure 9.42, which shows the cumulative yields obtained at different dates, thus shows maturity in one sense but not in the sense of the degree of completion of development of individual fruits. In the latter sense, potassium fertilization may be observed to increase the proportion of mature bolls and fibers (see, for example, Nelson, 1949); however, it is not known to what extent this was the case in the experiment described in Fig. 9.42 (see Fig. 9.38 and the associated text for further discussion).

Resistance to Disease

One of the most pronounced recorded effects of potassium supply on plant disease is on wilt of cotton caused by the fungus *Fusarium oxysporum* f. *vasinfectum.* Table 9.13 gives the average results of field experiments on potassium fertilization of cotton in Mississippi in which counts of wilted plants were made. If, as in the instance shown in Table 9.13, the severity of the disease increases as the deficiency of potassium becomes more pronounced, the disease accentuates the loss of crop due to the original deficiency. Effects of potassium on resistance to plant diseases are characteristically associated with effects of potassium on yield. Most frequently the effect is qualitatively similar to that shown in Table 9.13, although plants well supplied with potassium are most severely damaged by certain virus diseases (Cheo et al., 1952; Pound and Weathers, 1953). The usual effect of potassium thus is opposite to that of nitrogen.

Table 9.13. Yield of Cotton and Percentage of Plants Manifesting Infection with Wilt (*Fusarium oxysporum* f. *vasinfectum*) with Different Applications of Potassium in Fertilizer in Mississippi (Miles, 1936)

Potassium Applied per Hectare, kg	Yield of Seed Cotton per Hectare, kg	Proportion of Plants Wilted, %
0	561	32
22	975	10
45	1005	8

The variability of effects of potassium in different combinations of disease and host may be illustrated by Gassner and Hassebrauk's (1934) findings on infection of different varieties of cereals by different rusts. In most of the combinations of cereal-variety with rust-race tested, the varieties were either resistant or susceptible to infection regardless of the potassium supply. Varieties that were moderately susceptible to a particular rust with high potassium usually became more susceptible with low potassium but sometimes were not appreciably affected. In a few instances, varieties that were resistant with high potassium became moderately to strongly susceptible at low potassium levels.

The association of disease incidence with potassium supply does not seem to result from a favorable effect of potassium on the disease organisms (which require potassium) because the severity of diseases is usually least where the concentration of potassium in the plant is relatively high, nor does it seem to result from a bacteriostatic or fungistatic effect of high concentrations of potassium in plants because the concentrations are not high enough to be toxic to microorganisms. Rather, the effect of potassium is apparently exerted indirectly through some influence of potassium on the susceptibility of the plant to infection, on the spread of the disease after infection has occurred, or on the escape of the plant from the disease.

An effect of potassium supply on resistance to infection is evident in results of experiments by Allington and Laird (1954) on infection of *Nicotiana glutinosa* by tobacco mosaic virus. They applied the inoculum by dipping the leaves in an aqueous suspension of the virus different lengths of time after leaves of low-potassium and high-potassium plants had been wounded artificially. Where the inoculation was performed immediately, the greater number of lesions generally developed on the low-potassium leaves, which showed initial symptoms of potassium deficiency. With increasing time between wounding and inoculation, fewer lesions developed on high-potassium leaves; but the number of lesions on low-potassium leaves remained about the same or increased. Inoculation an hour after wounding resulted in six times as many lesions per low-potassium leaf as per high-potassium leaf in four experiments. In research on *Phytophthora infestans,* a fungus responsible for a rot of potato foliage and tubers, Alten and Orth (1941) found that germination of the fungal spores was inhibited by the amino acid arginine in low concentrations; this amino acid occurred in the tubers in concentrations that increased with the potassium supply. In a comparison of four varieties of potatoes, the one with the lowest concentration of arginine (below the toxic range) was relatively susceptible to the fungus; but another variety having a low concentration of arginine was more resistant to infection than the remaining two varieties, which had a concentration of arginine high enough to produce some inhibition of spore germination. Indications from this work therefore are that a change in concentration of arginine under the influence of potassium supply is one factor in resistance of potatoes to infection by the rot organism but that more is involved.

An instance in which potassium affected the spread of a disease was reported by Riley (quoted by Last, 1956). He found that increasing applications of potassium for tobacco reduced the lesions due to *Alternaria longipes* from 21 to 12 mm in diameter.

As an instance of an effect of potassium on disease escape, Last (1956) described observations on attack of cotton by *Alternaria* sp. on potassium-deficient soils in South Africa. The organism is only weakly parasitic and attacks senescent plants. Fertilization of the plants with sufficient potassium delayed senescence and hence delayed development of the disease. Eventually plants with or without the supplemental potassium fertilizer were equally attacked.

Resistance to Low Temperatures

Plants deficient in potassium are more likely to be damaged by exposure to low temperatures than are plants that are well supplied with potassium. Figure 9.43 shows the difference in appearance of low- and high-potassium potato plants after an unseasonably late frost in June in Michigan. In this instance, the frost affected the productivity of the plants during an important part of the growing season; and the damage was much greater to plants low in potassium than to those well supplied with potassium. Damage from early autumn frosts affects plant productivity for less time but still may vary with the potassium supply. Boysen's (1933) observations on frost damage to potatoes are pertinent in this connection. Plants that received no potassium fertilizer showed definite

Figure 9.43. Comparative injury caused by a late June frost to foliage of potato plants differing in supply of potassium on a potassium-deficient soil in Michigan. Most of the leaves on the low-potassium plant were killed and had turned brown by the time the picture was taken. Visible damage to the high-potassium plant was confined almost entirely to the terminal leaves. A few of these had turned brown by the time the picture was taken. The high-potassium plants recovered from the frost damage much more rapidly and completely than did the low-potassium plants. (Shickluna et al., 1965)

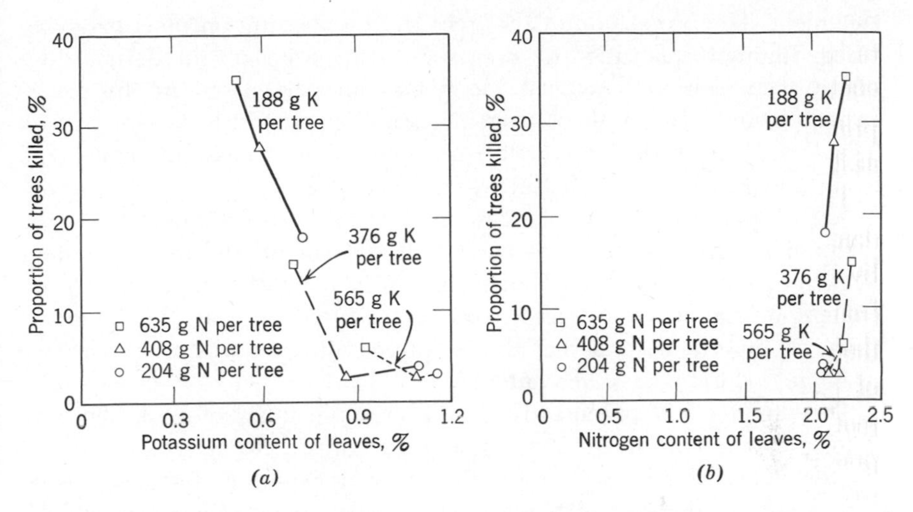

Figure 9.44. Percentage of initial stand of tung trees killed in consequence of an early autumn freeze in Mississippi versus average potassium and nitrogen percentages in the dry matter of the leaves over a 5-year period including the year in which the freeze occurred. The tung trees were in an experiment with different annual applications of nitrogen and potassium, average values for which are given in the figure. (Merrill et al., 1955; and unpublished data)

symptoms of potassium deficiency. Plants that received 33 and 66 kg of potassium per hectare showed no deficiency symptoms, but the plants with the greater application were somewhat the larger. The first frost on August 31 killed practically all the foliage of the unfertilized plants, about one-third of that on plants with 33 kg of potassium, and none on plants with 66 kg of potassium. A somewhat more severe frost on September 4 killed over half the foliage on the plots receiving 33 kg of potassium and about one-third the foliage on the plots receiving 66 kg of potassium. On September 13, a more severe frost killed all the remaining foliage. The heaviest application of potassium fertilizer thus lengthened the growing period 13 days. Final yields of tubers were 15, 22, and 27 metric tons/hectare on plots treated with 0, 33, and 66 kg of potassium per hectare.

Potassium exerts an effect also on low-temperature damage to plants that characteristically withstand freezing temperatures. Most higher plants that are winter-hardy do not withstand freezing temperatures at all times but only when they are properly conditioned. An ample supply of potassium aids in the conditioning process in the autumn. Figure 9.44 gives results from an experiment in Mississippi in which losses of tung trees from an unduly early freeze varied with the fertilizer treatment. The proportion of the trees lost by freezing on plots receiving three different quantities of potassium with each of three quantities of nitrogen is plotted against the potassium percentage in the leaves in part *a* and against the nitrogen percentage in the leaves in part *b*. The loss of trees decreased with an increase in potassium percentage in the leaves, whether the latter was associated with an increase in application of potassium or a decrease in application of nitrogen; but the range of nitrogen percentages was small and not related in a consistent fashion to the loss of trees. Hence the results suggest that the potassium content of the leaves or of the plants as a whole was the factor of primary importance and that the supply of nitrogen was of importance as it affected the potassium content.

The direct effect of potassium supply on the resistance of plants to damage from exposure to low temperatures must be small, as illustrated by the values in Table 9.14 for the freezing point of fluids expressed from roots of alfalfa grown with different levels of soil potassium. Although the lower freezing point was associated with the greater supply of potassium and with the greater concentration of potassium in the root tissue, the magnitude of the effects could hardly have much direct practical significance in the area in northern Wisconsin where the plants were grown. Winter temperatures there fall far below the freezing points observed. And, of course, the depression of the freezing point below zero was not attributable to potassium alone but rather to all the solutes. The effect of potassium supply on cold resistance is considered to be exerted for the most part indirectly through the physiological condition of plants.

Table 9.14. Winter-Killing, Potassium Content, and Carbohydrate Content of Alfalfa Roots, and Freezing Point of the Root Sap, as Determined in Late Autumn on Plants on a Silt Loam Soil in Northern Wisconsin (Wang et al., 1953)

Approximate Content of Exchangeable Potassium per Gram of Soil, μg	Winter-Killing of Alfalfa,[1] %	Potassium Content of Dry Roots, %	Carbohydrate Content of Dry Roots, %		Freezing Point of Root Sap, °C
			Sugars	Starch	
50	50	0.44	9.5	8.1	−1.06
150	20	0.63	12.6	9.4	−1.24

[1] Estimated on second-year hay in May over a period of 8 years.

According to Levitt's (1956) theory of the mechanism of low-temperature resistance, growing plants are not winter-hardy. Winter-hardiness involves accumulation of soluble carbohydrates in high concentrations, which is followed by binding of water by the cell proteins. As shown in Table 9.14, the roots of alfalfa grown on soil well supplied with potassium contained more sugar and starch than those of alfalfa grown on soil with relatively low potassium. The beneficial effect of a relatively high potassium supply on resistance to injury at low temperatures thus may have resulted from the effect of potassium supply on carbohydrate production. In terms of Levitt's theory, the damage to the tung trees in the early autumn freeze, mentioned previously, may be explained on the basis that the trees had not developed the required physiologically resistant condition. High-potassium trees were more nearly in a resistant condition than low-potassium trees, presumably because the high-potassium trees contained the more soluble carbohydrate.

Plant Competition

Figure 9.45 shows the yield of different components of a uniformly seeded, mixed meadow that was fertilized with different quantities of potassium. The clover component of the herbage was evidently far more

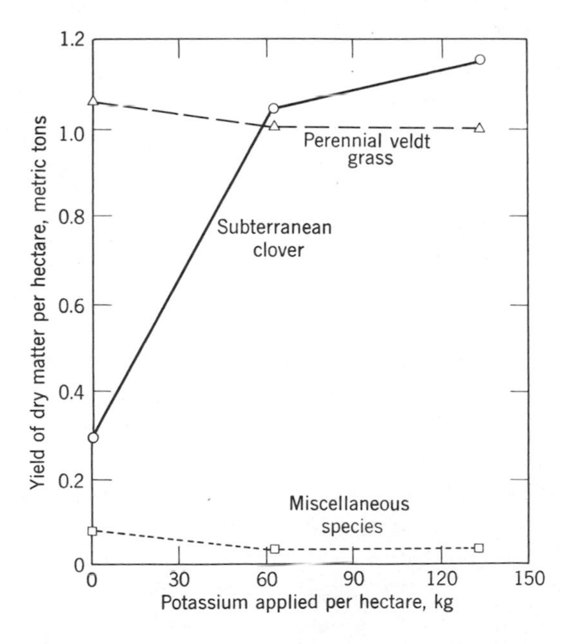

Figure 9.45. Yield of botanical components of a mixed meadow with different applications of potassium fertilizer to a sandy soil in Australia. (Rossiter, 1947)

productive on plots receiving much potassium than on control plots, but the grass made about the same growth at all levels of potassium. This relative behavior seems to be generally characteristic of the legumes and grasses grown together in meadow mixtures. From the standpoint of grass-legume competition, therefore, one may say that the competitive position of legumes is usually better where the supply of potassium is high than where it is low.

In attempting to analyze a situation such as the one illustrated in Fig. 9.45, one wishes to know which growth factors are limiting the yields of each species. In the experiment in question, the yield of grass was probably limited by the supply of nitrogen and not by potassium. The yield of clover, on the other hand, was evidently limited by potassium. If the supply of nitrogen had been ample for the grass, the yield of grass probably would have been greater; and response to potassium fertilization then might well have occurred. Greater growth and response of grass to potassium would in turn repress the clover because the grass would then compete more strongly for growth factors other than nitrogen.

As regards competition between the grass and legume for potassium, the data in Fig. 9.45 provide no information. All that can be said is that the grass seemed to have enough potassium but the clover did not. Perhaps the best information on competition for potassium may be derived from the ratio of the potassium percentage in a species in mixed culture to the potassium percentage in the same species in pure culture. If the ratio is unity, the species is competing equally well for potassium in mixed culture as in pure culture. If the ratio is greater than unity, the species is competing better, and, if less than unity, the species is competing poorer in mixed culture than in pure culture. As an example of use of this sort of evaluation, Warmke et al. (1952) grew Merker grass and kudzu in pure culture and in mixed culture in the field in Puerto Rico and analyzed both species for potassium. For Merker grass, the ratio of potassium percentages was $1.53/1.66 = 0.92$; for kudzu, the ratio was $0.72/1.16 = 0.62$. From these results, one may infer that neither species competed as well for potassium in associated growth as in pure culture but that kudzu was at the greater disadvantage in mixed culture.

The only theory that has attained prominence as an explanation for differences in competitive ability of different species for soil potassium is that the ratio of potassium to divalent cations held in exchangeable form on the root is greater with species having roots with a low cation-exchange capacity (e.g., grasses) than in species having roots with a high cation-exchange capacity (e.g., legumes). In view of findings relative to ion uptake that were discussed in Chapter 4, the applicability of this theory to potassium uptake seems questionable. Accordingly, the micro aspect of competition will be ignored, and attention will be directed instead to the macro aspect of the source of the potassium. Three different circumstances may be conceived in theory: withdrawal of potassium (1) from a common source by all species, (2) in part from a common source and in part from sources that may differ among species, and (3) from sources that differ among species. These will be considered in turn.

The first situation, in which all species withdraw potassium from a common source, may be exemplified by a tank of well-stirred nutrient solution in which all species are grown. Uptake of potassium by one species is promptly reflected as a reduction in the quantity and concentration of potassium in the solution from which all species are absorbing potassium. Consequently, for quantities and concentrations of potassium that are not excessive, the total uptake of potassium by a given species and the percentage content of potassium in it will decrease with an increase in total uptake of potassium by the competing species. For plants growing in soil, the first situation is approached if the species are grown together in a small container of soil because then the roots will be intermingled and confined to the same total volume. Figure 9.46 gives some of the results of an experiment of this kind in which ladino clover was grown in pure culture and in mixed culture with each of three grasses. Analyses of the clover showed that both the percentage content and total uptake of potassium decreased with an increase in potassium uptake by the associated grass.

The second situation, in which competing species withdraw potassium in part from a common source and in part from sources that may differ among species, is probably the prevailing one where agricultural plants of different species are grown together in meadows and pastures in the field. The roots of all species intermingle to some extent; but some, characteristically those of grasses, are concentrated in the upper part of the soil; and others may extend more deeply. The deeper extension of roots of certain species could be significant if subsoil potassium were high enough but probably is not of much importance in influencing competition for potassium in soils where subsoil potassium is relatively low and potassium in the surface is relatively high. The principal competition under the latter circumstances may be expected to take place in the upper part of the soil where the intermingling of roots occurs. Thus, in the case of associated growth of kudzu and Merker grass mentioned previously, kudzu would be expected to be the more effective competitor on the basis of depth of rooting; nevertheless, Merker grass was found

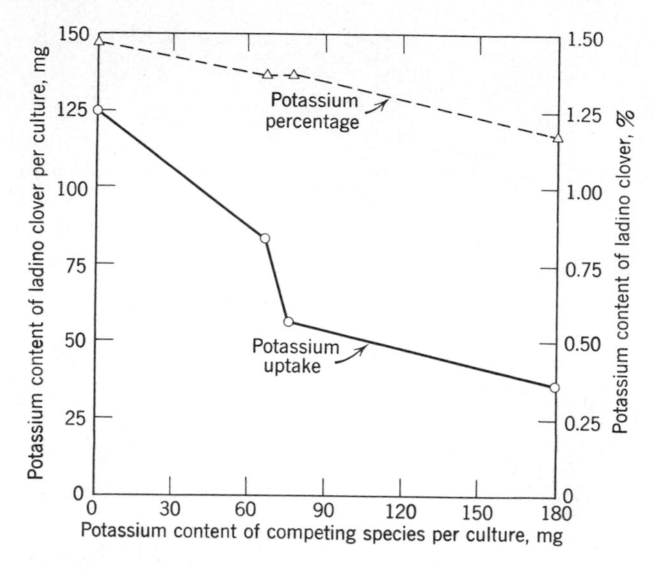

Figure 9.46. Potassium percentage in and total uptake of potassium by ladino clover grown alone and in competition with three species of grass in a fine sandy loam soil in greenhouse cultures. From left to right the competing species are none, smooth bromegrass, bluegrass, and bentgrass. Values are totals for three cuttings. (Gray et al., 1953)

to be the better competitor for potassium in field experiments by both Warmke et al. (1952) and Samuels and Landrau (1953).

There are perhaps no practical situations in which different species grown in association do not compete to some extent for the same source of potassium. Associations of herbaceous and tree crops, however, sometimes approach the theoretical concept of completely distinct sources. Tree crops normally have deeper root systems than the grasses and other herbaceous plants often grown as cover crops; and the presence of vigorous, shallow-rooted, cover crops would be expected to reduce the incidence of tree roots near the surface and hence to accentuate the segregation of root zones.

Observations made by Walsh and Clarke (1943) in a study of soil properties in relation to the condition of apple orchards in Ireland are pertinent. These investigators found that the presence or absence of symptoms of potassium deficiency on apple foliage was more closely associated with the content of exchangeable potassium in the subsoil than in the surface soil. Deficiency symptoms were never found if the subsoil was well supplied with potassium, but they were sometimes found if the surface soil was well supplied with potassium. In fact, symptoms of potassium deficiency on the apple leaves occasionally persisted despite application of potassium fertilizer to the surface of the soil. Their interpretation of these findings, verified by the relatively deep penetration of apple roots, was that the apple trees were absorbing potassium principally from the subsoil.

Cullinan and Batjer (1943) described a seemingly analogous situation in an experiment in a peach orchard in Maryland. In this experiment, various cover crops were grown between the trees. The cover crops were fertilized with potassium and made excellent growth, but the peach leaves remained low in potassium content and manifested symptoms of potassium deficiency. Application of potassium fertilizer in a ring beneath the trees and in quantities much higher than those used on the cover crops increased the potassium content of the leaves and produced normal growth of leaves and twigs.

Literature Cited

Acker, W. (1932) Untersuchungen über die Wirkung der Kalidüngung auf den Bau und das mechanische Verhalten des Gerstenhalmes. *Die Ernährung der Pflanze* **28**:401–403.

Agarwal, R. R. (1960) Potassium fixation in soils. *Soils and Fertilizers* **23**:375–378.

Allen, M. B., and D. I. Arnon. (1955) Studies on nitrogen-fixing blue-green algae. II. The sodium requirement of *Anabaena cylindrica. Physiol. Plantarum* **8**:653–660.

Allington, W. B., and E. F. Laird, Jr. (1954) The infection of *Nicotiana glutinosa* with tobacco mosaic virus as affected by potassium nutrition. *Phytopathology* **44**:297–299.

Allison, F. E., E. M. Roller, and J. E. Adams. (1959) Soil fertility studies in lysimeters containing Lakeland sand. *U.S. Dept. Agr. Tech. Bul. 1199.*

Alten, F., and G. Goeze. (1936) Der Einfluss der Düngung auf den Ertrag und die Güte der Flachsfaser. *Die Ernährung der Pflanze* **32**:1–14.

Alten, F., G. Goeze, and H. Fischer. (1938) Mineralsalzernährung, Kohlensäureassimilation und Stickstoffhaushalt junger Weizenpflanzen. *Die Ernährung der Pflanze* **34**:73–75.

Alten, F., and R. Gottwick. (1939) Unterschungen über den Einfluss der Düngung auf das Wachstum und die Entwicklung der Sojabohne. *Die Ernährung der Pflanze* **35**:277–284.

Alten, F., and H. Orth. (1941) Untersuchungen über den Aminosäurengehalt und die Anfälligkeit der Kartoffel gegen die Kraut- und Knollenfäule (*Phytophthora infestans* de By.). *Phytopath. Zeitschr.* **13**:243–271.

Anderson, M. S., M. G. Keyes, and G. W. Cromer. (1942) Soluble material of soils in relation to their classification and general fertility. *U.S. Dept. Agr. Tech. Bul.* 813.

Anderson, P. J., and T. R. Swanback. (1929) Report of the Tobacco Substation at Windsor for 1928. *Connecticut* (New Haven) *Agr. Exp. Sta. Bul.* 299.

Anderson, P. J., T. R. Swanback, O. E. Street, and others. (1930) Tobacco Substation at Windsor report for 1929. *Connecticut* (New Haven) *Agr. Exp. Sta. Bul. 311.*

Appling, E. D., and J. Giddens. (1954) Differences in sodium and potassium content of various parts of the cotton plant at four stages of growth. *Soil Sci.* **78**:199–203.

Arland, A., and R. Zwicker. (1959) Anwelktranspiration und Hydratur verschieden mit Kalium ernährter Haferpflanzen unter besonderer Berücksichtigung des Jahresrhythmus. *Zeitschr. Acker- u. Pflanzenbau* **108**:449–472.

Arnold, P. W. (1960) Nature and mode of weathering of soil-potassium reserves. *Jour. Sci. Food Agr.* **11**:285–292.

Arnold, P. W. (1960a) Potassium-supplying power of some British soils. *Nature* **187**:436–437.

Ayres, A. S., M. Takahashi, and Y. Kanehiro. (1947) Conversion of nonexchangeable potassium to exchangeable forms in a Hawaiian soil. *Soil Sci. Soc. Amer. Proc.* (1946) **11**:175–181.

Barber, S. A. (1961) The influence of moisture and temperature on phosphorus and potassium availability. *Trans. 7th Internat. Congr. Soil Sci.* (Madison, 1960) **3**:435–442.

Barber, S. A. (1962) A diffusion and mass-flow concept of soil nutrient availability. *Soil Sci.* **93**:39–49.

Barber, S. A., and R. P. Humbert. (1963) Advances in knowledge of potassium relationships in the soil and plant. *In* M. H. McVickar, G. L. Bridger, and L. B. Nelson (Eds.) *Fertilizer Technology and Usage,* pp. 231–268. Soil Science Society of America. Madison, Wis.

Barber, S. A., J. M. Walker, and E. H. Vasey. (1962) Principles of ion movement through the soil to the plant root. *Trans. Joint Meeting Comm. IV & V, Internat. Soc. Soil Sci., New Zealand,* **1962**:121–124.

Barshad, I. (1954) Cation exchange in micaceous minerals: I. Replaceability of the interlayer cations of vermiculite with ammonium and potassium ions. *Soil Sci.* **77**:463–472.

Baumeister, W. (1939) Der Einfluss mineralischer Düngung auf den Ertrag und die Zusammensetzung des Kornes der Sommerweizenpflanze. *Bodenk. u. Pflanzenernähr.* **12**:175–222.

Baumeister, W. (1958) Hauptnährstoffe. *In* W. Ruhland (Ed). *Handbuch der Pflanzenphysiologie.* IV. *Die mineralische Ernährung der Pflanze,* pp. 482–557. Springer Verlag, Berlin.

Bennett, O. L., R. D. Rouse, D. A. Ashley, and B. D. Doss. (1965) Yield, fiber quality and potassium content of irrigated cotton plants as affected by rates of potassium. *Agron. Jour.* **57**:296–299.

Biebl, R. (1958) Der Einfluss der Mineralstoffe auf die Transpiration. *In* W. Ruhland (Ed.) *Handbuch der Pflanzenphysiologie.* IV. *Die mineralische Ernährung der Pflanze,* pp. 382–426. Springer Verlag, Berlin.

Black, S. (1951) Yeast aldehyde dehydrogenase. *Arch. Biochem. Biophysics* **34**:86–97.

Bolle-Jones, E. W. (1955) The interrelationships of iron and potassium in the potato plant. *Plant and Soil* **6**:129–173.

Boswell, F. C., and W. L. Parks. (1957) The effect of soil potassium levels on yield, lodging, and mineral composition of corn. *Soil Sci. Soc. Amer. Proc.* **21**:301–305.

Bowling, J. D., and D. E. Brown. (1947) Role of potash in growth and nutrition of Maryland tobacco. *U.S. Dept. Agr. Tech. Bul.* 933.

Boyd, D. A. (1956) The effect of potassium on crop yield. *Potassium Symposium* **1956**:143–154.

Boysen, H. (1933) Kalidüngung und Nachtfrostgefahr. *Die Ernährung der Pflanze* **29**:270.

Bray, R. H., and E. E. DeTurk. (1939) The release of potassium from nonreplaceable forms in Illinois soils. *Soil Sci. Soc. Amer. Proc.* (1938) **3**:101–106.

Breemhaar, R. (1956) Toetsing van grond- en bladonderzoek bij aardappelen op rivierkleigrond. *Landbouwk. Tijdschr.* **68**:1004–1011.

Brownell, P. F., and J. G. Wood. (1957) Sodium as an essential micronutrient element for *Atriplex vesicaria,* Heward. *Nature* **179**:635–636.

Brummer, V. (1959) Lannoituksen vaikutuksesta sokerijuurikkaan satoon. *Acta Agralia Fennica* **94**:201–239.

Bruns, W. (1935) Untersuchungen über Nährstoffaufnahme und Wasserhaushalt der Ackerbohne. *Jour. f. Landw.* **83**:285–325.

Burghardt, H. (1962) Über die Bedeutung des Chlors für Pflanzenernährung unter besonderer Berücksichtigung des Chlorid/Sulfat-Problems. *Angew. Bot.* **36**:203–257.

Bussler, W. (1964) *Comparative Examinations of Plants Suffering from Potash Deficiency.* (Translated from the German edition by C. L. Whittles) Verlag Chemie, G.m.b.H. Weinheim/Bergstr.

Chaminade, R. (1936) La rétrogradation du potassium dans les sols. *Ann. Agron.* (N.S.) **6**:818–830.

Cheo, P.-C., G. S. Pound, and L. G. Weathers. (1952) The relation of host nutrition to the concentration of cucumber virus 1 in spinach. *Phytopathology* **42**:377–381.

Coleman, R. G., and F. J. Richards. (1956) Physiological studies in plant nutrition. XVIII. Some aspects of nitrogen metabolism in barley and other plants in relation to potassium deficiency. *Ann. Bot. N.S.* **20**:393–409.

Collander, R. (1941) Selective absorption of cations by higher plants. *Plant Physiol.* **16**:691–720.

Cook, M. G., and T. B. Hutcheson, Jr. (1960) Soil potassium reactions as related to clay mineralogy of selected Kentucky soils. *Soil Sci. Soc. Amer. Proc.* **24**:252–256.

Cooke, G. W. (1954) Recent advances in fertilizer placement. II. Fertilizer placement in England. *Jour. Sci. Food Agr.* **5**:429–440.

Correns, C. W., and W. von Engelhardt. (1938) Neue Untersuchungen über die Verwitterung des Kalifeldspates. *Chemie der Erde* **12**:1–22.

Cullinan, F. P., and L. P. Batjer. (1943) Nitrogen, phosphorus, and potassium interrelationships in young peach and apple trees. *Soil Sci.* **55**:49–60.

Danielson, R. E., and M. B. Russell. (1957) Ion absorption by corn roots as influenced by moisture and aeration. *Soil Sci. Soc. Amer. Proc.* **21**:3–6.

DeMumbrum, L. E., and C. D. Hoover. (1958) Potassium release and fixation related to illite and vermiculite as single minerals and in mixtures. *Soil Sci. Soc. Amer. Proc.* **22**:222–225.

Denison, I. A., W. H. Fry, and P. L. Gile. (1929) Alteration of muscovite and biotite in the soil. *U.S. Dept. Agr. Tech. Bul. 128.*

DeTurk, E. E., L. K. Wood, and R. H. Bray. (1943) Potash fixation in corn belt soils. *Soil Sci.* **55**:1–12.

Eckstein, O., A. Bruno, and J. W. Turrentine. (1937) *Potash Deficiency Symptoms.* Verlagsgesellschaft für Ackerbau M.B.H., Berlin.

Eddy, A. A., and C. Hinshelwood. (1951) Alkali-metal ions in the metabolism of *Bact. lactis aerogenes.* II. Connexion with viability, growth rate and enzyme activity. *Proc. Royal Soc. London* **138B**:228–237.

Eijk, M. van. (1939) Analyse der Wirkung des NaCl auf die Entwicklung, Sukkulenz und Transpiration bei *Salicornia herbacea,* sowie Untersuchungen über den Einfluss der Salzaufnahme auf die Wurzelatmung bei *Aster tripolium. Recueil trav. bot. néerl.* **36**:559–657.

Epstein, E., D. W. Rains, and O. E. Elzam. (1963) Resolution of dual mechanisms of potassium absorption by barley roots. *Proc. Natl. Acad. Sci.* **49**:684–692.

Failyer, G. H., J. G. Smith, and H. R. Wade. (1908) The mineral composition of soil particles. *U.S. Dept. Agr., Bur Soils Bul. 54.*

Food and Agriculture Organization of the United Nations. (1965) *Fertilizers.* An annual review of world production, consumption and trade. 1964. Food and Agriculture Organization of the United Nations, Rome.

Franck, Olle. (1931) Gödsling, mognadstid och kärnkvalitet. *Nordisk Jordbrugsforskning* **13**:282–290.

Fried, M., and R. E. Shapiro. (1961) Soil-plant relationships in ion uptake. *Ann. Rev. Plant Physiol.* **12**:91–112.

Gardner, E. H. (1960) Potassium release from several western Oregon soils and its relationship to crop growth and soil mineralogy. *Diss. Abstr.* **21**:408–409.

Garman, W. L. (1957) Potassium release characteristics of several soils from Ohio and New York. *Soil Sci. Soc. Amer. Proc.* **21**:52–58.

Gassner, G., and K. Hassebrauk. (1934) Der Einfluss der Mineralsalzernährung auf das Anfälligkeitsverhalten der zur Rassenbestimmung von Getreiderosten dienenden Standardsortimente. *Phytopath. Zeitschr.* **7**:63–72.

Gray, B., M. Drake, and W. G. Colby. (1953) Potassium competition in grass-legume associations as a function of root cation exchange capacity. *Soil Sci. Soc. Amer. Proc.* **17**:235–239.

Greenway, H., and M. G. Pitman. (1965) Potassium retranslocation in seedlings of *Hordeum vulgare. Australian Jour. Biol. Sci.* **18**:235–247.

Grim, R. E., R. S. Dietz, and W. F. Bradley. (1949) Clay mineral composition of some sediments from the Pacific Ocean off the California coast and the Gulf of California. *Bul. Geol. Soc. Amer.* **60**:1785–1808.

Grimes, D. W. (1966) An evaluation of the availability of potassium in crop residues. Ph.D. Thesis, Iowa State Univ., Ames.

Hanway, J. J., S. A. Barber, R. H. Bray, A. C. Caldwell, L. E. Engelbert, R. L. Fox, M. Fried, D. Hovland, J. W. Ketcheson, W. M. Laughlin, K. Lawton, R. C. Lipps, R. A. Olson, J. T. Pesek, K. Pretty, F. W. Smith, and E. M. Stickney. (1961) North central regional potassium studies. I. Field studies with alfalfa. *North Central Reg. Publ. 124 (Iowa Agr. Home Econ. Exp. Sta. Res. Bul. 494).*

Hanway, J. J., S. A. Barber, R. H. Bray, A. C. Caldwell, M. Fried, L. T. Kurtz, K. Lawton, J. T. Pesek, K. Pretty, M. Reed, and F. W. Smith. (1962) North central regional potassium studies. III. Field studies with corn. *North Central Reg. Publ. 135 (Iowa Agr. Home Econ. Exp. Sta. Res. Bul. 503).*

Harley, G. T. (1953) Production and processing of potassium materials. *Adv. Agron.* **3**:287–322.

Harmer, P. M., and E. J. Benne. (1941) Effects of applying common salt to a muck soil on the yield, composition, and quality of certain vegetable crops and on the composition of the soil producing them. *Jour. Amer. Soc. Agron.* **33**:952–979.

Harmer, P. M., and E. J. Benne. (1945) Sodium as a crop nutrient. *Soil Sci.* **60**:137–148.

Heimann, H. (1958) Irrigation with saline water and the ionic environment. *Potassium Symposium* **1958**:173–220.

Hellriegel, H., H. Wilfarth, H. Römer, and G. Wimmer. (1898) Vegetationsversuche über den Kalibedarf einiger Pflanzen. *Arb. deut. Landw. Gesell.*, Heft 34.

Hemingway, R. G. (1963) Soil and herbage potassium levels in relation to yield. *Jour. Sci. Food Agr.* **14**:188–195.

Hewitt, E. J. (1963) The essential nutrient elements: requirements and interactions in plants. *In* F. C. Steward (Ed.) *Plant Physiology, a Treatise.* Vol. III. *Inorganic Nutrition of Plants.* pp. 137–360. Academic Press, New York.

Hoagland, D. R., and T. C. Broyer. (1936) General nature of the process of salt accumulation by roots with description of experimental methods. *Plant Physiol.* **11**:471–507.

Hoffer, G. N., and R. H. Carr. (1923) Accumulation of aluminum and iron compounds in corn plants and its probable relation to rootrots. *Jour. Agr. Res.* **22**:801–823.

Hoffer, G. N., and J. F. Trost. (1923) The accumulation of iron and aluminium compounds in corn plants and its probable relation to root rots. II. *Jour. Amer. Soc. Agron.* **15**:323–331.

Hood, J. T., N. C. Brady, and D. J. Lathwell. (1956) The relationship of water soluble and exchangeable potassium to yield and potassium uptake by ladino clover. *Soil Sci. Soc. Amer. Proc.* **20**:228–231.

Hsiao, T. C.–T. (1964) Some aspects of ribonucleic acid and protein in maize. I. Characteristics of isolated ribosomes. II. Effects of potassium nutrition. *Diss. Abstr.* **25**:804–805.

Humbert, R. P. (1958) Potash fertilization in the Hawaiian sugar industry. *Potassium Symposium* **1958**:319–344.

Hunter, A. H., and P. F. Pratt. (1957) Extraction of potassium from soils by sulfuric acid. *Soil Sci. Soc. Amer. Proc.* **21**:595–598.

Ignatieff, V., and H. J. Page (Eds.) (1958) Efficient use of fertilizers. Revised ed. *FAO Agr. Studies 43.*

Itallie, T. B. van. (1938) Cation equilibria in plants in relation to the soil. *Soil Sci.* **46**:175–186.

Kachmar, J. F., and P. D. Boyer. (1952) Kinetic analysis of enzyme reactions. II. The potassium activation and calcium inhibition of pyruvic phosphoferase. *Jour. Biol. Chem.* **200**:669–682.

Kaila, A. (1965) Fixation of potassium in Finnish soils. *Maat. Aikak.* **37**:116–126.

Kapusta, E. C., and N. E. Wendt. (1963) Advances in fertilizer potash production. *In* M. H. McVickar, G. L. Bridger, and L. B. Nelson (Eds.) *Fertilizer Technology and Usage*, pp. 189–230. Soil Science Society of America, Madison, Wis.

Kellner, K. (1955) Die Adaptation von *Ankistrodesmus braunii* an Rubidium und Kupfer. *Biochem. Zentbl.* **74**:622–691.

Kelly, J. B. (1941) Composition of the displaced soil solution and exchangeable potassium content of high-lime soils in relation to potassium deficiency in corn. M.S. Thesis, Iowa State University, Ames.

Kidson, E. B. (1943) An ash skeleton method for the diagnosis of magnesium and potassium deficiencies in apple leaves and for the determination of their distribution in the leaf. *New Zealand Jour. Sci. Technol.* **24B**:140–145.

Kono, M., and J. Takahashi. (1962) Study on the effect of potassium on the breaking strength of paddy stem. *Soil Sci. Plant Nutr.* **8**, No. 2:39–40.

Kramer, S. (1963) Untersuchungen über die Wirkung chloridhaltiger Kalidünger auf Ertrag, Ascorbinsäuregehalt der Früchte und vegetative Entwicklung der Erdbeere. *Arch. für Gartenbau* **11**:175–190.

Lagatu, H., and L. Maume. (1932) Un cas d'absolue nécessité d'engrais potassique. *Le Progrès Agricole et Viticole* **97**:576–581.

Lardy, H. A. (1951) The influence of inorganic ions on phosphorylation reactions. *In* W. D. McElroy and B. Glass (Eds.) *Phosphorus Metabolism. A Symposium on the Role of Phosphorus in the Metabolism of Plants and Animals,* pp. 477–499. The Johns Hopkins Press, Baltimore, Md.

Last, F. T. (1956) The effect of potassium on parasitic plant diseases. *Potassium Symposium* **1956**:179–188.

Latzko, E. (1959) Die Funktion des Kaliums im Stoffwechsel der energiereichen Phosphate pflanzlicher und tierischer Organismen. *Agrochimica* **3**:148–164.

Lawton, K., and R. L. Cook. (1954) Potassium in plant nutrition. *Adv. Agron.* **6**:253–303.

Lehr, J. J. (1953) Sodium as a plant nutrient. *Jour. Sci. Food Agr.* **4**:460–471.

Levitt, J. (1956) *The Hardiness of Plants. Agronomy* **6**.

Levy, J. F. (1964) Exchangeable soil potassium, potassium uptake by plants, soil texture. (Translated title.) *Potasse* **38**:9–14. (*Soils and Fertilizers* **27**, Abstr. 1315. 1964).

Liebhardt, W. C., and J. T. Murdock. (1965) Effect of potassium on morphology and lodging of corn. *Agron. Jour.* **57**:325–328.

Lipman, J. G., and A. B. Conybeare. (1936) Preliminary note on the inventory and balance sheet of plant nutrients in the United States. *New Jersey Agr. Exp. Sta. Bul. 607.*

Lubin, M. (1964) Intracellular potassium and control of protein synthesis. *Federation Proc.* **23**:994–1001.

Luebs, R. E., G. Stanford, and A. D. Scott. (1956) Relation of available potassium to soil moisture. *Soil Sci. Soc. Amer. Proc.* **20**:45–50.

Luttkus, K., and R. Bötticher. (1939) Über die Ausscheidung von Aschenstoffen durch die Wurzeln. I. *Planta. Arch. f. wiss. Bot.* **29**:325–340.

MacDonald, R. A. (1960) Potash: occurrences, processes, production. *In* V. Sauchelli (Ed.) *Chemistry and Technology of Fertilizers,* pp. 367–402. Reinhold Publ. Corp., New York.

Mackenzie, R. C. (1955) Potassium in clay minerals. *Potassium Symposium* **1955**:123–143.

MacLean, A. J. (1961) Water-soluble K, per cent K-saturation, and pK-$\frac{1}{2}$p(Ca + Mg) as indices of management effects on K status of soils. *Trans. 7th Internat. Congr. Soil Sci.* (1960) **3**:86–91.

MacLean, A. J. (1961a) Potassium-supplying power of some Canadian soils. *Canadian Jour. Soil Sci.* **41**:196–206.

MacLean, A. J., R. L. Halstead, A. R. Mack, and J. J. Jasmin. (1964) Comparison of procedures for estimating exchange properties and availability of phosphorus and potassium in some eastern Canadian organic soils. *Canadian Jour. Soil Sci.* **44**:66–75.

Mann, C. E. T. (1924) The physiology of the nutrition of fruit trees. I. Some effects of calcium and potassium starvation. *Bristol Univ. Agr. Hort. Res. Sta. Ann. Rept.* **1924**:30–45.

Marel, H. W. van der, and J. T. N. Venekamp. (1955) Onderzoek naar het verschijnsel der kalifixatie in de Nederlandse gronden. *Versl. Landbouwk. Onderzoek.* No. 61.8.

McEwen, H. B., and B. C. Matthews. (1958) Rate of release of non-exchangeable potassium by Ontario soils in relation to natural soil characteristics and management practices. *Canadian Jour. Soil Sci.* **38**:36–43.

Mederski, H. J., and J. Stackhouse. (1961) Relation of plant growth and ion accumulation to soil moisture level stabilized with a divided root technique. *Trans. 7th Internat. Congr. Soil Sci.* (Madison, 1960) **3**:467–474.

Mengel, K. (1964) Influx und Efflux bei der Kaliumaufnahme junger Gerstenwurzeln. *Zeitschr. Pflanzenernähr., Düng., Bodenk.* **106**:193–206.

Merrill, S., Jr., G. F. Potter, and R. T. Brown. (1955) Effects of nitrogen, phosphorus, and potassium on mature tung trees growing on Red Bay fine sandy loam. *Proc. Amer. Soc. Hort. Sci.* **65**:41–48.

Merwin, H. D., and M. Peech. (1951) Exchangeability of soil potassium in the sand, silt, and clay fractions as influenced by the nature of the complementary exchangeable cation. *Soil Sci. Soc. Amer. Proc.* (1950) **15**:125–128.

Middelburg, H. A. (1955) Potassium in tropical soils: Indonesian archipelago. *Potassium Symposium* **1955**:221–257.

Miles, L. E. (1936) Effect of potash fertilizers on cotton wilt. *Mississippi Agr. Exp. Sta. Tech. Bul. 23.*

Moore, J. H., and W. H. Rankin. (1937) Influence of "rust" on quality and yield of cotton and the relation of potash applications to control. *North Carolina Agr. Exp. Sta. Bul. 308.*

Mortland, M. M. (1961) The dynamic character of potassium release and fixation. *Soil Sci.* **91**:11–13.

Mortland, M. M., K. Lawton, and G. Uehara. (1956) Alteration of biotite to vermiculite by plant growth. *Soil Sci.* **82**:477–481.

Moss, P. (1964) A potassium relationship between soil solution and cacao bark. *Nature* **201**:729–730.

Munson, R. D., and W. L. Nelson. (1963) Movement of applied potassium in soils. *Jour. Agr. Food Chem.* **11**:193–201.

Nash, V. E., and C. E. Marshall. (1957) Cationic reactions of feldspar surfaces. *Soil Sci. Soc. Amer. Proc.* **21**:149–153.

Nelson, W. L. (1949) The effect of nitrogen, phosphorus, and potash on certain lint and seed properties of cotton. *Agron. Jour.* **41**:289–293.

Nightingale, G. T., L. G. Schermerhorn, and W. R. Robbins. (1930) Some effects of potassium deficiency on the histological structure and nitrogenous and carbohydrate constituents of plants. *New Jersey Agr. Exp. Sta. Bul. 499.*

Olsen, C. (1948) The mineral, nitrogen and sugar content of beech leaves and beech leaf sap at various times. *Compt. Rend. Lab. Carlsberg, Ser. Chim.* **26**:197–230.

Olsen, C. (1948a) Adsorptively bound potassium in beech leaf cells. *Physiol. Plantarum* **1**:136–141.

Paauw, F. van der. (1958) Relations between the potash requirements of crops and meteorological conditions. *Plant and Soil* **9**:254–268.

Page, A. L., F. T. Bingham, T. J. Ganje, and M. J. Garber. (1963) Availability and fixation of added potassium in two California soils when cropped to cotton. *Soil Sci. Soc. Amer. Proc.* **27**:323–326.

Page, A. L., and T. J. Ganje. (1964) The effect of *p*H on potassium fixed by an irreversible adsorption process. *Soil Sci. Soc. Amer. Proc.* **28**:199–202.

Page, A. L., and J. P. Martin. (1964) Growth and chemical composition of citrus seedlings as influenced by Na additions to soils low in exchangeable K. *Soil Sci.* **98**:270–273.

Page, J. B., and L. D. Baver. (1939) Ionic size in relation to fixation of cations by colloidal clay. *Soil Sci. Soc. Amer. Proc.* **4**:150–155.

Parker, F. W., J. R. Adams, K. G. Clark, K. D. Jacob, and A. L. Mehring. (1946) Fertilizers and lime in the United States: resources, production, marketing, and use. *U.S. Dept. Agr., Misc. Publ. 586.*

Parks, W. L., and W. M. Walker. (1966) The effect of potassium fertilization rate and placement upon corn yields at different soil potassium levels. (Manuscript submitted to *Agronomy Journal*).

Place, G. A., and S. A. Barber. (1964) The effect of soil moisture and rubidium concentration on diffusion and uptake of rubidium-86. *Soil Sci. Soc. Amer. Proc.* **28**:239–243.

Porter, C. L. (1927) A study of the fungous flora of the nodal tissues of the corn plant. *Phytopathology* **17**:563–568.

Pound, G. S., and L. G. Weathers. (1953) The relation of host nutrition to multiplication of turnip virus 1 in *Nicotiana glutinosa* and *N. multivalvis*. *Phytopathology* **43**:669–674.

Pratt, P. F. (1951) Potassium removal from Iowa soils by greenhouse and laboratory procedures. *Soil Sci.* **72**:107–117.

Pratt, P. F., W. W. Jones, and F. T. Bingham. (1957) Magnesium and potassium content of orange leaves in relation to exchangeable magnesium and potassium in the soil at various depths. *Proc. Amer. Soc. Hort. Sci.* **70**:245–251.

Prummel, J. (1957) Fertilizer placement experiments. *Plant and Soil* **8**:231–253.

Rankama, K., and T. G. Sahama. (1950) *Geochemistry*. University of Chicago Press, Chicago, Ill.

Rathje, W. (1961) Zur ertragssteigernden Wirkung einer Natriumdüngung zu Zuckerrüben. *Zeitschr. Pflanzenernähr., Düng., Bodenk.* **94**:174–180.

Reed, J. F. (1953) Potash resources in the United States in relation to world supplies. *Agronomy* **3**:257–286.

Reitemeier, R. F., I. C. Brown, and R. S. Holmes. (1951) Release of native and fixed nonexchangeable potassium of soils containing hydrous mica. *U.S. Dept. Agr. Tech. Bul. 1049.*

Rich, C. I. (1958) Muscovite weathering in a soil developed in the Virginia Piedmont. *Natl. Conf. Clays and Clay Minerals* (1956) **5**:203–212.

Richards, F. J. (1954) Potassium deficiency in relation to putrescine production. *Huitième Congr. Internat. Bot.* (Paris, 1954), *Rapports et Commun.*, Sec. 11-12: 101–102.

Richards, F. J. (1956) Some aspects of potassium deficiency in plants. *Potassium Symposium* 1956:59–73.

Richards, F. J., and R. G. Coleman. (1952) Occurrence of putrescine in potassium-deficient barley. *Nature* **170**:460.

Richards, L. A., and C. H. Wadleigh. (1952) Soil water and plant growth. *Agronomy* **2**:73–251.

Rossiter, R. C. (1947) The effect of potassium on the growth of subterranean clover and other pasture plants on Crawley sand. 2. Field-plot experiments. *Jour. Council Sci. Indust. Res.* (Australia) **20**:389–401.

Russell, E. J., and D. J. Watson. (1938) The Rothamsted field experiments on barley 1852–1937. Pt. II. Effects of phosphatic and potassic fertilizers. Deterioration under continuous cropping. *Empire Jour. Exptl. Agr.* **6**:293–314.

Russell, E. W. (1954) The effect of soil forming factors on the potassium content of soils. *Potassium Symposium* **1954**:199–205.

Salmon, R. C. (1964) Potassium in different fractions of some Rhodesian soils. *Rhodesian Jour. Agr. Res.* **2**:85–90.

Samuels, G., and P. Landrau, Jr. (1953) Potassium competition in a kudzu-Merker grass association. *Jour. Agr. Univ. Puerto Rico* **37**:273–282.

Samuels, G., and P. Landrau, Jr. (1954) The influence of potassium on the yield and sucrose content of sugarcane. *Jour. Agr. Univ. Puerto Rico* **38**:170–178.

Schachtschabel, P. (1937) Aufnahme von nicht-austauschbarem Kali durch die Pflanzen. *Bodenk. u. Pflanzenernähr.* **3**:107–133.

Schachtschabel, P. (1937a) Mikroskopische und röntgenographische Untersuchungen von Böden. *Bodenk. u. Pflanzenernähr.* **5**:375–389.

Schachtschabel, P., and W. Köster. (1960) Chemische Untersuchungen an Marschen. II. Böden des Ems- und Weser-Jade-Gebietes. 3. Kaliumfixierung und Kaliumnachlieferung. *Zeitschr. Pflanzenernähr., Düng., Bodenk.* **89**:148–159.

Scheffer, F., E. Welte, and H. G. v. Reichenbach. (1960) Über den Kaliumhaushalt und Mineralbestand des Göttinger E-Feldes. *Zeitschr. Pflanzenernähr., Düng., Bodenk.* **88**:115–128.

Schmehl, W. R., and R. P. Humbert. (1964) Nutrient deficiencies in sugar crops. *In* H. B. Sprague (Ed.) *Hunger Signs in Crops,* 3rd Ed., pp. 415–450. David McKay Co., New York.

Schuffelen, A. C. (1954) The absorption of potassium by the plant. *Potassium Symposium* **1954**:169–181.

Schuffelen, A. C., and H. W. van der Marel. (1955) Potassium fixation in soils. *Potassium Symposium* **1955**:157–201.

Schulte, E. E., and R. B. Corey. (1965) Extraction of potassium from soils with sodium tetraphenylboron. *Soil Sci. Soc. Amer. Proc.* **29**:33–35.

Scott, A. D., and S. J. Smith. (1966) Susceptibility of interlayer potassium in micas to exchange with sodium. *Natl. Conf. Clays and Clay Minerals* **14**:69–81.

Scott, G. T., and H. R. Hayward. (1953) The influence of temperature and illumination on the exchange of potassium ion in *Ulva lactuca. Biochim. Biophys. Acta* **12**:401–404.

Semb, G., and G. Uhlen. (1954) A comparison of different analytical methods for the determination of potassium and phosphorus in soil based on field experiments. *Acta Agr. Scand.* **5**:44–68.

Shaw, W. M., and B. Robinson. (1960) Reaction efficiencies of liming materials as indicated by lysimeter leachate composition. *Soil Sci.* **89**:209–218.

Shepherd, L. N., J. C. Shickluna, and J. F. Davis. (1959) The sodium-potassium nutrition of sugar beets produced on organic soil. *Jour. Amer. Soc. Sugar Beet Technol.* **10**:603–608.

Shickluna, J. C., J. F. Davis, and R. E. Lucas. (1965) Why potatoes and onions need phosphorus and potassium on a virgin organic soil. *Better Crops with Plant Food* **49**, No. 5:20–24.

Smith, T. A. (1963) L-arginine carboxy-lyase of higher plants and its relation to potassium nutrition. *Phytochem.* **2**:241–252.

Sprague, H. B. (Ed.) (1964) *Hunger Signs in Crops.* 3rd Ed. David McKay Co., New York.

Stadtman, E. R. (1952) The purification and properties of phosphotransacetylase. *Jour. Biol. Chem.* **196**:527–534.

Stanford, G., and W. H. Pierre. (1947) The relation of potassium fixation to ammonium fixation. *Soil Sci. Soc. Amer. Proc.* (1946) **11**:155–160.

Stauffer, R. S. (1942) Runoff, percolate, and leaching losses from some Illinois soils. *Jour. Amer. Soc. Agron.* **34**:830–835.

Strasman, A., P. Quidet, and R. Blanchet. (1958) Valeur comparée des divers tests analytiques relatifs au potassium du sol, d'après la réaction des plantes aux engrais potassiques. *Compt. Rend. Acad. Agr. France* **44**:639–642.

Takahashi, T., and D. Yoshida. (1960) Occurrence of putrescine in the tobacco plant with special reference to the nutrient deficiency. *Soil and Plant Food* **6**:93.

Tinker, P. B. (1964) Studies on soil potassium. IV. Equilibrium cation activity ratios and responses to potassium fertilizer of Nigerian oil palms. *Jour. Soil Sci.* **15**:35–41.

Tinker, P. B. H. (1965) The effects of nitrogen, potassium and sodium fertilizers on sugar beet. *Jour. Agr. Sci.* **65**:207–212.

Titus, J. S., and D. Boynton. (1953) The relationship between soil analysis and leaf analysis in eighty New York McIntosh apple orchards. *Proc. Amer. Soc. Hort. Sci.* **61**:6–26.

Tobler, F. (1929) Zur Kenntnis der Wirkung des Kaliums auf den Bau der Bastfaser. *Jahrb. wiss. Bot.* **71**:26–51.

Turner, J. H., Jr. (1944) The effect of potash level on several characters in four strains of upland cotton which differ in foliage growth. *Jour. Amer. Soc. Agron.* **36**:688–698.

Van Ruymbeke, M. (1963) Le problème de la richesse inhérente des sols poldériens en potasse. *Pedologie* **13**:15–24.

Vermà, G. P. (1963) Release of non-exchangeable potassium from soils and micaceous minerals during short periods of cropping in the greenhouse. Ph.D. Thesis, Iowa State Univ., Ames.

Volk, G. M., and C. E. Bell. (1945) Some major factors in the leaching of calcium, potassium, sulfur and nitrogen from sandy soils. A lysimeter study. *Florida Agr. Exp. Sta. Bul. 416.*

Walker, W. M., and W. L. Parks. (1966) The effect of potassium fertilization rate and placement upon lodging in corn at different soil potassium levels. (Manuscript submitted to *Agronomy Journal.*)

Wallace, T. (1958) Potassium uptake in relation to soil moisture. *Potassium Symposium* **1958**:141–147.

Wallace, T. (1961) *The Diagnosis of Mineral Deficiencies in Plants by Visual Symptoms.* 2nd Ed., enlarged. Chemical Publ. Co., New York.

Walsh, T. (1954) The determination of potassium requirements by visual symptoms and plant analysis. *Potassium Symposium* **1954**:327–352.

Walsh, T., and E. J. Clarke. (1943) Characteristics of some Irish orchard soils in relation to apple tree growth. *Éire Dept. Agr. Jour.* **40**:61–122.

Walsh, T., and S. J. Cullinan. (1945) The effect of wetting and drying on potash-fixation in soils. *Empire Jour. Exptl. Agr.* **13**:203–212.

Wang, L. C., O. J. Attoe, and E. Truog. (1953) Effect of lime and fertility levels on the chemical composition and winter survival of alfalfa. *Agron. Jour.* **45**:381–384.

Warmke, H. E., R. H. Freyre, and J. Garcia. (1952) Evaluation of some tropical grass-legume associations. *Tropical Agr.* **29**:115–121.

Watson, D. J. (1956) The physiological basis of the effect of potassium on crop yield. *Potassium Symposium* **1956**:109–119.

Weaver, C. E. (1958) The effects and geologic significance of potassium "fixation" by expandable clay minerals derived from muscovite, biotite, chlorite, and volcanic material. *Amer. Mineral.* **43**:839–861.

Webster, G. C., and J. E. Varner. (1954) Peptide-bond synthesis in higher plants. II. Studies on the mechanism of synthesis of γ-glutamylcysteine. *Arch. Biochem. Biophys.* **52**:22–32.

Welte, E., and E. A. Niederbudde. (1963) Die unterschiedliche Bindungsfestigkeit von fixiertem Kalium und ihre mineralogische Deutung. *Plant and Soil* **18**:176–190.

Wilfarth, H., and G. Wimmer. (1902) Die Wirkung des Kaliums auf das Pflanzenleben nach Vegetationsversuchen mit Kartoffeln, Tabak, Buchweizen, Senf, Zichorien und Hafer. *Arb. deut. Landw. Gesell.*, Heft 68.

Williams, D. E. (1961) The absorption of potassium as influenced by its concentration in the nutrient medium. *Plant and Soil* **15**:387–399.

Williams, M. C. (1960) Effect of sodium and potassium salts on growth and oxalate content of halogeton. *Plant Physiol.* **35**:500–505.

Wit, C. T. de. (1953) A physical theory on placement of fertilizers. *Versl. Landbouwk. Onderz.* **59.4**.

Index